Power Systems

More information about this series at http://www.springer.com/series/4622

Naser Mahdavi Tabatabaei
Ali Jafari Aghbolaghi · Nicu Bizon
Frede Blaabjerg
Editors

Reactive Power Control in AC Power Systems

Fundamentals and Current Issues

Springer

Editors
Naser Mahdavi Tabatabaei
Electrical Engineering Department
Seraj Higher Education Institute
Tabriz
Iran

Ali Jafari Aghbolaghi
Zanjan Electric Energy Distribution Company
Zanjan
Iran

Nicu Bizon
Faculty of Electronics, Communication and Computers
University of Piteşti
Piteşti
Romania

Frede Blaabjerg
Department of Energy Technology
Aalborg University
Aalborg
Denmark

ISSN 1612-1287 ISSN 1860-4676 (electronic)
Power Systems
ISBN 978-3-319-51117-7 ISBN 978-3-319-51118-4 (eBook)
DOI 10.1007/978-3-319-51118-4

Library of Congress Control Number: 2017930608

Printed on acid-free paper

This Springer imprint is published by Springer Nature
The registered company is Springer International Publishing AG
The registered company address is: Gewerbestrasse 11, 6330 Cham, Switzerland

Dedicated to
all our teachers and colleagues
who enabled us to write this book,
and our family and friends
for supporting us all along

Foreword

Electric power systems will be operated in reliable and efficient situation considering reactive power control and voltage stability management. Reactive power margins are related to the voltage stability. The aspects are satisfied by designing and operating of right voltages limits, maximizing utilization of transmission systems and minimizing of reactive power flow. Therefore, controlling reactive power and voltage is one of the major challenges of power system engineering.

Reactive power as the dissipated power is affected by capacitive and inductive phenomena that they drop voltage and draw current in the form of heat or waste energy. Reactive power is generated by the capacitors and generators, whereas it is consumed by the inductors and is essential in the parallel connection circuits as power factor controlling and power transmission lines.

Reactive power control and voltage stability aspects are effective in reliability of electric power networks. Voltage instability commonly occurs as a result of reactive power deficiency. The trends are to reduce reactive power and increase voltage stability to improve efficiency and operation of power systems. There is a direct relation between reactive power and voltage behavior which serves the voltage collapse and rising effects in power systems.

Regulating the reactive power and voltage control should be done according to flexible and fast controlled devices. Placement and adjustment of reactive power play important roles in operation of reactive power compensation and voltage control. Therefore, the operations of reactive power resources in the power systems such as automatic transformer tap changer, synchronous condenser, capacitor banks, capacitance of overhead lines and cables, static VAR compensators and FACTS devices are very significant.

Reactive power control and voltage stability management are considered as regional challenges to meet, which otherwise can cause the scale of blackouts increase in the power systems. Theoretical and application issues in these areas help us to identify problems related to reliability and stability of the power systems and prevent the system degradation.

The above aspects are illustrated in this book by the editors and authors, in the following topics: electrical power systems operation and control, reactive power and voltage stability in power systems, reactive power control in transmission lines, reactive power compensation and optimal placement, reactive power in renewable resources, reactive power optimization and software applications, optimal reactive power dispatch, induction generator operation and analysis, communication networks and standards in power systems, power systems SCADA applications, and geomagnetic storms effects in electric networks.

The book chapters and materials are very efficient in theoretical and application issues and are highly recommended for studying and considering in educational and research fields.

November 2016

Academician Arif M. Hashimov
Institute of Physics
Azerbaijan National Academy of Sciences
Baku
Azerbaijan

Preface

The modern electric power systems are more expanded worldwide and include more energy resources and critical parts based on the requirements of the twenty-first century. General parts of electric power systems as generation, transmission and consumption are important to be analyzed and well operated for the development of industry and life.

The engineers and scientists need applicable and renewable methods for analyzing and controlling each part of the electric power systems and to overcome complicated actions which occur in the systems due to their operational and interconnection behaviors. The objective of the analysis is minimizing the losses of the networks and increasing the overall efficiency and economic advantages.

The central and distributed generation of electric power networks connect to more loads, transmission lines, transformers and energy sources together including nonlinear equipment such as power semiconductor devices. The engineers and scientists are interested in analyzing the power systems operations to control and develop the AC/DC networks including high voltage transmission lines and equipment.

Flexible and fast power flow control and transmission are expected to raise the network effective operation, power wheeling requirement and transmission capability as well as voltage stability. Computational intelligence methods are applied to electric power analysis to facilitate the effective analysis techniques and solve several power system problems especially in power transmission and voltage stability.

Reactive Power Control in AC Power Systems: Fundamentals and Current Issues is a book aimed to highlight the reactive power control and voltage stability concepts and analysis to provide understanding on how they are affected by different criteria of available generations, transmissions and loads using different research methods.

A large number of specialists joined as authors of the book chapters to provide their potentially innovative solutions and research related to reactive power control and voltage stability, in order to be useful in developing new ways in electric power analysis, design and operational strategies. Several theoretical researches, case analysis, and practical implementation processes are put together in this book that aims to act as research and design guides to help graduates, postgraduates and researchers in electric power engineering and energy systems.

The book, which presents significant results obtained by leading professionals from industries, research and academic fields, can be useful to a variety of groups in specific areas. All works contributed to this book are new, previously unpublished material or extended version of published papers in the proceedings of international conferences and transactions on international journals. The book consists of 16 chapters in three parts.

Part I Fundamentals of Reactive Power in AC Power Systems

The six chapters in the first part of this book present the fundamentals of reactive power in AC power systems considering different operating cases. The topics in this part include the advanced methods and applications in electric power systems and networks related to the fields of fundamentals of reactive power in AC power systems, reactive power role in AC power transmission systems, reactive power compensation in energy transmission systems with sinusoidal and nonsinusoidal currents, reactive power importance in wind power plants, and fundamentals and contemporary issues of reactive power control in AC power systems.

Chapter 1 describes the general overview of electric power systems including power generation, transmission and distribution systems, linear AC circuits in steady state conditions, flow of power between generator and customers is studied by using the active, reactive, apparent and complex power, electric power system quality, measurement and instrumentation methods of power systems parameters, and general standards in energy generation, transmission and marketing. The importance of reactive power in AC power systems and its various interpretations are also discussed in this chapter.

The basic theory of AC circuits, behavior of two-port linear elements and analysis methods of AC circuits are given in Chap. 2. The physical interpretation of electric powers in AC power systems, fundamental problems of reactive power consumption automated management in power systems, equipment for power factor correction, designing simple systems for compensating of reactive power for different levels of installation, the overall harmonic distortion of voltage and current, and qualitative and quantitative aspects related to active and reactive power circulation in AC power systems including several examples and case studies referring to classical linear AC circuits under sinusoidal and nonsinusoidal conditions are also the topics of this chapter.

Chapter 3 presents basic principles of power transmission operation, equipment for reactive power generation, shunt/series compensation, control of reactive power in power transmission system. The chapter describes the capacitive and inductive properties of power transmission lines and also reactive power consumption by transmission lines which increases with the square of current. The chapter states the sources, effects and limitations of the reactive power and flowing in transmission lines and transformers as well as control of reactive power should satisfy the bus voltages, system stability and network losses in the power systems.

The definition of reactive power under nonsinusoidal conditions in nonlinear electric power systems is described in Chap. 4. This chapter discusses and simulates the reactive power compensation for sinusoidal and nonsinusoidal situations, where nonlinear circuit voltages and currents contain harmonics and also the control algorithms of automatic compensators. The main aim of the chapter is based on the dissipative systems and cyclodissipativity theories for calculation of compensation elements for reactive power compensation by minimizing line losses. The chapter is also including the examples and computer simulations to show the mathematical framework for analyzing and designing of compensators for reactive power compensation in general nonlinear loads.

Chapter 5 deals with the rate of reactive power absorption or injected by the wind units and also the key role of reactive power generation and consumption in large-scale wind farms. The chapter describes requirements of reactive power compensation, voltage stability and also power quality improvement in the electric grid of wind turbine to reduce the power losses and control of voltage level. The units of wind turbines of types 1 to 4 are also categorized and discussed in the chapter considering their construction, generation, converters, reactive power and voltage control abilities. The coordination related to reactive power adjustment in the wind turbines is also discussed in this chapter.

The concept of power quality and voltage stability improvement based on the reactive power control is introduced in Chap. 6. The chapter describes the impact of reactive power flow in the power system and defines the power components of electrical equipment that produces or absorbs reactive power. Then the reactive power control and relations between reactive power and voltage stability are presented. The chapter also contains reactive power control methods for voltage stability and presents voltage control management based on case studies.

Part II Compensation and Reactive Power Optimization in AC Power Systems

The second part of this book tries to highlight in six chapters the concepts of reactive power optimization and compensation. The topics in this part include optimal reactive power control for voltage stability improvement, reactive power compensation, optimal placement of reactive power compensators, reactive power optimization in classic methods and also using MATLAB and DIgSILENT, and multi-objective optimal reactive power dispatch.

Chapter 7 is entirely focused on the voltage stability control using three main techniques of reactive power management, active power re-dispatch, and load shedding. The chapter discusses about determining the location of VAR sources and their setting and installation, online and offline reactive power dispatch, and optimal reactive power flow (ORPF). The reactive power flow and voltage magnitudes of generator buses, shunt capacitors/reactors, output of static reactive power compensators, transformer tap-settings are considered as the control parameters and are used for minimizing the active power loss and improving of the voltage profile in ORPF. This chapter also confers the reactive power dispatch as a nonlinear and nonconvex problem with equality and inequality constraints.

The reactive power compensators based on advanced industrial applications are highlighted in Chap. 8. The basic theoretical background of reactive power compensation as well as conventional compensators and improved FACTS are introduced in the chapter. The compensation devices including shunt, series and shunt-series configurations for transmission lines regarding their characteristics and also analytical expressions are presented in the chapter. The power flow control, voltage and current modifications as well as stability issues are also analyzed and compared for similar compensation devices and emerging technologies.

Chapter 9 provides a framework and versatile approach to develop a multi-objective reactive power planning (RPP) strategy for coordinated handling of reactive power from FACTS devices and capacitor banks. This chapter deals with power system operators for determining the optimal placement of FACTS devices and capacitor banks should be injected in the network to improve simultaneously the voltage stability, active power losses and cost of VAR injection. A formulation and solution method for reactive power planning, and voltage stability based on cost functions are also presented in the chapter.

Chapter 10 presents the reactive power optimization using artificial optimization algorithms as well as the formulations and constraints to implement reactive power optimization. The classic method of reactive power optimization and basic principles and problem formulation of reactive power optimization using artificial intelligent algorithms are discussed in the chapter. In addition, this chapter focuses on the particle swarm optimization algorithm and pattern search method application in reactive power optimization including the case studies.

The efficient approach using parallel working of MATLAB and DIgSILENT software with the intention of reactive power optimization is discussed in Chap. 11. This chapter presents the toolboxes, functions and flexibility powers of MATLAB and DIgSILENT in electrical engineering calculation and implementation. Also it provides the advantages of parallel calculations of MATLAB and DIgSILENT and relation of two software to carry out the heuristic algorithms as fast, simple and accurate as possible to optimize reactive power in AC power systems.

In Chap. 12, the reactive power compensation devices are modeled using deterministic multi-objective optimal reactive power dispatch (DMO-ORPD) and two-stage stochastic multi-objective optimal reactive power dispatch (SMO-ORPD) in discrete and continuous studies. They are formulated as mixed integer nonlinear program (MINLP) problems, and solved by general algebraic modeling system (GAMS). A case study for evaluation of the performance of different proposed MO-ORPD models is also shown in the chapter. This chapter presents the MO-ORPD problem taking into account different operational constraints such as bus voltage limits, power flow limits of branches, limits of generators voltages, transformers tap ratios and the amount of available reactive power compensation at the weak buses.

Part III Challenges, Solutions and Applications in AC Power Systems

The final part of this book consists of four chapters and considers some applications and case studies in AC power systems related to the issues of active and reactive power concepts. The topics in this part include self-excited induction generator, communications for electric power systems, SCADA applications for electric power systems and effect of geomagnetic storms on electrical networks.

Chapter 13 discusses about a three-phase self-excited induction generator in an autonomous power generation mode. The chapter presents generator operating points and control strategies to maintain the frequency at quasi-constant values and to use it as power converter such as a simple dimmer to control the reactive power. The frequency analysis in steady state and transient cases is studied in this chapter using a single-phase equivalent circuit as well as theoretical and numerical results are also validated on a laboratory test bench.

Chapter 14 describes communications applied for electric power systems including communication standards and infrastructure requirements for smart grids. The chapter presents three primary functions of smart grids to accomplish in real time requests of both consumers and suppliers based on communications technologies. The most usual communication systems including fiber optic communication, digital subscriber line/loop, power line communications, and wireless technologies for using the power system control for smart grids architecture are highlighted in the chapter. The case studies related to communication systems of electric power system are also carried out in this chapter.

The SCADA systems and applications in electric power networks are studied in Chap. 15. The chapter explains the role and theory of SADA systems, security, real-time control and data exchange between remote units and central units. The SCADA systems are also applied for optimization and realization of reactive power in AC power systems. Some disadvantages of dispatching systems such as graphical information and interface are explained in the chapter and the rules of improving them are also carried out. The flexibility designing of the systems for small and large networks are also explained.

Chapter 16 introduces the effect of geomagnetic fields called as storms on electric power systems. This chapter discusses about the physical nature of earth's magnetic field and its measurements in geomagnetic observatories and shows that the variation of geomagnetic field affect the operation of various distracting electronic devices, such as electrical transmission systems. An algorithm for calculating induced currents in the power transmission lines and also the violation of stability of the system considering the illustrative example are also derived in this chapter.

The editors recommend this book as suitable for an audience professional in electric power systems, as well as researchers and developers in the field of energy and power engineering. It is anticipated that the readers have sufficient knowledge in electric power engineering and also advanced mathematical background.

In total, the book includes theoretical background and case studies in reactive electric power and voltage stability concepts. The editors have made efforts to cover the essential topics of reactive electric power to balance theoretical and applicative aspects in the chapters of this book. The book has been written by a team of researchers from which use the dedicated intensive resources for achieving certain mental attitudes for interested readers. At the same time, the application and case studies are intended for real understanding and operation.

Finally, the editors hope that this book will be useful to undergraduate and graduate students, researchers and engineers, trying to solve reactive electric power problems using modern technical and intelligent systems based on theoretical aspects and application case studies.

Tabriz, Iran — Naser Mahdavi Tabatabaei
Zanjan, Iran — Ali Jafari Aghbolaghi
Piteşti, Romania — Nicu Bizon
Aalborg, Denmark — Frede Blaabjerg

Contents

List of Figures

List of Tables

Part I
Fundamentals of Reactive Power in AC Power Systems

Chapter 1
Electrical Power Systems

Horia Andrei, Paul Cristian Andrei, Luminita M. Constantinescu, Robert Beloiu, Emil Cazacu and Marilena Stanculescu

Abstract The general description of architecture and classification of power systems are presented at the beginning of the first chapter. Afterwards the basic concepts and analysis methods of electrical power systems are given, being accompanied by many examples of calculation. Also the power flow between the generators and consumers has been considered. In order to characterize the parameters of power systems next section of the first chapter is dedicated to measurement methods of power systems parameters. Electrical energy needed to power industrial or household electric consumers, like any other product, should satisfy specific quality requirements. In this respect the power quality aspects and the standards for power systems parameters are presented. The flow of reactive power and energy in power systems produces significant effects on the optimal functioning of suppliers and customers which are connected. Therefore in the last section of first chapter is defined the importance of reactive power in AC power systems and on its various understandings. Usually the first chapter is closed with a large list of bibliographic references.

H. Andrei (✉)
Doctoral School of Engineering Sciences, University Valahia of Targoviste, Targoviste, Dambovita, Romania
e-mail: hr_andrei@yahoo.com

P.C. Andrei · E. Cazacu · M. Stanculescu
Department of Electrical Engineering, University Politehnica Bucharest, Bucharest, Romania
e-mail: paul.andrei@upb.ro

E. Cazacu
e-mail: emil.cazacu@upb.ro

M. Stanculescu
e-mail: marilena.stanculescu@upb.ro

L.M. Constantinescu · R. Beloiu
University of Pitesti, Pitesti, Romania
e-mail: lmconst2002@yahoo.com

R. Beloiu
e-mail: robertbeloiu@yahoo.com

N. Mahdavi Tabatabaei et al. (eds.), *Reactive Power Control in AC Power Systems*, Power Systems, DOI 10.1007/978-3-319-51118-4_1

1.1 Chapter Overview

First part of this chapter is dedicated to an overview over all the topics presented in its sections. Power generation, transmission and distribution systems—that can be usually called electrical power systems—use almost exclusively AC circuits due to economic and technical advantages they offer. In this respect the second section of this chapter makes a general description of power systems. Generalities about the linear AC circuits in steady state conditions when the parameters as currents through and the voltages across the branches of circuits are sinusoidal are described in section three. Also the flow of power between generator and customers is studied by using the active, reactive, apparent and complex power like other important energetically parameters of AC circuits in sinusoidal state.

In the complex and interconnected power system a large variety of the electromagnetic field occurrences are present. They influence, in any time and location, the system parameters particularly his currents and voltages, one of the most important parameters being the power quality (*PQ*). Therefore all those who are interconnected and use the power system like power suppliers, distributors and customers are interested to preserve the *PQ* quality in their nominal values or in other words the electrical power as clean as possible.

Generally speaking the power quality term includes all the parameters which are defined for power systems. In order to check the *PQ* state, the parameters of power systems must be measured as accurately as possible. Herein in Sect. 4 of the chapter is presented an overview of the most important measurement and instrumentation methods of power systems parameters. In addition the parameters of *PQ* must be compared with a set of nominal (technical reference) values, namely standards. Particularly each country holds its own *PQ* standards and *PQ* instrumentation. On the other hand the globalization of the energy generation, transmission and market imposes general standards for common *PQ* parameters. A review of these international standards represents another purpose of this chapter which is allocated in Sect. 5.

The circulation of reactive power in power systems produces significant effects on the optimal functioning of suppliers and customers which are connected. Therefore about the importance of reactive power in AC power systems and on its various understandings refers the last section of the chapter.

This chapter is closed with a specific list of bibliographic references.

1.2 Introduction in Electrical Power Systems

Nowadays electrical power, together with natural resources, is becoming one of the most technical, economic and political factors. Often the stability and development of a region of the world depends on the respective countries' energy systems and resources among which the electricity plays a key role.

The traditional architecture of Electrical Power Systems (EPS) is based on power generation, transmission, distribution and usage interconnected subsystems [1, 2]. For the past century the rate of change about energy production, distribution and customers is significant: on one hand the depletion of natural resources and environmental protection measures have led to the use of renewable energy sources (RES) such as wind turbines (WT), photovoltaic (PV) modules, geo-thermal (GT) or bio-mass (BM) systems and to the penetration of sophisticated distributed generation (DG) and Micro and Smart Grids (MG) systems; on the other hand through the development of industry production is resulted diverse range of much sensitive industrial and residential equipment [3, 4].

More efficient operation modes of distribution subsystems have been implemented with the increasing of DG penetration rate. Thus the presence of a multitude of energy sources leads to improving the continuity of power supply to the industrial and residential consumers. One the other hand there are several technical and economic issues that can be exceeded by introducing of new operation and control concepts as MG paradigm. The MG is a complex flexible and system control system of power flow between the generators and consumers. MG provides real-time decisions and auxiliary services to networks (relieves congestions, aiding restoration after faults e.a.). Also MG can provide to customers their thermal and electric energy needs, enhance the PQ, reduce the pollution and the costs of consumed energy e.a.

Nowadays complex and complicated technical, economic and political processes of industry and residential consumers development influence the dynamics of EPS [5–7]:

- globalization means the inter-country or inter-continental networks integration, energy market and investment combination and technological integration;
- liberalization is associated with the development of regional or inter-regional energy markets;
- decentralization assumes the development of small and large units power, together with upgrading and renewal of transmission and distribution networks, and introducing of DG concept;
- diversification means the increasing of multitude of energy sources (fuels, renewable energy sources e.a.) and of types of power plants;
- modernization results from the development of old technologies and the implementation of new and efficient ones.

Considering that EPS have different functions and nominal voltages, respectively various constructive types, there are some criteria for their classification [8–10]:

- criteria of nominal voltage is important because with its help is determine the power and distance that can be transmitted, the cost of the transmission line and its equipment e.a. EPS classification taking into account the nominal voltage is presenting in Table 1.1.

Table 1.1 Classification of EPS

Class of EPS	Nominal voltage
Low voltage (LV)	50–1000 V
Medium voltage (MV)	1–35 kV
High voltage (HV)	35–275 kV
Very high voltage (VHV)	>300 kV

The nominal voltage means the rms value of a line voltage between phases; its standardized values are recommended by the International Electrotechnical Commission (CEI). For example the standardization values in kV are: 3; 3.3; 6; 10; 11; 15; 20; 22; 30; 45; 47; 66; 69; 110; 115; 132; 138; 150; 161; 220; 230; 287; 330; 345; 380; 400; 500; 700, and each country can adopt specific several such values. The LV is used in indoor electrical installations to supply directly the low voltage customers as well as the small urban and industrial networks, with power up to tens of kVA. The MV is used in urban and industrial networks for supply transformers with powers between tens and hundreds of kVA and also can supply directly medium voltage equipment. Transmission and distribution lines for powers between tens of MVA and 1–2 hundreds of MVA are carried out with HV. The VHV is used for transmission lines of powers between hundreds and thousands of MVA.

- criteria of functions classify EPS in usage (utilization), distribution and transmission networks.
- criteria of topology classify EPS in: (i) radial networks that means each customer can be supplied from one side (source) only; (ii) meshed networks that means each customer can be supplied from two sides; (iii) complex meshed networks that means each customer can be supplied from of more than the two sides.
- criteria of adopted power system classify EPS in AC respectively DC. Starting from historical scientific insights of M. Dolivo-Dobrovolski and N. Tesla now the widespread EPS system consists in 3 and 2-phases electrical AC networks. But from struggle between AC and DC transmission systems is possible in the near future that very high voltage DC systems to win.

1.3 Basic Concepts and Analysis Methods of Electrical Power Systems

From theoretical point of view EPS are considering as symmetrical three-phase systems, condition that is ensured through the symmetrization of transmissions and distribution lines, and transformers. Starting from this ideal assumption the symmetrical three-phase transmission and distribution networks, including equipment and transformers, can be analyzed by using symmetrical components method and the decomposition in single phase circuits. The equivalent circuits of EPS contain non-linear passive elements, as resistances, inductances and capacitances. Nonlinearity of these elements are neglected in most frequent calculations, considering that their values change relatively small according to the low and expected limits of change of the voltage, current or frequency. For this reason the equivalent circuits of EPS are considered linear and are disposed longitudinal and transversal as in Γ, T or Π scheme types [11–13].

A three-phase symmetrical and positive phase-sequence system in instantaneous values is expressed as [14, 15]

$$\begin{aligned} v_{d1}(t) &= V\sqrt{2}\sin(\omega t+\gamma) \\ v_{d2}(t) &= V\sqrt{2}\sin(\omega t+\gamma-\frac{2\pi}{3}) \\ v_{d3}(t) &= V\sqrt{2}\sin(\omega t+\gamma-\frac{4\pi}{3}) = V\sqrt{2}\sin(\omega t+\gamma+\frac{2\pi}{3}) \end{aligned} \tag{1.1}$$

respectively by the complex vectors

$$\begin{aligned} \underline{V}_{d1} &= Ve^{j\gamma} = \underline{V} \\ \underline{V}_{d2} &= Ve^{j(\gamma-\frac{2\pi}{3})} = \underline{V}e^{-j\frac{2\pi}{3}} = a^2\underline{V} \\ \underline{V}_{d3} &= Ve^{j(\gamma+\frac{2\pi}{3})} = \underline{V}e^{j\frac{2\pi}{3}} = a\underline{V} \end{aligned} \tag{1.2}$$

where $a = e^{j\frac{2\pi}{3}} = e^{-j\frac{4\pi}{3}} = -\frac{1}{2} + j\frac{\sqrt{3}}{2}$ is called phase operator. The representation of instantaneous values (1.1) in time domain and of complex vectors (1.2) in complex Cartesian coordinates are presented in Fig. 1.1a, b.

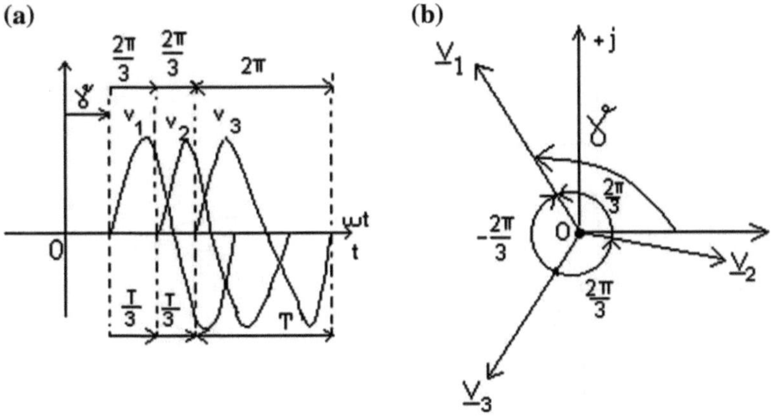

Fig. 1.1 Representation of three-phase symmetrical and positive phase-sequence system: **a** Time domain, **b** Cartesian coordinates

A three-phase symmetrical and negative phase-sequence system in instantaneous values is expressed as

$$\begin{aligned} v_{i1}(t) &= V\sqrt{2}\sin(\omega t + \gamma) \\ v_{i2}(t) &= V\sqrt{2}\sin(\omega t + \gamma + \frac{2\pi}{3}) \\ v_{i3}(t) &= V\sqrt{2}\sin(\omega t + \gamma - \frac{2\pi}{3}) \end{aligned} \quad (1.3)$$

respectively by the complex vectors

$$\begin{aligned} \underline{V}_{i1} &= Ve^{j\gamma} = \underline{V} \\ \underline{V}_{i2} &= Ve^{j(\gamma + \frac{2\pi}{3})} = \underline{V}e^{j\frac{2\pi}{3}} = a\underline{V} \\ \underline{V}_{i3} &= Ve^{j(\gamma - \frac{2\pi}{3})} = \underline{V}e^{-j\frac{2\pi}{3}} = a^2\underline{V} \end{aligned} \quad (1.4)$$

which have the representations in time domain and complex Cartesian coordinates illustrated in Fig. 1.2a, b.

A three-phase symmetrical and zero phase-sequence system in instantaneous values is expressed as

$$v_{01}(t) = v_{02}(t) = v_{03}(t) = V\sqrt{2}\sin(\omega t + \gamma) \quad (1.5)$$

and by the complex vectors

$$\underline{V}_{o1} = \underline{V}_{o2} = \underline{V}_{o3} = Ve^{j\gamma} = \underline{V} \quad (1.6)$$

The representation of values (1.5) and (1.6) is shown in Fig. 1.3a, b.

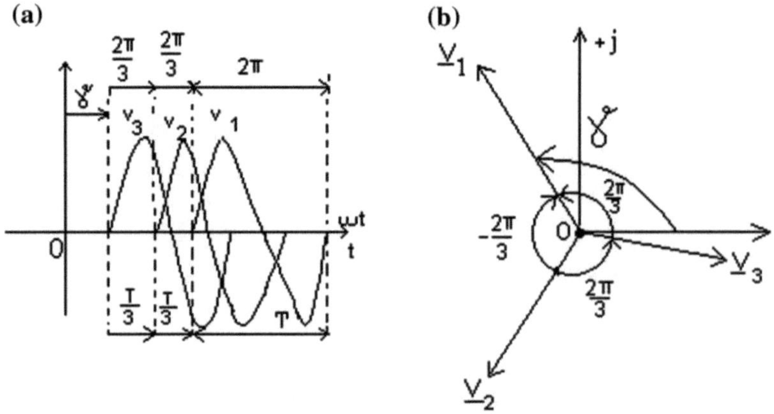

Fig. 1.2 Representation of three-phase symmetrical and negative phase-sequence system: **a** Time domain, **b** Cartesian coordinates

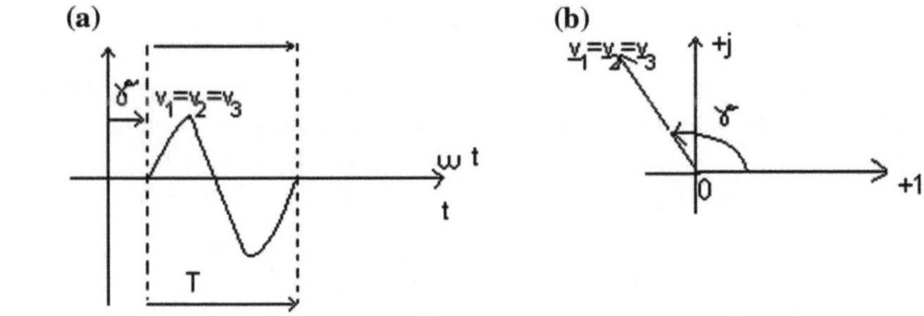

Fig. 1.3 Representation of three-phase symmetrical and zero phase-sequence system: **a** Time domain, **b** Cartesian coordinates

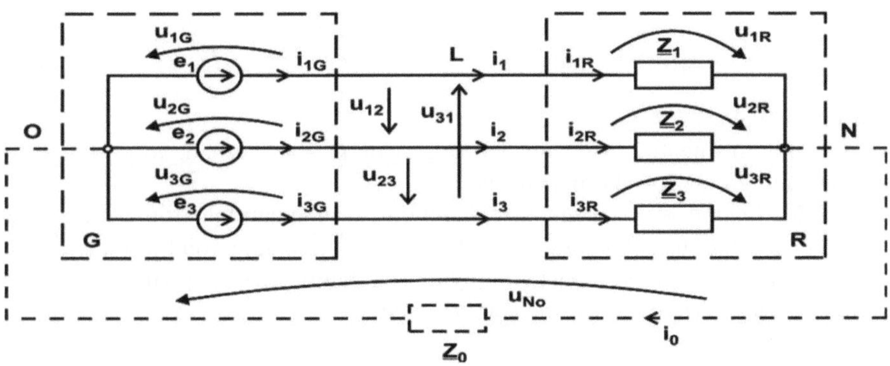

Fig. 1.4 Star connection

Practically, two general connections are considered in EPS: star and delta [16–18].

(i) *Star connection.* Let us consider an elementary EPS in which the generator (G) and consumer (load, receptor—R) are arranged into a star connection. The transmission line (L) links G with R. The star connection illustrated in Fig. 1.4 is defined in that all the phase terminals are connected together to form the neutral point. The both neutral nodes of the generator "O" and of the receptor "N" is linked by neutral wire. The electromotive voltages of generators e_1, e_2, e_3, the voltages across the terminals of generators u_{1_G}, u_{2_G}, u_{3_G}, the voltages between the transmission lines u_{12}, u_{23}, u_{31}, the currents in the transmission lines i_1, i_2, i_3 and the voltages across the terminals of receptor u_{1_R}, u_{2_R}, u_{3_R} there are three-phase systems.

For an adequate systematization of knowledge following definitions are useful:

- the triplets (e_1, u_{1_G}, i_{1_G}), (e_2, u_{2_G}, i_{2_G}) and (e_3, u_{3_G}, i_{3_G}) are called phase signals of generators;

- the doublets (u_{1_R}, i_{1_R}), (u_{2_R}, i_{2_R}) and (u_{3_R}, i_{3_R}) are called phase signals of receptors (consumers);
- the voltages (u_{12}, u_{23}, u_{31}) between of the three line wires are called line voltages, also the line impedances have not been taken into account;
- the currents (i_1, i_2, i_3) across the line wires are called line currents;
- the impedances $\underline{Z}_1$, $\underline{Z}_2$, $\underline{Z}_3$ are called phase impedances of receptor.

Generally speaking the set signals $(e_1, u_{1_G}, i_{1_G}, u_{1_R}, i_{1_R}, \underline{Z}_1)$ makes up phase 1 of star connection and similar definitions are used for the other two phases. For star connection the line currents are equal to the phase currents

$$i_{1G} = i_1 = i_{1R};\ i_{2G} = i_2 = i_{2R};\ i_{3G} = i_3 = i_{3R} \tag{1.7}$$

respectively in complex values

$$\underline{I}_{1G} = \underline{I}_1 = \underline{I}_{1_R};\ \underline{I}_{2G} = \underline{I}_2 = \underline{I}_{2_R};\ \underline{I}_{3G} = \underline{I}_3 = \underline{I}_{3_R} \tag{1.8}$$

and in rms values $I_l = I_{ph}$.

Also for the neutral wire are defined: i_0 or i_n—the current, u_{NO} or u_n—the voltage, and $\underline{Z}_0$ or $\underline{Z}_N$—the impedance.

Considering the positive reference directions for line currents from generators to receptors and for current across the neutral wire from neutral point of the receptors to that of the generators, then by applying Kirchhoff laws following relations in complex values are described the star connection

$$\begin{aligned}
&\underline{I}_1 + \underline{I}_2 + \underline{I}_3 = \underline{I}_0\\
&\underline{U}_{12} = \underline{U}_{1_R} - \underline{U}_{2_R}\\
&\underline{U}_{23} = \underline{U}_{2_R} - \underline{U}_{3_R}\\
&\underline{U}_{31} = \underline{U}_{3_R} - \underline{U}_{1_R}\\
&\underline{U}_{12} + \underline{U}_{23} + \underline{U}_{31} = 0\\
&\underline{U}_{1_R} + \underline{U}_{NO} + \underline{U}_{10} = 0\\
&\underline{U}_{2_R} + \underline{U}_{NO} + \underline{U}_{20} = 0\\
&\underline{U}_{3_R} + \underline{U}_{NO} + \underline{U}_{30} = 0
\end{aligned} \tag{1.9}$$

Another set of relations express Ohm's law for receptor phases and for neutral wire in complex values are expressed as

$$\begin{aligned}
&\underline{U}_{1_R} = \underline{Z}_1 \cdot \underline{I}_1\\
&\underline{U}_{2_R} = \underline{Z}_2 \cdot \underline{I}_2\\
&\underline{U}_{3_R} = \underline{Z}_3 \cdot \underline{I}_3\\
&\underline{U}_{NO} = \underline{Z}_0 \cdot \underline{I}_0
\end{aligned} \tag{1.10}$$

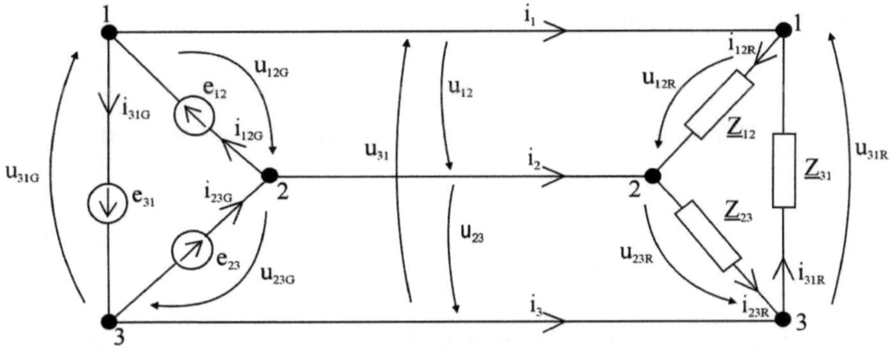

Fig. 1.5 Delta connection

(ii) *Delta connection.* Let us consider an elementary EPS in which the generator and receptor (consumer) are arranged into a delta (mesh) connection (Fig. 1.5). The transmission line links generator with receptor. Symmetrical cycling between the pairs of impedances leads so that the start terminal of one phase impedance is connected to the finish terminal of another are used for delta connection.

For an adequate systematization of knowledge following definitions are useful:

- the triplets $(e_{12}, u_{21_G}, i_{21_G})$, $(e_{23}, u_{23_G}, i_{23_G})$ and $(e_{31}, u_{31_G}, i_{31_G})$ are called phase signals of generators;
- the doublets (u_{12_R}, i_{12_R}), (u_{23_R}, i_{23_R}) and (u_{31_R}, i_{31_R}) are called phase signals of receptors (consumers);
- the voltages (u_{12}, u_{23}, u_{31}) between of the three line wires are called line voltages, also the line impedances have not been taken into account;
- the currents (i_1, i_2, i_3) across the line wires are called line currents;
- the impedances $\underline{Z}_{12}$, $\underline{Z}_{23}$, $\underline{Z}_{31}$ are called phase impedances of receptor.

Generally speaking the set signals $(e_{12}, u_{12_G}, i_{12_G}, u_{12_R}, i_{12_R}, \underline{Z}_{12})$ makes up phase 1 of delta connection and similar definitions are used for the other two phases. If the line impedances are not considering, for star connection the line voltages of generators are equal to the line voltage of receptor

$$u_{12_G} = u_{12_R} = u_{12}, u_{23_G} = u_{23_R} = u_{23}, u_{31_G} = u_{31_R} = u_{31} \tag{1.11}$$

and in rms values $V_l = V_{ph}$.

Based on Kirchhoff's current law (KCL) the following relations are true

$$\begin{aligned} \underline{I}_1 &= \underline{I}_{12} - \underline{I}_{31} \\ \underline{I}_2 &= \underline{I}_{23} - \underline{I}_{12} \\ \underline{I}_3 &= \underline{I}_{31} - \underline{I}_{23} \end{aligned} \tag{1.12}$$

By summing (1.12) results

$$\underline{I}_1 + \underline{I}_2 + \underline{I}_3 = 0 \tag{1.13}$$

Also applying Ohm's law on each phase of the receptor are obtained

$$\begin{aligned} \underline{U}_{12} &= \underline{Z}_{12} \cdot \underline{I}_{12} \\ \underline{U}_{23} &= \underline{Z}_{23} \cdot \underline{I}_{23} \\ \underline{U}_{31} &= \underline{Z}_{31} \cdot \underline{I}_{31} \end{aligned} \tag{1.14}$$

A three-phase system is called symmetrical if it is supplied by a symmetrical voltages system (1.1), (1.3) or (1.5). Otherwise it is considered unsymmetrical. If all the complex phase impedances of star or delta connection are equal the system is balanced, otherwise is non-balanced. A symmetrical and balanced system produces symmetrical lines and phases currents systems. In symmetrical and balanced systems all neutral points are the same potential and the current across the neutral wire is null. On the other hand in symmetrical and balanced conditions for star connection the rms value of line voltage is $\sqrt{3}$ times the rms value of phase voltage $V_l = \sqrt{3}V_{ph}$, respectively for delta connection the rms value of line current is $\sqrt{3}$ times the rms value of phase current $I_l = \sqrt{3}I_{ph}$.

The overall complex powers absorbed by a receptor in star (including the neutral wire) and delta connections are given by

$$\begin{aligned} \underline{S}_{R,star} &= \underline{U}_{1_R}\underline{I}_1^* + \underline{U}_{2_R}\underline{I}_2^* + \underline{U}_{3_R}\underline{I}_3^* + \underline{U}_{NO}\underline{I}_0^* = \underline{Z}_1 I_1^2 + \underline{Z}_2 I_2^2 + \underline{Z}_3 I_3^2 + \underline{Z}_0 I_0^2 \\ &= P_{R,star} + jQ_{R,star} \end{aligned} \tag{1.15}$$

respectively

$$\begin{aligned} \underline{S}_{R,delta} &= \underline{U}_{12_R}\underline{I}_{12}^* + \underline{U}_{23_R}\underline{I}_{23}^* + \underline{U}_{31_R}\underline{I}_{31}^* = \underline{Z}_{12} I_{12}^2 + \underline{Z}_{23} I_{23}^2 + \underline{Z}_{31} I_{31}^2 + \underline{Z}_0 I_0^2 \\ &= P_{R,delta} + jQ_{R,delta} \end{aligned} \tag{1.16}$$

where the real part represents the overall active absorbed power (P_R—dissipated in the resistors) and the imaginary part represents the overall absorbed reactive power (Q_R—dissipated in the inductors minus in the capacitors). When symmetrical and balanced conditions are verified then for any form of interlinkage (star or delta) the overall active and reactive absorbed powers are given by

$$\begin{aligned} P_R &= 3V_{ph}I_{ph}\cos\varphi = \sqrt{3}V_l I_l \cos\varphi \\ Q_R &= 3V_{ph}I_{ph}\sin\varphi = \sqrt{3}V_l I_l \sin\varphi \end{aligned} \tag{1.17}$$

where the angle φ is the phase displacement between the voltage and current phases, or in other word, is the phase angle of the complex phase impedance. Also the apparent (overall) power for star and delta connections is defined as

$$S_{R,star} = \sqrt{P_{R,star}^2 + Q_{R,star}^2}; \quad S_{R,delta} = \sqrt{P_{R,delta}^2 + Q_{R,delta}^2} \tag{1.18}$$

According to the Eq. (1.17) in symmetrical and balanced conditions for both connections the following expression is true

$$S_R = 3V_{ph}I_{ph} = \sqrt{3}V_lI_l \tag{1.19}$$

The calculation methods of EPS are based on the properties of three-phase circuits with different types of supplying system's voltages and on some properties of various phase connections [19, 20]. Two elementary methods are presented below.

(α) *Calculation of symmetrical and balanced three-phase circuits.* In this case we assume that the voltage generators are symmetrical and know, and also the receptor is balanced. Symmetrical three-phase circuits consisting of many star or delta receptors are solved by the use of single-phase circuit, for example of phase 1. Based on the principles of this elementary method have been developed software programs for the analysis of three-phase circuits. This method contains following steps [21, 22]:

(α1) The coupled inductances in phase 1 are replaced with equivalent inductance $-\underline{Z}_m$. Based on three-phase symmetrical currents property $\underline{I}_1 + \underline{I}_2 + \underline{I}_3 = 0$, then the voltage across the coupled inductance in phase 1 can be expressed as $\underline{U}_{m,1} = \underline{Z}_m\underline{I}_2 + \underline{Z}_m\underline{I}_3 = \underline{Z}_m(\underline{I}_2 + \underline{I}_3) = -\underline{Z}_m\underline{I}_1$, where the complex mutual impedance is $\underline{Z}_m = \pm j\omega M$, and M is the mutual inductance which can be considered in aiding (+) or opposite (−) direction. Analogous equivalences are applied to couple inductances of other two phases.

(α2) All the delta connections are replaced with equivalent star connections, by using the relation $\underline{Z}_Y = \underline{Z}_\Delta/3$, where $\underline{Z}_Y$ and $\underline{Z}_\Delta$ are the phase impedances of star respectively delta connections.

(α3) In a symmetrical and balanced three-phase star circuits all neutral points have the same potential and $\underline{I}_1 + \underline{I}_2 + \underline{I}_3 = \underline{I}_N = 0$. Therefore all neutral points can be connected each other through a null resistance wire without modifying the three-phase circuit voltages and currents. Such the single-phase circuit becomes a closed-loop.

(α4) The voltages and currents of phase 1 are calculated by using the Kirchhoff's laws in single-phase circuit. Finally on determine the other two phases voltages and currents through the use of phase operator a, according to the type of phase-sequences of the voltages generator system.

(α5) By using Eqs. (1.17) and (1.19) the overall active, reactive, apparent and complex absorbed power are calculated.

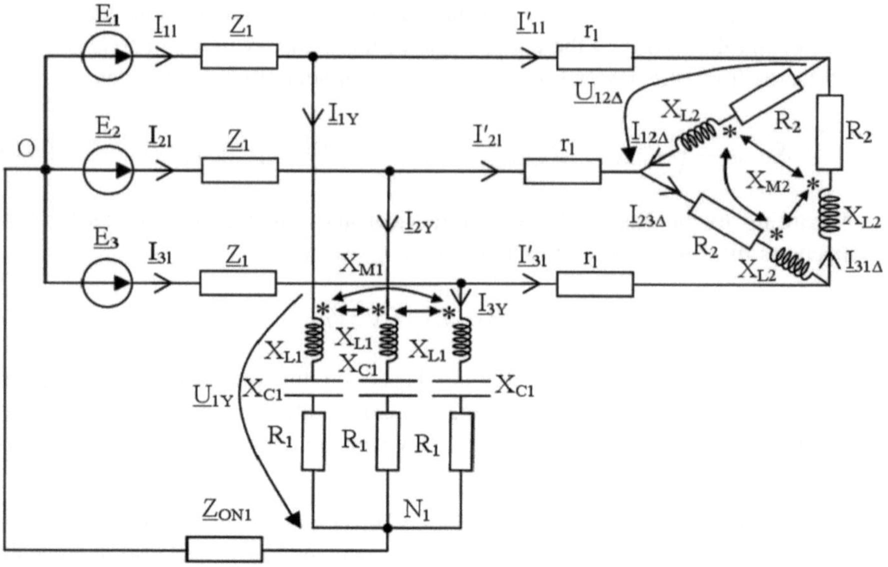

Fig. 1.6 Symmetrical and balanced three-phase system

Example 1.1 Let us consider a symmetrical three-phase generator in star connection shown in Fig. 1.6. Its positive phase-sequence voltages system are $\underline{E}_1 = 400\sqrt{2}$, $\underline{E}_2 = 400\sqrt{2}e^{-j\frac{2\pi}{3}}, \underline{E}_3 = 400\sqrt{2}e^{+j\frac{2\pi}{3}}$ supplying through a transmission line impedance $Z_1 = 40 + 80j$ first star connection receptor $R_1 = 40\ \Omega$, $X_{L1} = X_{C1} = X_{M1} = 40\ \Omega$, neutral wire impedance $\underline{Z}_{ON1} = 10(1-j)\ \Omega$, and through another transmission line impedance $r_1 = 10\ \Omega$, the second delta connection receptor $R_2 = 90\ \Omega$, $X_{L2} = 150\ \Omega$ and $X_{M2} = 30\ \Omega$.

In order to calculate the generator line currents ($\underline{I}_{11}$, $\underline{I}_{21}$, $\underline{I}_{31}$), the phase currents ($\underline{I}_{1Y}$, $\underline{I}_{2Y}$, $\underline{I}_{3Y}$) and voltages ($\underline{U}_{1Y}$, $\underline{U}_{2Y}$, $\underline{U}_{3Y}$) of star receptor, the phase currents ($\underline{I}_{12\Delta}$, $\underline{I}_{23\Delta}$, $\underline{I}_{31\Delta}$) and voltages ($\underline{U}_{12\Delta}$, $\underline{U}_{23\Delta}$, $\underline{U}_{31\Delta}$) of delta receptor and the overall absorbed powers, the above mentioned elementary method is applied:

(α1) The equivalences of coupled inductances of star and delta connections are presented in Fig. 1.7;
(α2) Replacement of delta connection with equivalent star connection whose neutral point is N_2 is shown in Fig. 1.8. There the values of equivalent impedances are

$$\underline{Z}_{1Y} = R_1 + j(X_{L1} - X_{M1} - X_{C1}) = 40(1-j)\Omega$$

$$\underline{Z}_{2Y_e} = \frac{\underline{Z}_{2\Delta}}{3} = \frac{R_2 + j(X_{L2} - X_{M2})}{3} = \frac{90 + 120j}{3} = 30 + 40j\Omega$$

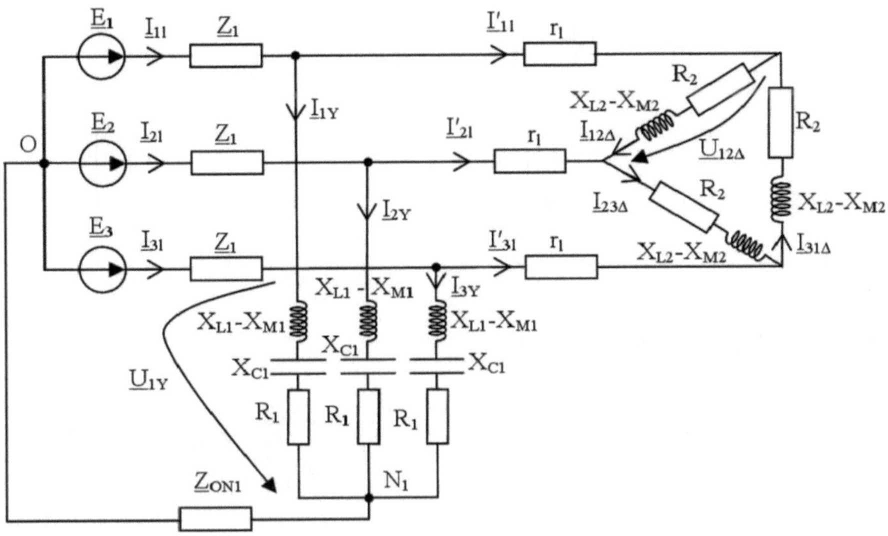

Fig. 1.7 Equivalent three-phase circuit

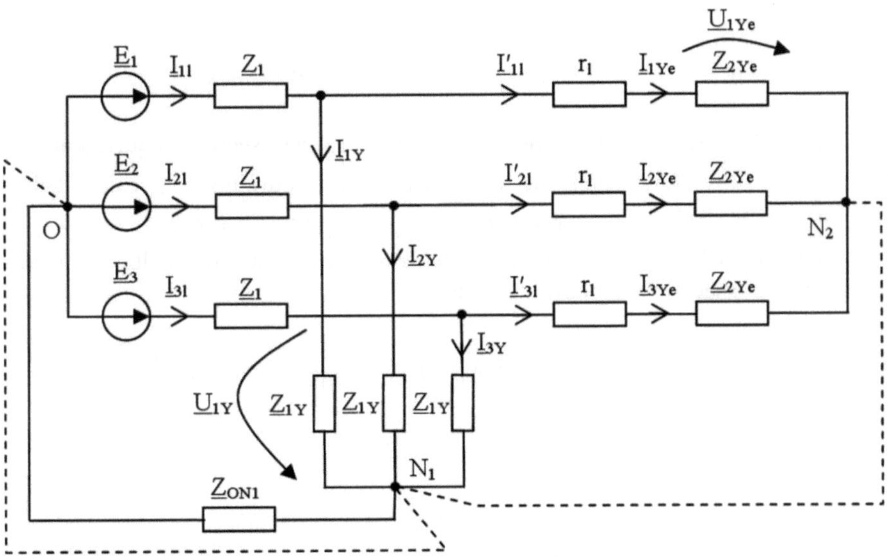

Fig. 1.8 Equivalence delta—star

(α3) A null resistance wire is introduced in order to connect all the neutral points O, N and N_1 (dash line in Fig. 1.8);
(α4) The single-phase circuit of phase 1 is shown in Fig. 1.9. By using Kirchhoff's laws one obtains

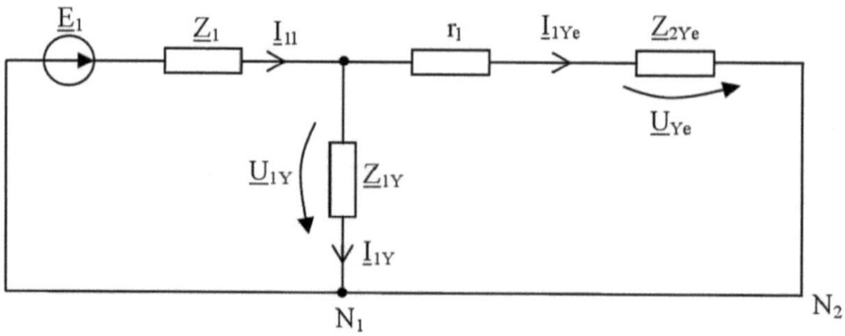

Fig. 1.9 Single-phase circuit of phase 1

$$\underline{I}_{11} = \frac{\underline{E}_1}{\underline{Z}_1 + \frac{\underline{Z}_{1Y}(r_1 + \underline{Z}_{2Y_e})}{\underline{Z}_{1Y} + r_1 + \underline{Z}_{2Y_e}}} = 5e^{-j\frac{\pi}{4}}\ \text{A}; \quad \underline{I}_{1Y} = \underline{I}_{11} \cdot \frac{r_e + \underline{Z}_{1Ye}}{\underline{Z}_{1Y} + r_e + \underline{Z}_{2Ye}} = \frac{5\sqrt{2}}{2} A$$

The currents of other two phases of first star receptor are obtained by using the phase operator

$$\underline{I}_{2Y} = \underline{I}_{1Y}e^{-j\frac{2\pi}{3}} = \frac{5\sqrt{2}}{2}e^{-j\frac{7\pi}{6}}\ \text{A}; \quad \underline{I}_{3Y} = \underline{I}_{1Y}e^{j\frac{2\pi}{3}} = \frac{5\sqrt{2}}{2}e^{j\frac{\pi}{6}}\ \text{A}$$

From KCL results the line 1 current of the second equivalent star $\underline{I}'_{11} = \underline{I}_{1Ye} = \frac{5\sqrt{2}}{2}e^{-j\frac{\pi}{2}}\text{A}$. Then the other two phase currents are calculated as $\underline{I}'_{21} = \underline{I}_{2Ye} = \underline{I}'_{11}e^{-j\frac{2\pi}{3}} = \frac{5\sqrt{2}}{2}e^{-j\frac{7\pi}{6}}\text{A}$, $\underline{I}'_{31} = \underline{I}_{3Ye} = \underline{I}'_{11}e^{j\frac{2\pi}{3}} = \frac{5\sqrt{2}}{2}e^{j\frac{\pi}{6}}\text{A}$.

If the properties of symmetrical systems are used thus the delta connection receptor is crossed by the phase currents

$$\underline{I}_{12\Delta} = \underline{I}_{1Ye}\sqrt{3}e^{-j\frac{\pi}{6}} = \frac{5}{2}\sqrt{6}e^{-j\frac{2\pi}{3}}\text{A}, \quad \underline{I}_{23\Delta} = \underline{I}_{2Ye}\sqrt{3}e^{-j\frac{\pi}{6}} = \frac{5\sqrt{6}}{2}e^{-j\frac{4\pi}{3}}\text{A},$$
$$\underline{I}_{31\Delta} = \underline{I}_{3Ye}\sqrt{3}e^{-j\frac{\pi}{6}} = \frac{5\sqrt{6}}{2}\text{A}$$

First star receptor has the phase voltages

$$\underline{U}_{1Y} = \underline{Z}_{1Y}\underline{I}_{1Y} = 200e^{-j\frac{\pi}{4}}\text{V}, \quad \underline{U}_{2Y} = \underline{U}_{1Y}e^{-j\frac{2\pi}{3}} = 200e^{-j\frac{11\pi}{12}}\text{V},$$
$$\underline{U}_{3Y} = \underline{U}_{1Y}e^{j\frac{2\pi}{3}} = 200e^{j\frac{5\pi}{12}}\text{V}$$

and also the phase voltages of delta receptor are given by

$$\underline{U}_{12\Delta} = \underline{Z}_{1\Delta}\underline{I}_{12\Delta} = 375\sqrt{6}e^{j\left(-\frac{2\pi}{3}+\arctan\frac{4}{3}\right)}\text{V}, \quad \underline{U}_{23\Delta} = \underline{U}_{12\Delta}e^{-j\frac{2\pi}{3}} = 375\sqrt{6}e^{j\left(-\frac{4\pi}{3}+\arctan\frac{4}{3}\right)}\text{V}$$

$$\underline{U}_{31\Delta} = \underline{U}_{12\Delta}e^{j\frac{2\pi}{3}} = 375\sqrt{6}e^{j\left(\arctan\frac{4}{3}\right)}\text{V}$$

(α5) The overall absorbed active, reactive, apparent and complex powers by receptors and transmission lines are respectively

$$\begin{aligned}
P_{abs} &= 3\text{Re}[\underline{Z}_e]\underline{I}_{11}^2 + 3\text{Re}\left[\underline{Z}_{1y}\right]\underline{I}_{1y}^2 + 3r_e\underline{I}_{11}^2 + 3\text{Re}[\underline{Z}_{2\Delta}]\underline{I}_{12\Delta}^2 = 6000\text{ W}\\
Q_{abs} &= 3\text{Im}[\underline{Z}_1]\underline{I}_{11}^2 + 3\text{Im}\left[\underline{Z}_{1y}\right]\underline{I}_{1y}^2 + 3\text{Im}[\underline{Z}_{2\Delta}]\underline{I}_{12\Delta}^2 = 6000\text{ VAr}\\
S_{abs} &= 3|\underline{Z}|_1\cdot\underline{I}_{11}^2 + 3|\underline{Z}|_{1y}\cdot\underline{I}_{1y}^2 + 3r_e\cdot\underline{I}_{11}'^2 + 3|\underline{Z}|_{12\Delta}\cdot\underline{I}_{12\Delta}^2 = \sqrt{P_{abs}^2 + Q_{abs}^2} = 6000\sqrt{2}\text{ VA}\\
\underline{S}_{abs} &= P + jQ = 6000 + 6000j
\end{aligned}$$

These powers are received from the generator whose overall complex power is

$$\underline{S}_{gen} = \underline{E}_1\underline{I}_{11}^* + \underline{E}_2\underline{I}_{21}^* + \underline{E}_3\underline{I}_{31}^* = 3\underline{E}_1\underline{I}_{11}^* = 6000 + 6000j$$

hence the conservation of active and reactive powers is proved

$$P_{gen} = \text{Re}\left[\underline{S}_{gen}\right] = 6000\text{ W} \equiv P_{abs};\ Q_{gen} = \text{Im}\left[\underline{S}_{gen}\right] = 6000\text{ VAr} \equiv Q_{abs}$$

Due to the losses in transmission lines, the active power transmission efficiency can be calculated as

$$\eta = \frac{3\text{Re}[\underline{Z}_{1Y}].\underline{I}_{1Y}^2 + 3\text{Re}[\underline{Z}_{2\Delta}].\underline{I}_{12\Delta}^2}{P_{abs}} = 0.68$$

Also the power factor (PF, also λ or $\cos\varphi$) of the considered EPS is defined as

$$\lambda = \cos\varphi = \frac{P}{S} = 0.706$$

(β) *Calculation of asymmetrical three-phase circuits by using the method of symmetrical components.* In EPS two kinds of asymmetry transverse and longitudinal may occur. Transverse asymmetry occurs when an unbalanced receptor is connected to a symmetrical three-phase network. Such an unbalanced load may take the form of asymmetrical short-circuits as line-to-line, one or two line-to-earth short-circuits. Longitudinal asymmetry occurs when the phases of a transmission line contain un-equal impedances (unsymmetrical section of transmission line) or when an open-circuit occurs in one or two phases. By the elementary method of symmetrical components an asymmetrical three-phase set of currents or voltages can be decomposed into three symmetrical systems positive, negative and zero

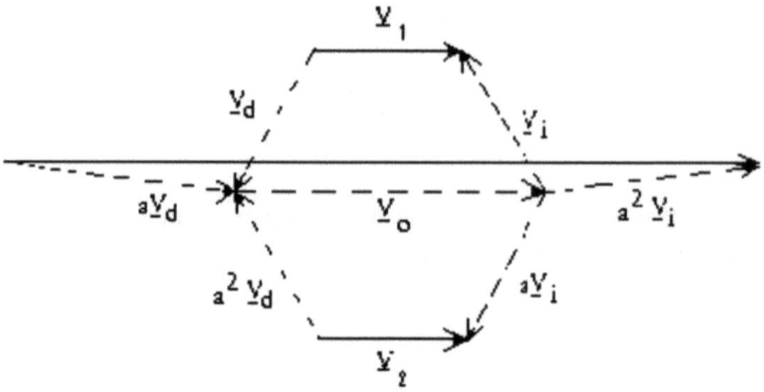

Fig. 1.10 Decomposition of an unsymmetrical system in three symmetrical systems

phase-sequence which are called symmetrical components [23]. Since it is based on superposition theorem the decomposition can be applied only to linear circuits.

It is demonstrates that any asymmetrical three-phase system $\underline{V}_1, \underline{V}_2, \underline{V}_3$ can be explained as the sum of three symmetrical systems: positive, negative and zero phase-sequences [24, 25], so

$$\begin{aligned} \underline{V}_1 &= \underline{V}_{o1} + \underline{V}_{d1} + \underline{V}_{i1} \\ \underline{V}_2 &= \underline{V}_{o2} + \underline{V}_{d2} + \underline{V}_{i2} \\ \underline{V}_3 &= \underline{V}_{o3} + \underline{V}_{d3} + \underline{V}_{i3} \end{aligned} \tag{1.20}$$

where

$$\underline{V}_{d1} = \underline{V}_d \ , \ \underline{V}_{d2} = \underline{V}_d \cdot e^{-j\frac{2\pi}{3}} = a^2 \underline{V}_d \ , \ \underline{V}_{d1} = \underline{V}_d \cdot e^{j\frac{2\pi}{3}} = a \underline{V}_d \tag{1.21}$$

$$\underline{V}_{i1} = \underline{V}_i \ , \ \underline{V}_{i2} = \underline{V}_i \cdot e^{j\frac{2\pi}{3}} = a \cdot \underline{V}_i \ , \ \underline{V}_{i1} = \underline{V}_i \cdot e^{-j\frac{2\pi}{3}} = a^2 \cdot \underline{V}_i \tag{1.22}$$

$$\underline{V}_{o1} = \underline{V}_{o2} = \underline{V}_{o3} = \underline{V}_o \tag{1.23}$$

Taking into account the relations (1.21)–(1.23), the decomposition (1.21)—shown in Fig. 1.10—can be rewritten in the form

$$\begin{aligned} \underline{V}_1 &= \underline{V}_o + \underline{V}_d + \underline{V}_i \\ \underline{V}_2 &= \underline{V}_o + a^2 \cdot \underline{V}_d + a \cdot \underline{V}_i \\ \underline{V}_3 &= \underline{V}_o + a \cdot \underline{V}_d + a^2 \cdot \underline{V}_i \end{aligned} \tag{1.24}$$

If it is considered know the asymmetrical system $\underline{V}_1, \underline{V}_2, \underline{V}_3$, then zero ($\underline{V}_o$), positive ($\underline{V}_d$) and negative-sequences ($\underline{V}_i$) are calculated as

$$\begin{aligned}\underline{V}_o &= \frac{1}{3}(\underline{V}_1 + \underline{V}_2 + \underline{V}_3)\\ \underline{V}_d &= \frac{1}{3}(\underline{V}_1 + a \cdot \underline{V}_2 + a^2 \cdot \underline{V}_3)\\ \underline{V}_i &= \frac{1}{3}(\underline{V}_1 + a^2 \cdot \underline{V}_2 + a \cdot \underline{V}_3)\end{aligned} \tag{1.25}$$

In order to evaluate the state of asymmetry, are defined two dimensionless parameters: the coefficient of dissymmetry $\varepsilon_d = \frac{V_i}{V_d}$, and the coefficient of asymmetry $\varepsilon_a = \frac{V_o}{V_d}$, where V_o, V_d, V_i are the rms values of zero, positive and negative-sequences. In real applications a system is considered symmetrical if ε_d and ε_a have values lower than 0.05.

The calculation of three-phase systems operating under asymmetrical conditions can be made based on the superposition theorem: are calculated separately each symmetrical component, and finally these components gather.

Example 1.2 Let us consider a balanced star receptor supplied by an asymmetrical system of phase voltages $\underline{U}_{10}$, $\underline{U}_{20}$, $\underline{U}_{30}$. The calculation of phase currents $\underline{I}_1$, $\underline{I}_2$, $\underline{I}_3$ is made by method of symmetrical components. Asymmetrical system shown in Fig. 1.11a, is decomposed in three symmetrical systems positive (Fig. 1.11b), negative (Fig. 1.11c) and zero-sequences (Fig. 1.11d).

Since the receiver is balanced, the current through the neutral wire is zero for positive and negative component, and for zero-sequence component is 3 times the

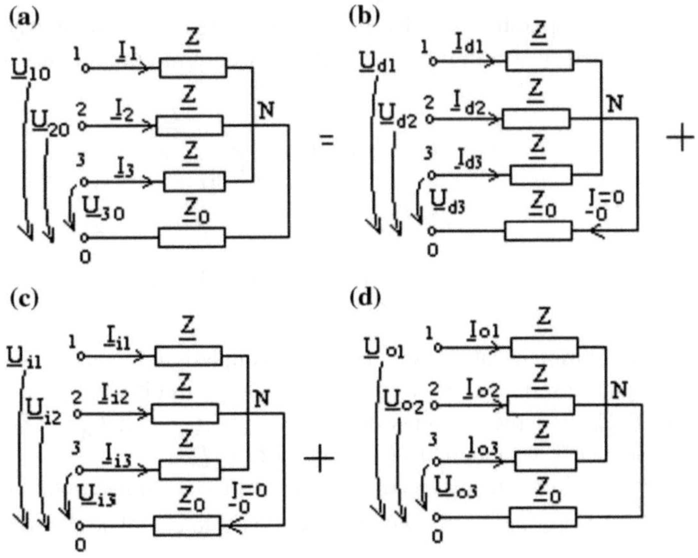

Fig. 1.11 Decomposition of the **a** unsymmetrical system, **b** positive, **c** negative, and **d** zero-sequences

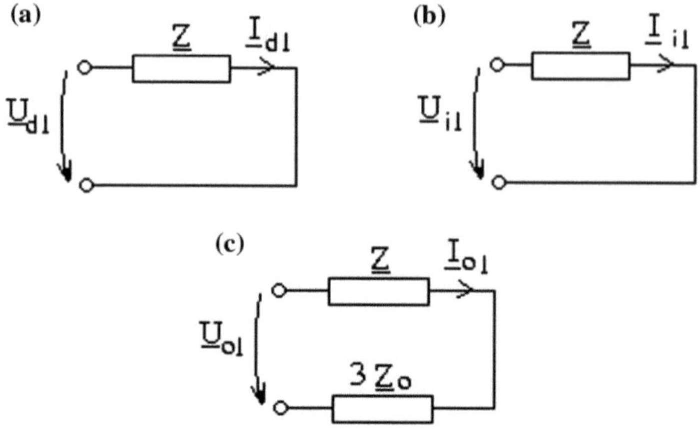

Fig. 1.12 Single phases of **a** positive, **b** negative, and **c** zero-sequence

phase current because the phase currents are equal. The reducing of the system to a single phase is made by using the equivalent circuit shown in Fig. 1.12: single phase of positive (Fig. 1.12a), negative (Fig. 1.12a), and zero-sequence (Fig. 1.12c).

Phase voltages corresponding to each circuit are respectively

$$\underline{U}_{d_1} = \underline{Z} \cdot \underline{I}_{d_1}; \quad \underline{U}_{i_1} = \underline{Z} \cdot \underline{I}_{i_1}; \quad \underline{U}_{o_1} = (\underline{Z} + 3\underline{Z}_0) \cdot \underline{I}_{o_1} \tag{1.26}$$

The two other phase currents for positive and negative sequence are calculated by using the phase operator. Finally the phase currents of initial asymmetrical system are calculated by summing of symmetrical current components as

$$\begin{aligned} \underline{I}_1 &= \underline{I}_{o_1} + \underline{I}_{d_1} + \underline{I}_{i_1} = \frac{\underline{U}_{o_1}}{\underline{Z} + 3\underline{Z}_0} + \frac{\underline{U}_{d_1}}{\underline{Z}} + \frac{\underline{U}_{i_1}}{\underline{Z}} \\ \underline{I}_2 &= \underline{I}_{o_2} + \underline{I}_{d_2} + \underline{I}_{i_2} = \frac{\underline{U}_{o_1}}{\underline{Z} + 3\underline{Z}_0} + a^2 \cdot \frac{\underline{U}_{d_1}}{\underline{Z}} + a \cdot \frac{\underline{U}_{i_1}}{\underline{Z}} \\ \underline{I}_3 &= \underline{I}_{o_3} + \underline{I}_{d_3} + \underline{I}_{i_3} = \frac{\underline{U}_{o_1}}{\underline{Z} + 3\underline{Z}_0} + a \cdot \frac{\underline{U}_{d_1}}{\underline{Z}} + a^2 \cdot \frac{\underline{U}_{i_1}}{\underline{Z}} \end{aligned} \tag{1.27}$$

If the initial asymmetrical system contains coupled inductances $\underline{Z}_m$ then for positive and negative-sequence is used the equivalent circuit where the equivalent impedance is $\underline{Z}$–$\underline{Z}_m$ and the voltages are $\underline{U}_d = (\underline{Z} - \underline{Z}_m) \cdot \underline{I}_d$; $\underline{U}_i = (\underline{Z} - \underline{Z}_m) \cdot \underline{I}_i$. Also for zero-sequence is obtained $\underline{U}_o = (\underline{Z} + 2\underline{Z}_m) \cdot \underline{I}_o$.

The delta connections are replaced by equivalent star connections considering relation $\underline{Z}_Y = \frac{\underline{Z}_\Delta}{3}$. Equations (1.15) and (1.16) are used to calculate the overall P and Q powers.

In conclusion, the elementary method of symmetrical components comprises the following steps:

(i) asymmetrical supplied voltage system is decomposed in three symmetrical components;
(ii) after the equivalence of coupled inductances and of delta-star connections, are calculated the impedances of positive, negative and zero-sequences;
(iii) symmetrical components of currents system are calculated by using single phase circuit and phase operator;
(iv) asymmetrical currents of initial system are calculated by adding the symmetrical current components.

1.4 State of the Art: Measurement Methods of Power Systems Parameters

In EPS systems it is important to measure its electrical characteristics [26]. In this section are described and illustrated measurement methods of voltage, current, active power and frequency.

Regardless of what parameter is measured, before the actual operation is done there are some precautions to be considered:

- the level of the measured parameter
- the scale of the measuring apparatus

In order to have a safe measuring operation, the maximum value of the measured parameter should not be higher than the maximum indication of the measuring device. If this precaution is not carefully considered, it could lead to permanent damage of the measuring instrument.

1.4.1 Voltage Measurement

(i) *Mono-phase circuits.* In mono-phase circuits, the voltage measurement consists in connecting a voltmeter at the terminals of the voltage source [27] as indicated in Fig. 1.13. Voltmeter V indicates the rms value of the voltage source.
(ii) *Three-phase circuits* contain two types of voltages, as are described in Sect. 1.3: line voltage—that is measured between two lines of the supply system, and phase voltage—that is measured between one line and the common point.

In Fig. 1.14 is shown the measurement principle of line and phase voltages. There voltmeter V_1 measures the line voltage between the phases U and V, while

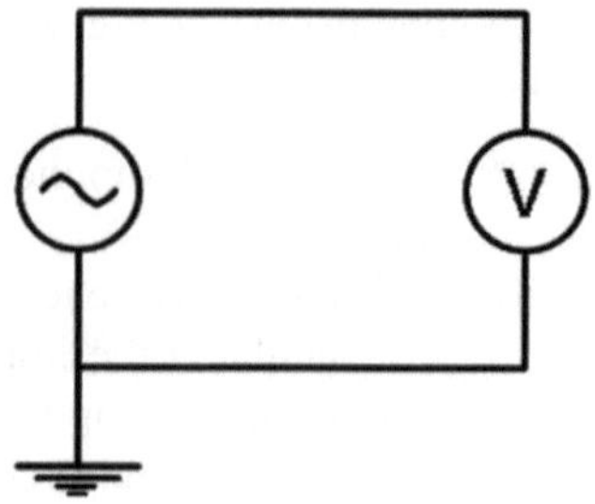

Fig. 1.13 Voltage measurement in mono-phase AC circuits

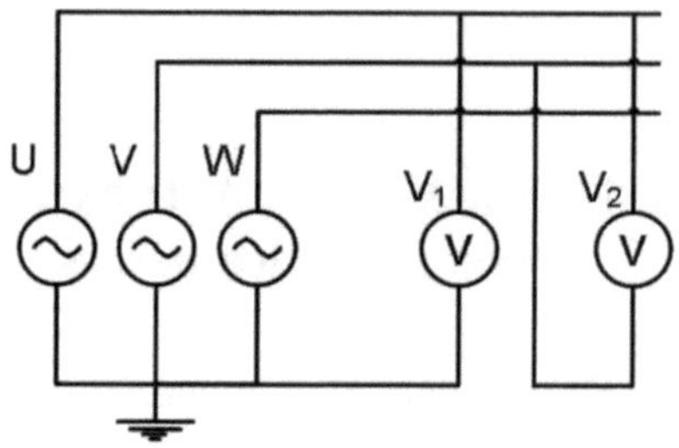

Fig. 1.14 Phase and line voltage measurements

voltmeter V_2 measures the phase voltage between the phases U and the common point.

1.4.2 Current Measurement

The current appears in an electrical circuit when it is closed on a load. If the circuit is open, so that no load is connected to the source's terminals, there is no current flow. The current is measured by connecting an ammeter in series with the load [28].

(i) *Mono-phase circuits.* In mono-phase circuits the current in measured by connecting in series an ammeter with the load. In Fig. 1.15 is indicated the procedure to measure the rms value of current across the load Z.

(ii) *Three-phase circuits.* In three-phase circuits the current is measured in the same way as in mono-phase circuits. The measuring of the current for each phase of the circuit there will be used three ammeters, one for each circuit phases. In Fig. 1.16 the ammeters A_1, A_2 and A_3 measure the rms values of current through the loads Z_1, Z_2 and Z_3, respectively.

If the loads are identical on all the three phases, they constitute a balanced circuit, it is enough to connect in series only one ammeter on only one phase, as the other rms values of currents are equals. This situation is displayed in Fig. 1.17.

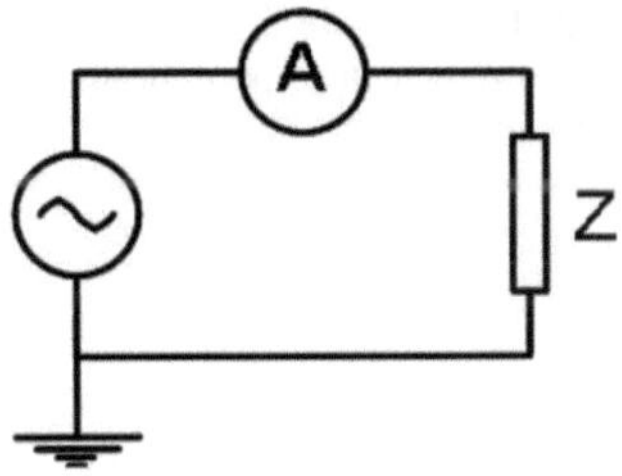

Fig. 1.15 Current measurement in mono-phase circuit

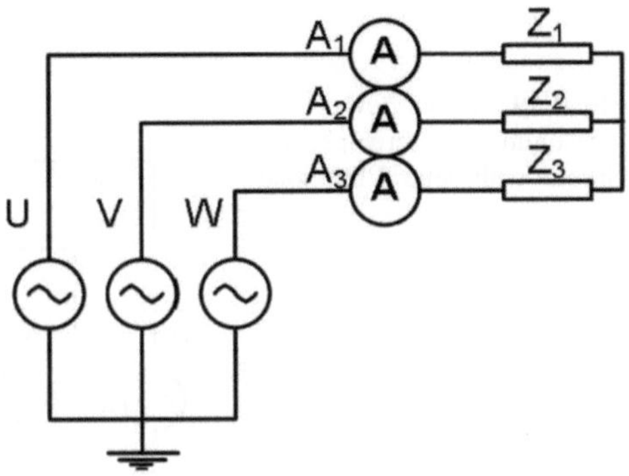

Fig. 1.16 Current measurement in three-phase circuit

Fig. 1.17 Current measurement for a three-phase balanced circuit

1.4.3 Active Power Measurement

(i) *Active power measurement in mono-phase circuits* can be done with the wattmeter. The wattmeter has two coils [29]: current and voltage. The voltage coil of the wattmeter is connected in parallel with the load, while the current coil is connected in series with the load, as displayed in Fig. 1.18.

(ii) *Active power measurement in three-phase circuits* is quite similar regardless of load configuration as far as procedure [30]. This consists in connecting a wattmeter in the circuit with the current coil in series with the load and with

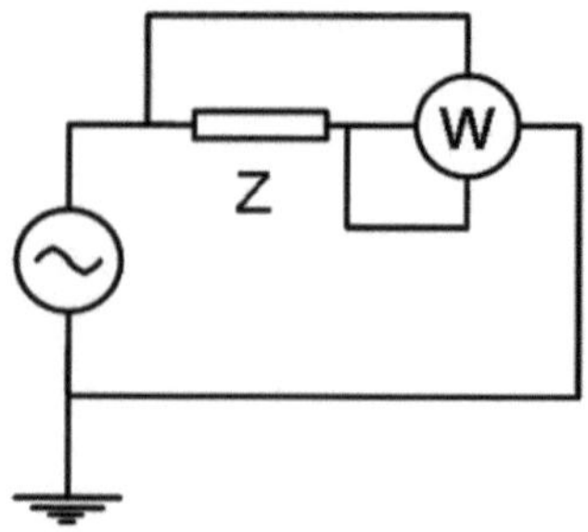

Fig. 1.18 Active power measurement in mono-phase circuits

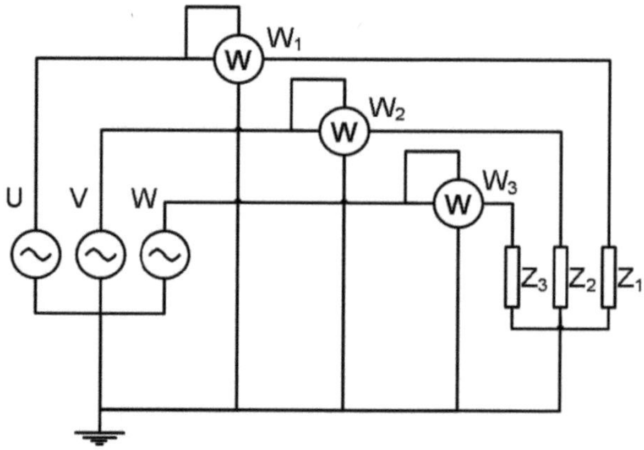

Fig. 1.19 Active power measurement in three-phase circuits with neutral line load

the voltage coil in parallel with the load. The three-phase circuits have some particularities due to load configurations.

If the load has a neutral line, such as star connected loads, then the power can be measured by using three wattmeters. The voltage coils of the wattmeters are connected between each phase and the neutral point of the load and voltage source [4]. The current coils are connected in series with the load on each phase, as displayed in Fig. 1.19.

If the load does not have an accessible neutral point or is delta connected, in order to measure the power using three wattmeters, these are connected such as to construct an artificial neutral point for them. This situation is indicated in Fig. 1.20. Thus the active power absorbed by the load can be expressed by

$$P = P_1 + P_2 + P_3 \tag{1.28}$$

If a neutral point is created, it is preferred that it has the same potential as one of the source phase as indicated in Fig. 1.21. In this situation the indication of the third wattmeter would be zero as its voltage coil would be connected at the same

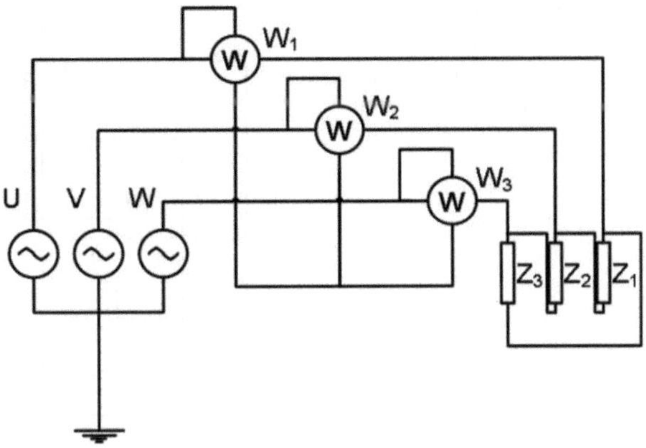

Fig. 1.20 Power measurement in three-phase circuits without neutral line load

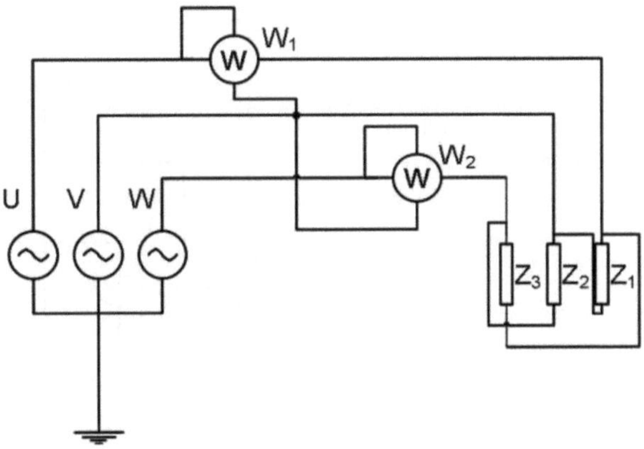

Fig. 1.21 Power measurement in three-phase circuits without neutral line load using two wattmeters

potential. Because of this, it is eliminated from the circuit as not being useful. The measured active power is the sum of the indications of each wattmeter, expressed as

$$P = P_1 + P_2 \tag{1.29}$$

A different situation is the case of balanced loads and symmetrical voltages on each source phase. In this case it is enough to use one wattmeter as indicated in Fig. 1.22. In this situation, the total power is determined by multiplying the indication of the wattmeter by three, as in expressed in Eq. (1.30)

$$P = 3 \cdot P_1 \tag{1.30}$$

The situation presented in Fig. 1.22 is valid if the load has an accessible neutral point. In case that the load does not have an accessible neutral point, or is delta connected, then it is created an artificial neutral point by using two additional

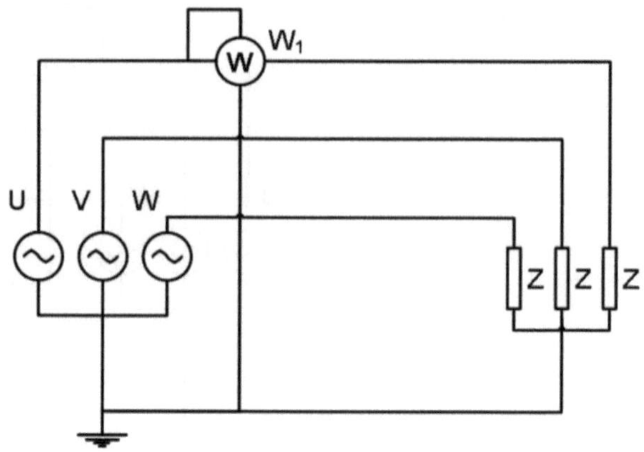

Fig. 1.22 Power measurement for balanced load and symmetrical source voltage with natural neutral point

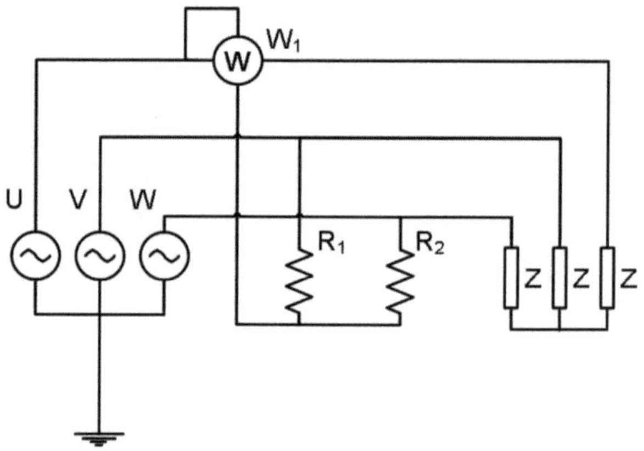

Fig. 1.23 Power measurement for balanced load and symmetrical source voltage with an artificial neutral point

resistors as indicated in Fig. 1.23. The values of the additional resistors R_1 and R_2 are equal to the value of the resistance of the voltage circuit of the wattmeter.

1.5 State of the Art: Standards for Power Systems Parameters

Electrical energy needed to power industrial or household electric consumers, as all the products, should satisfy same quality demands. These demands, like good *PQ* ensure the electrical equipment is to operate without errors and this is a responsively of the supplier [17, 31].

On the other hand, an important part of the equipment in use today, in particular electronic and computer devices generates distortion of the voltage supply in the installation, because of its non-linear characteristics, i.e. it draws a non-sinusoidal current with a sinusoidal supply voltage. In this case the *PQ* is a responsively of the electricity user.

In consequence, maintaining satisfactory *PQ* is a joint responsibility for the supplier and the electricity user.

Standard IEC 60038 distinguishes "two different voltages in electrical networks and installations" [32]:

- supply voltage, which "is the line-to-line or line-to-neutral voltage at the point of common coupling (PCC), i.e. main supplying point of installation";
- utility voltage, which "is the line-to-line or line-to-neutral voltage at the plug or terminal of the electrical device".

The main document dealing with demands concerning the supplier's side is *standard EN 50160* [33] which characterize voltage parameters of electrical energy in public distribution systems.

According with *Standard EN 50160*, "the main voltage characteristics of public distribution systems are":

Supply voltage—"the root mean square (rms) value of the voltage at a given moment at the PCC, measured over a given time interval".

Nominal voltage of the system (U_n)—"the voltage by which a system is designated or identified and to which certain operating characteristics are referred".

Declared supply voltage (U_c)—"is normally the nominal voltage U_n of the system. If, by agreement between the supplier and the user, a voltage different from the nominal voltage is applied to the terminal, then this voltage is the declared supply voltage U_c".

Normal operating condition—"the condition of meeting load demand, system switching and clearing faults by automatic system protection in the absence of exceptional conditions due to external influences or major events".

Voltage variation—"is an increase or decrease of voltage, due to variation of the total load of the distribution system or a part of it".

Flicker—"impression of unsteadiness of visual sensation induced by a light stimulus, the luminance or spectral distribution of which fluctuates with time".

Flicker severity—"intensity of flicker annoyance defined by the UIE-IEC flicker measuring method and evaluated by the following signals":

- *Short term severity* (P_{st}) "measured over a period of ten minutes";
- *Long term severity* (P_{lt}) "calculated from a sequence of 12 P_{st}—values over a two-hour interval, according to the following expression":

$$P_{lt} = \sqrt[3]{\sum_{i=1}^{12} \frac{P_{sti}^3}{12}} \tag{1.31}$$

Supply voltage dip—"a sudden reduction of the supply voltage to a value between 90 and 1% of the declared voltage U_c, followed by a voltage recovery after a short period of time". Conventionally, the duration of a voltage dip is between "10 ms and 1 min". *The depth of a voltage dip* is defined as the "difference between the minimum rms voltage during the voltage dip and the declared voltage".

Supply interruption—is a condition in which "the voltage at the supply terminals is lower than 1% of the declared voltage U_c". A supply interruption is classified as:

- "*prearranged* in order to allow the execution of scheduled works on the distribution system, when consumers are informed in advance";
- "*accidental*, caused by permanent (a long interruption) or transient (a short interruption) faults, mostly related to external events, equipment failures or interference".

Temporary power frequency over-voltages—"have relatively long duration, usually of a few power frequency periods", and originate mainly from switching operations or faults, e.g. sudden load reduction, or disconnection of short circuits.

Transient over-voltages—"are oscillatory or non-oscillatory, highly damped, short over-voltages with a duration of a few milliseconds or less, originating from lightning or some switching operations", e.g. at switch-off of an inductive current.

Harmonic voltage—"a sinusoidal voltage with a frequency equal to an integer multiple of the fundamental frequency of the supply voltage". Harmonic voltages can be evaluated:

- "*individually* by their relative amplitude U_h related to the fundamental voltage U_1", where h is the order of the harmonic;
- "*globally*, usually by the total harmonic distortion factor THD_u", calculated using the following expression:

$$THD_u = \sqrt{\sum_{h=2}^{40} (U_h)^2 \Big/ U_1} \tag{1.32}$$

Inter-harmonic voltage—"is a sinusoidal voltage with frequency between the harmonics", i.e. the frequency is not an integer multiple of the fundamental.

Voltage unbalance—"is a condition where rms value of the phase voltages or the phase angles between consecutive phases in a three-phase system is not equal".

Standard EN 50160 gives the main voltage parameters and their permissible deviation ranges at the customer's PCC in public low voltage (LV) and medium voltage (MV) electricity distribution systems, under normal operating conditions (Fig. 1.24). In this circumstances, LV signify that the phase to phase nominal rms voltage does not exceed 1000 V and MV signify that the phase-to-phase nominal rms value is between 1 and 35 kV.

The comparison of the EN 50160 requirements with those of the EMC standards EN 61000, listed in Table 1.2 show significant differences in various parameters [33–35].

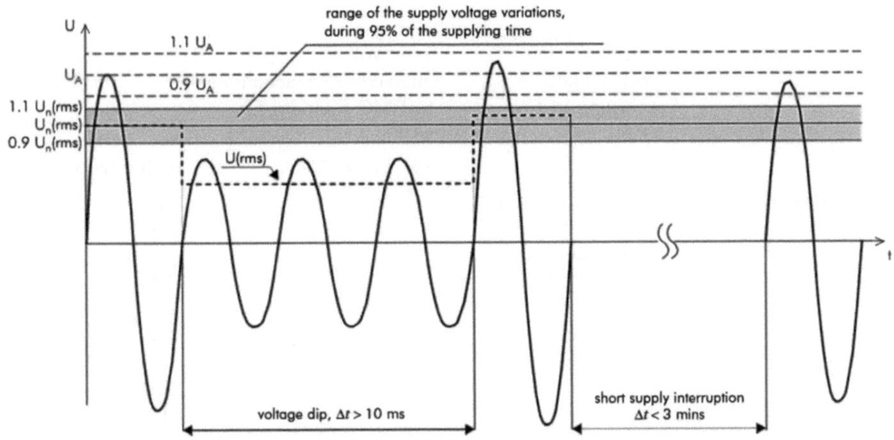

Fig. 1.24 Exemplification of a voltage dip and a short supply interruption, classified according to EN 50160; U_n—nominal voltage of the supply system (rms), U_A—amplitude of the supply voltage, U (rms)—the actual rms value of the supply voltage [3]

Table 1.2 Comparison of supply voltage requirements according to EN 50160 and the EMC standards EN 61000

No	Parameter	Supply voltage characteristics according to EN 50160	Low voltage characteristics according to EMC standard EN 61000	
			EN 61000-2-2	Other parts
1	Power frequency	LV, MV: mean value of fundamental measured over 10 s ±1% (49.5–50.5 Hz) for 99.5% of week –6%/+4% (47–52 Hz) for 100% of week	2%	
2	Voltage magnitude variations	LV, MV: ±10% for 95% of week, mean 10 min rms values (Fig. 1.1)		±10% applied for 15 min
3	Rapid voltage changes	LV: 5% normal 10% infrequently $P_{lt} \leq 1$ for 95% of week MV: 4% normal 6% infrequently $P_{lt} \leq 1$ for 95% of week	3% normal 8% infrequently $P_{st} < 1.0$ $P_{lt} < 0.8$	3% normal 4% maximum $P_{st} < 1.0$ $P_{lt} < 0.65$ (EN 61000-3-3) 3% (IEC 61000-2-12)

(continued)

Table 1.2 (continued)

No	Parameter	Supply voltage characteristics according to EN 50160	Low voltage characteristics according to EMC standard EN 61000	
			EN 61000-2-2	Other parts
4	Supply voltage dips	Majority: duration <1 s, depth <60% Locally limited dips caused by load switching on: LV: 10–50%, MV: 10–15% (Fig. 1.1)	urban: 1–4 months	up to 30% for 10 ms up to 60% for 100 ms (EN 61000-6-1, 6-2) up to 60% for 1000 ms (EN 61000-6-2)
5	Short interruptions of supply voltage	LV, MV: (up to 3 min) few tens—few hundreds/year Duration 70% of them <1 s		95% reduction for 5 s (EN 61000-6-1, 6-2)
6	Long interruption of supply voltage	LV, MV: (longer than 3 min) <10–50/year		
7	Temporary, power frequency over-voltages	LV: <1.5 kV rms MV: 1.7 U_c (solid or impedance earth) 2.0 U_c (unearthed or resonant earth)		
8	Transient over-voltages	LV: generally <6 kV, occasionally higher; rise time: ms—μs. MV: not defined		±2 kV, line-to-earth ±1 kV, line-to-line 1.2/50(8/20) T_r/T_h μs (EN 61000-6-1, 6-2)
9	Supply voltage unbalance	LV, MV: up to 2% for 95% of week, mean 10 min rms values, up to 3% in some locations	2%	2% (IEC 61000-2-12)
10	Harmonic voltage	LV, MV: (Table 1.2)	6%-5th, 5%-7th, 3.5%-11th, 3%-13th, *THD* <8%	5% 3rd, 6% 5th, 5% 7th, 1.5% 9th, 3.5% 11th, 3% 13th, 0.3% 15th, 2% 17th (EN 61000-3-2)
11	Inter-harmonic voltage	LV, MV: under consideration	0.2%	

Harmonic emissions are subject to various standards and regulations [36, 37]:

- "Compatibility standards for distribution networks";
- "Emissions standards applying to the equipment causing harmonics";
- "Recommendations issued by utilities and applicable to installations".

Currently, a triple system of standards and regulations is in force based on the documents listed below.

"Standards governing compatibility between distribution networks and products" determine the necessary compatibility between distribution networks and products:

- The harmonics caused by a device must not disturb the distribution network beyond certain limits;
- Each device must be capable of operating normally in the presence of disturbances up to specific levels;
- Standard IEC 61000-2-2 is applicable for public low-voltage power supply systems;
- Standard IEC 61000-2-4 is applicable for LV and MV industrial installations.

"Standards governing the quality of distribution networks" contain:

- Standard EN 50160 stipulates the characteristics of electricity supplied by public distribution networks;
- Standard IEEE 519 presents a joint approach between utilities and customers to limit the impact of non-linear loads. What is more, utilities encourage preventive action in view of reducing the deterioration of *PQ*, temperature rise and the reduction of power factor. They will be increasingly inclined to charge customers for major sources of harmonics.

"Standards governing equipment" contain

- Standard IEC 61000-3-2 for low-voltage equipment with rated current under 16 A;
- Standard IEC 61000-3-12 for low-voltage equipment with rated current higher than 16 A and lower than 75 A.

Maximum permissible harmonic levels

An estimation of typical harmonic contents often encountered in electrical distribution networks and the levels that should not be exceeded is presents in Table 1.3.

Example 1.3 For an assessment of supply voltage, three-phase network, 240 V/400 V are presented the measurement values which was done with PowerQ Plus MI 2392, a portable multifunction instrument for measurement and analysis of three-phase power systems shown in Fig. 1.25. The basic measurement time interval for: voltage, current, harmonics is a 10-cycle time interval. The 10-cycle measurement is resynchronized on each interval tick according to the IEC 61000-4-30 Class B [38, 39]. Measurement methods are based on the digital

Table 1.3 Maximum admissible harmonic voltages and distortion (%), *h*—order of harmonic [36, 37]

	LV	MV	HV	
Odd harmonics non-multiples of 3	5	6	5	2
	7	5	4	2
	11	3.5	3	1.5
	13	3	2.5	1.5
	$17 < h < 49$	$2.27\frac{17}{h} - 0.27$	$1.9\frac{17}{h} - 0.2$	$1.2\frac{17}{h}$
Odd harmonics multiples of 3	3	5	4	2
	9	1.5	1.2	1
	15	0.4	0.3	0.3
	21	0.3	0.2	0.2
	$21 < h < 45$	0.2	0.2	0.2
Even harmonics	2	2	1.8	1.4
	4	1	1	0.8
	6	0.5	0.5	0.4
	8	0.5	0.5	0.4
	$10 <= h <= 50$	$0.25\frac{10}{h} + 0.25$	$0.25\frac{10}{h} + 0.22$	$0.19\frac{10}{h} + 0.16$
THD_u	8	6.5	3	

(a) (b)

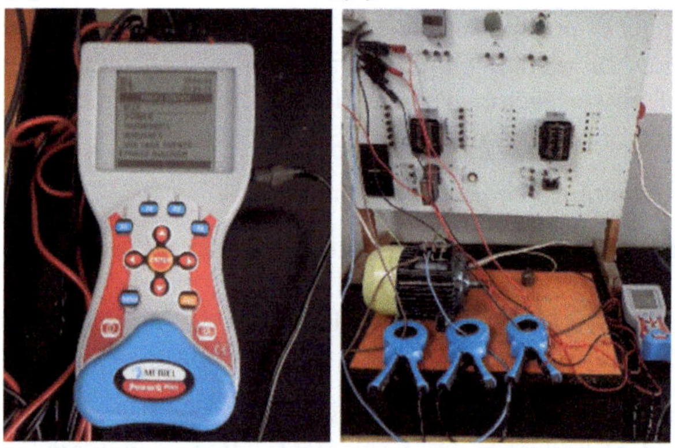

Fig. 1.25 **a** Power *Q*-meter, **b** Measurement stand

sampling of the input signals, synchronized to the fundamental frequency. Each input (3 voltages and 3 currents) is simultaneously sampled 1024 times in 10 cycles (Fig. 1.26).

The supply voltage, which is the line-to-line or line-to-neutral voltage at the PCC [38] i.e. main supplying point of installation. The measured values of supply voltage parameters without and with load (a three-phase asynchronous motor) are presented in Tables 1.4 and 1.5, respectively.

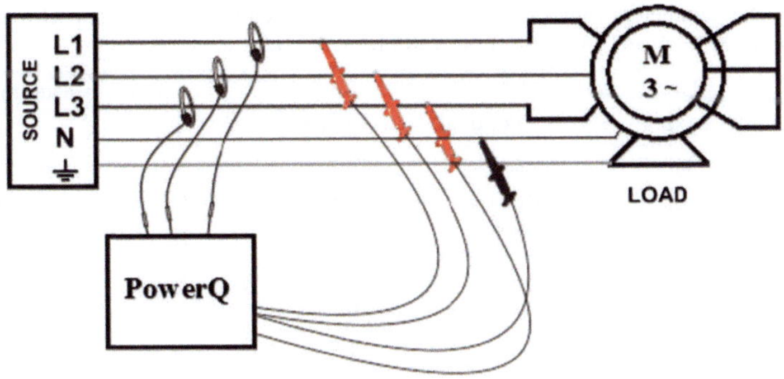

Fig. 1.26 Wiring diagram for U, I, f measurements

Table 1.4 The parameters values of supply voltage—without load

4 W	L1	L2	L3	Unit of measurement
U_L	235.8	235.3	234.4	V
THD_u	2.5	2.6	2.3	%
h_1	100	100	100	%
h_2	0	0	0	%
h_3	0.3	0.3	0.2	%
h_4	0	0	0	%
h_5	2.0	2.1	2.0	%
h_6	0	0	0	%
h_7	1.2	1.3	1.2	%
h_8–h_{10}	0	0	0	%
h_{11}	0.1	0.2	0.1	%
h_{12}	0	0	0	%
h_{13}	0	0.1	0	%
h_{14}–h_{18}	0	0	0	%
h_{19}	0	0	0.1	%
h_{20}–h_{50}	0	0	0	%
I_L	0	A		
THD_i	0	%		
f	49.98	Hz		
4 W	L12	L23	L31	Unit of measurement
U_{LL}	410.3	407.7	405.7	V
THD_u	2.5	2.4	2.3	%
I_0	0.5	A		
f	49.98	Hz		

Table 1.5 The parameters values of supply voltage—with load three-phase asynchronous electric motor 240 V/400 V, 6.6A/3.8 A, 1.5 kW, 1405 rpm, 50 Hz

4 W	L1	L2	L3	Unit of measurement
U_L	234.8	234.9	233.5	V
THD_u	2.3	2.4	2.3	%
h_1	100	100	100	%
h_2	0	0	0	%
h_3	0.2	0.2	0.2	%
h_4	0	0	0	%
h_5	1.8	2.0	1.9	%
h_6	0	0	0	%
h_7	1.2	1.1	1.2	%
h_8-h_{14}	0	0	0	%
h_{15}	0.1	0.0	0.0	%
h_{16}-h_{50}	0	0	0	%
I_L	2.3	2.4	2.5	A
f	49.98	Hz		
4 W	L12	L23	L31	Unit of measurement
U_{LL}	409.3	407.0	404.0	V
THD_u	2.3	2.6	2.5	%
I_0	0.6	A		
f	49.98	Hz		

In concordance with the measured values presented in Tables 1.4 1.5, then:

- voltage magnitude variations: ±10%, means range [216−264 V]/[360–440 V];
- harmonic voltage: 6%-5th, 5%-7th, 3.5%-11th, 3%-13th, THD_u <8%;
- frequency mean value of fundamental measured over 10 s, ±1% (49.5–50.5 Hz).

Thus the electric parameters of public distribution systems, in PCC and the utility voltage satisfy the requirements of standards set.

1.6 Importance of Reactive Power

Reactive energy and reactive power represent basic signals which are present in all alternating voltage installations, due to their nature and specificity, although they do not produce directly useful effects (light, heat, mechanical work etc.). Most of the time, the realization of useful effects is not possible without reactive energy consumption, taking into account the magnetization processes that take place in motor drives iron cores and in electrical transformers. Also, the leakage fluxes corresponding to the electric lines and to the coils determine reactive power consumption. The resistors, used as electrical energy receivers, consume only active energy, but its transfer through the installations placed upstream reactive determines energy and reactive power losses.

In general, one can say that, in nowadays electrical systems, the active energy and active power management take place only in the presence and with a big consumption of reactive energy and reactive power. From here it results also the concern that, the problems corresponding to producing and management of reactive power and reactive energy to be examined simultaneously with those related to active power and under higher efficiency taking into account their corresponding specificity. At the same time, one should notice also the signal that expresses the correlation between the two powers and energies categories that is the power factor PF, as a main signal for guiding the corresponding problems analysis.

As a general aspect for reactive power and reactive energy management in electrical installations, one considers as necessary, to underline the essential difference between the process required to produce active power and active energy that takes place, in principal, in electric power plants, belonging to the energetic system and the one corresponding to producing reactive energy and reactive power. This is due to that active energy production implies primary energy consumption, which is not the case of the reactive energy, excepting the case for covering some active energy losses, reduced as signal.

The necessity to have a reactive power flow control arises from the fact that, practically, in each station, for solving sinusoidal states, the adopted solution is that of using capacitor banks. These are installed, usually, on LV mains power supply of the transformer power, where are connected the consumers from all or from a part of the enterprise. The purpose, with priority, is to obtain an overall power factor, at least equal to the neutral value (imposed by the electrical energy provider contract), in the point used for measuring the active and reactive energy consumption of the electrical installations placed downstream, avoiding the overcompensation situations. In general, for the energy provided to the consumers, the values obtained for the overall power are in 0.93 … 0.97 range.

Many factories use also automatic multi-step power factor correction systems (MSPFC), to maintain constant value for the power factor. This is the situation for long-term functioning in industrial installations, on the occasion of the current jobs. One can affirm this solution contributes to a rational management of the reactive energy and reactive power, representing a priority in the electrical energy sector.

The analysis of this aspect is done, in general, at the enterprises level, because the electrical energy consumers from this one, and especially the operation, with very large nominal power range, sometimes due to over-sizes, as well as the corresponding low-loaded functioning, have the biggest weight from the general reactive power consumption, sometimes up to 60–70%.

The transformers from the electroenergetical system take around 20% out from the reactive power, the rest of the consumption belonging to the other installations.

The solution of placing a capacitor bank in the transformer power substation under the name of global compensation, presents advantage only for upstream installations, exterior to the enterprise, respectively for generating, transport and distribution installations. Using this type of compensation, the advantage represented by reactive and active power and energy losses, does not appear also for enterprise's installations. From this point of view the installations from the enterprise do not benefit from the advantage given by reactive power local production.

Taking into account the importance of the contribution, given by reactive power compensation measures, also to the losses reduction in distribution installation inside the enterprise, is indicated that, for on-going processes, to examine also the possibilities to de-centralize the banks in proper installations, that is to use with more economical compensation types, respectively, the individual or by sector consumers.

It is necessary to notice that, a special importance for consumers and electroenergetical system reactive power management, the apparition of new types of receivers, with non-linear electrical characteristics critically influence the reactive power management. The presence of the harmonic state, in many cases, determines practically the total revision of the compensation used currently, by exclusively installing the capacitor banks. This fact is determined by the reciprocal influence of these two aspects, the reactive power compensation and the harmonic, implying the application of a unique measure, respectively that of using filters that include also the existing capacitor banks.

Due to the complexity of the two problems and especially to the necessity and of the urgency of solving a phenomena assembly, one will examine, as follows, the most important theoretical and technical-economical specific aspects, that can be used to establish the on-going processes. In all cases, one should analyze, together, the two problems, taking into account that practically all modern electrical energy consumers present a nonlinear characteristic.

1.6.1 Reactive Power Flow Effects Evaluation

Electric devices are designed at a certain apparent power S that is proportional to the product between the rms values corresponding to the voltage U and to the current I. The power flow in the electroenergetical system is accompanied, function of the electric energy consumer structure, by the active power flow P, reactive power flow Q and distorted power flow D. The only useful one is the active power flow and its corresponding share from the apparent power necessary is computed using the *power factor* PF defined as [40]:

$$PF = \frac{P}{S} = \frac{P}{\sqrt{P^2 + Q^2 + D^2}} \tag{1.33}$$

The weights corresponding to the reactive and distorted powers are estimated using the *reactive factor* ρ and *the distorted factor* τ for the permanent harmonic state, according to the relationships

Fig. 1.27 Significance of phase-shifts φ and ψ

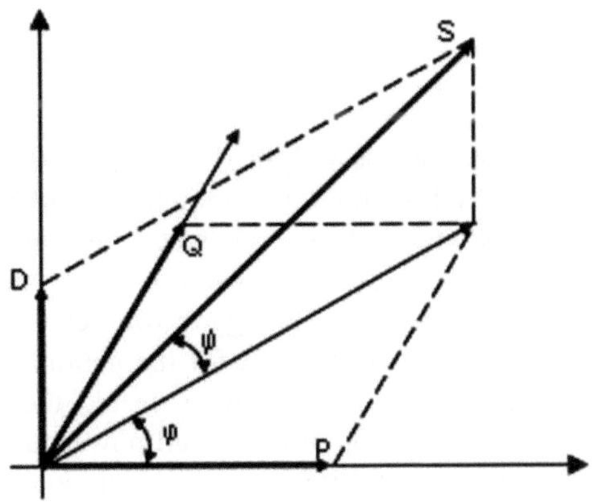

$$\rho = \frac{Q}{P} = \tan \varphi; \quad \tau = \frac{D}{\sqrt{P^2 + Q^2}} = \tan \psi \tag{1.34}$$

where the phasors P, Q, D form a three-orthogonal reference system, and the phase-shifts φ and ψ have the significance given in Fig. 1.27.

So, one obtains a new expression for the power factor $k = \cos \phi \cdot \cos \psi$. If one considers a sinusoidal permanent single-phase circuit, then: $D = 0 (\psi = 0)$ and $k = \cos \varphi$ so the power factor is numerically equal to the cosine of phase-shift angle between the voltage and the current.

For three-phase equilibrium circuits, linear and sinusoidal voltage powered, the power factor has exactly the same mathematical expression and significance as for single-phase circuit case. If the electric circuits are slightly asymmetrical, then the voltage-current phase-shifts are different from a phase to another $\varphi_1 \neq \varphi_2 \neq \varphi_3$ and the power factor will be:

$$PF = \frac{P_1 + P_2 + P_3}{\sqrt{(P_1 + P_2 + P_3)^2 + (Q_1 + Q_2 + Q_3)^2}} \tag{1.35}$$

where P_j and Q_j are the active, respectively the reactive powers corresponding to each one of the phases (j = 1, 2, 3).

The relationships above define the *instantaneous power factor* that corresponds to a certain moment form the consumer's installation functioning. Because the load presents fluctuations, the current legislation recommends the determination of the *weighted overall power factor* based on the active and reactive energy consumption E_a, E_r from a certain period, in the hypothesis that the receivers behave as a linear three-phase load, in equilibrium, working in sinusoidal permanent state:

$$\cos\varphi = \frac{E_a}{\sqrt{E_a^2 + E_r^2}} \tag{1.36}$$

Related to the weighted overall power factor, this can be *neutral* when is being determined without taking into consideration the reactive power compensation installations, and general, when one takes into account the powers provided by these installations. The value of the general weighted overall power factor from which the reactive energy power consumption is no longer charged, is called *neutral power* factor and for the national energetic system is $\cos_n^* = 0,92$.

For electrical installations inside the consumer: the power factor decreasing causes, the effects of a reduced power factor, means and methods for power factor mitigation, a technical-economical computation for placing the reactive power sources etc.

The irrational and big reactive power consumption, that generates a reduced power factor, presents a series of disadvantages for the electric installation, among which we mention [25]: the increase of active power losses in the passive elements of the installation, the increase of voltages' losses, installation' oversize and the its electric energy diminished transfer capacity.

1.6.2 Harmonic State Indicator—Specific PQ Parameters

Electrical installations that contain non-linear consumers absorb non-sinusoidal electric current even in the theoretical case of a perfect sinusoidal voltage supply. The non-sinusoidal electric current flows through the impedances corresponding to these electrical installations determine non-sinusoidal voltage drops that, superposed over the initial sinusoidal voltage signals, determine their distortion (shown in Fig. 1.28). This determines a supplementary solicitation of the electrical installations.

According to Fourier theory [3, 7] the periodic distorted signals can be decomposed in sinusoidal components whose frequencies are integer multiple of the frequency corresponding to the analyzed signal period. The sinusoidal signal with reference frequency (that corresponds to the analyzed period signal) is called *fundamental signal*, and the sinusoidal signals with frequency integer multiple of the fundamental frequency are called *harmonics*.

$$\begin{aligned} u &= U_0 + \sum_{k=1}^{\infty} \sqrt{2} U_k \sin(k\omega_1 t + \alpha_k); \\ i &= I_0 + \sum_{k=1}^{\infty} \sqrt{2} I_k \sin(k\omega_1 t + \beta_k) \end{aligned} \tag{1.37}$$

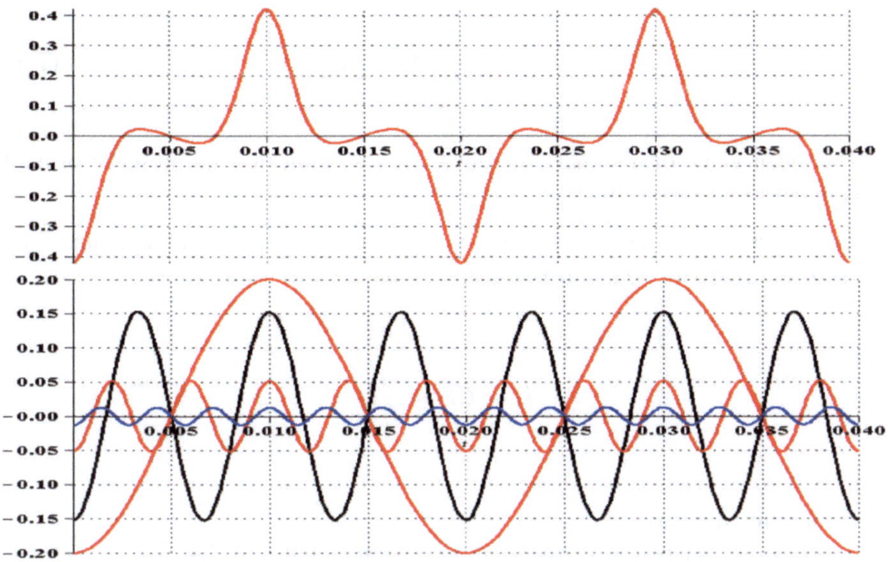

Fig. 1.28 Fourier series decomposition of a periodic distorted voltage signal

where: U_0 and I_0 represents the DC components of the input voltage and current, U_k and I_k are the rms values of harmonic voltages, respectively of the harmonic currents, ω_1 pulsation corresponding to the fundamental frequency, α_k, β_k—phase-shifts corresponding to the harmonic voltages signals, respectively to harmonic currents signals with respect to a reference axis.

To define the main indicators of harmonic state, we consider, as a reference periodic signal, the electric wave shape through the load *i*(*t*) developed in Fourier series. The following definitions are valid for any periodic signal [41, 42].

- *Harmonic level*—is defined as the ratio between the rms value of the harmonic of rank *k* and the rms value of the fundamental harmonic (rank 1):

$$\gamma_k[\%] = \frac{I_k}{I_1} \cdot 100[\%] \tag{1.38}$$

This factor is an important indicator for evaluating the distortion level, its maximum admitted values is being indicated in the voltage or current signal.

- DC−component of the signal is defined as the integral on a variation signal period:

$$\langle I\rangle = DC = \frac{1}{T}\int_{t_0+T}^{T} i(t)dt = I_0 \tag{1.39}$$

The mean value of a signal different from zero indicates the presence of a DC component in its spectrum.

- *Crest factor–CF*—defined as the ratio between the maximum value (amplitude i_M of the periodic non-sinusoidal signal) and its rms value I_{RMS}:

$$k_v = CF_i = \frac{i_M}{I_{RMS}} \tag{1.40}$$

- for a pure sine wave, $k_v = 1.41$;
- for periodic "shaped-top" distorted waveform signal $k_v > 1.41$;
- for a periodic "flat-top" distorted waveform signal $k_v < 1.41$.

The voltage signals characterized by a crest factor $k_{vi} > \sqrt{2}$ can present in time dangerous thermal solicitations.

- *Form Factor*—is the ratio between the rms value of the signal and the mean value corresponding to half a period $I_{med1/2}$:

$$k_f = \frac{I}{I_{med1/2}} = k_f = \frac{I_{RMS}}{\frac{1}{T}\int_{t_0}^{t_0+T} |i(t)|dt} \tag{1.41}$$

For the signals met in electrical installations the form factor can take the values:

- for a sinusoidal signal, $k_f = 1.11$;
- for a periodic signal shaper than a sinusoid $k_f > 1.11$;
- for a sinusoid flatter than a periodic signal $k_f < 1.11$.

- *Root Mean Square—rms Value:*

$$I_{RMS} = \sqrt{\frac{1}{T}\cdot\int_0^T i^2\cdot dt} \text{ or } I_{RMS} = \sqrt{I_0^2 + I_1^2 + I_2^2 + \ldots} = \sqrt{\sum_{k=0}^{n} I_k^2} \tag{1.42}$$

Remarks:

- For signals containing harmonics one uses the terminology *True rms Value*. To compute it, one should take into also the high-order harmonics (usual up to 50!).
- Not all measuring equipment can correctly measure the rms value of a non-sinusoidal signal (True rms Value), many of them displaying only the rms value corresponding to the fundamental (first harmonic)!
- *Residual component* (I_d)—is computed as the rms value of the harmonics corresponding to the analyzed signal:

$$I_d = \sqrt{I_{RMS}^2 - I_1^2} \cong \sqrt{I_0^2 + I_2^2 + I_3^2 + \ldots} \tag{1.43}$$

The distortion residue is a measure if the thermal effect determined by the distorted signal harmonic components.

- *Total Harmonic Distortion−THD*—is the ratio between the distortion residue of the signal and the fundamental rms value:

$$THD_i = \frac{I_d}{I_1} = \frac{\sqrt{I_{RMS}^2 - I_1^2}}{I_1} = \frac{\sqrt{I_2^2 + I_3^2 + I_3^2 + \cdots}}{I_1} = \frac{\sqrt{\sum_{k=2}^{n} I_k^2}}{I_1} \cdot 100\,[\%] \tag{1.44}$$

The *THD* is one of the indicators used to evaluate the distortion level, the maximum admitted values in the electrical network nodes being tabled [43].

- *Distortion Factor−DF*—is defined as the ratio between the distorted residue of the signal and the rms value of the signal:

$$DF_i = \frac{I_d}{I_{RMS}} = \frac{\sqrt{I_{RMS}^2 - I_1^2}}{I_{RMS}} = \frac{\sqrt{I_2^2 + I_3^2 + I_3^2 + \cdots}}{I_{RMS}} = \frac{\sqrt{\sum_{k=2}^{n} I_k^2}}{I_{RMS}} \cdot 100\,[\%] \tag{1.45}$$

The *DF* is much closed as value to *THD* and it is used also as a principal indicator for appreciating the distortion level.

Remark:

- In literature [44, 45], *THD* is also denoted by *THD-F* (distortion related to the fundamental) and DF can be found denoted by *THD-R* (distortion related to the rms value)
- One can easily prove the relationship: $\frac{I_1}{I_{rms}} = \frac{1}{\sqrt{1 + THDi^2}}$

- *The derating factor KF of the transformer's apparent power*—is defined (only for electric current) as follows:

$$KF = \frac{\sum_{k=1}^{n} (kI_k)^2}{\sum_{k=1}^{n} I_k^2} \qquad (1.46)$$

where I_k is the rms value of the harmonic of rank k of the electric current that flows through the transformer's windings. KF (defined only for the current) allows an evaluation of the supplementary heating of the transformers circulated by distorted currents that imply derating its installed power [46, 47].

In order to evaluate the *energetic indicators of harmonic state*, one defines the following types of powers:

- *Single-phase active power P* (*active power*) for harmonic state is:

$$P = \frac{1}{m \cdot T} \cdot \int_{\tau}^{\tau + m \cdot T} pdt = \frac{1}{mT} \int_{\tau}^{\tau + m \cdot T} uidt = \sum_{h=1}^{\infty} U_k I_m \cos \varphi_k \; [\mathrm{W}] \qquad (1.47)$$

where $\varphi_k = \alpha_k - \beta_k$ is the phase-shift between the signals corresponding to the voltage and to the current, in the plan of harmonic of rank k.

- *Active energy* W_a in harmonic state is:

$$W_a = \int_{\tau}^{\tau + t} P \cdot dt \; [\mathrm{Ws}] \qquad (1.48)$$

- *Single-phase reactive power Q* (*reactive power*) is given by the expression:

$$Q = \sum_{k=1}^{\infty} U_k I_k \sin \varphi_k \; [\mathrm{VAr}] \qquad (1.49)$$

- *Single-phase reactive energy* W_r in harmonic state is:

$$W_r = \int_{\tau}^{\tau+t} Q \cdot dt \,[\text{VArs}] \tag{1.50}$$

- *Single-phase apparent power S* (*apparent power*) is defined by the expression:

$$\begin{aligned} S &= U_{RMS} I_{RMS} = \sqrt{U_0^2 + U_1^2 + U_2^2 + \ldots} \cdot \sqrt{I_0^2 + I_1^2 + I_2^2 + \ldots} \\ &= \sqrt{\sum_{k=0}^{n} U_k^2} \cdot \sqrt{\sum_{k=0}^{n} I_k^2} \end{aligned} \tag{1.51}$$

- *Single-phase apparent energy W* is defined by:

$$W = \int_{\tau}^{\tau+t} S \cdot dt \,[\text{VAs}] \tag{1.52}$$

- *Distorted power*—is defined as a complement of active and reactive power related to the apparent power:

$$D = \sqrt{S^2 - P^2 - Q^2} \quad [\text{VAd}] \tag{1.53}$$

Remark:

- Most modern measuring equipment designed for measuring electric energy and electric power can measure instantly each of the 4 defined powers (energies).
- *Power Factor–PF*—is defined as the ratio between the active power *P* and the total apparent power *S:*

$$\lambda = PF = \frac{P}{S} = \frac{\sum_{k=1}^{\infty} U_k I_k \cos \varphi_k}{U_{RMS} I_{RMS}} \tag{1.54}$$

- *Displacement Power Factor–DPF*—is defined as the ratio between the active power P_1 and the apparent power S_1, corresponding to the fundamental harmonic (k = 1):

$$\lambda_1 = DPF = \frac{P_1}{S_1} = \frac{U_1 I_1 \cos \varphi_1}{U_1 I_1} = \cos \varphi_1 \tag{1.55}$$

Remarks:

- The two definitions for *PF* and *DPF* leads to different values displayed separately by modern measuring equipment. Moreover, the value of the angle φ_1 and its tangent tan φ_1 are also displayed.
- The equality $\lambda = \frac{P}{S} = \cos \varphi \ (PF = DPF)$ is valid only for a single-phase circuit and only in a *pure sinusoidal* state (ideal case), so the interpretation of the power factor function of voltage-current phase-shift should be avoided!
- If, it's necessary, the dimensioning of a capacitor bank for reactive power compensation can be done *only based on power factor* cosφ_1 = DPF and *not* on PF!
- One can easily prove the relationship: $PF = \frac{\cos \varphi_1}{\sqrt{1 + THDi^2}} = \frac{DPF}{\sqrt{1 + THDi^2}}$ (valid in the case of negligible voltage distortion $THD_u < 5\%$).

Most of the above mentioned parameters are nowadays commonly indicated by high accuracy measurement devices such *PQ* analyzers. They are able to perform a complete Fourier analysis for the three-phase current and voltage signals up to the 50 harmonic. Figure 1.29 illustrates these parameters measured with a Chauvin Arnoux 8335 device for a low-voltage industrial load.

General solutions for reducing the reactive power consumption are [48–50]:

- adopting, if possible, some technological processes, aggregates and technological and functioning schemes, characterized by a high power factor;
- judicious choice of the type and powers for the electric motor drives, for the transformers, avoiding over-sizing; introducing the synchronous motor instead of the asynchronous one will be justified from economical point of view considering the asynchronous motor individually compensated with derivation capacitors.

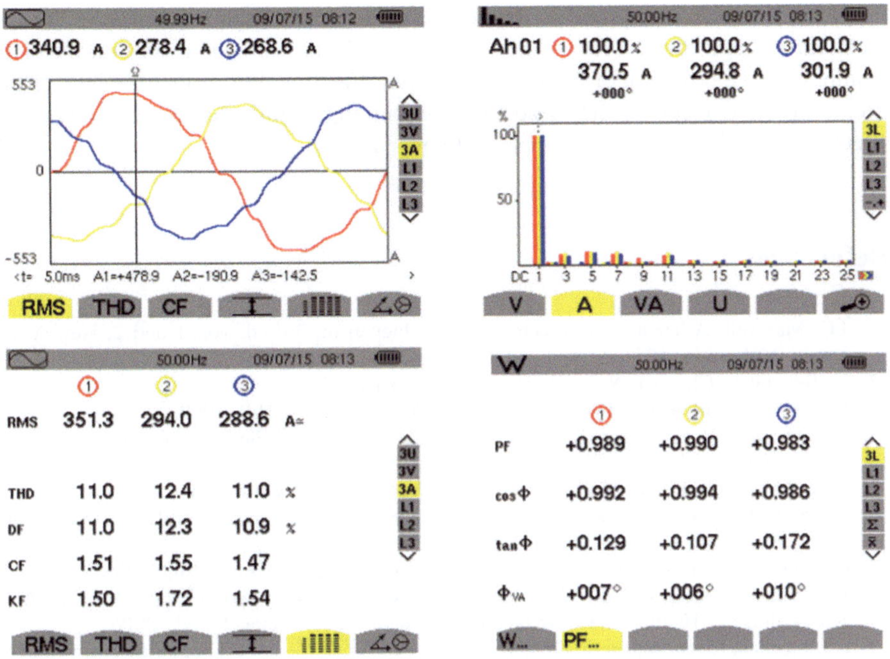

Fig. 1.29 Current signals, harmonics and various *PQ* parameters measured for a low-voltage industrial load

The principal means used especially in existing installations are the following:

- power transformers running in parallel, following the minimum reactive losses graph, when the exploitation conditions allow it;
- exploitation of synchronous motors at the maximum limit of producing reactive power;
- limiting the unloaded running of asynchronous motors, of special transformers, if the technological process allows it;
- using Y-Δ switches for low-voltage asynchronous motors, systematically loaded under 40% from its nominal load, for a long-term functioning in Y connection;
- replacing asynchronous motors and of the over-sized transformers, based on technical-economic analysis using the updated total spending method.

The main specialized reactive power sources are synchronous compensators and derivation capacitors.

Reactive power control has a significant role in role in supporting the power transmission systems.

Reactive power is required to deliver active power through transmission lines at a certain voltage level. Motors, transformers and other power loads, intrinsically require reactive power in order to maintain their functionality. Either the lake or the excess of reactive power may lead to numerous PQ disturbances, which adversely affect both the loads and the supply network [16].

References

1. J.C. Maxwell, A Treatise on Electricity and Magnetism, 3rd ed. vol. 1 and 2, New York, Dover Publication, 1954.
2. V. Bellevitch, Classical Network Theory, Holden-Day, 1968.
3. L.O. Chua, Nonlinear Network Theory, New York, McGraw Hill, 1969.
4. P. Penfield Jr, R. Spence, S. Duinker, Tellegen's Theorem and Electrical Networks, Research monograph no. 58, Massachusetts, M.I.T. Press, 1970.
5. R.W. Newcomb, Linear Multiport Synthesis, New York, McGraw-Hill, 1966.
6. M. Preda, P. Cristea, Fundamentals of Electrotechnics (in Romanian: Bazele Electrotehnicii), vol. 2, Bucharest, ed. Didactica si Pedagogica, 1980.
7. C.A. Desoer, E.S. Kuh, Basic Circuit Theory, New York, Mc Graw-Hill, 1969.
8. J.C. Willems, Dissipative and Dynamical Systems, Eur. J. Contr., vol. 13, pp. 134–151, 2007.
9. M. Vasiliu, I.F. Hantila, Electromagnetics, Bucharest, ed. Electra, 1st ed., 2006.
10. C.I. Mocanu, Electromagnetic Field Theory (in Romanian: Teoria Campului Electromagnetic, Bucharest, ed. Didactica si Pedagogica, 1st ed., 1981.
11. C.I. Mocanu, Electric Circuits Theory (in Romanian: Teoria circuitelor electrice) Bucharest, ed. Didactica si Pedagogica, 1st ed., 1979.
12. H. Andrei, Modern Electrical Engineering (in Romanian: Inginerie Electrica Moderna), vol. 1, ed. Bibliotheca, Targoviste, 2010.
13. J. Arrillage, N. Watson, S. Chen, Power System Quality Assessment, John Wiley & Sons, New York, 2001.
14. L.F. Beites, J.G. Mayordomo, A. Hernandes, R. Asensi, Harmonics, Inter Harmonic, Unbalances of Arc Furnaces: A New Frequency Domain Approach, IEEE Transactions on Power Delivery, 16(4), 2001, 661–668.
15. C. Cepisca, H. Andrei, S. Ganatsios, S.D. Grigorescu, Power Quality and Experimental Determinations of Electrical Arc Furnaces, Proc. 14th IEEE Mediterranean Electrotechnical Conference, MELECON, vol. 1 and 2, pp. 546–551, Ajaccio France, May 5–7, 2008.
16. A.E. Emmanuel, On the Assessment of Harmonic Pollution, IEEE Transaction on Power Delivery, vol. 10(3), 1995, 1693–1698.
17. A. Eberhard, Power Quality, InTech, Vienna, 2011.
18. A.E. Emmanuel, Apparent Power Definition for Three-Phase systems, IEEE Transactions on Power Delivery, 14 (3), 1999, 762–772.
19. J. Zhu, Optimization of Power System Operation, Willey & Sons, 2009.
20. P.S. Filipski, Y. Baghzouz, M.D. Cox, Discussion of Power Definitions Contained in the IEEE Dictionary, IEEE Transaction on Power Delivery, vol. 9(3), 1994, 1237–1244.
21. E.F. Fuchs, M.A.S. Masoum, Power Quality in Power Systems and Electrical Machines, Elsevier Academic Press, Amsterdam, 2006.
22. R. Hurst, Power Quality and Grounding Handbook, The Electricity Forum, Toronto, Canada, 1994.
23. V. Katic, Network Harmonic Pollution - A Review and Discussion of International and National Standards and Recommendations, Proc. of Power Electronic Congress - CIEP, pp. 145–151, Paris, France, October 24–26, 1994.

24. T.S. Key, J.S. Lai, IEEE and International Harmonic Standard Impact on Power Electronic Equipment Design, Proc. of International Conference Industrial Electronics, Control and Instrumentation - IECON, pp. 430–436, London, England, May, 25–27, 1997.
25. C. Sankaran, Power Quality, CRC Press, London, 2008.
26. P. Arpaia, Power Measurement, CRC Press LLC, 2000.
27. B. Gatheridge, How to Measure Electrical Power, EDN Network, 2012.
28. S. Humphrey, H. Papadopoulos, B. Linke, S. Maiyya, A. Vijayaraghavan, R. Schmitt, Power Measurement for Sustainable High-Performance Manufacturing Processes, Procedia CIRP, vol. 14, pp. 466–471, 2014.
29. M.A. Lombardi, Fundamentals of Time and Frequency, http://tf.nist.gov/general/pdf/ 1498. pdf, Accessed July 2015.
30. Oscilloscope, https://en.wikipedia.org/wiki/Oscilloscope.
31. H. Markiewicz, A. Klajn, Power Quality Application Guide, Wroclaw University of Technology, July 2004.
32. IEC 6038 - International Electrotechnical Commission Standard Voltages, 1999.
33. EN 50160 - European Norm Voltage Characteristics of Electricity Supplied by Public Distribution Systems, 1999.
34. IEC 61000-4-7, International Electrotechnical Commission.
35. IEC 61000-4-15, International Electrotechnical Commission.
36. F.C. De La Rosa, Harmonics and Power Systems, New York, Taylor & Francis, 2006.
37. Electrical Installation Guide, Schneider Electric, 2015.
38. Technische Anschlussbedingungen (Technical requirements of connection), VDEW.
39. Rozporzadzenie Ministra Gospodarki z dnia 25 wrzesnia 2000, w sprawie szczególowych warunków przylaczania podmiotów do sieci elektroenergetycznych, obrotu energia elektryczna, swiadczenia uslug przesylowych, ruchusieciowego i eksploatacji sieci oraz standardów jakosciowych obslugi odbiorców. Dziennik Ustaw Nr 85, poz. 957 (Rules of detailed conditions of connection of consumers to the electrical power network and quality requirements in Poland).
40. A.E. Emanuel, Power Definitions and the Physical Mechanism of Power Flow, John Wiley & Sons, 2011.
41. A. Baggini, Handbook of Power Quality, John Wiley & Sons, 2008.
42. IEEE 100-1996, The IEEE Standard Dictionary of Electrical and Electronics Terms, 6th Edition, 1996.
43. IEEE-WG, IEEE Working Group on Nonsinusoidal Situations, Practical Definitions for Powers in Systems with Nonsinusoidal Waveforms and Unbalanced Loads, IEEE Transactions on Power Delivery, vol. II, no. 1, 1996, 79–101.
44. R.C. Dugan, M.F. McGranaghan, S. Santoso, H.W. Beaty, Electrical Power Systems Quality, McGraw Hill Professional, 2012.
45. A. Kusko, M. Thomson, Power Quality in Electrical Systems, McGraw Hill Professional, 2007.
46. IEC 61000-4-30, Testing and Measurement Techniques-Power Quality Measurement Method, 1999.
47. IEEE Recommended Practice for Monitoring Electric Power Quality, IEEE 1159, 1995.
48. Romanian Norm for Limitation of Harmonic Pollution and Unbalance in Electrical Networks, PE 143, 2004.
49. Characteristics of Supplied Voltage in Public Distribution Networks, SREN 50160, October 1998.
50. Power Quality Application Guide, Voltage Disturbance, Standard EN50160, July 2004.

Chapter 2
Fundamentals of Reactive Power in AC Power Systems

Horia Andrei, Paul Cristian Andrei, Emil Cazacu and Marilena Stanculescu

Abstract The fundamentals of reactive power in AC power systems are discussed in the second chapter. The chapter presents basic theory of AC circuits including two-ports linear elements, basic equations and definition of powers in AC circuits. The phasor diagrams and power measurement techniques in AC networks are also presented. The chapter also investigates the effects of reactive power as well as power factor compensation in consumers. The end part of the chapter is related to minimum active and reactive absorbed power in linear AC circuits and also non-sinusoidal conditions. All of the parts include some practical examples and case studies. The chapter is closed with a large list of bibliographic references.

2.1 Chapter Overview

The chapter opens with an overview. Starting on the Kirchhoff's laws expressed in terms of symbolic (complex) form the basic theory of AC circuits is summarized in first section of chapter two. The impedance and admittance are used in order to characterize the behavior of two-ports linear elements. Also a review about the analysis methods of AC circuits is absolutely necessary to emphasize the equations systems that are commonly used. Then are introduced the definitions and physical interpretation of powers in AC power systems: the active, reactive and apparent

H. Andrei (✉)
Doctoral School of Engineering Sciences, University Valahia of Targoviste, Targoviste, Dambovita, Romania
e-mail: hr_andrei@yahoo.com

P.C. Andrei · E. Cazacu · M. Stanculescu
Department of Electrical Engineering, University Politehnica Bucharest, Bucharest, Romania
e-mail: paul.andrei@upb.ro

E. Cazacu
e-mail: cazacu@upb.ro

M. Stanculescu
e-mail: marilena.stanculescu@upb.ro

N. Mahdavi Tabatabaei et al. (eds.), *Reactive Power Control in AC Power Systems*, Power Systems, DOI 10.1007/978-3-319-51118-4_2

power, the power, reactive and deforming factor, the overall harmonic distortion of voltage and current, the harmonic level. Several examples put in evidence the theoretical aspects presented below. General problems about energy and power in AC power systems is intuitive and lead to a better understanding of the definitions of these energetic parameters. Accompanied by presentation of some modern measurement methods of these parameters and calculation examples for studied cases of power systems, all of them are the subjects exposed in the third section of this chapter.

The next section has as a main objective the reader's assimilation of the fundamental problems of reactive power consumption automated management in power systems and of some methods used to limit the negative effects due to a reduced power factor. On this line, there has been elaborated a selection of the equipment for power factor correction based on the analysis of the customer's electric energy quality. There are also presented the notions needed to design some simple systems used to compensate the reactive power for different levels of the installation. Several examples have been studied separately in this section.

The qualitative and quantitative aspects related to the active and reactive power circulation in AC networks are presented in section four of this chapter. In this way the recent Principles of Minimum Absorbed Active and Reactive Power (PMARP) in AC Power Systems are demonstrated and formulated. For AC circuits under sinusoidal and non-sinusoidal conditions the PMARP proves, on one hand, that the active power absorbed by all the resistances of the AC power system is minimum and, on the other hand, that the reactive power absorbed (generated) by all the resistive-inductive (resistive-capacitive) elements is minimum.

Also one demonstrates that (i) this principle is verified by the currents which satisfy Kirchhoff current law (KCL) and nodal method (NM), and (ii) the co-existence (CEAPP) of PMARP and of maximum active power transfer theorem (MPTT). Several examples presented hereinafter demonstrate the PMARP and CEAPP for classical linear and reciprocal AC circuits under sinusoidal and non-sinusoidal signals and prove the originality of the new theoretical concepts introduced by authors. The second chapter is closed with a specific list of bibliographic references.

2.2 Basic Theory of AC Circuits

2.2.1 Two-Port Linear Elements

(i) **Sinusoidal signals—Characterization, symbolic representation**

By definition, a sinusoidal signal is that signal whose time variation is described by an expression of the following form [1–4]

$$x(t) = X_{\max} \sin(\omega t + \varphi) = X\sqrt{2} \sin(\omega t + \varphi) \tag{2.1}$$

In Eq. (2.1) the signals have the following significance:

- $X_{\max}$—is the *amplitude* or the peak value of the sinusoidal signal and it represents the maximum positive value of $x(t)$ variation during one period.
- X—is *the effective* (*root mean square–rms*) value of the sinusoidal signal. Between amplitude and rms value there is a relationship as it can be deduced from (2.1), which is the dependence: $X_{\max} = X\sqrt{2}$. The rms value X is the value indicated by the measuring equipment.
- ω—*angular frequency*. For a given signal, between the angular frequency and its frequency (or period—T) there is the following relationship

$$\omega = 2\pi f = \frac{2\pi}{T} \tag{2.2}$$

- $\alpha = \omega t + \varphi$—represents the phase at a given moment in time (t arbitrary). For $t = 0$ one obtains the initial phase φ of the sinusoidal signal.

To illustrate better the physical significance of these signals we represent graphically the time variation of a sinusoidal signal in Fig. 2.1.

By definition, *the mean value* of a periodic signal is given by:

$$\langle x \rangle = \frac{1}{T} \int_{t_0}^{t_0 + T} x(t)\mathrm{d}t = 0 \tag{2.3}$$

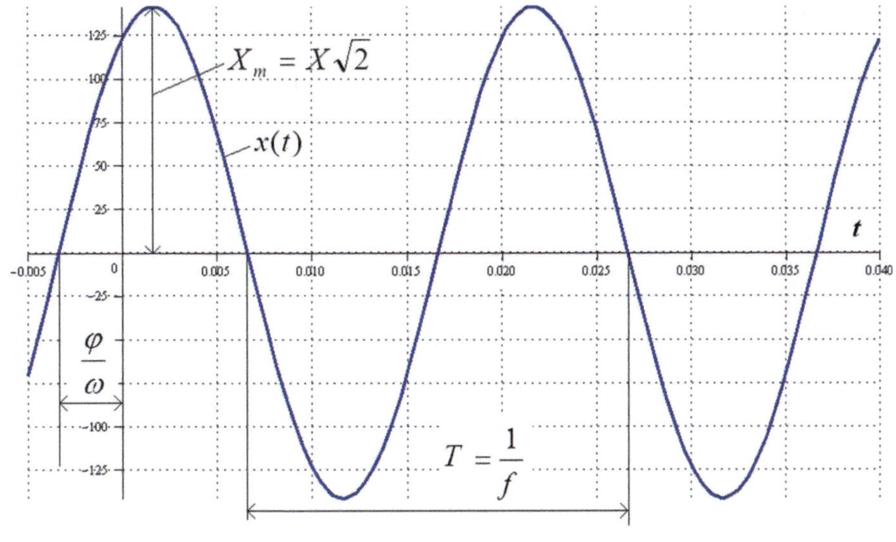

Fig. 2.1 Signals and values characteristic for a sinusoidal variation

From Eq. (2.3) for a sinusoidal signal its corresponding mean value is zero. A zero mean value periodic signal is called *alternative signal*. *The rms value* of a signal is the square root of the mean value of the square of its corresponding variation.

$$X = \sqrt{\langle x^2 \rangle} = \sqrt{\frac{1}{T}\int_{t_0}^{t_0+T} x^2(t)\mathrm{d}t} = \frac{X_{\max}}{\sqrt{2}} \tag{2.4}$$

For two signals having the same angular frequency ω one defines the *phase shift* φ as being the difference between the phases corresponding to the two sinusoidal signals—i.e. the difference between their initial phases.

$$\begin{aligned} x_1(t) &= X_1\sqrt{2}\sin(\omega t + \varphi_1) \\ x_2(t) &= X_2\sqrt{2}\sin(\omega t + \varphi_2) \\ \varphi &= (\omega t + \varphi_1) - (\omega t + \varphi_2) = \varphi_1 - \varphi_2 \end{aligned} \tag{2.5}$$

In Fig. 2.2, is presented the phase-shift between two signals having different amplitudes and different initial phases.

(ii) **Complex representation of sinusoidal signals**

For any sinusoidal signal *x*(*t*) of an angular frequency ω, one can bi-univocally associate a complex number $\underline{X}$ called *its complex* or *the complex image of x*(*t*), of

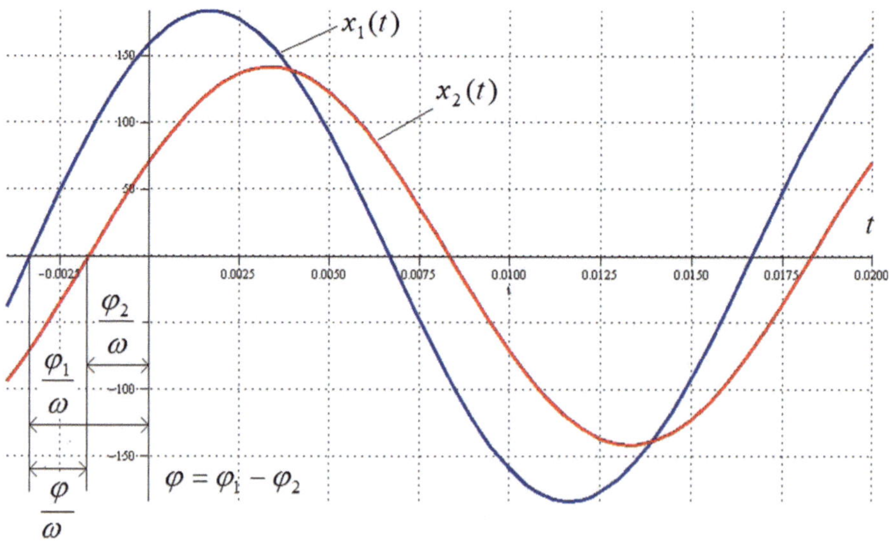

Fig. 2.2 The phase-shift between two sinusoidal signals

modulus equal to its rms value and the argument equal to the initial phase of the sinusoidal signa

$$x(t) = X\sqrt{2}\sin(\omega t + \varphi) \Leftrightarrow \underline{X} = Xe^{j\varphi} = X(\cos\varphi + j\,\sin\varphi) \qquad (2.6)$$

In Eq. (2.6) one denoted by $j = \sqrt{-1}$ the complex number of modulus equal to one and the phase equal to $\pi/2$. This analytical representation used for sinusoidal signals is called *complex representation*. This representation allows also a representation in the complex plan of the sinusoidal signals (Fig. 2.3).

This representation is very useful because it allows the computation of AC sinusoidal electrical circuits easier and allows also a better interpretation of the obtained results.

As a consequence, during the first step, the sinusoidal signals will be expressed using complex numbers. Then, after computation, using in principal the same equivalence theorems and the same solving methods as in DC, one will came back to time domain using bi-univocal properties of complex transformation.

(iii) **Two-ports linear circuits' elements**

Passive circuit elements

In principal, these circuit's elements are represented by: resistor, coil, capacitor and mutually connected coils, each of these elements being characterized by a single constant parameter: the resistance R, the inductivity L, the capacity C, respectively the mutual inductance M, that is a parameter apart from the parameters corresponding to the two magnetic coupled coils.

In Fig. 2.4 are presented the analytical equations characterizing each element. For magnetic coupled coils the sign between the two terms is + if i_1 and i_2 have the same direction with respect to the polarized terminals and the sign is − if i_1 enters the polarized terminal and i_2 exits the polarized terminal, or vice versa. As it is shown in the figure for the direction of the currents and for the terminals position marked in the figure, the sign is positive.

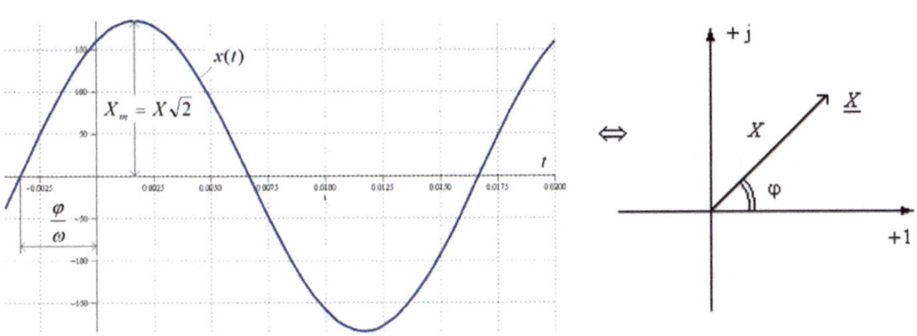

Fig. 2.3 Complex representation of sinusoidal and complex signals

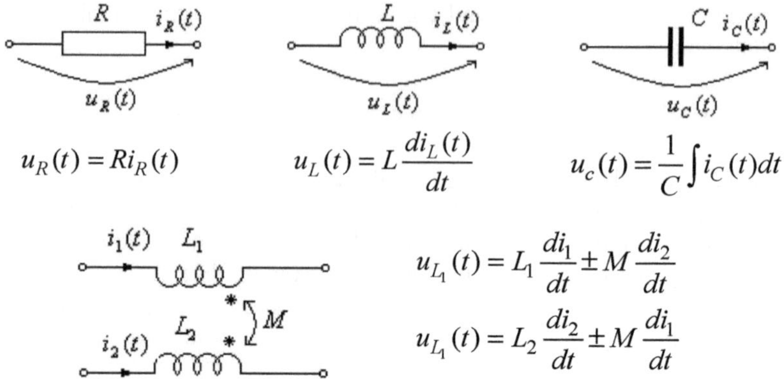

$$u_R(t) = Ri_R(t)$$

$$u_L(t) = L\frac{di_L(t)}{dt}$$

$$u_c(t) = \frac{1}{C}\int i_C(t)dt$$

$$u_{L_1}(t) = L_1\frac{di_1}{dt} \pm M\frac{di_2}{dt}$$

$$u_{L_1}(t) = L_2\frac{di_2}{dt} \pm M\frac{di_1}{dt}$$

Fig. 2.4 Passive circuit's elements

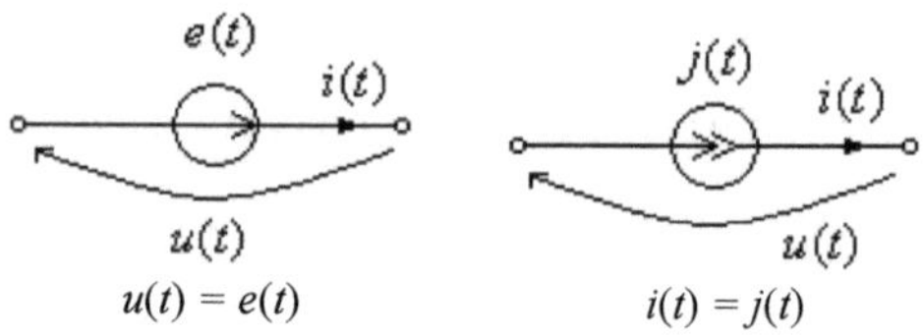

Fig. 2.5 Ideal current ad voltage generators

Active circuit elements

These elements are: the ideal voltage generator and the ideal current generator. The ideal voltage generator is characterized by that no matter what the value of the current intensity $i(t)$ is, this gives at its terminals a constant voltage $u(t)$ equal to the value of the voltage generator $e(t)$. The ideal current generator is characterized by that no matter what the value of the voltage $u(t)$ at its terminals is, this gives to the circuits a constant current $i(t)$ equal to the value given by the current generator $j(t)$.

The symbols and the functioning equations corresponding to the ideal voltage generator and to the ideal current generator are presented in Fig. 2.5.

For real voltage and current generators, there is also another component called inner resistance placed in series with the voltage generator and in parallel with the current generator.

(iv) **Complex imittances**

The computation of periodic AC electrical circuits can be done in a systematic way by using the notions of *complex impedance* respectively *complex admittance* named using the common name of *complex imittances*.

To do this, one considers a passive linear two-port system, whose constitutive inductive elements do not present magnetic couplings with the exterior.

The voltage and the current at its terminals have a sinusoidal variation Fig. 2.6.

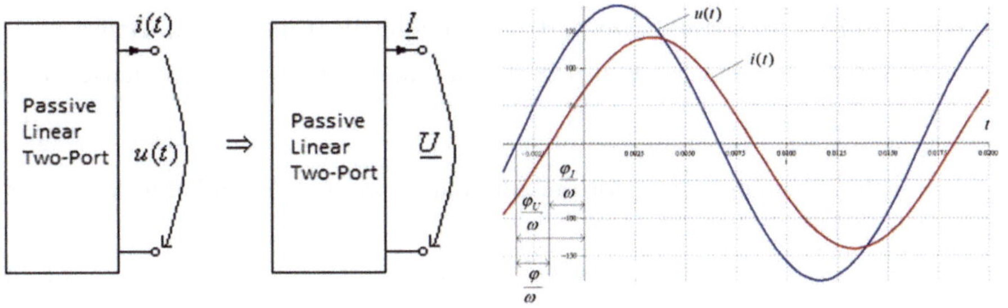

Fig. 2.6 Passive linear two-port system magnetically not coupled to the exterior

The time variation of the voltage and current at two-port system's terminals:

$$\begin{aligned} u(t) &= U\sqrt{2}\sin(\omega t + \varphi_U), \quad \underline{U} = Ue^{j\varphi_U} = U(\cos\varphi_U + j\sin\varphi_U) \\ i(t) &= I\sqrt{2}\sin(\omega t + \varphi_I), \quad \underline{I} = Ie^{j\varphi_I} = I(\cos\varphi_I + j\sin\varphi_I) \end{aligned} \tag{2.7}$$

By definition one calls *the complex impedance* corresponding to the two-port system as being the ratio between the complex images of the voltage at its terminals and the absorbed current (Ohm's law)

$$\underline{Z} = \frac{\underline{U}}{\underline{I}} = \frac{U}{I}e^{j(\varphi_U - \varphi_I)} = Ze^{j\varphi} = Z(\cos\varphi + j\sin\varphi) = R + jX \tag{2.8}$$

The modulus Z [Ω] is called the two-port system's real impedance and its argument $\varphi = \varphi_U - \varphi_I$ is called the two-port system's phase and

$$\begin{aligned} R &= \Re e\{\underline{Z}\} = Z\cos\varphi - \text{two-port system equivalent inner resistance } [\Omega] \\ X &= \Im m\{\underline{Z}\} = Z\sin\varphi - \text{two-port system equivalent inner reactance } [\Omega] \end{aligned} \tag{2.9}$$

Obviously one can determine the relationships

$$Z = \frac{U}{I} = \sqrt{R^2 + X^2} \tag{2.10}$$

By definition Y [S] one calls *complex admittance* of the two-port system the ratio between the complex images of the current and of the voltage at its terminals:

$$\underline{Y} = \frac{\underline{I}}{\underline{U}} = \frac{I}{U}e^{-j(\varphi_U - \varphi_I)} = Ye^{j\varphi} = Y(\cos\varphi - j\sin\varphi) = G - jB \tag{2.11}$$

In Eq. (2.11) one identifies G—the equivalent conductance and B—the equivalent susceptance as the real, respectively the changed sign coefficient of the imaginary part from $\underline{Y}$.

$$G = \Re e\{\underline{Y}\} = Y\cos\varphi - \text{two-port system equivalent inner conductance}[S]$$
$$B = \Im m\{Y\} = Y\sin\varphi - \text{two-port system equivalent inner susceptance } [S] \tag{2.12}$$

As for the impedance case, for the admittance we have the relationships

$$Y = \frac{I}{U} = \sqrt{G^2 + B^2}, \varphi = \arctan\frac{B}{G} \tag{2.13}$$

We notice that the admittance (real or complex) represents the inverse of the impedance (real or complex). As a consequence, between the parameters shown above one can determine a series of relationships

$$\begin{array}{lll} \underline{Y} = \dfrac{1}{\underline{Z}} & R = \dfrac{G}{Y^2} = GZ^2 & X = \dfrac{B}{Y^2} = BZ^2 \\ Y = \dfrac{1}{Z} & G = \dfrac{R}{Z^2} = RY^2 & B = \dfrac{X}{Z^2} = XY^2 \end{array} \tag{2.14}$$

Taking into account the Eqs. (2.10) and (2.13) we notice that it is possible to build two right triangles generically called the impedances' triangle and the admittances' triangle (Fig. 2.7).

The linear not-coupled with the exterior two-port system should compulsory satisfy the condition

$$\varphi \in \left[-\frac{\pi}{2}, \ \frac{\pi}{2}\right] \tag{2.15}$$

The condition (2.15) is equivalent to $\Re e\{\underline{Z}\} = R \geq 0$. If $\Im m\{\underline{Z}\} > 0$ or $\varphi > 0$ the regime is inductive. In this case we can equate the whole two-port system either in series or in parallel (depending on the way in which one works: either in impedance or in admittance) with a resistor in connection to an inductor.

$$\begin{array}{ll} \text{For series connection} & \text{For parallel connection} \\ R = \Re e\{\underline{Z}\} & G = \Re e\{\underline{Y}\} \\ X_L = \Im m\{\underline{Z}\} \Rightarrow L = \dfrac{X_L}{\omega} & B_L = \Im m\{\underline{Y}\} \Rightarrow L = \dfrac{1}{\omega B_L} \end{array} \tag{2.16}$$

Fig. 2.7 Imittances' triangles

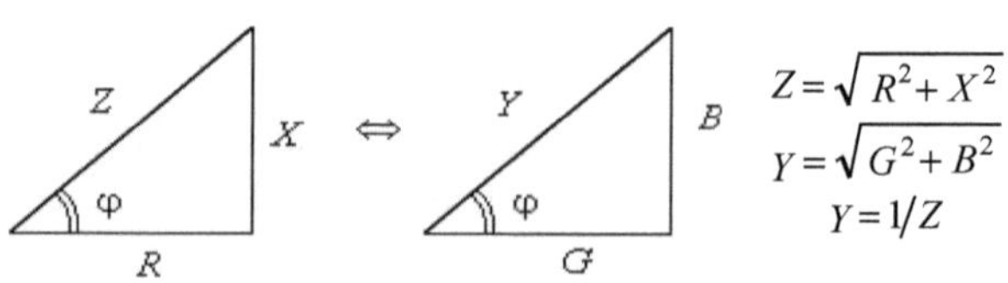

If $\Im m\{\underline{Z}\} < 0$ or $\varphi < 0$ the regime is capacitive. In this case we can equate the whole two-port system either in series or in parallel (depending on the way in which one works: either in impedance or in admittance) with a resistor in connection with a capacitor.

$$\begin{array}{ll} \text{For series connection} & \text{For parallel connection} \\ R = \Re e\{\underline{Z}\} & G = \Re e\{\underline{Y}\} \\ X_C = \Im m\{\underline{Z}\} \Rightarrow C = \dfrac{1}{\omega X_C} & B_C = \Im m\{\underline{Y}\} \Rightarrow C = \dfrac{B_C}{\omega} \end{array} \tag{2.17}$$

2.2.2 Basic Equations: Kirchhoff's Laws in Complex Representation [2–5]

For circuits containing resistors, coils, capacitors and ideal sinusoidal electromotive sources (*emfs*), the Kirchhoff's Currents (KCL) and Voltage Laws (KVL) in instantaneous values are given below

$$\sum_{l_k \in n_j} i_k = 0 \tag{2.18}$$

$$\sum_{l_k \in b_i} (R_k \cdot i_k + L_k \cdot \frac{di_k}{dt} + M_{kh} \cdot \frac{di_h}{dt} + \frac{1}{C_k} \int i_k dt) = \sum_{l_k \in bi} e_k \tag{2.19}$$

where: $j = 1, \ldots, n-1$ are the $n-1$ nodes of the circuit that can be expressed using $n-1$ independent KCL equations, $i = 1, \ldots, l-n+1$ are the $b = l-n+1$ loops of the circuit that can be expressed using $l-n+1$ KVL independent equations, and $k = 1, \ldots, l$ are the l branches of the circuit, eventually coupled with other $h = 1, \ldots, l,\ h \neq k$, branches of the circuit.

Using the properties of the symbolic method, the equations in complex form corresponding to Kirchhoff's laws are obtained

$$\sum_{l_K \in n_j} \underline{I}_k = 0, \quad j = 1, 2, \ldots, n-1 \tag{2.20}$$

$$\sum_{l_K \in b_j} (R_k \cdot \underline{I}_k + j\omega L_k \cdot \underline{I}_k + j\omega\, M_{kh} \cdot \underline{I}_h + \frac{1}{j\omega C_k} \cdot \underline{I}_k) = \sum_{l_k \in bi} E_k \tag{2.21}$$

with $i = 1, 2, \ldots, l - n + 1$

The solutions $\underline{I}_k$ of the complex Eqs. (2.20) and (2.21) are therefore the sinusoidal particular solutions of the integral-differential Eqs. (2.18) and (2.19).

The resistive, inductive and capacitive voltages from the left side of Eq. (2.21) can be written in complex form having a common representation that is $\underline{Z}.\underline{I}$ with the complex impedance defined according to its corresponding branch. Using this convention, KVL is compactly expressed as follows

$$\sum_{l_K \in b_j} \underline{Z}_k \cdot \underline{I}_k = \sum_{l_k \in b_j} \underline{E}_k \tag{2.22}$$

So, the statements of the two theorems are:

- KCL in complex form: "the sum of the currents' complex representations in a node n_j is zero";
- KVL in complex form: "the sum of the voltages' complex representations computed at the terminals of l_k branches along a loop b_j, is equal to the sum of the complex representation corresponding to the *emf* from the branches belonging to the same loop".

In this way, the Eqs. (2.20) and (2.21) or (2.22) represent—for a circuit in sinusoidal state having n nodes and l branches—the complete Kirchhoff independent equations system with a number of $l = (n–1) + (l–n + 1)$ equations and with l unknowns.

There is a formal analogy between AC (sinusoidal state) and DC circuits, both regarding the DC equations system and the AC circuits' complex equations systems, as well as regarding the signals that characterize and describe the DC circuits functioning and the AC circuits' complex signals. as the following

$$\begin{array}{cc} \text{DC} & \text{AC} \\ I \Leftrightarrow & \underline{I} \\ U \Leftrightarrow & \underline{U} \\ R \Leftrightarrow & \underline{Z} \\ E \Leftrightarrow & \underline{E} \end{array}$$

The only difference between the two functioning states (DC and AC), is represented by circuits with magnetic couplings in sinusoidal state that have no correspondent in DC state. That's why, the computation, analysis methods and the theorems established for DC state can be used without any modification for sinusoidal state circuits with no magnetic couplings.

The algorithm for AC circuit computation using Kirchhoff's laws is:

- compute the circuit's impedances, the complex *emf*s and the complex current sources;
- draw the equivalent complex scheme of the circuit using the corresponding complex;
- express Kirchhoff's equations in complex form and solve the system (either in currents unknowns, or in voltages unknowns);

– determine the corresponding instantaneous values form previously computed complex values.

2.2.3 Definitions of Powers in AC Circuits [1, 6, 7]

To define the powers in sinusoidal periodic regime we consider again the case of the linear, passive and inductive not-coupled with the exterior two-port system (Fig. 2.6).

The instantaneous power—p is defined as the received power at each instance of time at its terminals and is the product between the instantaneous values of the voltage and current, having the following expression

$$p(t) = u(t)i(t) = UI[\cos(\varphi_U - \varphi_I) - \cos(2\omega t + \varphi_U + \varphi_I)] \tag{2.23}$$

As one can notice from Eq. (2.23) the instantaneous power contains two terms: a constant term that characterizes the average power exchange of the two-port system and an alternative term that has the angular frequency twice as much as the frequency of the applied voltage.

Active power—P is by definition the average function of time of the instantaneous power

$$P = \langle p \rangle = \frac{1}{T} \int_{t_0}^{t_0+T} p(t)\mathrm{d}t = UI\cos(\varphi_U - \varphi_I) = UI\cos\varphi \quad [\mathrm{W}] \tag{2.24}$$

Taking into account the Eq. (2.24), the active power is always positive so it is received by the passive linear two-port system.

Taking into consideration the relationships stated for the linear two-port system case, the active power consumed by this one, can be also expressed function of its resistance, respectively its conductance

$$P = RI^2 = GU^2 \tag{2.25}$$

The active power is consumed by the active elements from a circuit (the resistances) its unit measure being the watt (W).

The reactive power—Q received by the two-port system is defined in a similar manner as the active power

$$Q = UI\sin\varphi \quad [\mathrm{VAr}] \tag{2.26}$$

This power changes its sign together with the phase shift φ between the voltage and the current, such that it can be both positive and negative, therefore consumed and generated by the two-port system. As in the active power case, the reactive power can be expressed function of reactances or susceptances

$$Q = XI^2 = BU^2 \tag{2.27}$$

The reactive power is "consumed" by the circuit's reactive elements (coils, capacitors and magnetic coupled coils), its unit measure being volt-ampere reactive (VAr).

The apparent power—S is by definition the product between the rms values of voltage and current

$$S = UI \quad [\mathrm{VA}] \tag{2.28}$$

As in the previous cases we can express the apparent power function of the passive linear two-port system's imittances as

$$S = ZI^2 = YU^2 \tag{2.29}$$

The apparent power is an indicator upon the circuit's functioning, being the maximum of the active power for $\varphi = 0$, respectively of the reactive power for $\varphi = \pi/2$. Its corresponding unit measure is (VA). Taking into account the definition procedure of these powers one can also introduce, as for the imittances' case, a triangle corresponding to the three powers: active, reactive ad apparent. In Fig. 2.8 there is represented the powers' triangle as well as the computation relationships for the active and reactive power, function of apparent power.

A very important signal from an energetic point of view is *the power factor* (*PF*) defined as the ratio between the two-port system's consumed active power and the apparent power

$$PF \equiv \cos\varphi \equiv \lambda = \frac{P}{S} \in [0 \;\; 1] \tag{2.30}$$

A synthesis of the powers defined above is *the complex power* $\underline{S}$, defined as the product between the complex image of the voltage applied to the two-port system and the complex conjugated complex image of the absorbed current

$$\underline{S} = \underline{U}\,\underline{I}^* = Se^{j\varphi} = S(\cos\varphi + j\sin\varphi) = P + jQ \tag{2.31}$$

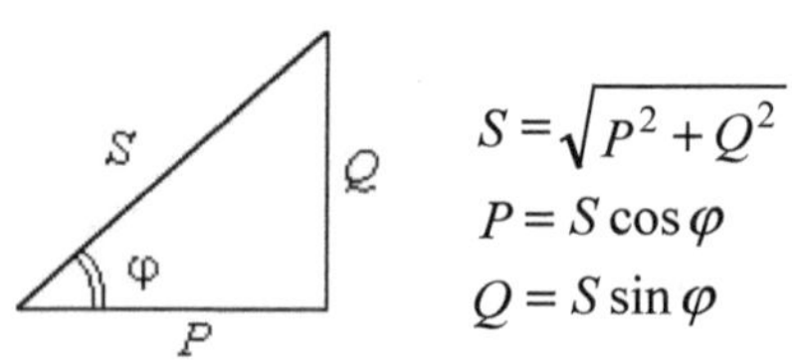

Fig. 2.8 Powers' triangle

As one can notice the modulus of the complex power is represented by the apparent power, its real part can be identified with the active power and the coefficient of the imaginary part can be identified with the reactive power defined for the two-port system. Equation (2.28) states these remarks.

$$\begin{aligned} |\underline{S}| &= S \\ P &= \Re e\{\underline{S}\} \\ Q &= \Im m\{\underline{S}\} \end{aligned} \tag{2.32}$$

For these reasons, when computing the powers, one proceeds directly to compute the complex power after which one identifies the active and reactive powers by separating its components.

Reactive circuits elements—energy sources (voltage sources, respectively current sources) that generate complex power in the circuit. For the voltage sources' case it is given by the product between the complex image of the voltage at its terminals and the complex conjugated image of the supplied current that circulates the source. For the current source, the complex apparent power is given by the product between the complex image of the voltage at its terminals and the complex conjugated image of the source current. For both sources the relationships are taken with the plus sign if the directions chosen for voltage and the current obeys the generator rule, otherwise the complex powers present minus sign in front of the above expressions (Fig. 2.5).

One should emphasize that the direction of the voltage at the current source terminals should be chosen from the extremity indicated by the arrow at the base. The total complex power for a circuit consists of a sum between all the complex powers corresponding to all energy sources (voltage and current) from that circuit; its real part should be equal to the active power, and its imaginary part should be equal to the reactive power of the circuit.

$$\begin{aligned} \underline{S} &= \sum_{k=1}^{n} \underline{E}_k \underline{I}_k^* + \sum_{l=1}^{n} \underline{U}_k \underline{J}_k^* = P + \mathrm{j}Q \\ P &= \sum_{k=1}^{n} R_k I_k^2 \\ Q &= \sum_{k=1}^{n} \omega L_k I_k^2 - \sum_{k=1}^{n} \frac{1}{\omega C_k} I_k^2 \pm \sum_{k=1}^{n} \sum_{l=1}^{m} 2M_k \Re e\{\underline{I}_k \underline{I}_l^*\} \end{aligned} \tag{2.33}$$

If one computes separately the active, respectively the reactive power, the following identities exist $P = \Re e\{\underline{S}\}$, $Q = \Im m\{\underline{S}\}$, respectively. The power balance mainly validates the computed currents values of a specific AC circuit.

2.2.4 Examples

Example 2.1 Let us consider the passive circuit from Fig. 2.9. The following parameters of the circuit elements are known: $R_3 = 2\,\Omega$, $f = 50\,Hz$, $L_2 = 40/\pi$ mH, $L_5 = 20/\pi$ mH, $L_6 = 60/\pi\,$mH, $C_1 = 5/\pi\,$mF, $C_4 = 10/\pi$ mF, $u_{AB}(t) = 128 \cdot \sin(\omega t + \pi/4)$ [V].
Find:

(1) the instantaneous voltages at the terminals of each passive element from the given circuit (using the voltage divider rule);
(2) the instantaneous electric currents for each passive element from the given circuit (using the current divider rule).

We build the equivalent scheme in Fig. 2.10a. The passive elements are characterized by impedances, and the source is characterized by a voltage phasor.
In the equivalent scheme we have

$$\underline{Z}_{C_1} = \frac{1}{j\omega \cdot C_1} = -2j, \underline{Z}_{C_4} = \frac{1}{j\omega \cdot C_4} = -j, \underline{Z}_{L_2} = j\omega \cdot L_2 = 4j$$
$$\underline{Z}_{L_5} = j\omega \cdot L_5 = 2j, \underline{Z}_{L_6} = j\omega \cdot L_6 = 6j, \underline{Z}_{R_3} = R_3 = 2$$
$$\underline{U}_{AB} = \frac{128}{\sqrt{2}} \cdot e^{j\cdot\frac{\pi}{4}} = 64\sqrt{2} \cdot \left[\cos\left(\frac{\pi}{4}\right) + j\,\sin\left(\frac{\pi}{4}\right)\right] = 64 \cdot (1 + j)$$

where $\omega = 2\pi f = 100\pi$.
(1) Determination of the electric voltages at the circuit's elements terminals. In order to use the relationships from the voltage divider rule one should have the impedances placed in series. Notice that the impedances $\underline{Z}_{C_4}$ and $\underline{Z}_{L_5}$ are connected in parallel and the equivalent impedance between these two is: $\underline{Z}_{e_1} = \frac{\underline{Z}_{C_4} \cdot \underline{Z}_{L_5}}{\underline{Z}_{C_4} + \underline{Z}_{L_5}}$. It results: $\underline{Z}_{e_1} = -2j$. The impedances $\underline{Z}_{R_3}$ and $\underline{Z}_{e_1}$ are connected in series (Fig. 2.10b), so the equivalent impedance between these two is: $\underline{Z}_{e_2} = \underline{Z}_{R_3} + \underline{Z}_{e_1}$. It results: $\underline{Z}_{e_2} = 2 - 2j$.

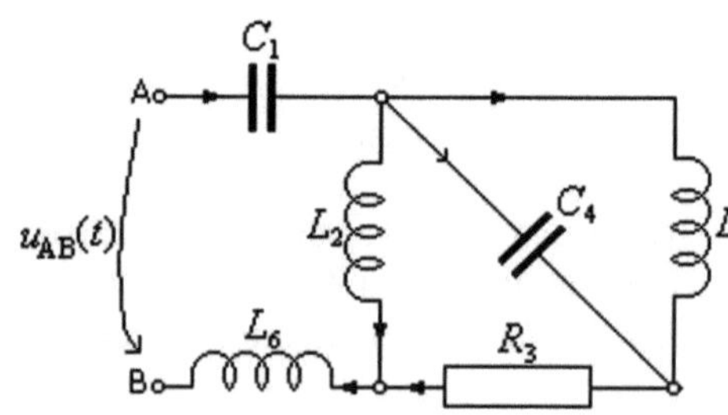

Fig. 2.9 AC circuit with 6 branches

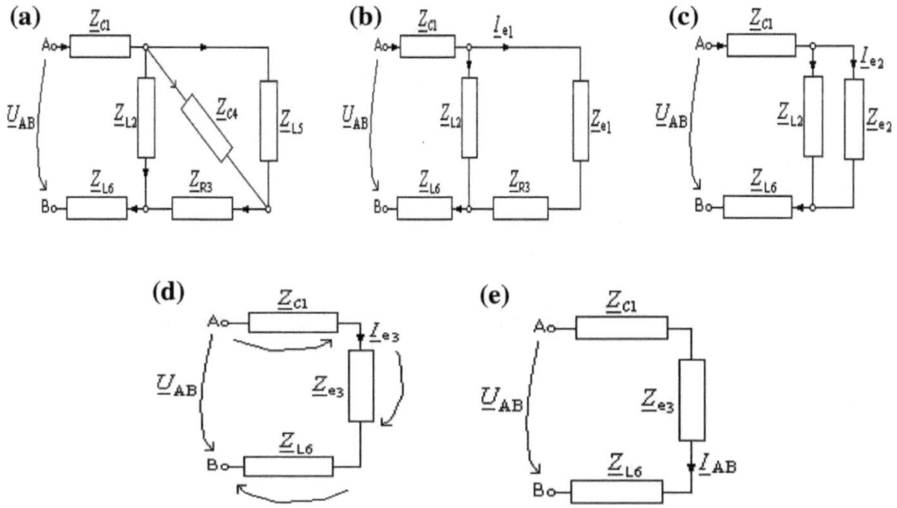

Fig. 2.10 The equivalent AC circuit

Notice the impedances $\underline{Z}_{e_2}$ and $\underline{Z}_{L_2}$ are connected in parallel (Fig. 2.10c), and the equivalent impedance between these two is: $\underline{Z}_{e_3} = \frac{\underline{Z}_{e_2} \cdot \underline{Z}_{L_2}}{\underline{Z}_{e_2} + \underline{Z}_{L_2}}$. It results: $\underline{Z}_{e_3} = 4 \cdot (1 - j)$.

We obtained three impedances in series connection. We can apply the relationships from the voltage divider rule (Fig. 2.10d).

$$\begin{cases} \underline{U}_{C_1} = \frac{\underline{Z}_{C_1}}{\underline{Z}_{C_1} + \underline{Z}_{e_3} + \underline{Z}_{L_6}} \cdot \underline{U}_{AB} \\ \underline{U}_{e_3} = \frac{\underline{Z}_{e_3}}{\underline{Z}_{C_1} + \underline{Z}_{e_3} + \underline{Z}_{L_6}} \cdot \underline{U}_{AB} \\ \underline{U}_{L_6} = \frac{\underline{Z}_{L_6}}{\underline{Z}_{C_1} + \underline{Z}_{e_3} + \underline{Z}_{L_6}} \cdot \underline{U}_{AB} \end{cases}$$

Results:

$$\begin{cases} \underline{U}_{C_1} = 32 \cdot (1 - j) = 32\sqrt{2} \cdot e^{-j\frac{\pi}{4}} \\ \underline{U}_{e_3} = 128 = 128 \cdot e^{j0} \\ \underline{U}_{L_6} = 96 \cdot (-1 + j) = 96\sqrt{2} \cdot e^{j\frac{3\pi}{4}} \end{cases}$$

Because $\underline{U}_{e_3}$ is the voltage from the equivalent impedances' terminals, we'll compute the voltages for each realized equivalence. The impedances $\underline{Z}_{e_2}$ and $\underline{Z}_{L_2}$ are connected in parallel. It results: $\underline{U}_{L_2} = \underline{U}_{e_2} = \underline{U}_{e_3}$. $\underline{U}_{L_2} = 128 = 128 \cdot e^{j0}$.

The impedances $\underline{Z}_{R_3}$ and $\underline{Z}_{e_1}$ are connected in series, so we can use the relationships corresponding to the voltage divider rule

$$\begin{cases} \underline{U}_{R_3} = \dfrac{\underline{Z}_{R_3}}{\underline{Z}_{R_3} + \underline{Z}_{e_1}} \cdot \underline{U}_{e_2} \\ \underline{U}_{e_1} = \dfrac{\underline{Z}_{e_1}}{\underline{Z}_{R_3} + \underline{Z}_{e_1}} \cdot \underline{U}_{e_2} \end{cases}$$

It results:

$$\begin{cases} \underline{U}_{R_3} = 64 \cdot (1 + j) = 64\sqrt{2} \cdot e^{j\frac{\pi}{4}} \\ \underline{U}_{e_1} = 64 \cdot (1 - j) \end{cases}$$

The impedances $\underline{Z}_{C_4}$ and $\underline{Z}_{L_5}$ are connected in parallel. It results: $\underline{U}_{e_1} = \underline{U}_{C_4} = \underline{U}_{L_5}$.

We obtain:

$$\begin{cases} \underline{U}_{C_4} = 64 \cdot (1 - j) = 64\sqrt{2} \cdot e^{-j\frac{\pi}{4}} \\ \underline{U}_{L_5} = 64 \cdot (1 - j) = 64\sqrt{2} \cdot e^{-j\frac{\pi}{4}} \end{cases}$$

Therefore, there are determined all the voltages at the terminals corresponding to each passive element in the complex domain. The instantaneous values of the voltages determined above are

$$\begin{cases} u_{C_1}(t) = 64 \cdot \sin\left(\omega t - \frac{\pi}{4}\right) [V]; \quad u_{L_2}(t) = 128\sqrt{2} \cdot \sin(\omega t) [V] \\ u_{R_3}(t) = 128 \cdot \sin\left(\omega t + \frac{\pi}{4}\right) [V]; \quad u_{C_4}(t) = 128 \cdot \sin\left(\omega t - \frac{\pi}{4}\right) [V] \\ u_{L_5}(t) = 128 \cdot \sin\left(\omega t - \frac{\pi}{4}\right) [V]; \quad u_{L_6}(t) = 192 \cdot \sin\left(\omega t + \frac{3\pi}{4}\right) [V] \end{cases}$$

(1) Determination of the currents across the circuit's elements. We should determine the current $\underline{I}_{AB}$ from the Fig. 2.10,e. Because the impedances $\underline{Z}_{C_1}$, $\underline{Z}_{e_3}$ and $\underline{Z}_{L_6}$ are in series connection the current across them is : $\underline{I}_{AB} = \frac{\underline{U}_{AB}}{\underline{Z}_{C_1} + \underline{Z}_{e_3} + \underline{Z}_{L_6}}$ and it results $\underline{I}_{AB} = \underline{I}_{C_1} = \underline{I}_{e_3} = \underline{I}_{L_6} = 16 \cdot (1 + j)$. So we have

$$\begin{cases} \underline{I}_{C_1} = 16 \cdot (1 + j) = 16\sqrt{2} \cdot e^{j\frac{\pi}{4}} \\ \underline{I}_{L_6} = 16 \cdot (1 + j) = 16\sqrt{2} \cdot e^{j\frac{\pi}{4}} \end{cases}$$

Because the impedances $\underline{Z}_{e_2}$ and $\underline{Z}_{L_2}$ are connected in parallel, we use the relationships from the current divider rule to determine the currents $\underline{I}_{L_2}$ and $\underline{I}_{e_2}$

$$\begin{cases} \underline{I}_{L_2} = \dfrac{\underline{Y}_{L_2}}{\underline{Y}_{L_2} + \underline{Y}_{e_2}} \cdot \underline{I}_{AB} \\ \underline{I}_{e_2} = \dfrac{\underline{Y}_{e_2}}{\underline{Y}_{L_2} + \underline{Y}_{e_2}} \cdot \underline{I}_{AB} \end{cases}$$

It results

$$\begin{cases} \underline{I}_{L_2} = 16 \cdot (1 - j) = 16\sqrt{2} \cdot e^{-j\frac{\pi}{4}} \\ \underline{I}_{e_2} = 32j \end{cases}$$

The impedances $\underline{Z}_{R_3}$ and $\underline{Z}_{e_1}$ are connected in series, so it results: $\underline{I}_{e_1} = \underline{I}_{R_3} = \underline{I}_{e_2}$. So: $\underline{I}_{R_3} = 32j = 32 \cdot e^{j\frac{\pi}{2}}$. Because the impedances $\underline{Z}_{C_4}$ and $\underline{Z}_{L_5}$ are connected in parallel, we determine the currents $\underline{I}_{L_5}$ and $\underline{I}_{C_4}$ using the relationships from the current divider rule

$$\begin{cases} \underline{I}_{L_5} = \dfrac{\underline{Y}_{L_5}}{\underline{Y}_{L_5} + \underline{Y}_{C_4}} \cdot \underline{I}_{e_1} \\ \underline{I}_{C_4} = \dfrac{\underline{Y}_{C_4}}{\underline{Y}_{L_5} + \underline{Y}_{C_4}} \cdot \underline{I}_{e_1} \end{cases}$$

We obtain:

$$\begin{cases} \underline{I}_{L_5} = -32j = 32 \cdot e^{-j\cdot\frac{\pi}{2}} \\ \underline{I}_{C_4} = 64j = 64 \cdot e^{j\cdot\frac{\pi}{2}} \end{cases}$$

The instantaneous values of the electric currents for each element are

$$\begin{cases} i_{C_1}(t) = 32 \cdot \sin\left(\omega t + \frac{\pi}{4}\right) [A]; \quad i_{L_2}(t) = 32 \cdot \sin\left(\omega t - \frac{\pi}{4}\right) [A] \\ i_{R_3}(t) = 32\sqrt{2} \cdot \sin\left(\omega t + \frac{\pi}{2}\right) [A]; \quad i_{C_4}(t) = 64\sqrt{2} \cdot \sin\left(\omega t + \frac{\pi}{2}\right) [A] \\ i_{L_5}(t) = 32\sqrt{2} \cdot \sin\left(\omega t - \frac{\pi}{2}\right) [A]; \quad i_{L_6}(t) = 32 \cdot \sin\left(\omega t + \frac{\pi}{4}\right) [A] \end{cases}$$

Example 2.2 In the circuit given in Fig. 2.11 we have: $R_1 = 2\,\Omega$, $\omega L_1 = \frac{1}{\omega C_1} = 3\,\Omega$, $\omega L_2 = 2\Omega$, $\frac{1}{\omega C_3} = 1\,\Omega$, $e_1(t) = 20 \cdot \sin\left(\omega t + \frac{\pi}{4}\right)$ [V], $j_3(t) = 10\sqrt{2} \cdot \sin\left(\omega t + \frac{\pi}{2}\right)$ [A].

(1) solve the circuit using Kirchhoff's laws;
(2) solve the circuit using the superposition principle;
(3) power balance.

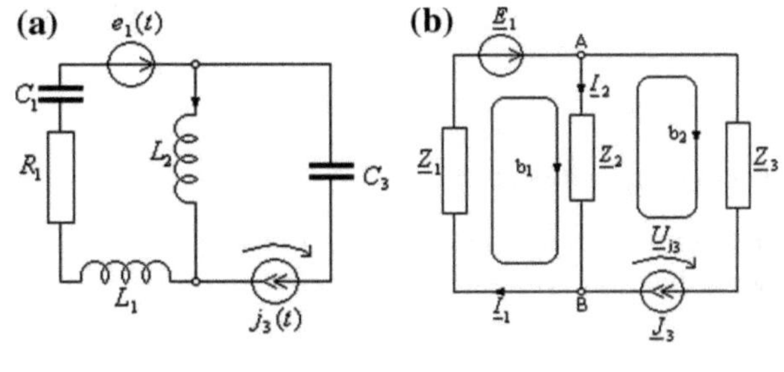

Fig. 2.11 AC circuit with two energy sources

The computation of AC circuits is done using the complex representation. In complex domain, the passive elements are characterized by impedances and the energy sources by phasors. For the passive circuit elements, the impedances are computed as follows

$$\underline{Z}_{R_1} = R_1 = 2,\ \underline{Z}_{L_1} = j\omega \cdot L_1 = 3j,\ \underline{Z}_{C_1} = \frac{1}{j\omega \cdot C_1} = -3j, \underline{Z}_{L_2} = j\omega \cdot L_2 = 2j,$$

$$\underline{Z}_{C_3} = \frac{1}{j\omega \cdot C_3} = -j$$

The voltage source is represented in complex domain by the phasor $\underline{E}_1 = \frac{20}{\sqrt{2}} \cdot e^{j\frac{\pi}{4}} = 10 \cdot (1+j)$ and the current source is represented in complex domain by the phasor $\underline{J}_3 = \frac{10\sqrt{2}}{\sqrt{2}} \cdot e^{j\frac{\pi}{2}} = 10j$. Grouping all the impedances from a branch in an equivalent impedance one obtains (Fig. 2.11b), in which: $\underline{Z}_1 = \underline{Z}_{R_1} + \underline{Z}_{L_1} + \underline{Z}_{C_1} = 2$, $\underline{Z}_2 = \underline{Z}_{L_1} = 2j$ and $\underline{Z}_3 = \underline{Z}_{C_1} = -j$.

(1) *Kirchhoff's equations method*

The topological elements are: 2 nodes, 3 branches and 2 loops. KCL and KVL are expressed as

$$\begin{cases} KCL(A) : -\underline{I}_1 + \underline{I}_2 + \underline{J}_3 = 0 \\ KVL(b_1) : \underline{Z}_1 \cdot \underline{I}_1 + \underline{Z}_2 \cdot \underline{I}_2 = \underline{E}_1 \\ KVL(b_2) : \underline{Z}_3 \cdot \underline{J}_3 - \underline{Z}_2 \cdot \underline{I}_2 - \underline{U}_{j_3} = 0 \end{cases}$$

We obtain

$$\begin{cases} \underline{I}_1 = 5j \\ \underline{I}_2 = -5j \\ \underline{U}_{j_3} = 0 \end{cases}$$

These values are transformed in time domain

$$i_1(t) = 5\sqrt{2} \cdot \sin\left(\omega t + \frac{\pi}{2}\right) \quad [\mathrm{A}]$$
$$i_2(t) = 5\sqrt{2} \cdot \sin\left(\omega t - \frac{\pi}{2}\right) \quad [\mathrm{A}]$$
$$u_{j_3}(t) = 0 \quad [\mathrm{V}]$$

(2) *Superposition principle*

The initial circuit (in complex domain) contains two ideal energy sources and so we'll have to compute two cases

Case 1

In this circuit the ideal voltage source is reduced to its inner resistance, and the ideal current source is characterized by the phasor $\underline{J}_3$ (Fig. 2.12, a). We determine the currents using the relationships from the current divider rule:

$$\begin{cases} \underline{I}_1' = \frac{\underline{Y}_1}{\underline{Y}_1+\underline{Y}_2} \cdot \underline{J}_3 = \frac{\frac{1}{\underline{Z}_1}}{\frac{1}{\underline{Z}_1}+\frac{1}{\underline{Z}_2}} \cdot \underline{J}_3 \\ \underline{I}_2' = \frac{\underline{Y}_2}{\underline{Y}_1+\underline{Y}_2} \cdot (-\underline{J}_3) = \frac{\frac{1}{\underline{Z}_2}}{\frac{1}{\underline{Z}_1}+\frac{1}{\underline{Z}_2}} . (-\underline{J}_3) \end{cases}$$

We obtain:

$$\begin{cases} \underline{I}_1' = 5 \cdot (-1+j) \\ \underline{I}_2' = 5 \cdot (-1-j) \end{cases}$$

To determine the voltage $\underline{U}_{j_3}'$ we apply KVL for the loop "b" such that: $(b) : \underline{Z}_3 \cdot \underline{J}_3 - \underline{Z}_2 \cdot \underline{I}_2' - \underline{U}_{j_3}' = 0$, and results: $\underline{U}_{j_3}' = 10j$.

Case 2

In this circuit the ideal current source is reduced to its inner resistance) and the ideal voltage source is characterized by the phasor $\underline{E}_1$ (Fig. 2.12b). Applying KVL for the loop "b" such that: $(b) : \underline{Z}_1 \cdot \underline{I}_1'' + \underline{Z}_2 \cdot \underline{I}_2'' = \underline{E}_1$. We take into account that $\underline{I}_1'' = \underline{I}_2''$, and results $\underline{I}_1'' = \underline{I}_2'' = 5$. One notices that $\underline{U}_{j_3}'' = -\underline{Z}_2 \cdot \underline{I}_2''$. We obtain: $\underline{U}_{j_3}'' = -10j$.

The final results are obtained by superposing the results obtained independently in each of the two cases. Taking into account the directions of the determined signals for each case compared to the directions from the initial circuit

$$\begin{cases} \underline{I}_1 = \underline{I}_1' + \underline{I}_1'' \\ \underline{I}_2 = \underline{I}_2' + \underline{I}_2'' \\ \underline{U}_{j_3} = \underline{U}_{j_3}' + \underline{U}_{j_3}'' \end{cases}$$

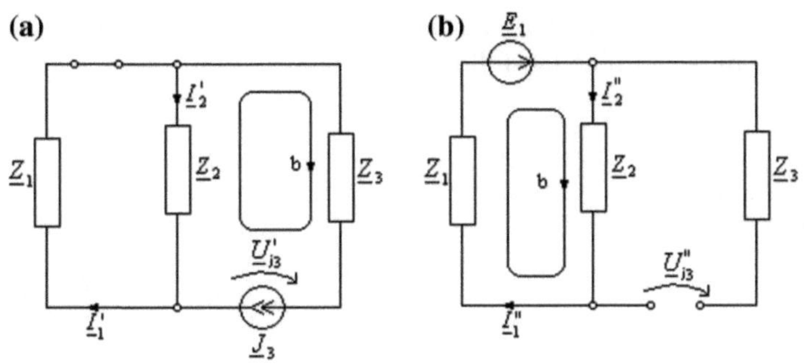

Fig. 2.12 Superposition principle

So, the final results are

$$\begin{cases} \underline{I}_1 = 5j \\ \underline{I}_2 = -5j \\ \underline{U}_{j_3} = 0 \end{cases}$$

The results obtained using the *superposition principle* are identical with the results obtained using the *Kirchhoff's method.*

(3) *Power balance*

The active power consumed in the circuit is: $P_c = R_1 \cdot I_1^2 = 2 \cdot 25 = 50\,\text{W}$. The reactive power consumed in the circuit is computed as follows:

$$\begin{aligned} Q_c &= \text{Im}(\underline{Z}_1) \cdot I_1^2 + \text{Im}(\underline{Z}_2) \cdot I_2^2 + \text{Im}(\underline{Z}_3) \cdot I_3^2 \\ &= 0 + X_{L_2} \cdot 25 - X_{C_3} \cdot 100 = -50 \ \text{VAr} \end{aligned}$$

The apparent complex power generated is computed as follows

$$\begin{aligned} \underline{S}_g &= \underline{E}_1 \cdot \underline{I}_1^* + \underline{U}_{j_3} \cdot \underline{J}_3^* = 10 \cdot (1 + j) \cdot (-5j) + 0 \cdot (-10j) \\ &= 50 - 50j = P_g + jQ_g \end{aligned}$$

We extract the generated active power as being: $P_g = \text{Re}\left\{\underline{S}_g\right\} = 50\text{W}$. In the same way we proceed for the generated reactive power: $Q_g = \text{Im}\left\{\underline{S}_g\right\} = -50\,\text{VAr}$.

We notice that $P_c = P_g$, $Q_c = Q_g$. Therefore, it is verified the power balance corresponding to the active and reactive powers consumed, respectively generated. Implicitly, the consumed, respectively the complex apparent powers balance is verified: $\underline{S}_c = \underline{S}_g$.

Example 2.3 A mono-phase receiver (AC electric drive shown in Fig. 2.13, has the following nominal data: the voltage U_n = 230 V, f = 50 Hz, the active power P_n = 1 kW and the power factor $\cos\varphi = 00.86$ (inductive). The consumer is connected using two copper conductors having the conductivity $\sigma = 56.87 \cdot 10^6$ S/m = 56.87 m/Ωmm^2 and the section A = 1.5 mm^2. The conductors have the length l = 50 m. Determine the following:

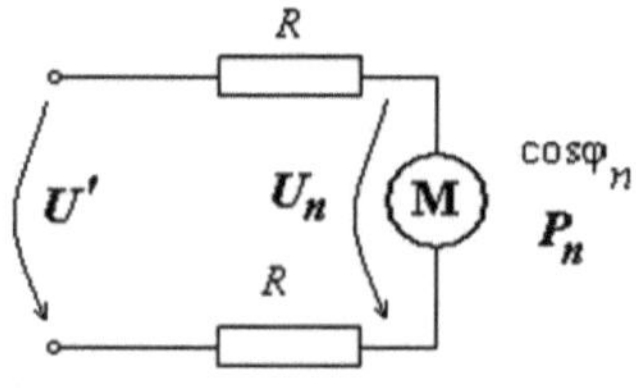

Fig. 2.13 A simple installation for an AC electric drive

(1) The current absorbed by the receiver (the corresponding rms and complex values) under nominal conditions.
(2) The reactive power, Q, the apparent power, S, corresponding to the receiver.
(3) The rms value U' of a voltage source such that at the consumers' terminals one finds its nominal voltage.

(1) From the expression of the active power one can determine the computation relationship for the absorbed current

$$I_n = I = \frac{P_n}{U_n \cos\varphi} = 5.05\ \text{A}$$

The complex value of the current (taking into account the inductive character) can be expressed as follows

$$\underline{I} = Ie^{j\varphi} = I(\cos\varphi - j\sin\varphi) = 4.34 - 2.57j$$
$$\sin\varphi_1 = \sqrt{1-\cos^2\varphi_1} = 0.51$$

(2) The reactive power is

$$Q = P\tan\varphi = P\frac{\sqrt{1-\sin^2\varphi}}{\cos\varphi} = 593.36\ \text{VAr}$$

The apparent power

$$S = \sqrt{P^2+Q^2} = \frac{P}{\cos\varphi} = 1162.79\ \text{VA} = 1.16\ \text{kVA}$$

(3) The rms value U' of the voltage needed to power the system, such that at the consumer one has the nominal value, can be determined, using the simplest mode, from the computed power balance. Therefore

$$S' = U'I = \sqrt{\left(P+2RI^2\right)^2+Q^2} \Rightarrow U' = \frac{\sqrt{\left(P+2RI^2\right)^2+Q^2}}{I}$$

where

$$R = \frac{l}{\sigma A} = 0.586\ \Omega$$

As a consequence, the rms value of the voltage at the end of the line is

$$U' = \frac{\sqrt{\left(P+2RI^2\right)^2+Q^2}}{I} = 235.45\ \text{V}$$

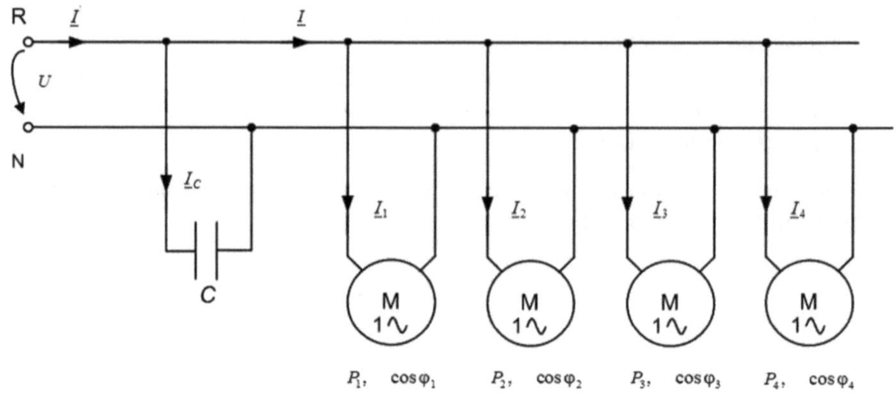

Fig. 2.14 The electric equipment in the mechanical work-place

Example 2.4 Let us consider a mechanical work-place, where there are mono-phase electric drives shown in Fig. 2.14, with the specifications given below (nominal electric power and the power factor (inductive)

$$P_1 = 1\text{ kW}, \cos\varphi_1 = 0.60, U = 220\ V$$
$$P_2 = 4\text{ kW}, \cos\varphi_2 = 0.70, U = 220\ V$$
$$P_3 = 5\text{ kW}, \cos\varphi_3 = 0.76, U = 220\ V$$
$$P_4 = 10\text{ kW}, \cos\varphi_4 = 0.80, U = 220\ V$$

All drives are powered in parallel with a sinusoidal voltage having the rms value $U = 220\ V$ and the frequency 50 Hz.
Find:

(1) The currents absorbed by the receivers and the total current for its corresponding power line (rms and complex values);
(2) The installation total power factor;
(3) The value of the capacitors' battery that should be placed in parallel with respect to the drives, such that the power factor of the entire installation to be $\cos\varphi' = 0.92$;
(4) The rms value of the current absorbed by the installation after introducing the capacitors' battery;
(5) The current drop and the losses on the power line after introducing the power factor compensation system (given in percentage).

Solution:

(1) From the expressions corresponding to the active power and to the power factor (inductive for all consumers), one can determine the electric currents absorbed by the drives (in complex and rms value), as well as the reactive power absorbed by these ones

$$I_1 = \frac{P_1}{U \cos\varphi_1} = 7.57 \text{ A}, \underline{I}_1 = I_1 e^{j\varphi_1} = I_1(\cos\varphi_1 - j\sin\varphi_1) = 4.54 - 6.06j$$

$$\sin\varphi_1 = \sqrt{1-\cos^2\varphi_1} = 0.8,\ Q_1 = P_1\tan\varphi_1 = P_1\frac{\sqrt{1-\sin^2\varphi_1}}{\cos\varphi_1} = 1.33 \text{ kVAr}$$

$$I_2 = \frac{P_2}{U \cos\varphi_2} = 25.97 \text{ A}, \underline{I}_2 = I_2 e^{j\varphi_2} = I_2(\cos\varphi_2 - j\sin\varphi_2) = 18.18 - 18.54j$$

$$\sin\varphi_2 = \sqrt{1-\cos^2\varphi_2} = 0.71,\ Q_2 = P_2\tan\varphi_2 = P_2\frac{\sqrt{1-\sin^2\varphi_2}}{\cos\varphi_2} = 4.08 \text{ kVAr}$$

$$I_3 = \frac{P_3}{U \cos\varphi_3} = 29.90 \text{ A}, \underline{I}_3 = I_3 e^{j\varphi_3} = I_3(\cos\varphi_3 - j\sin\varphi_3) = 22.72 - 19.43j$$

$$\sin\varphi_3 = \sqrt{1-\cos^2\varphi_3} = 0.64,\ Q_3 = P_3\tan\varphi_3 = P_3\frac{\sqrt{1-\sin^2\varphi_3}}{\cos\varphi_3} = 4.27 \text{ kVAr}$$

$$I_4 = \frac{P_4}{U \cos\varphi_4} = 56.81 \text{ A}, \underline{I}_4 = I_4 e^{j\varphi_4} = I_4(\cos\varphi_4 - j\sin\varphi_4) = 45.45 - 34.09j$$

$$\sin\varphi_4 = \sqrt{1-\cos^2\varphi_4} = 0.60,\ Q_4 = P_4\tan\varphi_4 = P_4\frac{\sqrt{1-\sin^2\varphi_4}}{\cos\varphi_4} = 7.50 \text{ kVAr}$$

The value of the total current absorbed by the installation is:

$$\underline{I} = \underline{I}_1 + \underline{I}_2 + \underline{I}_3 + \underline{I}_4 = 90.90 - 78.13j$$

with the rms value

$$I = 119.873 \text{ A}$$

An alternative method for determining the value of the rms current absorbed by the whole installation is used in the power balance formula. Therefore, the rms value of the total apparent absorbed power can be expressed function of the values corresponding to the voltage and to the current

$$S = UI = \sqrt{P^2 + Q^2}$$

where

$$\begin{aligned}
P &= P_1 + P_2 + P_3 + P_4 \\
Q &= Q_1 + Q_2 + Q_3 + Q_4 \\
S &= \sqrt{(20000)^2 + (17189.95)^2} = 26372.23\ VA \\
I &= \frac{S}{U} = \frac{26372.23}{220} = 119.873\,A
\end{aligned}$$

(2) The installation power factor is

$$\cos\varphi = \frac{P}{S} = \frac{P}{\sqrt{P^2+Q^2}} = \frac{20000}{236372.23} = 00.758$$

$$\tan\varphi = \frac{Q}{P} = 00.85$$

(3) The value of the capacitors' battery necessary for power factor correction at $\cos\varphi' = 0.92$ can be determined taking into account that the active power remains constant after connecting the capacitors. After connecting the capacitors, the reactive consumed power is

$$Q' = P\tan\varphi'$$

where

$$\tan\varphi' = \frac{\sqrt{1-\sin^2\varphi'}}{\cos\varphi'} = 0.48$$

On the other hand

$$Q' = P\tan\varphi' = P\tan\varphi + Q_C = P\tan\varphi - \omega CU^2 = 9.68\ \text{kVAr}$$

From where it results

$$C = \frac{P(\tan\varphi - \tan\varphi')}{\omega U^2} = 493.48\ \mu\text{F}$$

(4) The new value of the current absorbed by the installation after connecting the capacitors' battery can be determined by either applying the KCL or one takes into account the active power conservation.

– Applying the KCL, it results

$$\underline{I}' = \underline{I}_1 + \underline{I}_2 + \underline{I}_3 + \underline{I}_4 + \underline{I}_C$$

where

$$\underline{I}_C = j\omega CU = j34.10$$

with rms value $I' = 101.01$ A.

– If we take into account the active power conservation

$$P = UI \cos \varphi$$

respectively,

$$P = UI' \cos \varphi'$$

From where

$$I' = I \frac{\cos \varphi}{\cos \varphi'} = 119.873 \frac{0.758}{0.9} = 101.01 \text{ A}$$

(5) The percentage reduction of the current computed on the power line and of the active power losses can be evaluated as follows:

The current percentage reduction

$$\Delta I\% = \frac{I - I'}{I} \cdot 100 = \left(1 - \frac{I'}{I}\right) \cdot 100 = \left(1 - \frac{\cos \varphi}{\cos \varphi'}\right) \cdot 100 = 15.73\ \%$$

The percentage reduction of the active power losses

$$\begin{aligned} \Delta P\% &= \frac{\Delta P - \Delta P'}{\Delta P} \cdot 100 \\ &= \left(1 - \left(\frac{I'}{I}\right)^2\right) \cdot 100 = \left(1 - \left(\frac{\cos \varphi}{\cos \varphi'}\right)^2\right) \cdot 100 = 29.10\ \% \end{aligned}$$

where $\Delta P' = 2R_l\, I'^2$ and $\Delta P = 2R_l\, I^2$, represents the losses on the power line after and respectively before connecting the capacitors' battery, where the resistance corresponding to one power line is denoted by R_l.

2.3 Intuitive Understanding of Powers in AC Power Systems

2.3.1 *Energy and Power in AC Power Systems*

Let's consider a two-port circuits source (generator) and receptor—shown in Fig. 2.15-having at the terminals the voltage u_b and across it the current i. The *instantaneous power* defined as in Eq. (2.23) at the two-port terminals of the generator and receptor is:

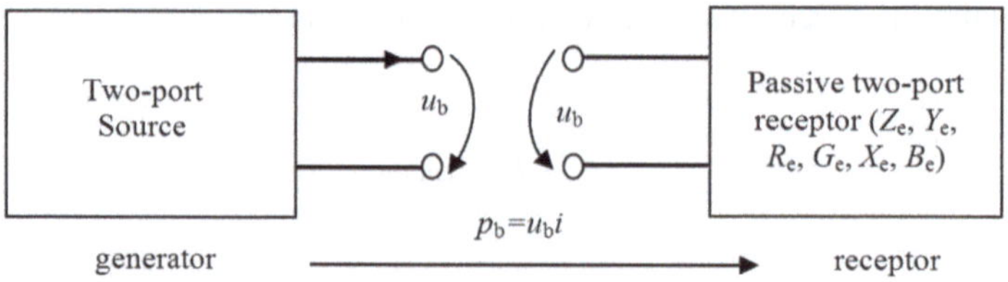

Fig. 2.15 Convention of the directions for defining the power at terminals

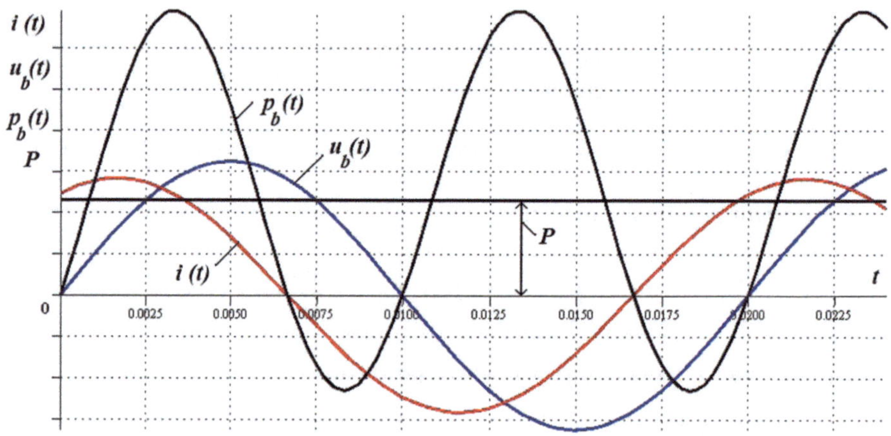

Fig. 2.16 The current and voltage signals of the instantaneous power

$$\begin{aligned} p_b &= u_b i = 2U_b I \sin(\omega t + \beta)\sin(\omega t + \gamma) \\ &= U_b I \cos\varphi - U_b I \cos(2\omega t + 2\beta - \varphi) \\ u_b &= U_b\sqrt{2}\sin(\omega t + \beta), \quad i = I\sqrt{2}\sin(\omega t + \beta - \varphi) \end{aligned}$$

This power is a periodic signal having a constant component and a sinusoidal one, of double frequency (Fig. 2.16).

The energy received (or generated) by a two-port system, in a time $\tau = nT$ or much bigger than a period $\tau \gg T$, is obtained by multiplying the active power with the corresponding time. Indeed, for the energy, we obtain the following successive expressions [2–4, 7]:

$$\begin{aligned} W_\tau &= \int_0^\tau p_b dt = \int_0^{nT} p_b dt = n\int_0^T p_b dt = nTP \\ &= \tau U_b I \cos\varphi = \tau R_e I^2 = \tau G_e U^2 \end{aligned} \tag{2.34}$$

where R_e and G_e are the equivalent resistance respectively conductance of the passive two-port receptor. The *active power* P received by a passive two-port receptor is always positive or at least zero (for non-dissipative circuits). The expression of the instantaneous power shows that even if the circuit is passive, so it is a receiver ($P > 0$), there are moments in time when it is negative, so the circuit gives energy. In those time moments, the energy accumulated in the coils' magnetic field or in the capacitors' electric field is partially returned to the power source.

The *apparent power* (2.28) of the considered passive two-port receptor can be expressed also using the equivalent impedance Z_e or admittance Y_e of passive two-ports receptor $S = U_b I = Z_e I^2 = \sqrt{R_e^2 + X_e^2} I^2 = Y_e U_b^2 = \sqrt{G_e^2 + B_e^2} U_b^2$, where X_e and B_e are the equivalent reactance respectively susceptance.

The apparent power does not have the same important energetic significance, as the active power, but it represents an important computation quantity, because it represents the maximum possible value of the active power at different not-variable values of the voltage and current and for variable phase. Because the machine drives, the electric transformers and the electric pieces of equipment are characterized by maximum admitted limit values of the current (such that the losses caused by Joule effect in conductors don't determine excessive heating) and of the voltage (such that the magnetic circuit doesn't saturate and the isolation is not damaged etc.), the apparent power characterizes their functioning limits and is usually written on the nameplate of the respective device [7, 8].

The *power factor* (at receivers) defined in (2.30) as a positive and subunitary ratio between the active power and the apparent power $\lambda = PF = P/S$, then for two-ports receptor in sinusoidal state is always $\lambda = \cos\varphi$. In order that a given apparent power installation to function with a maximum active power, the corresponding power factor should be maximum (or as close as possible to the unit) that is in sinusoidal state the phase-shift to be as small as possible.

The *reactive power* Q defined as in (2.26) is for the considered two-port receptor in sinusoidal state $Q = S\sin\varphi = U_b I\sin\varphi = X_e I^2 = B_e U_b^2$. The reactive power of the resistive circuit is zero, also the coils consume reactive power, and the capacitors generate reactive power.

The reactive power has been introduced based on the definition relationship built in analogy with the active power expression, parallelism that can be found also in other relationships. The reactive power does not correspond to an average energy contribution at the terminals. However it has a practical significance for the following reasons:

- the power factor (2.30) can be written as

$$\lambda = PF = \frac{P}{S} = \frac{\sqrt{S^2 - P^2}}{S} = \sqrt{1 - \left(\frac{P}{S}\right)^2} \tag{2.35}$$

- the last expression emphasizes the fact that power factor improvement is equivalent to the reactive power (compensation) reduction problem;

- as it will be shown, the reactive power has conservation properties and it will be used for computing the power balance, as if it will correspond to certain specific energy forms (different from the usual energy, that conditions the active power balance);
- the reactive power received by a passive network is proportional to the difference between the average value of the magnetic field energy of the network's coils and the average value of the electric field energy of the network's capacitors.
- the reactive power measures the un-compensation of the internal energy exchange between the magnetic and electric field.

2.3.2 Case Studies

Example 2.5 The instantaneous energetic balance of an *RLC* series circuit where the equation of the voltage of the circuit is $u = u_R + u_L + u_C$. Then by multiplying this equation with i, one obtains the equation corresponding to instantaneous power [6–9]

$$p_b = u_R i + u_L i + u_C i = Ri^2 + Li\frac{di_L}{dt} + Cu_c\frac{du_c}{dt} = Ri^2 + \frac{d(w_e + w_m)}{dt}$$

To simplify the expressions, one chooses the current as phase origin. Then it results

$$u_b = U_b\sqrt{2}\sin(\omega t + \varphi), \quad i = I\sqrt{2}\sin\omega t$$

The voltage can be written as the sum between the active and reactive components

$$u_b = U_b\sqrt{2}\cos\varphi\ \sin\omega t + U_b\sqrt{2}\sin\varphi\ \cos\omega t = u_R + u_X$$

With this decomposition, the instantaneous power can be written as the sum between two components

$$p_b = p_R + p_X = U_b I\ \cos\varphi(1 - \cos 2\omega t) + U_b I\ \sin\varphi\ \sin 2\omega t$$

The first term corresponds to current multiplication with the active component of the voltage and it represents the power developed in the circuit's resistance

$$p_R = u_R i = Ri^2 = RI^2 2\ \sin^2\omega t = UI\ \cos\varphi(1 - \cos 2\omega t)$$

This power is always positive (or zero), and it is called *pulsating* power and it has the average value equal to the active power.

The second term

$$p_X = u_X i = (u_L + u_C i) = \frac{d(w_m + w_e)}{dt} = p_b - p_R = UI \sin\varphi \sin 2\omega t$$

is the instantaneous power resulted from multiplying the current with the reactive component of the voltage and is called *harmonic instantaneous power*, having zero average value and the amplitude equal to the reactive power modulus. The harmonic instantaneous power is equal to the variation speed of the total instantaneous energy (electric w_e and magnetic w_m) of the circuit.

So, *the reactive* power (in modulus) is equal to the amplitude of the variation speed of the energy accumulated by the circuit electromagnetic field.

Another interpretation can be as follows. The reactive power is written in successive forms [10–14] as

$$Q = X_e I^2 = (X_C - X_L) I^2 = L\omega I^2 - \frac{I^2}{C\omega} = 2\omega\left(LI^2 - CU_C^2\right)$$

The magnetic energy average value is

$$\tilde{w}_m = \frac{1}{2T}\int_0^T L\, i^2 dt = \frac{1}{2}L\left\{\frac{1}{T}\int_0^T i^2 dt\right\} = \frac{1}{2}LI^2$$

and the electric energy average value is

$$w_e = \frac{1}{2T}\int_0^T C\, u_C^2 dt = \frac{1}{2}C\left\{\frac{1}{T}\int_0^T u_C^2 dt\right\} = \frac{1}{2}CU_C^2$$

Finally, we have the relationship

$$Q = 2\omega(\tilde{w}_m - \tilde{w}_e)$$

The reactive power is proportional to the difference between the average magnetic energy and average electric energy of the circuit. It results the reactive power is zero when $\tilde{w}_m - \tilde{w}_e = 0$. This relationship corresponds to a zero phase-shift $\varphi = 0$ and defines the resonance condition of the circuit.

2.4 Effects of Reactive Power on the Power System Parameters

The irrational and big reactive power consumption, that generates a reduced power factor, presents a series of disadvantages for the electric installation, among which we mention [15]: the increase of active power losses in the passive elements of the installation, the increase of voltage losses, installation oversize and its diminished electric energy transfer capacity.

(*a*) *Power losses* in the installation's conductors, having the resistance R are given by the following relationship:

$$\Delta P = 3RI^2 = \frac{R}{U^2} S^2 = \frac{RP^2}{U^2 \lambda^2}$$

We notice that it varies inverse proportional to the square of the *PF* at P = cte. and U = cte. Therefore, if the same active power P is transported using different power factors $\lambda_1 < \lambda_2$, then the power losses ΔP_1 and ΔP_2 are inter-dependent according to the relationship: $\Delta P_2 = \Delta P_1 \left(\frac{\lambda_1}{\lambda_2}\right)^2$ from where it results that, by improving the power factor, the power losses diminishes.

(*b*) *Rms value voltage alternations*

If *PF* is an inductive one, for the sinusoidal state, it takes place a voltage reduction on the power mains, and if the power factor is a capacitive one, the voltage in the installation increases. In the sinusoidal state, for an inductive power factor, the longitudinal voltage drop $\Delta U \cong U_1 - U_2$ is given by the relationship:

$$\Delta U = RI \cos \varphi + XI \sin \varphi = \frac{R \cdot P + X \cdot Q}{U_2} = \Delta U_a + \Delta U_r$$

where U_1 is the phase voltage at the power source terminals, U_2 is the phase voltage at the power bars, R and X—the electric resistance and, respectively, the reactance of the line that connects the source to the receiver, P and Q—active power and, respectively, the reactive power transported on a phase of the electric installation, and ΔU_a and ΔU_r—the voltages drops determined by the circulation of the active and, respectively reactive power. As a consequence, it results for $Q = 0$, that is without a reactive power circulation, one obtains: $\Delta U = \Delta U_{\min} = \Delta U_a$.

(*c*) *Diminishing the installation's active power loading capacity* due to a reduced power factor.

So, for the same apparent power S_n it corresponds many active powers $P_1 = S_n \lambda_1$, $P_2 = S_n \lambda_2$ function of power factor's value. If $\lambda_2 < \lambda_1$ then we have: $P_2 = \frac{\lambda_2}{\lambda_1} P_1$ from where it results active power reduction $P_2 < P_1$ increasing the reactive power consumption.

(*d*) *Electric installations oversize* (implies supplementary investment) that functions at a low power factor is explained by that the energy conductors (the electric line) are dimensioned function of the admissible voltage loss and is being verified for heating in a long enough time state. Therefore, if we take into account the admissible voltage losses expression (in a sinusoidal state) $\Delta U_{ad} = \Delta U_a + \Delta U_r$, whose quantity is normalized, then for given P and Q it results

$$\Delta U_r = X \cdot Q/U_n = \text{cte}$$

that leads to

$$\Delta U_a = \Delta U_{ad} - \Delta U_r = \rho \frac{L}{s} \frac{P}{U_n} = \text{cte.} \quad \text{or} \quad s = \rho \frac{L}{\Delta U_a} \frac{P}{U_n}$$

where

L the length of the line;
s the phase conductor's section;
ρ the resistivity of the conducting material;
P the active power flow.

For a given active power, the investment in electric energy sources is inverse proportional to the square of the power factor, and the installed apparent power varies inverse proportional to the power factor.

2.4.1 Investigating the Powers Flow Process in AC Systems

In AC electric installations, characterized from the electric point of view, by a scheme containing active and reactive elements, in general, it takes place a transfer of *active power P* from the source to the receiver, in correlation with the consumers' requirements, as well as a transfer of *reactive power Q* and of *distortion power D*.

An objective energetic characterization of a consumer is done using *the apparent power S* that connects the above mentioned powers [16, 17].

Lately, due to the large-scale use in all domestic and industrial installations of power electronics and due to numerous non-linear receivers, practically we cannot talk about receivers for which the voltage signals and especially the signal of the absorbed current to be pure sinusoids. So, the weight of the distortion power becomes more significant. Therefore, one defines the following: *apparent, active, reactive and distortion power*. For a single-phase consumer, these become

$$\begin{aligned} S &= UI = \sqrt{\sum_{k=0}^{n} U_k^2} \cdot \sqrt{\sum_{k=0}^{n} I_k^2} \\ P &= \sum_{k=0}^{\infty} U_k I_k \cos \varphi_k \\ Q &= \sum_{k=1}^{\infty} U_k I_k \sin \varphi_k \\ D &= \sqrt{S^2 - P^2 - Q^2} \end{aligned} \tag{2.36}$$

where U, I represent the rms value of the voltage respectively of the current and φ_k is the phase shift between the voltage and the current, corresponding to the harmonic of rank k, with $k = 1 \div n$. The unit measures for the four types of powers are given in their definition order: VA, W, VAr and VAd. Most of modern measuring pieces of equipment designed for measuring the electric power and energy can measure instantly each of the 4 powers (energies) previously defined.

The reactive power of the total load of an electric energy consumer has, usually, an inductive character, the load current being phase-shifted behind the voltage; in this case, one considers, conventionally, that the reactive power is positive ($Q_L > 0$) and the receivers represent *reactive power consumers*. For other receiver, the absorbed current is phase-shifted before the voltage; these receivers are considered, conventionally, *reactive power sources*, and the corresponding power in taken in computation with minus sign ($Q_C < 0$). The total reactive power is: $Q = Q_L - Q_C$.

In electric installations, *the Power Factor* (*PF*) defined in (2.30) can be rewritten as:

$$\lambda \equiv PF = \frac{P}{S} = \frac{\sum_{k=0}^{\infty} U_k I_k \cos \varphi_k}{UI} \tag{2.37}$$

In a similar manner, the *Displacement Power Factor* (*DPF*) is defined as the ratio between the active power P_1 and the apparent power S_1, corresponding to the fundamental harmonic ($k = 1$)

$$\lambda_1 \equiv DPF = \frac{P_1}{S_1} = \frac{U_1 I_1 \cos \varphi_1}{U_1 I_1} = \cos \varphi_1 \tag{2.38}$$

We mention that the two definitions for the power factor (*PF* and *DPF*) lead, in general, to different values and the equality $\lambda = \frac{P}{S} = \cos \varphi$ ($PF = DPF$) is valid only in single-phase circuits and only for pure *sinusoidal state* (ideal case). As a consequence, the interpretation of the power factor function of the phase-shift between the current and the applied voltage signals should be carefully taken into consideration.

To characterize the high order harmonics loading degree corresponding to the voltage wave shape or to the consumer's absorbed current, one can use several indicators, the most used one being the *Total Harmonic Distortion* (*THD*) defined in percentage as the ratio between the distortion residue of the signal and the rms value of the fundamental harmonic. So, for current

$$THD_i = \frac{1}{I_1} \sqrt{\sum_{k=1}^{n} I_k^2} \cdot 100\,[\%] \tag{2.39}$$

where I_k represents the rms value corresponding to harmonic k and I_1 is the one corresponding to the fundamental harmonic. Using the value of the total harmonic distortion, one can easily prove the relationship

$$PF = \frac{\cos \varphi_1}{\sqrt{1 + THD_i^2}} = \frac{DPF}{\sqrt{1 + THD_i^2}} \tag{2.40}$$

valid for the case when the voltage distortion is negligible $THD_u < 5\%$ (frequently enough for low-voltage installations). Therefore, if $DPF = 0.92$ but $THD_i = 30\%$ the value of the power factor is only $PF = 0.88$.

The previous relationships define the *instantaneous power factor* that corresponds to a certain moment from the consumer's installations functioning.

Because the electric charge presents fluctuations, the current legislation recommends the determination *of the weighted mean power factor* based on the consumption of:

– active energy

$$W_a = \int_0^t P \cdot dt$$

– reactive energy

$$W_r = \int_0^t Q \cdot dt \tag{2.41}$$

– distortion energy

$$W_d = \int_0^t Q \cdot dt$$

– for a certain period t (month, year)

$$PF_{med} = \frac{W_a}{\sqrt{W_a^2 + W_r^2 + W_d^2}}$$

From an energetic point of view, this is useful to characterize the consumer's installation and to charge its consumed electric energy. The weighted average power factor can be *natural*—when is determined without taking into consideration the reactive power compensation and *general*—when considering its evaluation one takes into account the losses corresponding to this installation. The value of the general weighted mean power factor from which the reactive energy consumption is no further charged is called *neutral power factor*. This value is being determined using technical-economical computation for minimizing the active power losses and is a local quantity that depends on the position of the consumers in the electric network. At international energetic system level the established value is equal to 0.92 [18, 19].

2.4.2 Reactive Power Consumers

The main reactive power consumers are *the asynchronous motors and the electrical transformers* which consume around 60% respectively 25% from the installation's total reactive power due to producing alternating magnetic fields [20–25]. At industrial consumers' level the weight is around 70% for asynchronous motors and around 20% for transformers. The difference between the reactive power consumptions for asynchronous motors and transformers, at the same active power and the same magnetic stress, comes from the fact that the magnetization reactive power, that constitute the most important component of the reactive power, depends on the volume of the magnetic circuit to which, for asynchronous motors' case, one adds also the volume of the air-gap (not existent for transformers). Another component of the reactive power for the asynchronous motors and for transformers is the reactive power dependent on load, called also dispersion reactive power.

(A) For *asynchronous motors* case, the magnetization reactive power (or the no load power) represents the most part from the motor's reactive power, function of motor's loading and function of the air-gap. Taking into account the exploitation average loading: $\beta < 0.5$ (evaluated using the loading factor as the ratio between the mechanical shaft power P and mechanical nominal power P_n of the motor $\beta = P/P_n$), one can approximate the reactive power of an asynchronous motor as being constant and independent of the load, while the active power depends on the motor' load. Also, the motors' active power remains practically constant for small deviations of the voltage compared to the nominal voltage, while the reactive power depends essentially on the voltage's variation. *Asynchronous motors* functioning with a load factor $\beta < 1$ due to an inappropriate technological exploitation, determines the power factor reduction under its nominal value.

The reactive power Q absorbed by asynchronous motor for any load P, is determined using the relationship [26, 27]

$$Q = Q_0 + Q_d = Q_n\left[\alpha + (1 - \alpha)\beta^2\right] = Q_0 + (Q_n - Q_0)\beta^2$$

where $\alpha = Q_0/Q_n$ is the ratio between the reactive power Q_0 for unloaded running $(\beta = 0)$ and the reactive power Q_n absorbed at nominal load $(\beta = 1)$; $Q_d = (1 - \alpha) \cdot \beta^2 Q_n$—dispersion reactive power. Considering the definition of PF in sinusoidal state, for a symmetrical loading of the three-phase network, it results

$$\lambda = \cos\varphi = \frac{\beta}{\sqrt{\beta^2 + \left[\alpha + (1-\alpha)\beta^2\right]^2 \tan^2\varphi_n}} \tag{2.42}$$

For values $\beta < 0.5$ the reduction of the *PF* under its nominal value is highly accentuated. If in exploitation, the power voltage of the asynchronous motors increases, it results an increase of the absorbed reactive power, having undesired consequences upon *PF*—Fig. 2.17a. This is due to the increase of the magnetization current in the saturated area. In Fig. 2.17b is presented, as a rough guide, the dependence of the absorbed reactive power function of the nominal one (Q/Q_n) for asynchronous motors of low and high power function of the relative value of the power voltage for different loading factor's values [28, 29].

(B) For *transformers* the absorbed reactive power is computed using the relationship [26, 27, 30, 31]

$$Q = Q_0 + Q_d = \frac{S_n}{100} \cdot \left(i_0 + k_f \cdot \beta^2 \cdot u_{sc}\right)$$

where S_n is the nominal apparent power, i_0—the current for unloaded running (expressed in percentage function of nominal current), k_f—the form factor of the load signal (defined as the ratio between the mean square value and the mean value, of the load current, computed for a given time interval), $\beta = S/S_n$—the transformer load factor, u_{sc}—the voltage short-circuit voltage (expressed in percent).

As for asynchronous motors case, the transformers functioning at a power under the nominal one determines the reduction of *PF*. In real exploitation conditions the

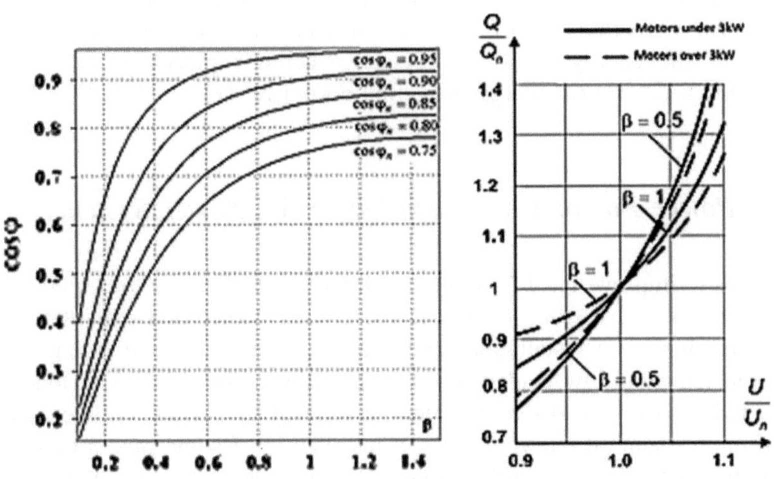

Fig. 2.17 Power factor variation function of asynchronous motor's loading and the reactive power consumption for small and big power motors function of the relative power voltage

transformers' total reactive power can be evaluated to 10% of the nominal power (8% unloaded running power and 2% dispersion power).
(C) *Electric ovens* installations (arc or induction) consume reactive power from the oven's power supply transformer, the oven adjustable autotransformer and the power supply circuit of the oven from the transformer. Therefore, for a three-phase arc oven, the *PF* of the installation is, usually, sufficiently big (0.8 ÷ 0.9) but the absolute reactive power consumption is big, compared to the oven's power (that can reach a value of 80 MW for a capacity of 400 t).
(D) *Electric Distribution Lines* can determine reactive power consumption Q_L that can be computed using the relationship $Q_L = \omega LI^2$ where L is the line's equivalent inductivity, ω—the voltage pulsation on the line, and I—the electric current that circulates the line. On the other hand, the electric lines generate reactive power Q_C due to their capacity C against the ground: $Q_C = \omega CU^2$. For electric lines case, the resulting reactive power can have positive or negative values function of the values of the two components Q_L and $Q_{C,}$ the first one depending on the square of the electric current that circulates the line and the second one depending on the square of the voltage.
(E) *Power Controlled rectifying installations* supplied by a transformer represent also an important reactive power source. The schemes adopted for controlled rectifiers can lead also to a small power factor [32–37]. Lately, the consumer containing power electronic circuits represent, due to their large scale use, the main distortion power source from installations. As a consequence, the power factor of the whole installation containing such receivers is diminished.
(F) A contribution to the reactive power flow is given by *electric discharge lamps* in metallic vapours when they are connected in an uncompensated inductive ballast schemes.

The *PF* monitoring is very important for the producer, the transporter, the distributer, the provider and the final user of electric energy, because it influences the performance characteristics of all these services, it determines the availability capacity of the energy transfer for electro-energetic pieces of equipment and imposes the final electric energy providing costs [29–31].

2.4.3 Power Factor Compensation in AC Power Systems

The means to improve the *PF* are grouped into *natural means and special means. Natural means* derive from a rational and correct choice and exploitation of the existing machinery in installations and consists of technical and managerial measures (replacing the low-loaded transformers and motors, nonlinear loads aggregation, reducing the unloaded functioning time of the machineries etc.) and *the special* ones assume the introduction in installation of some pieces of equipment generating reactive power and/or limiting the distorted state (installation of some capacitor banks, of active filters networks conditioners etc.).

As a consequence, *PF* improvement includes, on a first stage, *operations for mitigation the high harmonics' attendance from the wave shapes* of the voltage and the absorbed current, and, on the second stage, *the limitation of reactive power flow*. Any method dedicated to reducing the reactive power absorbed by a consumer is efficient only if the power voltage is practically sinusoidal.

(1) *The use of capacitors banks as a reactive power source*

This way for improving the power factor, even it's highly spread, it's efficient only if the signals corresponding to the current and to the voltage are close to a sinusoid (the contribution of the distorted state is modest). The placement of the capacitors in a low voltage installation represents the compensation procedure that can be [36–38]: global (placement in a single point for the entire installation), by sector (group with sector), local or individual (at each equipment) or a combination between the last two. Generally speaking the ideal compensation is applied at the consumption place and it has the level in concordance with the instantaneous power values, but in practice the choice is decided by economical and technical factors—Fig. 2.18.

(2) *Individual compensation* is applied, firstly, in the big reactive power loads (asynchronous motors, electric ovens etc.) and with continuous functioning, ensuring the reactive power compensation at the consumption place and unloading the remaining of the network from the reactive power flow, with all the advantages deriving from this one—Fig. 2.19.

For individual compensation of *asynchronous motors*, the capacitors bank is usually directly connected to the motors' terminals and the decoupling from the network taking place at the same time as the motor, using the same switching device. To avoid the overcompensation that occur for the no loaded and the auto-excitation (when the motor' breaks), around 90% from the unloaded functioning power is compensated, ensuring a power factor of around 0.9 at normal loading and approx. 0.95 at incomplete loading or unloaded functioning. The reactive power Q_c necessary to compensate three-phase asynchronous motors is determined by the motors' unloaded operating current value I_0 (given in theirs catalogue data)

$$Q_C = Q_0 = 0.9\sqrt{3}U_n I_0 \tag{2.43}$$

For motors having powers bigger than 30 kW one can choose a covering value $Q_C \cong 0.35P_n$ [39].

For power distribution transformers' case, the compensation bank is placed on the low voltage part and very often consists of a fixed step (necessary to compensate the unloaded absorbed reactive power) followed by a step that covers the necessary reactive power at nominal load: the bank's dimensioning takes place using the relationship

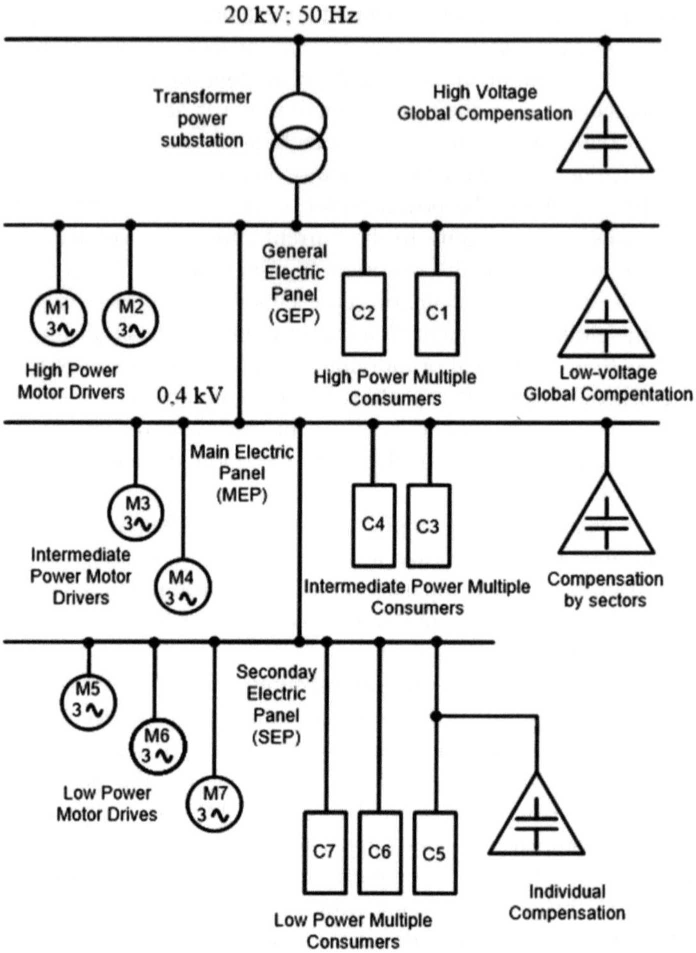

Fig. 2.18 Examples for placing the capacitors bank

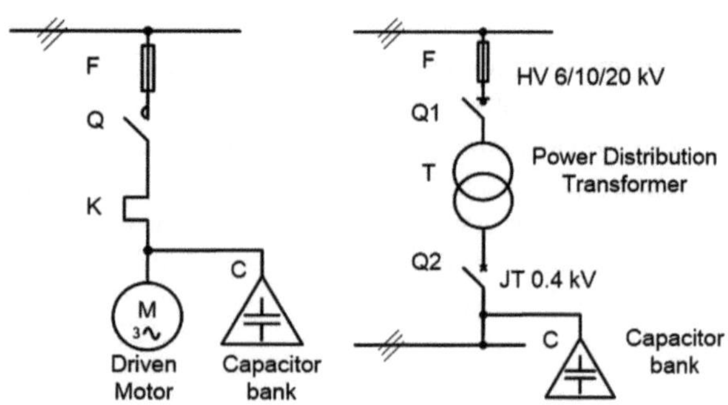

Fig. 2.19 Individual compensation for asynchronous motors and for transformers

$$Q_C \cong Q_0 + Q_d = \frac{i_0[\%]}{100} S_n + \frac{u_{sc}[\%]}{100} S_n \cdot \beta^2 \tag{2.44}$$

where i_0 [%] respectively u_{sc} [%] represent the percent values corresponding to the unloaded running current and to the short circuit voltage of the transformer (given in its catalogue data) and $\beta = S/S_n$ is the load factor rate. Usually, one can consider: $Q_C = (0.1 \div 0.2)S_n + (0.5 \div 0.6)S_n$.

The individual compensation is applied also to illumination lamps' with metallic vapours electric discharge, the capacitors' values is being mentioned by the producer as a function of the lamp's rated power.

(3) *Compensation on receivers' groups* (by sector) is applied when the reactive power consumers are grouped, the capacitors' banks being connected to the mains from the distribution panels corresponding to the receivers' groups. The bank's power and the operating mode are established as a function of the receivers' uncompensated reactive power signal. This compensation mode limits also the reactive power circulation in the network above the setting place.

(4) *Centralized compensation* can be realized by connecting the capacitors bank at the mains (coupling points) from the general electric distribution panel from the transformer power substation. The bank is executed in commutation steps, usually automated, function of the reactive power that should be compensated, corresponding to powered receivers running. The bank can be connected also to the intermediate high voltage of the transformer power substation. For a centralized compensation, intermediate high voltage, the positive technical effects are not present at low voltage installations belonging to the consumer (downstream to the installing place). But, the big consumers powered directly in intermediate high voltage, use the centralized compensation that provides a general power factor bigger than the neutral power factor value such that the electrical energy consumption to be billed only for active energy.

(5) *Mixed compensation* uses all the procedures presented before for reactive power compensation. The solution is applied in steps or, when there are certain conditions specific to the respective consumer.

The power Q_b corresponding to the capacitors bank is determined such that, to a certain given active power P, absorbed under a power factor $\cos\varphi_1$, to obtain an improved power factor $\cos\varphi_2 > \cos\varphi_1$. The reactive power consumed by these capacitors will be: $Q_b = -3\omega C U_c^2$. The receiver consumes a reactive power before adding the bank, $Q_1 = P\mathrm{tg}\varphi_1$. The necessary power from the network after adding the capacitors will be $Q_2 = Q_1 + Q_b$, from where

$$Q_b = Q_1 - Q_2 = P(\tan\varphi_1 - \tan\varphi_2) \tag{2.45}$$

where $Q_2 = P\tan\varphi_2$ represents the reactive power received from the network after introducing the capacitors bank—the active power P remains constant, and $\cos\varphi_2$ is the new power factor that should be realized. As a consequence, the value of the capacity for one capacitor from the triangle bank will be

$$C = \frac{P(\tan\varphi_1 - \tan\varphi_2)}{3\omega U_c^2} \tag{2.46}$$

In Fig. 2.20 the connection possibilities for the capacitors that form the bank in single and three phase installations are presented. If one takes into account the power loss ΔP_C in the capacitors' dielectric the power necessary for the bank becomes

$$Q_b = P \cdot \tan\varphi_1 - (P + \Delta P_C) \cdot \tan\varphi_2 = Q_1 - Q_2 \text{ with } \Delta P_c = \frac{p_d}{100} \cdot Q_b \tag{2.47}$$

where $p_d = 0.25 \div 0.35\%$, represents the losses from the dielectric expressed as a percent function of the capacitors bank power Q_b. These depend on the tangent of the material's losses angle $tg\delta$ after the relationship: $\Delta P_C = \omega C U^2 \tan\delta$ and can be considered (especially in low voltage) negligible.

One can note that for Y bank's connection result capacitors having a capacity three times bigger, because instead of U_c is the voltage $U_f = U_l/\sqrt{3}$. Results $C_\Delta/C_Y = (U_f/U_l)^2 = 1/3$ and, as a consequence, the power factor compensation is an economical and technical problem, taking into account that, in low voltage, the cost of the capacitors is proportional to their capacity. This is the reason why Δ connection is preferred. The Y assembly is advantageous for intermediate high voltage networks, because nominal voltage of the capacitors is then reduced. One can notice that the maximum voltage for Δ connection capacitors is over 580 V compared to Y connection where the voltage is only 320 V. The capacitors bank power depends on the network voltage variation. So, if the network voltage varies from U_n to U_{n2} the capacitors bank power modifies from Q_b to the value $Q_{b2} = Q_b(U_{n2}/U_n)^2$.

The time variation signal corresponding to the reactive power consumption Q in an electrical installation is, in general, not a linear one. The necessary reactive power depends on the way in which the electric energy is being used in various technological processes, fact emphasised by the load variation. Therefore, the absorbed reactive power can vary slowly or rapidly in very large limits. In this

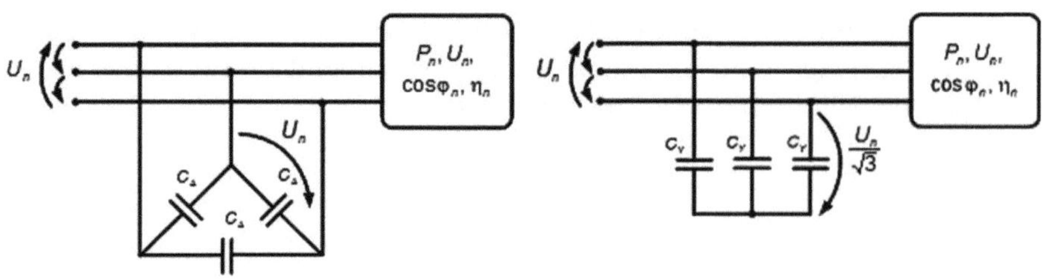

Fig. 2.20 Connections possibilities for capacitors bank

situation, the adoption of some capacitor banks with fix value can determine over and under compensations regimes. To avoid such a major inconvenient, the problem is solved by realizing a capacitor bank from several sections (stages) of fixed power. Each section consists of automatic controlled commutation pieces of equipment, following several criteria, for example, function of the mains voltage, the load current, and the direction of the reactive power change with the energetic system or function of the running time corresponding to different loads—Fig. 2.21.

The selection of the number and of the values for the capacitors bank sections should take into account that the control efficiency increases with the number of steps, making possible to follow very closely the load reactive power signal. On the other hand, an excessive fractioning of the bank becomes at a certain point not viable from economical point of view because it implies the use of a complex commutation apparatuses. Usually, the capacitor banks use a number of 4 ÷ 12 sections having the powers between 5 ÷ 25 kVAr [34, 38, 39].

For the scheme presented in Fig. 2.21, the control block CB establishes the value of the power factor (cos φ) and knowing the capacities corresponding to the bank's sections, connects the elements necessary to realize the prescribed power factor. Exceeding the prescribed power factor leads to disconnect the capacitors bank sections.

Systems' selection for reactive power compensation in non-sinusoidal state. To select a system for power factor correction in an installation where there are present high-order harmonics in the current and voltage waves shape, function of the user's data, one uses the following methods [35–40]:

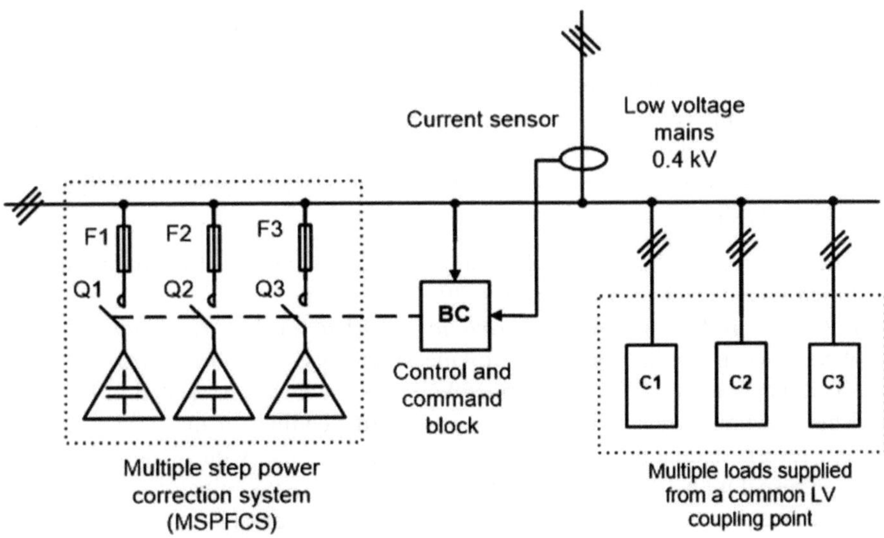

Fig. 2.21 The principle scheme for an automated controlled power factor in an installation

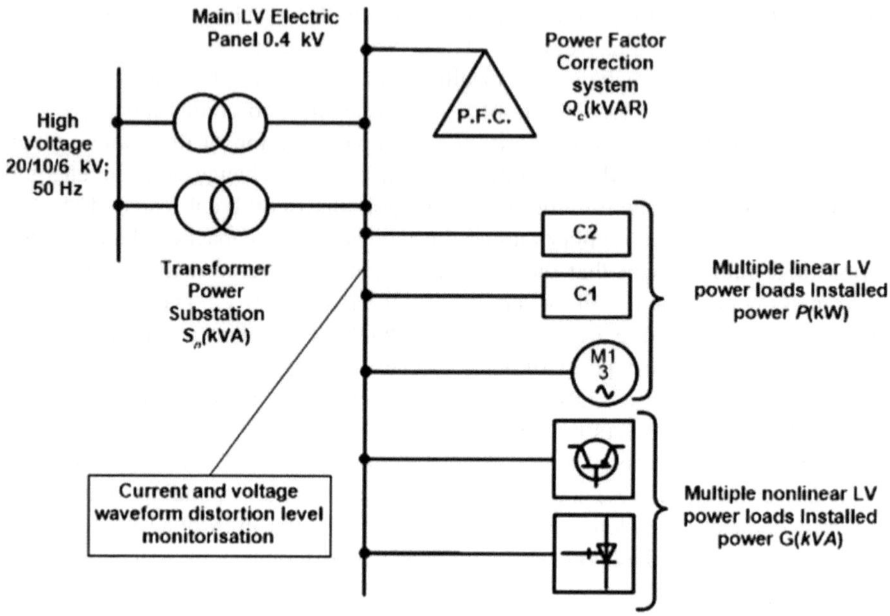

Fig. 2.22 Connecting the power factor correction system in a non-sinusoidal state system

(A) *Distorted installed power procedure*, whose principle is described in Figs. 2.22 and 2.23. The following notations are used:

- G is the sum of apparent powers taken for all pieces of equipment that generate harmonics (power static converters, inverters variable speed drives, etc.) connected to the mains where the capacitors bank is also connected. If, for some equipment the active power is given, to compute the apparent power one should take into consideration the covering *PF*.
- S_n is the sum of apparent powers taken for all system's power transformers to which the distribution mains belong.

The above method can be applied only for limited voltage and current harmonic distorted level ($THD_u < 5\%$) ($THD_i < 30\%$).
(B) *Voltage and current total harmonic distorted level determination method*, applicable when one knows exactly (by measurements) the values corresponding to THD_u respectively to THD_i. The principle of this method is presented in Fig. 2.24.

The selection methods indicate the use of (depending on the harmonic's distortion level): standards capacitor banks, capacitor banks with increased nominal voltage (usual with 10%), capacitor banks with increased nominal voltage and with detuned reactors or harmonic mitigation filters (active, passive or hybrid).

One should mention that in actual conditions, where the installation's load with voltage or especially current high-order harmonics are very frequent, it is used the

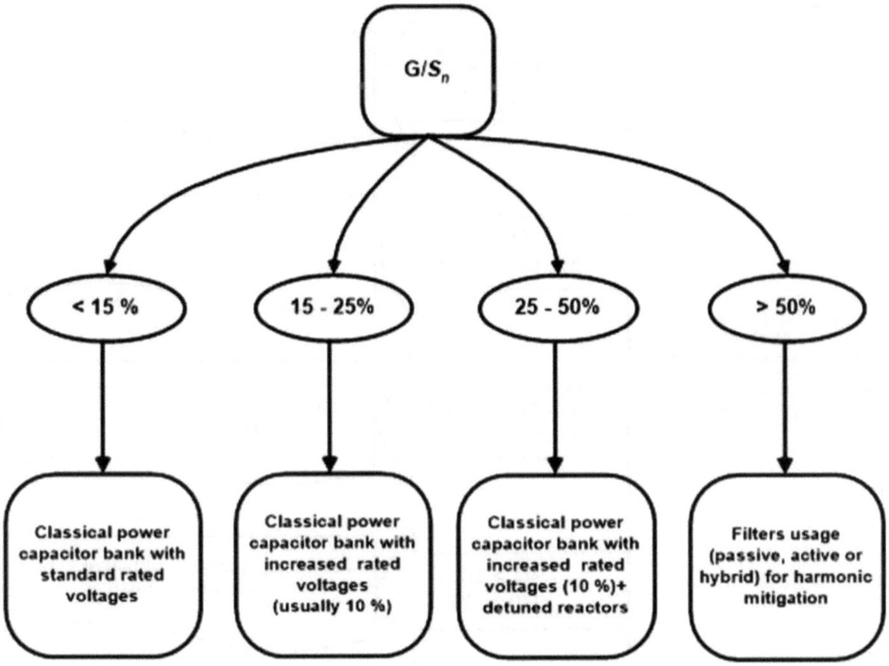

Fig. 2.23 Selecting the reactive power compensation possibility function of installation's nonlinear receivers' weight

solution (cheaper than that with active filters) of capacitor banks placed in series with certain coils also called "detuned reactors" shown in Fig. 2.25 [38–40]. The reactive power corresponding to these detuned reactors is chosen usually as a percent, called detuned reactor factor, of $p = 5$, 7 or 11% from the compensation reactive power of the bank. Therefore, by a correct selection of the detuned reactor factor, one may avoid the resonance on the dominant installation's harmonics and at the fundamental frequency the compensation function is satisfactory maintained. Moreover, the detuned reactors, that have values of mH order, together with the connecting conductors' inductivity, limit the commutation inrush currents in the case of banks from automatic power factor correction system (APFC). Moreover, the detuned reactors supplementary stress the capacitors' dielectric, reason for which these should be dimensioned at voltages at least 10% bigger than the nominal ones.

The reduction of the connecting current value is done by placing in series some very small inductivity coils (a few μH). For low-voltage banks where the harmonics do not have a consistent weight ($THD_{u,i} < 10\%$) also the connecting cable, out of which several turns are coiled, realizes a current limitation. So, only the conductivity of a cable of 10 m is sufficient to limit the current necessary to connect a bank up to 25 kVAr [29–31].

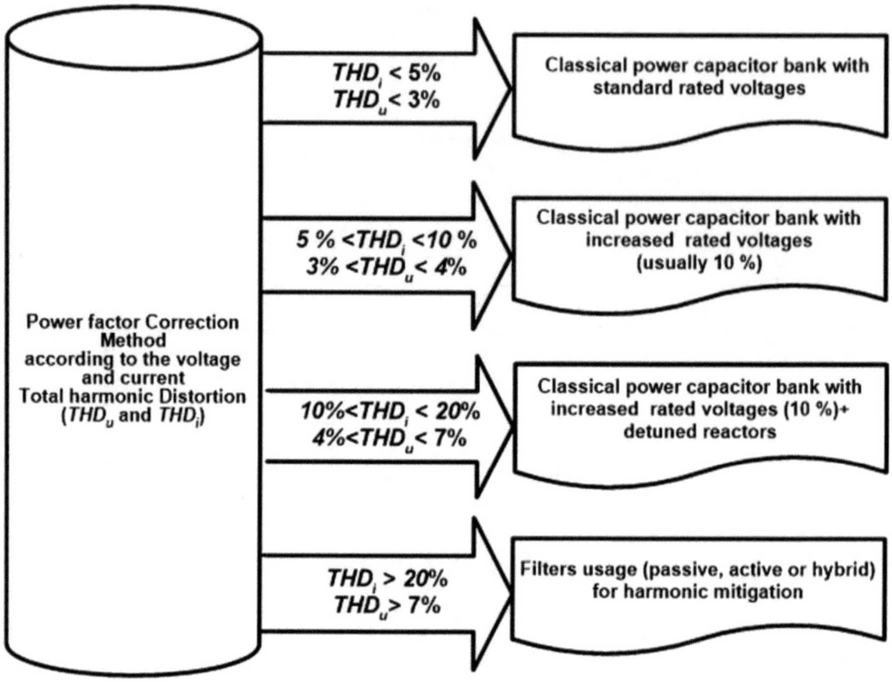

Fig. 2.24 Selection of reactive power compensation functions of installation's nonlinear loads' weight

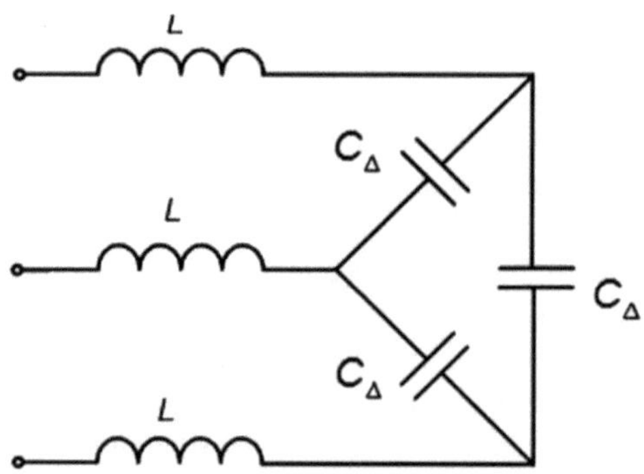

Fig. 2.25 Connecting the detuned reactors in a Δ connection capacitors bank

These measures should determine the consumers to function with a power factor in exploitation at values very close to the nominal one and finally to lead to an increased average weighted power factor at least up to the neutral power factor value (usually 0.92).

2.4.4 Case Studies

Example 2.6 Let's consider an industrial load that contains linear power consumers (especially asynchronous single and three-phase power motors) for which the electric power quality parameters are presented in Fig. 2.26. The figures are captured by using a Fluke 435—a "class A" power quality analyzer instrument. As one can notice, from the main parameters analysis (current and voltage signals, harmonics distortions etc.), the system allows the connection of a capacitor bank.

Its power is evaluated taking into consideration both the value of the actual measured power factor ($PF_1 = \cos\varphi_1$), as well as that of the desired power factor (mostly is the value of the neutral factor value: $PF_2 = \cos\varphi_2 = 0.92$ or even 0.95). Knowing the active power $P = 223.7\,kW$ and choosing for $\cos\varphi_2 = 0.95$ we can determine the reactive capacitor bank value

$$Q_{bat} = P \cdot (\tan\varphi_1 - \tan\varphi_2) = 223.7 \cdot (0.619 - 0.203) \cong 93.2 \text{ kVAr}$$

The three-phase capacitors used to improve the *PF* can be placed in Δ or Y connection (Fig. 2.20). The bank's capacity on a phase results from the following relationship:

$$C_\Delta = \frac{Q_C}{\omega \cdot \left(U/\sqrt{3}\right)^2} = \frac{93.2 \cdot 10^3}{2 \cdot \pi \cdot 50 \cdot \left(400/\sqrt{3}\right)^2} = 5.56 \text{ mF}$$

$$C_\Delta = \frac{C_\Delta}{3} = \frac{Q_C}{\omega \cdot U^2} = \frac{93.2 \cdot 10^3}{2 \cdot \pi \cdot 50 \cdot (400)^2} = 1.85 \text{ mF}$$

So, the for the analyzed system case, one recommends an automatic power factor correction system (APFC) (in steps taking into consideration that energy can vary in large limits, function of the number of loads connected at the same time) of power value: $Q_{bat} = 95$ kVAr. A solution of the automatic capacitor bank can be realized in three steps of 30, 30 respectively 55 kVAr. With the new power factor (0.95), the current absorbed by the consumer becomes

$I_2 = I_1 \frac{PF_1}{PF_2} = 402 \cdot \frac{0.86}{0.95} = 363$ A, (the measured current one being $I_1 = 402$ A)
With this value, the losses in a three-phase copper power cable 3 × CYY 2 × 120 mm^2, can be computed. The resistance on the Cu cable per unit length of section 120 mm^2 is $r_l = 0.154$ mΩ/m. But on each phase two such cables of length $l = 100$ m are used. Thus the total specific resistance is given by the parallel equivalent resistance of the cables

$$r_t = \frac{r_l}{2} = \frac{0.154}{2}\frac{\text{m}\Omega}{\text{m}} = 0.077\frac{\text{m}\Omega}{\text{m}}$$

$$R_0 = r_t \cdot l = 0.077 \cdot 100 = 7.7 \text{ m}\Omega$$

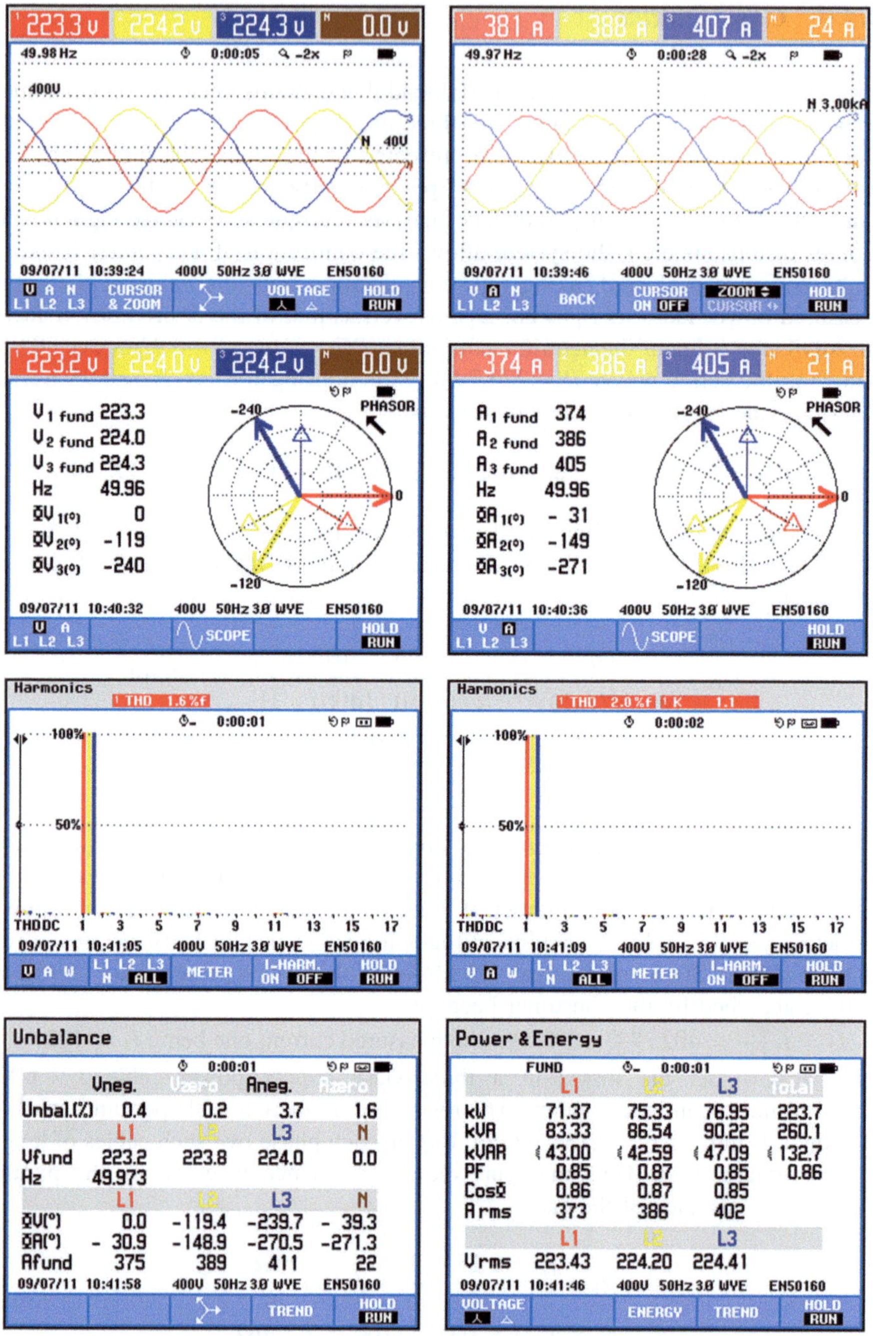

Fig. 2.26 The energy quality parameters of the industrial consumer under investigation

Considering the average working temperature of the cable θ_2 = 42 °C, (resulted eventually from analyzing the thermal image), we will compute the value of the conductors' resistance

$$R_L = R_0(1 + \alpha \cdot (\theta_2 - \theta_1)) = 7.7 \cdot (1 + 0.00339 \cdot (42 - 20)) = 8.27\ m\Omega$$

The power losses before and after compensating the reactive energy flow are:
Before

$$\Delta P_1 = 3 \cdot I_1^2 R_L = 3 \cdot 402^2 \cdot 8.27 \cdot 10^{-3} = 4\ \text{kW}$$

and after

$$\Delta P_2 = 3 \cdot I_2^2 R_L = \Delta P_1 \left(\frac{PF_2}{PF_1}\right)^2 = 4\left(\frac{0,86}{0,95}\right)^2 = 3.27\ \text{kW}$$

Results a consumer's power cable loss

$$\Delta P_1 - \Delta P_2 = 730\ \text{W}$$

or given in percent

$$\varepsilon = \frac{\Delta P_1 - \Delta P_2}{\Delta P_1} = 18.25\%$$

The power cable consumed energy can be correct evaluated if one knows the form factor of the current variation (on the most loaded phase), that is represented in Fig. 2.27.

With this information one will compute the average value I_m and the square average value I_{mp} of the current, and its corresponding form factor, respectively

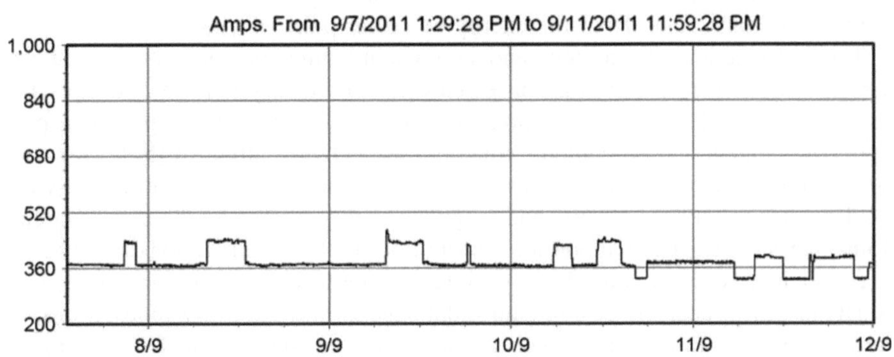

Fig. 2.27 The current variation on the consumer's most loaded phase taken on a monitor interval

$$I_m = 363\,\text{A},\ I_{mp} = 368\,\text{A},\ k_f = I_{mp}/I_m = 1.063$$

The energy for a 30 days period (720 h) before and respectively after compensation will be

$$\Delta E_1 = 3 \cdot k_f I_m^2 R_L \tau_f 10^{-3} = 3 \cdot 1.063 \cdot (363)^2 (8.27 \cdot 10^{-3}) \cdot 720 \cdot 10^{-3} = 2.502\ \text{MWh}$$

$$\Delta E_2 = 3 \cdot k_f \left(I_m \frac{PF_1}{PF_2} \right)^2 R_L \tau_f 10^{-3} = \left(\frac{PF_1}{PF_2} \right)^2 \Delta E_1 = 2.049\ \text{MWh}$$

It results a consumer's power cable losses reduction

$$\Delta E_1 - \Delta E_2 = 453\ \text{kWh}$$

or given in percent

$$\delta = \frac{\Delta E_1 - \Delta E_2}{\Delta E_1} = 18.10\%$$

2.5 New Principle of Minimum Active and Reactive Absorbed Power (PMARP) in AC Power Systems

In different real systems such as mechanics, thermodynamics, Earth climate system, hydrology, scientific studies have been carried out to express the balanced state of the systems in energy and/or power terms [41–50]. These studies are mostly based on the concept that for conservative systems it is possible to define suitable functionals in Hilbert space related to power and/or energy. An important step forward in order to define the functional optimization problems is to calculate their limits. On the other hand the studies cited below report that the steady state in mechanic, thermodynamic, hydrology or climate conservative system is an edge state, generally a minimum state, from the energy or power point of view.

The introduction of "content and co-content" function applied to non-linear resistors represented a milestone in electric circuits theory [51–53]. Authors like Penfield in [54] demonstrate for nonlinear resistive circuits the "variational principle" of "content and co-content" function, respectively Smith in [55], [56] analyses for linear passive circuits the average energy storage, the minimum energy and the dependence between electric and magnetic energy storage to one-port of circuit. Romanian scientists have important contributions to the minimum/maximum theory of power in electric circuits. For example, the application of Hilbert space functionals to power analysis in non-sinusoidal state is introduced by Ionescu in [57], the theorem of minimum power absorbed by resistances in DC circuits is demonstrated by Mocanu in [58], and the analysis and solving of electromagnetic field based on energy functional are defined by Hantila in [59] and Fireteanu in [60].

This section refers to new theorems (principles) of minimum active and reactive absorbed power (PMARP) for linear electric circuits in quasi-stationary regime which are based on power functionals and Hilbert space properties such in the recent contributions of authors [61–69]. On the other hand, based on analogy between linear electric and magnetic circuits in quasi-stationary regime, similar principles are described by authors in [70–74]. In this work for AC circuits, including non-sinusoidal regime, is demonstrated a complete set of principles of minimum absorbed active and reactive power, are defined the specific conditions and limits of their applications and finally are provided suggestive examples for each circuit. For all the types of AC circuits the principles demonstration are similar. As a starting point is the definition of power functionals and their expressing depending on the potential of circuit nodes which are considered variable.

In this respect the Kirchhoff voltage law (KVL) and the topology of circuit are used. Then by imposing the conditions for obtaining the minimum point of functionals (when existing), it is demonstrated that these equations are similar with Kirchhoff current law (KCL) and nodal method (NM). Consequently by using the power flow theory, either a complete system of Kirchhoff equations (KVL and KCL) is obtained from which one determines the circuit currents and voltages or a complete system of NM from which one computes the potentials of nodes. Or in other words, KCL equations can be substituted with those arising from PMARP. Likewise in this section is demonstrated the coexistence (CEAPP) of PMARP and maximum power transfer theorem (MPTT) for active power in AC circuits. Therefore PMARP can be considered as part of basic concept of circuit theory.

2.5.1 *PMARP for Linear AC Circuit*

Let us consider the common branch of a linear and reciprocal AC circuit under sinusoidal signal shown in Fig. 2.28. The circuit contains *N* nodes and *K* branches with voltage sources and RLC passive elements. In order to put in evidence the

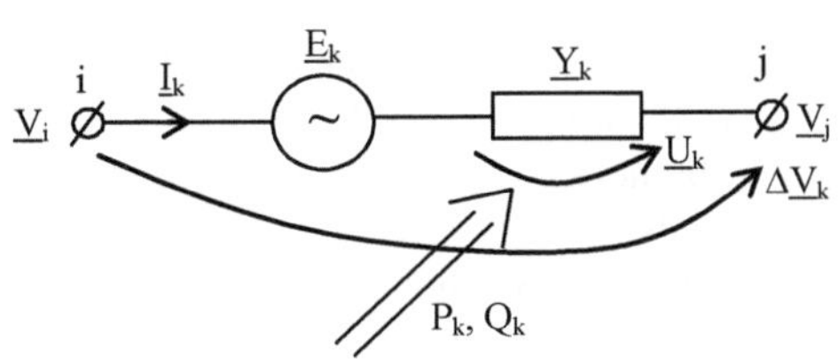

Fig. 2.28 Representation of an AC circuit branch

power flow between the sources and passive elements, one defines in $\Re^{2(N-1)}$—dimensional Hilbert space, two functionals. The first one is the active power Eq. $\Im_P : \Re^{2(N-1)} \to \Re$, and the second one is the reactive power Eq. $\Im_Q : \Re^{2(N-1)} \to \Re$ defined as

$$\Im_P = \mathrm{Re}[\underline{u}^T \underline{i}^*] \tag{2.48}$$

$$\Im_Q = \mathrm{Im}[\underline{u}^T \underline{i}^*] \tag{2.49}$$

where $\underline{u}$ represents the K-dimensional vector in C^K of complex voltages at the branch admittances terminals and $\underline{\boldsymbol{i}}$ represents the K-dimensional vector in C^K of the branch complex currents. In this section the superscript * denotes the conjugate operator. Related to the branch voltages and current baseline direction presented in Fig. 2.28, the power Eqs. (2.48) and (2.49) mean: (i) $\Im_P$ is the *active power absorbed by all the passive elements' conductances (resistances) of the circuit,* respectively (ii) $\Im_Q$ is the *reactive power absorbed* or *generated by all the reactive elements (inductors and capacitors) of the circuit.* By using KCL equation for each k-branch $\Delta \underline{V}_k = \underline{V}_{i,k} - \underline{V}_{j,k} = \underline{U}_k - \underline{E}_k = \frac{\underline{I}_k}{\underline{Y}_k} - \underline{E}_k$, for $k = 1, \ldots, K$ and considering as variables the complex potentials of circuit nodes $\underline{V}_i = x_i + jy_i$, for $i = 1, \ldots, N-1$ $x_i, y_i \in \Re$, then $\Im_P$ and $\Im_Q$ can be expressed as

$$\begin{aligned} \Im_P(x, y) &= \mathrm{Re}[\underline{u}^T \underline{i}^*] = \mathrm{Re}[(\Delta \underline{v} + \underline{e})^T \underline{Y}^* (\Delta \underline{v} + \underline{e})^*] \\ &= \mathrm{Re}[(C\underline{V} + \underline{e})^T \underline{Y}^* (C\underline{V} + \underline{e})^*] \\ &= \sum_{\substack{k=1,K \\ i,j=1,N-1 \\ i \neq j}} G_k[(x_i - x_j + a_k)^2 + (y_i - y_j + b_k)^2] \end{aligned} \tag{2.50}$$

$$\begin{aligned} \Im_Q(x, y) &= \mathrm{Im}[\underline{u}^T \underline{i}^*] = \mathrm{Im}[(\Delta \underline{v} + \underline{e})^T \underline{Y}^* (\Delta \underline{v} + \underline{e})^*] \\ &= \mathrm{Im}[(C\underline{V} + \underline{e})^T \underline{Y}^* (C\underline{V} + \underline{e})^*] \\ &= \sum_{\substack{k=1,K \\ i,j=1,N-1 \\ i \neq j}} B_k[(x_i - x_j + a_k)^2 + (y_i - y_j + b_k)^2] \end{aligned} \tag{2.51}$$

where $\Delta \underline{v}$ is the K-dimensional vector in C^K of branch complex voltages, $\underline{e}$ is the K—dimensional vector in C^K of branch voltage sources expressed as $\underline{E}_k = a_k + jb_k$, where a_k, b_k are considered constant, $\underline{Y} = \mathrm{diag}(\underline{Y}_1, \underline{Y}_2, \ldots, \underline{Y}_K)$ is the $K \times K$—dimensional diagonal matrix of branch admittances, where $\underline{Y}_k = G_k - jB_k$, $k = 1, \ldots, K$ with $G_k > 0$, $B_k > 0$ for inductive branches respectively $B_k \langle 0$ for capacitive branches, $\underline{V}$ is the reduced (N-1)-dimensional vector in C^{N-1} of nodes potentials, and $C = [c_{1,n}]$ is the $K \times (N–1)$-dimensional reduced branch-to-node incidence matrix whose elements $c_{1,n}$ can take the values $-1, 0$ or 1. The superscript T is used to indicate the transposition.

The next step of PMARP demonstration is to analyze the sign functionals defined in (2.50) and (2.51), and as a consequence of the nature, minimum or maximum, of thereof extreme point.

(α) Evidently if $G_k > 0$, then the active power Eq. is a quadratic form (strictly positive) $\Im_P(x, y)\rangle 0$, and its *extreme point is a minimum*. Therefore one can affirm that the *active power absorbed by all the passive elements (conductances) of the AC circuit is minimum*. By imposing the concurrent conditions $\partial \Im_P/\partial x_i = 0$ and $\partial \Im_P/\partial y_i = 0$, for $i = 1, \ldots, N-1$, then is obtained the minimum point of $\Im_P$. Related to the potential of the i-node of the circuit these derivative conditions are expressed as

$$\begin{aligned}\frac{\partial \Im_P}{\partial x_i} &= \frac{\partial}{\partial x_i}\mathrm{Re}[(C\underline{V}+\underline{e})^T \underline{Y}^*(C\underline{V}+\underline{e})^*] \\ &= 2\sum_{l_k \in n_i} c_{l_k,n_i} G_k(x_i - x_j + a_k) = 0\end{aligned} \tag{2.52}$$

$$\begin{aligned}\frac{\partial \Im_P}{\partial y_i} &= \frac{\partial}{\partial y_i}\mathrm{Re}[(C\underline{V}+\underline{e})^T \underline{Y}^*(C\underline{V}+\underline{e})^*] \\ &= 2\sum_{l_k \in n_i} c_{l_k,n_i} G_k(y_i - y_j + b_k) = 0\end{aligned} \tag{2.53}$$

for $k = 1, \ldots, K$, $i,j = 1, \ldots, N-1, i \neq j$.

(β) The sign of the reactive power Eq. depends on the sign of B_k: (β1) when all the branches of the AC circuit have an equivalent resistive-inductive admittance $B_k > 0$, $k = 1, \ldots, K$, then $\Im_Q(x, y) > 0$, so $\Im_Q$ is strictly positive and its *extreme point is a minimum*. Therefore one can affirm that the *reactive power absorbed by all the reactive elements of AC circuit is minimum*; (β2) when all the branches of AC circuit have an equivalent resistive-capacitive admittance $B_k < 0$, $k = 1, \ldots, K$, then $\Im_Q(x, y) < 0$. In this case reversing the sign of $\Im_Q$ and changing its significance from absorbed in "generated", then $-\Im_Q > 0$ is strictly positive and its *extreme point is a minimum*. Therefore one can affirm that the *reactive power generated by all the reactive elements of the AC circuit is minimum*; (β3) when the branch admittances of the circuit verify the particular case of resonance condition, then $\Im_Q(x, y) = 0$; (β4) when the circuit includes inductive and capacitive branches and is not possible to conclude about the sign of B_k then the extreme of $\Im_Q$ can not be fixed.

Related only to the cases (β1) and (β2) by imposing the concurrent conditions $\partial \Im_Q/\partial x_i = 0$ and $\partial \Im_Q/\partial y_i = 0$, for $i = 1, \ldots, N-1$ then is obtained the minimum point of $\Im_Q$. With respect to the potential of the i-node these derivative conditions are expressed as

$$\frac{\partial \Im_Q}{\partial x_i} = \frac{\partial}{\partial x_i}\mathrm{Im}[(C\underline{V}+\underline{e})^T \underline{Y}^*(C\underline{V}+\underline{e})^*] = 2\sum_{l_k \in n_i} c_{l_k,n_i} B_k(x_i - x_j + a_k) = 0 \tag{2.54}$$

$$\frac{\partial \Im_Q}{\partial y_i} = \frac{\partial}{\partial y_i} \mathrm{Im}[(C\underline{V} + \underline{e})^T \underline{Y}^* (C\underline{V} + \underline{e})^*] = 2 \sum_{l_k \in n_i} c_{l_k, n_i} B_k (y_i - y_j + b_k) = 0 \quad (2.55)$$

Finally, the minimum points of $\Im_P$ and $\Im_Q$ are the solutions of concurrent conditions (2.52)–(2.55) which means the below system with 4(*N*–1) Eqs.

$$\begin{aligned}
&\frac{\partial \Im_P}{\partial x_i} = \sum_{l_k \in n_i} c_{l_k, n_i} G_k (x_i - x_j + a_k) = 0; \quad \frac{\partial \Im_P}{\partial y_i} = \sum_{l_k \in n_i} c_{l_k, n_i} G_k (y_i - y_j + b_k) = 0 \\
&\frac{\partial \Im_Q}{\partial x_i} = \sum_{l_k \in n_i} c_{l_k, n_i} B_k (x_i - x_j + a_k) = 0; \quad \frac{\partial \Im_Q}{\partial y_i} = \sum_{l_k \in n_i} c_{l_k, n_i} B_k (y_i - y_j + b_k) = 0
\end{aligned} \quad (2.56)$$

If we calculate the algebraic sum of Eq. (2.56), with the derivatives $\partial \Im_P / \partial y_i$ and $\partial \Im_Q / \partial x_i$ multiplied by *j* respectively –*j*, we will get

$$\sum_{l_k \in n_i} c_{l_k, n_i} \underline{Y}_k (\underline{V}_i - \underline{V}_j + \underline{E}_k) = 0 \quad (2.57)$$

Equation (2.57) represent for $i, j = 1, \ldots, N - 1, i \neq j$ the Eq. of *NM* applied for $N - 1$ nodes of the considered AC circuit. Withal by rewriting the Eq. (2.57) we will obtain

$$\sum_{l_k \in n_i} c_{l_k, n_i} \underline{I}_k = 0 \quad (2.58)$$

that represents even the *KCL Eqs*. for *N*–1 nodes of the considered AC circuit.

Taking into account those shown so far one can formulate the following Principle of Minimum Active and Reactive Power (PMARP): *"In reciprocal and linear AC circuits the condition of minimum active absorbed power in the resistances and minimum reactive absorbed (generated) power in the inductances (capacitances) is consistent with the NM and KCL"*. In more specific terms, *"in reciprocal and linear AC circuits: (a) the branch currents and voltages are distributed such as the absorbed active power and the absorbed (generated) reactive power in the resistive-inductive (resistive-capacitive) branches of the circuit are minimum; (b) PMARP along with KVL constitute the basic forms of linear and independent Eqs. system of AC circuits analysis; (c) PMARP is consistent with NM*".

Also it is possible to analyze together PMARP and MPTT in what concerns the active power flow in AC circuits [75]. First according to MPTT let's consider that certain AC circuit branches—named load—verify the conditions for maximum active power transfer. In terms of PMARP that means the case (β3) when $\Im_Q(x, y) = 0$ and the active power Eq. $\Im_P(x, y)$ has a minimum. For these reasons can be formulated the Co-Existence of Active Power Principles (CEAPP) for linear

AC circuits as: "*in linear and reciprocal AC circuits the branches currents and voltages comply simultaneously PMARP and MPTT when one of the branch admittance verifies the MPTT condition*".

2.5.2 PMARP for Linear AC Circuits Under Non-sinusoidal Conditions

The concepts illustrated above can be extended to the case of AC circuits operating in non-sinusoidal regime. A similar linear circuit with K branches and N nodes is considered but in this case operating under non-sinusoidal conditions with radian frequency ω at the fundamental harmonic. For each branch of the circuit, the voltage and the current are expressed by using the Fourier series

$$u_k(t) = \sum_{h=0}^{H} U_k^{(h)} \sqrt{2} \sin(h\omega t + \varphi_k^{(h)}) \tag{2.59}$$

$$i_k(t) = \sum_{h=0}^{H} I_k^{(h)} \sqrt{2} \sin(h\omega t + \varphi_k^{(h)} - \gamma_k^{(h)}) \tag{2.60}$$

where $k = 1, \ldots, K$ is the number of branches, H is the finite number of harmonics considered, $U_k^{(h)}$ and $I_k^{(h)}$ represent the rms values of the hth harmonic of voltage and current, respectively, while $\varphi_k^{(h)}$ and $\varphi_k^{(h)} - \gamma_k^{(h)}$ represent the phase angle of the hth harmonic of voltage and current, respectively. Considering branch k of the circuit (Fig. 2.29), by using the complex representation for the hth harmonic of voltage and current, with the same conventional circuit representation adopted in Sect. 2.5.1 for AC circuits in sinusoidal regime, we get

$$\underline{\Delta V}_k^{(h)} = \underline{V}_{i,k}^{(h)} - \underline{V}_{j,k}^{(h)} = \underline{U}_k^{(h)} - \underline{E}_k^{(h)} = \frac{\underline{I}_k^{(h)}}{\underline{Y}_k^{(h)}} - \underline{E}_k^{(h)} \tag{2.61}$$

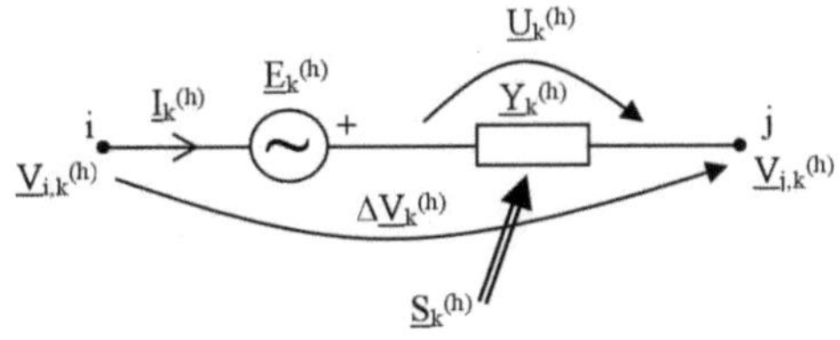

Fig. 2.29 AC circuit under non-sinusoidal conditions

In the Hilbert space we can define, similarly with Eqs. (2.50) and (2.51), the functionals corresponding to the *h*th harmonic of active and reactive absorbed power by all the *K* branches. We obtain the Eqs. $\Im_R^{(h)} : \Re^{2(N-1)} \to \Re$ and $\Im_Q^{(h)} : \Re^{2(N-1)} \to \Re$ such that, respectively:

$$\begin{aligned} \Im_R^{(h)}(x_1^{(h)}, \ldots, x_{N-1}^{(h)}, y_1^{(h)}, \ldots, y_{N-1}^{(h)}) &\equiv \mathrm{Re}[\bar{S}^{(h)}] \\ &= \sum_{k=1}^{K} G_k^{(h)}[(x_{i,k}^{(h)} - x_{j,k}^{(h)} + a_{E,k}^{(h)})^2 + (y_{i,k}^{(h)} - y_{j,k}^{(h)} + b_{E,k}^{(h)})^2] \end{aligned} \tag{2.62}$$

$$\begin{aligned} \Im_Q^{(h)}(x_1^{(h)}, \ldots, x_{N-1}^{(h)}, y_1^{(h)}, \ldots, y_{N-1}^{(h)}) &\equiv \mathrm{Im}[\bar{S}^{(h)}] \\ &= \sum_{k=1}^{K} B_K^{(h)}[(x_{i,k}^{(h)} - x_{j,k}^{(h)} + a_{E,k}^{(h)})^2 + (y_{i,k}^{(h)} - y_{j,k}^{(h)} + b_{E,k}^{(h)})^2] \end{aligned} \tag{2.63}$$

These equations are a function class C^2 in $\Re^{2(N-1)}$. Considering $G_k^{(h)} > 0$ for $k = 1, \ldots, K$ the real equation of the complex power is always positive defined, that is, $\Im_R^{(h)}(x_1^{(h)}, x_2^{(h)}, \ldots, x_{N-1}^{(h)}, y_1^{(h)}, y_2^{(h)}, \ldots, y_{N-1}^{(h)}) > 0$, for all the pairs $(x_i^{(h)}, y_i^{(h)})$ with $i = 1, \ldots, N-1$ and $h = 1, \ldots, H$; then, *the active absorbed power has a minimum.* The reactive power equation can be analyzed, for all the pairs $(x_i^{(h)}, y_i^{(h)})$, $i = 1, \ldots, N-1$, and for each harmonic $h = 1, \ldots, H$ as single-phase AC circuits described in Sect. 2.5.1 for AC circuits. Similarly, such as in case (β1) with resistive-inductive branches, the Eq. (2.63) is *positive defined*, thus the *total reactive power absorbed by the inductive components has a minimum.* Furthermore, likewise case (β2) with resistive-capacitive branches by changing the sign and significance of the imaginary functional, in order to obtain a *positive defined* functional and to conclude that the *total reactive power generated by the capacitive components has a minimum.* For the general case in which reactive components of different nature—inductive and capacitive—coexist at the given harmonic $h = 1, \ldots, H$, the sign of Eq. $\Im_Q^{(h)}$ cannot be fixed and the extreme of $\Im_Q^{(h)}$ cannot be stated.

Related only to the cases (β1) and (β2), the minimum point of $\Im_Q$ is obtained from the below 4(*N*–1) Eqs. system that results by imposing to the Eqs. (2.62) and (2.63) the derivative concurrent conditions for extreme

$$\begin{aligned} \frac{\partial \Im_R^{(h)}}{\partial x_i^{(h)}} = 0, \quad \frac{\partial \Im_R^{(h)}}{\partial y_i^{(h)}} = 0 \\ \frac{\partial \Im_Q^{(h)}}{\partial x_i^{(h)}} = 0, \quad \frac{\partial \Im_Q^{(h)}}{\partial y_i^{(h)}} = 0 \end{aligned} \tag{2.64}$$

where $i = 1, \ldots, N-1$ and $h = 1, \ldots, H$.

By calculating the algebraic sum of the solutions, with the derivatives taken with respect to the imaginary parts multiplied by $(-j)$, in the same fashion as seen in AC examples, we obtain the expressions

$$\sum_{l_k \in n_i} c_{l_k,n_i} \underline{Y}_k^{(h)} (\underline{V}_i^{(h)} - \underline{V}_j^{(h)} + \underline{E}_k^{(h)}) = 0 \tag{2.65}$$

$i,j = 1,\ldots,N-1, i \neq j$, $k = 1,\ldots,K$ that represent NM Eqs. for hth harmonic expressed at $N-1$ nodes of the circuit. On the other hand by using in (2.65) Ohm's law for each circuit branch then results

$$\sum_{l_k \in n_i} c_{l_k,n_i} \underline{I}_k^{(h)} = 0 \tag{2.66}$$

which are the KCL equations for the hth harmonic, expressed for $N-1$ nodes of the circuit.

Taking into account those shown so far, one can formulate the following *Principle of Minimum Active and Reactive Power (PMARP)*: "*In reciprocal and linear AC circuits the condition of minimum active absorbed power in the resistances and minimum reactive absorbed (generated) power in the inductances (capacitances) is consistent with the NM and KCL*". In more specific terms, "*in reciprocal and linear AC circuits: (a) the branch currents and voltages are distributed such as the absorbed active power and the absorbed (generated) reactive power in the resistive-inductive (resistive-capacitive) branches of the circuit are minimum; (b) PMARP along with KVL constitute the basic forms of linear and independent equations system of AC circuits analysis; (c) PMARP is consistent with NM*".

Consequently, we may state the *Principle of Minimum Active and Reactive absorbed Power in linear circuits under Non-sinusoidal conditions (PMARPN)*: "*In linear and reciprocal circuits under non-sinusoidal conditions, for each* h*th harmonic, the minimum of the active absorbed power and minimum of the reactive absorbed (generated) power by the inductances (capacitances) is consistent with the NM and KCL*". Another description of the same principle can be stated as follows: "*In the linear and reciprocal circuits under non-sinusoidal periodic signals for each* h*th harmonic: (a) the branch currents and voltages are distributed such as the absorbed active power and the absorbed (generated) reactive power in the resistive-inductive (resistive-capacitive) branches of the circuit are minimum; (b) PMARPN along with KVL constitute the basic forms of linear and independent Eqs. system of AC circuits analysis; (c) PMARPN is consistent with NM*".

It is essential to remark that the same observations regarding the limits of applicability of the minimum power principle indicated in Section for AC circuits hold at every harmonic order. Thus, the minimum power principle applies when no mixture of inductive and capacitive components appears in the circuit at any frequency. Also a similar CEAPP can be stated for each frequency of AC circuits in non-sinusoidal conditions.

The existence of the minimum power condition is verified if, starting from all inductive (or capacitive) reactive components, no resonance conditions is reached in the circuit at any frequency in the observed range, since the reactive nature of the circuit would change above the resonance frequency [76]. This could be the case of filters containing both inductive and capacitive components, whose equivalent admittance exhibits capacitive behavior at relatively low frequency up to the resonance conditions, and inductive behavior at frequencies higher than the one corresponding to the resonance conditions.

2.5.3 Practical Examples

Example 2.7 demonstration of PMARP for the AC circuit shown in Fig. 2.30 with $N = 2$ nodes and $K = 3$ resistive-inductive branches. The admittances contain only resistive and inductive elements, that is, $G_k > 0$ and $B_k > 0$ for $k = 1, 2, 3$, satisfying the conditions imposed in case β1. The real and imaginary parts of nodes potentials are considered variables, while the voltage source is constant. Then, by applying KVL the expressions of the complex conjugated currents of the circuit branches are

$$\begin{aligned}
&\underline{V}_1 = x_1 + jy_1;\ \underline{V}_2 = x_2 + jy_2;\ \underline{E}_1 = a + jb \\
&\underline{I}_1^* = (G_1 + jB_1)[(x_2 - x_1 + a) - j(y_2 - y_1 + b)] \\
&\underline{I}_2^* = (G_2 + jB_2)[(x_1 - x_2) - j(y_1 - y_2)] \\
&\underline{I}_3^* = (G_3 + jB_3)[(x_1 - x_2) - j(y_1 - y_2)]
\end{aligned}$$

The active and reactive absorbed powers of the circuit branches elements expressed by the Eqs. (2.50) and (2.51) are positive defined $\Im_R > 0$, $\Im_Q > 0$ in the forms

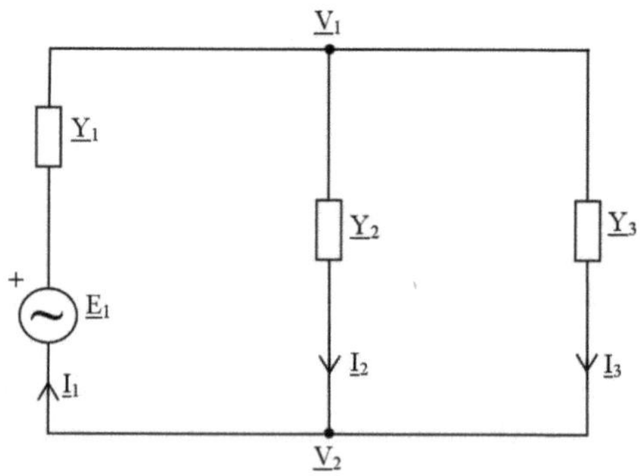

Fig. 2.30 Resistive-inductive AC circuit

$$\begin{aligned}
\Im_R = & \Big\{G_1\left[(x_2 - x_1 + a)^2 + (y_2 - y_1 + b)^2\right] + G_2\left[(x_1 - x_2)^2 + (y_1 - y_2)^2\right] \\
& + G_3\left[(x_1 - x_2)^2 + (y_1 - y_2)^2\right]\Big\} \\
\Im_Q = & \Big\{B_1\left[(x_2 - x_1 + a)^2 + (y_2 - y_1 + b)^2\right] + B_2\left[(x_1 - x_2)^2 + (y_1 - y_2)^2\right] \\
& + B_3\left[(x_1 - x_2)^2 + (y_1 - y_2)^2\right]\Big\}
\end{aligned}$$

Related only to x_1 and y_1 as variables, the minimum of the Eqs. $\Im_R$ and $\Im_Q$ is obtained by imposing the derivative concurrent conditions for extreme as the solution of the 4-Eqs. system

$$\begin{aligned}
\frac{\partial \Im_R}{\partial x_1} &= -G_1(x_2 - x_1 + a) + G_2(x_1 - x_2) + G_3(x_1 - x_2) = 0 \\
\frac{\partial \Im_R}{\partial y_1} &= -G_1(y_2 - y_1 + b) + G_2(y_1 - y_2) + G_3(y_1 - y_2) = 0 \\
\frac{\partial \Im_Q}{\partial x_1} &= -B_1(x_2 - x_1 + a) + B_2(x_1 - x_2) + B_3(x_1 - x_2) = 0 \\
\frac{\partial \Im_Q}{\partial y_1} &= -B_1(y_2 - y_1 + b) + B_2(y_1 - y_2) + B_3(y_1 - y_2) = 0
\end{aligned}$$

On the other hand, by summing up the first equation of the system and the third equation multiplied by $(-j)$, and summing up the second equation with the fourth equation multiplied by $(-j)$, we obtain KCL expressed at node 1, with $L^{(1)} = \{1, 2, 3\}$

$$-\underline{I}_1^* + \underline{I}_2^* + \underline{I}_3^* = 0$$

Withal the system above represents the NM expressed in node 1.

A similar result is obtained if the variables are considered the real part x_2 and imaginary part y_2 of the potential $\underline{V}_2$.

Example 2.8 Demonstration of the PMARP for a resistive-capacitive AC circuit similar to that shown in Fig. 2.30. Unlike the previous circuit, the admittances of this circuit contain only resistive and capacitive elements, that is $G_k > 0$ and $B_k < 0$ for $k = 1, 2, 3$, satisfying the conditions imposed in case β2. The expressions of the potentials, voltage source, and complex conjugated currents of the circuit are the same as in example 2.7. Also the Eqs. (2.50) and (2.51) have the same relations as in the precedent example, where the real functional is always positive defined $\Im_R > 0$ but the imaginary functional is negative $\Im_Q < 0$. Then, if changing the sign of the reactive power functional, one achieves "positive" sign for $-\Im_Q$ and its significance is changing in *reactive generated power by all the capacitive elements which have a minimum.* Assuming x_1 and y_1 as variables, the minimum of the equation $\Im_R$, and of the equation $\Im_Q$ with reversed sign, is the solution of the system

$$\frac{\partial \Im_R}{\partial x_1} = -G_1(x_2 - x_1 + a) + G_2(x_1 - x_2) + G_3(x_1 - x_2) = 0$$
$$\frac{\partial \Im_R}{\partial y_1} = -G_1(y_2 - y_1 + b) + G_2(y_1 - y_2) + G_3(y_1 - y_2) = 0$$
$$\frac{\partial(-\Im_Q)}{\partial x_1} = -[-B_1(x_2 - x_1 + a) + B_2(x_1 - x_2) + B_3(x_1 - x_2)] = 0$$
$$\frac{\partial(-\Im_Q)}{\partial y_1} = -[-B_1(y_2 - y_1 + b) + B_2(y_1 - y_2) + B_3(y_1 - y_2)] = 0$$

Similarly as in the previous example, we obtain the KCL and NM expressed at node 1, with $L^{(1)} = \{1, 2, 3\}$

$$-\underline{I}_1^* + \underline{I}_2^* + \underline{I}_3^* = \sum_{k \in L^{(1)}} \underline{I}_k^* = 0$$

Example 2.9 Demonstration of PMARP for a three-phase star circuit. Let us consider the star circuit with neutral point n shown in Fig. 2.31, where the branch impedances are modeled as series-connected resistances and inductances, and can be transformed into equivalent admittances $\underline{Y}_k = 1/\underline{Z}_k = G_k - jB_k$, for $k = 0, 1, 2, 3$. For the sake of representation, the values of $B_k > 0$ for $k = 1, 2, 3$ satisfy the conditions imposed in case β1. We start from known three-phase symmetrical voltages of the system

$$\underline{U}_{10} = U; \quad \underline{U}_{20} = -a - jb; \quad \underline{U}_{30} = -a + jb$$

and if we consider the neutral point displacement voltage $\underline{U}_{n0} = x + jy$ as variable, then the Eqs. (2.50) and (2.51) power can be expressed as

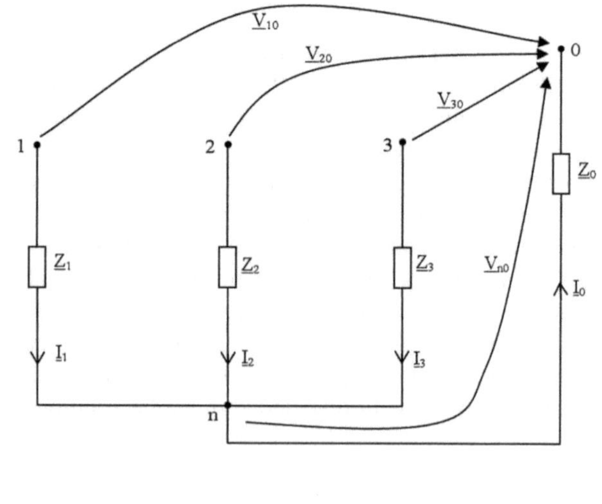

Fig. 2.31 Three-phase circuit with star connection

$$\Im_R = \left\{ G_1[(U-x)^2+y^2]+G_2[(-a-x)^2+(-b-y)^2]+G_3[(-a-x)^2+(b-y)^2]+G_0(x^2+y^2)\right\}$$

respectively

$$\Im_Q = \left\{ B_1[(U-x)^2+y^2]+B_2[(-a-x)^2+(-b-y)^2]+B_3[(-a-x)^2+(b-y)^2]+B_0(x^2+y^2)\right\}$$

The minimum of the Eqs. $\Im_R$ and $\Im_Q$ is given by the solution of the system containing 4-Eqs.

$$\frac{\partial \Im_R}{\partial x}=0, \frac{\partial \Im_R}{\partial y}=0, \frac{\partial \Im_Q}{\partial x}=0, \frac{\partial \Im_Q}{\partial y}=0$$

so that

$$\begin{aligned}
\frac{\partial \Im_R}{\partial x} &= -G_1(U-x)-G_2(-a-x)-G_3(-a-x)+G_0x=0;\\
\frac{\partial \Im_R}{\partial y} &= G_1y-G_2(-b-y)-G_3(b-y)+G_0y=0\\
\frac{\partial \Im_Q}{\partial x} &= -B_1(U-x)-B_2(-a-x)-B_3(-a-x)+B_0x=0;\\
\frac{\partial \Im_Q}{\partial y} &= B_1y-B_2(-b-y)-B_3(b-y)+B_0y=0
\end{aligned}$$

From this system, by summing up the first equation and third equation multiplied by (−*j*), respectively by summing up second equation with fourth equation multiplied by (−*j*), is obtained

$$x=\frac{(G_1-jB_1)\,U-(G_2-jB_2)\,a-(G_3-jB_3)\,a}{(G_0-jB_0)+\sum\limits_{k=1}^{3}(G_k-jB_k)}$$

and

$$y=\frac{-(G_2-jB_2)\,b+(G_3-jB_3)\,b}{(G_0-jB_0)+\sum\limits_{k=1}^{3}(G_k-jB_k)}$$

The outcomes of above equations reproduce the well-known relation

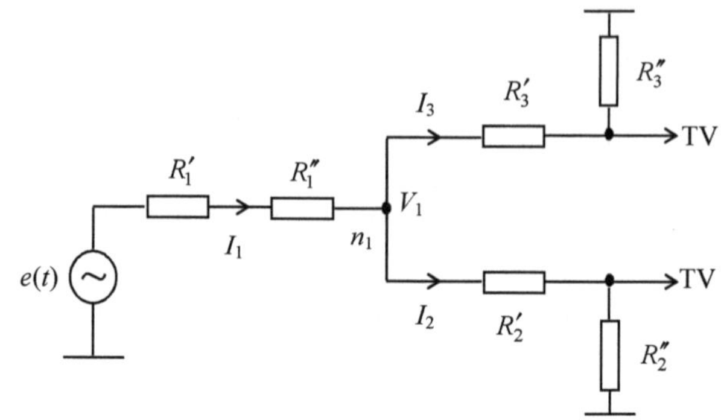

Fig. 2.32 The "splitter"

$$\underline{U}_{n0} = x + jy = \frac{\sum_{k=1}^{3} (G_k - jB_k)\underline{U}_{k0}}{(G_0 - jB_0) + \sum_{k=1}^{3} (G_k - jB_k)}$$

that incorporates KCL expressed at node *n*.

Example 2.10 Let's consider a linear circuit—"splitter"—which is used in TV cables as power amplifiers for signals, shown in Fig. 2.32. The instantaneous values of non-sinusoidal voltage source have the practical form:

$$e(t) = 0.27386 + 0.27386\sqrt{2}\sin(2\pi 40 \cdot 10^6 t) + 0.27386\sqrt{2}\sin(2\pi 900 \cdot 10^6 t)\ \text{V}$$

If one considers the potential V_1 variable, for each harmonic of the voltage source we can calculate the minimum of the absorbed power equations.
(a) The absorbed power equation for DC regime of splitter can be expressed by

$$\Im_R^{(0)} = \frac{1}{2}\left[G_1(-V_1^{(0)} + E^{(0)})^2 + G_2 V_1^{(0)^2} + G_3 V_1^{(0)^2}\right]\rangle 0, \forall V_1^{(0)} \in \Re$$

where $G_1 = \frac{1}{R_1' + R_1''}$, $G_2 = \frac{1}{R_2' + R_2''}$, $G_3 = \frac{1}{R_3' + R_3''}$ do not depend on the frequency. Then, the minimum point of functional is obtained when

$$\frac{\partial \Im_R^{(0)}}{\partial V_1^{(0)}} = -G_1(-V_1^{(0)} + E^{(0)}) + G_2 V_1^{(0)} + G_3 V_1^{(0)}) = 0$$

relation that is identical with KCL expressed in node 1.
(b) If we consider the potential $\underline{V}_1^{(p)} = x + jy^{(p)}$, for the two harmonics, $p_1 = 8{\cdot}10^5$ and $p_2 = 18{\cdot}10^6$, then for the AC regime of the splitter the functional of the complex absorbed power is

$$\Im_R^{(p)} = G_1\left[(-x^{(p)} + E^{(p)})^2 + y^{(p)2}\right] + G_2(x^{(p)2} + y^{(p)2}) + G_3(x^{(p)2} + y^{(p)2})\rangle 0, \forall(x^{(p)}, y^{(p)}) \in \Re$$

Then, the minimum point of the active power Eq. $\Im_R$ is the solution of the system

$$\frac{\partial \Im_R^{(p)}}{\partial x^{(p)}} = -G_1(-x^{(p)} + E^{(p)}) + G_2 x^{(p)} + G_3 x^{(p)} = 0$$
$$\frac{\partial \Im_R^{(p)}}{\partial y^{(p)}} = G_1 y^{(p)} + G_2 y^{(p)} + G_3 y^{(p)} = 0$$

and, if we calculate the algebraic sum of the solutions of the system above, with the second equation multiplied with $-j$, results KCL expressed in node 1

$$-\underline{I}_1^{(p)^*} + \underline{I}_2^{(p)^*} + \underline{I}_3^{(p)^*} = 0$$

The imaginary part of power equation is zero, $\Im_Q^{(p)} = 0$. By using a Multisim measurement procedure, shown in Fig. 2.33a–c, the absorbed active power for different values of the resistances is calculated. The results are shown in Table 2.1. The first line contains the nominal values of conductance, and the second line of the table contains the adapted values of all the resistances, i.e. each resistance absorbs from the circuit the maximum active power. In this case, the value of the absorbed active power is big compared with the other values.

Example 2.11 Demonstration of CEAPP for AC circuit presented in Fig. 2.34, where the numerical values of elements are $\underline{E}_1 = 100$ V, $\underline{Z}_1 = \underline{Z}_2 = 10(1 + j)\,\Omega$. The load impedance $\underline{Z}_L = 5(1 + j)\,\Omega$ was selected so as to verify MPTT.

Assuming that the potential of node 1, $\underline{V}_1 = x_1 + jy_1$ is variable, then the power Eqs. (2.50) and (2.51) are expressed as follows

$$\Im_P = \frac{1}{20}\left[(-x_1 + 100)^2 + (-y_1)^2\right] + \frac{1}{20}\left[(-x_1)^2 + (-y_1)^2\right] + \frac{1}{10}\left[(x_1)^2 + (y_1)^2\right]$$
$$\Im_Q = \frac{1}{20}\left[(-x_1 + 100)^2 + (-y_1)^2\right] + \frac{1}{20}\left[(-x_1)^2 + (-y_1)^2\right] + \frac{1}{10}\left[(x_1)^2 + (y_1)^2\right]$$

By applying the extreme conditions for $\Im_R$ and $\Im_Q$ then results

$$\frac{\partial \Im_P}{\partial x_1} = -\frac{-x_1 + 100}{20} - \frac{-x_1}{20} + \frac{x_1}{10} = 0; \; \frac{\partial \Im_P}{\partial y_1} = -\frac{-y_1}{20} - \frac{-y_1}{20} + \frac{y_1}{10} \equiv 0$$
$$\frac{\partial \Im_Q}{\partial x_1} = -\frac{-x_1 + 100}{20} - \frac{-x_1}{20} + \frac{x_1}{10} = 0; \; \frac{\partial \Im_Q}{\partial y_1} = -\frac{-y_1}{20} - \frac{-y_1}{20} + \frac{y_1}{10} \equiv 0$$

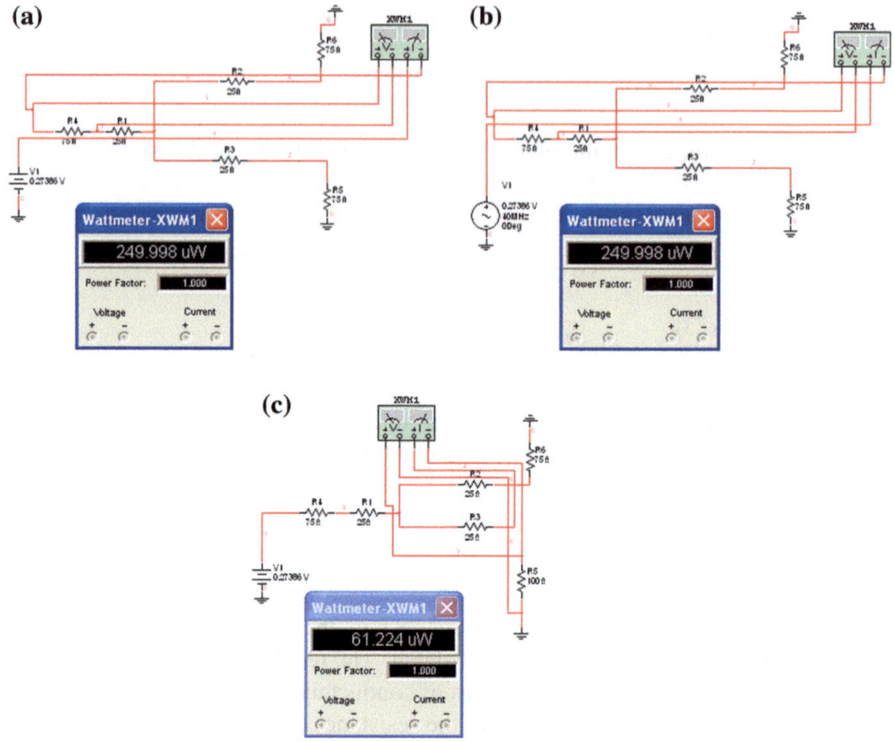

Fig. 2.33 Multisim analysis of power absorbed by, **a** R_1' in DC regime, **b** R_1'' in AC regime, **c** R_2'' in DC regime

Table 2.1 Absorbed active power for different values of the resistances

Values of conductances (S)	Total absorbed power in DC and AC (mW)
$G_1 = G_2 = G_3 = 0.01$	499.593
$G_1 = 0.01$; $G_2 = 0.013$; $G_3 = 0.01$	524.818
$G_1 = 0.01$; $G_2 = 0.02$; $G_3 = 0.025$	1.662
$G_1 = 0.01$; $G_2 = 0$; $G_3 = 0.01$	374.832
$G_1 = 0.01$; $G_2 = 0.04$; $G_3 = 0.01$	624.810
$G_1 = 0.02$; $G_2 = 0.01$; $G_3 = 0.01$	749.664

Fig. 2.34 AC circuit with CEAPP

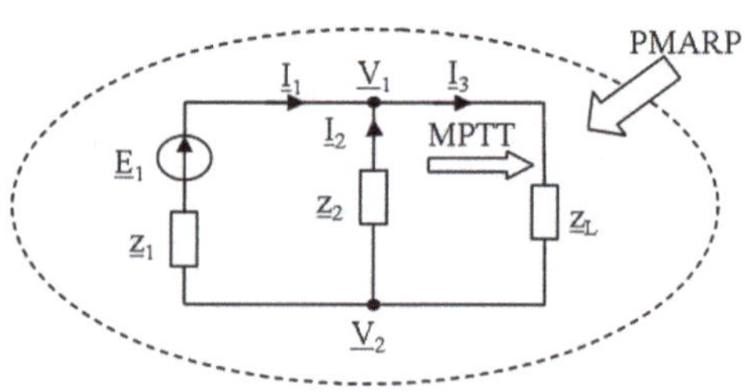

From the above system by summing up the first equation and the second equation multiplied by (j), and the third equation multiplied by ($-j$) and the fourth equation results the equation of *NM* expressed at node 1

$$-\frac{1-j}{20}[(-x_1+100)+j(-y_1)]-\frac{1-j}{20}[(-x_1)+j(-y_1)]+\frac{1+j}{10}[x_1+jy_1]=0$$

Furthermore, if we rewrite the NM relation we'll obtain KCL at node 1

$$-\underline{I}_1-\underline{I}_2+\underline{I}_3=0$$

For the numerical value $\underline{V}_1 = 25\text{V}$ which results from KVL, the total absorbed active power is $P_{abs,tot}$ = 125 W. This numerical value coincides with the minimum value of active power functional (PMARP) by solving the above system. Also for these numerical values of the circuit elements as consequences of the MPTTT it results: (i) the reactive power is null (resonant conditions, case β3), (ii) the load absorbs maximum active power $P_{L\max}$ = 62.5 W, and (iii) the efficiency of the circuit in what concerns the flow of active power from circuit to load is 50%. Thus the conditions of CEAPP are verified.

2.6 Conclusion

The Principle of Minimum absorbed Power (PMP) is extended and demonstrated from DC to AC circuits (PMARP) under sinusoidal and non-sinusoidal signals. Also from active power flow is proved the co-existence of PMARP and MPTT for AC circuits. In other works as [61–74] for magnetic circuits in stationary and quasi-stationary regime the authors introduced and proved the Principle of Minimum absorbed Energy (PME).

From a theoretical point of view, the introduction of these principles for AC linear and reciprocal circuits in quasi-stationary state is extremely useful in order to formulate their basic system of equations related to the flow of active and reactive power. On the other hand, all these principles are formulated for natural distribution of the potentials, currents and voltages of circuit that verify the fundamental laws KCL and KVL, which approaches the proposed principles to the basic theory of circuits.

The examples presented above that refer to linear and reciprocal AC circuits, prove the introduced theoretical concepts and reveal the direct dependence between the power flow with the classical theory of electric circuits.

References

1. C. Alexandre, M.O. Sadiku, Fundamentals of Electric Circuits, McGraw-Hill, 5th Edition, 2012.
2. R.C. Dorf, J.A. Svoboda, Introduction to Electric Circuits, Wiley, 8th Edition, 2010.
3. W. Hayt, J. Kemmerly, S. Durbin, Engineering Circuit Analysis, McGraw-Hill, 8th Edition, 2011.
4. J.W. Nilsson, S. Riedel, Electric Circuits Prentice Hall, 9th Edition, 2010.
5. M. Nahvi, J. Edminster, Schaum's Outline of Electric Circuits, McGraw-Hill, 5th Edition, 2011.
6. A.E. Emanuel, Power Definitions and the Physical Mechanism of Power Flow, John Wiley & Sons, 2011.
7. C. Sankaran, Power Quality, New York: CRC Press, 2002.
8. F.C. de la Rosa, Harmonics and Power Systems, New York: Taylor & Francis, 2006.
9. A. Baggini, Handbook of Power Quality, John Wiley & Sons, 2008.
10. R.C. Dugan, M.F. Mc Granaghan, S. Santoso, W.H. Beaty, Electrical Power Systems Quality, McGraw Hill Professional, 2012.
11. A. Kusko, M. Thomson, Power Quality in Electrical Systems, McGraw Hill Professional, 2007.
12. E.F. Fuchs, M.A.S. Masoum, Power Quality in Power Systems and Electrical Machines, Elsevier Inc., 2008.
13. IEEE Standard 1159, IEEE Recomandated Practices for Monitoring Electric Power Quality, 1995.
14. EN Standard 50160, Characteristics of Voltage at a Network User's Supply Terminals: Limits and Values, 2011.
15. W. Hofmann, J. Schlabbach, W. Just, Reactive Power Compensation: A Practical Guide, John Wiley & Sons, 2012.
16. ABB, 2008 Power Factor Correction and Harmonic Filtering in Electrical Plants, Technical Application Paper.
17. W. Hofmann, W. Just, 2003, Reactive Current Compensation in the Operating Practice: Design, Energy Saving, Harmonics, Voltage Quality, VDE Verlag, 4, Over Edition.
18. Schneider Electric, 2001, Power Quality, Cahier Technique no., 1999.
19. G.G. Seip, Electrical Installations Handbook - Siemens, Wiley, 3rd Edition, 2000.
20. J.M. Broust, Industrial Electrical Equipment and Installations - Conception, Coordination, Implementation, Maintenance, Editor, Dunod, 2013.
21. K. Tkotz, (Management of the work group), Electrical Engineering, European Teaching, 28. Revised and Expanded Edition, 2012.
22. B.D. Jenkins, M. Coates, Electrical Installation Calculations, Blackwell Science, 3rd Edition, 2003.
23. C. Linsley, Advanced Electrical Installation Work, Newnes Publishing House, 2008.
24. A.J. Watkins, C. Kitcher, Electrical Installation Calculations, vol. 1+2, Elsevier and Newnes Publishing, 6th Edition, 2006.
25. S. Christopher, Electrical Installation, Nelson Thornes Ltd., 3rd Edition, 2005.
26. S. Gunter, Electrical Installations Handbook, John Wiley and Sons- VCH; 3rd Edition, 2000.
27. B. Atkinson, R. Lovegrove, Electrical Installation Designs, Wiley, 4th Edition, 2013.
28. I. Kasikci, Analysis and Design of Low-Voltage Power Systems: An Engineer's Field Guide, John Wiley and Sons – VCH, 1st Edition, 2004.
29. Schneider Electric, Electric Installation Guide, 2013.
30. J. Schonek, Y. Nebon, LV Protection Devices and Variable Speed Drives (Frequency Converters), Schneider Electric, Cahier Technique no. 204, 2002.
31. D. Clenet, Electronic starters and variable speed drives, Schneider Electric, Cahier Technique no. 208, 2003.

32. E. Aucheron, Electric Motors and How to Improve their Control and Protection, Schneider Electric, Cahier Technique no. 207, 2004.
33. G. Baurand, V. Moliton, The protection of LV motors, Schneider Electric, Cahier Technique no. 211, 2007.
34. ABB, Three-Phase Asynchronous Motors - Generalities and ABB Proposals for Coordination of Protective Devices, Technical Application Papers, vol. 7, 2004.
35. E. Bettega, J.N. Fiorina, Active Harmonic Conditioners and Unity Power Factor Rectifiers, Schneider Electric, Cahier Technique no. 183, 1999.
36. J.N. Fiorina, Inverters and Harmonics (Case Studies of Non-Linear Loads, Schneider Electric, Cahier Technique no. 159, 1993.
37. Ph. Ferraci, Power Quality, Schneider Electric, Cahier Technique no. 199, 2001.
38. C. Collombet, J.M. Lupin, J. Schonek, Harmonic Disturbances in Networks, and their Treatment, Schneider Electric, Cahier Technique no. 152, 1999.
39. R. Calvas, Electrical Disturbances in LV, Schneider Electric, Cahier Technique no. 141, 2000.
40. J. Schonek, The Singularities of the Third Harmonic, Schneider Electric, Cahier Technique no. 202, 2001.
41. R.C. Dewar, Information Theory Explanation of the Fluctuation Theorem, Maximum Entropy Production, and Self-Organization Criticality in Non-Equilibrium Stationary States, Journal of Physics, 2003, A36, 631–641.
42. H. Ozawa, A. Ohmura, R.D. Lorenz, T. Pujol, The Second Law of Thermodynamics and the Global Climate System - A Review of the Maximum Entropy Production Principle, Revue of Geophysics, 2003, 41, 1018.
43. I. Rodriguez Iturbe, A. Rinaldo, Fractal River Basins: Chance and Self-Organization, New York, USA: Cambridge University Press, 1997.
44. G.H. North, R.F. Cahalan, J.A. Coakley, Energy Balance Climate Model, Revue of Geophysics, 1981, 19, 91–121.
45. C.D. Rodgers, Minimum Entropy Exchange Principle-Reply, Journal of Meteorology Society, 1976, 102, 455–7.
46. J. Clemente Gallardo, J.M.A. Scherpen, Relating Lagrangian and Hamiltonian Formalism of LC Circuits, IEEE Transactions on Circuits and Systems, 2003, 50, no. 10, 1359–1363.
47. E.N. Lorenz, Generation of Available Potential Energy and the Intensity of the General Circulation, Oxford, UK: Pergamon Press, 1960, 86–92.
48. A. Kleidon, Testing the Effect of Life on Earth's Functioning: How Gaian is the Earth System?, Clim Change, 2002, 52, 383–389.
49. A. Kleidon, K. Fraedrich, T. Kunz, F. Lunkeit, The Atmospheric Circulation and States of Maximum Entropy Production, Geophys. Res. Lett., 2003, 30, 2223.
50. A. Bejan, Shape and Structure from Engineering to Nature, Cambridge, UK: Cambridge University Press, 2000.
51. J.C. Maxwell, A Treatise on Electricity and Magnetism, vol. I and II, 3rd Ed., New York, USA: Dover Publications, Inc., 1954.
52. W. Millar, Some General Theorems for Nonlinear Systems Possessing Resistance, Phil. Mag., 1951, ser. 7, vol. 42, no. 333, 1150–1160.
53. T.E. Stern, On the Equations of Nonlinear Networks, IEEE Transactions on Circuits Theory, 1996, vol. CT-13, no. 1, 74–81.
54. P. Jr. Penfield, R. Spence, S. Duinker, Tellegen's Theorem and Electrical Networks, Research monograph, no. 58, Massachusetts, USA: M.I.T. Press, 1970.
55. W.E. Smith, Electric and Magnetic Energy-Storage in Passive Non-Reciprocal Networks, Electronics Letters, 1967, 3, 389–391.
56. W.E. Smith, Average Energy Storage by an One-Port and Minimum Energy Synthesis, IEEE Transactions on Circuit Theory, 1970, 5, 427–430.
57. V. Ionescu, Hilbert Space Applications to Distorted Signal Analysis, (in Romanian: Aplicatii ale spatiilor Hilbert la studiul regimului deformant), Electrotehnica, 1958, 6, 280–286.

58. C.I. Mocanu, Electric Circuits Theory (in Romanian: Teoria circuitelor electrice), Bucharest, Romania: Editura Didactica si Pedagogica, 1979.
59. I.F. Hantila, N. Vasile, E. Demeter, S. Marinescu, M. Covrig, The Stationary Electromagnetic Field in Non-Linear Media (in Romanian: Campul electromagnetic stationar in medii neliniare), Bucharest, Romania, Editura ICPE, 1997.
60. V. Fireteanu, M. Popa, T. Tudorache, L. Levacher, B. Paya, Y. Neau, Maximum of Energetic Efficiency in Induction through-Heating Processes, Proceedings of HES Symposium, Padua, Italy, 2004, 80–86.
61. F. Spinei, H. Andrei, Energetical Minimum Solution for the Resistive Network, Proceedings of National Symposium of Theoretical Electrotechnics SNET, Politehnica University of Bucharest, Romania, 2004, 443–450.
62. H. Andrei, F. Spinei, C. Cepisca, Theorems about the Minimum of the Power Functional in Linear and Resistive Circuits, Proc. of Advanced Topics in Electrical Engineering ATEE'04, Politehnica University of Bucharest, Romania, 2004, 22–18.
63. H. Andrei, F. Spinei, C. Cepisca, N. Voicu, Contributions Regarding the Principles of the Minimum Dissipated Power in Stationary Regime, Proc. of IEEE Conference on Circuits and Systems CAS, Dallas, USA, 2006, 143–47.
64. H. Andrei, F. Spinei, I. Caciula, P.C. Andrei, The Systematic Analysis of the Absorbed Power in DC Networks with Modifiable Parameters Using a New Mathematic Algorithm, Proc. of IEEE-AQTR, Cluj Napoca, Romania, 2008, 121–124.
65. H. Andrei, F. Spinei, P. Andrei, U. Rohde, M. Silaghi, H. Silaghi, Evaluation of Hilbert Space Techniques and Lagrange's Method for the Analysis of Dissipated Power in DC Circuits, Proc. of IEEE-European Conference on Circuit Theory and Design ECCTD'09, Antalya, Turkey, 2009, 862–865.
66. H. Andrei, F. Spinei, C. Cepisca, P.C. Andrei, Mathematical Solution to Solve the Minimum Power Point Problem for DC Circuits, Proc. of IEEE-AQTR Conference, May 28–30, Cluj Napoca, 2010, 322–328.
67. H. Andrei, G. Chicco, F. Spinei, Minimum Dissipated Power and Energy - Two General Principles of the Linear Electric and Magnetic Circuits in Quasi-Stationary Regime, pp. 140–205, chapter 5 of the book Advances in Energy Research: Distributed Generations Systems Integrating Renewable Energy Resources, Edited by N. Bizon, Nova Science Publishers, New York, 2011.
68. H. Andrei, P.C. Andrei, Matrix Formulations of Minimum Dissipated Power Principles and Nodal Method of Circuits Analysis, Proc. of IEEE-Advanced Topics in Electrical Engineering - ATEE, 23–25 May, 2013, Bucharest, Romania, paper ELCI 1.
69. H. Andrei, P.C. Andrei, G. Oprea, B. Botea, Basic Equations of Linear Electric and Magnetic Circuits in Quasi-stationary State Based on Principle of Minimum Absorbed Power and Energy, Proc. of IEEE-ISFEE, Bucharest, 28–29 Nov, 2014, 1–6.
70. H. Andrei, F. Spinei, An Extension of the Minimum Energetical Principle in Stationary Regime for Electric and Magnetic Circuits, Proc. of Advanced Topics in Electrical Engineering ATEE'06, Politehnica University of Bucharest, Romania, 2006, 210–213.
71. H. Andrei, F. Spinei, The Minimum Energetical Principle in Electric and Magnetic Circuits, Proc. of IEEE European Conference on Circuit Theory and Design ECCTD, Sevilla, Spain, 2007, 906–909.
72. H. Andrei, F. Spinei, An Extension of the Minimum Energy Principle in Stationary Regime for Electric and Magnetic Circuits, Romanian Journal of Technical Sciences Series in Electrotechnique and Energetic, 2007, 52, 419–427.
73. H. Andrei, G. Chicco, F. Spinei, C. Cepisca, Minimum Energy Principle for Electric and Magnetic Circuits in Quasi-Stationary Regime, Journal of Optoelectronics and Advanced Materials, 2008, 10 (5), 1203–1207.
74. H. Andrei, P.C. Andrei, G. Mantescu Matrix Formulation of Minimum Absorbed Energy Principle and Nodal Method of Magnetic Circuits Analysis, Proc. of 14th IEEE International Conference on Optimization of Electrical and Electronic Equipment - OPTIM 2014, 22–24, May 2014, Brasov, 59-64.

75. L.O. Chua, Introduction to Nonlinear Network Theory: Part I, New York, USA: Mc Graw Hill Book Company, 1969.
76. H. Lev-Ari, A. Stankovic, Hilbert Space Techniques for Modeling and Compensation of Reactive Power in Energy Processing System, IEEE Transactions on Circuits and Systems 2003, 50 (4), 540–56.

Chapter 3
Reactive Power Role and Its Controllability in AC Power Transmission Systems

Esmaeil Ebrahimzadeh and Frede Blaabjerg

Abstract This chapter is a general introduction to the reactive power role in the voltage control and the stability in power transmission systems. It starts with a brief overview of the potential limitations related to the transmission system loading and also different ways that the reactive power can affect the power system operation. Different reactive power generation technologies based on capacitors and power electronic converters are reviewed as possible sources for the reactive power compensation. Also, voltage and reactive power control methods of the mentioned technologies are briefly explained for use in AC power transmission systems.

3.1 Introduction

Reactive power is generated when the current waveform is not in phase with the voltage waveform because of inductive or capacitive components. Only the component of current in phase with voltage generates active power that does the real work. Reactive power is required for producing the magnetic and electric fields in capacitors and inductors. Power transmission lines have both capacitive and inductive properties. A typical transmission line can be presented by a PI equivalent model as shown in Fig. 3.1.

E. Ebrahimzadeh (✉) · F. Blaabjerg
Department of Energy Technology, Aalborg University, Aalborg, Denmark
e-mail: ebb@et.aau.dk

F. Blaabjerg
e-mail: fbl@et.aau.dk

N. Mahdavi Tabatabaei et al. (eds.), *Reactive Power Control in AC Power Systems*,
Power Systems, DOI 10.1007/978-3-319-51118-4_3

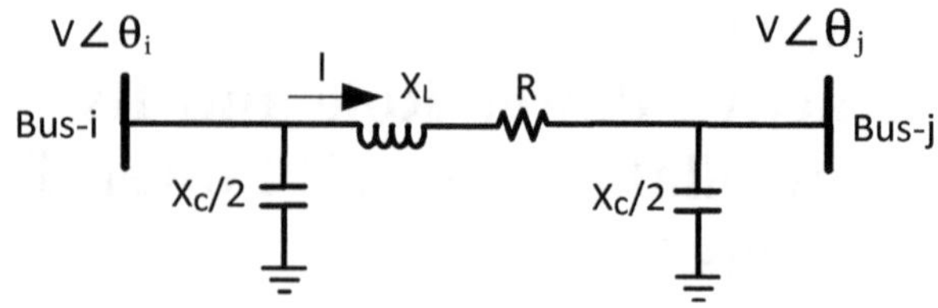

Fig. 3.1 Transmission line connecting two buses (i, j) presented by a PI equivalent model

The line capacitance supplies the reactive power ($Q_{produced}$) and the line inductance consumes the reactive power ($Q_{consumed}$), which can be calculated in an ideal line (R = 0) as:

$$Q_{produced} = \frac{V^2}{X_C} \tag{3.1}$$

$$Q_{consumed} = I^2 X_L \tag{3.2}$$

where V, X_C, X_L, and I are the bus voltage, the line's capacitive reactance, the line's inductive reactance and the line current, respectively. Therefore, the amount of reactive power consumed by a line is related to the current flow in the line; the amount of reactive power supplied by a line is related to the line voltage. At a Surge Impedance Loading (SIL), the supplied reactive power is equal to the absorbed reactive power:

$$Q_{Produced} = Q_{Consumed} \tag{3.3}$$

By substituting (3.1) and (3.2) in (3.3), the surge impedance (Z_0) can be obtained:

$$Z_0 = \frac{V}{I} = \sqrt{\frac{X_L}{X_C}} \tag{3.4}$$

The surge impedance loading (SIL) is equal to the voltage squared divided by the surge impedance:

$$SIL = \frac{V^2}{Z_0} \tag{3.5}$$

The above expression shows that an ideal line (with zero resistance) loaded at its surge impedance loading does not produce or consume reactive power, so it will have the same voltage at both ends. When a line is loaded above its SIL, it acts like a shunt reactor which are absorbing the reactive power from the system, and when a line is loaded below its SIL, it acts like a shunt capacitor which are supplying the reactive power to the system [1]. Balance of both consumption and production of reactive power at a particular loading level results into a flat voltage profile along the line.

The consumption of the reactive power by transmission lines increases with the square of current. Thus, reactive power is difficult to transport along long lines. In a power system, the goal is to maximize the utilization of the transmission system but some factors limit the loading capability of the transmission systems, which are as discussed in Sect. 3.2. Equation (3.5) shows that the transmitted power through a long transmission line can be increased by increasing the value of the line voltage (V) or by reducing the surge impedance (Z_0). This illustration demonstrates that there are two main variables that can be directly controlled for improving the performance of the power system. These are [2]:

- Voltage
- Impedance

Increasing the line voltage is the most common method for increasing the power limit under heavy loading conditions. But there are some economical and practical limitations. The surge impedance can be decreased by either increasing the capacitance of the line or by reducing the inductance of the line. With the establishment of "which" variables can be controlled in a power system, the next question is "how" these variables should be and can be controlled [2]. The answer is reactive power generation equipment, which is discussed in Sects. 3.3 and 3.4. For example, series capacitors or shunt capacitors can be used to reduce the value of the surge impedance.

3.2 Basic Principles of Power Transmission Operation

A transmission system is a complicated network of the transmission lines which connect all power substations to the loads. The AC systems can be connected together by the transmission lines to create a large power system for exchanging electrical energy. In a power system, the goal is to use the transmission lines with the least possible power losses and to maximize its loading capability by considering emergency conditions all the time. But some factors limit the loading capability of transmission systems, which are as follows:

3.2.1 Thermal Limit

The thermal limit of an overhead transmission line is reached when the current flow heats the conductor material up to a temperature above which the conductor material gradually looses mechanical strength [3]. In fact, the thermal capability of an overhead transmission line is a function of environment temperature, wind conditions, conductor conditions and its distance from the ground. Excessive heat causes that the transmission lines loose its mechanical resistance and reduce its expected useful life time. Flowing current over the heat capability is allowed only for a short and limited time. According to the definition, normal and nominal current capacity of a transmission line is a current that can be flown over the line for an unlimited time.

The line current (I in Fig. 3.1) can be divided into two components:

$$I = I\,\cos\theta + I\,\sin\theta \tag{3.6}$$

where θ is the phase difference between the line voltage and the line current. So the line losses can be obtained as follow:

$$P = R(I\,\cos\theta + I\,\sin\theta)^2 \cong R(I\,\cos\theta)^2 + R(I\,\sin\theta)^2 = P + P' \tag{3.7}$$

where $P' = R(I\,\sin\theta)^2$ is the reactive power losses. Therefore, by reducing the reactive power, the line losses are decreased and consequently the loading capacity is increased. The following methods can potentially help to increase the loading capacity of a transmission line [4]:

- Phase shifting transformers
- Series capacitors or series reactors to adjust the impedances of the lines
- FACTS elements to control the reactive power flows using power electronics.
- The phase shifting transformers and series capacitors or series reactors are usually the less expensive options while FACTS are very flexible and also more costly.

3.2.2 Voltage Limit

Voltage limits normally require that the voltage level within a transmission system to be maintained within a specified interval, for instance $\pm 5\%$ of the nominal voltage. The voltage in the transmission line can be changed by the change of the load or occurrence of the fault in transmission and distribution lines or other equipment. In these cases, it should be noted that the dynamic and transient voltages should be remained within a given range. If the line voltage exceeds more than the maximum rated value, it can result in a short circuit and may cause damage to transformers and other equipment in the substations. The voltage in AC transmission line is almost related to the level of reactive current of the line as well as line's reactance. Capacitors and reactors can be installed on the lines to control the voltage changes along the line.

3.2.3 Stability Limit

According to the definition, power system stability is the ability of the power system to remain in a balanced condition during normal operation of the system and to bring back balanced conditions within minimum possible time after the occurrence of disturbance. In general, in the literature, four different types need to be dealt with the steady state stability, dynamic stability, transient stability and voltage stability [5].

3.2.3.1 Steady-State Stability

Steady state stability refers to system power stability in response to small disturbances and continuous changes in the load. Steady state stability can be improved by

- Increasing the voltage level of the network
- Adding new lines to the transmission systems
- Reducing the series reactance of the line with bundling the lines, with installation of series capacitors along the line.

3.2.3.2 Transient Stability

In a power system, transient stability is the ability of the system in damping the oscillations due to severe disturbances [6], for instance, the reaction of the voltage to faults in the transmission system caused by events such as lightning. Transient stability of the system can be improved by increasing the system voltage and increasing the *X*/*R* ratio of the power system. An increase in the system voltage profile and *X*/*R* ratio implies an increase in the power transfer ability. Thus it helps to improve the stability.

3.2.3.3 Dynamic Stability

The ability of a power system to maintain stability under sudden small disturbances is investigated under the name of dynamic stability (also known as small-signal stability) [7]. For instance, power oscillations occurring from disconnection of large amounts of generation or load, or switching of some of the lines.

3.2.3.4 Voltage Stability

Voltage Stability is the ability of the system to maintain steady state voltages at all the system buses when subjected to a disturbance [6]. The system voltage might be unstable, if the load demand suddenly increases, or a disturbance occurs. One of the important factors that plays a significant role in the voltage instability is the inability of the system to provide the required reactive power. Voltage instability causes voltage collapse in which the buses' voltage begins to drop progressively and uncontrollably [8, 9]. Placement of series and shunt capacitors and reactive power controllers can prevent voltage instability. Such compensation has the purpose of injecting reactive power to maintain the voltage magnitude in the buses close to the nominal values, as well as to reduce the line currents and therefore the total system losses.

In brief, reactive power generation technologies can provide remedies for all of the above voltage and stability issues, and create the possibilities to run the transmission system closer to its thermal limit by controlling two main variables of the power system: Voltage and impedance.

3.3 Equipment for Reactive Power Generation in Power System

Reactive Power can be generated by power plants, capacitors, static compensators and synchronous condensers. Reactive power generation by power plant has two problems: first, the reactive power generation capacity of a power plant is limited and secondly, this huge power occupies the capacity of transmission line, transformers, and imposes some losses in the system. The presence of reactive power sources near to the consumption not only reduce the costs, but also increase the capacity of the transmission line.

3.3.1 Parallel Capacitor

Based on the amount of voltage drop, some parallel capacitor banks are connected to the network and provide the required reactive power. It increases the load power factor and finally the active power capacity of transformers.

3.3.2 Series Capacitor

Series capacitors at the network are used to reduce the impedance of the transmission lines which increases the power transmission capacity and reduces the voltage drop. Nowadays, it is also used for dynamic stability and prevention of Sub-Synchronous Resonance (SSR). In other words, the series capacitor is used to compensate the series inductive effect of lines by creating a negative reactance. Capacitive reactance should always be less than inductive reactance, and this condition is considered to determine the capacity of the series capacitor. Lack of attention to this condition may result in over-compensation at the end of the line, which is not desired.

3.3.3 Reactor

Reactors are reactive power consumers which are mostly installed in substations and at the end of long transmission lines in parallel. Basically, a circuit breaker is installed with reactors to connect them to the network, when it is needed. Usually, the reactor is switched on when the network load is minimum, and it is switched off when the network load is high.

3.3.4 Synchronous Condenser

A synchronous condenser is a synchronous motor, which is running at no load and can be operated as an inductor or a capacitor by controlling its excitation current. The machine has three operating states, which are dependent on the power factor: under-excited state, normal-excited state, and over-excited state. An under-excited synchronous motor draws both active and reactive power from the network. An over-excited synchronous motor draws active power from the network, while delivers reactive power to the network. A normal-excited motor draws only active power.

Because of the energy storage capability, synchronous condensers with automatic control has faster and smoother response to reactive power consumers. Therefore they are more effective than the parallel capacitor banks. These reactive power suppliers improve the voltage and frequency stability and they are able to supply energy in transients caused by short-circuit fault. Automatic excitation control of synchronous condensers can improve the system stability by generating lagging kVar at low loads and leading kVar at high loads. Also, they can prevent over-voltage phenomena at low loads, which is called the Ferranti effect. Relatively high losses are the main disadvantage of this device.

3.3.5 *Reactive Power Control Transformer*

Reactive power control transformers adjust secondary voltage by tap changer and as a result, keep the reactive power within a specified range.

3.3.6 *Static Reactive Power Generators with Variable Impedance*

Static reactive power generators with variable impedance change the reactive power amount by switching the capacitor banks and reactors. The aim of the approach is to provide variable impedance for compensation of the transmission system.

3.3.6.1 Thyristor-Controlled Reactor (TCR)

A single-phase Thyristor-Controlled Reactor (TCR) is shown in the Fig. 3.2a. The device consists of a fixed reactor with inductance L and a thyristor based AC switch. The effective reactance of the inductor can be changed continuously by controlling the firing angle of the thyristor.

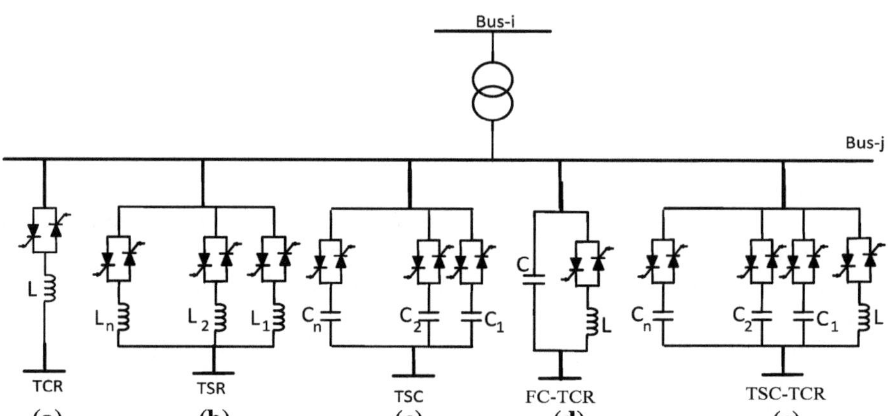

Fig. 3.2 Parallel compensators for reactive power control, **a** Thyristor-Controlled Reactors, **b** Thyristor-Switched Reactor, **c** Thyristor-Switched Capacitor, **d** Fixed Capacitor Thyristor-Controlled Reactor, **e** Thyristor-Switched Capacitor-Thyristor-Controlled Reactor

3.3.6.2 Thyristor-Switched Reactor (TSR)

A Thyristor-Switched Reactor (TSR) consists of several parallel inductors and switches which has no firing angle control and is shown in Fig. 3.2b. Using the switches without firing angle control leads to lower losses, but the control of reactive power is not continuous.

3.3.6.3 Thyristor-Switched Capacitor (TSC)

In a Thyristor-Switched Capacitor (TSC), thyristor based ac switches (without control of firing angle) are used to switch on or off the parallel capacitor units which provide the required reactive power of the system (Fig. 3.2c). Unlike the parallel reactors, the parallel capacitors cannot be switched based on the firing angle to control the reactive power continuously.

3.3.6.4 Fixed Capacitor Thyristor-Controlled Reactor (FC-TCR)

One of the main arrangements to provide reactive power is to use a Fixed Capacitor and a Thyristor-Controlled Reactor (FC-TCR), which is shown in Fig. 3.2d. The capacitive fixed reactive power along with variable reactive power of the TCR generates an output reactive power.

3.3.6.5 Thyristor-Switched Capacitor-Thyristor-Controlled Reactor (TSC-TCR)

A basic configuration of a single phase Thyristor-Switched Capacitor-Thyristor-Controlled Reactor (TSC-TCR) is shown in Fig. 3.2e. For a given range of the power output, the arrangement consists of n TSC branches and a TCR. The output capacitive reactive power is changed by the TSCs in steps and a relatively small output of the inductive reactive power is used to eliminate the excess reactive power for providing the required reactive power.

3.3.6.6 Thyristor-Switched Series Capacitor (TSSC)

The main element of a Thyristor-Switched Series Capacitor (TSSC) is a capacitor which is connected in parallel to a thyristor based ac switch as shown in Fig. 3.3a. A TSSC can only play a role of a discrete capacitor for compensation and there is no continuous control over it.

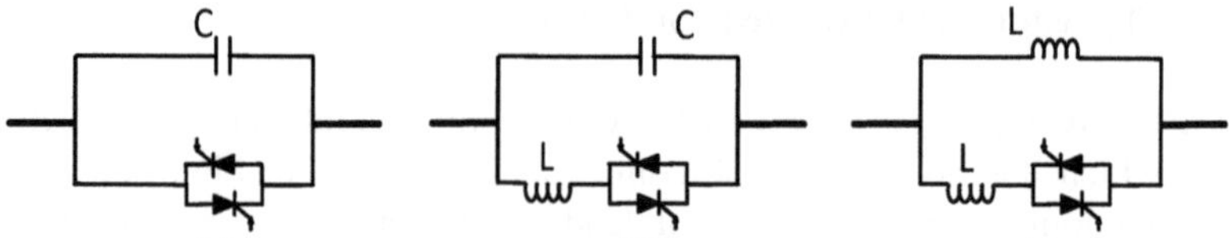

Fig. 3.3 Series compensators for reactive power control, **a** Thyristor-Switched Series Capacitor (TSSC), **b** Thyristor-Controlled Series Capacitor (TCSC), **c** Thyristor-Controlled Series Reactor (TCSR)

3.3.6.7 Thyristor-Controlled Series Capacitor (TCSC)

Thyristor-Controlled Series Capacitor (TCSC) consists of a capacitor bank in parallel with a thyristor-controlled reactor to provide the series capacitive reactance with the smoothly changes (Fig. 3.3b). The impedance of reactor is designed to be much less than the impedance of the series capacitor. At the firing angle of 90°, the TCSC helps to limit the fault current. The TCSC may consist of several smaller capacitors with different size in order to achieve a better performance.

3.3.6.8 Thyristor-Controlled Series Reactor (TCSR)

A Thyristor-Controlled Series Reactor (TCSR) is an inductive reactance compensator, which includes a series reactor along with a controlled-reactor as shown in Fig. 3.3c.

3.3.7 Static Reactive Power Generators Based on the Power Electronic Converters

More recently, interruptible (self-commutated) thyristors and other power semiconductors are used to generate and to absorb reactive power, without the use of ac capacitors or reactors, in which the output voltage is controlled for generating the required reactive power. So, a static reactive power compensator with a power electronic convertor is a system, which can provide controlled reactive current from an AC power source.

3.3.7.1 STATic Synchronous COMpensator (STATCOM)

A STATic synchronous COMpensator (STATCOM) is a static synchronous generator, which operates as a parallel reactive power compensator and can control the output capacitive or inductive current as shown in Fig. 3.4 [10].

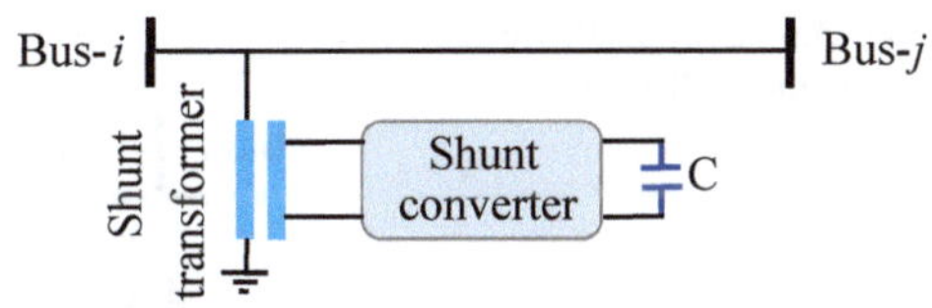

Fig. 3.4 STATic synchronous COMpensator (STATCOM)

3.3.7.2 Static Synchronous Series Compensator (SSSC)

Static Synchronous Series Compensator (SSSC) is a static synchronous generator based on power electronics without an external energy source, and works as a series compensator (Fig. 3.5) [11]. It is used to increase or to decrease the reactive voltage drop along the transmission line and consequently control the transferred electrical power.

3.3.7.3 Unified Power Flow Controller (UPFC)

A Unified Power Flow Controller (UPFC) is a combination of STATCOM and SSSC, which are connected through a dc link, to compensate active and reactive power simultaneously without external energy source (Fig. 3.6) [12]. It is a complete compensator for controlling the active and reactive power as well as the network voltage [13, 14].

3.3.8 Interline Power Flow Controller (IPFC)

An Interline Power Flow Controller (IPFC), as shown in Fig. 3.7, can draw the active power from one side of a line and injected to the other side of the line [15]. Therefore, unlike the SSSC, this compensator is able to control both the phase and amplitude of the injected voltage into the line.

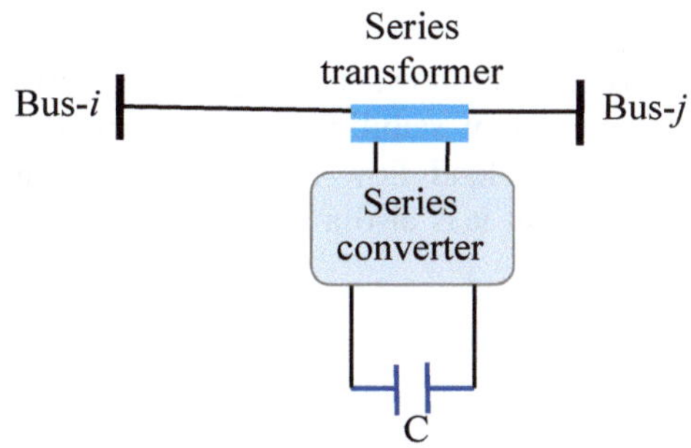

Fig. 3.5 Static Synchronous Series Compensator (SSSC)

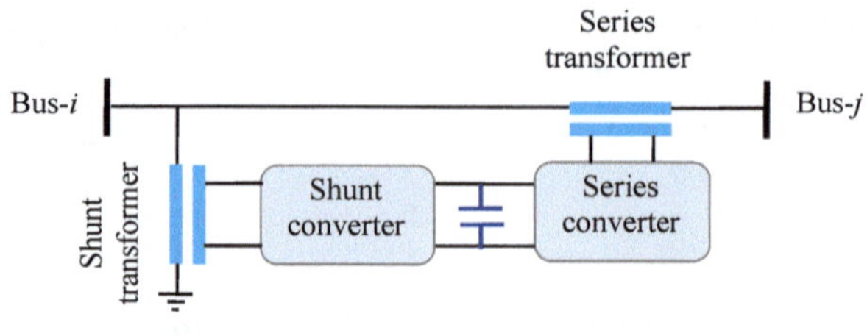

Fig. 3.6 Unified Power Flow Controller (UPFC)

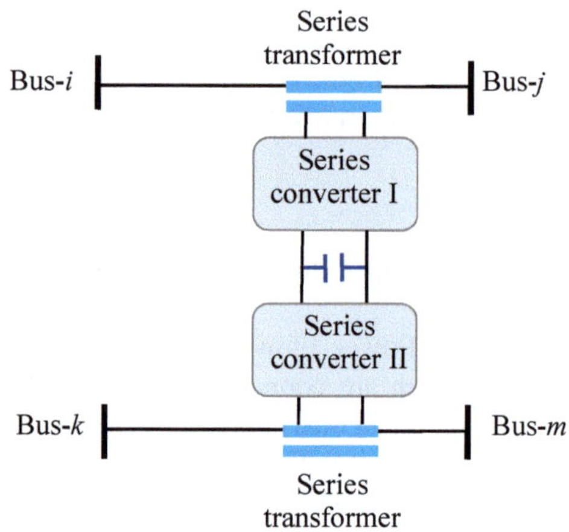

Fig. 3.7 Interline Power Flow Controller (IPFC)

Table 3.1 Comparison between reactive power sources for power system stability enhancement [16]

Reactive power controller	Stability enhancement	Load flow	Voltage control	Transient stability	Dynamic stability	Required time (s)
UPFC	Y	High	High	Medium	Medium	0.6
TCSC	Y	Medium	Low	High	Medium	1.5
FC-TCR	Y	Low	high	Low	Medium	7
SSSC	Y	Low	High	Medium	Medium	11

Table 3.1 shows a comparison between various reactive power sources for power system stability enhancement. It is found that UPFC is a more effective device for load flow, voltage control and stability enhancement of the power system, but it is also a more expensive solution.

3.4 Control of Reactive Power in a Power Transmission System

Figure 3.8 shows a simplified model of a power transmission system, where X_L is the reactance of the transmission line, $V_i \angle \theta_i$ and $V_j \angle \theta_j$ are voltage phasors of the grid buses. The purposes of the reactive power control in power transmission system are to

(a) transmit as much power as feasible on a line of the specified voltage and
(b) to control the voltage along the line within the limits [17].

The active and reactive power at buses *i* and *j* can be obtained by Eqs. (3.8) and (3.9), respectively:

$$P_i = \frac{V_i V_j}{X_L} \sin(\theta_i - \theta_j), \quad Q_i = \frac{V_i (V_i - V_j \cos(\theta_i - \theta_j)}{X_L} \tag{3.8}$$

$$P_j = \frac{V_i V_j}{X_L} \sin(\theta_i - \theta_j), \quad Q_j = \frac{V_j (V_j - V_i \cos(\theta_i - \theta_j)}{X_L} \tag{3.9}$$

It can be seen from (3.8) and (3.9) that the active and reactive power can be controlled by the voltages, phase angles and line impedance of the transmission system. Reactive power compensation can be implemented by reactive power generators, which are connected to the transmission line in parallel or in series [18]. The principles of shunt and series reactive power controllers are described below.

3.4.1 Shunt Compensation

Shunt reactive compensation is used in transmission systems to adjust the voltage magnitude, improve the voltage quality and the system stability. Shunt-connected reactors reduce the line over-voltages by consuming the reactive power, while shunt-connected capacitors maintain the voltage levels by compensating the reactive power. Instead of inductors or capacitors, there are reactive power generators based on the power electronic converters, which can generate the reactive power independent of the voltage at the point of connection.

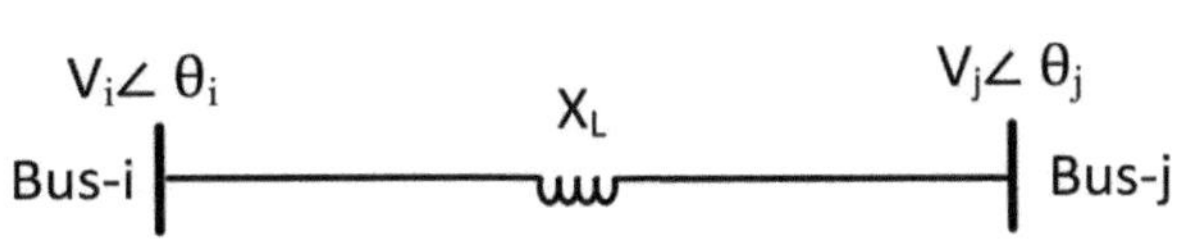

Fig. 3.8 Model of a lossless power transmission system

3.4.1.1 Shunt-Connected Capacitors

Figure 3.9 shows a simplified model of a compensated transmission line, in which the voltage magnitudes of the buses are assumed as $V \angle \theta_i$ and $V \angle 0$. An ideal controlled shunt-connected capacitor C is expected to regulate the voltage at the connection point as $V \angle \theta_i/2$ [19].

By using the above assumptions, (3.8) and (3.9), the reactive powers at buses i and j can be obtained as

$$Q_i = Q_j = 2\frac{V^2}{X_L}(1 - \cos(\frac{\theta_i}{2})) \tag{3.10}$$

Therefore, the injected reactive power by the capacitor to adjust the mid-point voltage can be calculated as:

$$Q_c = -(Q_i + Q_j) = -4\frac{V^2}{X_L}(1 - \cos(\frac{\theta_i}{2})) \tag{3.11}$$

3.4.1.2 Shunt Compensation Based on the Power Electronic Converters

Figure 3.10 depicts an equivalent circuit of the system, which is compensated by a shunt power electronic converter. There are different configurations available for the shunt compensator such as modular multi-level inverter, six-pulse three phase

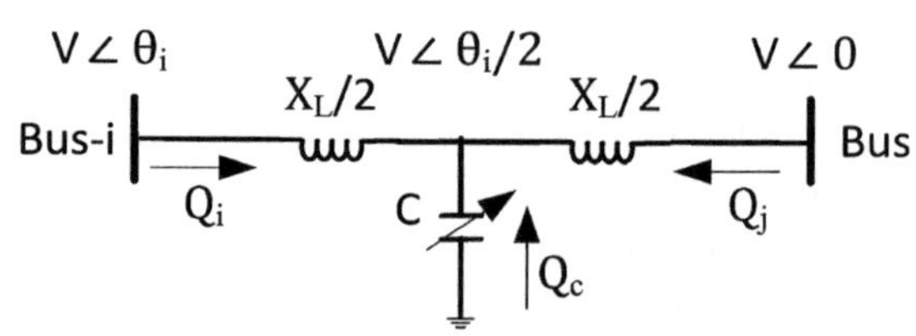

Fig. 3.9 Simplified model of a compensated transmission line by a shunt-connected capacitor

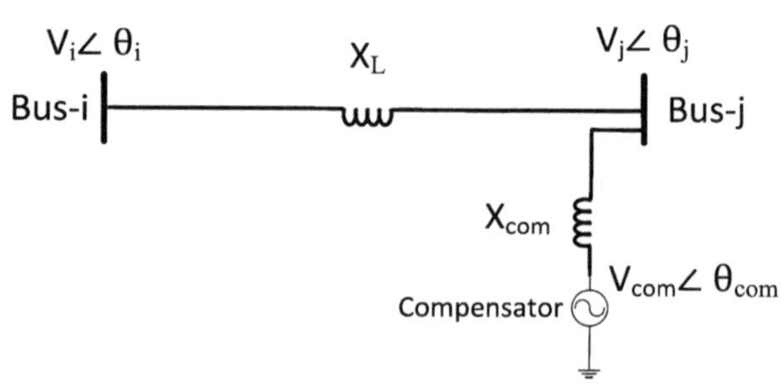

Fig. 3.10 Shunt compensation based on the power electronic converters

inverter, cascaded H-bridge converter, Neutral Point Clamped (NPC) inverter etc. [20].

Since the phase difference between the converter voltage and grid bus-j voltage (θ_{com}–θ_j) is small under normal operation, then $\sin(\theta_{com}-\theta_j) \approx (\theta_{com}-\theta_j)$ and $\cos(\theta_{com}-\theta_j) \approx 1$ [21]. Therefore, based on (3.8) and (3.9), the active power P_{com} and reactive power Q_{com} flowing out of the parallel converter are

$$P_{com} = \frac{V_j V_{com}}{X_{com}} \sin(\theta_{com} - \theta_j) \approx \frac{V_j V_{com}}{X_{com}} (\theta_{com} - \theta_j) \tag{3.12}$$

$$Q_{com} = \frac{V_j V_{com}}{X_{com}} \cos(\theta_{com} - \theta_j) - \frac{V_j^2}{X_{com}} \approx \frac{V_j}{X_{com}} (V_{com} - V_j) \tag{3.13}$$

From (3.13), the reactive power Q_{com} can be controlled by changing the voltage difference between the converter and grid bus (V_{com}–V_j). When the reactive power Q_{com} is changed, the voltage V_j changes slightly as well. This can be used to regulate the voltage at the PCC. Hence, a shunt reactive power controller mainly has two different operation modes: one is called the direct Q control mode, which provides the desired amount of reactive power, and the other is called the voltage regulation mode, which regulates the PCC voltage. Equation (3.12) shows the relationship between the active power P_{com} and the phase difference of the converter voltage and bus voltage. The real power flowing in or out forces the DC-link voltage to increase or decrease. As a result, it can be regulated by controlling the phase angle of the voltage generated by the converter. When the parallel compensator is operated in voltage regulation mode, it implements the following *V-I* characteristic [22].

Figure 3.11 shows as long as the reactive current stays within the minimum and maximum current values ($-I_{max}$, I_{max}) imposed by the converter rating, the voltage is regulated at the reference voltage V_{ref}.

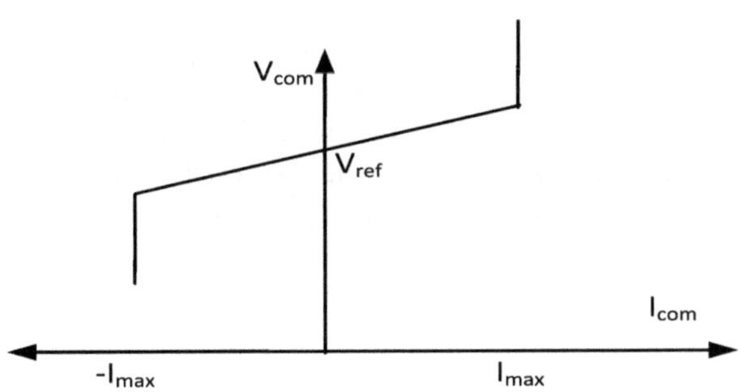

Fig. 3.11 *V-I* characteristic of the shunt compensator

3.4.2 Series Compensation

Series compensation controls the series impedance of the transmission line. Based on Eqs. (3.8) and (3.9), the AC power transmission is basically limited by the series reactive impedance of the transmission line. Series compensation with capacitors is the most common strategy to cancel the reactance part of the line. Like shunt compensation, series compensation may also be implemented with power electronic converters.

3.4.2.1 Series Capacitors

A simplified model of a series-compensated transmission line is shown in Fig. 3.12.

The transmission line is assumed ideal and it is represented by the reactance X_L. A series controlled capacitor is connected in the transmission line. The overall series inductance of the compensated transmission line is:

$$X_{total} = X_L - X \tag{3.14}$$

Therefore, a series capacitor can cancel the reactance part of the line. This increases the maximum power, reduces the transmission angle at a given level of power transfer, and increases the surge impedance loading. Based on Eq. (3.8), the transmitted active power in the compensated line is calculated as:

$$P_i = \frac{V_i V_j}{X_L - X} \sin(\theta_i - \theta_j) \tag{3.15}$$

3.4.2.2 Series Compensation Based on the Power Electronic Converters

Like shunt compensation, series compensation may also be implemented with voltage source converters. Under compensation conditions, the converter injects a voltage vector between two buses in series. The equivalent circuit to explain the injected voltage vector by the converter is shown in Fig. 3.13. As it can be seen, bus-j voltage vector can be expressed as follow:

$$\vec{V}_j = \vec{V}_{inj} + \vec{V}_i \tag{3.16}$$

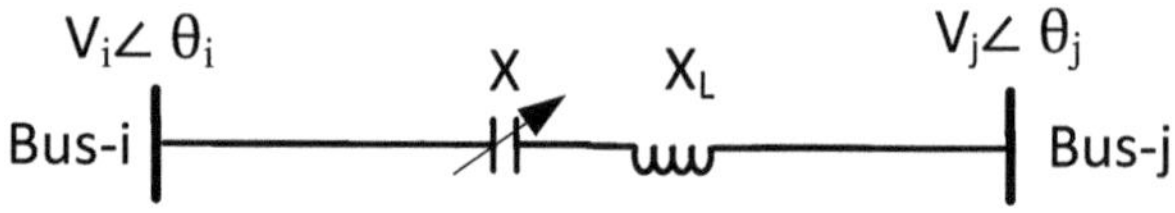

Fig. 3.12 Simplified model of a series compensated transmission line

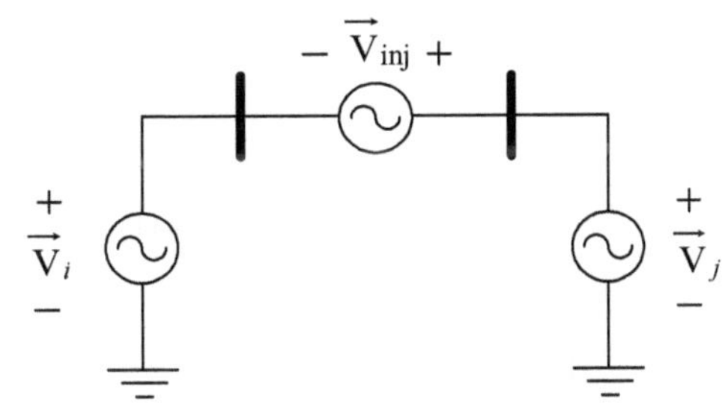

Fig. 3.13 Series compensation based on the power electronic converter

where $\vec{V}_i$, $\vec{V}_{inj}$ and $\vec{V}_j$ are bus-*i* voltage vector, injected voltage vector by the series converter and bus-j voltage vector, respectively. By injecting an appropriate voltage vector by the converter, the voltage can be regulated.

The control scheme is one of the important parts of the series compensator and has four basic functions:

- First, the grid buses voltages must be estimated
- After estimation, using an appropriate compensation method, the control scheme generates the voltage references.
- After producing the voltage references, the switching commands are generated by the appropriate modulation technique.
- When the current magnitude exceeds the rated converter range, the control scheme will generate the appropriate commands to the protection devices.

The two following methods have been usually used in the literature for compensation and voltage control:

1. After transformation of the three-phase voltages to the synchronous reference frame, dq-components of the voltages are controlled using PI controllers as shown in Fig. 3.14 [23, 24].
2. By using a phasor estimation method such as Kalman filter, Discrete Fourier Transform, or Least Error Squares, phasor parameters of the sensed voltages and currents are estimated separately for each phase. Then the control scheme generates the voltage references for each phase [25].

Figure 3.15a depicts the vector diagram of the voltages and current during the compensation and Fig. 3.15b shows the control scheme based on phasor estimation method. Here $\vec{V}_i = V_i\angle\theta_i, \vec{V}_j = V_j\angle\theta_j, \vec{V}_{inj} = V_{inj}\angle\theta_{inj}$ and $\vec{I} = I\angle\theta$ are the vectors of bus-i voltage, bus-j voltage, injection voltage by the converter and the line transmission current, respectively. V_{nom} is the compensated voltage magnitude. ϕ is the phase difference between the current and voltage of the bus-*j*. γ is phase difference between the injected voltage and bus-*i* voltage. According to the Fig. 3.15, the injected power is calculated as

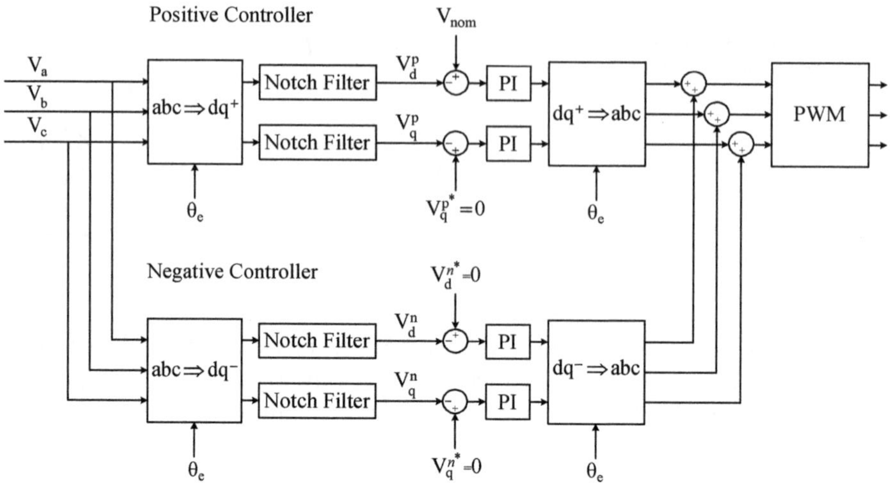

Fig. 3.14 Voltage control block diagram in dq reference frame

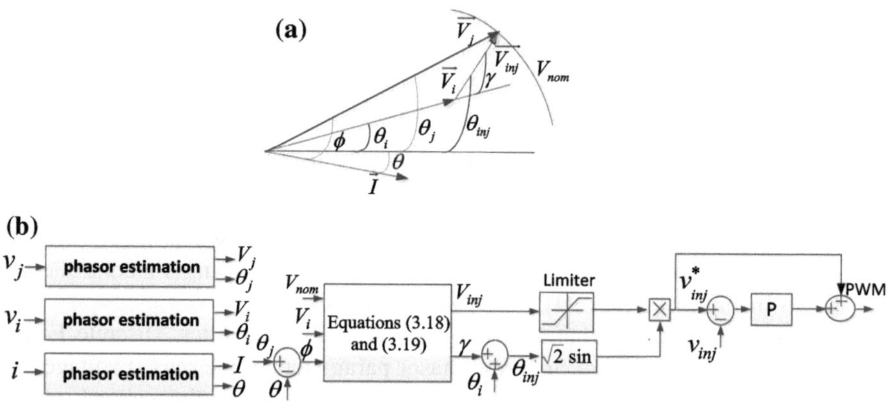

Fig. 3.15 Voltage control based on phasor estimation by series compensator, **a** vector diagram of the voltages and current during compensation, **b** block diagram of the control scheme

$$P_{inj} = P_{out} - P_{in} = V_j I \cos(\phi) - V_i I \cos(\phi - \theta_j + \theta_i) \tag{3.17}$$

where P_{out} is the output power and P_{in} is the input power. Since the compensation should be done by exchanging only reactive between the series converter and the network, so $P_{inj} = 0$ is assumed. In other words, the injected voltage should be perpendicular to the current, so no active power is exchanged between the converter and the network. By doing a small amount of computations and trigonometric relations, the injected voltage magnitude (V_{inj}) and γ are obtained as follows:

$$\gamma = \arcsin\left(\frac{V_{nom}\cos(\phi)}{V_i}\right) \tag{3.18}$$

$$V_{inj} = \sqrt{V_{nom}^2 + V_i^2 - 2V_{nom}\,V_i\,\sin(\gamma + \phi)} \tag{3.19}$$

After producing the voltage references, the inverter output voltages are injected in series by three single-phase transformers. To eliminate the switching frequency harmonics, a low-pass filter for each phase is used, which consists of the leakage inductance of the series transformer and the filter capacitor. Also, a parallel switch for each phase is used to bypass the series converter in fault conditions.

3.5 Conclusion

An overview of the basic principles of the power transmission operation and the reactive power role in the transmission system has been presented. The flow of reactive power causes additional heating of the lines and voltage drops in the network. High reactive power consumption by heavily loaded transmission line lead to voltage dips in the system and limit the generation of the active power. Voltage and reactive power control causes that a stable, efficient, and reliable operation of the power system is achieved and utilization of the transmission system is maximized.

References

1. V. Patel, J.G. Jamnani, Techniques to Increase Surge Impedance Loading Level of EHV AC Transmission Lines for Improving Power Transfer Capability, IEEE International Conference on Computation of Power, Energy Information and Communication, (ICCPEIC), 2015, pp. 518–523.
2. A.K. Mohanty, A.K. Barik, Power System Stability Improvement Using FACTS Devices, Int. J. Mod. Eng. Res., vol. 1, no. 2, pp. 666–672, 2009.
3. F.S. Moreira, T. Ohishi, J.I.D.S. Filho, Influence of the Thermal Limits of Transmission Lines in the Economic Dispatch, IEEE Power Engineering Society General Meeting, 2006, pp. 1–6.
4. R. Mathur, R. Varma, Reactive Power Control in Electrical Power Transmission Systems, 2002, Wiley-IEEE Press.
5. P. Kundur, Power System Stability and Control, 1994, McGrawHill.
6. D. Biggar, M. Hesamzadeh, Introduction to Electric Power Systems, 2014, Wiley-IEEE Press.
7. X. Yu, S. Cao, H. Jia, P. Zhang, Impact of the Exciter Voltage Limit to Power System Small Signal Stability Region, Proc. of IEEE Power Engineering Society General Meeting, 2007, pp. 1–7.

8. K. Bhattacharya, J. Zhong, Reactive Power as an Ancillary Service, IEEE Trans. Power Syst., vol. 16, no. 2, pp. 294–300, 2001.
9. T. Zabaiou, L.A. Dessaint, I. Kamwa, Preventive Control Approach for Voltage Stability Improvement Using Voltage Stability Constrained Optimal Power Flow based on Static Line Voltage Stability Indices, IET Gener., Transm. Distrib., vol. 8, no. 5, pp. 924–934, 2014.
10. Z. Chen, F. Blaabjerg, Y. Hu, Stability Improvement of Wind Turbine Systems by STATCOM, Proc. of IEEE Industrial Electronics, 2006, pp. 4213–4218.
11. F.A.L. Jowder, Influence of Mode of Operation of the SSSC on the Small Disturbance and Transient Stability of a Radial Power System, IEEE Trans. Power Syst., vol. 20, no. 2, pp. 935–942, 2005.
12. X. Jiang, J.H. Chow, A.A. Edris, B. Fardanesh, E. Uzunovic, Transfer Path Stability Enhancement by Voltage-Sourced Converter-Based FACTS Controllers, IEEE Trans. Power Deliv., vol. 25, no. 2, pp. 1019–1025, 2010.
13. M.H. Haque, Evaluation of First Swing Stability of a Large Power System with Various FACTS Devices, IEEE Trans. Power Syst., vol. 23, no. 3, pp. 1144–1151, 2008.
14. S. Kannan, S. Jayaram, M. Salama, Real and Reactive Power Coordination for a Unified Power Low Controller, IEEE Trans. Power Syst., vol. 19, no. 3, pp. 1454–1461, 2004.
15. V. Azbe, R. Mihalic, The Control Strategy for an IPFC Based on the Energy Function, IEEE Trans. Power Syst., vol. 23, no. 4, pp. 1662–1669, 2008.
16. G.M. Vireshkumar, F.R. Basangouda, H.J. Suresh, Review on Comparison of FACTS Controllers for Power System Stability Enhancement, International Journal of Scientific and Research Publications, vol. 3, no. 3, pp. 57–58, 2013.
17. K.R. Padiyar, FACTS Controllers in Transmission and Distribution, 2007, New Age Intenational.
18. E. Acha, V.G. Agelidis, O. Anaya-Lara, T.J.E. Miller, Power Electronic Control in Electrical Systems, 2002, Newnes.
19. T.J.E. Miller, Reactive Power Control in Electric Systems, 1982,Wiley-Interscience.
20. P.L. Nguyen, Q.C. Zhong, F. Blaabjerg, J. M. Guerrero, Synchronverter-Based Operation of STATCOM to Mimic Synchronous Condensers, Proc. of IEEE Conference on Industrial Electronics and Applications, (ICIEA), 2012, no. 2, pp. 942–947.
21. L. Liu, H. Li, Y. Xue, W. Liu, Reactive Power Compensation and Optimization Strategy for Grid-Interactive Cascaded Photovoltaic Systems, IEEE Trans. Power Electron., vol. 30, no. 1, pp. 188–202, 2015.
22. X. Jiang, X. Fang, J.H. Chow, A.A. Edris, E. Uzunovic, M. Parisi, L. Hopkins, A Novel Approach for Modeling Voltage-Sourced Converter-Based FACTS Controllers, IEEE Trans. Power Deliv., vol. 23, no. 4, pp. 2591–2598, 2008.
23. J.G. Nielsen, M. Newman, H. Nielsen, F. Blaabjerg, Control and Testing of a Dynamic Voltage Restorer (DVR) at Medium Voltage Level, IEEE Trans. Power Electron., vol. 19, no. 3, pp. 806–813, 2004.
24. Y. Shokri, E. Ebrahimzadeh, H. Lesani, S. Afsharnia, Performance Improvement of DFIG-Based Wind Farm Using Multilevel Cascaded Hbridge Converter under Unbalanced Grid Voltage Conditions, Proc. of 14th Int. Conf. on Environ. and Elec. Eng. (EEEIC), 2014, pp. 158–163.
25. E. Ebrahimzadeh, S. Farhangi, H. Iman-eini, F.B. Ajaei, R. Iravani, Improved Phasor Estimation Method for Dynamic Voltage Restorer Applications, IEEE Trans. Power Deliv., vol. 30, no. 3, pp. 1467–1477, 2015.

Chapter 4
Reactive Power Compensation in Energy Transmission Systems with Sinusoidal and Nonsinusoidal Currents

Milan Stork and Daniel Mayer

Abstract A standout amongst the most noteworthy current talks in electrical designing is the meaning of the responsive power under nonsinusoidal conditions in nonlinear electric frameworks. New meanings of responsive power have been talked about in the most recent years. Despite the fact that the component of electric energy stream for nonsinusoidal conditions is all around depicted today, so toll is not yet accessible a summed up power hypothesis, and hypothetical figuring for the configuration of such gadgets as dynamic channels or element compensators. Thusly the undertaking of planning compensators for advance energy transmission with nonlinear time-fluctuating loads in nonsinusoidal administrations is, a long way from clear. Voltage and current harmonics created by nonlinear loads increment power losses in transmission frameworks and, in this manner, negatively affect effectivity of appropriation frameworks and parts. While a few harmonics are brought on by framework nonlinearities, for example, transformer immersion, most harmonic are delivered by power electronic loads, for example, flexible velocity drives and diode span rectifiers. In this chapter the reactive power compensation for sinusoidal and nonsinusoidal circumstances, where nonlinear circuit voltages and streams contain harmonic are explained and reenacted. The results can be used for control algorithms of automatic compensators which are also described. The main aim of this chapter is based on the dissipative systems theory and therefore theory of cyclodissipativity which can be used for calculation of compensation elements (capacitors, inductors) for reactive power compensation. The compensation elements are determined by minimizing line losses. It will show that approach base on dissipative systems theory provides an important mathematical framework for

M. Stork (✉)
Department of Applied Electronics and Telecommunications, University of West Bohemia, Plzen, Czech Republic
e-mail: stork@kae.zcu.cz

D. Mayer
Department of Theory of Electrical Engineering, University of West Bohemia, Plzen, Czech Republic
e-mail: mayer@kte.zcu.cz

N. Mahdavi Tabatabaei et al. (eds.), *Reactive Power Control in AC Power Systems*, Power Systems, DOI 10.1007/978-3-319-51118-4_4

analyzing and designing of compensators for reactive power compensation even for general nonlinear loads. The presented theory is supplemented by a series of examples and simulations.

4.1 Introduction

A standout amongst the hugest current examinations in electrical designing is the meaning of the reactive power (RP) under nonsinusoidal conditions in nonlinear electric frameworks. New meanings of RP have been talked about in the most recent years. Despite the fact that the component of electric energy stream for nonsinusoidal conditions is all around portrayed today, so admission is not yet accessible a summed up power hypothesis, and hypothetical computations for the configuration of such gadgets as dynamic channels or element compensators. In this manner, the undertaking of planning compensators for improve energy transmission with nonlinear time-changing loads in nonsinusoidal administrations is, a long way from clear. Voltage and current harmonics created by nonlinear loads increment power loss in transmission frameworks and, in this manner, negatively affect effectivity of conveyance frameworks and parts. While a few harmonics are created by framework nonlinearities, for example, transformer immersion, most harmonic are delivered by power electronic loads, for example, notice adjustable speed drives and diode span rectifiers. In this chapter the reactive power compensation for sinusoidal and nonsinusoidal circumstances, where voltages and streams contain harmonic are contemplated and mimicked. The outcomes can be utilized for control calculations of programmed compensators. The fundamental point of this section depends on the dissipative frameworks hypothesis and along these lines hypothesis of cyclodissipativity which can be utilized for estimation of compensation components (capacitors, inductors) for RP pay. It will demonstrate that approach base on dissipative frameworks hypothesis gives a numerical structure to breaking down and planning of compensators.

Reactive power investigation and compensation are significant for conveyed era and deregulation, which are the two primary procedures reshaping electric energy frameworks today. Conveyed era requires a portrayal of complex loads with bidirectional power streams, both in enduring state and in incessant homeless people. Deregulation requires exact evaluation of power streams, as well as of pay endeavors and different administrations that incorporates numerous members and chiefs.

Our inspiration is to think about inert power and probability of pay for energy frameworks. Today, the exchanging power converters can working at time interims much shorter than the key (50 or 60 Hz) period. Numerous broadly utilized outline instruments (e.g., those taking into account the idea of "quick reactive power") are described solely in the time space. Their application once in a while results in presentation of harmonic that were not present in the first waveforms. This, thusly, might energize unmodeled progress, and bring about unsatisfactory drifters. The specialized means for online pay of polyphase frameworks are simply getting to be accessible. This is valid for power electronic converters (e.g., voltage-sourced inverters with quick exchanging and short reaction times) furthermore for control equipment (e.g., reasonable miniaturized scale controllers with sign preparing ability).

Improving energy exchange from an air conditioner source to a load is an established issue in electrical designing. The outline of power frameworks is such that the exchange happens at the key frequency of the source. Practically speaking, the proficiency of this exchange is typically diminished because of the phase shift in the middle of voltage and current at the major frequency. The phase shift emerges to a great extent because of energy streams portraying electric engines. The power variable (PF), characterized as the proportion between the genuine or dynamic power (normal of the immediate power) and the clear power (the result of rms root mean square estimations of the voltage and current), then communicates the energy transmission proficiency for a given load. The standard way to deal with diminishing the responsive power is to utilize a compensator between the source and the load. Theoretical configurations of the compensator regularly expect that the identical source comprises of a perfect generator having zero output impedance and delivering an altered, simply sinusoidal voltage. On the off chance that the load is straight time invariant (LTI), the subsequent enduring state mutt rent is a stage moved sinusoid, and the PF is the cosine of the phase shift point. Accessible compensation innovations incorporate e.g. the pivoting hardware as well as exchanged capacitors and inductors and power electronic converters, for example, dynamic channels and adaptable air conditioning transmission frameworks.

4.2 Reactive Power Review

Energy handling frameworks have experienced significant changes as of late. This procedure is principally determined by a longing to build proficiency, with diminishment in energy costs, energy losses (and cooling necessities), and part measure. For productivity changes have been created the new advancements, (for example, power electronics), and use of quicker actuators and controllers.

It is imperative that there exist useful impediments applicable for displaying and control of dormant power. For instance, because of limited exchanging frequency there exists a scope of frequencies where a dynamic channel (regularly a voltage sourced inverter) can work viably. The execution criteria are themselves innovation subordinate—numerous instruments accessible in the field can just gauge amounts in the scope of roughly 1.5 kHz, however the most astounding harmonic utilized for metering reasons for existing is frequently the 23rd. Another for all intents and purposes essential class of requirements is powered by material science of voltage sourced inverters, for example, limited energy storage, with constrained permitted voltage variety of the primary capacitor and conceivably lacking voltage extent for following of higher current harmonics.

(a) Harmonic Standards: Standards and suggestions that recommend execution criteria for electrical hardware are advanced by various associations around the world (more than 50 national guidelines are said in [1]). The most powerful ones originate from the International Electrotechnical Commission (IEC) and by the IEEE, [1–3]. While the IEEE-519 prescribed practice concentrates on the client/framework interface (the purported purpose of regular coupling) and on higher power levels, the IEC-1000 standard (additionally embraced as the European standard) incorporates limits for individual bits of gear, and approaches to gauge the harmonic. These archives commonly list the extent of particular (low) current harmonics (both in relative and in outright terms), and an aggregate amount known as the total harmonic distortion (*THD*), that confines a potentially weighted total of squares of consonant sizes. Watch that all significant global measures are indicated in the frequency area. While these benchmarks can be deciphered as outside imperatives powered on the specialist outlining an energy handling framework, it is critical to recall the real framework indications of expanded harmonics. These incorporate hardware impacts, (for example, disappointments of wires, power component redress capacitors, transformer and engine overheating, breaker stumbling) and framework wide issues (squinting of glowing lights and glint of glaring lights, impedance on correspondence frameworks) [3]. A late study of North American utilities [4] recommended that while the quantity of loads that create current harmonic when supplied with a (near) sinusoidal voltage is expanding powerfully, there exists an essential absence of comprehension of energy streams including nonsinusoidal waveforms and of approaches to decipher benchmarks.

A word about classification while everybody concurs that dynamic (or genuine) power is the dc part of the prompt power (every one of these amounts will be characterized right away), a few creators save the name responsive power for the supplement of dynamic power (the dormant power) at major frequency, or for the component that can be repaid with direct latent circuit component; we utilize the assignments reactive and idle reciprocally (for additional on a proposed brought together phrasing see [5]).

(b) Power Systems: A spearheading chip away at this subject was attempted in Europe in the seventies, and brought about the Fryze-Buchholz-Depenbrock (FBD) strategy, compressed by the originator in [6]. This reference brings up the likelihood of figuring ideas in the frequency space. Late down to earth uses of this time-area approach are exhibited in [7, 8].

Different references by Czarnecki, e.g. [9–13] together with here and there warmed running with trade [14], tended to the issue from a power structure perspective, and portrayed diverse miracles of utilitarian energy for circuit-based compensation in unbalanced three-stage systems. A repeat space presentation of Czarnecki breaking down of three-stage systems is presented in [9]; we later give a complete explanation in the general case. Note that other orthogonal weakenings are doable for the sections of the inactive present, for instance, the one showed in

[15] that disconnects the part compensable by direct shunt components. A related responsibility by an IEEE Power-Engineering Society Working Group is shown in [16]. An instructive and sensible power system test is given in [17].

The unmistakable power is one of the key thoughts while describing diverse sums in polyphase structures, and a material science based comprehension is given in [18]; a former study on the subject is depicted in [19]. A novel power system organized presentation of the thought responsive power with respect to vector spaces and repeat parts is given in [20]. From the theoretical perspective, our investigation heading contrasts from the strategy depicted in [20] in our emphasis on projections in vector spaces where a polyphase sign is identified with by a singular part (point) of a suitable Hilbert space. This new vector space licenses us to decide a united treatment of each and every genuine class of responsive power compensators as projections to suitable sub-spaces, and structures a reason for fruitful numerical procedure.

(c) Power Electronics: The exploration on polyphase frameworks in nonsinusoidal operation has been reinvigorated by [21] that drew closer the issue from a down to earth stance, and proposed a helpful definition for the "quick" reactive power. (An undifferentiated from idea was presented before by Depenbrock as pointed out in [6], yet remained basically obscure outside Germany.) The idea was introduced in the time space, and in the exceptional instance of three-stage frameworks, so that the thought of vector (external) results of vectors could be utilized. An augmentation to four-conductor frameworks that can represent the zero-sequence components was presented in [22, 23]. Note that the momentary responsive power relates to the part of the inert power that can be repaid without energy stockpiling. A compensator construct only with respect to this idea won't just neglect to totally dispose of the idle power, yet might likewise bring extra harmonic [24] into the present waveform, as we show later. This is recognized by one of the originators of the quick responsive power idea in [25]. Hypothetical ideas proposed by Depenbrock and Akagi-Nabae have been broadly tried tentatively, for the most part in dynamic channel applications. Some illustrative studies are [26–31]. On account of three-stage three-conductor frameworks, it is some of the time gainful to express applicable amounts regarding space vectors. The change is clear, and a trial confirmation is introduced in [32].

4.3 Theoretical Background—Properties of Homogenous Operators

In this part the properties of homogenous operators are presented [33, 34]. They are important for next parts of this chapter.

For any given periodic signal $x(t)$ of period T the homogenous integral operator is defined as

$$\widehat{x}(t) = \omega\left(\psi(t) - \overline{\psi}\right) = \frac{2\pi}{T}\left(\psi(t) - \overline{\psi}\right) \tag{4.1}$$

where integral of voltage $x(t)$ is

$$\psi(t) = \int_0^t x(\tau)d\tau \tag{4.2}$$

and the average value over period T is

$$\overline{\psi} = \int_0^T \psi(\tau)d\tau \tag{4.3}$$

Similarly, the homogenous differential operator is

$$\breve{x}(t) = \frac{1}{\omega}\frac{dx(t)}{dt} \tag{4.4}$$

Note that $\widehat{x}$ and $\breve{x}$ are dimensionally homogenous to basic quantity x [33]. The internal product of two periodic variables $x(t)$ and $y(t)$ is defined as

$$\langle x, y\rangle = \frac{1}{T}\int_0^T x(t)y(t)dt \tag{4.5}$$

and norm of variable $x(t)$ (X is rms—root mean square) is

$$\|x\| = \sqrt{\langle x, x\rangle} = \sqrt{\frac{1}{T}\int_0^T x^2(t)dt} = X \tag{4.6}$$

The homogenous operators have following properties [from (4.7) to (4.17)]

$$\left\langle x, \widehat{x}\right\rangle = \left\langle x, \breve{x}\right\rangle = 0 \tag{4.7}$$

$$\left\langle \widehat{x}, \breve{x}\right\rangle = -\|x\|^2 \tag{4.8}$$

$$x = \breve{y} \Leftrightarrow \widehat{x} = y \tag{4.9}$$

$$\left\langle \widehat{x}, y \right\rangle = -\left\langle x, \widehat{y} \right\rangle \tag{4.10}$$

$$\left\langle \breve{x}, y \right\rangle = -\left\langle x, \breve{y} \right\rangle \tag{4.11}$$

$$\left\langle \widehat{x}, \breve{y} \right\rangle = \left\langle \breve{x}, \widehat{y} \right\rangle = -\langle x, y \rangle \tag{4.12}$$

Moreover, if x and y are sinusoidal quantities with rms values respectively equal to X and Y and phase difference equal to φ, the following properties hold:

$$\|x\| = \left\|\widehat{x}\right\| = \left\|\breve{x}\right\| = X \tag{4.13}$$

$$\widehat{x} + \breve{x} = 0 \tag{4.14}$$

$$x^2 + \widehat{x}^2 = x^2 + \breve{x}^2 = x^2 - \widehat{x}\breve{x} = 2\|x\|^2 = 2X^2 \tag{4.15}$$

$$xy - \widehat{x}\breve{y} = xy - \breve{x}\widehat{y} = 2XY \cos \varphi \tag{4.16}$$

$$\widehat{x}y - x\widehat{y} = x\breve{y} - \breve{x}y = 2XY \sin \varphi \tag{4.17}$$

Moreover, expressing variable $x(t)$ by its Fourier series:

$$x(t) = \sum_{k=1}^{\infty} x_k(t) = \sum_{k=1}^{\infty} \sqrt{2}X_k \sin(k\omega t + \varphi_k) \tag{4.18}$$

For homogenous variables

$$\widehat{x}(t) = \sum_{k=1}^{\infty} \widehat{x}_k(t) = -\sum_{k=1}^{\infty} \sqrt{2}\frac{X_k}{k} \cos(k\omega t + \varphi_k) \tag{4.19}$$

$$\breve{x}(t) = \sum_{k=1}^{\infty} \breve{x}_k(t) = \sum_{k=1}^{\infty} \sqrt{2}kX_k \cos(k\omega t + \varphi_k) \tag{4.20}$$

4.4 Cyclodissipativity Approach

It will be reflected the typical energy transmission from an AC generator to a load. The voltage and current of the supplier are functions of time and are stand for the column vectors v_S, $i_S \in \Re^q$. The load is defined by a (conceivably nonlinear and time-varying) q–port system Σ [35, 36].

There are will the two following assumptions:

(**A1**) All the signs in the system are periodic with fundamental period T and fit into the space

$$L_2[0,T) : \left\{ x : [0,T) \rightarrow \Re^q \left| \|x\|^2 := \frac{1}{T} \int_0^T |x(\tau)|^2 d\tau < \infty \right. \right\} \tag{4.21}$$

where $\|.\|$ *is named the RMS* value of b *and* $|.|$ *is the Euclidean standard.*

(**A2**) *The supplier is ideal, in the sense that* v_S *stays unaffected for all loads.*

Assumption (A1) captures the essentially rational method that the organism functions in a periodic, though not unavoidably sinusoidal, steady-state system. This is the case of the vast majority of applications of attention for the problem at hand [36].

Assumption (A2) is the similar to saying that the supplier has zero impedance and is defensible by the fact that most AC device function at a specified voltage, with the actual drained current being stated by the load [36].

The active power supplied by the source is defined as

$$P := \langle v_S, i_S \rangle = \frac{1}{T} \int_0^T v_S^T(t) i_S(t) dt \tag{4.22}$$

and $\langle . \rangle$ means the internal creation in $L_2[0, T)$. From (4.22) and the Cauchy-Schwarz inequality [36] is

$$P \leq \|v_S\| \cdot \|i_S\| =: S \tag{4.23}$$

where is defined the apparent power S. From the inequality above we achieve that, under Assumption (2), S is the uppermost average power supplied to the load for all loads that have the identical rms current $\|i_S\|$. The characteristics holds if and only if v_S and i_S are collinear. If this is not the case $P < S$ and compensation schemes are presented to decrease this incongruity [36]. That is, to maximize the proportion P/S - that is called the *PF*. A typical compensation formation (which can be easily prolonged to n-phase system) is shown in Fig. 4.1 where, to preserve the rated

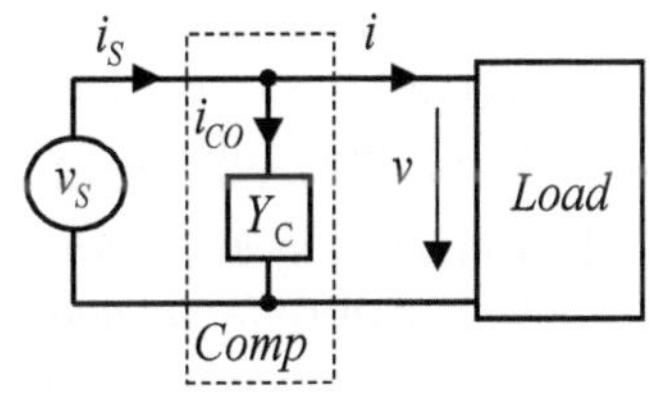

Fig. 4.1 Principle of typical reactive power compensation

voltage at the load stations the compensator Y_C is located in parallel. Similarly, to evade power indulgence, Y_C is restricted to be lossless, that is

$$\langle v, i_{CO}\rangle = 0 \tag{4.24}$$

where i_{CO} is the compensator current and notification that $v_S = v$ (Fig. 4.1). From the former can be result that the *PF* compensation problem is accurately equivalent to the problem of minimization of $\|i_S\|$ subject to the restriction (4.24). From

$$\|i_S\|^2 = \|i\|^2 + \|i_C\|^2 + 2\langle i_C, i\rangle \tag{4.25}$$

it is clear that an obligatory situation to decrease the rms value of i_S is

$$\langle i_C, i\rangle < 0 \tag{4.26}$$

It turns out that the latter circumstance and the limitation of compensator losslessness can be recognized using the perception of cyclo-dissipativity and its connected abstract energy [36–41].

Definition 1 Assume a dynamical system, with input $u \in L_2[0,T)$ and output $y \in L_2[0,T)$, acknowledges a state–space picture with state vector $x \in X$ [42]. The system is cyclo-dissipative with respect to the supply rate $w(u, y)$, where w: $L_2[0, T) \times L_2[0,T) \rightarrow \Re$, if and only if

$$\int_0^T w(u(t), y(t))dt \geq 0 \tag{4.27}$$

For all u: $[0, T) \rightarrow L_2[0,T)$ such that $x(T) = x_0$ where $x(0) = x_0$. It is said to be cyclo-lossless if the inequality holds with identity. *In other words, a system is cyclo-dissipative when it cannot create (abstract) energy over closed paths in the state-space.* It might, though, produce energy alongside some preliminary share of such a trajectory; if so, it would not be dissipative. *On the other hand, every dissipative system is cyclo-dissipative.* For instance, (possibly nonlinear) RLC circuits with input and output their terminal currents and voltages, respectively, are cyclo-dissipative with supply rate $w(u, y) = u^T y$ provided that all resistances are passive (This can be easily proven using Tellegen's Theorem [42–44]). Note that it is not presumed that inductors and capacitors are passive-that is, that their stored energy is non–negative-if so, the circuit is in addition passive. It has been shown in [42–44] that, correspondingly to dissipative systems, one can use storage functions and dissipation differences to characterize cyclo-dissipativity.

Theorem 1 *A system with state depiction is cyclo-dissipative if, for all* $x \in X$ *which are both controllable and reachable, there exists a virtual storage function* ϕ: $X \rightarrow R$ [42]. *That is, a function that satisfies*

$$\phi(x_0) + \int_0^T w(u(t), y(t))dt \geq \phi(x_1) \tag{4.28}$$

for $u \in L_2[0,T)$ such that $x(0) = x_0$ and $x(T) = x_1$. In following part, the instances of reactive power compensation based on cyclo-dissipativity method are presented.

4.5 Reactive Power Compensation

In this part on the first, the simple example concerning linear circuit and nonharmonic power source is presented. The second example is devoted to nonlinear circuit. Booth examples are simply extended to q-phase version.

Example 4.1 [36]: Suppose linear system, power source, compensator and serial connection of *RLC* and nonharmonic power source (Fig. 4.2) [36]. Supply voltage v_S is

$$v_s(t) = \sqrt{2}(A_1 \sin(\omega_0 t) + A_3 \sin(3\omega_0 t) + A_5 \sin(5\omega_0 t)) \tag{4.29}$$

where A_1 = 360, A_3 = 144, A_5 = 42, ω_0 = 100π [rad/s] and R = 15 [Ω], L = 0.08 [H], C = 0.0212 [mF]. Load impedances Z_n (for n = 1, 3, 5) are given

$$Z_n = R + j\omega_0 nL + \frac{1}{j\omega_0 nC} \quad n = 1, 3, 5 \tag{4.30}$$

For calculated values Z_1 = 15 – 125.14j; Z_3 = 15 + 24.97j; and Z_5 = 15 + 95j, the effective currents are

$$i_n = A_n / Z_n \quad n = 1, 3, 5 \tag{4.31}$$

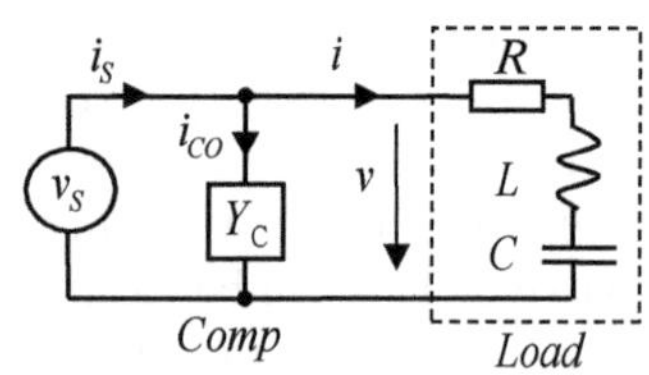

Fig. 4.2 Example of RP compensation for linear *RLC* load and nonharmonic source

Numerical values of currents are $i_1 = 0.34 + 2.8361j$; $i_3 = 2.54 - 4.24j$; $i_5 = 0.0681 - 0.4313j$.

Total effective current through the load is

$$i = \|i_n\| = \sqrt{\sum_n |i_n|^2} = \sqrt{|i_1|^2 + |i_3|^2 + |i_5|^2}; \quad n = 1, 3, 5 \tag{4.32}$$

therefore $i = 5.7257$ [A]. For *uncompensated circuit* ($Y_C \to 0$) $i_S = i$ and apparent power is

$$S = \|v_S\| \cdot \|i_S\| = \left(\sqrt{|A_1|^2 + |A_3|^2 + |A_5|^2}\right) \cdot \left(\sqrt{|i_1|^2 + |i_3|^2 + |i_5|^2}\right) \tag{4.33}$$

Result is $S = 2233$ [VA] and active power (average power) is

$$P = \langle v_S, i_S \rangle = \frac{1}{T} \int_0^T v_S^T(t) i_S(t) dt = R i_S^2 \tag{4.34}$$

$P = 491.8$ [W]. *PF* for uncompensated circuit ($Y_C \to 0$) is

$$PF = \frac{P}{S} = \frac{491.76}{2233} = 0.22 \tag{4.35}$$

For circuit compensated by capacitor

$$\|i_S\|^2 = \|i + i_{CO}\|^2 = \|i\|^2 + \|i_{CO}\|^2 + 2\langle i, i_{CO} \rangle \tag{4.36}$$

and current i_{CO} for compensation capacitor

$$i_{CO}(t) = C_{CO} \frac{dv_S(t)}{dt} \tag{4.37}$$

therefore

$$i_{CO} = C_{CO} \dot{v}_S \tag{4.38}$$

and substitution of Eq. (4.18) into (4.16)

$$\|i_S\|^2 = \|i + i_{CO}\|^2 = \|i\|^2 + \|C_{CO} \dot{v}_S\|^2 + 2\langle i, C_{CO} \dot{v}_S \rangle t \tag{4.39}$$

after some manipulations

$$\|i_S\|^2 = \|i + i_{CO}\|^2 = \|i\|^2 + C_{CO}^2\|\dot{v}_S\|^2 + 2C_{CO}\langle i, \dot{v}_S\rangle \tag{4.40}$$

For compensated circuit $\|i_S\|^2 < \|i + i_{CO}\|^2$ Eq. is

$$\|i_S\|^2 < \|i + i_{CO}\|^2 = \|i\|^2 + C_{CO}^2\|\dot{v}_S\|^2 + 2C_{CO}\langle i, \dot{v}_S\rangle \tag{4.41}$$

Using the property (4.42)

$$\langle \dot{x}, y\rangle = -\langle x, \dot{y}\rangle \tag{4.42}$$

in Eq. (4.41) results

$$\|i_S\|^2 < \|i + i_{CO}\|^2 = \|i\|^2 + C_{CO}^2\|\dot{v}_S\|^2 - 2C_{CO}\left\langle \dot{\hat{i}}, v_S\right\rangle \tag{4.43}$$

Minimal value of i_S (or minimum of apparent power *S*) can be found for

$$\frac{d}{dC_{CO}}\left(\|i\|^2 + C_{CO}^2\|\dot{v}_S\|^2 - 2C_{CO}\left\langle \dot{\hat{i}}, v_S\right\rangle\right) = 0 \tag{4.44}$$

therefore

$$2C_{CO}\|\dot{v}_S\|^2 - 2\left\langle \dot{\hat{i}}, v_S\right\rangle = 0 \tag{4.45}$$

Optimal value of compensation capacitor is

$$C_{CO} = \frac{\left\langle \dot{\hat{i}}, v_S\right\rangle}{\|\dot{v}_S\|^2} \tag{4.46}$$

where

$$\left\langle \dot{\hat{i}}, v_S\right\rangle = \mathrm{Re}\{j\omega_0 i_1 A_1\} + \mathrm{Re}\{j3\omega_0 i_3 A_3\} + \mathrm{Re}\{j5\omega_0 i_5 A_5\} = 282800 \tag{4.47}$$

and

$$\|\dot{v}_S\|^2 = (A_1\omega_0)^2 + (A_3 3\omega_0)^2 + (A_5 5\omega_0)^2 = 3.5563\cdot 10^{10} \tag{4.48}$$

The optimal capacitor for compensation of reactive power [according (4.46)] is

$$C_{CO} = 282800/\left(3.5563\cdot 10^{10}\right) = 7.9521[\mu\mathrm{F}]$$

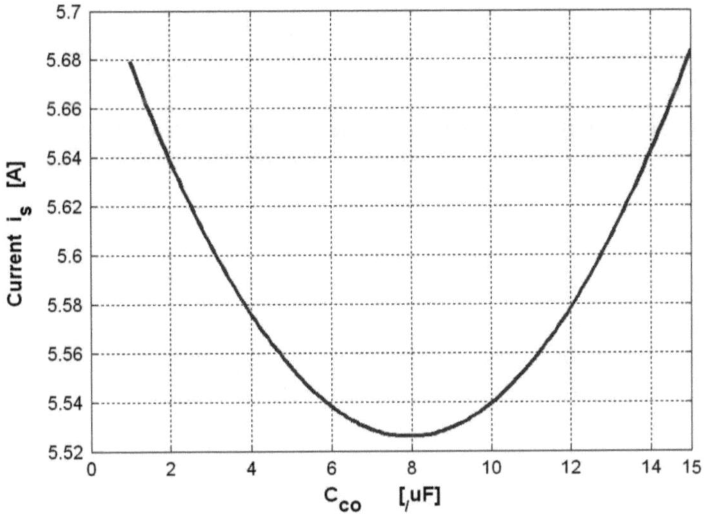

Fig. 4.3 Current i_S versus value of compensation capacitor C_{CO} (Example 4.1)

This result was confirmed by simulation (searching for optimal value) as Fig. 4.3. Currents through the compensation capacitor

$$i_{COn} = jn\omega_0 C_{CO} A_n;\ n = 1, 3, 5 \tag{4.49}$$

i_{CO1} = 0 + 0.899*j*; i_{CO3} = 0 + 1.079*j*; i_{CO5} = 0 + 0.525*j* and currents from source after compensation i_{S1} = 0.34 + 3.735*j*; i_{S3} = 2.545 – 3.158*j*; i_{S5} = 0.0681 + 0.0933*j*, and total effective current from source for circuit with compensation capacitor is i_S = 5.5259 (without compensation was 5.72 A) and apparent power is S = 2155. PF for compensated circuit is $PF = P/S = 491.8/2155 = 0.228$.

The circuit can be compensated also by inductor where compensation current is given as

$$i_L = \frac{1}{L_C}\int v_S dt \tag{4.50}$$

For compensation by means of inductor, Eq. (4.43) is changed

$$\|i_S\|^2 < \|i + i_L\|^2 = \|i\|^2 + \left\|\frac{1}{L_C}\int v_S\right\|^2 + 2\left\langle\frac{1}{L_C}\int v_S, i\right\rangle \tag{4.51}$$

For compensated circuit

$$\|i_S\|^2 < \|i + i_L\|^2 = \|i\|^2 + \frac{1}{L_C^2}\left\|\int v_S\right\|^2 + \frac{2}{L_C}\left\langle i, \int v_S\right\rangle \tag{4.52}$$

After some manipulation, optimal value of compensation inductor is

$$L_C = \frac{-\left\|\int v_S\right\|^2}{\left\langle i, \int v_S\right\rangle} \tag{4.53}$$

where

$$\left\|\int v_S\right\|^2 = \left(\left|\frac{A_1}{\omega_0}\right|\right)^2 + \left(\left|\frac{A_3}{3\omega_0}\right|\right)^2 + \left(\left|\frac{A_5}{5\omega_0}\right|\right)^2 = 1.34 \tag{4.54}$$

and

$$\left\langle i, \int v_S\right\rangle = \mathrm{Re}\left\{j\frac{A_1}{\omega_0}i_1\right\} + \mathrm{Re}\left\{j\frac{A_3}{3\omega_0}i_3\right\} + \mathrm{Re}\left\{j\frac{A_5}{5\omega_0}i_5\right\} = -2.59 \tag{4.55}$$

Therefore optimal compensation inductor [according Eq. (4.53)] is

$$L_C = -1.337/-2.5909 = 0.516\ [\mathrm{H}] \tag{4.56}$$

This result was confirmed by simulation (Fig. 4.4). Current through the compensation inductor

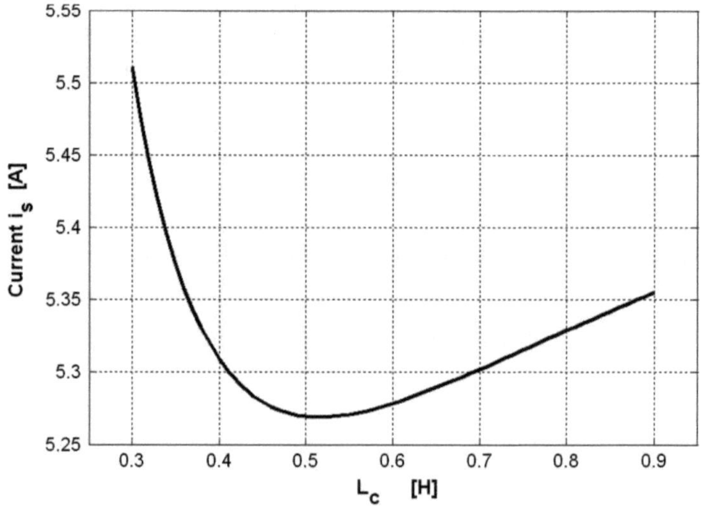

Fig. 4.4 Current i_S versus value of compensation inductor L_C (Example 4.1)

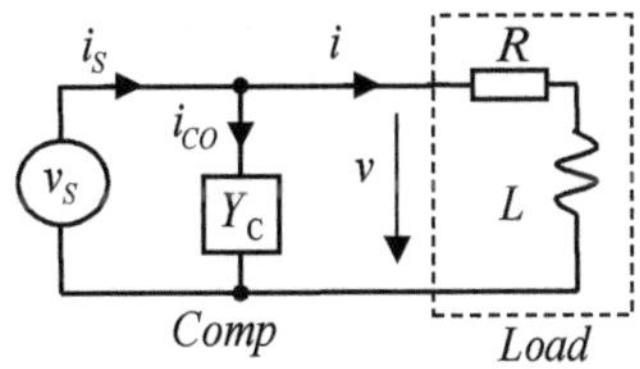

Fig. 4.5 Example of RP compensation for linear *RL* load and nonharmonic source

$$i_{Ln} = A_n / jn\omega_0 L_C; \quad n = 1, 3, 5\ldots \tag{4.57}$$

Calculated values are $i_{L1} = 0 - 2.22j$; $i_{L3} = 0 - 0.296j$; $i_{L5} = 0 - 0.052j$. Currents from source after compensation are $i_{S1} = 0.34 + 0.616j$; $i_{S3} = 2.545 - 4.534j$; $i_{S5} = 0.068 - 0.483j$ and total effective current from source for circuit with compensation inductor is $i_S = 5.27$, and apparent power is $S = 2055$, and *PF* for compensated circuit is $PF = P/S = 491.8/2055 = 0.24$.

This result was confirmed by simulation (searching for optimal value) as Fig. 4.4.

Example 4.2 [36]: Suppose linear system, power source, compensator and serial connection of *RL* and nonharmonic power source shown in Fig. 4.5 [36], instead of nonlinear inductor the nonharmonic power source is used. Supply voltage v_S is

$$v_s(t) = \sqrt{2}(A_1 \sin(\omega_0 t) + A_3 \sin(3\omega_0 t)) \tag{4.58}$$

where $A_1 = 220$, $A_3 = 70$, $\omega_0 = 100\pi$ [rad/s] and $R = 10$ [Ω], $L = 0.2$ [H]. Similarly as in Example 4.1, the load impedances Z_n (for n = 1, 3) are given

$$Z_n = R + j\omega_0 nL \quad n = 1, 3 \tag{4.59}$$

For calculated values $Z_1 = 10 + 62.8j$, $Z_2 = 10 + 188.5j$, the effective currents are $i_1 = 0.54 - 3.4j$; $i_3 = 0.02 - 0.37j$;

For *uncompensated circuit* ($Y_C \to 0$), total effective current through the load according Eq. (4.22) is $i_S = i = 3.48$ [A] and apparent power (4.23) is $S = 802.9$ [VA] and active power (4.24) is $P = 121$ [W]. For uncompensated circuit is $PF = 0.15$. By the same way as in previous Example 4.1, using (4.40), (4.45) and (4.46) calculated values are

$$\left\langle \hat{\dot{i}}, v_S \right\rangle = 260450 \tag{4.60}$$

and

$$\|\dot{v}_S\|^2 = 9.13 \cdot 10^9 \tag{4.61}$$

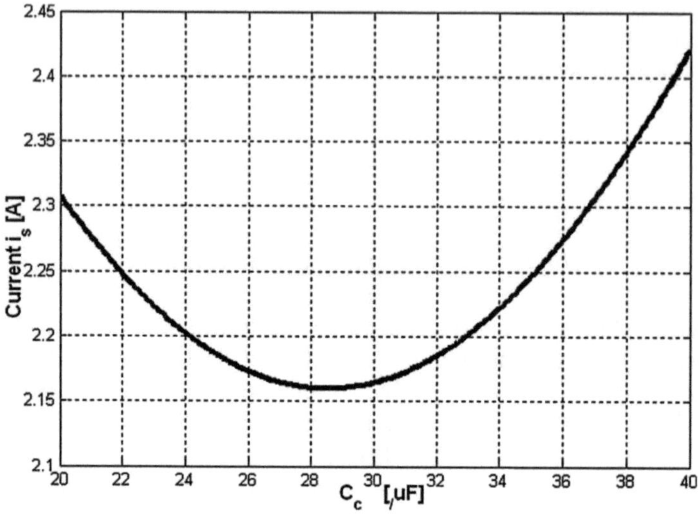

Fig. 4.6 Current i_S versus value of compensation capacitor C_{CO} (Example 4.2)

Optimal value of compensation capacitor is

$$C_{CO} = \frac{\left\langle \hat{\dot{i}}, v_S \right\rangle}{\|\dot{v}_S\|^2} = \frac{260450}{9.13 \cdot 10^9} = 28.5 \cdot 10^{-6} \tag{4.62}$$

This result was confirmed by simulation (searching for optimal value) as Fig. 4.6.

For circuit compensated by capacitor $i = 2.16$ [A] and apparent power is $S = 498.6$ [VA], active power is $P = 121$ [W] and. $PF = 0.22$. The compensation by means of inductor is not possible because result according Eq. (4.53) has negative sign. Simulation results are shown in Fig. 4.7, where compensation capacitor is connected in time ≥ 0.44 s.

Example 4.3 [36]: Suppose nonlinear circuit with triac (Fig. 4.8) and non-sinusoidal power source $v_S(t) = 220\sin(100\pi t) + 30\sin(300\pi t)$, $R = 10\ \Omega$ and $\alpha = \pi/2$. By the same way as in previous Example 4.1, using (4.60), (4.61) and (4.62) calculated values are

$$\left\langle \hat{\dot{i}}, v_S \right\rangle = -\langle i, \dot{v}_S \rangle = -3.6 \cdot 10^5 \tag{4.63}$$

and

$$\|\dot{v}_S\|^2 = 5.58 \cdot 10^9 \tag{4.64}$$

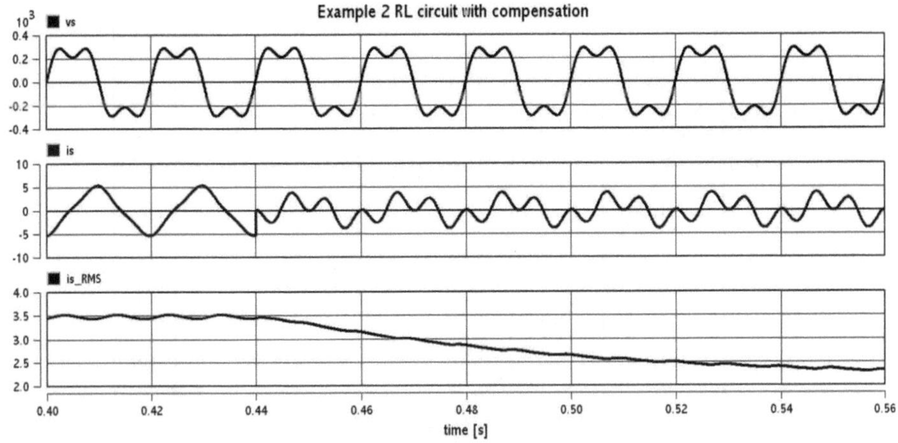

Fig. 4.7 Time evolution of voltage (*top*), current from source (*middle*) and effective current from source for uncompensated (time <0.44) and compensated circuit (time ≥ 0.44)

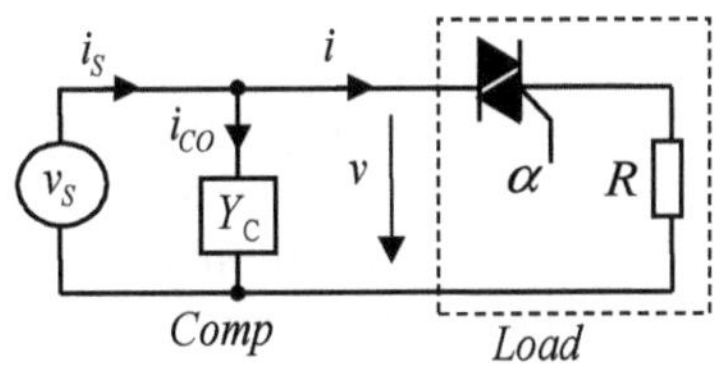

Fig. 4.8 Nonlinear circuit with triac

Optimal value of compensation capacitor can be calculated from

$$C_{CO} = \frac{-\langle i, \dot{v}_S \rangle}{\|\dot{v}_S\|^2} = \frac{-(-3.6 \cdot 10^5)}{5.58 \cdot 10^9} = 64.7 \cdot 10^{-6} \tag{4.65}$$

Optimal value of compensation capacitor (C_{CO} = 65 [μF]) was confirmed by simulation as Fig. 4.9. For uncompensated circuit effective current is

$$i_S = 15.7\,[\text{A}],\ P = 2462[\text{W}],\ S = 3484\,[\text{VA}],\ PF = 0.71$$

After compensation effective current from source is

$$i_S = 14.92\,[\text{A}],\ P = 2462\,[\text{W}],\ S = 3315\,[\text{VA}],\ PF = 0.743$$

The time evolutions of signals in nonlinear circuit are shown in Fig. 4.10. It is important to note that inductor is not possible for compensation because result according Eq. (4.53) has negative sign.

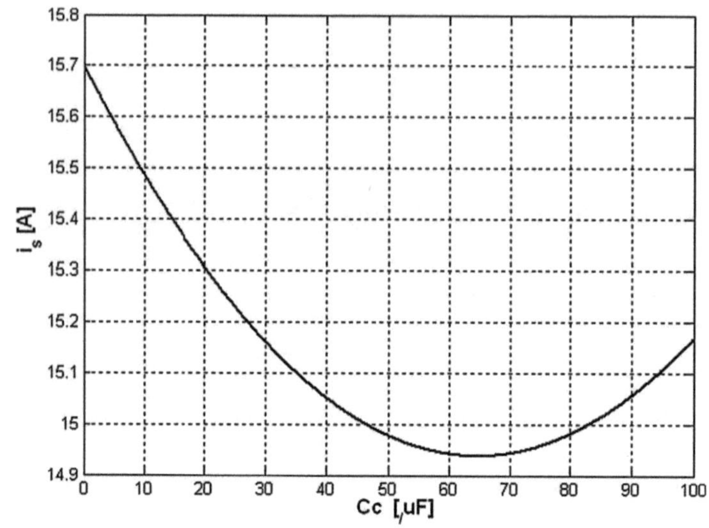

Fig. 4.9 Current i_S versus value of compensation capacitor C_{CO} for nonlinear circuit (Example 4.3)

Previous results can be used for automatic compensation of RP. The time evolution of signals (for circuit in Fig. 4.8) is shown in Fig. 4.11. In time of $t_1 = 0.2$ [s] start automatic compensation [repeated evaluation of Eq. (4.64)] and changing value of compensation capacitor according condition in circuit ($C_{CO} = f(i_S(t), v_S(t))$).

From the beginning, until t_2 the power source voltage is $v_S(t) = 220\sin(100\pi t) + 30\sin(300\pi t)$ after t_2 voltage is sinusoidal $v_S(t) = 220\sin(100\pi t)$.

The effective value of supply current i_i is repeatedly calculated and smoothed by low-pass filter. From the Fig. 4.11 can be seen decreasing supply current value after start of compensation and once more decreasing after changing supply voltage to pure sinusoidal. In Fig. 4.12, the Value of compensation capacitor versus triac fire angle alpha is displayed.

4.6 Extension to Three-Phase Networks

The recently proposed networks were devoted to the analysis of single phase systems. In this chapter it is extended to three-phase networks.

The extension of the above theory to poly-phase systems is straightforward for 4 wire Y connection, but for 3 wire network with unbalanced terms, delta connection and compensation of reactive power opens several questions [45–47]. In this part the compensation of reactive power and unbalanced currents of three-phase network will described [48–52].

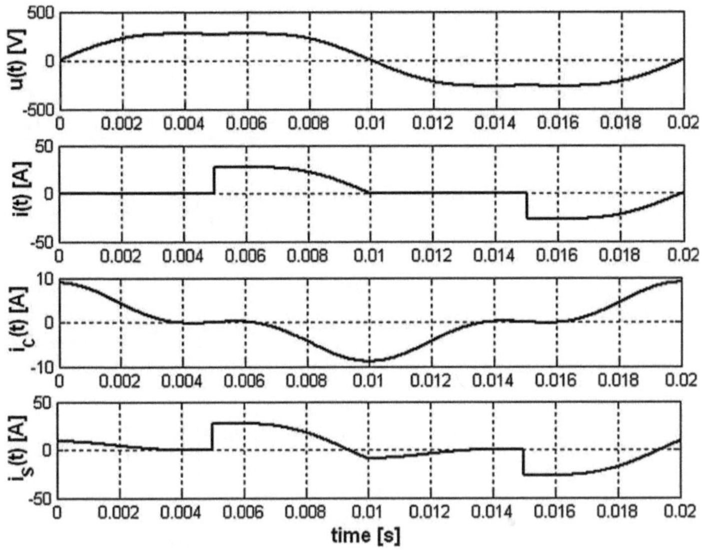

Fig. 4.10 Voltage and currents in nonlinear circuit. From *top* to *bottom*: Power source voltage, current through the load, current through compensation capacitor and current from source after compensation (Example 4.3)

Considerations in this part are devoted to three-phase, three-wire circuits, shown in Fig. 4.13 with linear, time-invariant loads supplied with a sinusoidal symmetrical voltage [53–55]. For any such loads there exist equivalent resistive and balanced loads shown in Fig. 4.14 that at the same voltage has the same active power *P*, as the original load. The Three-phase source with line resistance, transformer *TR* (Δ to Y connection) and load Z_R, Z_S, Z_T is shown in Figs. 4.15 (without compensation) and 4.16 (with compensation).

Suppose three-phase, sinusoidal line-to-ground voltages u_R, u_S and u_T and line currents i_R, i_S and i_T. The active and reactive powers is defined as

$$P = \frac{1}{T}\int_0^T (u_R i_R + u_S i_S + u_T i_T)dt = \sum_{f=R,S,T} U_f I_f \cos(\varphi_f) \tag{4.66}$$

$$Q = \sum_{f=R,S,T} U_f I_f \sin(\varphi_f) \tag{4.67}$$

There are several definitions of apparent power *S*. The definition according (4.68) is used

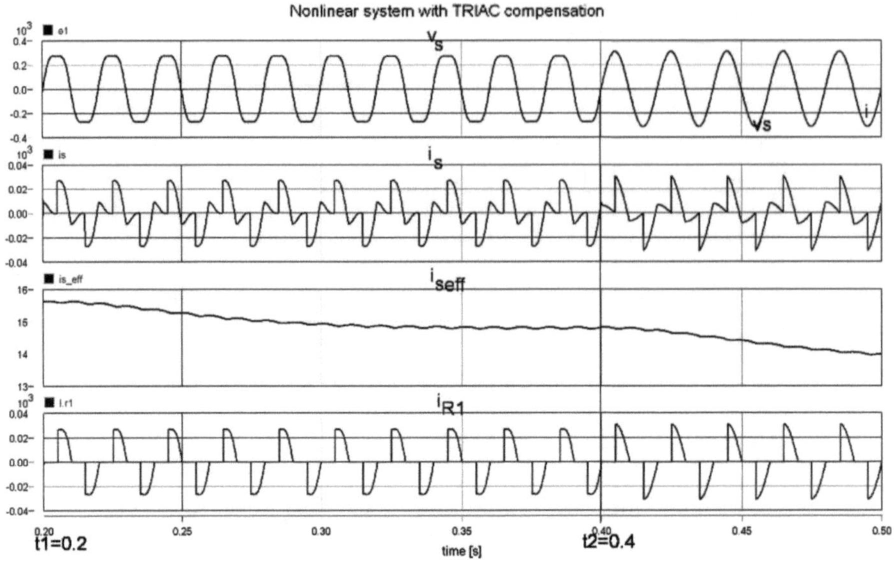

Fig. 4.11 Time diagram of voltage and currents in nonlinear circuit during the automatic compensation. From *top* to *bottom*: Voltage of power source, supply current i_S, effective value of supply current, current through the load. Compensation start in time $t = 0.2$. Source voltage is changed in $t = 0.4$ to pure sinusoidal. The time slice is $\langle 0.20 \div 0.50 \rangle$ (Example 4.3)

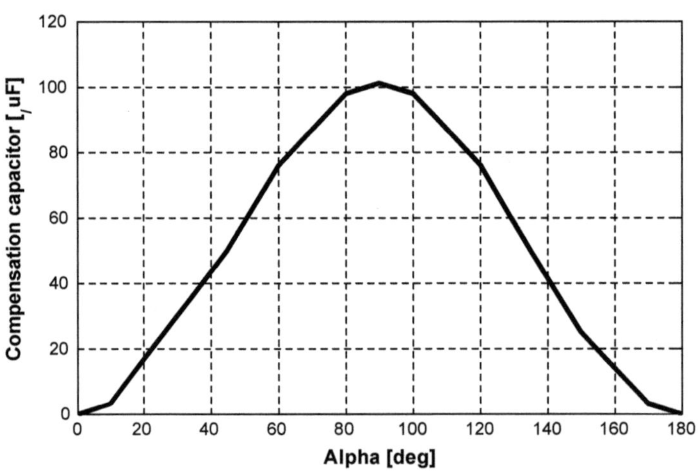

Fig. 4.12 Value of compensation capacitor versus alpha (Example 4.3)

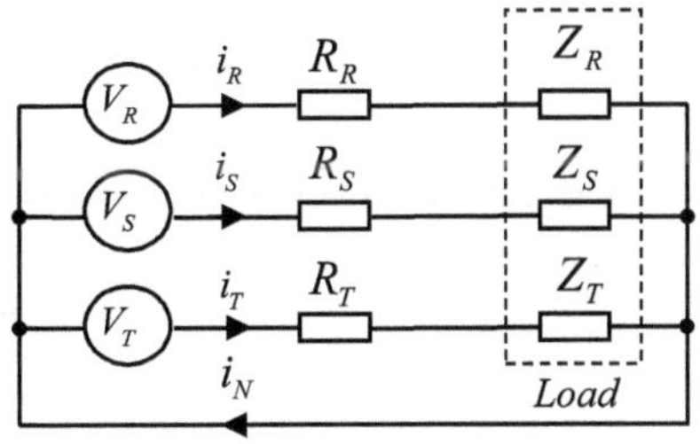

Fig. 4.13 Three-phase star (Y) network with line resistances R_R, R_S, R_T, loads Z_R, Z_S, Z_T and neutral wire

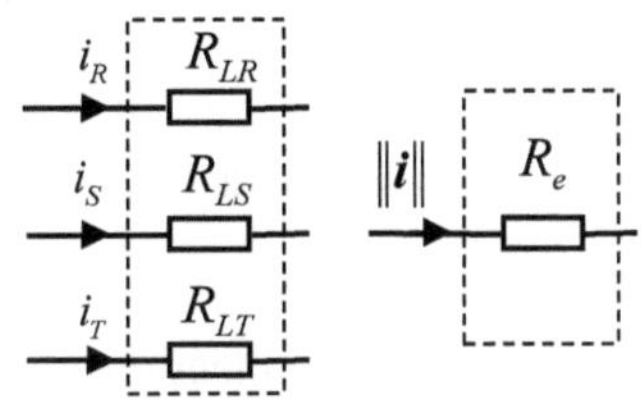

Fig. 4.14 Three-phase resistive load R_{LR}, R_{LS}, R_{LT} and its equivalent R_e

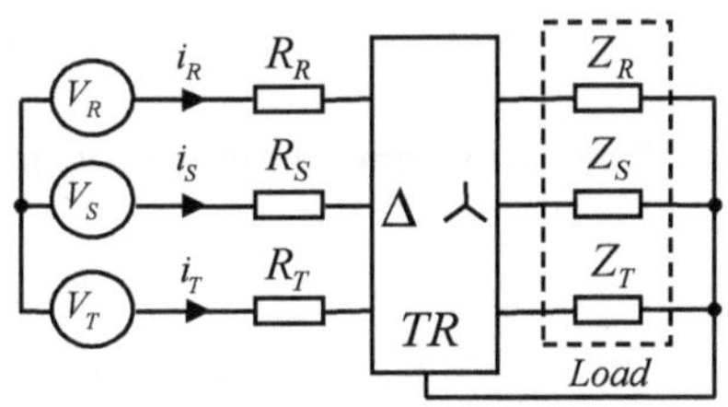

Fig. 4.15 Three-phase source with line resistance, transformer *TR* (Δ to Y) and load Z_R, Z_S, Z_T

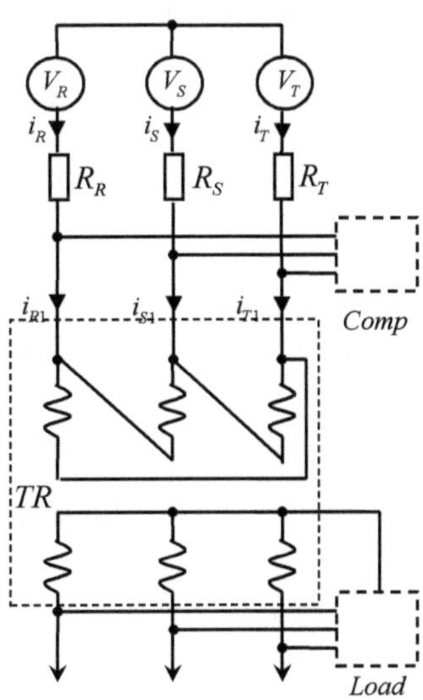

Fig. 4.16 Three-phase network with transformer *TR* (Δ to Y connection), load and compensation

$$S = \sqrt{U_R^2 + U_S^2 + U_T^2} \cdot \sqrt{I_R^2 + I_S^2 + I_T^2} = \|u\| \cdot \|i\| \tag{4.68}$$

The active power in three-phase system in frequency domain is

$$P = (u, i) = \mathrm{Re} \sum_{n \in N} U_n^T I_n^* \tag{4.69}$$

where asterisk denotes a conjugate number. The RMS value of a three-phase vector is defined as:

$$\|x\| = \sqrt{\sum_{n \in N} X_n^T \cdot X_n^*} = \sqrt{\sum_{n \in N} [X_{Rn}, X_{Sn}, X_{Tn}] \begin{bmatrix} X_{Rn}^* \\ X_{Sn}^* \\ X_{Tn}^* \end{bmatrix}} \tag{4.70}$$

For serial connection of R_Z and L_Z between phase R-S admittance is

$$Y_{RS} = \frac{1}{R_Z + j\omega L_Z} = \frac{1}{R_Z + X_Z} \tag{4.71}$$

The equivalent admittance Y_e for three admittances is

$$Y_e = Y_{RS} + Y_{ST} + Y_{TR} \tag{4.72}$$

The equivalent admittance G_e and susceptance B_e are given by

$$G_e = \mathrm{Re}(Y_e); \quad B_e = \mathrm{Im}(Y_e) \tag{4.73}$$

Unbalanced admittance A is

$$A = -(Y_{ST} + Y_{TR} e^{j2\pi/3} + Y_{RS} e^{j4\pi/3}) \tag{4.74}$$

Voltage v across phase is

$$v = \sqrt{3} v_f \tag{4.75}$$

Active current i_a, reactive current i_r unbalanced current i_u and total current i are

$$\begin{aligned} &i_a = G_e v = G_e \sqrt{3} v_f; \; i_r = |B_e| v; \; i_u = |A \cdot v| \\ &\|i\| = \sqrt{\|i_a\|^2 + \|i_r\|^2 + \|i_u\|^2} \end{aligned} \tag{4.76}$$

Active power P

$$P = G_e v_s^2 = G_e \|v\|^2 \tag{4.77}$$

Reactive power Q

$$Q = \|v\| \cdot \|i_r\| = -\mathrm{Im}\left\{ Y_{RS}\|u_{RS}\|^2 + Y_{ST}\|u_{RS}\|^2 + Y_{TR}\|u_{RS}\|^2 \right\} = -B_e \|v\|^2 \tag{4.78}$$

Unbalanced power D

$$D = \|v\| \cdot \|i_u\| = |A| \|v\|^2 \tag{4.79}$$

Apparent power S is calculated by

$$S = \|v\| \cdot \|i\| = \sqrt{P^2 + Q^2 + D^2} \tag{4.80}$$

Power Eq. (4.80) shows that reactive power and unbalanced power (reactive current and unbalanced current) increase apparent power and decrease power factor [Eq. (4.35)]. Thus reduction of booth currents leads to PF improvement. These currents can be reduced by shunt compensator. Compensator is composed from reactive passive parts e.g. inductors and capacitors, or three-phase inverter with control systems. The structure of reactive compensator is shown in Fig. 4.18 (*Comp*), susceptances W_{RS}, W_{ST} and W_{TR}, connected in delta configuration (it is supposed that W_{RS}, W_{ST} and W_{TR} are ideal loss-less parts).

From unbalanced admittance A and equivalent susceptance B_e is possible calculate the compensation susceptances W_{XY} from following equations. The reactive current is compensated if:

$$B_e + W_{RS} + W_{ST} + W_{TR} = 0 \tag{4.81}$$

The unbalanced current is compensated if

$$\begin{aligned} &A - j\left(W_{ST} + e^{j2\pi/2} W_{TR} + e^{j4\pi/2} W_{RS}\right) = 0 + j0 \\ &A - j\left(W_{ST} + \left(-\frac{1}{2} + j\frac{\sqrt{3}}{2}\right) W_{TR} + \left(-\frac{1}{2} - j\frac{\sqrt{3}}{2}\right) W_{RS}\right) = 0 + j0 \end{aligned} \tag{4.82}$$

The Eq. (4.82) is split on real and imaginary parts. Real part:

$$\mathrm{Re}\left\{ A - j\left(W_{ST} + \left(-\frac{1}{2} + j\frac{\sqrt{3}}{2}\right) W_{TR} + \left(-\frac{1}{2} - j\frac{\sqrt{3}}{2}\right) W_{RS}\right)\right\} = 0 \tag{4.83}$$

Imaginary part:

$$\mathrm{Im}\left\{A - j\left(W_{ST} + \left(-\frac{1}{2} + j\frac{\sqrt{3}}{2}\right)W_{TR} + \left(-\frac{1}{2} - j\frac{\sqrt{3}}{2}\right)W_{RS}\right)\right\} = 0 \qquad (4.84)$$

Therefore real part

$$\mathrm{Re}\{A\} = \frac{\sqrt{3}}{2}W_{RS} - \frac{\sqrt{3}}{2}W_{TR} \qquad (4.85)$$

Imaginary part

$$\mathrm{Im}\{A\} = -\frac{1}{2}W_{RS} - \frac{1}{2}W_{TR} + W_{ST} \qquad (4.86)$$

Add Eqs. (4.86) and (4.81) multiplied by −1

$$\left.\begin{aligned} \mathrm{Im}\{A\} &= -\tfrac{1}{2}W_{RS} - \tfrac{1}{2}W_{TR} + W_{ST} \\ B_e &= -W_{RS} - W_{TR} - W_{ST} \end{aligned}\right\} + \qquad (4.87)$$

Result is

$$\mathrm{Im}\{A\} + B_e = -\frac{3}{2}W_{RS} - \frac{3}{2}W_{TR} \qquad (4.88)$$

Now add Eqs. (4.88) and (4.85) multiplied by $\sqrt{3}$

$$\left.\begin{aligned} \mathrm{Im}\{A\} + B_e &= -\tfrac{3}{2}W_{RS} - \tfrac{3}{2}W_{TR} \\ \sqrt{3}\mathrm{Re}\{A\} &= \tfrac{3}{2}W_{RS} - \tfrac{3}{2}W_{TR} \end{aligned}\right\} + \qquad (4.89)$$

Result is

$$\sqrt{3}\mathrm{Re}\{A\} + \mathrm{Im}\{A\} + B_e = -3W_{TR} \qquad (4.90)$$

After manipulation

$$W_{TR} = \left[-\sqrt{3}\cdot\mathrm{Re}(A) - \mathrm{Im}(A) - B_e\right]/3 \qquad (4.91)$$

and also

$$W_{RS} = \left[\sqrt{3}\mathrm{Re}(A) - \mathrm{Im}(A) - B_e\right]/3 \qquad (4.92)$$

$$W_{ST} = [2\cdot\mathrm{Im}(A) - B_e]/3 \qquad (4.93)$$

Values of compensation capacitors C_{XY} or inductors L_{XY} are calculated by

$$\text{if}\begin{cases} W_{XY} > 0 & \text{then} \quad C_{XY} = \frac{W_{XY}}{\omega} \\ W_{XY} < 0 & \text{then} \quad L_{XY} = -\frac{1}{\omega W_{XY}} \end{cases} \tag{4.94}$$

The previous derivation and results will used in next two examples for three-phase network compensation.

Example 4.4 [50, 51]: Suppose three-phase network with loss-less transformer *TR* (1:1) in Δ−Y connection and non-symmetrical resistive load $Z_L = R_L = 3\ \Omega$, $V_R = 277\angle 0°$, $V_S = 277\angle 120°$, $V_T = 277\angle 240°$ V, $\omega = 2\pi f = 314$, $R_R = R_S = R_T = 0.01\ \Omega$ (Figs. 4.17, 4.18, 4.19 and 4.20).

$$Y_{RS} = \frac{1}{R_Z + j\omega L_Z} = \frac{1}{R_Z + X_Z} = \frac{1}{3 + j0} \tag{4.95}$$

The equivalent admittance Y_e is

$$Y_e = Y_{RS} + Y_{ST} + Y_{TR} = \frac{1}{3 + j0} + 0 + 0 = 0.33 \tag{4.96}$$

The equivalent admittance G_e and susceptance B_e are given by

$$G_e = \mathrm{Re}(Y_e) = \mathrm{Re}(Y_{RS}) = 0.33; \quad B_e = \mathrm{Im}(Y_e) = \mathrm{Im}(Y_{RS}) = 0 \tag{4.97}$$

Unbalanced admittance A is

$$A = -(Y_{ST} + Y_{TR}e^{j2\pi/3} + Y_{RS}e^{j4\pi/3}) = -(0.33e^{j4\pi/3}) = 0.167 + 0.289j \tag{4.98}$$

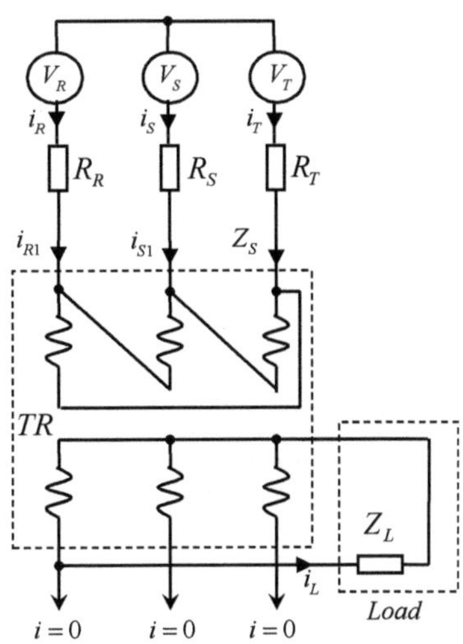

Fig. 4.17 Three-phase network with transformer *TR* (Δ to Y connection) with unbalanced load

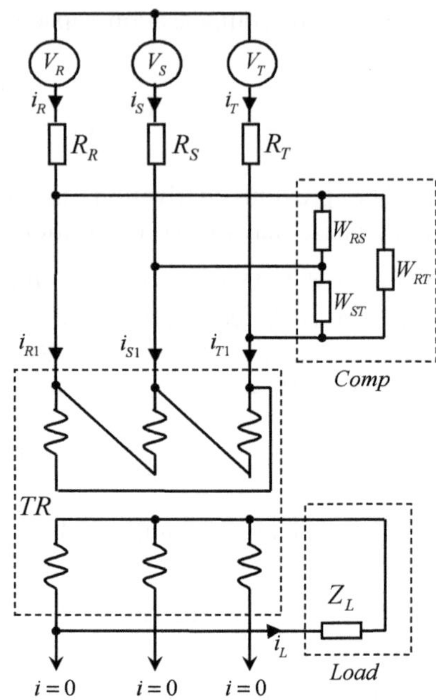

Fig. 4.18 Three-phase network with transformer *TR* (Δ to Y connection) with unbalanced load and compensation circuit

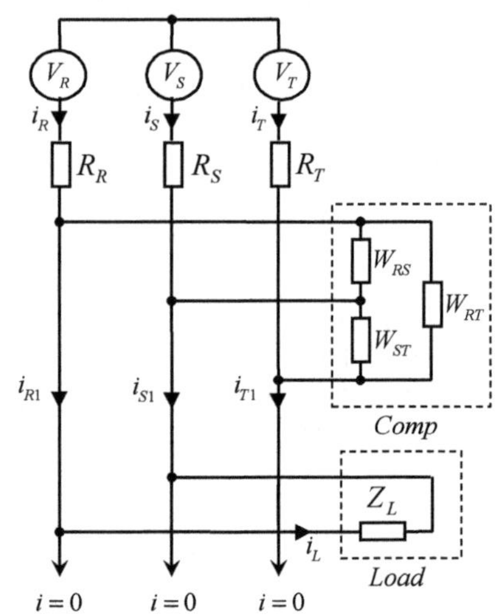

Fig. 4.19 Three-phase equivalent network with unbalanced load and compensation circuit. Avoid of transformer (Figs. 4.17 and 4.18) was possible by means of Y-load to Δ-load transformation

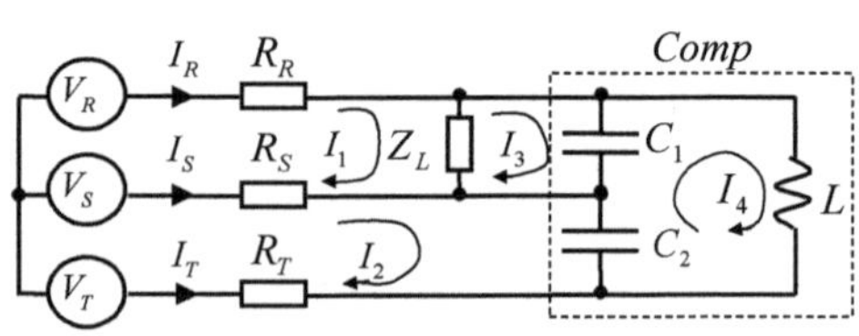

Fig. 4.20 Three-phase network with load Z_L and compensation parts C_1, C_2 and L

Voltage across phase v is

$$v = \sqrt{3}v_f = \sqrt{3} \cdot 277 = 480 \tag{4.99}$$

Active current i_a, reactive current i_r unbalanced current i_u and total current i are

$$\begin{aligned} & i_a = G_e v = G_e \sqrt{3} v_f = 160;\ i_r = |B_e| v = 0;\ i_u = |A \cdot v| = 160 \\ & \|i\| = \sqrt{\|i_a\|^2 + \|i_r\|^2 + \|i_u\|^2} = \sqrt{160^2 + 0 + 160^2} = 226 \end{aligned} \tag{4.100}$$

Active power

$$P = G_e v_s^2 = G_e \|v\|^2 = 76729 \tag{4.101}$$

Reactive power Q

$$Q = -\mathrm{Im}\left\{ Y_{RS}\|u_{RS}\|^2 + Y_{ST}\|u_{RS}\|^2 + Y_{TR}\|u_{RS}\|^2 \right\} = -B_e v_s^2 = -B_e \|v\|^2 = 0 \tag{4.102}$$

Unbalanced power D

$$D = \|v\| \cdot \|i_u\| = |A| \|v\|^2 = 479.7 \cdot 160 = 76.7 \cdot 10^3 \tag{4.103}$$

Phase currents for uncompensated circuit $i_R = i_S = 160$ A, $i_T = 0$. Apparent power S and power factor PF for uncompensated circuit

$$S = \|v\| \cdot \|i\| = 479.7 \cdot 226 = 1.1 \cdot 10^5;\ PF = \frac{P}{S} = \frac{76729}{110000} = 0.7 \tag{4.104}$$

From unbalanced admittance A and equivalent susceptance B_e is possible calculate the compensation susceptances W_{xy} according following equations

$$W_{RS} = \left[\sqrt{3}\mathrm{Re}(A) - \mathrm{Im}(A) - B_e\right]/3 = \frac{0.289 - 0.289}{3} = 0 \tag{4.105}$$

$$W_{ST} = [2 \cdot \mathrm{Im}(A) - B_e]/3 = \frac{2 \cdot 0.289 - 0}{3} = 0.192 \tag{4.106}$$

$$W_{TR} = \left[-\sqrt{3} \cdot \mathrm{Re}(A) - \mathrm{Im}(A) - B_e\right]/3 = \frac{-0.289 - 0.289 - 0}{3} = 0.192 \tag{4.107}$$

Values of compensation capacitors and inductors are calculated according Eq. (4.84)

$$
\begin{aligned}
C_1 &= \frac{W_{RS}}{\omega} = \frac{0}{314} = 0 \\
C_2 &= \frac{W_{ST}}{\omega} = \frac{0.192}{314} = 6.13 \cdot 10^{-4} \\
L &= -\frac{1}{\omega W_{TR}} = -\frac{1}{314 \cdot (-0.192)} = 0.016
\end{aligned}
\tag{4.108}
$$

Phase currents for compensated circuit $\|i_R\| = \|i_S\| = \|i_T\| = 92.4$ A. Apparent power S and power factor PF for compensated circuit is

$$
S = \|v\| \cdot \|i\| = 479.7 \cdot \sqrt{3 \cdot 92.4^2} = 7.9 \cdot 10^3;\ PF = \frac{P}{S} = \frac{76700}{79000} = 0.97
\tag{4.109}
$$

The circuit diagram for network with compensation is shown in Fig. 4.20. Solving the Eqs. (4.85)–(4.89) gives the same results (phase currents) as previous calculation.

$$
\begin{aligned}
& R_R I_1 + Z_L(I_1 - I_3) + R_S(I_1 - I_2) = U_R\angle 0 - U_S\angle - 120 \\
& R_S(I_2 - I_1) - \frac{j}{\omega C_2}(I_2 - I_4) + R_T I_2 = U_S\angle - 120 - U_T\angle 120 \\
& Z_L(I_3 - I_1) - \frac{j}{\omega C_1}(I_3 - I_4) = 0 \\
& j\omega L I_4 - \frac{j}{\omega C_2}(I_4 - I_2) - \frac{j}{\omega C_1}(I_4 - I_3) = 0
\end{aligned}
\tag{4.110}
$$

Matrix M derived from (4.110) is

$$
M = \begin{bmatrix}
R_R + R_S + Z_L & -R_S & -Z_L & 0 \\
-R_S & R_S + R_T + \frac{-j}{\omega C_2} & 0 & \frac{j}{\omega C_2} \\
-Z_L & 0 & Z_L + \frac{-j}{\omega C_1} & \frac{j}{\omega C_1} \\
0 & \frac{j}{\omega C_2} & \frac{j}{\omega C_1} & j\omega L + \frac{-j}{\omega C_1} + \frac{-j}{\omega C_2}
\end{bmatrix}
\tag{4.111}
$$

Matrix equation is

$$
M \cdot \begin{bmatrix} I_1 \\ I_2 \\ I_3 \\ I_4 \end{bmatrix} = \begin{bmatrix} U_R\angle 0 - U_S\angle - 120 \\ U_S\angle - 120 - U_T\angle 120 \\ 0 \\ 0 \end{bmatrix}
\tag{4.112}
$$

Currents I_1 to I_4 can are calculated from

$$\begin{bmatrix} I_1 \\ I_2 \\ I_3 \\ I_4 \end{bmatrix} = M^{-1} \begin{bmatrix} U_R\angle 0 - U_S\angle -120 \\ U_S\angle -120 - U_T\angle 120 \\ 0 \\ 0 \end{bmatrix} \quad (4.113)$$

Phase currents are

$$I_R = I_1;\ I_S = I_2 - I_1;\ I_T = -I_2 \quad (4.114)$$

The theoretical derivations were supported by simulation. The network was simulated without compensation part and with compensation part, connected by switch SW (Fig. 4.21). The time evolution of V_R, I_R, V_S, I_S and V_T, I_T before compensation is shown in Fig. 4.22. The same time evolution of phase's voltages and currents after compensation is shown in Fig. 4.23. The Fig. 4.24 presents time evolution of effective values of phase currents before and after connection of compensation circuit (compensation circuit was connected in time = 0.44 s)

Example 4.5 [50, 51]: Suppose three-phase network with loss-less transformer *TR* (1:1) in Δ–Y connection and non-symmetrical *RL* load $Z_L = R_L + j\omega L_Z = 30 + j10$, $V_R = 6000\ \angle\ 0°$, $V_S = 6000\ \angle\ 120°$, $V_T = 6000\ \angle\ 240°$ V, $\omega = 2\pi f = 314$ [rad/sec], $R_R = R_S = R_T = 0.1\ \Omega$ (Fig. 4.17).

$$Y_{RS} = \frac{1}{R_Z + j\omega L_Z} = \frac{1}{R_Z + X_Z} = \frac{1}{30 + j10} = 0.03 - 0.01j \quad (4.115)$$

The equivalent admittance Y_e is

$$Y_e = Y_{RS} + Y_{ST} + Y_{TR} = 0.03 - 0.01j \quad (4.116)$$

The equivalent admittance G_e and susceptance B_e are given by

$$G_e = \mathrm{Re}(Y_e) = \mathrm{Re}(Y_{RS}) = 0.03;\ B_e = \mathrm{Im}(Y_e) = \mathrm{Im}(Y_{RS}) = -0.01 \quad (4.117)$$

Unbalanced admittance A is

$$A = -(Y_{ST} + Y_{TR}e^{j2\pi/3} + Y_{RS}e^{j4\pi/3}) = 0.0237 + 0.021j \quad (4.118)$$

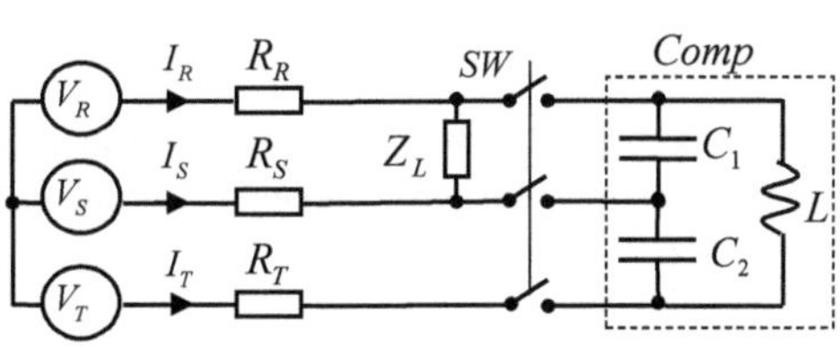

Fig. 4.21 Three-phase network with load Z_L, compensation parts C_1, C_2 and *L* and switch *SW* which is used for connection compensation parts

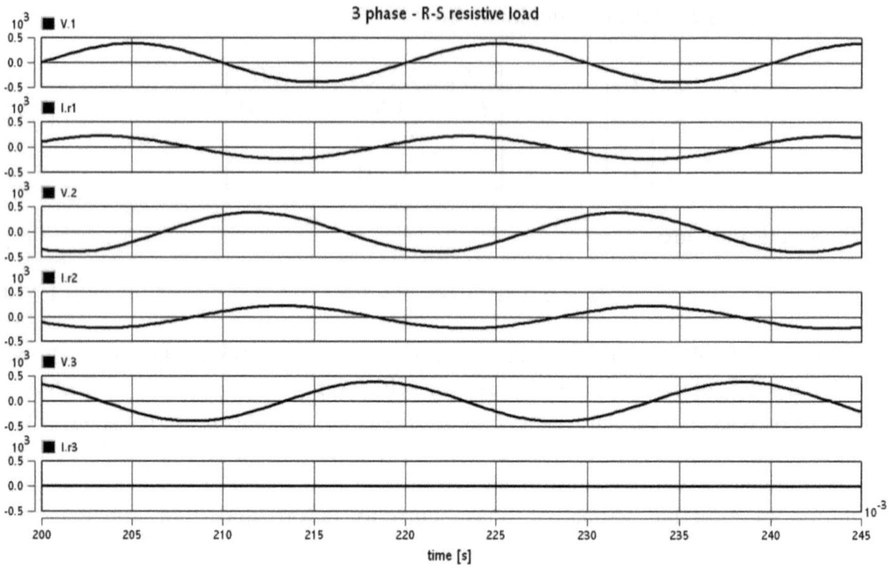

Fig. 4.22 Time evolution of voltages and currents in circuit according Example 4.4, before compensation. From *top* to *bottom*: Voltage V_R, current i_R, voltage V_S, current i_S, voltage V_T, current i_T

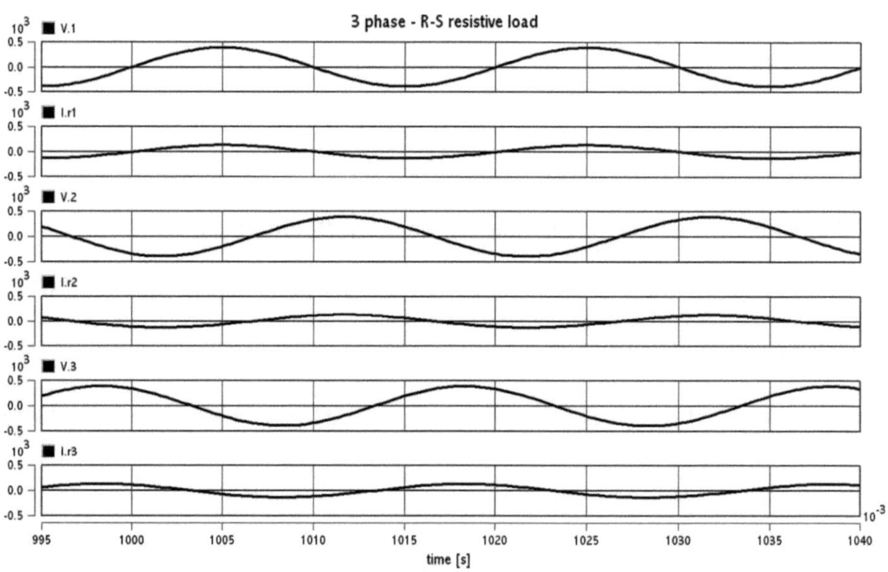

Fig. 4.23 Time evolution of voltages and currents in circuit according Example 4.4, after compensation. From *top* to *bottom*: Voltage V_R, current i_R, voltage V_S, current i_S, voltage V_T, current i_T

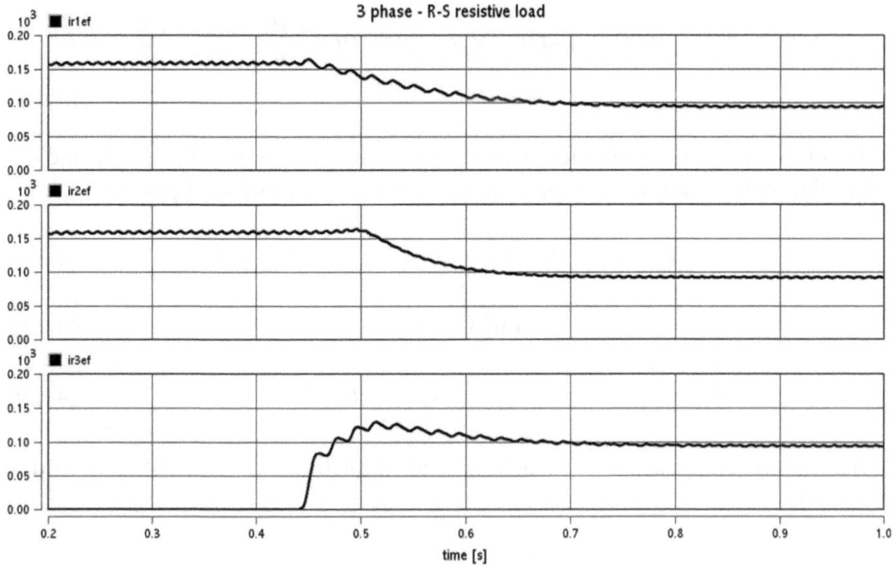

Fig. 4.24 Time evolution of RMS currents in circuit according Example 4.4, before and after compensation. Compensation parts are connected in time ≥ 0.44. From *top* to *bottom*: RMS currents i_R, i_S, i_T

Active current i_a, reactive current i_r unbalanced current i_u and total current i are

$$
\begin{aligned}
& i_a = G_e v = G_e \sqrt{3} v_f = 311.8;\ i_r = |B_e| v = 104;\ i_u = |A \cdot v| = 328.6 \\
& \|i\| = \sqrt{\|i_a\|^2 + \|i_r\|^2 + \|i_u\|^2} = 464.7
\end{aligned}
\tag{4.119}
$$

Active power

$$P = G_e v_s^2 = G_e \|v\|^2 = 3.24 \cdot 10^6 \tag{4.120}$$

Reactive power Q

$$Q = -B_e v_s^2 = -B_e \|v\|^2 = 1.08 \cdot 10^6 \tag{4.121}$$

Unbalanced power D

$$D = \|v\| \cdot \|i_u\| = |A| \|v\|^2 = 10392 \cdot 328.6 = 3.4 \cdot 10^6 \tag{4.122}$$

Phase currents for uncompensated circuit $i_R = i_S = 328.6$ A, $i_T = 0$. Apparent power S and power factor PF for uncompensated circuit

$$S = \|v\| \cdot \|i\| = 10392 \cdot 464.7 = 4.83 \cdot 10^6; \; PF = \frac{P}{S} = \frac{3.24 \cdot 10^6}{4.83 \cdot 10^6} = 0.67 \tag{4.123}$$

From unbalanced admittance A and equivalent susceptance B_e is possible calculate the compensation susceptances W_{xy} according following equations:

$$W_{RS} = \left[\sqrt{3}\mathrm{Re}(A) - \mathrm{Im}(A) - B_e\right]/3 = 0.01 \tag{4.124}$$

$$W_{ST} = [2 \cdot \mathrm{Im}(A) - B_e]/3 = 0.0173 \tag{4.125}$$

$$W_{TR} = \left[-\sqrt{3} \cdot \mathrm{Re}(A) - \mathrm{Im}(A) - B_e\right]/3 = -0.0173 \tag{4.126}$$

Values of compensation capacitors and inductors are calculated according Eq. (4.84)

$$\begin{aligned} C_1 &= \frac{W_{RS}}{\omega} = \frac{0.01}{314} = 3.18 \cdot 10^{-5} \\ C_2 &= \frac{W_{ST}}{\omega} = \frac{0.0173}{314} = 5.52 \cdot 10^{-5} \\ L &= -\frac{1}{\omega W_{TR}} = -\frac{1}{314 \cdot (-0.0173)} = 0.184 \end{aligned} \tag{4.127}$$

Phase currents for compensated circuit $i_R = i_S = i_T = 180.4$ A. Apparent power S and power factor PF for compensated circuit is

$$\begin{aligned} S &= \|v\| \cdot \|i\| = 10392 \cdot \sqrt{3 \cdot 180.4^2} = 3.25 \cdot 10^6 \\ PF &= \frac{P}{S} = \frac{3.24 \cdot 10^6}{3.25 \cdot 10^6} = 0.99 \end{aligned} \tag{4.128}$$

The theoretical results were confirmed by simulation. The time evolution of V_R, I_R, V_S, I_S and V_T, I_T before compensation is shown in Fig. 4.25. The same time evolution of phase's voltages and currents after compensation is shown in Fig. 4.26. The Fig. 4.27 presents time evolution of effective values of phase currents before and after connection of compensation circuit (compensation circuit was connected in time = 0.44 s)

It past hypothesis and samples was demonstrated, that direct loads with low power component, (for example, prompting engines) and unbalance can be amended with a detached system of capacitors or inductors. Nonlinear loads, for example, rectifiers, contort the current drawn from the framework. In such cases, dynamic or aloof power variable remedy might be utilized to check the twisting and raise the power element. The gadgets for redress of the power variable might be at a focal substation, spread out over a circulation framework, or incorporated with

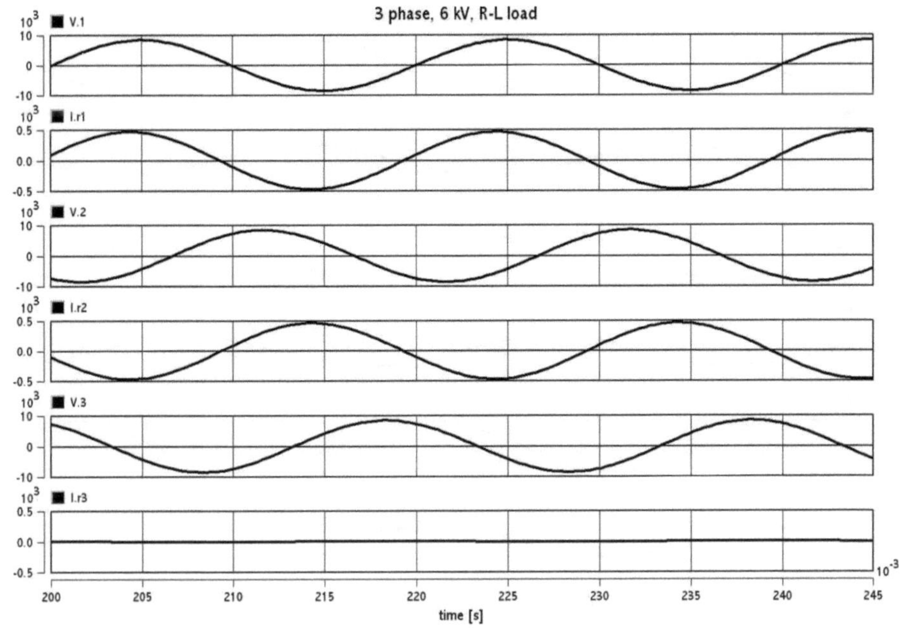

Fig. 4.25 Time evolution of voltages and currents in circuit according Example 4.5, before compensation. From *top* to *bottom*: Voltage V_R, current i_R, voltage V_S, current i_S, voltage V_T, current i_T

power devouring gear. For nonlinear load the higher harmonic are produced. These harmonics can be e.g. stifled by serial reverberation circuits.

Rather than utilizing an arrangement of exchanged capacitors for pay, an emptied synchronous engine can supply reactive power. The reactive power drawn by the synchronous engine is an element of its field excitation. This is alluded to as a synchronous condenser. It is begun and joined with the electrical system. It works at a main power variable and puts VARS (reactive energy) onto the system as required to bolster a framework's voltage or to keep up the framework power factor at a predefined level. The condenser's establishment and operation are indistinguishable to substantial electric engines. Its essential leverage is the simplicity with which the measure of remedy can be balanced; it acts like an electrically variable capacitor. Not at all like capacitors, the measure of reactive power supplied is relative to voltage (not the square of voltage), this enhances voltage security on extensive systems. Synchronous condensers are regularly utilized as a part of association with high-voltage direct-current transmission ventures or in expansive mechanical plants, for example, steel factories.

For power variable remedy of high-voltage power frameworks or huge, fluctuating modern loads, power electronic gadgets, for example, the Static VAR compensator [56, 57] or STATCOM [58, 59] are progressively utilized. These frameworks can remunerate sudden changes of power element a great deal more

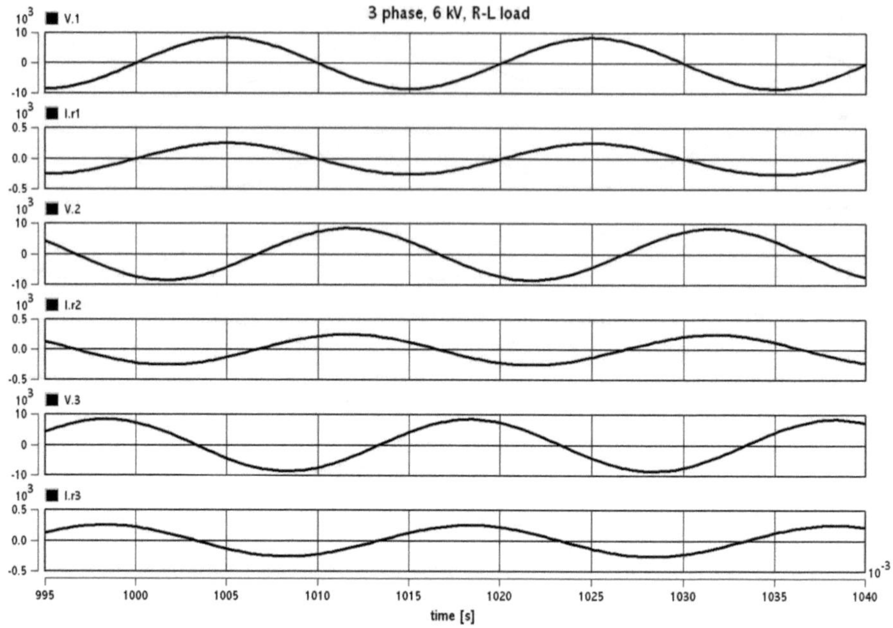

Fig. 4.26 Time evolution of voltages and currents in circuit according Example 4.5, after compensation. From *top* to *bottom*: Voltage V_R, current i_R, voltage V_S, current i_S, voltage V_T, current i_T

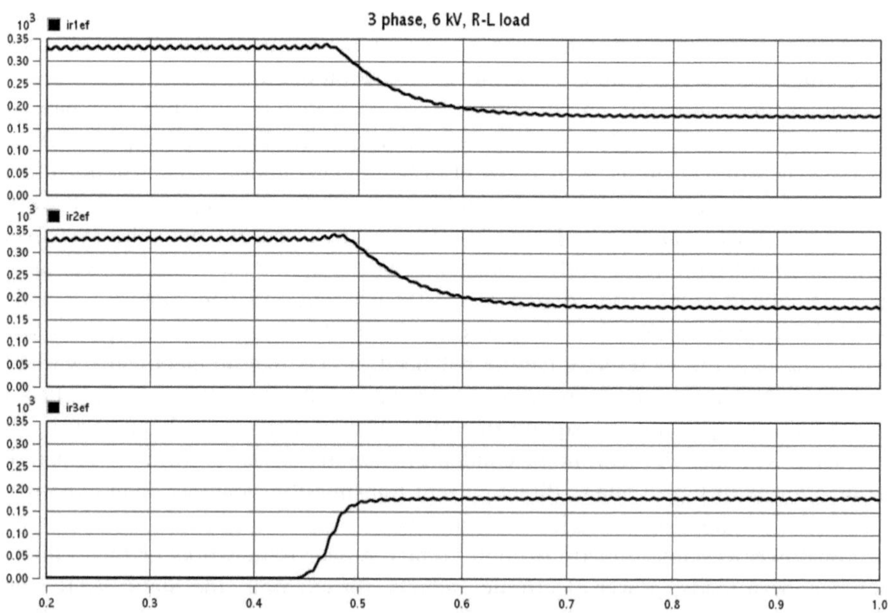

Fig. 4.27 Time evolution of RMS currents in circuit according Example 4.4, before and after compensation. Compensation parts are connected in time ≥ 0.44. From *top* to *bottom*: RMS currents i_R, i_S, i_T

quickly than contactor exchanged capacitor banks, and being strong state require less upkeep than synchronous condensers.

4.7 Quantifying of Reactive Power in Energy Meters

The sum and unpredictability of family unit electrical gear has expanded immensely in the course of the most recent couple of years. Electronic counterbalance lighting, PC screens and ventilation systems are welcome options to our homes however accompanied extra weights. One of these is on the power matrix, as these machines produce more signal harmonic.

This adjustment at last buyer profile is a hindrance for energy wholesalers which charge energy construct just with respect to dynamic power. With the use of nonlinear loads to electrical cables the dynamic energy no more speaks to the aggregate energy conveyed. As a reaction to enhance charging, the estimation of reactive energy is increasing interest.

Electromechanical meters have set a point of reference in responsive energy charging. Despite the fact that they are data transfer capacity restricted and can't consider harmonic of the line frequency, they are upheld by the worldwide standard for rotating current static VAR-hour meters for responsive energy (IEC-1268) [60]. The standard characterizes reactive energy estimations at the crucial line frequency, which suggests that it is not compulsory to incorporate harmonics. It likewise indicates extra testing conditions to check the vigor of the estimations against the third consonant, the dc counterbalance in the present data, and the line frequency variety. The different responsive power estimation routines exhibited in this chapter are assessed against these basic tests of the IEC-1268 (Table 4.1) [61].

4.7.1 Reactive Power IEEE Definition

The reactive power is well-defined in the IEEE Standard Dictionary 100-1996 under the energy "magner" as:

$$RP = \sum_{n=1}^{\infty} V_n I_n \sin(\varphi_n) \tag{4.129}$$

where V_n and I_n are respectively the voltage and current RMS values of the nth harmonics of the line frequency, and φ_n is the phase difference between the voltage and the current nth harmonics. An agreement is also adopted stating that the reactive energy should be positive for inductive load.

In an electrical framework containing absolutely sinusoidal voltage and current waveforms at a settled frequency, the estimation of reactive power is simple and can

Table 4.1 Error benchmark of different reactive energy calculation methods

Test		Power triangle	Time delay	Low pass filter
IEC 1268-reference test	Voltage and current Input *f*: $PF = 0$	Negligible	Negligible	Negligible
IEC 1268-frequency variation test	Reference test $f \pm 2\%$ and $PF = 0.87$	Negligible	5.4%	Negligible
IEC 1268-harmonic test	Reference test +10% of the third harmonics on the current signal	0.5%	Negligible	Negligible
IEC 1268-DC component test	Reference test with half way rectified sin wave on the current input	Negligible	Negligible	Negligible
Reference test +10% of the third harmonics on voltage input and 20% of the third harmonics on current input ($\varphi_1 = \varphi_2 = 30°$)		1.9%	4%	1%

be expert utilizing a few systems without blunders. In any case, in the vicinity of nonsinusoidal waveforms, the energy contained in the harmonic causes estimation blunders.

By Fourier hypothesis any occasional waveform can be composed as an aggregate of sin and cosine waves. As energy meters manage occasional signs at the line frequency both current and voltage inputs of a solitary stage meter can be portrayed by:

$$v(t) = \sum_{n=1}^{\infty} \sqrt{2} V_n \sin(n\omega_0 t) \tag{4.130}$$

$$i(t) = \sum_{n=1}^{\infty} \sqrt{2} I_n \sin(n\omega_0 t + \varphi_n) \tag{4.131}$$

where V_n, I_n are and φ_n are defined as in Eq. (4.129) and ω_0 is frequency in rad.s^{-1}.

4.7.2 Reactive Power Calculation

Method 1—Power Triangle

The Power triangle method is based on the assumption that the three energies, apparent, active and reactive, form a right-angle triangle. The reactive power can then be processed by estimating the active and apparent energies and applying:

$$RP = \sqrt{Apparent\_power^2 - Active\_power^2} \tag{4.132}$$

Despite the fact that this strategy gives amazing results with immaculate sinusoidal waveforms, perceptible mistakes show up in vicinity of harmonic (Table 4.1).

Method 2—Time Delay

A period deferral is acquainted with movement one of the waveforms by 90° at the crucial frequency and duplicate the two waveforms:

$$RP = \frac{1}{T}\int_0^T v(t) \cdot i\left(t + \frac{T}{4}\right) dt \tag{4.133}$$

where T is the period of the fundamental frequency. In an electronic DSP system, this technique can be applied by delaying the samples of one input by the quantity of samples representing a quarter-cycle of the fundamental line frequency f_{line} (Fig. 4.28).

This technique presents disadvantages if the line frequency changes and the quantity of tests no more speaks to a quarter-cycle of the crucial frequency. Huge mistakes are then acquainted with the outcomes (Table 4.1).

Method 3—Low-Pass Filter

A steady 90° phase shift over frequency with a lessening of 20 dB/decade is presented. This arrangement, which has been executed by Analog Devices, can be acknowledged with a solitary pole low-pass channel on one channel information (Fig. 4.29). On the off chance that the cut-off frequency of the low-pass channel is much lower than the central frequency, this arrangement gives a 90° phase shift at any frequency higher than the principal frequency. It likewise constricts these frequencies by 20 dB/decade. The square chart of the framework utilizing first request low-pass channel technique is appeared in Fig. 4.30.

Also to technique 2, this arrangement is defenseless to varieties of the line frequency. Be that as it may, a dynamic compensation of the addition constriction with the line frequency can be accomplished by assessing the line time of the sign (Table 4.1). As clarified above, various techniques can be utilized to figure the reactive power. The hypothetical meaning of the responsive power is hard to actualize in an electronic framework at a sensible expense. It requires a committed DSP to prepare the Hilbert change important to get a consistent stage movement of 90° at every frequency. A few arrangements have been created to beat this constraint.

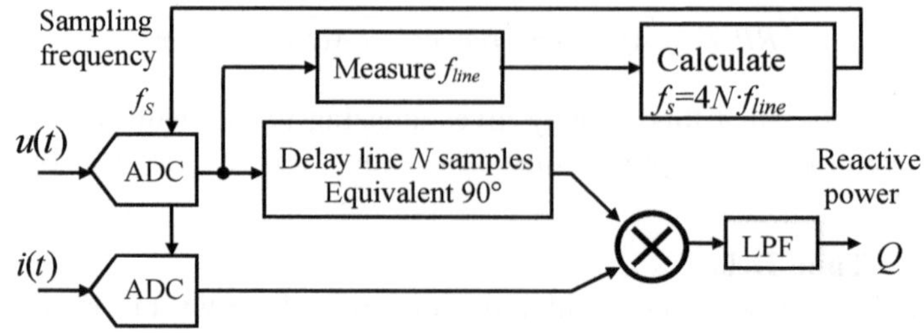

Fig. 4.28 Block diagram of the system using frequency changing compensated $T/4$ delay algorithm

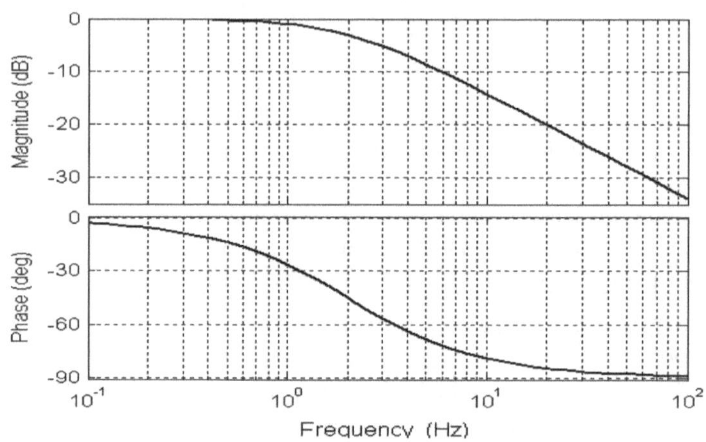

Fig. 4.29 Magnitude and phase response of low-pass filter 1st order with cutoff frequency 2 Hz

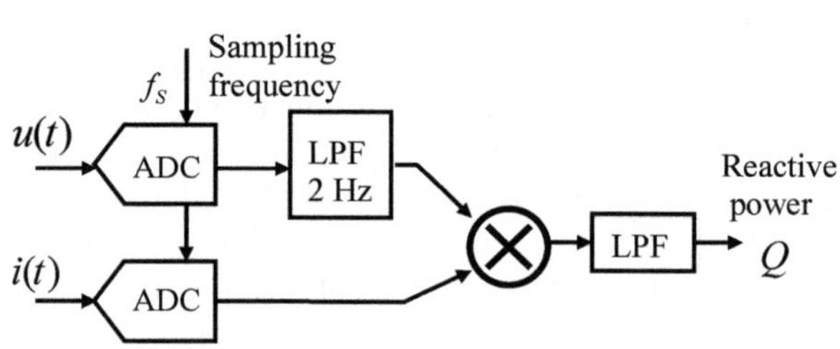

Fig. 4.30 Block diagram of the system using first order low-pass filter method

4.7.3 Example of Integrated Circuits for Reactive Power Measuring

The ADE7753 [62] is single-phase and ADE7854A/ADE7858A/ ADE7868A/ADE7878A [63–67] are high accuracy, 3-phase electrical energy measurement ICs with serial interfaces and three flexible pulse outputs. The devices incorporate second-order Σ–Δ analog-to-digital converters (ADCs), a digital integrator, reference circuitry, and all signal processing required to perform total (fundamental and harmonic) active, reactive (ADE7858A, ADE7868A, and ADE7878A), and apparent energy measurement and rms calculations. The ADE7878A can also perform fundamental-only active and reactive energy measurement and rms calculations.

A fixed function digital signal processor (DSP) executes the signal processing. The DSP program is stored in the internal ROM memory. The ADE7854A/ADE7858A/ADE7868A/ADE7878A can measure active, reactive, and apparent energy in various 3-phase configurations, such as wye or delta services, with both three and four wires. Aside from regular rms measurements, which are updated every 8 kHz, these devices measure low ripple rms values, which are averaged internally and updated every 1.024 s. The devices provide system calibration features for each phase, that is, rms offset correction, phase calibration, and gain calibration.

Harmonic Power Measurement with Electronic Energy Meters
Measuring Harmonic power is an important performance measure of new energy meters. Electro-mechanical meters typically measure only up to the 5th harmonic. IEC specification does not require measurement above 5th harmonic. Harmonic power has been measured in the field at up to 9.3% of the total active power up to the 50th harmonic. Accurate measurement of harmonics can be limited by the characteristics of the ADCs.
Minimum requirements for electronic meters:
Analog input bandwidth ≥ 4 kHz
ADC Sampling frequency $\geq$ 8ksps
With an analog input bandwidth of >4 kHz, Signals up to the 63rd harmonic will be processed correctly (in 50 and 60 Hz systems)

Tests Description and Test Results

Harmonic components in the voltage and current
Test Conditions
Channel 1 input: Sine wave at 50 Hz (140 mV) + Sine wave at 250 Hz (56 mV)
Channel 2 input: Sine wave at 50 Hz (250 mV) + Sine wave at 250 Hz (25 mV)

Reference Conditions
Channel 1 input: Sine wave at 50 Hz (140 mV)
Channel 2 input: Sine wave at 50 Hz (250 mV)

Measurement
Energy Accumulated over 4095 half line cycles (40.95 s)
Result
0.09% Error in Active Energy Measurement (IEC Limit is 0.8%)
0.46% Error in Reactive Energy Measurement

DC and even harmonics in the AC current

Half-Wave rectified waveform test (Fig. 4.31)

Test Conditions

Channel 1 input: Half wave rectified waveform at 50 Hz (300 mV)

Channel 2 input: Sine wave at 50 Hz (250 mV)

Reference Conditions

Channel 1 input: Sine wave at 50 Hz (150 mV)

Channel 2 input: Sine wave at 50 Hz (250 mV)

Measurement

Energy Accumulated over 4095 half line cycles (40.95 s)

Result

0.28% Error in Active Energy Measurement (IEC Limit is 3%)

0.27% Error in Reactive Energy Measurement (IEC Limit is 3%)

Odd harmonics in the AC current

Phase-Fired waveform test (Fig. 4.32)

Test Conditions

Channel 1 input: Phase fired rectified waveform at 50 Hz (80 mV)

Channel 2 input: Sine wave at 50 Hz (250 mV)

Reference Conditions

Channel 1 input: Sine wave at 50 Hz (40 mV)

Channel 2 input: Sine wave at 50 Hz (250 mV)

Measurement

Energy Accumulated over 4095 half line cycles (40.95 s)

Result

0.34% Error in Active Energy Measurement (IEC Limit is 3%)

0.62% Error in Reactive Energy Measurement

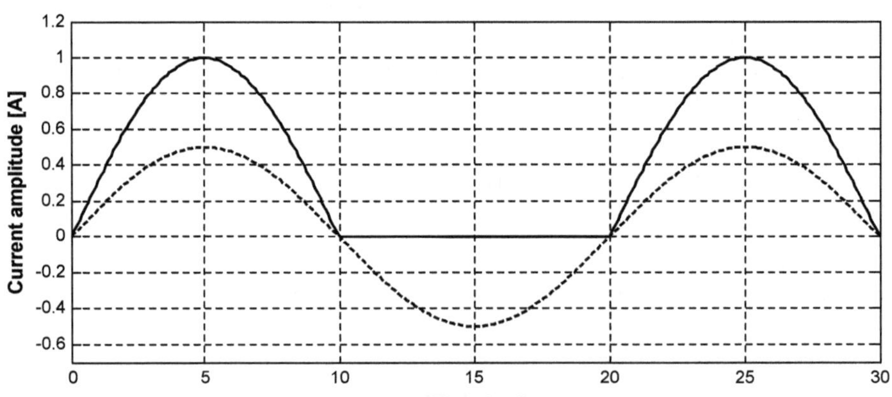

Fig. 4.31 Half-Wave rectified waveform test. *Solid line*—test waveform, *dash line*—reference waveform

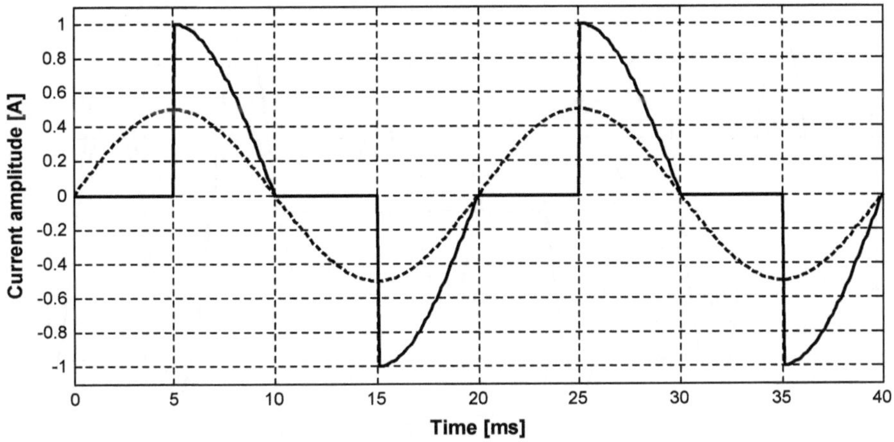

Fig. 4.32 Phase-Fired waveform test. *Solid line*—test waveform, *dash line*—reference waveform

Sub-harmonics in the AC current circuit

Burst-Fired waveform test (Fig. 4.33)

Test Conditions

Channel 1 input: Phase fired rectified waveform at 50 Hz (80 mV)

Channel 2 input: Sine wave at 50 Hz (250 mV)

Reference Conditions

Channel 1 input: Sine wave at 50 Hz (40 mV)

Channel 2 input: Sine wave at 50 Hz (250 mV)

Measurement

Energy Accumulated over 4095 half line cycles (40.95 s)

Result

0.28% Error in Active Energy Measurement (IEC Limit is 3%)

0.29% Error in Reactive Energy Measurement

4.7.4 Summary

With more nonlinear loads in family unit machines, measuring reactive energy precisely turns into a key issue for energy wholesalers. Conventional estimation techniques like the Power triangle and the Time delay com-employ with universal gauges yet indicate restrictions in the vicinity of harmonics or line frequency variety. With the most recent headways in incorporated circuit advancement, as proposed e.g. by Analog Devices, energy meter planners can now effectively actualize more exact responsive energy estimations and consequently, fulfill developing necessities from energy suppliers. The ADCs of Electronic Energy Meters should have wide enough bandwidth (>4 kHz) to measure harmonic power.

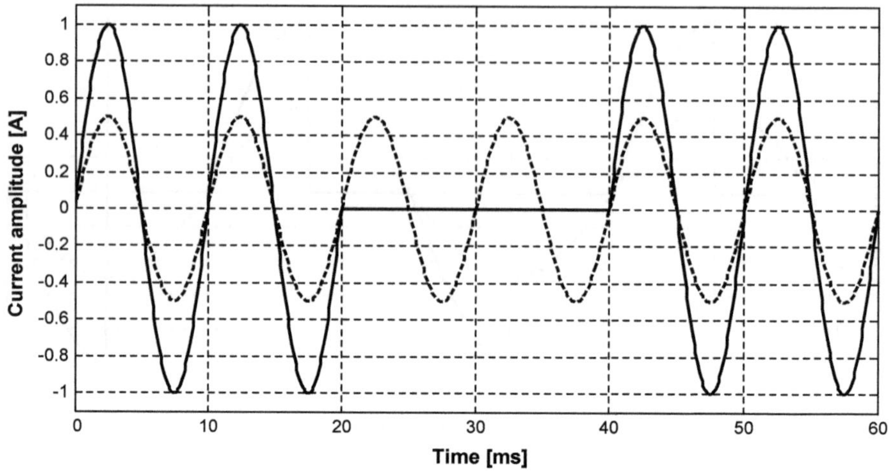

Fig. 4.33 Burst-Fired waveform test. *Solid line*—test waveform, *dash line*—reference waveform. Test waveform is on for 2 cycles and off for 2 cycles

The ADC sampling frequency should be high enough (>8 ksps) to prevent distortion of the sampled waveform. ADE products exceed the Watt-hour and VAR-hour IEC specifications for harmonic influence. The rms measurement of the ADE7753 and ADE7854A/ADE7858A/ADE7868A/ADE7878A has been proven accurate over a wide range of harmonics.

4.8 Active Power Factor Correction

An ideal Power Factor Corrector (PFC) takes from the supply V_i a current I_L which is proportional to the supply voltage as pure resistance load R_L [67–69]

$$I_L = \frac{V_i}{R_L} \tag{4.134}$$

Power factor definition—for sinusoidal input voltage

$$PF = \frac{V_1 I_1 \cos(\varphi_1)}{V_1 I_{iRMS}} = \frac{I_1}{I_{iRMS}} \cos(\varphi_1) \tag{4.135}$$

Where I_1/I_{iRMS} is distortion factor (*DF*) and $\cos(\varphi_1)$ is displacement factor and *THD* is total harmonic distortion.

$$THD = \frac{\sqrt{I_{iRMS}^2 - I_1^2}}{I_1} \quad (4.136)$$

The ideal $PF = 1$ implies:

(a) Zero phase shift between voltage and current fundamental component ($\varphi_1 = 0$)
(b) Zero current harmonic content (Fig. 4.34).

Reasons for power factor correction

(a) Increased source efficiency

- lower losses on source impedance
- lower voltage distortion (cross-coupling)
- higher power available from a given source

(b) Reduced low-frequency harmonic pollution
(c) Compliance with limiting standards (IEC 555-2, IEEE 519 etc.)

Power factor correction techniques
Passive methods: LC filters

- power factor not very high
- bulky components
- high reliability
- suitable for very small or high power levels

The easiest approach to control the harmonic current is to utilize a channel that passes current just at line frequency (50 or 60 Hz), as Fig. 4.35 [70–72]. The channel comprises of capacitors or inductors, and makes a non-direct gadget look more like a straight load. A sample of aloof PFC is a valley-fill circuit.

A detriment of latent PFC is that it requires bigger inductors or capacitors than an identical power dynamic PFC circuit [73–75]. Likewise, by and by, detached PFC is frequently less viable at enhancing the power variable [76].

Dynamic techniques: High-frequency converters

- high power factor (drawing nearer solidarity)
- probability to present a high-frequency insulating transformer
- design subordinate high-frequency harmonics era (EMI issues)
- suitable for little and medium power levels

Active PFC is the use of power electronics to adjust the waveform of current drawn by a load to progress the power factor [76]. Some kinds of the active PFC are buck, boost, buck-boost and synchronous condenser. Active power factor adjustment can be single-stage or multi-stage. In the case of a switched-mode power supply, a boost converter is placed in between the connection rectifier and the core input capacitors. The boost converter endeavors to preserve a constant DC bus voltage on its output while drawing a current that is continuously in phase with and at the same frequency as the line voltage. Another switched-mode converter inside

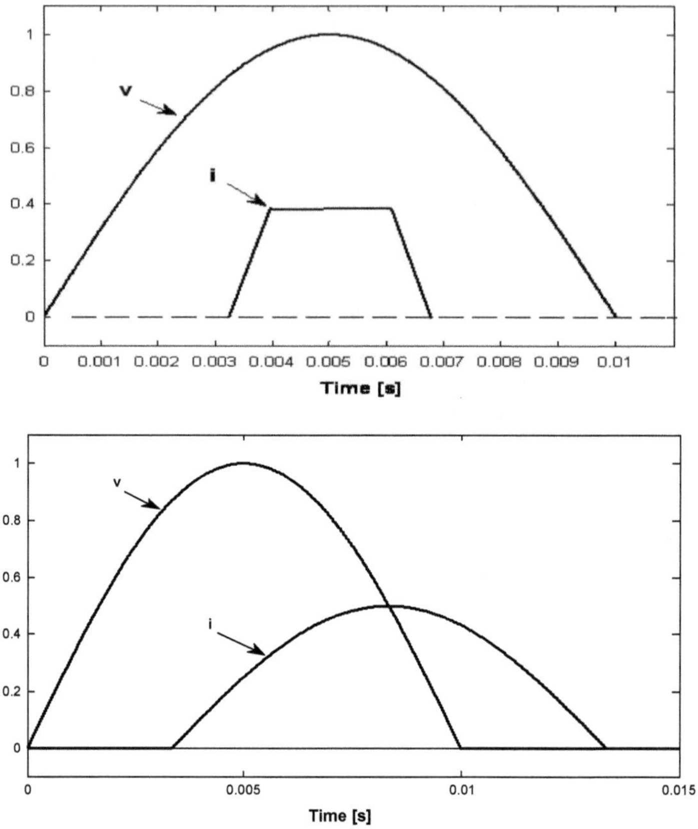

Fig. 4.34 Example of non-ideal *PF*. Zero displacement between voltage and current fundamental component (*top*), but higher harmonic in current ($\cos(\varphi_1) = 0$, $DF \neq 0$). Zero current harmonic content (*bottom*), but nonzero phase shift ($\cos(\varphi_1) \neq 0$, $DF = 0$)

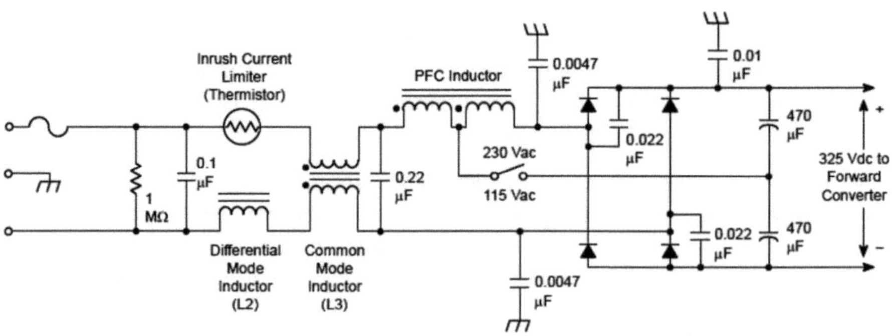

Fig. 4.35 Example of passive PFC [70, 71]

the power source produces the anticipated output voltage from the DC bus. This method necessitates extra semiconductor switches and control electronics, but permits inexpensive and minor passive components. It is regularly used in practice. Switched-mode power sources (SMPS) with passive PFC can accomplish power factor of approximately 0.7–0.75, SMPSs with active PFC, up to 0.99 power factor, while a SMPSs without any power factor rectification have a power factor of only about 0.55–0.65 [77].

Due to their very extensive input voltage range, many power supplies with active PFC can spontaneously regulate to function on AC power from about 100 V (Japan) to 230 V (Europe). That feature is principally welcome in power supplies for laptops [78–83].

Active power factor correction (DPFC), occasionally mentioned to as "real-time power factor correction," is used for electrical steadying in cases of fast load changes (e.g. at large manufacturing sites). DPFC is suitable when normal power factor correction would affect over or under correction [83] DPFC uses semiconductor switches, normally thyristors, to rapidly connect and disconnect capacitors or inductors from the grid in order to progress power factor.

Active power factor corrector

Standard conformation—Two stage PFC, Cascade connection, shown in Fig. 4.36 [68]. More detailed of block diagram of the classic PFC circuit is displayed in Fig. 4.37.

There are two ways of PFC operation; intermittent and continuous mode. Intermittent way is when the boost converter's MOSFET is turned on when the inductor current touches zero, and turned off when the inductor current encounters the anticipated input reference voltage as shown in Fig. 4.38. In this mode, the input current waveform tracks that of the input voltage, consequently reaching a power factor of near to 1 [84, 85].

Intermittent mode can be used for SMPS that have power levels of 300 W or less. Compared to continuous type equipment, discontinuous ones use greater cores and have upper RI^2 and skin effect losses due to the greater inductor current fluctuates. With the amplified swing a bigger input filter is also obligatory. On the positive cross, since discontinuous type equipment change the boost MOSFET on when the inductor current is at zero, there is no opposite retrieval current (I_{RR}) specification required on the boost diode. This means that fewer expensive diodes can be used.

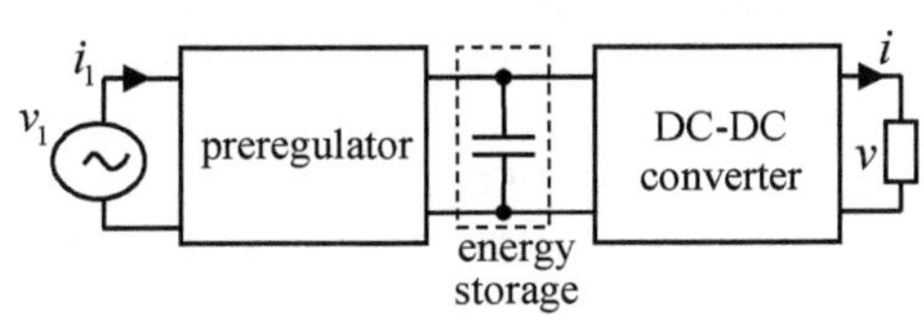

Fig. 4.36 The block diagram of two stage PFC cascade connection

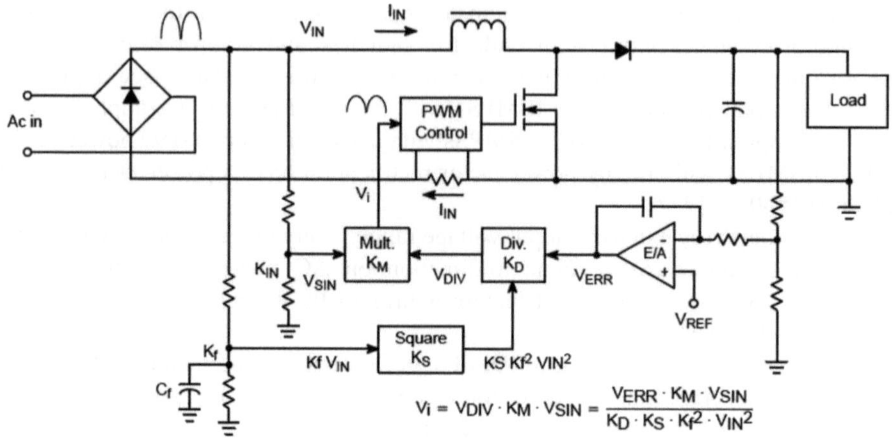

Fig. 4.37 Block diagram of the classic PFC circuit [68]

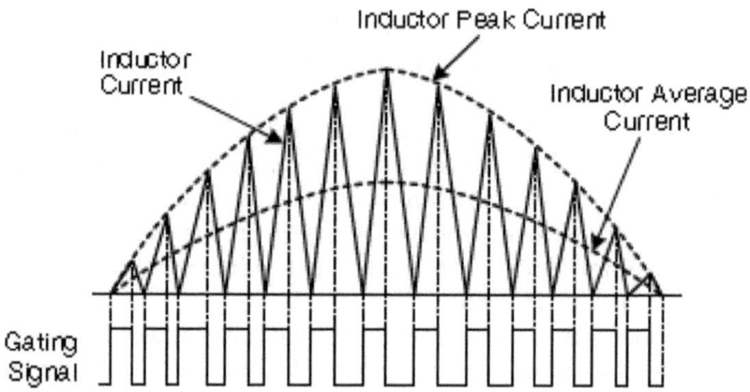

Fig. 4.38 Discontinuous mode of PFC operation

Continuous mode naturally suits SMPS power levels bigger than 300 W. This is where the boost converter's MOSFET does not switch on when the boost inductor is at zero current, in its place the current in the energy transmission inductor at no time grasps zero during the switching sequence (Fig. 4.39).

Leading Edge Modulation/Trailing Edge Modulation (LEM/TEM) versus Trailing Edge Modulation/Trailing Edge Modulation (TEM/TEM)

Leading edge/trailing edge modulation is a patented Fairchild technique to synchronize the PFC controller to the pulse-width modulation (PWM) controller [85]. Naturally TEM/TEM is used in PFC/PWM controllers which outcomes in an extra stage as well as a larger PFC bulk capacitor (as shown below).

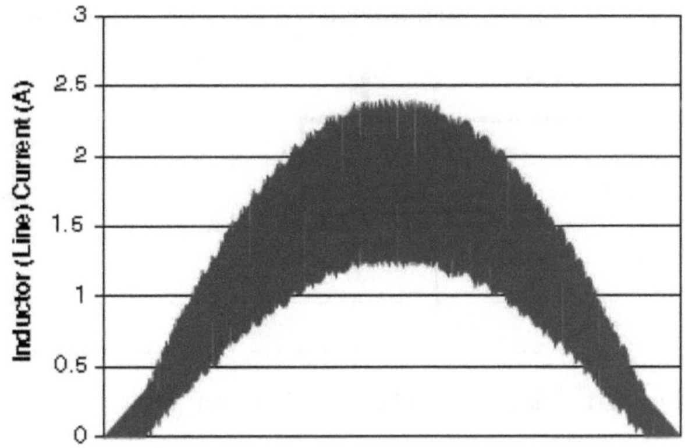

Fig. 4.39 Inductor current in continuous mode of PFC operation

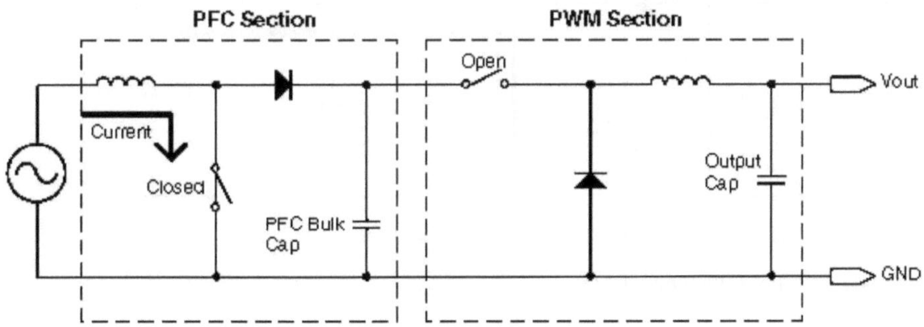

Fig. 4.40 Energizing the PFC Inductor [85]

Figure 4.40 illustrates the PFC inductor being energized. Figure 4.41 shows the energy from the inductor being transferred into the PFC bulk capacitor. When the PWM switch is shut, as shown in Fig. 4.42, the energy kept within the PFC bulk capacitor is used to drive the load. Every time this cycle is recurred, the PFC bulk capacitor has to be completely charged since it is fully discharged when the PWM switch is shut.

Fairchild Patented Leading Edge Modulation/Trailing Edge Modulation (LEM/TEM) Technique

In LET/TEM the PFC and PWM switches are tied together, but opening and closing 180° out of phase, so when the PFC switch is open the PWM switch is shut and vice versa [85]. Originally while the PFC switch is shut, the PFC inductor is energized, once the PWM switch is shut, both the output and the PFC bulk capacitor are energized. Figures 4.43 and 4.44 show that upon frequency of this

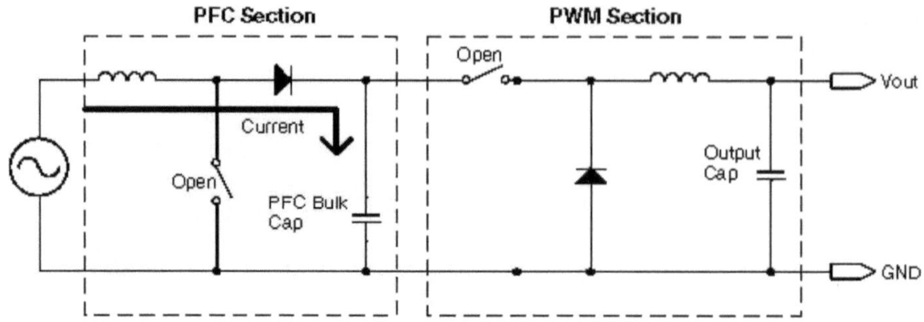

Fig. 4.41 Charging the PFC Bulk Capacitor [85]

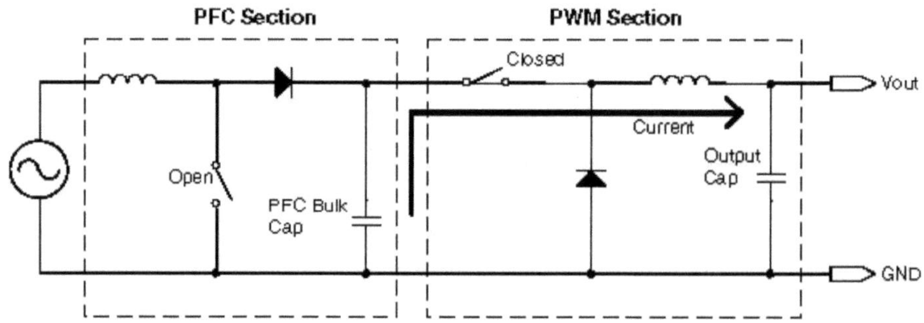

Fig. 4.42 Powering the Output [85]

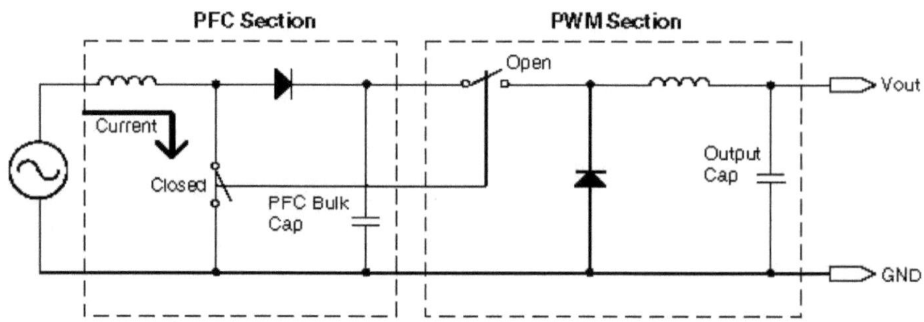

Fig. 4.43 Energizing the PFC Inductor [85]

cycle, the PFC bulk capacitor does not have to be that huge because it is not driving the output all by itself, the PFC inductor is serving out as well.

There are numerous standards in place to drive power consumption to a power factor of 1 and preserve total harmonic alteration to a minimum. Depending on the output power and the designer's requirements, a SMPS can be designed with either a discontinuous or continuous method standalone PFC controller, or a continuous

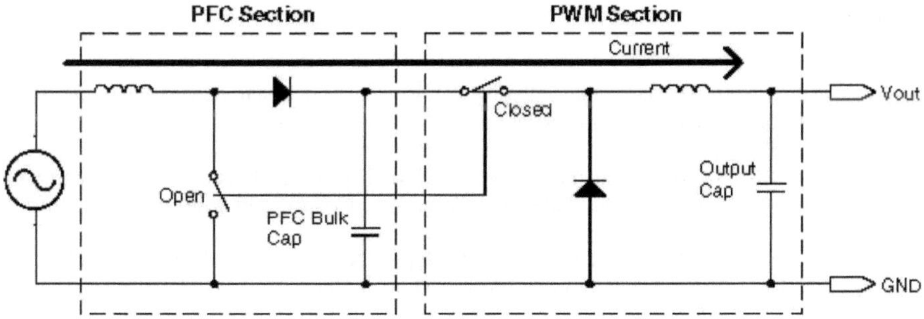

Fig. 4.44 Charging the PFC Bulk Capacitor and Powering the Output [85]

PFC/PWM mode device can be used. PFC controllers are predicted to quickly grow in near future, and standards are dropping the smallest power restrictions on systems that necessitate PFC, more and more PFC controllers will be used.

4.9 Conclusion

In this chapter the theory and principle reactive power compensation for one-phase and three-phase networks were described. On the beginning the cyclodissipativity used for reactive power compensation was presented. Results of this theory were used in several examples. It was demonstrated that the results can be used for control algorithms of automatic compensator. In the first parts of chapter the reactive power decreasing (or power factor increasing) was based on passive compensation (capacitor, inductor). The one subchapter was also devoted to digital power and energy meter description. In the last subchapter the active power filters were also described.

The main aim of this chapter was to show different principles of reactive power compensation, theory and examples which can be used for effective calculation of compensation elements.

Acknowledgements Milan Stork's participation was supported by Department of Applied Electronics and Telecommunications, University of West Bohemia, Plzen, Czech Republic and by the European Regional Development Fund and the Ministry of Education, Youth and Sports of the Czech Republic under the Regional Innovation Centre for Electrical Engineering (RICE), project No. No. LO1607, the Internal Grant Agency of University of West Bohemia in Pilsen, the project SGS-2015-002 and GA15-22712S.

References

1. V.A. Katic, Network Harmonic Pollution - A Review and Discussion of International and National Standards and Recommendations, Proc. Power Electronics Congress - CIEP, 1994, pp. 145–151.
2. T.S. Key, J.S. Lai, IEEE and International Harmonic Standard Impact on Power Electronic Equipment Design, Int. Conf. Industrial Electronics, Control and Instrumentation IECON, 1997, pp. 430–436.
3. A. Rash, Power Quality and Harmonics in the Supply Network: A Look at Common Practices and Standards, Proc. Mediterranean Electrotechnical Conf. MELECON, 1998, pp. 1219–1223.
4. IEEE PES Working Group on Nonsinusoidal Situations, A Survey of North American Electric Utility Concerns Regarding Nonsinusoidal Waveforms, IEEE Trans. Power Delivery, vol. 11, pp. 73–78, Jan. 1996.
5. A. Ferrero, Definitions of Electrical Quantities Commonly Used in Nonsinusoidal Conditions, Eur. Trans. Electric Power, vol. 8, pp. 235–240, 1998.
6. M. Depenbrock, The FBD Method, a Generally Applicable Tool for Analyzing Power Relations, IEEE Trans. Power Systems, vol. 8, pp. 381–387, May 1993.
7. M. Depenbrock, V. Staudt, The FBD Method as a Tool for Compensating Total Nonactive Currents, Proc. Int. Conf. Harmonics and Quality of Power, 1998, pp. 320–324.
8. M. Depenbrock, V. Staudt, Stability Problems if Three-Phase Systems with Bidirectional Energy Flow Are Compensated Using the FBD Method, Proc. Int. Conf. Harmonics and Quality of Power, 1998, pp. 325–330.
9. L.S. Czarnecki, Minimization of Unbalanced and Reactive Currents in Three-Phase Asymmetrical Circuits with Nonsinusoidal Voltage, Proc. Inst. Elect. Eng. B, vol. 139, pp. 347–354, 1992.
10. L.S. Czarnecki, Physical Reasons of Currents RMS Value Increase in Power Systems with Nonsinusoidal Voltage, IEEE Trans. Power Delivery, vol. 8, pp. 437–447, Jan. 1993.
11. L.S. Czarnecki, Misinterpretations of Some Power Properties of Electric Circuits, IEEE Trans. Power Delivery, vol. 9, pp. 1760–1769, Oct. 1994.
12. L.S. Czarnecki, Power related Phenomena in Three-Phase Unbalanced Systems, IEEE Trans. Power Delivery, vol. 10, pp. 1168–1176, July 1995.
13. L.S. Czarnecki, Comments on Active Power Flow and Energy Accounts in Electrical Systems with Nonsinusoidal Waveforms and Asymmetry, IEEE Trans. Power Delivery, vol. 11, pp. 1244–1250, July 1996.
14. L.S. Czarnecki, Comments, with a Reply on a New Control Philosophy for Power Electronic Converters as Fictitious Power Compensators, IEEE Trans. Power Electron., vol. 5, pp. 503–504, Oct. 1990.
15. P. Mattavelli, P. Tenti, Design Aspects of Hybrid Compensation Systems, Eur. Trans. Electric Power, vol. 8, pp. 375–382, 1998.
16. IEEE PES Working Group on Nonsinusoidal Situations, Practical Definitions for Powers in Systems with Nonsinusoidal Waveforms and Unbalanced Loads: A Discussion, IEEE Trans. Power Delivery, vol. 11, pp. 79–101, Jan. 1996.
17. E.B. Makram, S. Varadan, Definition of Power Components in the Presence of Distorted Waveforms Using Time-Domain Technique, Proc. Southeastern Symp. System Theory, 1991, pp. 536–539.
18. A. Emanuel, Apparent Power: Components and Physical Interpretation, Proc. Int. Conf. Harmonics and Quality of Power, 1998, pp. 1–13.
19. W. Shepherd, P. Zand, Energy Flow and Power Factor in Nonsinusoidal Circuits, Cambridge, U.K.: Cambridge Univ. Press, 1979.
20. N. LaWhite, M. Ilic, Vector Space Decomposition of Reactive Power for Periodic Nonsinusoidal Signals, IEEE Trans. Circuits Syst. I, vol. 44, pp. 338–346, Apr. 1997.

21. H. Akagi, Y. Kanazawa, A. Nabae, Instantaneous Reactive Power Compensators Comprising Switching Devices without Energy Storage Components, IEEE Trans. Ind. Application, vol. 20, pp. 625–630, May/June 1984.
22. F.Z. Peng, J.S. Lai, Generalized Instantaneous Reactive Power Theory for Three-Phase Power Systems, IEEE Trans. Instrum. Measur., vol. 45, pp. 293–296, Feb. 1996.
23. F.Z. Peng, G.W. Ott, D.J. Adams, Harmonic and Reactive Power Compensation Based on the Generalized Instantaneous Reactive Power Theory for Three-Phase Four-Wire Systems, IEEE Trans. Power Electron., vol. 13, pp. 1174–1181, Nov. 1998.
24. J.L. Willems, Critical Analysis of the Concepts of Instantaneous Power Current and Active Current, Eur. Trans. Electric Power, vol. 8, pp. 271–274, 1998.
25. A. Nabae, T. Tanaka, A Quasi Instantaneous Reactive Power Compensator for Unbalanced and Nonsinusoidal Three-Phase Systems, Proc. IEEE Power Electronics Specialists Conf., 1998, pp. 823–828.
26. J.H.R. Enslin, J.D. van Wyk, Measurement and Compensation of Fictitious Power under Nonsinusoidal Voltage and Current Conditions, IEEE Trans. Instrum. Measurement, vol. 37, pp. 403–408, Sept. 1988.
27. F.P. Venter, J.D. van Wyk, L. Malesani, A Comparative Evaluation of Control Strategies for Current-Fed Converters as Filters for Nonactive Power in Networks, Proc. IEEE IAS Annual Meeting, 1992, pp. 829–836.
28. P. Salmeron, J. C. Montano, Instantaneous Power Components in Polyphase Systems under Nonsinusoidal Conditions, Proc. Inst. Elect. Eng. Sci. Meas. Technol., vol. 143, pp. 151–155, 1996.
29. W. le Roux, J.D. van Wyk, Correspondence and Difference between the FBD and Czarnecki Current Decomposition Methods for Linear Loads, Eur. Trans. Electric Power, vol. 8, pp. 329–336, 1998.
30. W. le Roux, J.D. van Wyk, Evaluation of Residual Network Distortion during Compensation According to the Instantaneous Power Theory, Eur. Trans. Electric Power, vol. 8, 1998, pp. 337–344.
31. R. Gretsch, M. Neubauer, System Impedances and Background Noise in the Range 2 kHz to 9 kHz, Eur. Trans. Electric Power, vol. 8, pp. 369–374, 1998.
32. G. Blajszczak, Space Vector Control of a Unified Compensator for Nonactive Power, Proc. Inst. Elect. Eng. C, vol. 141, pp. 207–211, 1994.
33. P. Tenti, P. Mattavelli, A Time-Domain Approach to Power Term Definitions under Non-Sinusoidal Conditions, Sixth International Workshop on Power Definitions and Measurements under Non-Sinusoidal Conditions Milano, October 13–15, 2003
34. E. Tedeschi, P. Tenti, Cooperative Design and Control of Distributed Harmonic and Reactive Compensators, International School on Nonsinusoidal Currents and Compensation Lagow, Poland, 2008.
35. D. Jeltsema, E. Garcia Canseco, R. Ortega, J.M.A. Scherpen, Towards a Regulation Procedure for Instantaneous Reactive Power, 16th IFAC World Congress, Prague, Czech, July 4–8 2005.
36. E.G. Canesco, R. Grino, R. Ortega, M. Salichs, A.M. Stankovic, Power-Factor Compensation of Electrical Circuits, IEEE Control Systems Magazine, April 2007, pp. 46–59.
37. D. Luenberger, Optimization by Vector Space Methods, New York: Wiley & Sons Inc., 1969.
38. R. Ortega, E. Garcia Canseco, Inter-Connection and Damping Assignment Passivity Based Control: A Survey, European Journal of Control: Special Issue on Lagrangian and Hamiltonian Systems, vol. 10, pp. 432–450, December 2004.
39. A. Astolfi, R. Ortega, R. Sepulchre, Stabilization and Disturbance Attenuation of Nonlinear Systems Using Dissipativity Theory, European Journal of Control, vol. 8, no. 5, pp. 408–433, 2002.
40. D. Mayer, J. Hrusak, M. Stork, On State-Space Energy Based Generalization of Brayton-Moser Topological Approach to Electrical Network Decomposition, Springer, Computing, 2013.

41. D.J. Hill, P.J. Moylan, Dissipative Dynamical Systems: Basic Input-Output and State Properties, Journal of the Franklin Institute, vol. 309, no. 5, pp. 327–357, May 1980.
42. P. Penfield, R. Spence, S. Duinker, Tellegen's Theorem and Electrical Networks, MIT Press, 1970.
43. D. Jeltsema, Modeling and Control of Nonlinear Networks: A Power Based Perspective, Ph. D. dissertation, Delft University of Technology, The Netherlands, May 2005.
44. D. Jeltsema, R. Ortega, J.M.A. Scherpen, On Passivity and Power-Balance Inequalities of Nonlinear RLC Circuits, IEEE Trans. on Circuits and Systems-Fund. Theory and Appl., vol. 50, no. 9, 2003, pp. 1174–1178.
45. L.S. Czarnecki, What is Wrong with the Budeanu Concept of Reactive and Distortion Power and Why it Should Be Abandoned, IEEE Trans. on Instrumentation and Measurement, vol. IM-36, no. 3, September 1987, pp. 834–837.
46. L.S. Czarnecki, Powers in Nonsinusoidal Networks: Their Interpretation, Analysis, and Measurement, IEEE Trans. on Instrumentation and Measurement, vol. 39, no. 2, April 1990, pp. 340–345.
47. L.S. Czarnecki, Power Related Phenomena in Three-Phase Unbalanced Systems, IEEE Trans. on Power Delivery, vol. 10, no. 3, pp. 1168–1176, 1995.
48. L.S. Czarnecki, Closure on Instantaneous Reactive Power p-q Theory and Power Properties of Three-Phase Systems, IEEE Trans. on Power Delivery, vol. 23, no. 3, 2000.
49. L.S. Czarnecki, Powers and Compensation in Circuits with Periodic Voltages and Currents, Part 8: Balancing and Reactive Power Compensation of Three-Phase Linear Loads, Journal on Electrical Power Quality and Utilization, vol. 7, no. 1, 2001, pp. 9–15.
50. L.S. Czarnecki, Currents' Physical Components (CPC) in Circuits with Nonsinusoidal Voltages and Currents, Part 1: Single-Phase Linear Circuits, Electrical Power Quality and Utilization Journal, vol. XI, no. 2, 2005, pp. 27–48.
51. L.S. Czarnecki, Currents' Physical Components (CPC) in Circuits with Nonsinusoidal Voltages and Currents, Part 2: Three-Phase Three-Wire Linear Circuits, Electrical Power Quality and Utilization Journal, vol. XII, no. 1, 2006, pp. 3–13.
52. P. Tenti, J.L. Willems, P. Mattavelli, E. Tedeschi, Generalized Symmetrical Components for Periodic Non-Sinusoidal Three Phase Signals, Seventh International Workshop on Power Definition and Measurements under Nonsinusoidal Conditions, Cagliari, Italy, July 2006.
53. L.S. Czarnecki, Could Power Properties of Three-Phase Systems be Explained in Terms of the Poynting Vector?, IEEE Trans. on Power Delivery, vol. 21, no. 1, Jan. 2006, pp. 339–344.
54. E. Tedeschi, P. Tenti, P. Mattavelli, Cooperative Operation of Active Power Filters by Instantaneous Complex Power Control, Proc. of the 7th Int. Conf. on Power Electronics and Drive Systems (PEDS 07), Bangkok, Nov. 2007.
55. L.S. Czarnecki, Powers and Compensation in Circuits with Nonsinusoidal Voltages and Currents, Part 2: Current Physical Components and Compensation in LTI Circuits, On-Line Journal on Control and Electrical Engineering, vol. 1, no. 2, 2010, pp. 71–79
56. D.J. Sullivan, Improvement in Voltage and Dynamic Performance of Power Transmission System Using Static Var Compensators, BSEET, Pennsylvania State University, 1995.
57. N. Sabai, H.N. Maung, T. Win, Voltage Control and Dynamic Performance of Power Transmission System Using Static Var Compensator, World Academy of Science, Engineering and Technology, 42, 2008
58. K.K. Sen, STATCOM-Static Synchronous Compensator Theory, Modelling and Applications, IEEE PES Winter Meeting 2, 1999, pp. 1177–1183.
59. G.A. Adepoju, O.A. Komolafe, Analysis and Modelling of Static Synchronous Compensator (STATCOM): A comparison of Power Injection and Current Injection Models in Power Flow Study, International Journal of Advanced Science and Technology, vol. 36, November 2011.
60. J. Radatz, The IEEE Standard Dictionary of Electrical and Electronics Terms, Sixth Edition Standards Coordinating Committee 10, Terms and Definitions.
61. E. Moulin, Measuring Reactive Power in Energy Meters, Metering International, Issue 1, 2002, http://libvolume3.xyz/electrical/btech/semester3/electricalandelectronic-measurements/measurementofpowerandenergy/measurementofpowerandenergytutorial1.pdf

62. ADE7753 - Single-Phase Multifunction Metering IC with di/dt Sensor Interface, Analog Devices, http://www.analog.com/media/en/technical-documentation/data-sheets/ADE7753.pdf
63. ADE7854A/ADE7858A/ADE7868A/ADE7878A - Polyphase Multifunction Energy Metering IC, Analog Devices, http://www.analog.com/media/en/technicaldocumentation/data-sheets/ADE7854A_7858A_7868A_7878A.pdf
64. H. Mani, Differences Between the ADE7854A/ADE7858A/ADE7868A/ADE7878A and ADE7854/ADE7858/ADE7868/ADE7878 Products, Analog Devices, AN-1272, http://www.analog.com/media/en/technical-documentation/application-notes/AN-1272.pdf
65. Evaluating the ADE7854A/ADE7858A/ADE7868A/ADE7878A Energy Metering ICs, Analog Devices, http://www.analog.com/media/en/technical-documentation/user-guides/EVAL-ADE7878AEBZ_UG-545.pdf
66. ADE7880 - Polyphase Multifunction Energy Metering IC with Harmonic Monitoring, Analog Devices, http://www.analog.com/media/en/technical-documentation/data-sheets/ADE7880.pdf
67. ADE7816 - Six Current Channels, One Voltage Channel Energy Metering IC, Analog Devices, http://www.analog.com/media/en/technical-documentation/data-sheets/ADE7816.pdf
68. P. Tenti, G. Spiazzi, Harmonic Limiting Standards and Power Factor Correction Techniques, 6th European Conference on Power Electronics and Applications - EPE'95, 1995
69. IEEE 519 Recommended Practices and Requirements for Harmonic Control in Electrical Power Systems, IEEE Industry Applications Society/ Power Engineering Society. 1993.
70. N. Mohan, et al., Power Electronics: Converters, Applications, and Design, New York: NY, USA, John Wiley & Sons, Inc., 1995.
71. O. Garcia, J.A. Cobos, R. Prieto, P. Alou, J. Uceda, Single Phase Power Factor Correction: A Survey, IEEE Trans. Power Electron., vol. 18, no. 3, pp. 749–755, May 2003.
72. Power Factor Correction Handbook, http://www.onsemi.com/pub_link/Collateral/HBD853-D.PDF, ON Semiconductor, 2007.
73. B. Schramm, Power Supply Design Principles: Techniques and Solutions, Part 3, http://www.nuvation.com/corporate/news/newsletter/fall2006/.
74. W.H. Wolfle, W.G. Hurley, Quasi-Active Power Factor Correction with a Variable Inductive Filter: Theory, Design and Practice, IEEE Transactions on Power Electronics, vol. 18, No. 1, January 2003.
75. W.H. Wolfle, W.G. Hurley, Quasi-Active Power Factor Correction: The Role of Variable Inductance, Power Electronics, http://www.nuigalway.ie/power_electronics/projects/quasi_active.html.
76. I. Sugawara, Y. Suzuki, A. Takeuchi, T. Teshima, Experimental Studies on Active and Passive PFC Circuits, INTELEC 97, 19th International Telecommunications Energy Conference, pp. 57178, 19–23 Oct 1997, doi:10.1109/INTLEC.1997.
77. C. Chavez, J.A. Houdek, Dynamic Harmonic Mitigation and Power Factor Correction, http://dx.doi.org/10.1109/EPQU.2007.4424144, IEE Electrical Power Quality, 2007.
78. D. Adar, G. Rahav, S. Ben Yaakov, A Unified Behavioral Average Model of SEPIC Converters with Coupled Inductors, IEEE PESC'97, vol. 1, pp. 441–446, 1997.
79. V. Vorperian, Simplified Analysis of PWM Converters Using Model of PWM Switch, Parts I (CCM) and II (DCM), IEEE Trans. on Aerospace Electronic Systems, vol. 26, 1990, pp. 497–505.
80. B. Keogh, Power Factor Correction Using the Buck Topology - Efficiency Benefits and Practical Design Considerations, Texas Instruments Power Supply Design Seminar SEM1900, Topic 4 TI Literature Number: SLUP264, 2010.
81. J.P.R. Balestero, F.L. Tofoli, R.C. Fernandes, G.V. Torrico, F.J.M. de Seixas, Power Factor Correction Boost Converter Based on the Three-State Switching Cell, IEEE Trans. Ind. Electron., vol. 59, no. 3, pp. 1565–1577, Mar. 2012.
82. C. Qiao, K.M. Smedley, A Topology Survey on Single-Stage Power Factor Corrector with a Boost Type Input-Current-Shaper, IEEE Trans. Power Electron., vol. 16, no. 3, pp. 360–368, May 2011.

83. A. El Aroudi, R. Haroun, Cid-Pastor, L.M. Salamero, Suppression of Line Frequency Instabilities in PFC AC-DC Power Supplies by Feedback Notch Filtering the Pre-Regulator Output Voltage, IEEE Trans. on Circuit and Systems-I: Regular Papers, vol. 60, no. 3, March 2013.
84. Design of Power Factor Correction Using FAN7527, Application Note AN4107, Fairchild Semiconductor, http://www.fairchildsemi.com/an/AN/AN-4107.pdf, 2004.
85. Power Factor Correction (PFC) Basics, Application Note 42047, Fairchild Semiconductor, http://www.fairchildsemi.com/an/AN/AN-42047.pdf, 2004.

Chapter 5
Reactive Power Control in Wind Power Plants

Reza Effatnejad, Mahdi Akhlaghi, Hamed Aliyari, Hamed Modir Zareh and Mohammad Effatnejad

Abstract Studies in this chapter have been performed on the interaction between wind farm, reactive power compensation, and the power system network. The fluctuation of the loads and the output of wind turbine units in power system have made the reactive power compensation an effective procedure. Considering the wind turbine power plant as a distributed generation unit, there would be some positive effect on the network, i.e. distributed system and upper hand grid reliability improvement, improving the environmental issues and development of power grid planning. In order to achieve better condition of reactive power in the network the existing conventional Asynchronous Induction motor (Constant Speed) should be replaced by Wound Rotor Synchronous Induction motor (variable speed), namely, Doubly Fed Induction Generator (DFIG). The control system of a DFIG wind turbine is usually comprised of two parts: electrical and mechanical control. The former includes the control of converter in the rotor side and control of converter in the grid side and the latter includes the control of the angel of turbine blade. The standard IEEE 30-bus System is Consider as the test system. Three methods are applied. Newton-Raphson algorithm, using second generation of smart genetic

R. Effatnejad (✉) · M. Akhlaghi · H. Modir Zareh
Department of Electrical Engineering, Karaj Branch, Islamic Azad University, Karaj, Alborz, Iran
e-mail: reza.efatnejad@kiau.ac.ir

M. Akhlaghi
e-mail: me_akhlaghi@hotmail.com

H. Modir Zareh
e-mail: hamed64.modirzare@hotmail.com

H. Aliyari
Department of Electrical Engineering, Faculty of Electrical Engineering, Qazvin Branch, Islamic Azad University, Qazvin, Iran
e-mail: hamedaliyary@ut.ac.ir

M. Effatnejad
Department of Electrical Engineering, Civil Aviation Technology College, Tehran, Iran
e-mail: m.effatnejad@yahoo.com

N. Mahdavi Tabatabaei et al. (eds.), *Reactive Power Control in AC Power Systems*, Power Systems, DOI 10.1007/978-3-319-51118-4_5

algorithm with non-dominated sorting without any power plant, and the last is using second generation of smart genetic algorithm with the non-dominated sorting with the assumption of the presence of wind power plants. Results show that the presence of wind power plant is effective in improving the reactive power in the grid.

5.1 Introduction

The daily increase of need to electrical energy has been converted into a serious problem and taking into consideration the environmental problems of electricity generation by the fossil fuel power plants, the development of renewable energy resources seems to be the only logic way.

Today, concurrent with the development of the new technologies, electricity world is facing noticeable changes from the viewpoint of issues such adjustment, social and environmental regulations. The prevailing power plant is equipped in particular with special instruments for the control of grid frequency, in order to guarantee the system balance. A similar controlling structure has been developed in the area of voltage control, though its use has been less prevailing.

With regard to the increase of the share of renewable energy systems from energy generation, such as photovoltaic systems and wind farms, the grid management has been converted into one of the very important necessities in creation of a balance between generation and consumption in an ideal and highly output form [1].

In addition, the energy generation based on renewable energy systems causes to make the process of forecasting and control of load distribution in the grid to be much more difficult. In such a scenario, the load distribution management is changed into a determining factor in the connection of generated energy by non-centralized renewable resources to the grid.

At present, renewable energy resources exist in a broad but non-centralized form at the sub-distribution and distribution systems. They are very frequent and their accessibility is predictable but they are not fully reliable. However, these resources have not been utilized fully yet to be able to detect their weak points and strengths. In a separate form, these resources cause the reduction of ability to control the voltage. These units are mostly managed by distribution system operators and have a difficult coordination with the transmission system operators.

With the fast expansion of the technology of wind energy and considering the national policies on renewable energy resources, wind energy has been noticed vastly in energy generation. The wind units in a useful and noticeable rate of energy are available. They have a zero fuel cost and the energy generation by these units is very ideal from the environmental viewpoint.

The wind farms at a large scale will have a great impact on the reliability and stable performance of the power system for two main reasons. First, the areas which are under the blowing of rich resources of wind energy and can be used in order to develop the generation of energy from wind power. In most of the cases, they are located in a region that the power grid faces problems and weakness for connection. Secondly, it is the nature of the wind energy which has been identified as an unstable energy. So, the active power from the generator connected to the wind turbine changes with the wind speed. The serious problem is the intensive reduction of voltage quality of the local electricity grid in the place or near wind farm. The power grid causes fluctuation of reactive power and in the continuation; it will have impact on system voltage and even causes a global blackout.

The adjustment of the reactive power is one of the important issues in connection with the large wind farms which should receive a specific notice. The rate of the reactive power absorption or injected by the wind units in the farm and electricity grid is changing constantly and its reason is the constant changes of power due to change in wind speed. By the way, considering the fact that the size and number of wind farms which contribute in the generation of electrical energy is increasing, it is not possible to ignore the reactive power which is generated in wind farms in large scales.

With the development of use of wind energy resources in larger scales, all effects which this energy has on the power quality and tolerated by the electricity grid, is not economic and even sometimes seems to be unbearable. So, proper compensation of reactive power besides the wind turbine units seems to be an essential issue. The compensation of reactive power can cause the increase of adjustable ability of the reactive power which may lead to the voltage stability in wind farms.

Optimization and control of reactive power in the power system through suitable allocation of reactive power resources and rational compensation of the reactive loads is the best effective method to reduce the losses of power in the grid and control of voltage level at power grid.

The units of wind turbines in power station scales could be able to adjust the voltage by the dynamic supply of reactive power. From the viewpoint of utilization of power system, wind farms should have the control ability in agreement with other resources of electricity generation. The ability of voltage adjustment of wind turbines units is different depending on the technology used in the construction of generation and also its manufacturer. The units of wind turbines of the type of 1 and 2 which are based on induction generator, do not have any control ability of voltage in nature. Type 3 and 4 of wind turbine units include power electronic converters, so they have the ability to adjust the reactive power and consequently they have the ability of voltage control. For some reasons, in most of the cases, this ability is not used in the type 3 units, but mostly they are used in the mode of unity power factor. When the adjustment of reactive power is employed, coordination is made among the reactive adjustment points of wind turbine units usually by a central controller which determines an ideal program for all existing units in the wind farm [2].

5.2 Indices Affecting the Wind Turbine Units from the Reactive Power Viewpoint

The increase of wind power share and larger wind turbines have made the evaluation of voltage conditions of connection to the wind turbine grids with further details and using the load flow analysis, an interesting issue. In load flow analysis, voltage of each knot is determined by the given load. In a radius grid consisting of wind units and consumers, the minimum voltage occurs mainly when there is zero generation and maximum consumption and the maximum of voltage occurs in minimum load and maximum generation. If the minimum and maximum of the voltage, both are in an acceptable level, it can be said that the situation of the grid is confirmed, otherwise, the grid should be reinforced.

Having precise and confirmed information on the maximum output power and reactive power and other indices related to that are the calculation requirements which have been mentioned before.

- **Maximum of Output Power**

The maximum of output power of wind turbine is the necessary feature in determining the strength necessary for the grid in the connection point of the place where the wind turbine unit is installed. On this basis, the following data seems to be necessary.

The reference power which based on definition is the highest point of curve for the power according to IEC 1400-12/2 standard.

The maximum of continues power includes the maximum of continues power which the system of turbine control permits and the output power irrespective of atmospheric conditions (wind and air density) and the grid should not surpass that. In practice, it means that the wind turbine should be equipped with a mechanism to control its own performance, so that this continues output power will never exceed a certain limit. The maximum of continues power can be obtained with the evaluation of the controlling system of turbine or calculations in accordance with the IEC 1400-12/12 standard [3, 4].

The maximum instantaneous power is the maximum of instantaneous output power from turbine in the normal operating condition and standard air density. Measuring this power is also taken place according to IEC 1400-12/2 standard.

Wind turbines with fixed speed, with the controlling system of pitch and stall have the ability to produce the peak power output, i.e. higher than nominal power and at the same time, the wind turbines of flexible speed and Optislip, due to speed or flexible slip, have certain limits in their own moment output power.

- **Reactive Power**

The reactive power of wind turbines (generation or consumption) has also important and necessary features in determining the rate of strength of the grid in the joint connection point of the place where the turbine is installed. The reactive power (or power factor) can be determined through measuring special functions which are related to the output of the wind turbines power.

The reactive power consuming in the wind turbine units which are connected to frequency converter (variable speed wind turbines) is usually zero. Whereas the consumption of reactive power in prevailing types of wind turbine with induction generations varies according to a function of their generation active power. Wind turbines with induction generator are usually used along with compensators. So the coefficient of their power depends on the size and type of generator design is variable from 1 in zero generations to 0.98 in nominal generation. In the event of need, it is possible to achieve the coefficient of unity power factor through connection of larger capacitor bank to the wind turbine [3].

- **Flicker**

Flicker is a description of the fast changes of voltage on incandescent lamps. Fast changes of voltage are created due to change in the consumption of grid loads which leads to the creation of fluctuation in the active and reactive power. Wind turbines are also sources of power fluctuations which mainly are due to the impact of disruption (turbulence) of wind and tower shade which leads to the periodical fluctuation of power in a frequency in which the rotor blades passes through vertical pivots of the tower. Flicker also depends on the ratio of *X*/*R* and level of short circuit (fault level) in the Point of common coupling (PCC) [3, 5].

- **Harmonic**

Voltage deviation from the 50 Hz full sinusoidal shape curve leads to creation of harmonic and noise in the network. Harmonic and noise causes the increase of losses in the power system. In some cases, they can lead to the creation of overload on batteries, transformers and other electrical equipment. The creation of disruption in telecommunication systems and fault of controlling equipment are other outcomes of these two phenomena. Since the frequency converters create a current with an incomplete sinusoidal wave shape, the of the variable speed wind units equipped with the frequency converter can lead to the harmonic generation in the power network and the current harmonic leads to the creation of harmonic in the voltage waveform. The amplitude of voltage harmonic is related to the current harmonic amplitude and impedance of the grid in the current frequency [3, 5].

5.3 Types of Wind Turbine Connection to the Power Grid

Wind Turbine Structure

The main parts of wind turbine unit contain the main tower, blades, gearbox, generator and axel of the turbine. In order to control the speed of rotating shaft of the generator, the gearbox is used.

In order to gain the maximum energy from the wind, it is necessary that the angle of blades changes with the change in the wind speed, and this function is done by controlling the angle of the blade. Also after measuring the wind direction, a small engine called Yaw, turns the whole upper part of the turbine tower to be placed in proper line with the wind blowing direction.

Types of Wind Turbines

- *Fixed Speed Wind Turbines*

By the early 1990, the standard of installation and utilization was based on wind turbines with fixed speed. In these types of turbines, irrespective of wind speed, the speed of turbine rotor (shaft) is fixed. This speed depends on the frequency of the generator construction grid and also the ratio of gears in the gearbox.

These types of turbines have Squirrel-cage induction generator or with the wounded rotor which are directly connected to the power grid. These generators are equipped with a soft starter and capacitor banks for the reactive power compensation. These generators have been designed such that in a specific speed of wind could have the greatest output. In order to increase the generation power of generator, these wind turbines have two types of adjustment on the stator windings.

One is used in low speeds of wind (mainly 8 poles) and the other is used in the average or high speed (4 poles or 6 poles). These types of turbines have advantages such as simplicity, reinforcement and high reliability and many scientific and research works have been made on them. The price of electrical parts and their drive is also low. The important disadvantages of these turbines also include the uncontrollable consuming reactive power, mechanical stress and limited control of power quality. Due to the performance of their fixed speed, all fluctuations in the wind speed emerge in form of fluctuations in the mechanical moment and thereby in form of fluctuations in the electrical power of the grid. As for the weak grids, the power fluctuations can lead to large voltage fluctuations which cause the considerable losses in the transmission lines [6].

- *Variable Speed Wind Turbines*

In recent years, wind turbines with variable speed have formed the dominant majority (among the installed turbines). In this status, it is possible to adjust the rotor rotating speed (with the increase or reduction of acceleration). The fixed speed is kept and stabilized in a fixed and pre-determined rate to be able to achieve a high power factor. In this status, the generator torque is kept relatively fixed and changes in the wind blowing leads to changes in the generator speed.

The electrical system of wind turbines with variable speed is more complex than turbines with fixed speed. These types of turbines mainly have induction generator or Synchronous one and are connected to the grid through a power converter. The advantages of these types of turbines include: the increase of energy obtained from the wind, improvement of the power quality and reduction of mechanical stress. Their disadvantages also include: losses in the electronic drives equipment, using more devices and increase of cost resulting from the equipment of electronic systems. In these types of turbines, the power fluctuations resulting from wind fluctuations mainly appear in form of changes in the rotor speed of turbine and generator [6].

- *Power Control Concepts*

The most simple, strong and cheap control method is the inactive control in which the blades are screwed inside the ball with a fixed angle. The rotor design is such that in the event that the wind speed exceeds a certain limit, rotor could loss the wind power. So the aerodynamic powers of blades are limited. Such an adjustment of aerodynamic power causes to have less power fluctuations proportional to the adjustment of power in a fast steps. Some of the deficiencies of this method are the low output in low speeds of wind, lack of auxiliary start and changes in the maximum power of the stable status as a result of changes in air density and network frequency.

Other type of control is the step control (active control) in which the blades can be twisted in the times when the output power is very low or high towards the wind direction or opposite to it accordingly.

In general, the advantages of this type of control are the good control of power, auxiliary restart and emergency stoppage. From electrical viewpoint, the good control means that in high speeds, the main rate of output power to be close to the nominal rate of generator. Some of its disadvantages are the additional complexity resulting from the step mechanism and more fluctuations of the power in high speeds. During storm and limited speeds of the step mechanism, the instantaneous power fluctuates around the nominal rate [6].

The third controlling strategy is active fatigue control. In the low speeds of wind, in order to have access to the maximum output, the blades are turned like a step controlled wind turbine. In the high speeds, blades are into a deep and slow fatigue and in the direction opposite to the controlled turbine with pace. With this type of control, a clearer limited power (without high fluctuations of controlled wind turbines with pace) is obtained. This type of control has advantages such as ability to compensate the air density changes. Combination with the step mechanism has eased the stoppages of the emergency status and start of wind turbines [6].

5.4 Types of Modern Generators

Type A: Fixed Speed

As the SCIG constantly extracts the reactive power from the grid, this arrangement uses a capacitor bank to compensate the reactive power. The soft, smooth and clear connection to the grid is obtained through a soft starter. As it was mentioned, on the case of weak grids, the wind fluctuations are converted into voltage fluctuations. These types of turbines can absorb variable quantities of power from the grid and this also increases the voltage fluctuation and line losses. So, some of the most important deficiencies of this generator and related systems include:

- Lack of control over speed
- A need to having a strong grid
- A need to having a strong mechanical structure to be able to bear the high mechanical stresses.

Type B: Limited Variable Speed

That category of wind turbines which are limited with the variable speed have variable resistance in the rotor and are identified with the name of Opt slip. In this type of arrangement, the wounded rotor induction generation is used. Generator is directly connected to the grid, and in order to have a better connection, a soft starter is used. The variable resistance of rotor of these types of generators can be adjusted with the controlled converter by optic pulses which have been installed above the rotor axel. This optic equipment is in need of expensive slip rings (which demands brush and maintenance).

In this case, the rotor resistance can be changed and thus control the slip in this way. By this means, the output power of the system is controlled. The range of dynamic speed control depends on the size of the rotor variable resistance. Typically, the speed limit is zero to 10% over the Synchronous speed. The energy resulting from the conversion of additional energy is lost in form of heat. In 1998, two engineers, namely Wallen and Oliner introduced the concept of replacement in which, passive parts has been put forth instead of power electronic converters. This concept has a 10% slip but unfortunately had no control on slip [6].

Type C: Variable Speed with Frequency Converter with Fractional Capacity

This type of generator is known as doubly fed induction generator (DFIG). The converter capacity (frequency converter with fractional capacity) is about 30% of the nominal power of the generator. This converter does the compensation of reactive power and by the way, at the time of connection to the power grid, it gives a clear and smoother state. This type of arrangement has a broader limit as compared with Optislip in the dynamic control of voltage. The limit of their speed is usually in a broader limit as compared with Synchronous speed. Smaller frequency converter is more economic from the economic point of view. The main deficiency of this method is the use of slip rings and their protection against the grid faults [6].

Type D: Variable Speed with Frequency Converter with Full Capacity

In this method, generator is connected to the grid through a converter (frequency converter with full capacity). The frequency converter does the compensation of reactive power and at the time of connection to the grid has a clear and smoother feature. This type of generator can be motivated in electrical form like WRIG, WRSG's or to be incited through a permanent magnet (PMSG). Some of the wind turbines of full variable speed do not have any kind of gearboxes. In these cases, the multi-polar generators which receive commands directly (in a large diameter) are used [6].

Asynchronous Generators (Induction)

Asynchronous generator is the most prevailing generator which is used in wind turbines. The mechanical strength and simplicity and cheap price are some of its advantages and need to magnetizing current is among its disadvantages. Its consuming reactive power is supplied either by capacitor bank or power electronic devices. The created magnetic field in it is rotating with a speed which is determined by the windings poles and grid frequency and is called Synchronous speed. If rotor turns with a speed more than Synchronous speed, in that magnetic field, voltage is induced and the current is flowing in its windings.

In SCIG generators whose speed changes only about many percent due to changes of wind speed, so that they are mostly used along with the turbines of type "A" (fixed speed). SCIG generator has a sharp torque-speed characteristic, so that the vibrations of wind power are directly transferred to the power grid. These transient states especially at the time of connection of wind turbine to the grid are critical because they give rise to the creation of inrush current of seven to eight times of the nominal flow. In the weak grids, this rate of inrush current can cause intensive disruption. So, connection of SCIG generators to the grid is done slowly and by a soft starter to limit the inrush current.

In SCIG generators, there is a liner relation between the rate of necessary capacitor bank, active power, terminal voltage and rotor speed. That means, in the event of intensive winds, the turbine can produce a more active power if it could absorb more reactive power from the grid. In these types of generators, if a fault occurs, due to the unbalance of mechanical and electrical powers, the rotor speed increases. So by removing the fault, SCIG extracts more reactive power from the grid which causes the greater reduction of the voltage [6].

In the wounded rotor induction generator, the electrical specifications of the rotor can be controlled from outside and affects its voltage. The rotor winding can be connected through a brush and slip rings to the outside or through power electronic devices which might need to slip rings and brushes or might not need, generator can become magnetic through stator or rotor. There is a possibility for retrieval of slip energy from rotor circuit and rotating to the stator outlet. In wind turbines, most of the following arrangements which are related to WRIG are used.

1. Induction generator of (Optslip) which is used in type (OSIGB).
2. Doubly fed Induction Generation which is used in type C.

In Optislip generators, converter is controlled in optic form (The converter which changes the rotor resistance), so that there is no need to slip rings. Stator is also connected directly to the power grid. The range of operating speed in this case as compared to SCIG generators is greater and more advanced from the viewpoint of system. For a specific spectrum, this concept can reduce the mechanical loads and power vibrations resulting from storms. The disadvantages of this method include:

1. Speed limit. Mainly it is about zero to 10% and this is independent from the rate of variable resistance of rotor.
2. It has a weak control on active and reactive powers.
3. Slip power against flexible power is annihilated in form of loss.

The doubly fed induction generator (DFIG) is an interesting option or an increasing market and with demand. DFIG is a WRIG in which its stator is directly connected to the three phase grid with fixed frequency and rotor winding is fed through a voltage source converter with back to back single direction IGBT switches. The doubly fed phrase refers to this reality that stator gives voltage for the loads fed to the power grid and rotor voltage is created by the power converter. The performance speed of this system is in a vast scope but limited. Converter compensates the difference of electrical and mechanical frequency by injecting the current into the rotor with a flexible frequency. Both during the normal operating condition and during faults, the generator behavior is adjusted by the power converter and its controllers.

Power converter is comprised of two converters, the converter in the side of the grid and converter in the side of rotor, which are controlled independently. The main idea is that the converter of the side of rotor controls the powers of active and reactive power by controlling the rotor flow. Whereas the converter of the line side, controls the voltage of DC side to be assured of the performance of converter in the coefficient of unity power factor (zero reactive power). Depending on the operating conditions of drive, power is entering into the rotor or exited from it. In the over-Synchronous condition the power can be flowed towards the grid and in under-Synchronous conditions, power flows from grid towards rotor. In both cases, power goes from stator towards grid side. The advantages of DFIG generators include:

1. Having ability to control the reactive power
2. Having ability to make an independent control of active and reactive powers with rotor flow control
3. Ability of magnetism by rotor side
4. Ability to produce the reactive power which is delivered to stator.

As it was mentioned, converter in the side of grid operates at the unity power factor and does not involve itself in the exchange of reactive power. Of course on the case of weak grids, in the event of voltage fluctuations, DFIG generators might be commanded to exchange the reactive power with the grid, and the ability to control voltage is not related to the overall generator power but related with speed limit and slip power. Converter price is proportional to the range of speed around the Synchronous speed. The inevitable deficiency of DFIG generators is the use of slip rings [6].

Synchronous Generator

Synchronous generator in comparison with the induction generator with similar specifications and similar size is more expensive and complex. Its most important advantage is lack of need to a flow for magnetizing and thereby, through a converter (power electronic), it is connected to the main grid. Converter has the following two important objectives too:

1. For acting as a reinforce of power against the power vibrations (resulting from flexible energy of wind and storm and transit states of the side of power system
2. For the control of magnetic field to maintain Synchronism and prevent from Synchronization problems with the power grid.

The application of such a generator, gives permits to work with a variable speed to wind turbines. These types of generators are either of the type of winded rotors (WRSG) or of the type of permanent magnetic rotor (PMSG).

Due to known state of stable and transit performance of these types of generates, no much discussion are made on them. Other generators in use for wind turbines include [6]:

1. High voltage generators
2. Switched reluctance generators
3. Diagonal flux generators.

5.5 Wind Turbine Requirements During Connection to the Grid Considering Reactive Power Aspects

Deciding about the system of reactive power compensation in the area of wind turbine power plants design, many cases should be taken into consideration. Most of the wind power stations for connection to the grid are bound to observe certain principles which has been confirmed through a series of agreement under the supervision of the management of all-nation electricity grid and necessary parameters for this connection is defined in accordance with the agreement. These necessities can be taken from the world, local standards or cases necessary for the local grid. Some of the necessities can be the result of the study of effects of the system which had been conducted earlier. For example, in some standards, the minimum coefficient of power and Voltage ride through (VRT) for the connection of wind turbines to the grid is considered as the main requirements. The wind power plants designers should be aware that a reactive power compensator cannot solely be a guarantee for low voltage ride through in the wind turbine power plant [5].

After meeting the main conditions for the connection of wind turbine to the grid, other cases such as the time of responding, voltage control requirements, necessities for the control of power factor, stable susceptance requirements, low voltage ride through, high voltage ride through, post emergency requirements and voltage retrieval requirements are considered as another part of the requirements for the set of wind power plant and reactive power compensation system.

- **Power Factor**

The agreement of the wind power plant power factor usually should be measured with POI. Wind power is authorized to meet the condition of power factor through the abilities of wind turbine generator (WTG), fixed or parallel reactors/capacitors or a combination of both states. For example FERC institution makes wind power plants bound to observe the power factor from 0.95 lead to 0.95 lag, of course, provided that the system studies by the officials of the transmission system indicates that such a condition is necessary to guarantee the security of utilization. Wind power plant usually pursues the voltage planning (static voltage) imposed from the transmission system which following it, the power factor of wind power plant is determined in the operating state and desired time. It is possible that the wind power plant be unable to secure the all conditions of the power factor limitation in all possible operating scenarios [5].

- **Voltage Dynamic Backup Requirements**

Some institutions believe that the sufficiency of voltage dynamic supply, like stabilizer of power system and automatic voltage regulations in the conventional power plants units is one of the requirements of wind power plants. Such obligations can be imposed when the authorities of transmission system has come to this conclusion through system studies that the dynamic capability of wind turbine power plant is necessary for maintaining the security of system operating condition [5].

- **Low Voltage Ride through (LVRT) Requirements**

Low voltage ride through ability in wind power plants is one of its requirements for the connection to the power transmission grid. Wind turbine power plants are bound to remain in the state of connection to the grid during the occurrence of 3 phase faults with the time to remove the faults of 4 to 9 cycles. In the other words, the wind turbine unit should not be outage for the state in which the voltage in the side of high voltage of the wind unit transformer substation is up to the time of 0.15 s in zero quantity. Also, wind power plants should remain in the circuit during the phase to the ground faults with the removing delay time and retrieval the post fault voltage to the pre fault amount. LVRT requirements are not applied for the faults among the generator terminals of wind turbine and the side of high voltage of wind power plant transformer [5, 7].

- **High Voltage Ride Through (HVRT) Requirements**

Wind power plants are exposed to high voltages and these voltages might occur due to reasons such as removal and clearance of faults, losing large loads or other system transient states in the grid. Some institutes intends to include certain obligations about HVRT. In many European countries, wind power plants bound to remain in the generation circuit in situations in which the high voltage is up to 110% of nominal voltage at the POI. Also these power plants should remain in circuit for higher voltages if the fault time is less than the pre-determined time [5, 7].

5.6 Wind Turbine Power Plant from Distributed Generation Viewpoint

Considering the wind turbine units as the distributed generation at the distribution and sub-distribution levels, it is possible to achieve good results in different areas related to electricity energy and environmental effects. In the continuation of the discussion, some of these effects are pointed out briefly.

- **Effect of Wind Turbine Units in the Role of Distributed Generation Resources on Reliability of Distribution System**

One of the most important feature of distribution system is its level of reliability. As we know, one way to improve the distribution system reliability is to reduce the outage time or in the other words, the reduction of load retrieval time for a part of consumers which have been detached from the electricity grid as an incident. Distributed generation resources can improve the reliability in this way. When there is a failure in the main grid, they can supply part of the load in island form which is not in the failure area and has been in outage as a result of fault in the grid. Thus, the time of load retrieval in these consumers, from the time of the repair of out-of-service part of the grid, will reduce to the duration of switching and DG re-running.

This state of DG performance, is also called backup state. In this state, it is possible that DG acts prior to failure in parallel form with the main grid and after the fault occurs in the grid, enters the island form. In this event, it should be noted that after the fault occurs, first DG is detached from the grid and after detaching the failure cause, it is connected to the grid for supplying a part of the load once again. The reason for doing this process is to prevent from the lack of protective coordination in the grid and also formation of unwanted islands which in addition to technical problems will bring about the personnel security issues too. Another issue which should be noticed is that the DG should have required capacity for supplying the load inside the island [7–9].

- **The Impact of Wind Turbine Units in the Role of Distributed Generation on the Reliability of Upper Hand Grid**

As the presence of a distributed generation unit can have impact on the reliability of distribution grid, it could influence on the upper hand grid reliability, because with the installation of a unit in distribution grid, the load curve seen in the side of the distribution substation changes. As this curve contributes in the trend of reliability calculation, so that with the installation of each DG in the distribution grid, as much as the reduction of the substation load curve, the reliability index of upper hand grid will change [8, 10].

- **The Effect of Wind Turbine Units in Improving the Environmental Issues**

Employing the wind turbine units in the role of resources for distributed generation in the power grid can be effective in improving environmental issues in two ways:

1. Limiting the dissemination of greenhouse gases: Using the distributed generation resources in particular, the resources based on renewable energies such as wind and co-generation of heat and power (CHP), limits the dissemination of greenhouse gases which is itself one of the stimulants for the use of these resources in the power system.
2. Preventing from the construction of transmission lines and new power plants: One of the other important effects of the distributed generation resources from the environmental viewpoints is to prevent from construction of transmission lines and new power plants whose performance increases the public opposition because the construction of these power plants reduces the environment beauty and capture a large space too [8, 11].

- **The Effect of Installation of Wind Turbine Units in DG Role in Planning for the Development of Power Grid**

The effect of distributed generation resources on the planning for the development of power grid should be studied from some aspects. These cases include the development of power grid for the grid load supply or in other words the study of system sufficiency at two levels of distribution and generation.

One of the important effects of the presence of distributed generation resources in the distribution grid is the concurrent replacement or delay of investment for the development of distribution and sub-distribution of substations. As these resources by themselves as the small power plants deal with the supplying the grid loads, so the rate of electrical power received from the generation and transmission system is reduced and investment for the promotion of substations capacity or construction of new posts are postponed. The important point which should be taken into consideration on this case is to determine the level of sections of distribution cables.

As the rate of load growth is determined with the measurement of grid load in distribution substations, the presence of distributed generation resources might cause the misleading of system designers and some of the distribution lines face overload gradually. As the occurrence of this issue will cause the disconnection of distribution lines and reduction of reliability, more accurate assessments should be done on this issue. Another issue which is put forth at distribution level is to study the sufficiency of power system at the level of distribution grid. The issue is put forth in this way that the presence of wind turbine units in the role of distributed generation resources can to which extent be an alternative for the power purchased from the transmission grid. In order to respond to this question, certain studies have been done to study the sufficiency of the distribution system in the presence of distributed generation resources [8].

5.7 Doubly Fed Induction Machine

The presence of wind energy resources in the power grids is expanding and in many wind farms, the doubly fed induction machines are used for the generation of electrical energy. When the power grid faces a serious disruption, the wind turbine generations in the wind farms are not able to adjust the reactive power dynamically because they are under operation with the unity power factor. So, this status lead to the intensive fall of voltage to the extent that the wind turbine generators makes tripping and are outage and this issue has directly impact on the regular performance of wind farm and threaten the grid security. So, the study of the quality of contribution of wind farms in the voltage regulation of power grid and quality of full utilization from the adjustment ability of the wind turbines reactive power is an effective way to reduce the voltage fluctuations and control of voltage durability in the common connection point of the grid and assurance of the secure performance of the power grid system [12].

As the power generation in distributed and uncontrolled form increase the risk of power system operation and may lead to reduction of power quality (placing the electrical variables outside the limit of use, risk of de-functioning or exit of equipment, etc.), so the power generation companies try to control the reactive power of wind farms through doubly fed machines and on this basis, it is possible to achieve the control of voltage level at distribution grids. It is worth mentioning that in these types of generations, the control of active and reactive power is possible and consequently, they can be used as a source for the continuous reactive power in the electricity generation institutes [12, 13].

5.7.1 Doubly Fed Induction Generator (DFIG) Modeling

The main structure of a DFIG is shown in Fig. 5.1. Wind turbine is connected to DFIG through a mechanical system and this mechanical system includes a low speed shaft (axle) and a high speed shaft which have been detached from each other by a gearbox. The induction generator and wounded rotor in this structure, fed from both sides the rotor and stator of electrical grid. In this machine, stator has been directly connected to power grid whereas the rotor has been connected to grid through AC/DC/AC variable frequency power electronic converter. In order to generate the power in the fixed voltage and frequency for a power grid in the speed limit between upper Synchronous and lower-Synchronous speed, it is necessary to control the transfer power between rotor and grid both from the viewpoint of quantity and direction. The variable frequency converter is comprised of two pulse modulation width converter (PMW) which includes Rotor Side Converter and Grid Side Converter (GSC).

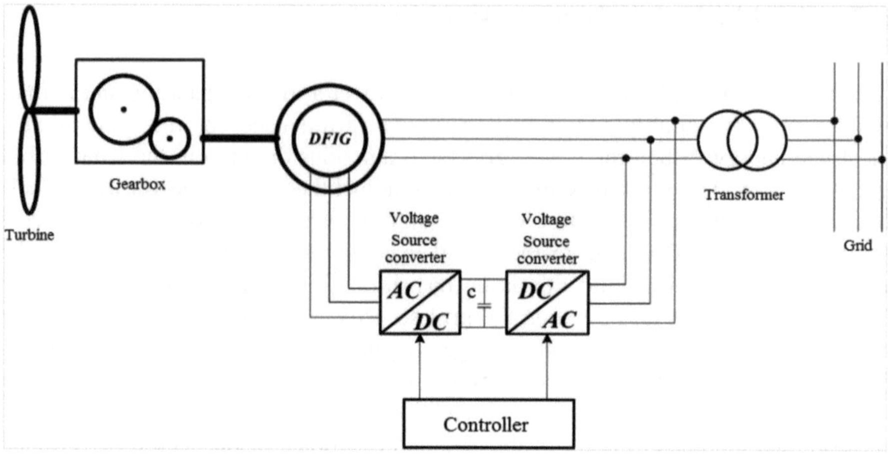

Fig. 5.1 DFIG structure

These converters have been connected back to back through a DC link capacitor. The Crow-bar system which exists in most DFIG is used to short circuit the rotor circuit at the time of fault in order to protect the rotor windings and RSC (Rotor Side Converter) against over-currents of rotor windings.

For a more precise discussion on this case, the following cases can be taken into consideration [14]. In DFIG system, the frequency of stator is fixed and equal to the grid frequency. As it was said, with the control of applied frequency to rotor, with regard to the change in the wind speed, it is possible to control the system.

In the event that rotor is fed with a variable frequency and voltage by the power electronic converter, the generator will provide a variable frequency and voltage in stator.

With regard to the explanations presented, it can be shown that in the remaining state, the mechanical speed of axle ω_{sh} is obtained from the following relation:

$$\omega_{sh} = \omega_s \pm \omega_r \tag{5.1}$$

where ω_{sh}, ω_s and ω_r, accordingly include the angle speed of stator voltage, angle speed of applied voltage on rotor and angle speed of rotor shaft.

Positive sign in the above relation is when the sequence of rotor and stator phase are similar and $\omega_{sh} < \omega_s$. This state is called sub-synchronous performance. The negative sign in this state is corresponding with the state in which the rotor phase sequence is negative (opposite to the sequence of stator phase) and $\omega_{sh} > \omega_s$. This state is called the over-synchronous mode.

With the assumption that DFIG function in the steady state, the existing relation between mechanical power, electrical power of rotor and electrical power of stator has been shown in Fig. 5.2. In this figure P_m indicates the mechanical power delivered to generator, P_r indicates the rotor power, P_{airgap} is the power of generator

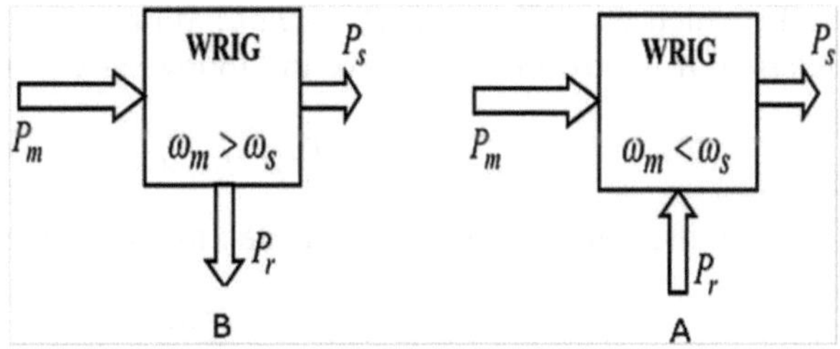

Fig. 5.2 DFIG power flow diagram

air distance and P_s is equal to the power delivered to the grid by stator. In this figure and relations of P_g indicate the total generating and delivered power to the grid.

$$P_m = P_s \pm P_r \tag{5.2}$$

Considering the DFIG equations, simply it can be shown that the following relation is established between stator power, rotor power and the power applied on generator shaft.

One system including DFIG can deliver the electrical power through both ways of stator to the grid but through rotor circuit can absorb energy and power from the grid too.

Delivery or absorption power by rotor depends on the speed of generator turn. If generator is in the over-Synchronous mode, power through rotor and through converters will be injected to the grid and if the generator is in sub-Synchronous mode, in this case rotor will absorb the power through converters. These two working modes are shown in Fig. 5.2 in which ω_s is equal to Synchronous speed and ω_r is equal to rotor speed.

With regard to the Fig. 5.2, it can be learned that in the event that the stator losses is ignored, the Eq. (5.3) and in the event of ignoring the rotor losses, the Eq. (5.4) is obtained.

$$P_{airgap} = P_s \tag{5.3}$$

$$P_{airgap} = P_m - P_r \tag{5.4}$$

From the above two and equation, the stator power is stated by using the following relation:

$$P_s = P_m - P_r \tag{5.5}$$

The above relation can be written in form of generator torque:

$$T\omega_s = T\omega_r - P_r \tag{5.6}$$

In which $P_s = T\omega_s$ and $P_m = T\omega_r$. With the right arrangement of the above relation, it is possible to achieve the following equation:

$$P_r = T(\omega_s - \omega_r) \tag{5.7}$$

Then the rotor power can be related to slip in the following form:

$$P_r = -sT\omega_s = -sP_s \tag{5.8}$$

In which slip means the same s, in terms of phrases of ω_s and ω_r are stated as follows:

$$s = \frac{\omega_s - \omega_r}{\omega_s} \tag{5.9}$$

With the combination of the above relations, the mechanical power i.e. P_m is stated as follows:

$$P_m = P_s - P_r = P_s - sP_s = (1 - s)P_r \tag{5.10}$$

Finally, the total delivered power to the grid, i.e. P_g is stated by using the following relation:

$$P_g = P_s + P_r \tag{5.11}$$

With regard to the equation $P_r = -sP_s$, it is seen that the direction of flow in rotor depends on the rotor performance speed. The direction of power includes to two states of sub-Synchronous and super-Synchronous performance. So, the rotor circuit in the generator state can both absorb the electrical power and inject the electrical power to the grid. Table 5.1 shows different working modes for the doubly fed induction generator [15].

The dynamic equation of a doubly fed induction generator with three phases in the Synchronous rotating reference frame of d–q is written as follows [15]:

$$V_{ds} = V_s I_{ds} - \omega\lambda_{qs} + \frac{d\lambda_{ds}}{dt} \tag{5.12}$$

Table 5.1 Possible states for the doubly fed induction generator

Under sync		Upper sync		
Motor	Generator	Motor	Generator	Functional status
0>	0<	0>	0<	P_m
>0	0>	0>	0<	P_r
0>	0<	0>	0<	P_s

$$V_{qs} = V_s I_{qs} - \omega \lambda_{ds} + \frac{d\lambda_{qs}}{dt} \tag{5.13}$$

$$V_{dr} = V_s I_{dr} - s\omega \lambda_{qr} + \frac{d\lambda_{dr}}{dt} \tag{5.14}$$

where ω_s shows the Synchronous reference frame rotating speed and $s\omega_s = \omega_s - \omega_r$ shows the slip frequency and the linking flow of DFIG is specified with the following relations.

$$\lambda_{ds} = L_{ls} i_{ds} + L_m (i_{ds} - i_{dr}) = L_s i_{ds} + L_m i_{dr} \tag{5.15}$$

$$\lambda_{qs} = L_{ls} i_{qs} + L_m (i_{qs} - i_{qr}) = L_s i_{qs} + L_m i_{qr} \tag{5.16}$$

$$\lambda_{dr} = L_{lr} i_{dr} + L_m (i_{ds} - i_{dr}) = L_m i_{ds} + L_r i_{dr} \tag{5.17}$$

$$\lambda_{qr} = L_{lr} i_{qr} + L_m (i_{qs} - i_{qr}) = L_m i_{qs} + L_r i_{qr} \tag{5.18}$$

where L_{lr}, L_m, L_{ls}, $L_r = L_m + L_{lr}$, $L_s = L_m + L_{ls}$ of the rotor and stator linking inductances and counterpart. DFIG electromagnetic torque is shown as follows:

$$T_e = \frac{3}{2}\frac{p}{2} L_m (i_{qs} i_{dr} - i_{ds} i_{qr}) \tag{5.19}$$

where p shows the number of poles of induction machines. Ignoring the losses of the power related to the stator resistance, active and reactive powers of stator are obtained in form of the following equations:

$$P_s = \frac{3}{2}(V_{ds} i_{ds} - V_{qs} i_{qs}) \tag{5.20}$$

$$Q_s = \frac{3}{2}(V_{qs} i_{ds} - V_{ds} i_{qs}) \tag{5.21}$$

And the active and reactive power of rotor is specified with the following equations:

$$P_r = \frac{3}{2}(V_{dr} i_{dr} - V_{qr} i_{qr}) \tag{5.22}$$

$$Q_r = \frac{3}{2}(V_{qr} i_{dr} - V_{dr} i_{qr}) \tag{5.23}$$

5.7.2 DFIG Shaft System Model

DFIG wind turbine shaft system has been formed in form of *an* integrated shaft or in form of two high speed and low speed axels which have been connected to each other by a gearbox. In the first model, the fix of total inertia of the system is specified as follows:

$$H_m = H_t + H_g \tag{5.24}$$

where H_g is generator inertia and H_t is turbine inertia. The electromechanical equation of DFIG wind turbine generator is shown as follows:

$$2H_m \frac{d\omega_m}{dt} = T_m + T_e - D_m\omega_m \tag{5.25}$$

which T_m shows the mechanical torque applied on turbine in pre-unit, T_e is the electromagnetic torque of machine and ω_m is the revolving speed in the per unit and m shows the mortal coefficient of shaft system. Due to the fact that there is a possibility of risk of rotating fluctuation of wind turbine and electrical quantities, in shaft system, mostly two models of shafts are used which one has a low speed and related to turbine and the other has a greater speed and related to generator and these two parts have been connected to each other by a gearbox. This type of model has been shown in Fig. 5.3.

Electromechanical equations of this system are obtained as follows:

$$2H_t \frac{d\omega_t}{dt} = T_m - D_t\omega_t - D_{tg}(\omega_t - \omega_r) - T_{tg} \tag{5.26}$$

$$2H_g \frac{d\omega_r}{dt} = T_{tg} - D_g\omega_r + D_{tg}(\omega_t - \omega_r) - T_e \tag{5.27}$$

$$\frac{dT_{tg}}{dt} = K_{tg}(\omega_t - \omega_r) \tag{5.28}$$

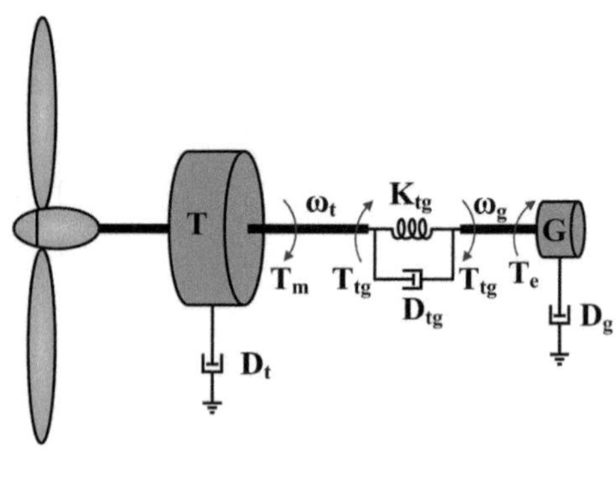

Fig. 5.3 DFIG wind turbine two part shaft system model

where ω_t and ω_r show the turbine and generator rotor speed in per unit and T_{tg} the local torque, D_{tg}, D_t and D_g show the turbine mortality coefficient, generator and gearbox and K_{tg} shows the strength of gearbox. Ignoring the mortality coefficient of turbine and generator, the function of transfer from electrical torque of the generator with the speed of ω_r rotor for DFIG wind turbine has been shown as follows. In Fig. 5.3, N_1/N_g specifies the ratio of gears of gearbox.

$$\frac{W_r}{T_e} = \frac{1}{2(H_t + H_g)s} \frac{2H_t s^2 + D_{tg}s + K_{tg}}{\frac{2H_tH_g}{H_t + H_g}s^2 + D_{tg}s + K_{tg}} \tag{5.29}$$

5.7.3 DFIG Wind Turbine Generator Control

The control system of a DFIG wind turbine usually is comprised of two parts: The part of electrical control and the part of mechanical control. The part of electrical control includes the control of converter in the rotor side and control of converter in the grid side.

The part of mechanical control includes the control of the angel of turbine blade. Figure 5.4 shows the system of simulated wind power plant in MATLAB software which is connected to a power grid.

Rotor Side Control Converter

The necessity of the rotor side control converter is to achieve the following objectives: [16]

1. Adjustment of DFIG rotor speed for maximum power absorption
2. Maintaining the output frequency and voltage of DFIG stator in fixed quantity
3. Control of DFIG reactive power.

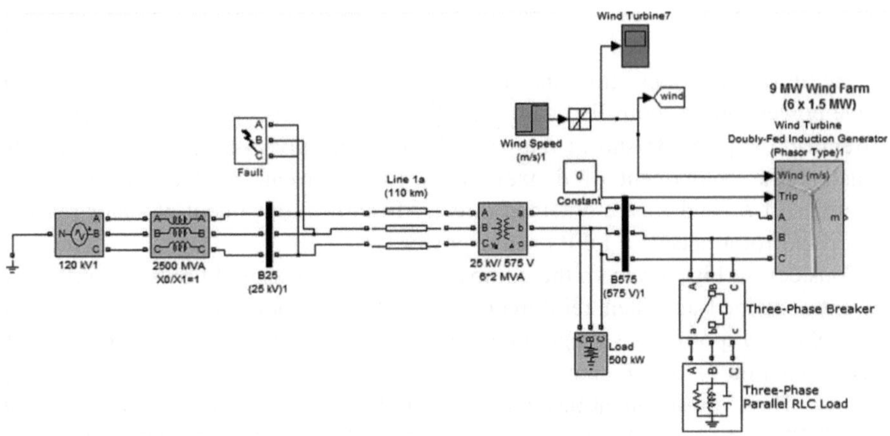

Fig. 5.4 Wind power plant system in MATLAB software

In DFIG, the mentioned objectives are usually obtained by adjusting the rotor current in the revolving frame of reference stator flux. In the rotating reference frame, the stator flux of d axel matches with the linking flux of stator Fig. 5.4 shows the design of RSC vector control [16].

In the d–q reference frame, there prevails a mathematical relation among the components of d and q axel which is shown as follows:

$$i_{qs} = \frac{-L_m i_{qr}}{L_s} \tag{5.30}$$

$$i_{ds} = \frac{L_m (i_{ms} - i_{dr})}{L_s} \tag{5.31}$$

$$T_e = \frac{-\frac{3}{2}\frac{p}{2} L_m^2 i_{ms} i_{qr}}{L_s} \tag{5.32}$$

$$Q_s = \frac{3}{2}\omega_s L_m^2 i_{ms}(i_{ms} - i_{dr})L_s \tag{5.33}$$

Such that

$$V_{dr} = r_r i_{dr} + \sigma L_r \frac{di_{dr}}{dt} - s\omega_s \sigma L_r i_{qr} \tag{5.34}$$

$$V_{qr} = r_r i_{qr} + \sigma L_r \frac{di_{qr}}{dt} - s\omega_s(\sigma L_r i_{dr} + \frac{L_m^2 i_{ms}}{L_s}) \tag{5.35}$$

$$i_{ms} = \frac{V_{qs} - r_s i_{qs}}{\omega_s L_m} \tag{5.36}$$

$$\sigma = 1 - \frac{L_m^2}{L_s L_r} \tag{5.37}$$

The above equations show that the DFIG (ω_r) rotor speed as a result of stator active power can be controlled with the adjustment of the component of q (I_{qr}) axel current and Eq. (5.23) shows that the Q_s stator reactive power is controlled by adjusting the component of d axel of I_{dr} rotor current. So that, the reference quantities of I_{qr} and I_{dr} are obtained directly from the adjustment of stator reactive power and rotor speed or DFIG stator power.

Considering Fig. 5.5 from the comparison of real measured flows and reference currents obtained and their reinforcement by PI controllers, the reference components of d and q of rotor voltage are obtained. In Fig. 5.5, V_{qr} and V_{dr} components only depend on I_{dr} and I_{qr} currents. So, they can be adjusted independently by I_{dr} and I_{qr}. Then, these quantities are compared with the simplified quantities in Eqs. (5.30) and (5.31). Then they are modulated by PWM for IGBT.

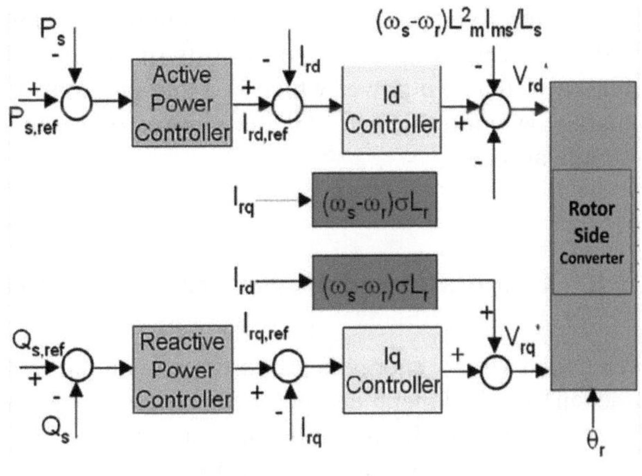

Fig. 5.5 RSC controlling circuit

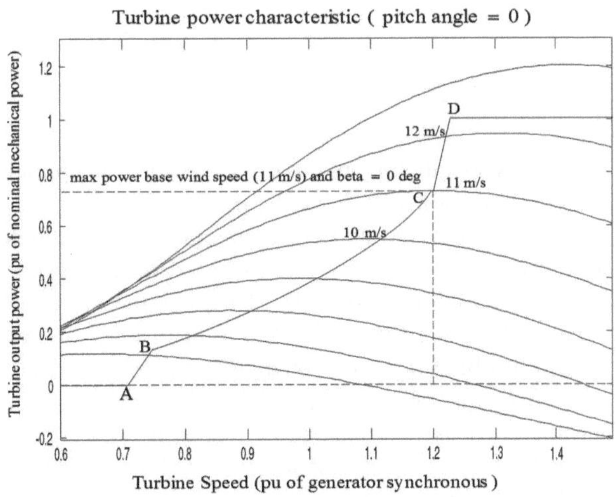

Fig. 5.6 Curve of feature of power absorption

The reactive power control can be used to achieve the coefficient of ideal power in the DFIG connection point. When WTG feeds a strong power system, the reference reactive power can be adjusted for simplicity in zero. For the control of active power, the curve of feature of power-speed is used which is known as the curve of feature of absorption. [Calculations for the wind turbine power].

This feature has been shown in Fig. 5.6 by ABCD curve. In this curve, first the turbine speed is measured. Then the mechanical power related to the same speed is used as the reference power for the control circle of active power P_s^*. The point between B and C is the geometrical place of the maximum of turbine power and power in D point and higher than that is equal to one per unit.

The reactive power causes changes in the systemized stabilized voltage and also indirectly causes the increase of costs of power system. Thus, the optimal distribution of reactive power which is a sub-problem of optimal load flow (OPF) and mainly is done through suitable control of reactive power resource is of great significance.

5.8 Calculation of Power Resulting from Wind

The wind kinetic energy is proportional with the square of its speed or when it hits the surface; its kinetic energy is converted into pressure (power) over that surface.

As we know, the multiplication of power in speed gives the power and as the wind power is proportional with the square of its speed, the wind power will be proportional with the cubic of its speed.

$$E_c = 0.5MV_w^2 \tag{5.38}$$

$$P = 0.5pAV_w^3 \tag{5.39}$$

As it is specified from the Eq. (5.38), the mechanical power of wind power has a direct connection with the cube power of wind speed. The high speeds of wind are not usually repeated and are non-economic to be able to put the accessible power and also the controlling systems based on using these types of speed, as a result, with the aerodynamic design of blades, the increase of the power in lieu of high speed will be prevented. The curve of Fig. 5.7 shows the changes of output power of wind turbine in lieu of changes of wind speed.

One of the most prevailing methods to stabilizing the output power of wind turbine is to use the change of angle of blades paces. Usually for modeling, the mutual effect of wind and turbine blades, a coefficient of power in Eq. (5.39) is used. Thus, the following relation is obtained:

$$P_m = 0.5pAV_w^3C(\theta, \lambda) \tag{5.40}$$

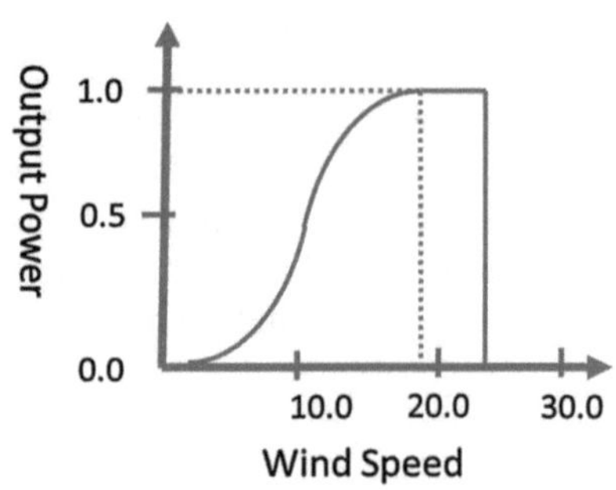

Fig. 5.7 Curve of output power in lieu of change of wind speed

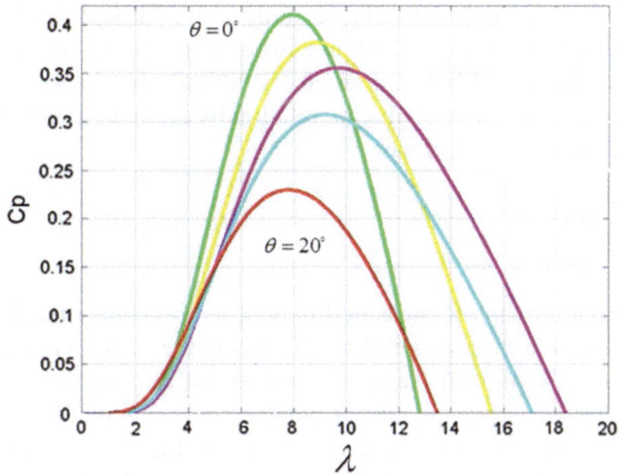

Fig. 5.8 Change of power coefficient in terms of λ variable

where $C(\theta, \lambda)$ is the power coefficient of θ angle of blades pace and λ is the ratio of speed of blade top to the wind speed. Figure 5.8 shows the method of change of power coefficient in terms of λ variable.

Theoretically, the maximum of wind power in a wind turbine which might be changed into mechanical energy, 16.27 is almost 3.59% of wind kinetic energy and this issue in 1972 was put by Betez. The aerodynamic deficiencies in practical machines and mechanical and electrical loses cause the real power to be less than the rate calculated through theory and in practice, the fixed coefficient of 593.0 cannot surpass 4.0.

5.9 Air Density Changes Proportional with the Height and Temperature Degree

Air density changes with the height changes and also with the changes of temperature degree.

Air density in regular conditions of 60 F is equal to 5.15 C near the sea 21.1 kg/m^3.

Air density in the height over the earth lowers. The ratio of air density in different heights to the air density in zero height of the sea surface is called the ratio of density in height.

Air density in different degrees of temperature changes such that in the degrees higher than 60 F, air density reduces. The ratio of air density in different temperature to the air density in 60 °F is called the ratio of density in temperature.

Changes of Wind Speed Proportional with the Height

Wind speed in the heights of about some thousands meters is basically related to the difference of atmosphere pressure. The closeness of wind to the earth makes the wind speed to be reduced noticeably. Though finding a precise relation between wind speed and its distance from the earth due to the situation of the earth from the viewpoint of ups and downs is difficult, but in this line, the experimental relations have been presented. One relation by Helman to obtain the speed in the inaccessible heights is as follows:

$$V_{h_2} = V_{h_1}\left(\frac{h_2}{h_1}\right)^a \tag{5.41}$$

where h_1 is the measured height, h_2 the concerned height, V_{h1} is the wind average speed in the measured height, V_{h2} is the wind average speed in the ideal height and a is the sign of Helman whose quantity depends on unsmooth statuses of heating classes and its quantity is obtained experimentally. For example, the quantity of a in the seashore areas is less than points far from the sea such that in the surrounding areas it is almost 1429.0 and in the forest and hills areas it is usually between 2.0 and 3.0.

As we see, the wind speed increases with the increase of height from the land surface. The 10 m height has been identified as a suitable height for the installation of average wind turbines in the world.

5.10 Load Control, Frequency and Voltage-Reactive Power in Diesel Generators

The main part of the system of frequency-load control is the system of adjustment of diesel engine speed which receives the frequency deviation or frequency or power and converts it into a suitable reaction to control the rate of input fuel into the diesel engine.

For adjusting the diesel engine speed to control load-frequency, three types of control can be employed which include:

- Speed manual changer
- By governor
- Through feedback.

For the control of voltage—power reactive, the simplest solution, is to control the Synchronous generator inciting which is point out at the following lines:

The current to incite the Synchronous generators is usually supplied by a DC generator which is the same shaft with turbine-generator.

The full control system is known as automatic voltage regulator (AVR). In the simplest way, the output voltage of Synchronous generator is sampled and after, it is compared with the reference voltage and the resulting error after reinforcement is applied on the inciting field of an Amplidyne which the Amplidyne also will incite and control the main field of generator.

In the large systems of wind power for the control of load-frequency and voltage-power control of reactive, it is possible to use the microprocessor methods to increase the easiness and accuracy of the job which is beyond the scope of the present discussion.

Traditionally, OPF issue is used in form of economic load flow. In this case, the object function is to minimize the fuel cost, but the functions of other objective can also use in this issue. In some cases, it is possible that a number of functions concurrently become optimal as the target functions which in this case, the issue takes the form of target multi-function. The formulation of the issue is as the following form:

$$\min/\max\left\{F(x,u) = \begin{bmatrix} f_1(x,u) \\ f_2(x,u) \\ \cdot \\ f_n(x,u) \end{bmatrix}\right\}; n = 1, 2, \ldots, N_{obj} \tag{5.42}$$

$$\text{s.t.}\begin{cases} h(s,u) \leq 0 \\ g(s,u) = 0 \end{cases} \tag{5.43}$$

where u is the vector of controlling or independent variables of grid including the generative powers apart from reference bass, generators voltages, transformer tap and injected reactive power by parallel elements and can be expressed in the following form.

$$u = (P_G, V_G, T, Q_{sh}) \tag{5.44}$$

$$x = (P_{Gref}, V, \delta, Q_G) \tag{5.45}$$

Also, x is the vector of status or dependent variables of the grid including load bass voltage, base voltage phase, generators reactive power, generative power in reference bass and can be expressed as follows:

$G(x, u)$ represents the equal constraints which indicates the equations of system load distribution. With the adjustment of u as the controlling variable in each stage and solving the non-linear equations of load flow, the corresponding x quantities are calculated.

$H(x, u)$ indicates the unequal constraints and includes the following cases:

Equal constraints:

$$P_{Gi} - P_{Di} = \sum_{j=1}^{N_{Buses}} V_i V_J Y_{ij}\, Cos(\theta_{ij} - \delta_i + \delta_j)]\ \forall i, j \in N_{Buses} \tag{5.46}$$

$$Q_{Gi} - Q_{Di} = \sum_{j=1}^{N_{Buses}} V_i V_J Y_{ij} \sin(\theta_{ij} - \delta_i + \delta_j)]\ \forall i, j \in N_{Load\ Buses} \tag{5.47}$$

$$\sum_{i}^{N_g} P_{Gi} = P_D + P_{Loss} \quad (5.48)$$

$$\sum_{i}^{N_g} Q_{Gi} = Q_D + Q_{Loss} \quad (5.49)$$

Unequal constraints:

(A) Capacity limits of generation units which includes the high and low limit of voltage rate, generation power of active and reactive. The output power of each generator should not be more than its nominal rate and also it should not be less than the quantity which is necessary for the durable use of steam boiler. So, the generation is limited such that it could place between the two predetermined limits of minimum and maximum:

$$V_{G_i}^{\min} \le V_{G_i} \le V_{G_i}^{\max} \quad (5.50)$$

$$P_{G_i}^{\min} \le P_{G_i} \le P_{G_i}^{\max} \quad (5.51)$$

$$\begin{gathered} Q_{G_i}^{\min} \le Q_{G_i} \le Q_{G_i}^{\max} \\ i = 1, 2, , \ldots, N_g \end{gathered} \quad (5.52)$$

(B) Compensational power limits by parallel elements

$$Q_{sh_i}^{\min} \le Q_{sh_i} \le Q_{sh_i}^{\max} \quad (5.53)$$

(C) Limit of tap transformer

$$t_i^{\min} \le t_{i_i} \le t_i^{\max} \quad (5.54)$$

(D) Limit of equipment use which includes the acceptable scope for the voltage and loading rate:

$$V_i^{\min} \le V_{i_i} \le V_i^{\max} \quad (5.55)$$

$$\begin{cases} P_{Li} \le P_{Li}^{\max} \\ Q_{Li} \le Q_{Li}^{\max} \end{cases} ; i = 1, 2, 3, \ldots, N_{Branches} \quad (5.56)$$

In this discussion, the optimization of reactive power is concerned which the objectives can be stated as follows [17–19].

5.10.1 Minimizing the Real Losses

One of important goals of optimal use of reactive power is to minimize the losses of real power in the transfer grid which in this chapter has been considered as the function of the goal of optimization problem. The rate of losses in the transfer grid can be calculated as follows [18, 20, 21]:

$$P_{Loss} = \sum_{k=1}^{N_L} g_k [V_i^2 + V_j^2 - 2V_i V_j \cos(\delta_i - \delta_j)] \tag{5.57}$$

where V_i and V_j are the primary and last bus bar voltage, θ_{ij} is the angle difference between bus bar i and j and g_{ij} is the conductive media between bus bar i and j.

5.10.2 Durability Indicator Calculation

There are many indicators for the analysis of the voltage improvement in power systems including *P–V* curve analysis, *Q–V* curve analysis and *L*-index indicator. In this research, the *L*-index indicator is used to analysis the durability and voltage sensitivity. For this purpose, one system of n bus bar is divided into two groups of generative and load bus bars. The bus bars 1 to g are the generative bus bars and bus bars $g + 1$ are load bus bars.

$$\begin{bmatrix} I_g \\ I_l \end{bmatrix} = \begin{bmatrix} Y_{gg} & Y_{gl} \\ Y_{lg} & Y_{ll} \end{bmatrix} \begin{bmatrix} V_g \\ V_l \end{bmatrix} \tag{5.58}$$

With regard to the admittance matrix, the L indicator for load bus bars is obtained from the following relation.

$$L_j = \left| 1 - \sum_{i=1}^{N_g} F_{ij} \frac{V_i}{V_j} \right|, \quad j = N_g + 1, \ldots, n \tag{5.59}$$

$[F_{ij}]$ can be calculated with regard to the admittance matrix in the following form.

$$[F_{ij}] = -[Y_{LL}]^{-1}[Y_{LG}] \tag{5.60}$$

L is an index between zero and one. To the extent that this index is closer to one, to the same extent it indicates the instability and disruption of voltage and the more this index is closer to zero, the more durability it has.

$$L = \max(L_j), \quad j \in \alpha_j \tag{5.61}$$

5.10.3 Voltage Profile Indicator

As voltage is one of the most important standards from the viewpoint of power quality in presenting services by the electricity companies, so that in the distribution networks, great attention has been made to study the impact of units on voltage. The optimization of voltage profile in power systems is of great significance.

In the power grids, there is an effort to minimize the voltage profile. In V_i^{ref} calculations, 1 per-unit is considered.

$$\Delta V_L = \sum_{i-1}^{N_{PQ}} \left| V_i - V_i^{ref} \right| \tag{5.62}$$

5.11 IEEE Standard 30-Bus Test System

The single line 30-bus bar standard system is shown in Fig. 5.9. As it is obvious in Fig. 5.9 it has 6 generator buses bar. The specifications of these generator units are shown in Table 5.2. As it is clear, the limits of active power generation and cost coefficients and coefficients of pollution rate of each of the units have been pointed out.

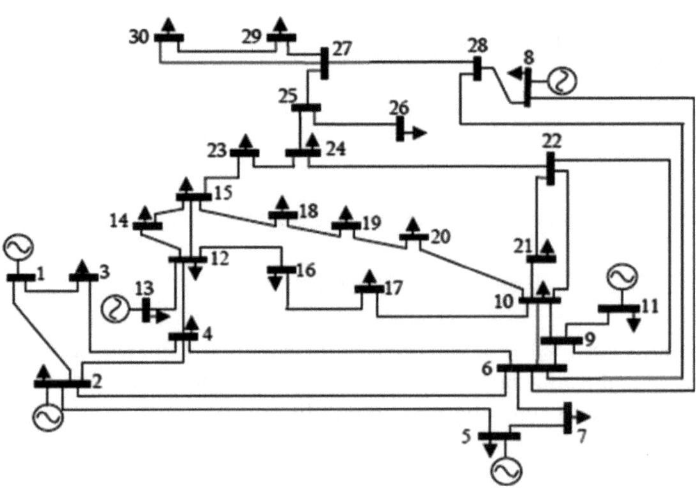

Fig. 5.9 IEEE standard 30-bus test system

Table 5.2 Specifications of the system generator units [22]

No.	λ	ξ	γ	β	α	c	b	a	P_{Gmin}	P_{Gmax}
P_{G1}	2.857	2e-4	0.0649	−0.05543	0.04091	100	200	10	5	150
P_{G2}	3.333	5e-4	0.05638	−0.06047	0.02543	120	150	10	5	150
P_{G5}	8	e-6	0.04586	−0.05094	0.04258	40	180	20	5	150
P_{G8}	2	2e-3	0.0338	−0.0355	0.05326	60	100	10	5	150
P_{G11}	8	e-6	0.04586	−0.05094	0.04258	40	180	20	5	150
P_{G13}	6.667	e-5	0.0515	−0.05555	0.06131	100	150	10	5	150

Table 5.3 N.R. results

Method	N.R.
Loss	4.6174
Voltage deviation	0.6783
Voltage stability	0.1171

The first method is to use the algorithm of N.R. whose results can be observed in Table 5.3.

The next step is to calculate the system calculations with the smart algorithm of the second generation of genetics with non-dominated sorting.

In this scenario, it is assumed that no wind power plant is placed in the grid and we have no limitation on this case and the problem is defined inform of a three-target problem to reduce the losses, voltage profile and voltage durability. It is clear that the reading of the problem in this state has three dimensions. With the implementation of algorithm of the best reading in the tridimensional space, they have formed a curve which totally none of the spots have a priority over the other one and the selection of the best response is done merely based on the indicators of losses, voltage profile and voltage durability and their preference is made with the employment of algorithm system.

Here it is worth pointing that from the viewpoint of Pareto superiority, the extreme points of Pareto front in fact is the optimal solution to the optimization issue, when each of the objectives are studied exclusively.

So, considering this issue, it is possible to study the quality of responses resulting from the suggested algorithm can be studied during the analysis of responses.

Figure 5.10 shows the set of optimal Pareto. After optimization, it is observed that the best reading of the last repetition of Algorithm has N-member which none of them has priority over the other one. The lack of existing continuity is due to the disrupted nature of optimization process. The resulting optimal point based on algorithm output is a suggestion. The best responses which optimize each of the three objectives in this state exclusively are displayed in Table 5.4.

The next step is to calculate the system with the smart algorithm of the second generation of genetics with the non-dominated sorting with the assumption of the presence of wind power plants with 10 MW active powers with power coefficient of 0.9.

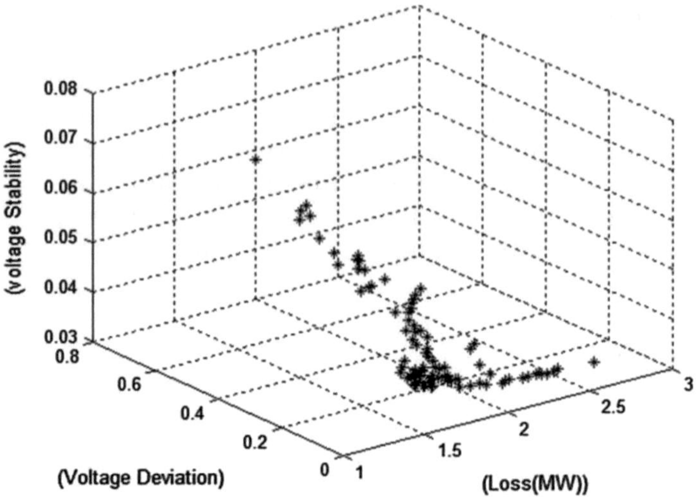

Fig. 5.10 Curves of the first Pareto front without wind turbine

Table 5.4 Results without wind turbine

Method	Best loss	Best voltage deviation	Best voltage stability
Loss	1.5198	4.1485	5.3467
Voltage deviation	0.5392	5.6788e–06	0.5073
Voltage stability	0.0486	0.0589	0.301

In this scenario, it is assumed that one power plant has been placed in the grid and the capacity generation limit of 10 MW active powers with power coefficient of 0.9 and the goal is to improve the reactive power. The problem is defined in for of a three-target problem to reduce the losses, voltage profile and voltage durability. It is clear that the reading of the problem in this state is three dimensional.

With the implementation of the best algorithm of reading in tridimensional space, they have formed a curve which totally none of the points is preferred to the others and the selection of the most suitable response is merely done based on the indicators of loss indictors, voltage profile and voltage durability and their preference is done by employing the algorithm system.

Figure 5.11 shows the optimal Pareto of this system. After optimization, it is observed that the best reading of the last repetition of Algorithm has N member and none of them has priority over the others. The resulting optimal point based on algorithm output is a proposal. The best response which optimizes exclusively each of the three objectives put forth in this state is displayed in Table 5.5.

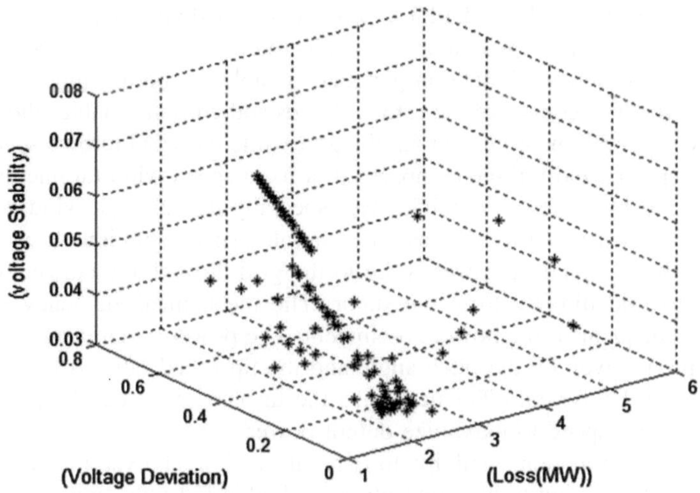

Fig. 5.11 Curves of the first Pareto front with wind turbine

Table 5.5 Results with wind turbine

Method	Best loss	Best voltage deviation	Best voltage stability
Loss	1.4503	1.7768	2.4338
Voltage deviation	0.5167	9.7316e–05	0.3332
Voltage stability	0.0709	0.0447	0.0305

Comparing the results, it is clearly learned that the presence of wind power plant is effective in improving the reactive power in the grid.

5.12 Conclusion

The role of wind power plant in today's environmental and energy dependable development is so crucial. Lots of studying and researches have been performed on wind power technologies and numbers of wind farm have been utilized around the world. The performance of overall wind power plant depends on the subsystem such as reactive power compensation and energy storage to maintain stability. However, with the increasing capacity of the wind power plant the cost and benefit of these subsystems became unfeasible. By increasing the wind farm capacity the cost for reactive power and energy storage increases. In future works, it could be feasible to test wind power plant using combined capacitor and reactive power compensation which could low the cost. The overall development of these subsystems in wind power plant depends on their cost.

It is also concluded that a DFIG is a wound-rotor doubly-fed electric machine (similar to a Synchronous generator), and as its rotor circuit is controlled by a power electronics converter, the induction generator is able to control import and export reactive power. The control of the rotor voltages and currents enables the induction machine to remain Synchronized with the grid while the wind turbine speed varies. A variable speed wind turbine compared to a fixed speed wind turbine utilizes the available wind resource more efficiently, especially during light wind conditions. The converter cost is not as high as other variable speed solutions because only a fraction of the Mechanical Power is fed to the grid through the converter, the rest being fed to grid directly from the stator. The mechanical efficiency in a wind turbine is dependent of the power co-efficient. The power co-efficient of a rotating wind turbine is given by the pitch angle and the tip speed ratio. Adjustable speed will improved the system efficiency since the turbine speed can be adjusted as a function of wind speed to maximize output power.

Performing simulations and by looking through real experiments it can be concluded that the wind turbine unit utilizing double fed induction generator is an important and effective tool from the voltage regulation point of view. So it plays a significant role in supplying the reactive power of the network.

References

1. A. Berizzi, C. Bovo, V. Ilea, M. Merlo, A. Miotti, F. Zanellini, Decentralized Reactive Power Control of Wind Power Plants, 2nd IEEE Energycon Conference & Exhibition, Future Energy Grids and Systems Symp, 2012.
2. D.F. Opila, A.M. Zeynu, I.A. Hiskens, Wind Farm Reactive Support and Voltage Control, IREP Symposium - Bulk Power System Dynamics and Control - VIII (IREP), Buzios, RJ, Brazil, August 1–6, 2010.
3. M.A. Nielsen, Power Quality and Grid Connection of Wind Turbines, Proc. of Solar '97 - Australian and New Zealand Solar Energy Society, Paper 154: Nielsen.
4. R. Jacobson, B. Gregory, Wind Power Quality Test for Comparison of Power Quality Standards, National Wind Technology Center, June 1999.
5. E.H. Camm, et al., Reactive Power Compensation for Wind Power Plants, IEEE PES Wind Plant Collector System Design Working Group, 2009.
6. A. Zare Bargabadi, Design of Power Oscillation Damping Controller Using Hybrid Fuzzy Logic and Computational Intelligence Base on Power System Stabilizer (PSS) and Supplementary Controller of Doubly Fed Induction Generator (DFIG) Wind Turbine, M.Sc. Thesis, Karaj Branch, Islamic Azad University, Summer 2014.
7. J.M. Garcia, M. Babazadeh, Control of Large Scale Wind Power Plants, IEEE Power and Energy Society General Meeting, 2012.
8. H. Modirzare, Analysis Impact of DG at Power System Planning Aspect of Reactive Power for Improvement Power Quality, M.Sc. Thesis, Karaj Branch, Islamic Azad University, Summer 2013.
9. H. Modirzare, P. Ramezanpour, R. Effatnejad, Optimal Allocation of DG Units and Var Compensators Suspect to GA Based Reactive Power for Power Losses Decreasing and Voltage Stability and Profile Improvements, International Journal on Technical and Physical Problems of Engineering (IJTPE), issue 18, vol. 6, no. 1, pp. 125–130, March 2014.

10. M. Akhlaghi, Improving Reliability Indices in HL2 System Through Reactive Power Management, M.Sc. Thesis, Karaj Branch, Islamic Azad University, February 2014.
11. M. Akhlaghi, P. Ramezanpour, R. Effatnejad, Weak Points Identification of HL2 Systems Using Contingency Analysis, The International Conference in New Research of Electrical Engineering and Computer Science, September 2015.
12. V. Karunakaran, R. Karthikeyan, Reactive Power Management for Wind Electric Generator, International Journal of Scientific & Engineering Research, vol. 2, issue 5, May 2011.
13. X. Chen, Reactive Power Compensation and Energy Storage in Wind Power Plant, A Major Qualifying Project Report Submitted to the Faculty of Worcester Polytechnic Institute in Partial Fulfillment of Requirements for the Degree of Bachelor of Science, January 2012.
14. H. Akagi, H. Sato., Control and Performance of a Doubly-Fed Induction Machine Intended for a Flywheel Energy Storage System, IEEE Trans. on Power Electronics, vol. 17, no. 1, pp. 109–116, 2002.
15. V. Akhmatov, Induction Generators for Wind Power, Multi-Science Publishing, Brentwood, 2007, ISBN 10: 0906522404.
16. Y. Mishra, S. Mishra, M. Tripathy, N. Senroy, Z. Dong, Improving Stability of a DFIG-Based Wind Power System with Tuned Damping Controller, IEEE Trans. on Energy Conversion, vol. 24, no. 3, pp. 650–660, 2009.
17. Z.X. Liang, J.D. Glover, A Zoom Feature for a Programming Solution to Economic Dispatch Including Transmission Losses, IEEE Trans. Power Syst., 7(2):544–50, 1992.
18. H. Aliyari, R. Effatnejad, A. Areyaei, Economic Load Dispatch with the Proposed GA Algorithm for Large Scale System, Journal of Energy and Natural Resources, 3(1):1–5, 2014.
19. R. Effatnejad, H. Aliyari, H. Tadayyoni, A. Abdollahshirazi, Novel Optimization Based on the Ant Colony for Economic Dispatch, International Journal on Technical and Physical Problems of Engineering (IJTPE), issue 15, vol. 5, no. 2, pp. 75–80, June 2013.
20. H. Shayeghi, A. Ghasemi, Application of MOPSO for Economic Load Dispatch Solution with Transmission Losses, International Journal on Technical and Physical Problems of Engineering (IJTPE), issue 10, vol. 4, no. 1, pp. 27–34, March 2012.
21. A. Khorsandi, S.H. Hosseinian, A. Ghazanfari, Modified Artificial Bee Colony Algorithm Based on Fuzzy Multi-Objective Technique for Optimal Power Flow Problem, Electric Power Systems Research, 95: 206–213, 2013.
22. H. Aliyari, Application of Meta-heuristic Algorithms for Multi-Objective Optimization of Reactive Power, Power Losses and Cost Function in Power System, M.Sc. Thesis, Science and Research Alborz Branch, Islamic Azad University, August 2014.

Chapter 6
Reactive Power Control and Voltage Stability in Power Systems

Mariana Iorgulescu and Doru Ursu

Abstract Reactive power control is sometimes the best way to enhance power quality and voltage stability. In the first part of chapter we describe the reactive power flow impact in the system starting from the definitions of power components and presentation of the electrical equipment that produces or absorbs the reactive power. Then we present the reactive power control and the relations between voltage stability and reactive power. The third part of chapter contains reactive power control methods for voltage stability. In the end of chapter we present the management of voltage control based on case studies.

6.1 Introduction

It is known that the energy system is a great system, complex and continuous change. The electro-energetically system is a system made up of an ensemble of interconnected sub-systems. These sub-systems consist of generators, transformers, electric lines and many types of consumers, those that actually constitute the load of this system [1, 2]. Every moment of time the system operates in and out of its elements causes changes in system parameters. The quality of electricity delivered to consumers is a very important requirement in the power system operation. For every moment in time, for a system with a stable functioning, there must be a very good balance between the quantity of energy produced by hydro-generators, turbo-generators, nuclear reactors and the energy consumed by the loads at a certain moment. For an optimal functioning of the system and for reducing the energy losses in the system, it is necessary to lower the reactive power circulation [2].

M. Iorgulescu (✉)
Faculty of Electronics, Communications and Computers, University of Pitesti,
Pitesti, Romania
e-mail: marianaiorgulescu@yahoo.com

D. Ursu
CEZ Distribute, Craiova, Romania
e-mail: doru.ursu@cez.ro

N. Mahdavi Tabatabaei et al. (eds.), *Reactive Power Control in AC Power Systems*,
Power Systems, DOI 10.1007/978-3-319-51118-4_6

Due to a permanent need for consuming electric energy, on the one hand, and the rise in the prices of the classic fuels, on the other, new sources of renewable energy have been developed [3, 4]. Their placement has led to the appearance of the problems related to the instability of the system and of the voltage in its nodes. The voltage stability is intimately related to the reactive power control, achieved by means of series capacitors, synchronous condensers, and the modern static compensator.

Another extremely important problem in the energetic system is the quality of the delivered electric energy as it conditions the proper functioning of the electro-energetic system and of its consumers.

Power quality parameters are:

- Voltage supply;
- Frequency;
- Total harmonic distortion of the voltage wave;
- Symmetry phase voltage system;
- Continuity of supply.

Voltage stability and frequency stability have lately become became the two important electric power quality parameters in the functioning of the power system. Equally important is the knowledge of how the power system elements that can cause system instability function. The voltage is a parameter of the electric energy which is different in the energy system nodes. This parameter depends on the values of the impedance of the elements which are in the network or exit the system and, implicitly, on the voltage drops on these impedances. Maintaining the imposed voltage level in a network node constitutes a problem of the area corresponding to that node and can be achieved by different means of adjusting the voltage. The ways the voltage can be adjusted in the power system is the focus of the specialists in the field of electric energy. The voltage stability within the nodes of the system is very important in coordinating and operating the system. Problems related to the voltage instability led to the fall of energy systems in countries like Japan in July 1987, on August 14th 2003 in US and Canada, on September 23rd 2003 in eastern Denmark and southern Sweden and days later in Italy and Central Europe following the cascade outage [5].

Considering the possibility of such blackouts appearing in energy systems, artificial intelligence based systems have been developed and tested, aimed at helping the adjustment of voltage and, implicitly, the system's stable functioning [6]. Such an example of information system was called STABTEN and it was developed and tested by using the CIGRES 32 scheme [7].

The method of using such a device for controlling voltage in an energy system pilot, electronic device which is called ASVR—Automatic Secondary Voltage Regulation—as it is presented in the final of the chapter, is a successfully used method for adjusting voltage by means of optimal reactive power control.

6.2 Impact of Reactive Power Flow in Power System

In order to study the reactive power impact in the system, it is important to start with the term—power—in the electrical system it refers to the energy-related quantities flowing in the transport and distribution network.

The instantaneous power is the product of voltage and current with the following equation:

$$p(t) = u(t)i(t) \tag{6.1}$$

Voltage and current can be in phase or not. When these are not in phase there are two components of power:

- active power P, that is measured in Watts

$$P = UI \cos \varphi \,[\mathrm{W}] \tag{6.2}$$

- reactive power Q (it is known as imaginary of apparent power), that is measured in VAr

$$Q = UI \sin \varphi \,[\mathrm{VAr}] \tag{6.3}$$

In the above equations, I and U are rms values of current and voltage, and φ is the phase angle by which the current out phasing the voltage.

The combination of active power with reactive is apparent power S, measured in VA.

$$S = P + jQ[\mathrm{VA}] \tag{6.4}$$

Reactive power—Q represents a part of apparent power—S. Reactive power is in opposition with active power—P. Reactive power is necessary to maintain voltage and to distribute active power through transmission lines. In this way different loads that use reactive power to convert the received power in mechanical, illumination and others will be in use. In case the presence of reactive power is under the acceptable level the voltage cracks down. The system can become unstable because the loads do not receive necessary power through the transmission line.

It is important to study reactive power flow in system because some equipment depends on these few factors can absorb or produce this type of power. Depending on the phase angle between current and voltage, the electrical equipment will consume an amount of reactive power.

In power system there are types of electrical equipment of the system that producing or absorbing reactive power, such as [10]:

- **Synchronous machines** are electrical machines with three operating conditions: motor, generator and condenser. In generator mode the electrical machines depending on the excitation level can generate or absorb reactive power together with active power. When the machine is overexcited it supplies reactive power, and when is under excited it absorbs reactive power.
- **Loads** are in a great number in the power system. The number of loads, their characteristics or power are dependent on the time of day, season or weather. In rated state every load absorbs active power and reactive power both leading to voltage variations. Some loads operate with low power factor (can be indeed 0.6), with important losses of active power in transmission network. In this case is mandatory for industrial consumers to improve the load power factor by natural means first and specialized devices second.
- **Transmission lines** are parts of the transport and distribution system. Across the lines the voltage and current parameters change in every moment, proving the series and shunt parameters. Based on equivalent circuit it is settled that a transmission line is characterized by the following circuit parameters:

  - longitudinal or series parameters: resistance R and reactance X;
  - and transversal or shunt parameters: conductance G and susceptance B.

Symbols of the parameters indicate their per-kilometer values.

Depending on the load current lines absorb or supply reactive power. In case of loads below the natural load, the lines produce net reactive power, if the opposite situation applies, at loads above natural load, the lines absorb reactive power.

The reactive power is given by [1]:

$$Q[\text{VAr}] = U^2[\text{V}]B[\text{Siemens/km}] \tag{6.5}$$

All parameters of the line are significantly different functions of the conductor size, material, spacing, height above ground, and temperature.

- The distribution network is built sometimes in *underground cables*. These cables are always loaded below their natural loads. The series parameters—especially reactance is lower but shunt parameter—capacitance is higher, affects by charging reactive power under all operating conditions. Based on these, it is very important to design the line with the proper length in order to respect the cable thermal capability.

  Reactive power flow in electrical network has a negative impact on the power system. In practice almost always the specialists work to reduce the level of reactive power in order to improve the system efficiency.

  Effects of the reactive power flow in network are:
- active power losses increase. In reactive power presence this losses will be:

$$\Delta P = \Delta P_a + \Delta P_r \tag{6.6}$$

- equipment oversize that increase the installation's cost. It is known that the equipment's proportion is according to:

$$S_n = P_n / \cos \varphi_n \tag{6.7}$$

Operating mode with:

$$\cos \varphi < \cos \varphi_n \tag{6.8}$$

leads to

$$S_{proportion} > S_n \tag{6.9}$$

- decrease of supporting capacity of electrical networks:

$$P = S \cos \varphi < P_n = S_n \cos \varphi_n \tag{6.10}$$

- increase of lost voltage:

$$\Delta U = U_1 - U_2 = RI \cos + XI \sin = \Delta U_a + \Delta U_r \tag{6.11}$$

6.3 Reactive Power Control in Electrical Networks

6.3.1 Reasons and Nature of Voltage Variations in Electrical Networks

Reactive power control is fulfilled according to reactive power variations. In steady state of electrical network the voltage is a parameter variable in time and space. Voltage variation in space is given by the different voltage nodes induced by the drops in voltage. These drops in voltage are brought by active and reactive power flow in electrical networks and power transformers.

Considering a power transformer or an electrical network with the following single phase equivalent diagram, Fig. 6.1—longitudinal impedance with equation [8, 9]:

$$Z = R + jX \tag{6.12}$$

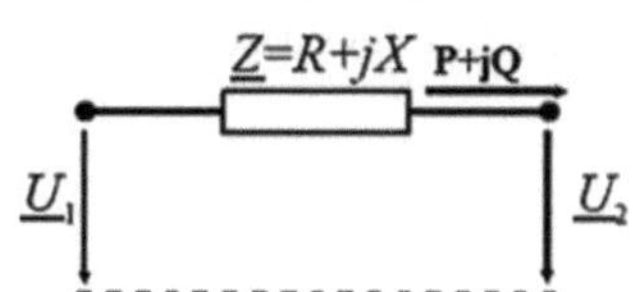

Fig. 6.1 Single phase equivalent circuit

It establishes the complex drop in voltage with two components:

- a longitudinal component ΔU, regularly named drop in voltage, with the equation:

$$\Delta U = U_1 - U_2 \cong \frac{PR + QX}{U_2} \cong \frac{PR + QX}{U_n} \tag{6.13}$$

- a transversal component δU, determination of angle θ value between voltage in the extreme of network, with the equation:

$$\delta U = \frac{PX - QR}{U_2} \cong \frac{PX - QR}{U_n} \tag{6.14}$$

Practical longitudinal impedances of transmission lines are characterized by the small value of resistance comparative with inductive reactance $R \ll X$.

Based on the phase diagram of drop in voltage, Fig. 6.2, and equations of drop in voltage, approximate equations of ΔU are

$$\Delta U \cong \frac{QX}{U_n} \tag{6.15}$$

$$\theta = \arcsin \frac{\delta U}{U_1} \cong \arcsin \frac{PX}{U_n^2} \tag{6.16}$$

These equations point out an important characteristic of the system, the level of voltage is determined by the reactive power flow and the θ value by the active power flow.

Time variations of voltage in network nodes are tie-in with the voltage drops variations in time. These variations have the following causes:

- active and reactive power flow variability in time as a consequence of variation of absorbed power by the consumers and produced power by the power plant, also.

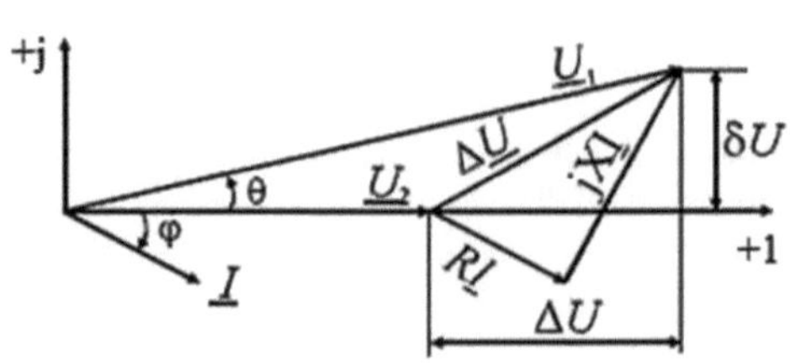

Fig. 6.2 Phasor diagram of drop in voltage on longitudinal impedance

– modification of the network topology, longitudinal impedance and transversal admittance. This change is imposed by the necessity for optimizing the technical indicators of loaded transmission lines.

As a result, in one node of the network the voltage is variable in time $U(t)$. Voltage variation in time in the node is named voltage deviation, measured as per cent at one moment in time, with the equation:

$$\Delta U(t) = \frac{U(t) - U_n}{U_n} 100[\%] \tag{6.17}$$

These voltage variations can be hard growing or fast.

Hard growing variations appear as a consequence of a flawed control or overloaded electric network. This variation can be automatically controlled through regulators. The regulators send an impulse to beakers in order to introduce the compensation equipment into the circuit, such as synchronous compensators, bank of capacitors or transversal coil.

Fast growing of reactive power can be controlled automatically trough regulators with static commutation (GTO) in order to introduce quickly compensation equipment in the circuit.

Synchronous machines, bank of capacitors, GTO and various types of other equipment are used to maintain voltages throughout the transmission system. Injecting reactive power in the system increases voltages, and absorbing reactive power decreases voltages.

6.3.2 Reactive Power Control Methods for Voltage Stability

The voltage is one parameter that fluctuates in time due to the modification of the total power absorbed by the receptor, together with operating specification of the power system (revisions, failures, etc.). Also, the voltage fluctuates in dimension because of lost voltage in network leads to different values of voltage in different nodes.

Level of voltage is indissolubly linked to a stable operating of the power system. Also the voltage adjustment is interrelated with the reactive power control possibilities. Reactive power control methods for voltage stability are presented in the next section.

6.4 Voltage Stability in Power System

6.4.1 Voltage Control by Reactive Power Flow Adjustment

The voltage adjustment by reactive power flow control can be continuous, used like a primary means of voltage regulation, or discrete used like a secondary means of adjustment.

The principle of voltage regulation by reactive power flow adjustment can be illustrated in Fig. 6.3.

In Fig. 6.3, k is the installation of reactive power compensation linked to consumer bus, producing Q_k reactive power. In this case reactive power transmitted on the line decreases to (Q_2–Q_k) value. As a consequence, dropping voltage decreases and the voltage value increases.

In another case, with the k installation absorbing reactive power, having inductive consumer operating, dropping voltage increases and U_2 decreases. The conclusion of this paragraph is that by modifying the reactive power, we can control voltage on consumer bus.

This type of control is based on the reactive power sources of the power system (generators, synchronous compensators, capacitors). The same role is played by the coil mounted in the high voltage electrical network. These are consumers of reactive power.

In the next section there are presented characteristics of important voltage control means: *generators, condensers and inductors.*

Generators: The main function of electric-power generators is to convert different type of energy into electric power. The generators have significant control on their terminal voltage and reactive power output. This can produce or absorb reactive power depending on the magnetizing current value.

The increasing of the magnetic field in the synchronous machine implies the raising of generator's terminal voltage in order to produced reactive power. The magnetic field increasing requires current increasing in the rotating field winding.

Absorption of reactive power is limited by the magnetic-flux design in the stator, which leads to over-heating of the stator-end iron. It is known that in every electrical machine there are the core-end heating limits [10].

Synchronous Compensators are synchronous motors used sometimes to provide dynamic voltage support to the power system as they provide mechanical

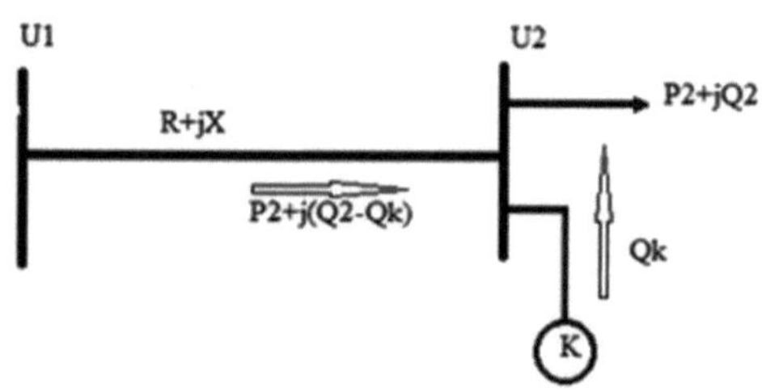

Fig. 6.3 Reactive power compensation on consumer bus

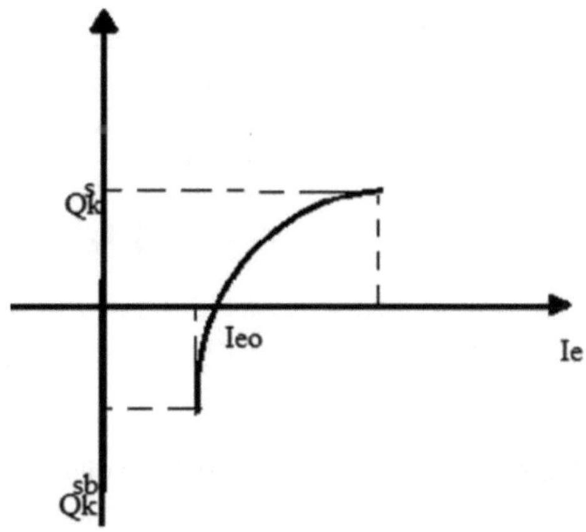

Fig. 6.4 Reactive power variation versus magnetizing current

power to their load. In some power plants the hydro units are designed to permit the generator to function without its mechanical power source simply to provide the reactive power capability to the power system [10]. These hydro units work as a compensators when the active power generation is not necessary, Fig. 6.4.

For a magnetizing current value $I_{e0,produced}$ reactive power is zero. In case $I_e > I_{e0}$ the synchronous condenser operates overexcited, with power factor capacitive conveying in electric network reactive power. When $I_e < I_{e0}$ the synchronous condenser operates under excited, with power factor inductive absorbing from electric network reactive power. Proportion of maximum reactive power absorbed Q_k^{sb} and produced power Q_k^s it is a very important characteristic of the condenser operation.

Advantages of using synchronous condenser are:

- operating possibilities as reactive power generator or consumer;
- continuous regulation of produced or absorbed reactive power by magnetization current;
- contribution to the system stability, with auto-regulation effect, once the decrease of voltage leads to an increase in reactive power supplied.

A disadvantage can be the limited applicability of synchronous condensers on a large scale. The disadvantage is due to the consumption of active power around 3% of machine's reactive power rating, or operating expenses.

Capacitors are passive devices that generate reactive power, with some advantages:

- significant real-power losses around 2–3% from the rated power;
- low operating expense;
- possibility to space out payment of investment by gradually developing the capacitor bank with new elements.

Capacitor banks are composed of individual capacitor connected in series and/or parallel in order to obtain the desired capacitor-bank voltage and capacity rating

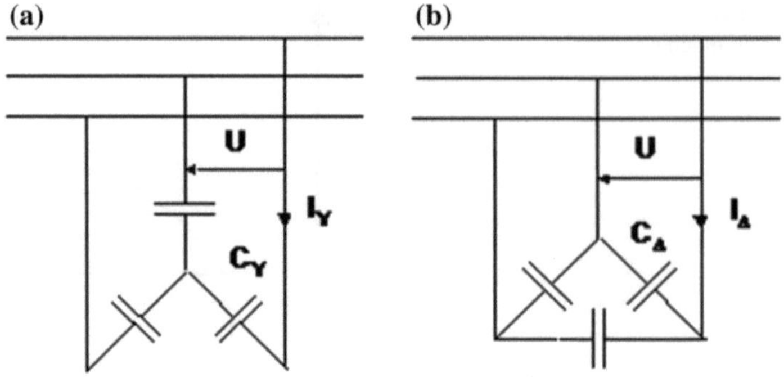

Fig. 6.5 **a** Star connection of bank-capacitor, **b** triangle connection

[10], Fig. 6.5 [13]. Banks capacitor are discrete devices but they are often configured with several steps to provide a limited amount of variable control, and are spread in small power [10].

Produced reactive power is given by

$$Q_k = m\omega CU^2\ [\text{MVar}] \tag{6.18}$$

where

- U—network voltage, in kV;
- C—bank capacity, in F;
- m—coefficient with value 1 in star connection and *3* in triangle connection.

The output of capacitors is proportional to U^2. In fact this effect is bad in voltage control and in operating conditions of the system, also. To avoid this, the bank capacitors should be input by autotransformer.

Inductors are passive devices that absorb reactive power. The inductors are shunt connected to high voltage transmission lines or to the tertiary autotransformer's winding with purpose to absorb capacitive power generate by these lines in low load operation. In this case there were reduced possibilities to have in the system overvoltage produced by capacitive currents flow.

Inductors are built in three phases or formed with three single phases. Frequently the principle of reactive power control consists in modifying the number of inductors in shunt connection.

Static VAR Compensators (SVCs)

By combining the banks capacitor with inductors we will obtain static VAR compensators. These compensators can be designed to absorb or produce reactive power. These devices have an operating principle like a synchronous condenser, the reactive power can be controlled continuously or step by step. In order to adjust in the continuous mode the power of installation is used for inductor's control a bias

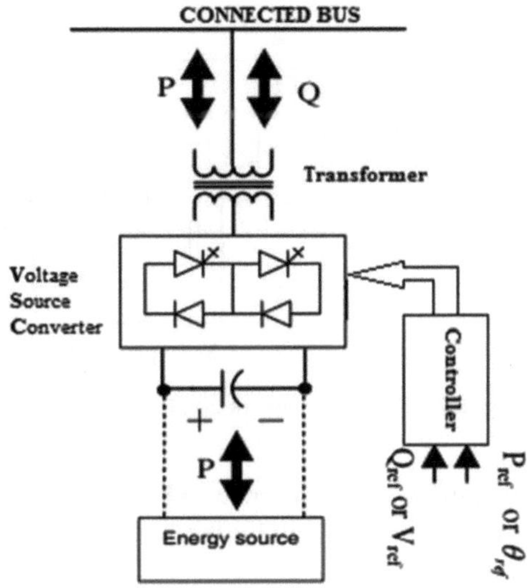

Fig. 6.6 STATCOM schematic diagram [16]

winding or thyristors control scheme. This control is very useful for the receivers, whose operations produce power surges in the electrical network.

Static Synchronous Compensators (STATCOMs)

The STATCOM (STATic synchronous COMpensator) is a shunt-connected reactive compensation equipment which generates and absorbs reactive power. Its output can be varied so as to maintain control of the power system specific parameters [14–16, 19].

The STATCOM belongs to the family of devices known as flexible AC transmission system (FACTS). The schematic diagram is presented in Fig. 6.6.

STATCOMs give the power system the following advantages:

- using capacitors and inductors combined with fast switches, finally, STATCOM uses power electronics to synthesize the reactive power output [10].
- output capability is commonly symmetrical, providing as much capability for production as absorption [10].
- controls are designed to provide a fast and effective voltage control
- STATCOMs capacity is not affected by degraded voltage.
- STATCOMs are current limited so their MVAr capability responds linearly to voltage, increasing STATCOMs' usefulness in preventing voltage collapse [10].

The voltage control in the medium voltage network can be made using a system called by its developers UPQC (Unified Power Quality Conditioner), Fig. 6.7 [17]. It has the role of compensating the voltage variations and of ensuring the user's clamps receive a set voltage, practically constant. This installation was tested within the medium voltage network which a photo-voltaic plant was connected to.

The installation has the role of correcting the i_r electric current loop absorbed by the user as the parallel converter introduces a i_f current with such a synthesized form

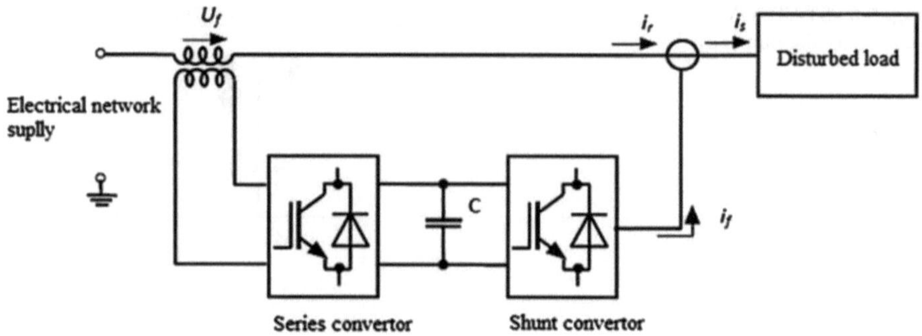

Fig. 6.7 Electrical network conditioner UPQC [17]

that the i_r electric current to be sinusoidal. The U_f voltage introduced in series with the network voltage by the serial converter maintains a constant value of the user supply voltage.

6.4.2 Voltage Control by Power Network Parameters Adjustment

This process of voltage adjustment involves changes in electrical network parameters, either by connecting or disconnecting circuits operating in parallel, or by compensating the inductive reactance power lines.

Some parts of the electrical network are built in parallel connection—two or three circuits, transmission lines or power transformers connected in parallel [8]. One of these can be activated or also stopped from operating, can modify longitudinal parameters (impedance) of network, leading to change voltage in consumption nodes.

In the analyzed Fig. 6.8 under the minimum load operating conditions or empty load voltage U_2 consumer bars may increase above the rated value. Disconnecting a circuit power line or a transformer in the network increase the longitudinal impedance of the circuit and voltage drop, respectively, so that the voltage level can lower the U_2 consumer bars.

Fig. 6.8 Electrical network to a consumer through two parallel circuits

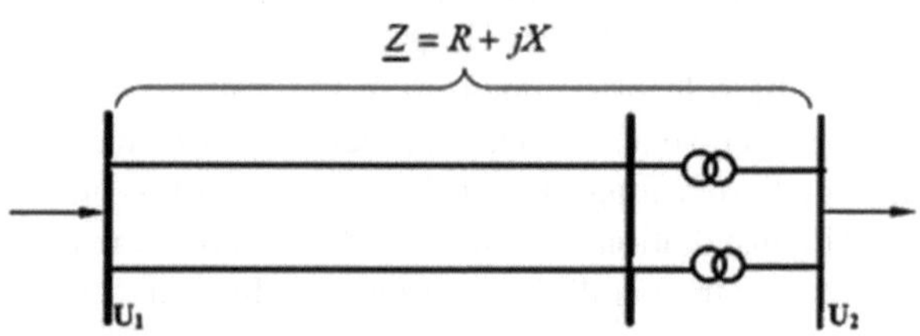

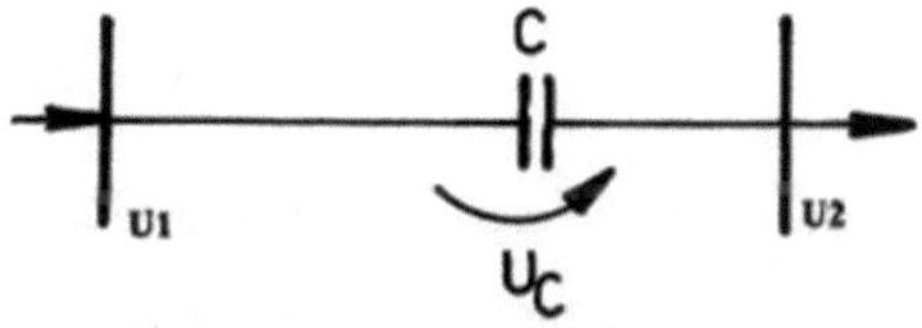

Fig. 6.9 Capacitor mounted in cascade on transmissions line

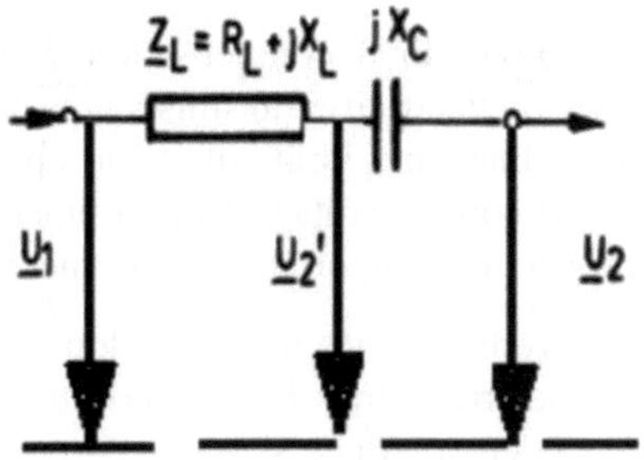

Fig. 6.10 Single phase equivalent circuit of compensated electrical network

In practice, disconnecting transformers operating in parallel, with voltage control purpose in low load operating case is a measure recommended in terms of reducing power losses and active energy in electricity network. Because the power transformer is a component of the electrical networks that presents a high level of safety in operation, it is generally accepted to operate transformer stations with a single transformer.

Regarding take off transmission lines, it is important to know that these are less reliable in operation and in practice, it is not recommended to disconnect transmission line circuits to control voltage in electrical networks.

Another procedure used for voltage control is used cascade circuit compensation of an electrical network by inductive reactance. This is accomplished by the capacitors mounted in cascade with reactance, Fig. 6.9, obtaining a reduction longitudinal reactance of the line and thus a decrease of voltage drops.

In Fig. 6.10 there is represented a single phase equivalent circuit of a compensated electrical network

$$X_L = \omega L \tag{6.19}$$

is transmissions line reactance, and

$$X_C = \frac{1}{\omega C} \tag{6.20}$$

is capacitive reactance series on line. Resulted reactance will be

$$X = X_L - X_C = \omega L - \frac{1}{\omega C} \tag{6.21}$$

Based on the capacitive reactance's value there are the following situations:

- under-compensated line, when $X_C < X_L$;
- total compensated line, when $X_C = X_L$;
- over compensated line, when $X_C > X_L$.

Longitudinal compensation with banks capacitor series connected with long lines of electricity transmission is done in order to increase the transport capacity of these lines and limits of stable operation of the power system.

To control voltage from a practical standpoint, series compensation with banks capacitor is important for the medium voltage lines in the following cases:

- overhead lines relatively long, slow voltage variations reduction to consumers;
- electrical lines feed to consumers that produce shock load in the network, in order to reduce shock occurring in the voltage corresponding to power supply.

On the other hand, it must be known that the compensation process causes some negative phenomenon: ferro-resonance, under-synchronous resonance or oscillation of the synchronous machines. For limiting them a research to develop possibilities and removing their appearance is required.

Measures that can be taken to avoid these are:

- restrict the compensation to between 1.5 and 2;
- using series or parallel resistors with banks capacitors.

Another drawback of the procedure- series bank capacitors for compensation of the inductive reactance of the power lines is the increase of the short-circuit current, symmetrical and non-symmetrical. The capacitors are protected by special circuits in the short circuit case within power system. This method is used in heavy growing and fast variation of reactive power in the system.

6.4.3 Node Voltage Set Up

The importance of voltage control in electrical network nodes is to change the level of voltage in admissible in them. This process seems to be the most natural, but it will not remove the cause of voltage fluctuations causing variations reactive loads transiting through electric transmission elements and distribution network. This voltage control method consists in injection of supplementary voltage in order to compensate longitudinal component of lost voltage. The injection voltage is realized by autotransformers or power transformer equipped with adjuster under longitudinal load or transversal. Often are used in power system voltage control by the automat voltage regulator.

Base on introducing additional voltage in an electrical network, there are the following types of control methods [18]:

- longitudinal adjustment, which is intended to compensate the longitudinal component of drop in voltage. Additional being introduced in phase with the line voltage at that point;
- transverse adjustment when additional voltage is in with quadrature voltage network at that point, and serves mainly to modify the power flow in the mesh network topology;
- adjustable long-transverse, when the longitudinal adjustment is used to adjust the transverse correlated.

In terms of structural and functional features of the used means to introduce additional voltage, there are to be distinguished two types of control:

- direct control, which is performed using power transformers and autotransformers fitted with sockets for adjusting the working windings, as well as with devices for switching them in the presence or absence of load;
- indirect control, which is performed by means of special transformers and special autotransformer-overvoltage transformer. These constitute complexes devices that can be associated to regulating power transformers non-adjustable or can be used separately. These also include the regulators induction.

6.4.4 Management of Voltage Control

Base of voltage adjustment consists of generating units by using their capability to absorb and debit reactive power (to modify the magnetization current and implicit electromotive force linked by synchronous reactance), within diagram *P-Q* available.

For basic voltage control use the following devices:

- Synchronous condenser;
- Synchronous generators operating in thermo and hydro plants;
- Wind power plants;
- Photovoltaic power plants.

All other means of voltage control, maintained by the consumers/network operators have a secondary role regardless of their number and capability and they come in support of the generating sets without replacing the first.

These are:

- Banks capacitors;
- Inductors;
- Control by power transformer sockets;
- Connect/disconnect of electrical lines transmission.

6.5 Case Studies

1. **Q–V voltage regulation approach in the pilot node using an automatic voltage control (ASVR—Automatic Secondary Voltage Regulation) developed in Romania**

The goal is to maintain the required voltage in the pilot node within the desired voltage domain. ASVR equipment consists of an independent regulator working on the principle of regulatory with negative feedback response.

ASVR are placed in the power plant or electrical station, and the desired voltage it is received by the voltage domain that must keep it in pilot node.

ASVR = Automatic Secondary Voltage Regulation in pilot node of 400 kV—automation which controls the voltage in the 400 kV pilot node from a transformation station.

Further on we will refer to such a system associated to a wind plant, but this system can be used for any kind of plant, photo-voltaic, for example.

The wind plants management system—which is capable of controlling in real time depending on the imposed voltage level, meaning the decrease of increase of produced reactive power, by controlling the voltage in the medium voltage network (MT) and in the 110 kV, through the modification of the plots at the transformers (power transformer 110/MT from the power transformer transformation stations and power transformer 400/110 kV from the system station), according to the set voltage value from the 400 kV pilot node, but also to the park's reactive power availability at that moment, increasing or decreasing the transit of active energy from the wind park, if the adjustment analysis takes into consideration the electric scheme from Fig. 6.11.

The automatic installation for the secondary adjustment of the ASVR voltage is destined to the voltage adjustment at the 400 kV level in the 400/110/MT station, as the reference is given the voltage value on the station's 400 kV bars.

Thus, in order to achieve this objective, there are used both the wind generator capacity to produce/consume reactive power, but also switching the plots of the MT/110 kV transformers and the 400/110/MT transformers.

The primary adjustment loop U–Q of the wind generator group has as reference the value of the voltage on the 110 kV bars so that the reactive power variation of the generators would follow the variation of this voltage, which, in turn—according to the ASVR logic—is proportional to the voltage variation on the 400 kV bars. Additionally, in order to extend the adjustment band, under the conditions of maintaining the inferior voltage of 110 kV, MT and 0.5 kV within the admitted limits, there are used the voltage regulators of the MT/110 kV transformers, respectively the 400/110/MT which act on the plot switches at the ASVR command.

From a practical standpoint, compared to the initial (primary) function of adjusting the voltage to the 110 kV through the variation of the wind generators

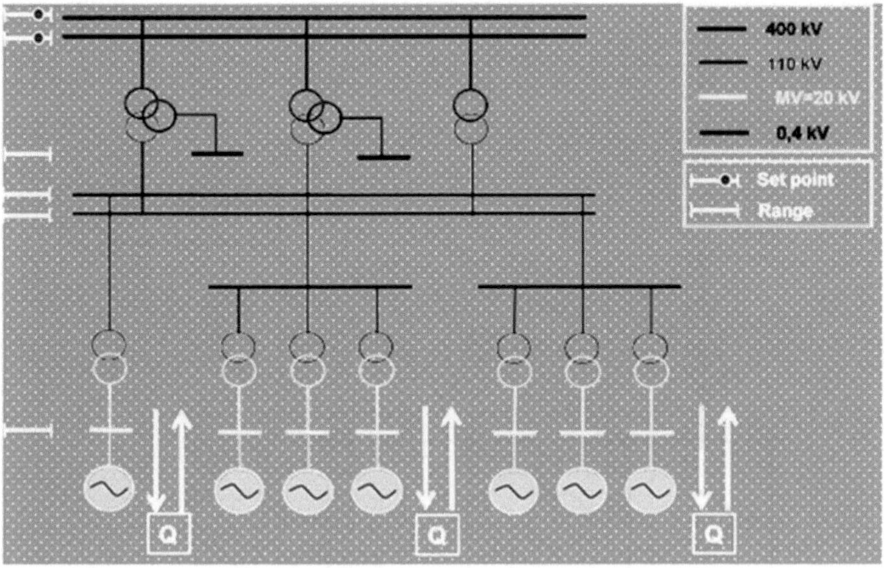

Fig. 6.11 The transformer station through which a central debits

reactive power, the ASVR installation is designed to perform the following supplementary functions:

- The voltage adjustment by modifying the reactive energy produced by the wind generators from the wind power plants by means of the command-control system WFMS, having as reference the voltage at the level of 400 kV bars;
- automatic switching of 400/110/MT transformer plots from the 400 kV station;
- automatic switching of MT/110 kV transformer plots from all stations corresponding to the plant;
- coordinating the three preious functions.
- observing the condition to maintain the voltage levels inferior to 400 kV (meaning 110 kV, MT and 0.4 kV) within the admitted limits.
- ASVR is composed of three adjustment loops, as follows:
- adjustment loop voltage 400 kV;
- adjustment loop voltage 110 kV;
- adjustment loop voltage MT.

The three adjustment loops function simultaneously, as they coordinated by the ASVR logic. Not including one of the loops in the ASVR scheme implies the ASVR functioning without the adjustment system corresponding to that loop (for example, not including the Mt adjustment loop determines the ASVR functioning without the MT/110 kV plot switching).

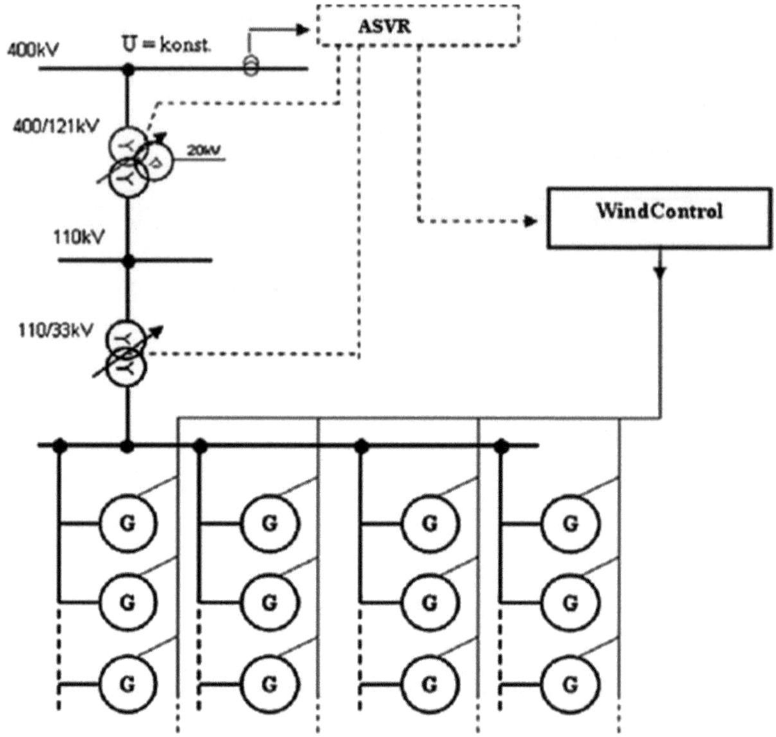

Fig. 6.12 ASVR integration in electrical network

The coordinating function of the three adjustment loops leads to a hierarchy of their actions, so that the adjustment process records the following hierarchy (priority):

1. loading/unloading with reactive energy the wind generator ensemble from the electric plant;
2. plot switching at the MT/110 kV transformers of the producer (from the plant stations);
3. plot switching at the 400/110/MT transformers from the 400/110 kV where the pilot node is considered.

The adjustment loop ensemble which ASVR consists of is schematically presented in Fig. 6.12.

Release period/start: The interval between two control interventions are usually set on 18 ÷ 20 s. Also every deviation must be eliminated by occurred adjusting 120 s of its release.

- The ASVR will command the power transformers 400/110 kV on the basis of criterion of minimal voltage difference between the sockets;
- In case of increase voltage 110 kV will be used by the transformer with the socket on lower voltage;
- When voltage decreases 110 kV the transformer is used with socket to the higher voltage;
- The criterion of reactive power transfer will be applied in voltage equality between the sockets.

2. ***Q–V* voltage control in wind power plant, with imposed *Q***

We have a wind power plant to which a ASVR is used in order to maintain voltage in the 110 kV, the consign the plant receives being of reactive power Q_{abs} or debited Q, according to the 'scale' chart from Fig. 6.13 and the adjustment itself consists in loading/unloading the wind generator group from the wind plant with reactive energy.

In the diagram presented in Fig. 6.13, imposed Q, as follows ASVR, Q absorbed or Q produced to be as close to the preset value. These values are indicated by Q measured which falls or how much it increases depending on the preset value Q. Also voltage of 110 kV aims at absorbing or dispensing Q, dropping respectively U with an increase for which we have in view the achievement of a certain preset Q.

Acquisition and execution of preset reactive power provided by ASVR controllers will be using the entire reactive power reserves that actions wind power plant, including the existing compensation means to them. In case $P = 0$ for a wind plant will provide for the adjustment $Q = 0$ or $Q \neq 0$, and the controller must know the limits of maximum and minimum reactive power that wind plant led so it can provide active power when it's generated or not.

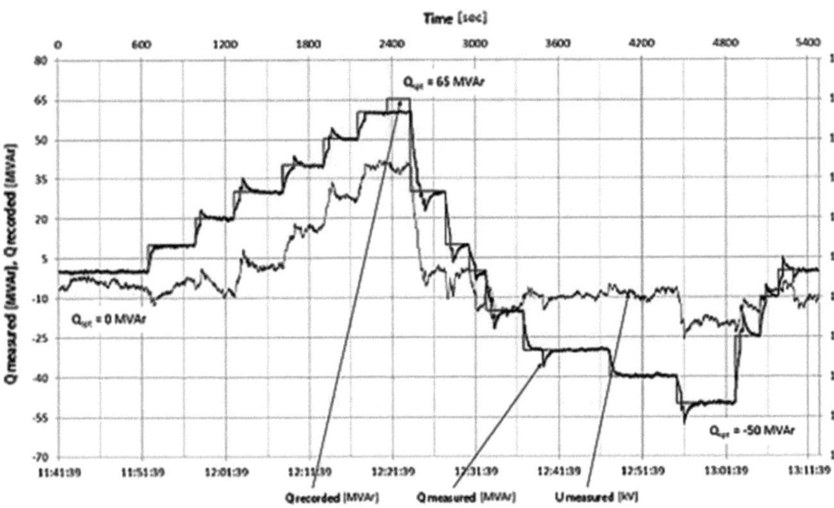

Fig. 6.13 *Q–V* diagram with imposed *Q*

3. ***Q–V* voltage control in photovoltaic power plant, with imposed *U***

We have a wind power plant to which a ASVR is used in order to maintain voltage in the 110 kV, the consign the plant receives being of voltage adjustment *U*, according to the 'scale' chart from Fig. 6.13 and the adjustment itself consists in the three loops' coordinated functioning, thus leading to a hierarchy of their actions so that the adjustment process can be performed.

In the diagram, in Fig. 6.14 can be seen as an imposed voltage, ASVR follows that voltage 110 kV to be as close to the preset value. These values are indicated by measured voltage that goes up and down depending on the value of voltage. Also instruction for reactive power *Q* is produced or absorbed in order to increase, decrease voltage respectively, as we have in mind to reaching any preset voltage.

According to the short-circuit on bar power at the voltage level for which the voltage is aimed to be adjusted, there are the results $\Delta Q/\Delta U$ [MVAr/kV] for the voltage level we want to maintain, meaning, for example, that, at a variation of 2 kV we approximate ±16 MVAr (Fig. 6.14), respectively at a de ±8 MVAr variation, as the voltage varies with ±1 kV/8MVAr (the mentioned values are just for exemplification, as they depend on the voltage level where the adjustment is made). The dU/dT value shows the response speed, the decrease/increase of voltage in a minute, indicating the system's capability to reach the values for *U* consign, according to the voltage rise.

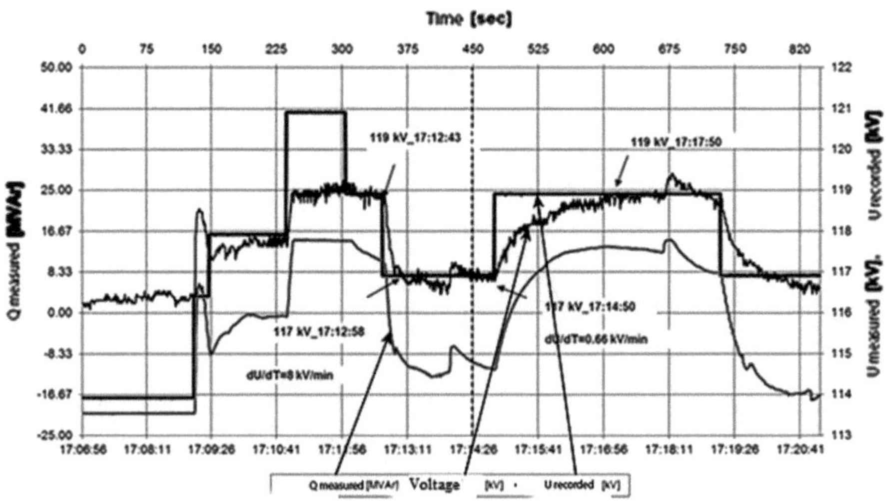

Fig. 6.14 *Q–V* diagram photovoltaic power plant—*U* preset

6.6 Conclusion

The wide development of renewable energy sources connected to the high medium voltage lines imposed the development of specific solutions for the voltage control in the point where these sources connect or in their vicinity.

Today's energy systems are facing the problem that, in some areas of the system there are is an exceeding production, while in other areas the production is lower, while energy is transited from the sources in order to balance the consumption. All these occur as result of the way the plants are located, as they, for different reasons, chosen for other criteria than the optimal location and, considering the degree to which the network in loaded, the power is transited to a consumption node which can be, in some situations, very far from the source, this leading to a decrease in the area's level of voltage. The phenomenon can also appear in reverse, meaning an increase in voltage, in the situations when the plant connection is made in nodes with reduced short-circuit power, thus appearing the limitation of the power debited by the plant in accordance with the voltage maximally admitted at the final consumers, especially in the cases of plants recorded at the MT lines.

For these reasons, it is important to control the voltage level, and this is why one must always analyze solutions to maintain the voltage within the limits which are adequate for the user's system.

For the transport network, one of the solutions is the ASVR system which maintains the voltage in pilot node chosen in the system, compensating the voltage variations in the system, with the three presented adjustment loops, thus using two of the adjustment methods, the reactive energy absorption and debit and the longitudinal and long-transversal voltage adjustment in the transformers by using the plots. The ASVR can be also used in the distribution network, with less control loops, two or even one.

Considering the voltage variation at the final user clamps, from each voltage level from the system, depends in each moment on production, consumption, network loading, using an automated voltage regulator for that voltage level allow for the necessary conditions of ensuring the quality of the energy from the network and, consequently, for all users.

References

1. C.W. Taylor, Power System Voltage Stability, McGraw Hill, 1994.
2. T.J.E. Miller, Reactive Power Control in Electric System, 1982, John Wiley & Sons, Inc., USA.
3. P.W. Sauer, Chapter 2, Reactive Power and Voltage Control Issues in Electric Power Systems.
4. J. Hossain, H.R. Pota, Robust Control for Grid Voltage Stability: High Penetration of Renewable Energy, Power Systems, DOI:10.1007/978-981-287-116-9_2, Springer Science-Business Media, Singapore, 2014.

5. IEEE Power Engineering Society Administrative Committee of the Power System Dynamic Performance Committee, Causes of the 2003 Major Grid Blackouts in North America and Europe, and Recommended Means to Improve System Dynamic Performance.
6. W. Nakawiro, Voltage Stability Assessment and Control of Power Systems Using Computational, Intelligence, http://duepublico.uniduisburgessen.de/servlets/DerivateServlet/Derivate-26570/Nakawiro_Worawat_Diss.pdf, 2011.
7. C. Bulac, M. Eremia, A.I. Bulac, I. Tristiu, Voltage stability in SEE Phenomenon Assessment and Control through Artificial Intelligence Techniques, Power Journal 47-1999, no. 2.
8. A. Poeata, et al., Transport and Distribution of Electricity, Bucharest, pp. 267–310.
9. M. Eremia, P. Simon, D. Petricica, D. Gheorghiu, Some Aspects of Hierarchical Voltage - Reactive Power Control, Power Engineering Society Summer Meeting, 2001, vol. 2, DOI: 10.1109/PESS.2001.970170, Publication Year: 2001, pp. 872–880, vol. 2.
10. http://electrical-engineering-portal.com/how-reactive-power-is-helpful-to-maintain-a-system-healthy.
11. F.A. Viawan, D. Karlsson, Coordinated Voltage and Reactive Power Control in the Presence of Distributed Generation, Power and Energy Society General Meeting - Conversion and Delivery of Electrical Energy in the 21st Century, 2008 IEEE, DOI:10.1109/PES.2008.4596855, Publication Year: 2008, pp. 1–6.
12. E. Demirok, P. Casado Gonzalez, K.H.B. Frederiksen, D. Sera, P. Rodriguez, R. Teodorescu, Local Reactive Power Control Methods for Overvoltage Prevention of Distributed Solar Inverters in Low-Voltage Grids, IEEE Journal of Photovoltaics, vol. 1, no. 2, December 2011.
13. S. Bisanovic, M. Hajro, M Samardzic, One Approach for Reactive Power Control of Capacitor Banks in Distribution and Industrial Networks, Electrical Power and Energy Systems, 60, pp. 67–73, 2014.
14. C.A. Canizares, Modeling and Implementation of TCR and VSI Based FACTS Controllers, Internal Report, ENEL and Politecnico di Milano, Milan.
15. H. Mori, Y. Goto, A Parallel Tabu Search Based Method for Determining Optimal Allocation of FACTS in Power Systems, Proc. of the International Conference on Power System Technology (PowerCon 2000), vol. 2, pp. 1077–1082, 2000.
16. S. Kamel, F. Jurado, D. Vera, A Simple Implementation of Power Mismatch STATCOM Model into Current Injection Newton-Raphson Power-Flow Method, Electrical Engineering, 96:135–144
DOI 10.1007/s00202-013-0288-4, Springer, 2014.
17. A. Marinescu, M. Radulescu, N. Golovanov, Control in Blood Medium Voltage Networks Are Connected Renewable Energy, VIIIth Edition of the ASTR Conference.
18. G. Georgescu, et al., Transport and Distribution of Electricity, Ghe. Asachi Publishing House, Iasi 2001.
19. E. Nasr Azadani, Optimal Placement of Multiple STATCOM for Voltage Stability Margin Enhancement Using Particles Warm Optimization, Electrical Engineering, September 2008.

Part II
Compensation and Reactive Power Optimization in AC Power Systems

Chapter 7
Optimal Reactive Power Control to Improve Stability of Voltage in Power Systems

Ali Ghasemi Marzbali, Milad Gheydi, Hossein Samadyar, Ruhollah Hoseyni Fashami, Mohammad Eslami and Mohammad Javad Golkar

Abstract The current power systems have works near to the marginal voltage stability due to the market performance as well as their weightier operation loadings along with consideration of environmental constraints of transmission as well as generation capacity enlargement. In other words, at the present time wind power has confirmed to be one of the most efficient and competitive renewable resources and therefore, its use is indeed continually growing. Little wind power infiltration planes are generally contained in the current grid networks in view of that it is passively controlled and operated. On the other hand, this statement is no more suitable for immediately after the wind power energy infiltration commences

A. Ghasemi Marzbali (✉)
Technical Engineering Department, University of Mohaghegh Ardabili, Ardabil, Iran
e-mail: ghasemi.agm@gmail.com

M. Gheydi · H. Samadyar
Young Researchers and Elites club, Science and Research Branch, Islamic Azad University, Tehran, Iran
e-mail: gheydi.m@srbiau.org

H. Samadyar
e-mail: hosseinsamadyar@yahoo.com

R.H. Fashami
Electrical Engineering Department, Damavand Branch, Islamic Azad University, Tehran, Iran
e-mail: rhfashami@gmail.com

M. Eslami
Department of Electrical and Computer Engineering, Nikshahr Branch, Islamic Azad University, Zahedan Branch, Nikshahr, Iran
e-mail: mohammad.eslami@chmail.ir

M.J. Golkar
Department of Electrical and Computer Engineering, Zahedan Branch, Islamic Azad University, Zahedan, Iran
e-mail: javad.golkar.1368@gmail.com

N. Mahdavi Tabatabaei et al. (eds.), *Reactive Power Control in AC Power Systems*,
Power Systems, DOI 10.1007/978-3-319-51118-4_7

growing, a broad scope of scientific issues can come out, namely: voltage rise, bi-directional power flow, improved power quality issues as well as distorted voltage stability. The additional improvement of electricity construction from renewable resources in a trustworthy as well as consistent system performance is driving transmission as well as distribution control utilizers to employ novel working models that are not presently extant. A serious subject of the demanding status described in the foregoing is the reactive power managing that involves the planning as well as operation deeds that are asked for to be executed to get better voltage profile as well as stability in the power networks. For this reason, voltage stability is a major issue of current power systems. It signifies the capableness of a power system to keep voltage when the required load is boosted. Researches about this kind of instability fact proceed with its control as well as evaluation. The first one designates if a power system runs in the safe operational area, while the second one will carry out essential control actions if a power system gets close to unsafe operational zone. Diverse approaches put forth in the chapter deal with offline and online purposes. The center of attention of this chapter is the second part; it means control of voltage stability. Three major methods have been utilized for voltage stability which are reactive power management, load shedding and active power re-dispatch. Reactive power management signifies the ways designating the place of novel VAR sources and/or settings of the VAR sources that are installed currently and the settings of facilities including on-load tap changers (OLTCs). Reactive power sources ordinarily consist of synchronous generators/condensers, reactor/capacitor banks, as well as flexible AC transmission systems (FACTS) controllers. It can be classified into two subjects as reactive source programming as well as reactive power dispatch. For reactive programming, the concerned temporal duration is the coming few months or years, and besides considering the optimum milieu of facilities that are installed currently, installation of novel reactive power sources is contemplated. It is performed in offline and online ways. The main purposes of offline reactive dispatch can be found in the duration of the coming few days or weeks, while, another model is carried out in the coming few minutes or hours. Opposing the reactive planning, both online and offline reactive power dispatches only designate the optimum settings of extant facilities. Optimal reactive power flow (ORPF) which is a specific instance of the optimal power flow (OPF) issue is an utterly significant instrument with regard to assured and gainful utilization of power systems. The OPF's control parameters have a proximate connection with the reactive power flow, including shunt capacitors/reactors, voltage magnitudes of generator buses, output of static reactive power compensators, transformer tap-settings. In the ORPF problem, the transmission power falloff is brought to a minimum and the voltage profile is modified and the operating and physical constraints are satisfied. Note that shunt capacitors/reactors and tap-settings of transformers are discrete variables while and except other variables are continuous. Hence, the reactive power dispatch issue is nonlinear, non-convex has equality and inequality limitations and has discrete and continuous variables.

7.1 Introduction

By the elevated exploitation as well as loading of the grid transmission system and besides because of refined enhanced operating conditions the issue of voltage stability and voltage collapse draws increasing consideration. A voltage drop could be taken in the power systems or subsystems and could emerge very suddenly [1]. Constant controlling of the system status is hence necessary. The reason of the 1977 New York blackout has been substantiated to be the reactive power issue. The 1987 Tokyo blackout was accepted to be because of reactive power deficit as well as a voltage drop at peak load in summer.

These facts have strongly shown that reactive power play an important role in the security of power systems as view of voltage stability. An appropriate compensation of system voltage profiles will improve the system securities in the operation and will decrease system losses [2]. The essential purpose of voltage regulation in the distribution system performance is to maintain the status voltage in the power system steady in the suitable scope. The desirable voltages could be acquired by either directly manipulating the voltage or reactive power flow which in its own right will influence the voltage collapse. The reactive tools usually employed for the voltage and reactive power control are on-load tap-changer (OLTC) transformers, steps voltage regulator and switched shunt capacitors [3, 4]. Such reactive tools are generally utilized on the basis of a presumption that power runs in just one direction and the voltage diminishes along the feeder, from the substation to the remote end.

An OLTC transformer is a transformer with automatically changeable taps. The OLTC is a section of most of HV/MV substation transformers [5]. A shunt capacitor produces reactive power to make up for the reactive power demand and hence increases the voltage. Shunt capacitors could be installed in the substation (hereinafter referred to as substation capacitors) or along the feeder (hereinafter referred to as feeder capacitors). A steps voltage regulator is an autotransformer with automatically adjustable taps that is ordinarily installed when the feeder is too long in such a manner which voltage regulation with OLTC and shunt capacitors is not enough. Voltage and reactive power control entails suitable coordination between the extant voltage and reactive power control equipment [6].

Many distribution network operators (DNOs) control these equipment locally via use of customary controllers to keep the voltages in the distribution system about approved range while bring to a minimum the voltage collapse and power falloffs. Various techniques have been presented in order to get better voltage and reactive power control in the distribution system for programming and operation stages. Within this time, many scholars have presented the trouble of voltage and reactive power control in distribution power systems through concentrating on automated distribution power system, with off-line setting control or real time control. The offline setting control intends to explore a dispatch program for OLTC movement and capacitor switching on the basis of day-ahead load prediction, in the meantime the real time control endeavors to control the capacitor and OLTC on the basis of real time surveying and trainings [7]. The major difficulty in utilization of the off-line setting control way is its affiliation to remote control and communication links to all capacitors. Nonetheless, a lot of DNOs do not communicate with links that are downstream to the feeder capacitor locations.

In the other hand, the nature of modern power systems has changed due to a variety of factors: the increased demand for sustainability, rises in the price of oil and the need for the reduction of greenhouse gases, all of which have driven a large increase in the level of wind generation in the power system. The intergovernmental panel on climate change has cited that wind energy will be the primary source of renewable generation in the electricity sector [8]. Wind generation in both Europe and the United States is the dominant renewable resource currently present in power systems. In Europe, wind energy is set to triple in penetration by the year 2020, with 15.7% of the continent's total energy supplied by wind generation [9]. In the United States, there is currently 42,432 MW of installed capacity providing 2.3% of the U.S. electricity mix, with the number set to rise to 25% by the year 2025 [10]. With wind generation set to become a significant generation resource in power systems around the world, it will become increasingly important to fully understand its impacts and interaction with the conventional elements in power systems. In fact, the real power unit output is generally restricted by radius (V_tI_a), as follows

$$P_G^2 + Q_G^2 \leq (V_tI_a)^2 \tag{7.1}$$

Constrain of field is circular (V_tE_f/X_s) at (0, $-V_t^2/X_s$). It can be defined as

$$P_G^2 + \left(Q_G + \frac{V_t^2}{X_s}\right)^2 \leq \left(\frac{V_tE_f}{X_s}\right)^2 \tag{7.2}$$

where, P_G, Q_G, V_t, I_a, E_f and X_s are active power, reactive power, terminal voltage, armature current, Excitation voltage and Synchronous reactance of the synchronous generator, respectively.

Fundamentally, power systems have been designed and operated around the concept of generation delivery from large synchronous machines. These machines have high levels of reliability and complex control systems that allow the system to maintain high levels of operational security. The correct operation and control of these machines across the full spectrum of time-frames is critical for maintaining reliable power system operation and stability [11–13].

Maintaining voltage stability requires that the various components and elements of the system can interact without issue across all of the timeframes of the stability spectrum. Wind generation will have significant impact across the power system stability time-frame and as wind generation becomes a more common source of generation in the system, new mitigation techniques will be necessary to continue operating the power system in an assured and stable way [14]. While the effect of reactive power reserves (RPRs) on system stability is widely acknowledged, few studies have been conducted to investigate how RPR levels could be used to indicate the amount of voltage stability margin (VSM) [15].

In contemporary years, certain papers with mathematical algorithms have been presented to think out the reactive power dispatch (RPD) problem [16–19]. These algorithms, including Non-linear Programming (NLP), Newton method, Gradient method, Linear Programming (LP), Jacobian matrix, Quadratic Programming (QP), interior point methods and so on, have been fruitfully used for thinking out the RPD problem. However, certain drawbacks are still linked to them. The RPD problem is non-linear, non-differential and non-convex problem with more than one local optimum, while these methods work based linearizing which make them less efficient in finding the global optimum. On the other hand, some of these methods suffer from special shortages, such as mathematical complexity and insecure convergence (NLP), piecewise quadratic cost approximation (QP), convergences to local optima (Newton method), a simplified piecewise linear estimate (LP), etc. Also, their optimization process mainly depends on the initial solution and can easily fall into local optima.

To overcome these disadvantages, different heuristic-based techniques are developed for solving the RPD problem. Population-based optimization techniques inspired by nature may be classed in two significant categories that are swarm intelligence and evolutionary algorithms. Many methods rooted in these techniques including fuzzy Adaptive Particle Swarm Optimization (FAPSO) [20], Real Genetic Algorithms (RGA) [21], Tabu Search (TS) [22], Particle Swarm Optimization (PSO) [23], Improved PSO (IPSO) [24], stochastic method [25], Hybrid Stochastic Search (HSS) [26], Differential Evolution (DE) [27], Artificial Bee Colony (ABC) [28] as well as other methods have been broadly employed in the problem of RPD. However, these methods appear to be proper approaches for the unravelment of the RPD variable optimization problem.

Attention to use optimization methods in distributed generation is increasing [29, 30]. In [31], an optimization method employed for wind power is offered, by a primal-dual predictor corrector interior point method, employed to explore the operating points of a single WT in a WF. The voltage stability based on reactive power control has been reviewed in rest of this chapter. The real transmission power loss minimization is consider as function, meantime the permitted transformer capacity, voltage range and conductor current capacity are added as the loading limitations. Moreover, OLTC operation and voltage fluctuation index are also analyzed. The reactive power in term of voltage stability is shown with reactive power control and local voltage in which timing of feeder capacitors are monitored and is still broadly employed.

7.2 Voltage Stability Based RPD Model

The different purposes of power system are sum of voltage deviations on load busses, system transmission falloffs, voltage stability, security, etc. Such purposes are contradictory in their essence and couldn't be dealt with by customary single purpose optimization techniques. Generally, the RPD model could be explained as follows in mathematical terms:
Problem Purposes

- *Purpose 1: Bringing to a minimum the total real power losses*

Transmission falloffs in the network can be stated as economic losses procuring no advantages. Thus, transmission falloffs are understood as a falloff in proceeds via the utility. The intensity of each falloff requires accurate estimation and applicable moves made to bring them to a minimum. When the transmission falloffs are stated with regard to bus voltages and associated angles, the falloffs could be stated based on Newton–Raphson as follows

$$J_1 = P_{loss}(x, u) = \sum_{k=1}^{N_L} g_k[V_i^2 + V_j^2 - 2V_iV_j\cos(\theta_i - \theta_j) \tag{7.3}$$

i and j end respectively if g_k is the conductance of the line i-j, V_i and V_j are line voltages and θ_i and θ_j are the line angles at the line. The kth network branch is k. It connects bus i to bus j. If N_D is the set of numbers of power demand bus and j = 1,2, …, N_j where, N_j is the set of numbers of buses in adjacency with bus j, i = 1,2, …, N_D. P_G is the active power in line i and j. x and u are the dependent variables vectors and vector of control variables, respectively.

- *Objective 2: Bringing voltage deviation to a minimum*

Satisfying user's demands with the smallest expense with a desirable continuity of supply and sufficiently little deviation in voltage is the second function of one RPD problem. The following is its expression

$$J_2 = VD(x, u) = \sum_{i=1}^{Nd} |V_i - 1.0| \tag{7.4}$$

where, N_d is number of load buses.

- *Objective 3: Minimization of L-index voltage stability*

Voltage stability and voltage drop problem draws increasing attention by increasing the power transmission system loading and exploitation, a voltage collapse could happen in systems or subsystems. And it could emerge very suddenly [15, 34]. *L-index*, L_j of the *j*th bus could be stated by means of below equation

$$\begin{cases} L_j = \left| 1 - \sum_{i=1}^{N_{PV}} F_{ji} \frac{V_i}{V_j} \right|, j = 1, 2, \ldots, N_{PQ} \\ F_{ji} = -[Y_1]^{-1}[Y_2] \end{cases} \tag{7.5}$$

where, N_{PV} and N_{PQ} are numbers of *PV* and *PQ* buses, respectively. Y_1 and Y_2 are the Y_{BUS} system sub-matrices acquired following the segregation of *PQ* and *PV* bus parameters as shown in the below equation

$$\begin{bmatrix} Y_1 & Y_2 \\ Y_3 & Y_4 \end{bmatrix} \begin{bmatrix} V_{PQ} \\ V_{PV} \end{bmatrix} = \begin{bmatrix} I_{PQ} \\ I_{PV} \end{bmatrix} \tag{7.6}$$

where *L-index* is calculated for the whole load buses. L_j shows there weren't load case and voltage drop circumstances of bus *j* in a feasible numerical range of [0, 1]. Thus, a representative *L* delineating the fixedness of the complete system is formulated as follows

$$L = \max(L_j), j = 1, 2, \ldots, N_{PQ} \tag{7.7}$$

In the optimal RPD problem, incorrect tuning of continuous and discrete control variable settings might boost the *L-index* value, which may reduce the system voltage fixedness outskirt. Let the maximum value of *L-index* be denoted as $L_{\max}$. Therefore, to improve the voltage fixedness and to keep the system remote from the voltage drop margin, the formula for the succeeding purpose function will be as follows

$$J_3 = VL(x, u) = L_{\max} \tag{7.8}$$

Objective Constraints

- *Limitations 1: Equality Limitations*

Power balance is equality limitations. To rephrase, the total power generation (P_G) must cover the total demand (P_D) as well as total real power losses in transmission lines. Equality constraints of real and reactive power in each bus can be expressed as below

$$\begin{cases} P_{G_i} - P_{D_i} = V_i \sum_{j=1}^{N_B} V_j [G_{ij} \cos(\theta_i - \theta_j) + B_{ij} \sin(\theta_i - \theta_j)] \\ Q_{G_i} - Q_{D_i} = V_i \sum_{j=1}^{N_B} V_j [G_{ij} \sin(\theta_i - \theta_j) - B_{ij} \cos(\theta_i - \theta_j)] \end{cases} \tag{7.9}$$

where; N_B and Q_{Gi} are the numbers of buses and the reactive power produced for ith bus, respectively; P_{Di} and Q_{Di} are real and reactive power at the ith load bus, respectively; G_{ij} and B_{ij} are the transfer conductance and susceptance between bus i and bus j, respectively; V_i and V_j are the voltage intensities at bus i and bus j, respectively; and θ_i and θ_j are the voltage angles at bus i and bus j, respectively. The equality limitations in (7.9) are nonlinear equations which could be thought out by employing Newton-Raphson method to create an answer to the load flow problem. Within the duration of answering, the real power output of one generator, titled slack generator, remains to fill in the real power losses and satisfy the equality limitation in Eq. (7.9). The load flow answer produces all bus voltage intensities and angles. Therefore, the real power losses in transmission lines could be obtained using Eq. (7.3).

- *Limitation 2: Generation Capacity Limitations*

For solid performance, the generator reactive power and bus voltage can be constrained through lower and upper limits as below

$$Q_i^{\min} \le Q_i \le Q_i^{\max}, v_i^{\min} \le v_i \le v_i^{\max} \tag{7.10}$$

where, $Q_i^{\min}$, $Q_i^{\max}$, $v_i^{\min}$ and $v_i^{\max}$ are the minimal and maximal value for reactive power and voltage magnitude of the ith transmission line, respectively. A clarified input/ output curve of the thermal unit understood as heat rate curve is indicated in Fig. 7.1.

- *Limitation 3: Line flow Limitations*

A significant limitation of RPD problem is the line limitation. Since any line has a constrained capacity for current power, the constraint should be checked following the power system load flow. For that reason, this section argued the answer for RPD problem with line flow limitations. The following is the modeling of this constraint

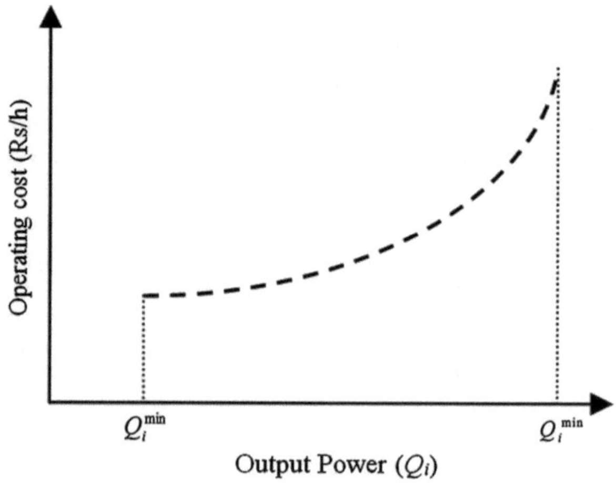

Fig. 7.1 Operating expenses curve for one generator

$$\left|S_{Lf,k}\right| \leq S_{Lf,k}^{\max}, k = 1, 2, \ldots, L \tag{7.11}$$

where $S_{Lf,k}$ is the real line k power flow; $S_{Lf,k}^{\max}$ is the upper limit of power flow of line k. L is the number of transmission lines [3].

- *Constraints 4: Discrete control variables*

The shunt susceptance (B_{sh}) and transformer tap settings (T_i) values are taken as discrete values. These must be constrained by their lower and upper limits as below

$$\begin{cases} T_i^{\min} \leq T_i \leq T_i^{\max} \\ B_{sh_i}^{\min} \leq B_{sh_i} \leq B_{sh_i}^{\max} \end{cases} \tag{7.12}$$

Problem formulation
Adding up the entire purpose performances and the equality and inequality limitations, a nonlinear limited multi-purpose optimization problem in mathematical terms could be the formula for the RPD problem, which can be represented by

$$\begin{aligned} & J_{Final} = \min_{P_G}[VL(x,u), P_{loss}(x,u), VD(x,u)] \\ & \text{s.t. :} \\ & g(x,u) = 0 \\ & h(x,u) \leq 0 \\ & x^T = [[V_L]^T, [S_L]^T, [Q_G]^T] \\ & u^T = [[V_G]^T, [Q_C]^T, [T]^T] \\ & J_{Final} = \min_{P_G}[VL(x,u), P_{loss}(x,u), VD(x,u)] \end{aligned} \tag{7.13}$$

where, g and h are the equality and inequality limitations, respectively. $[V_L]$, $[Q_G]$ and $[S_L]$ are the vectors of load bus voltages, generator reactive power outputs and the transmission line loadings, respectively. $[V_G]$, $[T]$ as well as $[Q_C]$ are the vectors of generator bus voltages, transformer taps and reactive compensation instruments, respectively.

7.3 Reactive Power Capacity and Control Options in Wind Farms

The requisite of the membership of WFs in grid control matters has increased the inclusion of power electronics and the expansion of new WT generation concepts, leading to variable speed wind turbines [30]. The most popular wind generation technology employed among them today is the DFIG. In this study, the DFIG

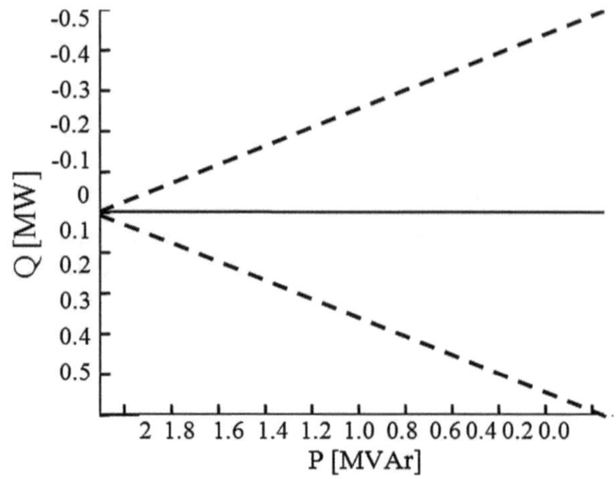

Fig. 7.2 *Q* property of an individual WT appertaining to G80-2.0 MW

Table 7.1 Properties of wind power

Parameter	Value	Parameter	Value
Diameter	80 m	PN	2.1 MW
Rotational speed	1680 rpm	Voltage	0.68 kV
Type	DFIG	Power factor	0.978 CAP–0.957 IND

technology is employed. The capacity of reactive power injection into the grid generally is associated to the used control approach, the active power generation and the converter size. The P-Q quality of the WTs employed in this study is given in Fig. 7.2. Specifically, it shows the business wind generator Gamesa WT G80-2.0 MW [26]; the useful data are given in Table 7.1. This WT acts with Power Factor (*PF*) 0.98 capacitive and 0.96 inductive. Consequently, the reactive power capability is bounded relying on the active power production. Its reactive power capacity is shown with the WT's trait and the influences of lines as well as cables. Thus, the WFP-Q trait is analogous to the same for the WTs. However, it's oriented to the capacitive side, as displayed in Fig. 7.3.

The green line shows the *PQ* trait of WF for $PF = 1$. For a power production less than 10 MW, the cable influence is larger than the transformer influence, since the transformer influence is paramount for large output power from WF. Thus, novel consumption and production of WF areas are modeled, for a range in which the WF alters the reactive power requisites exists. If the WF gets a capacitive reactive power in the small active power production range, the WTs set point will alter to be inductive. Voltage control can be performed via reactive power injection as well as transformer taps, like the way it is given in Table 7.2.

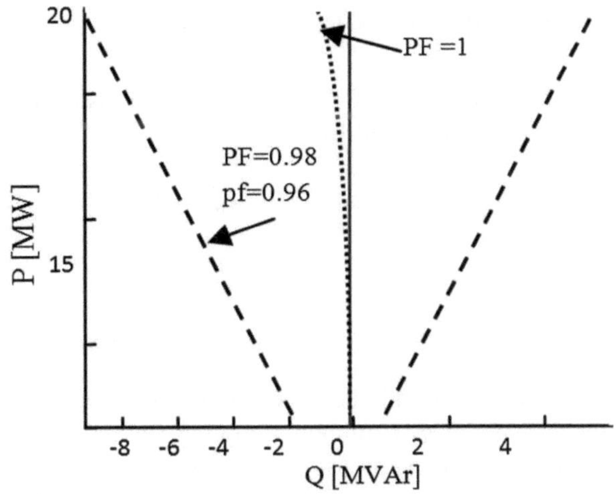

Fig. 7.3 *Q* property of the tested WF comprised of twelve G80-2.0 MWWTs, with no compensation gear

Table 7.2 Voltage and reactive power control alternatives in wind farms

Model	Control variable			
	Reactive power	Voltage	Discrete (Dis)/Continuous (Con)	Response time
Capacitor	•		Dis	Slow
STATCOM	•		Con	Fast
SVC	•		Con	Fast
DFIG	•		Con	Fast
Sync	•		Con	Fast
Tap		•	Dis	Slow

7.3.1 Objective Function

As the WF has to be capable to acquire reactive power from the grid, the optimum handle or reactive power resources in the WF must be handled. For optimizing the RPD among WTs and the control of various equipment like STATCOM or capacitor bank, one HBMO algorithm can be used. This fitness is given to bring the active power drops to a minimum along the WF cables or lines

$$\min J(Var_x, Var_y) = \min P_{losses} \tag{7.14}$$

where, Var_y denote the transformer tap location, the STATCOM reactive power setting and the capacitor bank state as independent variables, and Var_x represent the dependent variables that are the single WT reactive power outputs.

The control parameters are given by a *j*-dimensional vector, where *j* is the number of the enhanced variables. Meanwhile, any vector denotes one solution. *i* solutions exist and each one is a nominee answer.

7.3.2 Objective Constraints

The reactive power of each WT, tab of transformers and the STATCOM reactive power are constrained by

$$Var_{WT_i}^{\min} \le Var_{WT_i} \le Var_{WT_i}^{\max}, i = 1, 2, \ldots, N_G \tag{7.15}$$

$$T_i^{\min} \le T_i \le T_i^{\max} \tag{7.16}$$

$$Var_{Statcom}^{\min} \le Var_{Statcom} \le Var_{Statcom}^{\max} \tag{7.17}$$

Moreover, an additional equality limitation exists. The reactive power prerequisite in Point of Common Coupling (PCC) for the voltage control task can be modeled as an equality limitation by

$$Var_{PCC}^{*} = Var_{PCC}^{meas} \tag{7.18}$$

Within this study, the exploration of possible answers is employed to certificate an answer which meets the limitations that can be defined as follows

$$SO_i^{k+1} = \begin{cases} SO_i^k + v_i^{k+1}, & SO_i^{\min} \le SO_i^k + v_i^{k+1} \le SO_i^{\max} \\ SO_i^{\max}, & SO_i^k + v_i^{k+1} > SO_i^{\max} \\ SO_i^{\min}, & SO_i^{\min} > SO_i^k + v_i^{k+1} \end{cases} \tag{7.19}$$

With Eq. (7.9), the inequality restraints are met, the equality restraint (7.18) still needs to be answered. Because in the WF the main elements influencing the reactive power flow are the reactive power consumption of transformers, a scheme for meeting the equality constraint is introduced. Yet, bearing in mind the goal to

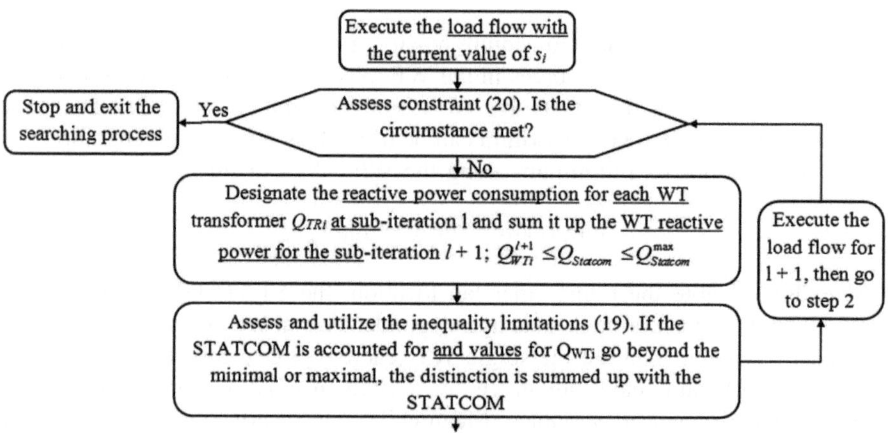

Fig. 7.4 Flowchart for the possible answer exploration process

decrease the searching CPU time to explore a possible answer, the equality constrain is improved and an error value ε is modeled.

$$\left|Var_{PCC}^{*} - Var_{PCC}^{meas}\right| < \varepsilon \tag{7.20}$$

This strategy is given in Fig. 7.4.

7.4 Voltage Stability Based RPP Model

Voltage deviation and stability limited VAr programming or Reactive Power Planning (RPP) is a significant demanding effect in power systems [32]. Investigations on voltage fixedness are fundamentally related to reactive power compensation sources. Having adequate reactive power compensation sources, principally in the shunt connection, the voltage stability boundary could be boosted a lot to make sure of system security [33]. Thus, appropriate programming of reactive power is a serious matter because of its specialized harness and the large expense of relocating a shunt compensator in economic terms when that is assembled. As an important demanding problem in power system research, RPP or VAr programming is a mixed integer nonlinear optimization programming with a huge scale of optimal variables [34]. The VAr programming is delineated in this part of applicable operational account limitations at various load levels. The mathematic formulation of optimization-based RPP model via accounting for voltage fixedness limitation could be stated as follows

$$\begin{aligned}
&\min \alpha \sum_{k=1,\ldots,N_L} P_{loss}^{(k)} + \beta \sum_{i=1,\ldots,N} Q_{ci} y_i + \gamma \sum_{i=1}^{Nd} \left|V_i - V_i^0\right|, \text{s.t.:} \\
&\sum_{i=1}^{N} y_i = N_c \\
&P_{G_i}^{(k)} - P_{L_i}^{(k)} - V_i^{(k)} \sum_{jw_i} V_j^{(k)} \times (G_{ij} \cos\theta_{ij}^{(k)} + B_{ij} \sin\theta_{ij}^{(k)}) = 0 \\
&Q_{G_i}^{(k)} + Q_{C_i}^{(k)}(y_i) - Q_{L_i}^{(k)} - V_i^{(k)} \sum_{jw_i} V_j^{(k)} \times (G_{ij} \sin\theta_{ij}^{(k)} - B_{ij} \cos\theta_{ij}^{(k)}) = 0 \\
&V_{i\min}^{(k)} \le V_i^{(k)} \le V_{i\max}^{(k)}, -S_{ij\max}^{(k)} \le S_{ij}^{(k)} \le S_{ij\max}^{(k)} \\
&P_{G_i\min}^{(k)} \le P_{G_i}^{(k)} \le P_{G_i\max}^{(k)}, i \in NG \\
&Q_{C_i\min}^{(k)} \le Q_{C_i}^{(k)} \le Q_{C_i\max}^{(k)}, i \in NG \\
&P_{tieline}^{(k)} \le TTC(Q_C^{(k)})
\end{aligned} \tag{7.21}$$

and,

$$P_{loss}^{(k)} = \sum g_{ij}((V_i^{(k)})^2 + (V_j^{(k)})^2 - 2V_i^{(k)} V_j^{(k)} \cos\theta_{ij}^{(k)}) \tag{7.22}$$

where, NG, N, N_L and N_c, are number of generators, number of buses, number of operation load levels and number of VAr sources which must be installed, respectively.

Subscript k indicates different load levels which k = 1,…, N_L. $T^{(k)}$, α, β and y_i are the time period of load level kth year, the energy expense per kWh, the calculated mean yearly preservation and assignment expense and binary variable (y_i = 1" if the VAr tool is installed at bus i, or else, "0"), respectively. PG_i, QG_i, PL_i, and QL_i are the generator active outputs, reactive power outputs, the load active and reactive power demands at bus i, respectively. S_{ij}^k is the line flow of line i-j where load level is k; other than for the value of zero, Q_{ci} is housed in a certain VAr capacity interval at bus i, Q_{ci}^k is the VAr capacity needed at load level k, and Q_{ci} is the ultimate VAr size at bus i; $P_{tieline}$ is the total active power flow with the tie lines from the source zone to sink zone; $TTC(Q_c^{(k)})$ is a piecewise linear interpolation function employed as static voltage fixedness limitation to be delineated later in this research, and Q_c is the Q_{ci} set at all nominee buses.

The Eq. (7.3) brings the yearly expense of power system, voltage deviation and real falloffs plus VAr tool to a minimum. Nonetheless, additional purposes could be accounted for namely bringing generation expense and VAr installation expense, etc. to a minimum However, such points do not alter the point of focus in this study, and other purpose performances could be accounted for if it is needed in a certain power system. The mentioned VAr programming plane is static voltage stability (SVS) limited OPF model. The important cause for the computational demand of the voltage fixedness limited OPF model is the requisite of two constraints and variables sets associated to the regular performance and drop points [32]. The two variables sets offer a demand to answer the optimization model, particularly for a huge power system having several conditions. To depict this demand, TTC could be employed to roughly offer the SVS limitation. At a stable VAr compensation model by accounting for certain plausible conditions that are clarified in advance, the security limited TTC optimization model is given by

$$
\begin{aligned}
&\min_{k}\left\{\max \sum_{\substack{i\,\in\, Source\,Area\\ j\,\in\, Sink\,Area}} \left(P_{ij}^{(k)}-P_{ij0}\right)\right\}, \text{s.t.:}\\
&\sum_{i=1}^{N} y_i = N_c\\
&P_{G_i}^{(k)}-P_{L_i}^{(k)}-V_i^{(k)}\sum_{jw_i} V_j^{(k)}\times\left(G_{ij}\cos\theta_{ij}^{(k)}+B_{ij}\sin\theta_{ij}^{(k)}\right)=0\\
&Q_{G_i}^{(k)}+Q_{C_i}^{(k)}(y_i)-Q_{L_i}^{(k)}-V_i^{(k)}\sum_{jw_i} V_j^{(k)}\times\left(G_{ij}\sin\theta_{ij}^{(k)}-B_{ij}\cos\theta_{ij}^{(k)}\right)=0\\
&V_{i\min}^{(k)}\le V_i^{(k)}\le V_{i\max}^{(k)},\ -S_{ij\max}^{(k)}\le S_{ij}^{(k)}\le S_{ij\max}^{(k)}\\
&P_{G_i\min}^{(k)}\le P_{G_i}^{(k)}\le P_{G_i\max}^{(k)},\ i\in \text{Source Area}\\
&Q_{C_i\min}^{(k)}\le Q_{C_i}^{(k)}\le Q_{C_i\max}^{(k)},\ i\in \text{Source Area}\\
&\frac{P_{Li}^{(k)}}{P_{Li}^{0}}=\frac{Q_{Li}^{(k)}}{Q_{Li}^{0}},\ i\in \text{Sink Area}
\end{aligned}
\tag{7.23}
$$

where, Q^0_{Li} and P^0_{Li} are the base case reactive and real power demands at load bus i, respectively. P_{ij0} expresses the base case power flow between line i-j; and $P^{(k)}_{ij}$ indicates the line flow of line i-j, having Var compensation. Meanwhile, the superscript $k = 1,\ldots, N_{cntg}$, characterizes various performance statuses with the regular performance, and $k > 0$ characterizing the post-condition statuses for the kth condition phenomenon.

7.5 Simulation

A. Voltage stability in wind farm

Wind Energy is one of the greatly encouraging renewable energy sources in Mongolia. WF within Inner Mongolia owns 200 wind turbines which are broken down into 20 sets, and there are 10 turbines in any set. The total installed capacity is as high as 300 MW. The wind turbine increases to 35 kV by box-type transformer T2, afterwards by 20 35 kV coupled transmission lines to the substation, which is comprised of two principal parts of the 220 kV transformer step-up substation.

There are 403 nodes in the wind farm model, because of its sophistication and restricted area [35], only 42 nodes were selected as shown in Fig. 7.5. To account for cable lines losses, Static VAr compensation (SVC) is installed in the nodes for simulation. By computing the wind farm in the network, the optimum reactive compensation could be answered. Reference power SB is 100 MW, reference voltage is 220 kV, and the largest reactive investment is 500 million.

Encrypt the SVC and on-load tap modifier. 50 is the highest code, and 100 is the highest number of disasters. Table 7.3 shows the compared *of* reactive power optimization between Traditional Genetic Algorithm (TGA), Improved Genetic Algorithm (IGA) and Honey Bee Mating Optimization (HBMO). The IGA reduces the network losses and financing of compensation gear. And it's more appropriate than TGA.

B. Voltage stability founded on RPP

The IEEE 118-bus system [37, 38] is employed for case study, as indicated in Fig. 7.6. This power system is altered by decreasing maximal generator reactive power output and boosting reactive load, as indicated in Tables 7.4 and 7.5. This adjustment is employed in order to build the system stressed enough in the manner that the reactive power compensation is required. Therefore, the voltage intensity at bus 30 in the altered system is less than 0.944 p.u. in the case with heavy loading. The altered system information are employed as the "heavy-load" case. Afterwards, loads are moved down by 0.78 to construct the "medium-load" case, and moved down more by 0.76 to make the "light-load" case. In the total 8760 h yearly, heavy-load and light-load cases are presumed to be 1200 h each condition, and therefore, the medium-load case gets 6360 h. α, β and χ are 40 $/MWh, 3600 $/year and 6400 kV for each VAr tool.

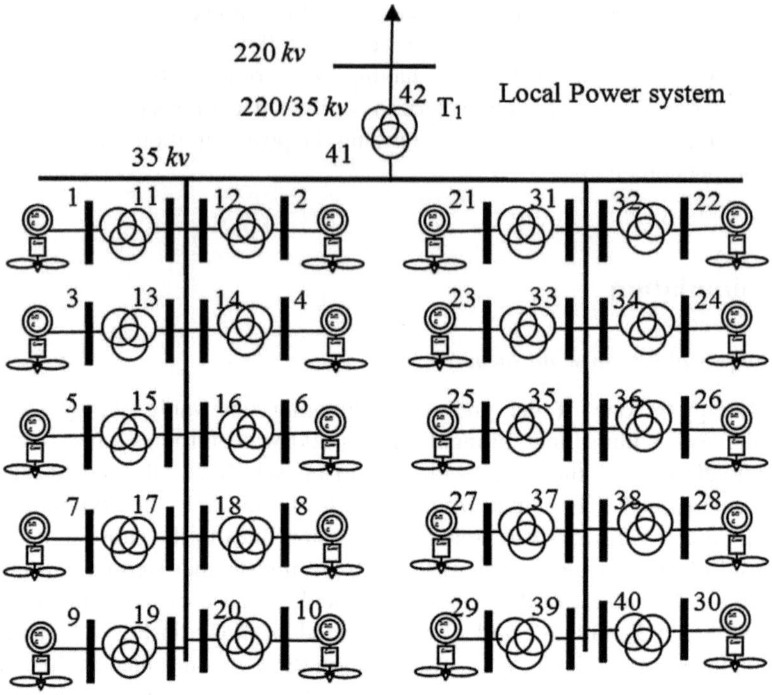

Fig. 7.5 The model of the wind farm having 42 nodes

Table 7.3 Contrasted conclusions of reactive power optimization in wind farms

Project	Financing of reactive power compensation (million Yuan)	The system falloff (kW)		
		$V = 4$ m/s	$V = 8$ m/s	$V = 12$ m/s
TGA [36]	338	1872	2480	3129
IGA [36]	336	1731	2292	2892
HBMO [1]	311	1721	2282	2671

In this section, certain indices founded on voltage fixedness are indicated to mention new nominee VAr tools. Furthermore, we must employ the indices that could provide data at all load buses. Hence, three indices constituted voltage and voltages fixedness associated information of any bus to appear in the index dataset matrix for fuzzy grouping. They are *L-index*, H/H_0 *Index*, and ultimately a voltage fluctuation index.

(1) H/H_0 *Index:* This index formulated in Eq. (7.24), is used to grant the test system delicate bus associated data.

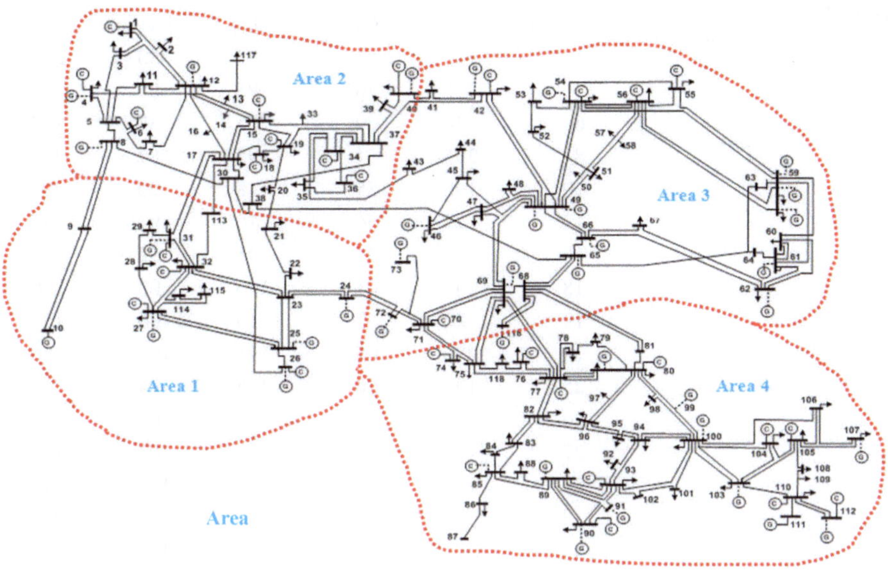

Fig. 7.6 IEEE 118-bus system

Table 7.4 Altered reactive load demand in the IEEE 118-bus system

PQ bus no	*Q* MVar	*PQ* bus no	*Q* MVar	*PQ* bus no	*Q* MVar
2	24	30	5	57	17
3	43	33	26	58	17
5	6	35	36	60	83
7	24	37	5	63	5
9	5	38	4	64	5
11	75	39	36	67	33
13	39	41	45	68	5
14	19	43	21	71	5
16	30	44	21	75	51
17	17	45	58	78	72
20	23	47	32	79	41
21	19	48	24	81	5
22	15	50	21	82	55
23	13	51	21	83	25
28	21	52	23	84	16
29	28	53	28	86	26

Table 7.5 Altered generator maximal reactive power output

Gen no	$Q_{G\max}$	Gen no	$Q_{G\max}$	Gen no	$Q_{G\max}$
1	12	42	295	80	277
4	295	46	95	85	19
6	46	49	205	87	997
8	295	54	295	89	297
10	195	55	18	90	297
12	115	56	10	91	97
15	25	59	175	92	5
18	45	61	295	99	96
19	20	62	15	100	151
24	296	65	195	103	36
25	136	66	195	104	19
26	996	69	295	105	19
27	296	70	27	107	197
31	296	72	97	110	19
32	38	73	97	111	997
34	21	74	5	112	997
36	21	76	17	113	197
40	295	77	67	116	997

$$\frac{H}{H_0} = \left[\frac{H_1}{H_{01}}, \ldots, \frac{H_k}{H_{0k}}\right] \tag{7.24}$$

where, H and H_0 are base case voltage intensity vectors at each bus and the voltage intensity with all loads adjusted to 0 for the power system [39].

(2) *L-Index:* With the boosted loading and utilization of the power transmission system and also because of enhanced optimized performance, the issue of voltage fixedness and voltage drop draw increasing attention. A voltage drop could happen in systems or subsystems and could emerge very suddenly. *L-index*, L_j of the jth bus could be stated by using the equation below

$$\begin{cases} L_j = \left|1 - \sum_{i=1}^{N_{PV}} F_{ji}\left(\frac{V_i}{V_j}\right)\right| \\ [Y_1] \times F_{ji} = -[Y_2], j = 1, 2, \ldots, N_{PQ} \end{cases} \tag{7.25}$$

where, N_{PV} and N_{PQ} are number of PV and PQ bus, respectively. Parameters Y_1 and Y_2 are the system Y_{BUS} sub-matrices acquired following segregating the PQ and PV bus bar parameters as shown in Eq. (7.26)

$$\left\{ \begin{bmatrix} Y_1 & Y_2 \\ Y_3 & Y_4 \end{bmatrix} = \begin{bmatrix} I_{PQ} \\ I_{PV} \end{bmatrix} \times \begin{bmatrix} V_{PQ} \\ V_{PV} \end{bmatrix}^{-1} \right\} \tag{7.26}$$

where *L-index* is computed for the entire load buses and L_j indicates no load case and voltage drop circumstances of bus j in a possible numerical range of (0, 1).

(3) *Bringing voltage deviation to a minimum:* The three index of an RPP problem is to satisfy consumer's demands with the lease expense with a desirable expectance of persistence of supply and adequately little voltage deviation. It can be stated as

Table 7.6 Three indices employed for fuzzy grouping algorithm

PQ bus no	*L*-index	H/H_0	Voltage fluctuation (%)	*PQ* bus no	*L*-index	H/H_0	Voltage fluctuation (%)
2	0.9554	1.0441	1.4252	57	0.9552	1.0222	1.6092
3	0.9577	1.0431	1.6084	58	0.9519	1.0356	1.0193
5	0.9543	1.0721	1.6271	60	0.9728	1.0523	1.3364
7	0.9540	1.0692	1.5476	63	0.9765	1.0223	1.1155
9	0.9762	1.0561	1.0672	64	0.9627	1.0522	1.6726
11	0.9672	1.0365	1.1797	67	0.9512	1.0623	1.4891
13	0.9663	1.0702	1.2301	68	0.9573	1.0111	1.3433
14	0.9540	1.0454	1.4635	71	0.9603	1.0112	1.3234
16	0.9751	1.0302	1.0932	75	0.9540	1.0228	1.0472
17	0.9684	1.0821	1.4954	78	0.9521	1.0091	1.0253
20	0.9605	1.0761	1.0731	79	0.9521	1.0022	1.7192
21	0.9654	1.0456	1.4490	81	0.9521	1.0435	1.0451
22	0.9620	1.0541	1.3393	82	0.9692	1.0074	1.3584
23	0.9522	1.0510	1.5352	83	0.9723	1.0234	1.0665
28	0.9571	1.0181	1.4912	84	0.9693	1.0692	1.5623
29	0.9535	1.0261	1.6224	86	0.9632	1.0024	1.5616
30	0.9555	1.0413	1.6123	88	0.9664	1.0806	1.4963
33	0.9572	1.0211	1.2296	93	0.9585	1.0646	1.1035
35	0.9625	1.0727	1.4875	94	0.9723	1.0425	1.4534
37	0.9513	1.0162	1.1354	95	0.9534	1.0527	1.3557
38	0.9771	1.0196	1.0218	96	0.9723	1.0225	1.6687
39	0.9782	1.0149	1.5113	97	0.9529	1.0392	1.4454
41	0.9647	1.0197	1.3435	98	0.9630	1.0833	1.5493
43	0.9641	1.0378	1.3297	101	0.9682	1.0473	1.3134
44	0.9601	1.0272	1.6215	102	0.9728	1.0452	1.2973
45	0.9772	1.0813	1.4145	106	0.9523	1.0229	1.5662
47	0.9611	1.0374	1.4273	108	0.9779	1.0422	1.0548
48	0.9533	1.0161	1.5944	109	0.9733	1.0541	1.0915
50	0.9730	1.0751	1.5528	114	0.9642	1.0590	1.1195
51	0.9610	1.0873	1.3926	115	0.9632	1.0342	1.2683
52	0.9570	1.0382	1.1283	117	0.9633	1.0321	1.5714
53	0.9620	1.0094	1.1636	118	0.9578	1.0864	1.5515

$$J_2 = VD(x, u) = \sum_{i=1}^{Nd} |V_i - V_i^{sp}| \tag{7.27}$$

where, N_d is number of load buses. The fuzzy grouping technique is first performed for the heavy load case by employing three various kinds of indices: *L-index*, H/H_0 index, and the voltage fluctuation index. Particularly, the voltage fluctuation manifests the relative deviation of the voltage intensity at the maximal TTC between zones 2 and 4 from the voltage intensity at the base operation case. Table 7.6 indicates the value of the indices at *PQ* buses, and Fig. 7.7 demonstrates post-standardization indices.

Because ordinarily more than one nominee location is available for installation of VAr (particularly in huge-scale power system), the least four values for any index are analyzed. The six weakest buses are {79, 78, 81, 37, 67, 23} on the basis of *L-index*, buses {79, 76, 82, 78, 53, 68} on the basis of H/H_0 index, and buses {79, 58, 38, 78, 81, 75} employing the voltage fluctuation index. Based on graphical review, the indices do not create identical assessment for the whole buses, which could similarly be observed from Fig. 7.7.

To explore an all-out exploitation of all the indices, it is mandatory to bring in the fuzzy theory to cluster the weak load buses. The whole PQ buses in the test system are contained in the fuzzy grouping procedure. The fuzzy clustering method categorizes the whole load buses into three sets that are: {79}; {78, 76, 81}; {other PQ buses}. From the perspective of voltage intensities, Bus 79 has both the least base-case voltage intensity and the maximum voltage fluctuation, therefore it finally

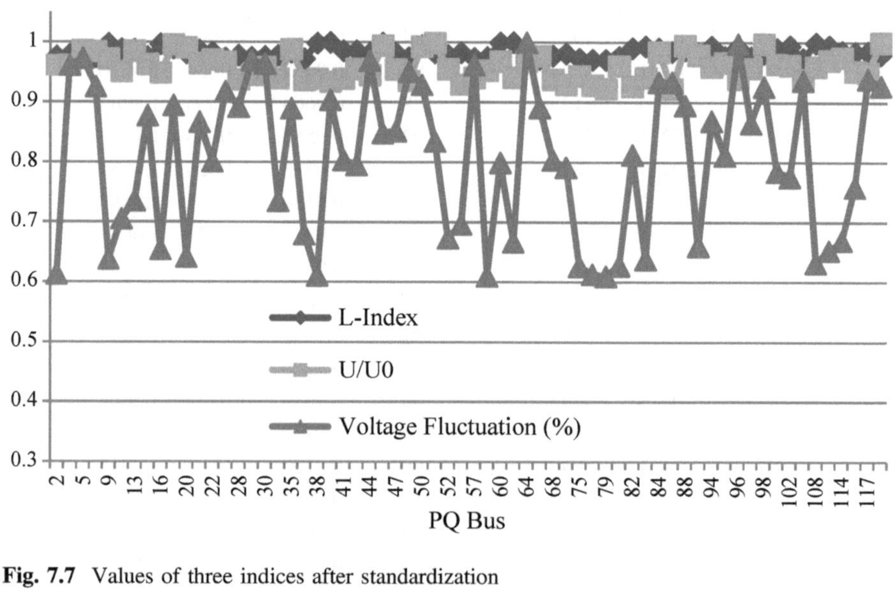

Fig. 7.7 Values of three indices after standardization

differentiates itself from the entire other buses. Furthermore, Buses 78, 76 and 81 are weak buses possessing resembling properties categorized in the second group. From the conclusions of the fuzzy grouping algorithm, Buses 30, 78, 76, and 81 are selected as nominee buses for novel VAr tools.

7.6 Conclusion

In this chapter, different models of reactive power dispatch/control system has been illustrated to minimize instability of bus voltage and generators using coordinate and optimal design of the generator voltages, transformers, switchable reactors and capacitors. Control of bus voltage has been shown to modify the reactive power aims and overtake them via the communication power system to the simulator. The RPD problem is implemented based on the active power aims place by active dispatch software.

Examinations of voltage fixedness are basically associated to reactive power compensation sources. Having sufficient reactive power compensation sources, mainly in the shunt connection, the voltage fixedness confine could be much booted to make sure of system security. This chapter solves the voltage stability as view of reactive power control problem by accounting for conditions and various load levels. In other words, this chapter offers voltage fixedness in customary distribution systems. Reactive power transmission, various voltage unstableness mechanisms and the static and dynamic reactive power sources role in various voltage unstableness mechanisms are examined. The chapter commences with a sketch of understood conclusions to update the audience's knowledge. Afterwards, the case study will demonstrate certain significant characteristics of the voltage fixedness concept in distribution systems. Moreover, reactive power dispatch in the wind farm is debated. It shows voltage stability with wind farm is better if correctly models its uncertainties.

References

1. A. Ghasemi, K. Valipour, A. Tohidi, Multiobjective Optimal Reactive Power Dispatch Using a New Multiobjective Strategy, Electrical Power and Energy Systems, vol. 57, pp. 318–334, 2014.
2. K. Ayana, U. Kilic, Artificial Bee Colony Algorithm Solution for Optimal Reactive Power Flow, Applied Soft Computing, vol. 12, pp. 1477–1482, 2012.
3. T.A. Short, Electric Power Distribution Handbook, CRC Press LLC, 2004.
4. T. Gonen, Electric Power Distribution System, McGraw-Hill Book Company, 1986.
5. M. Paramasivam, A. Salloum, V. Ajjarapu, V. Vittal, N.B. Bhatt, S. Liu, Dynamic Optimization Based Reactive Power Planning to Mitigate Slow Voltage Recovery and Short Term Voltage Instability, IEEE Trans. Power Syst., vol. 28, no. 4, pp. 3865–3873, Nov. 2013.
6. C.W. Taylor, Power System Voltage Stability, New York, NY, USA: McGraw-Hill, 1994.

7. B. Zhou, K.W. Chan, T. Yu, C.Y. Chung, Equilibrium-Inspired Multiple Group Search Optimization with Synergistic Learning for Multiobjective Electric Power Dispatch, IEEE Trans. Power Syst., vol. 28, no. 4, pp. 3534–3545, Nov. 2013.
8. M. Zare, T. Niknam, A New Multiobjective for Environmental and Economic Management of Volt/Var Control Considering Renewable Energy Resources, Energy, vol. 55, pp. 236–252, 2013.
9. D.S.B. Alencar, D.A.A. Marcio Formiga, Multiobjective Optimization and Fuzzy Logic Applied to Planning of the Volt/Var Problem in Distributions Systems, IEEE Trans. Power Syst., vol. 25, no. 3, pp. 1274–1281, 2010.
10. T. Niknam, M. Zare, J. Aghaei, Scenario Based Multiobjective Volt/Var Control in Distribution Networks Including Renewable Energy Sources, IEEE Trans Power Deliv., vol. 27, no. 4, pp. 2004–2019, 2012.
11. A. Ghasemi, H. Shayeghi, H. Alkhatib, Robust Design of Multimachine Power System Stabilizers Using Fuzzy Gravitational Search Algorithm, Electrical Power and Energy Systems, vol. 51, pp. 190–200, 2013.
12. G. Rogers, Power System Oscillations, 1st Ed, Springer, 1999.
13. H. Shayeghi, A. Ghasemi, A Multiobjective Vector Evaluated Improved Honey Bee Mating Optimization for Optimal and Robust Design of Power System Stabilizers, Electrical Power and Energy Systems, vol. 62, pp. 630–645, 2014.
14. M. Noshyar, H. Shayeghi, A. Talebi, A. Ghasemi, N.M. Tabatabaei, Robust Fuzzy-PID Controller to Enhance Low Frequency Oscillation Using Improved Particle Swarm Optimization, International Journal on Technical and Physical Problems of Engineering (IJTPE), vol. 5, no. 1, pp. 17–23, 2013.
15. A. Ghasemi, M.J. Golkar, A. Golkar, M. Eslami, Reactive Power Planning Using a New Hybrid Technique, Soft Comput., vol. 20, pp. 589–605, 2016.
16. K.Y. Lee, Y.M. Park, J.L. Ortiz, A United Approach to Optimal Real and Reactive Power Dispatch, IEEE Trans. Power Appar. Syst. PAS, vol. 104, no. 5, pp. 1147–1153, 1985.
17. S. Granville, Optimal Reactive Power Dispatch through Interior Point Methods, IEEE Trans. Power Syst., vol. 9, no. 1, pp. 98–105, 1994.
18. N.I. Deeb, S.M. Shahidehpour, An Efficient Technique for Reactive Power Dispatch Using a Revised Linear Programming Approach, Int. J. Electr. Power Syst. Res., vol. 15, pp. 121–134, 1988.
19. N. Grudinin, Reactive Power Optimization Using Successive Quadratic Programming Method, IEEE Trans. Power Syst., vol. 13, no. 4, pp. 1219–1225, 1998.
20. Z. Wen, L. Yutian, Multiobjective Reactive Power and Voltage Control Based on Fuzzy Optimization Strategy and Fuzzy Adaptive Particle Swarm, Int. J. Electr. Power Energy Syst., vol. 30, pp. 525–532, 2008.
21. Q.H. Wu, Y.J. Cao, J.Y. Wen, Optimal Reactive Power Dispatch Using an Adaptive Genetic Algorithm, Int. J. Electr. Power Energy Syst., vol. 20, no. 8, pp. 563–569, 1998.
22. D. Nualhong, S. Chusanapiputt, S. Phomvuttisarn, S. Jantarang, Reactive Tabu Search for Optimal Power Flow under Constrained Emission Dispatch, Proc. Tencon., pp. 327–330, 2004.
23. H. Yoshida, K. Kawata, Y. Fukuyama, S. Takayama, Y. Nakanishi, A Particle Swarm Optimization for Reactive Power and Voltage Control Considering Voltage Security Assessment, IEEE Trans. Power Syst., vol. 14, no. 4, pp. 1232–1239, 2000.
24. L.D. Arya, L.S. Titare, D.P. Kothari, Improved Particle Swarm Optimization Applied to Reactive Power Reserve Maximization, Int. J. Electr. Power Energy Syst., vol. 32, pp. 368–374, 2010.
25. Z. Hua, X. Wang, G. Taylor, Stochastic Optimal Reactive Power Dispatch: Formulation and Solution Method, Int. J. Electr. Power Energy Syst., vol. 32, pp. 615–621, 2010.
26. D.B. Das, C. Patvardhan, Reactive Power Dispatch with a Hybrid Stochastic Search Technique, Int. J. Electr. Power Energy Syst., vol. 24, pp. 731–736, 2002.
27. D. Karaboga, S. Okdem, A Simple and Global Optimization Algorithm for Engineering Problems: Differential Evolution Algorithm, Turk. J. Electr. Eng., vol. 12, no. 1, 2004.

28. K. Ayana, U. Kilic, Artificial Bee Colony Algorithm Solution for Optimal Reactive Power Flow, Appli Soft Compu., vol. 12, pp. 1477–1482, 2012.
29. P. Vovos, A. Kiprakis, A. Wallace, G. Harrison, Centralized and Distributed Voltage Control: Impact on Distributed Generation Penetration, IEEE Trans Power Syst., vol. 22, no. 1, pp. 476–83, 2007.
30. J.K. Kaldellis, K.A Kavadias, A.E. Filios, A New Computational Algorithm for the Calculation of Maximum Wind Energy Penetration in Autonomous Electrical Generation Systems, Appl. Energy, vol. 86, pp. 1011–23, 2009.
31. R. Almeida, E. Castronuovo, J. Lopes, Optimum Generation Control in Wind Parks When Carrying out System Operator Requests, IEEE Trans Power Syst., vol. 21, no. 2, pp. 718–25, 2006.
32. A. Monica, A. Hortensia, A.O. Carlos, A Multiobjective Approach for Reactive Power Planning in Networks with Wind Power Generation, Renewable Energy, vol. 37, pp. 180–191, 2012.
33. H. Amaris, M. Alonso, Coordinated Reactive Power Management in Power Networks with Wind Turbines and Facts Devices, Energy Conversion and Management, vol. 52, no. 7, pp. 2575–2586, 2011.
34. S. Ramesha, S. Kannan, S. Baskar, Application of Modified NSGA-II Algorithm to Multiobjective Reactive Power Planning, Applied Soft Computing, vol. 12, pp. 741–753, 2012.
35. J. Wiik, J O. Gjerde, T. Gjengedal, Steady State Power System Issues When Planning Large Wind Farms," IEEE Power Engin Soci Win Meeting, 2002, pp. 657–661.
36. Z. Xiang Jun, T. Jin, Z. Ping, P. Hui, W. Yuan Yuan, Reactive Power Optimization of Wind Farm based on Improved Genetic Algorithm, Energy Procedia, vol. 14, pp. 1362–1367, 2012.
37. A. Rabiee, M. Vanouni, M. Parniani, Optimal Reactive Power Dispatch for Improving Voltage Stability Margin Using a Local Voltage Stability Index, Energy Conversion and Management, vol. 59, pp. 66–73, 2012.
38. A. Ghasemi, H. Afaghzadeh, O. Abedinia, S.N. Mohammad, Artificial Bee Colony Algorithm Technique for Economic Load Dispatch Problem, Proceedings of EnCon2011 4th Engineering Conference Kuching, Sarawak, Malaysia, pp. 1–6, 2011.
39. Y. Wang, F. Li, Q. Wan, H. Chen, Reactive Power Planning Based on Fuzzy Clustering, Gray Code, and Simulated Annealing, IEEE Transactions on Power Systems, vol. 26, no. 4, pp. 2246–2255, 2011.

Chapter 8
Reactive Power Compensation in AC Power Systems

Ersan Kabalci

Abstract This chapter introduces most widely used reactive power compensators considering the recent advances seen in industrial applications. In order to provide better and deeper knowledge for authors, the basic principles of reactive power compensation and symmetrical systems are presented primarily. The theoretical backgrounds are discussed by comparing each approach and application types in detail. The remainder of the chapter is organized by considering the comprehensive figure that is illustrated in the third section. Thereby, the first generation conventional compensators and lately improved FACTS are introduced in the following sections. Furthermore, the compensation devices are also listed according to their integration to transmission line as shunt, series, and shunt-series devices. The circuit diagrams and control characteristics of each compensation device are presented with its analytical expressions. The power flow control, voltage and current modifications, and stability issues are illustrated with phasor diagrams in order to create further knowledge on operation principles for each device. The comparisons are associated with similar devices and emerging technologies.

8.1 Introduction

This chapter deals with reactive power definition, analytical background, and compensation methods applied for reactive power. The reactive power compensation is also known as VAR compensation in several textbooks. The VAR compensation implies the volt-ampere-reactive that is unit of the reactive power. The demands of lower power losses, faster response to parameter change of the system, and higher system stability have stimulated the development of the flexible ac transmission systems (FACTS) that stands for compensation systems connected to the transmission line in series or shunt. Besides the series and shunt connections of

E. Kabalci (✉)
Department of Electrical and Electronics Engineering, Faculty of Engineering and Architecture, Nevsehir HBV University, Nevsehir, Turkey
e-mail: kabalci@nevsehir.edu.tr

N. Mahdavi Tabatabaei et al. (eds.), *Reactive Power Control in AC Power Systems*, Power Systems, DOI 10.1007/978-3-319-51118-4_8

controllers, the most comprehensive compensators are implemented with semi-conductor converters based on multilevel topologies.

The static synchronous compensator (STATCOM) that is usually defined as self-commutated or static VAR compensator is a voltage source converter (VSC) based on controllable switches to control the reactive power Q continuously. Furthermore, STATCOM is shunt connected to the utility grid or system at the point of common coupling (PCC). The maximum value of Q is adjusted relatively to the voltage, and the maximum available Q is slightly decreased when the voltage is decreased. Thus, the instability of the voltage causes imbalances on Q. The load current consists of three components as active power (P), reactive power (Q), and harmonic contents where the demanded current should be purely sinusoidal and in phase with the line voltage. The STATCOM also eliminates harmonic contents of the generated voltage and current waveforms by using several control methods and topological configurations.

The multipulse converters are developed using the most widely known six-pulse configurations. The variations of multipulse converters are built by combining six-pulse converters via phase-shifting isolation transformers. On the other hand, multilevel converters were developed as an alternative to the multipulse configurations owing to their multi MVA switching capability that is inherited from series or parallel connection of modular cells. The most widely known topologies of multilevel converters are diode clamped, flying capacitor, and cascaded H-bridge configurations that are also introduced in this chapter. The multilevel converter topologies provide several advantages such as harmonic elimination, lower electromagnetic interference, better output waveforms, and increased power factor correction capabilities together. Furthermore, each switch can be controlled individually to robustly tackle the unbalanced load operations even in higher switching frequencies relatively to the multipulse configuration. The following sections introduce the basic principles of reactive power compensation, the state-of-art in compensator devices, conventional and FACTS compensators. The control and operation characteristics of converters are also surveyed in terms of the main topological issues.

8.2 Basic Principles of Photovoltaic Energy Conversion

8.2.1 Background

The increased energy demand of the power industry has caused several requirements since past a few decades. The distributed generation (DG) and microgrid opportunities that are additionally promoted by renewable energy source (RES) integration have accelerated the growth of conventional grid. Therefore, the number of power plants types and substations, and the length of transmission and distribution lines are increased rapidly than ever before. The most widely installed

Fig. 8.1 The power triangle

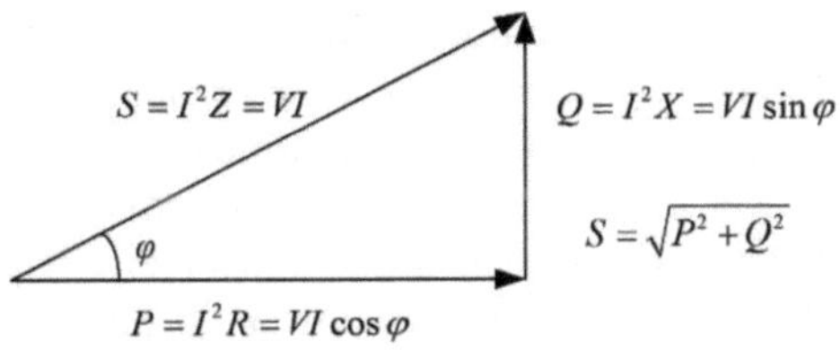

devices to control the conventional grid are mechanic circuit breakers where the response times are quite long to interact rapidly against the frequently changing loads. The loads require electrical or magnetic field to generate the reactive power. The reactive power consumers using inductive power are electrical motors, generators with lagging power factor, transformers, and transmission lines with heavy current loads. The transmission lines without any load or with low-level loads, rectifiers with capacitor filters, capacitors, generators operating at lagging power factor generate reactive power to meet the electrical field requirement where these kinds of loads presents capacitive power.

The electrical power is defined in three types as apparent power S, active power *P*, and reactive power *Q* in ac where the apparent power consist of real part P and imaginary part *Q* as seen in Fig. 8.1.

The equations seen in Fig. 8.1 are depended to impedances where apparent resistance and reactance compose it as follows;

$$Z = R + jX \tag{8.1}$$

where the reactance is depended on the particular frequency under consideration and can be calculated for capacitances or inductances as given in Eq. (8.2), respectively [1–3];

$$X_C = \frac{1}{\omega C}, \quad X_L = \omega L \tag{8.2}$$

The equations given above prove that the reactive power is depended on the reactance where the voltage and frequency are effective parameters. The ac power systems are expected to present constant voltage and frequency at any node under ideal conditions. Furthermore, the total harmonic distortion (*THD*) ratio should be zero and power factor (*PF*) should be unity in an ideal ac power system. The constant frequency is provided by sustaining the balance between generated and consumed power rates that is completely irrespective of voltage.

On the other hand, the voltage plays a vital role on system stability where it can be easily affected by varying reactive power. Thus, it clearly depicts that the focus of reactive power compensation is related to pursuing the system stability for all players in a transmission system, for both of generators and loads. The reactive power compensation is handled in two aspects as load compensation to improve the power quality for individual or particular loads, and transmission compensation that deals with long-distance and high voltage transmission lines [4].

The load compensation is expected to manage three main objectives that are listed as *power factor correction, load balancing,* and *voltage regulation*. The power factor correction minimizes the required reactive power that is met by central power stations. A great part of industrial loads operates at lagging PF that causes reactive power demand to consume. The power factor correction is used to provide the required reactive power locally instead of consuming from utility grid. Therefore, the *PF* of loads are increased up to unity value by decreasing the reactive power demand from utility, and the efficiency and capacity of generating stations are improved that support to sustain voltage stability on the transmission lines.

The *load balancing* is also performed in a similar way to power factor correction in order to decrease the required load current that tends to be higher than required active power rates. Since the higher load current increases joule losses in transmission and distribution lines, the load balancing allows minimizing line losses by this way. The *voltage regulation* is related to maintain utility supply at the allowed limits against rapidly and heavily changing load situations, that cause to voltage dips and flickers. It is based on installing a robust power system including high sized and number of generating units, and interconnecting to construct an intensive network. Since this approach seems high-cost and open to severe faults, the appropriate sizing and reactive power compensation of transmission and distribution lines can carry out another solution. The effect of *PF* on the transmission line losses can be expressed analytically as seen in Eq. (8.3) where the losses are quadratically increased against decreasing *PF*.

$$P_L = \frac{lP^2}{kAV^2(\cos\varphi)^2} \tag{8.3}$$

The parameters given in the equation are;

l = length of the conductor,
P = carried active power,
k = electrical conductivity,
A = cross-section of the conductor,
V = line voltage,
$\cos\varphi$ = fundamental frequency *PF*.

The transmission line losses can be decreased to k_{PL} factor by improving the $\cos\varphi$ fundamental frequency PF to $\cos\varphi_1$;

$$k_{PL} = 1 - \left(\frac{\cos\varphi}{\cos\varphi_1}\right)^2 \tag{8.4}$$

where the reactive power compensation in power systems provides to increase system stability by managing the *PF*. The reactive power compensation helps to increase available maximum load of any transmission line to the thermal limits under stability ranges without complex sizing requirements. This is obtained by using traditional reactive power compensations such as series or shunt capacitors, and variable compensators. On the other hand, the most recent compensation technologies under FACTS group enables to manage system stability relevant to voltage control, power demand control, and transient controls [1, 4].

8.2.2 The Theory of Reactive Power Compensation

The basic relations across the source and load should be realized to comprehend reactive power compensation theory. A pure resistive load as seen in Fig. 8.2a generates a phase difference δ between load voltage V and source voltage E while consuming power. The PF angle φ in a resistive loaded system is zero since that is between V and I. The voltage difference between source and load is expressed with jX_SI that is orthogonal to terminal voltage as shown in Fig. 8.2b.

The pure inductive loaded system and phasor diagram are illustrated in Fig. 8.3 referring to aforementioned approach. The pure inductive loads, i.e. shunt reactors used in tap-changing transformers and generation stations, do not draw power and δ between load voltage V and source voltage E is zero. Since the voltage drop jX_SI is in phase between V and E, the load voltage is easily affected by the inductive load current. This situation is the main reason of the shunt reactors usage to reduce the voltage.

The phasor diagram of the pure capacitive loaded system seen in Fig. 8.4a is opposite to pure inductive load. It also do not draw power and δ between load voltage V and source voltage E is zero as similar to pure inductive loaded system.

The capacitive loads, i.e. shunt capacitors, are used to increase the line voltage in the transmission and distribution lines, and this type of applications are used to

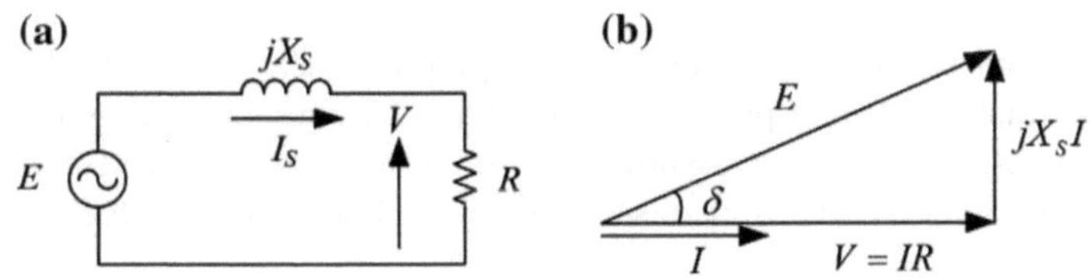

Fig. 8.2 Pure resistive loaded system, **a** circuit diagram, **b** phasor diagram

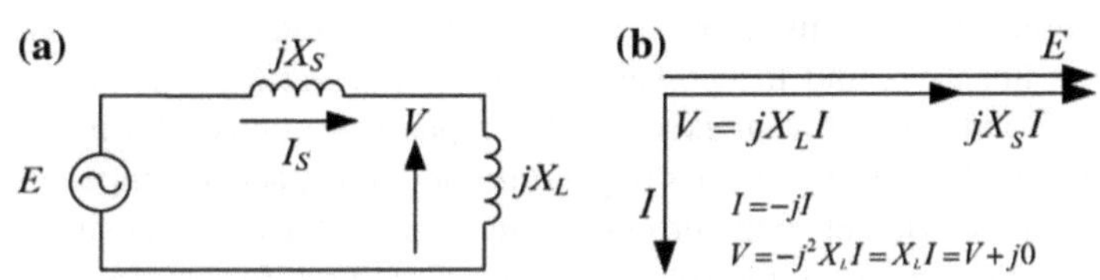

Fig. 8.3 Pure inductive loaded system, **a** circuit diagram, **b** phasor diagram

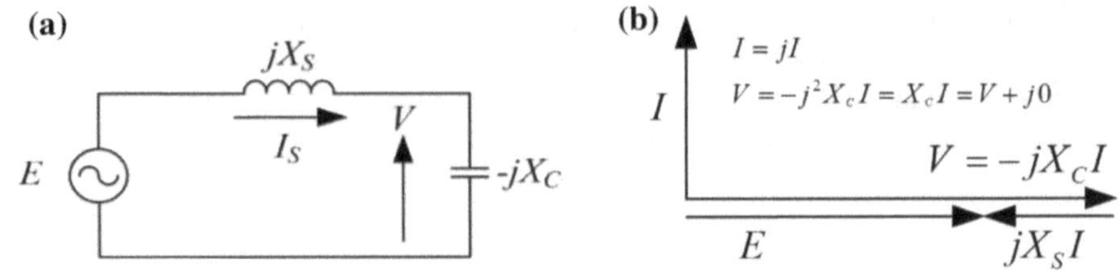

Fig. 8.4 Pure capacitive loaded system, **a** circuit diagram, **b** phasor diagram

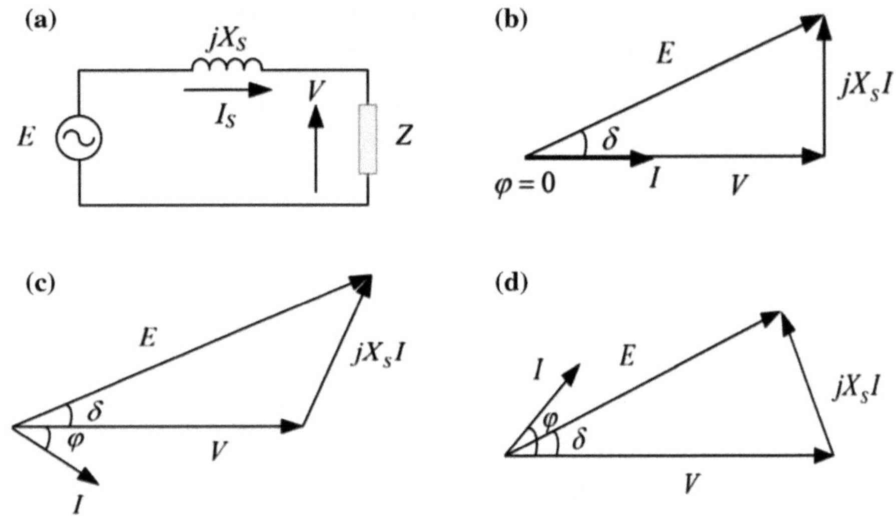

Fig. 8.5 *PF* analyses in an AC system, **a** circuit diagram, **b** resistive load phasor diagram, **c** inductive load phasor diagram, **d** capacitive load phasor diagram

tolerate the tendency to reduce the line voltage under heavy loads [4–6]. The load type defines that the current phasor to be in phase, lagging, or leading to the voltage phasor in an AC system. If we assume a power system installed as seen in Fig. 8.5a with terminal voltage E and transmission line impedance with jX_S, the PF is defined by load type.

In a resistive loaded application of this system, the current and terminal voltage will remain in same phase as seen in Fig. 8.5b where the system operates at unity *PF*. In an inductively loaded system, the current phasor is moved to clockwise as shown in Fig. 8.5c. In this case, the phase angle between current and terminal voltage φ is negative, and the power system operates at lagging PF that requires higher source voltage E for an exact load comparing to unity PF operation. In the third option, the leading PF occurs depending on the capacitive impedance of load where the phase angle between current and terminal voltage φ is positive as seen in Fig. 8.5d. At this operation, the source voltage E required for an exact load is lower than unity PF operation.

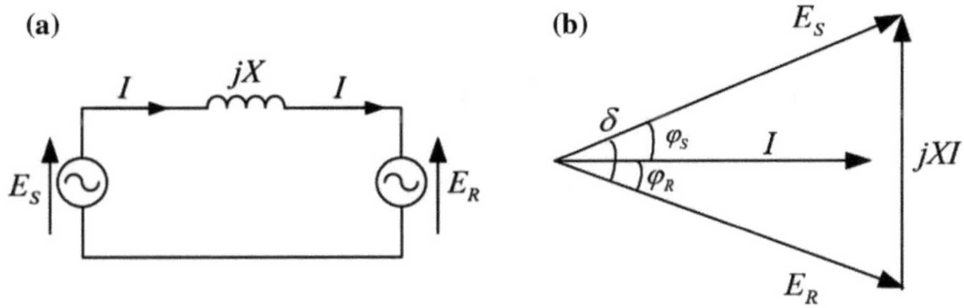

Fig. 8.6 Symmetrical system with source and receptor sections, **a** circuit diagram, **b** phasor diagram

The symmetrical systems where both sides of transmission line are comprised of generators or sources instead of source-load system are one of the most important application area of reactive power compensation. The most basic structure of such a system is illustrated in Fig. 8.6 similarly to pure resistive, inductive, and capacitive loaded systems.

The symmetrical system includes two sources at each side of transmission line where the basic definitions are done as source voltage E_S and receptor voltage E_R. However, this simple model can be assumed as generation stations connected over a line with jX impedance. Although the system may include several industrial loads, they are not considered in this power system since the main concern is directed to power exchange between generators. The bi-directional power flow is analyzed regarding to source voltage E_S, receptor voltage E_R, impedance X, and phase angle among voltage phasors δ as follows;

$$P = \frac{E_S E_R}{X} \sin \delta \tag{8.5}$$

Besides the phase angle δ, there are two *PF* angles for each terminal voltages where φ_S denotes the phase angle between source voltage E_S and I phasor while φ_R depicts the phase angle between receptor voltage E_R and I phasor [4, 6–8].

The basic idea of compensation is based on ideal shunt or ideal series compensators that constitute a lossless power system among the generators and loads connected over a transmission line as expressed in symmetrical system. The operation of an ideal shunt compensator consisting of capacitors compensates the power system by conditioning the *PF* angle where its operating principle is explained referring to Fig. 8.7. A symmetrical system and a shunt compensator integrated to the system at the middle of the transmission line is illustrated in the figure where source and receptor impedances (Z_S, Z_R), line parameters ($jX_{S/2}$), voltages at PCC with phase angles (E_S, E_R, δ_1, δ_2) are indicated. The shunt compensator connected in the middle of the transmission line is a voltage source that is continuously controlled where E_S and E_R sources supply same magnitude and phase angles are defined by the degrees of δ_1 and δ_2. Furthermore, voltage and phase

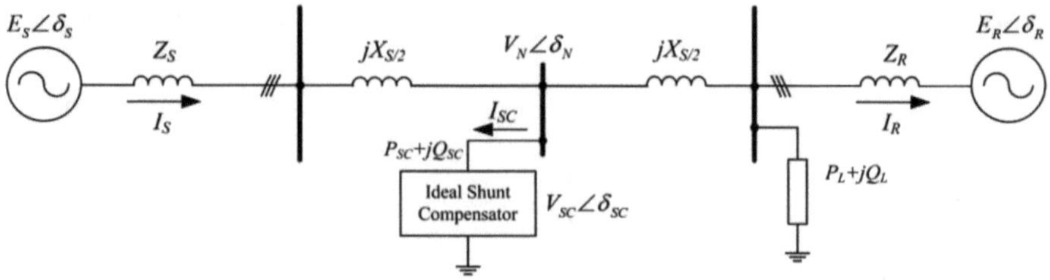

Fig. 8.7 Ideal shunt compensator connection to the symmetrical system

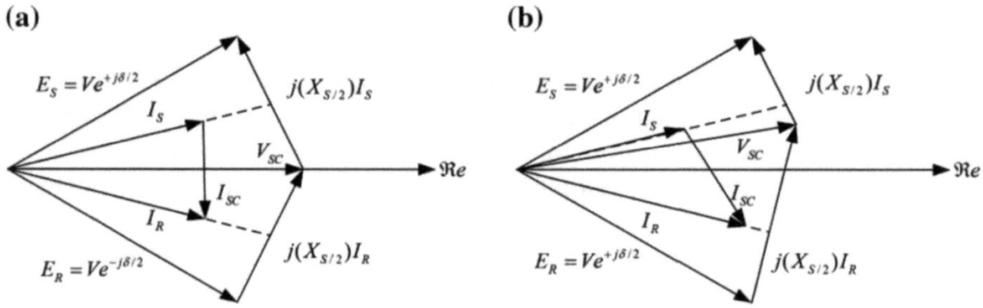

Fig. 8.8 Phasor diagrams of ideal shunt compensator, **a** reactive power compensation, **b** reactive and active power compensation

angles of the shunt compensator are expressed with V_{SC} and δ_2 as seen in the middle of the circuit diagram. The voltage and its phase angle at the node connection are depicted with V_N and δ_N.

The phasor diagrams of the symmetrical system with shunt compensation is depicted in Fig. 8.8 where the phase angles of E_S and E_R are assumed as (+δ/2) and (−δ/2), respectively. The phase angles indicate that I_S current flows from first source to the line while I_R flows from line to second source, namely receptor. The I_{SC} phasor that is orthogonal to the V_{SC} as seen in Fig. 8.8a shows the obtained current along the shunt compensator where it means the compensator does not consider the active power P for generation or consuming. In this case, the compensator only draws reactive power Q at the terminal connections. The power transferred from E_S to E_R is calculated as seen in Eq. (8.6) by assuming that there is not any P exchange between line and shunt compensator,

$$P_1 = \frac{2E^2}{X_S}\sin(\delta/2) \tag{8.6}$$

where the P_1 is the active power flowing from E_S source while E is the vector sum of E_S and E_R sources. The drawn power would be at the rate of calculated value of Eq. (8.7), if there is not any compensation is performed in the power system.

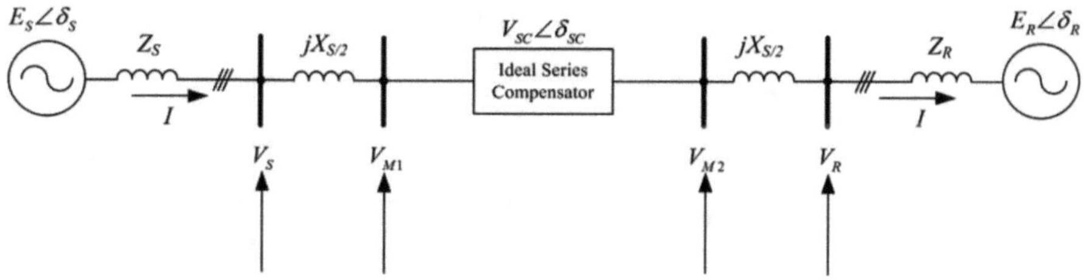

Fig. 8.9 Ideal series compensator connection to the symmetrical system

$$P_1 = \frac{V^2}{X_L}\sin(\delta) \tag{8.7}$$

Since [2 sin(δ/2)] is ever greater than sinδ, that is in the range of [0, 2π], the control capability of line is increased by the compensator. When the phase angles of E_S and E_R differ from δ/2, active and reactive power components are generated in the power exchange through sources as shown in Fig. 8.8b. In this situation, the shunt compensator should be operated to adjust any of P or Q owing to its power electronics based structure. Furthermore, the device structure also varies according to requirement of P or Q compensation since they have opposite phasor characteristic caused by the stored energy type [6–9].

The ideal series compensator that is connected to the transmission line in series is constituted by a voltage source V_{SC} as shown in Fig. 8.9. The current value that is delivered from source to receptor over transmission line is calculated as follows;

$$I = \frac{(E_{SR} - E_{SC})}{jX_S}, \quad E_{SR} = E_S - E_R \tag{8.8}$$

The series compensator utilizes capacitor banks to minimize the overall reactance of a transmission line at the line frequency where the reactance balance is arranged by the reactive power of capacitors. The reactive voltage that is generated by capacitors provides to improve phase angle and voltage stability in addition to load sharing optimization.

The phasor diagrams of the symmetrical system with ideal series compensation is depicted in Fig. 8.10 where the phase angles of E_S and E_R are assumed as (+δ/2) and ($-\delta$/2), respectively. The series compensator does not generate or consume any active power when the compensator voltage V_{SC} is generated orthogonal to the line current I. The power type of the series compensator is just reactive power and the reactance is balanced by using capacitive or inductive impedances. The equivalent impedance X_{eq} is calculated by

$$X_{eq} = X_S(1 + k) \tag{8.9}$$

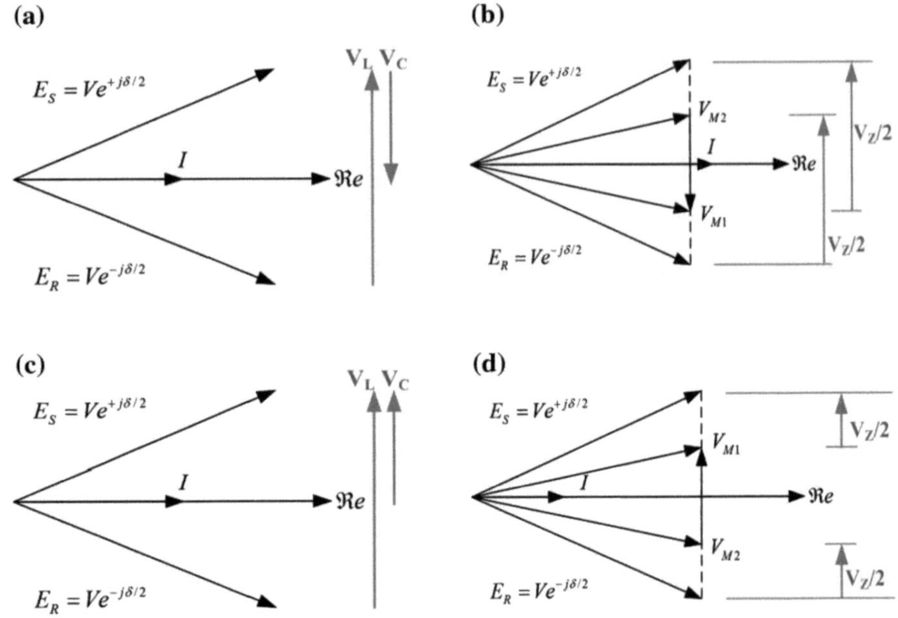

Fig. 8.10 Phasor diagram of ideal series reactive compensator; **a** capacitive operation without compensation, **b** capacitive operation with compensation, **c** reactive operation without compensation, **d** reactive operation with compensation

$$k = \frac{X_C}{X_S}, \quad 0 \le |k| \le 1 \tag{8.10}$$

where X_C is the series compensation reactance. The compensation factor k is positive at the inductive reactance and negative at the capacitive reactance. The compensation voltage V_{SC} and the delivered power are calculated as follows

$$V_{SC} = IX_{eq} \tag{8.11}$$

$$P_S = \frac{V^2}{X_S(1-k)} \sin\delta \tag{8.12}$$

where the equation depicts that the delivered power can be adjusted by selecting the proper k factor. Similarly, the reactive power delivered by series compensator to receptor is determined as seen in Eq. 8.13;

$$Q_S = \frac{2V^2 k}{X_S(1-k)^2}(1 - \cos\delta) \tag{8.13}$$

The phasor diagram of the symmetrical system seen in Fig. 8.9 is illustrated in Fig. 8.10a without any series compensation. The voltage phasor V_S corresponding to the line reactance X_S, and the compensation voltage phasor V_C are expressed on the right of Fig. 8.10b for an exact compensation value. The phasor of line current I leads to the compensation voltage V_C by 90° at this situation. In this case, the voltage drop occurred on the line $V_Z = E_C - E_S - V_C$ is greater than the regular voltage drop V_L. The compensation process increases the current flowing on the line according to the situation before compensation that proves the series compensation is essential to increase the delivered power.

The same symmetrical system without compensation is shown in Fig. 8.10c where the series inductive compensation is given in Fig. 8.10d. The phase difference of compensation voltage phasor V_C and drop voltage phasor V_L are 0° that provides generating lower equivalent voltage drop V_Z comparing to capacitive compensation seen in Fig. 8.10b in this case. This process causes to generating lower current to flow on the transmission line where the delivered power level is decreased. In any application of series compensation, capacitive or inductive, there is not any active power is generated or drawn.

The active power transfer characteristic according to power angle δ is illustrated in Fig. 8.11 in a transmission line where the cases are surveyed to compare the line without any compensation, with shunt compensation, with series compensation, and phase-shift compensation. The assumptions are performed regarding to Figs. 8.7 and 8.9, where the compensation factor of series compensator k is adjusted to 50% (0.5) regarding to Eq. (8.9).

The shunt compensator is not used to increase the power transfer capability in its regular operation angle that is originally lower than 90° and is usually around 30°. In case increasing the power is required the series compensation is the most appropriate selection as Fig. 8.11 depicts.

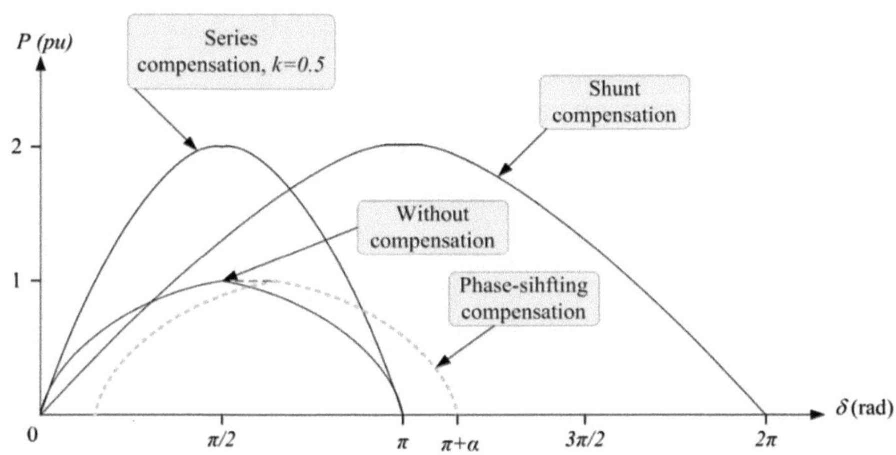

Fig. 8.11 Power transfer characteristics for several compensation cases

The most important supplement of shunt compensator is to increase the power stability of the system. The phase-shifting compensator has a great potential for symmetrical systems under exceeding or unpredictable phase differences. It is preferred to stabilize the phase fluctuations instead of increasing the power [6, 9–11].

8.2.3 Devices Used in Reactive Power Compensation

The FACTS originated from several reactive power compensators propose extensive use of power electronic devices and systems in ac power control. There are several FACTS devices are proposed and being studied since 1990. The first generation conventional compensators were constituted of inductive and capacitive reactive elements to build shunt, series, and shunt-series compensators as illustrated in Fig. 8.12. The improvements of power semiconductors promoted new device types and synthesizes of circuit topologies depending the capabilities of semiconductors. The aggregated terms of FACTS include several device types as seen on the right hand side of Fig. 8.12 where the commutation style is one of the most

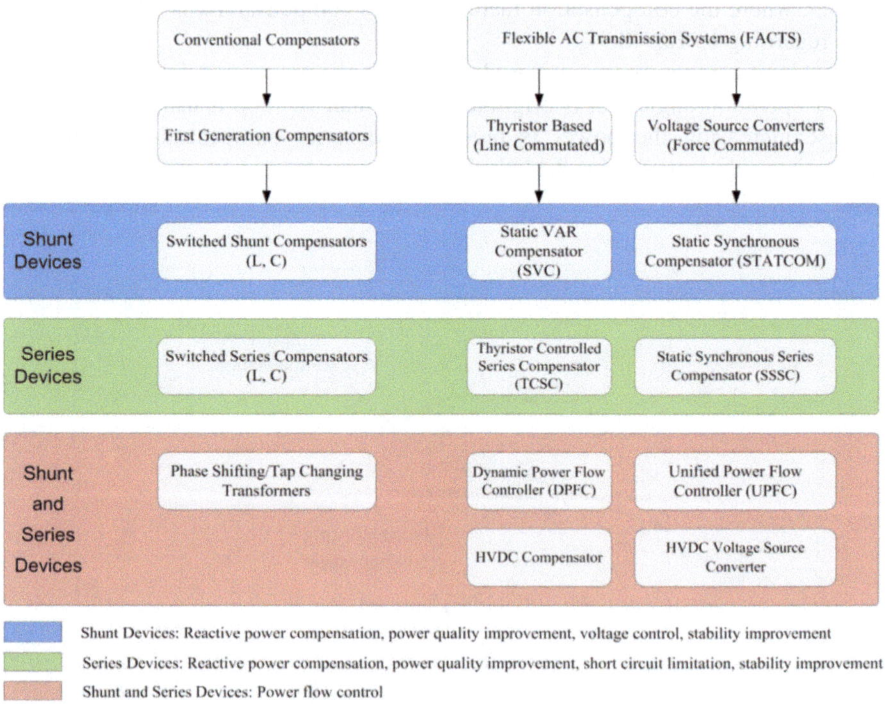

Fig. 8.12 A comprehensive list of FACTS devices

widely accepted indicator to classify device topologies. The connection type to the transmission line is another significant indicator to detect the usage of FACTS.

The essential application areas of FACTS are power flow control, voltage control, VAR compensation, stability improvement in terms of phase and voltage, power conditioning, and interconnection of several sources in distributed generation. The connection type to the transmission line and semiconductors of compensators involve several different operating characteristics. The conventional compensators are constructed with static components such as diodes, inductances, capacitances, and mechanic controllable devices such as tap-changing transformers.

The modern FACTS devices are installed by using more sophisticated components as thyristors or bi-directional semiconductors such as gate turn-off thyristors (GTOs), insulated gate bipolar transistors (IGBTs) or insulated gate commutated thyristors (IGCTs). The thyristor-based topologies are known as line commutated due to unidirectional switching property of the device is depended to the half-cycle polarity of the line. However, the advanced technology of bi-directional switches allow commutating the FACTS at the desired period of entire cycle [4, 6, 7, 12]. In this case, voltage source converters (VSCs) generate completely controllable output in terms of magnitude and phase angle owing to pulse width modulation (PWM) methods used to control IGBTs or IGCTs. The PWM frequency is freely adjusted to decrease THD ratio of the generated output voltage and compensate the power quality of the transmission line. Therefore, particular designs can be carried out in terms of device topology and commutation methods [6, 11–14].

The line commutated FACTS devices can be listed as thyristor-switched capacitor (TSC), thyristor-controlled reactor (TCR), static VAR compensator (SVC), thyristor-controlled series compensator (TCSC) that all are based on aforementioned ideal shunt or series compensator methodology as shown in Fig. 8.12. The right hand side of FACTS listed in Fig. 8.12 includes VSCs with advanced technologies of STATCOM, static synchronous series compensator (SSSC), unified power flow controller (UPFC)), dynamic power flow controller (DPFC) and HVDC VSCs. The FACTS improve the stability and power quality by adapting to rapidly changing line situations and dynamically responding even at the voltage dip and flicker cases.

The illustration given in the figure is also coloured in order to express application aims of compensators regarding to particular requests. The reactive power compensation and voltage control is primarily performed by selecting shunt devices that are shown in the first line of the figure. The SVCs are capable to present more accurate and smoother control comparing to mechanically switched shunt compensators. In this case, the STATCOM is the most recent and the most robust shunt compensator device comparing to others. The series devices shown in green box are preferred to compensate the line stability and reactive power depending to its adaptive impedance feature. These devices are implemented according to TSC, TCR, and SVCs where the main difference is their series connection to the transmission line. In the most cases, the series compensators are used a thyristor by-pass in order to protect the compensator.

The TCSC provides a controllable capacitance to perform power flow control and oscillation damping among the sources and loads. The TCSC is a quite effective device against the damping problems occurred during the integration of large power systems, and they can more accurately overcome similar problems than SSSC device that is not contemporarily being used in the transmission lines since their higher costs.

Furthermore, the TCSC prevents the of sub-synchronous resonance (SSR) problem that is seen as an interaction between heat based generating systems and series compensated systems. One of the most significant application areas of series compensators is seen in distribution lines of industrial plants against voltage dips and flickers. These special devices are known as dynamic voltage restorer (DVR) or static voltage restorer (SVR) [6, 11, 13–18].

The most extensively researched and improving compensation devices designed in shunt and series connection to perform power flow control. These are shown in red box of Fig. 8.12 as a new line. The increased energy demand requires dynamic and balanced power flow control between overloaded units and spare power systems of transmission line. The phase shifting or tap changing transformers (PST) are the first generation devices of shunt-series compensators.

As it is easily realized, the lower response speed, lower control ability, and periodic maintenance requirements of these devices are their drawbacks comparing to semiconductors. The unified power flow controller (UPFC) is a more sophisticated and convenient compensator owing to its fast and dynamic structure. It can be assumed as a composition of a STATCOM and an SSSC using common dc link for power flow control. The dc link is generated by a small capacitor connected between STATCOM and SSSC where the STATCOM draws active power and SSSC generates power at the same ratio to balance the transmission line.

The main drawback of this installation is its high cost caused by power semiconductors used in both devices. The dynamic power flow controller (DPFC) that is a combination of one traditional PST and a SSSC/thyristor switched series reactor (TSSR) is improved to tackle this situation by providing a simpler device to perform the same issue with low cost. The DPFC provides more dynamic performance than a traditional PST where it controls the reactance by a set of thyristors integrated to the transmission line serially. It is capable to control the active and reactive power delivered through the transmission line. Depending on the reactive power requirements of the system, the DPFC configuration is implemented by using a mechanically switched shunt capacitor (MSC) [6, 7, 11, 14].

The high voltage direct current (HVDC) compensators are associated with FACTS in industrial applications as shown in Fig. 8.12. HVDC and FACTS experienced researches and improvements primarily based on thyristors and recently based on bi-directionally controlled semiconductors. The power converters defined as back-to-back (B2B) systems can be built up with thyristors or bi-directional semiconductors such as IGBT, IGCT where they allow to the complete control of power flow while the thyristor based topologies controls only the active power. Although the drawbacks of thyristor devices, the VSCs based on HVDC B2B systems are capable to decouple frequency at the both sides of symmetrical systems. Furthermore, many PWM control methods decreasing THD and electromagnetic interference (EMI) are implemented owing to advantages of recent power semiconductors.

The HVDC and FACTS are fully or are partially integrated in many applications. It is noted that there are more than a hundred HVDC power plants are installed in worldwide, delivering more than 80 GW power. The current source converters (CSCs) alternatively to VSCs find application areas in power systems especially at 1000 MW and over. However, they are assumed as a mature technology since thyristor based configurations. The CSCs are naturally resistant against short circuits since the series inductor helps to limit inrush currents appearing at faulty operations while VSCs are weaker at line faults. The dc circuit breakers and cable are more attractive for VSC HVDCs in this case [7, 13–16].

8.3 Conventional Equipment for Reactive Power Compensation

The VAR compensators based on generating are classified into two groups as rotating and static generators. The antecedent applications of shunt capacitor and shunt inductor as power flow controllers (PFCs) are dated to 1914. The power semiconductors have attracted the improvement of static VAR compensators starting from last half of previous century. Then, the thyristor switched capacitors and reactors are improved to utilize rapid and dynamic response of power electronic devices. Afterwards, they are associated with tap-changing and PSTs that is the most reliable way to control the delivered power on the parallel transmission lines. The PFCs such as PSTs and series compensators were widely used in steady-state operation of transmission line that the technique is implemented by observing the effect of compensator on the power flow. The technique lying behind PST compensation is its unique and different design criteria comparing to regular power transformers. It can be produced as two-core or single-core symmetrical/asymmetrical constructions.

The power flow control based on PST can be performed by adjusting the phase-shift angle δ_C between source and receptor sides of symmetrical system as seen in power controller plane in Fig. 8.13. This method is used to determine steady-state conditions of the transmission line where the delivered active power is compensated by PFC [11, 19, 20–25].

The power controller plane is used to detect the power limits of the transmission line and the compensator. The bounded area is called the "operating area" depicting all operating points of the PFC to ensure operation is limited in the steady-state. The first generation compensators based on capacitor or inductors were arranged regarding to steady-state assumptions. However, the widely fluctuating loads expose the power system to large reactive power variations. In this case, the fixed capacitor banks lack to compensate the reactive power leading to

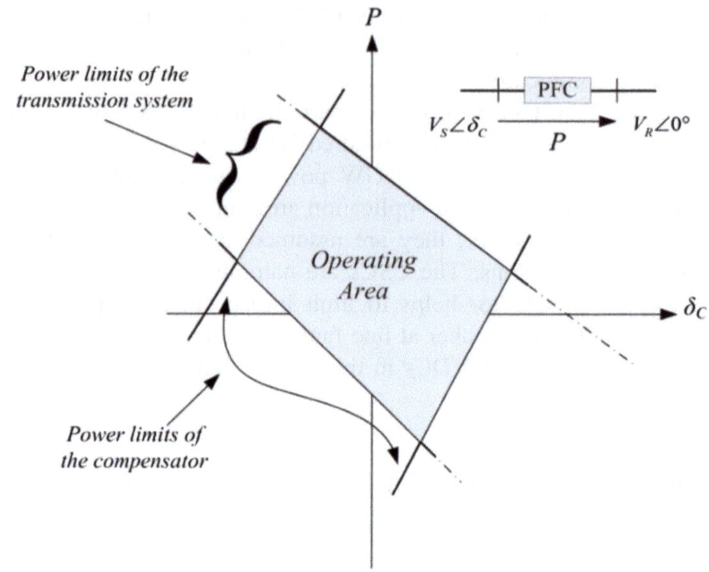

Fig. 8.13 Operating area of a power flow controller (PFC) in the power controller plane

over-compensation or under-compensation. The switched capacitor and reactors are proposed to tackle this drawback by providing variable compensation owing to variable switching angle. The primary switching applications were being performed using mechanical switches such as relays and circuit breakers that are replaced with power semiconductors lately [6, 9, 11].

8.3.1 Thyristor Switched Capacitor (TSC)

The TSC is shown in Fig. 8.14a that is constructed with a line transformer, bi-directional thyristor valve, and a capacitor where a small inductance can be added in series to limit the current. Either the SVCs are defined with switched or control as previously expresses TSC and TCR. The switching term is used to emphasize that the thyristors are turned on at the exact angles only when zero-voltage switching (ZVS) condition is achieved for TSC. The voltages at the thyristor terminals are kept at zero in ideal conditions. However, the required positive amplitude across anode to cathode for switching thyristors is provided in practice, and thyristor valve provides connection and disconnection of capacitors to the line. In the case of a small reactor is preferred to use, it should be located among the transformer and thyristor valve to prevent inrush currents under critical conditions.

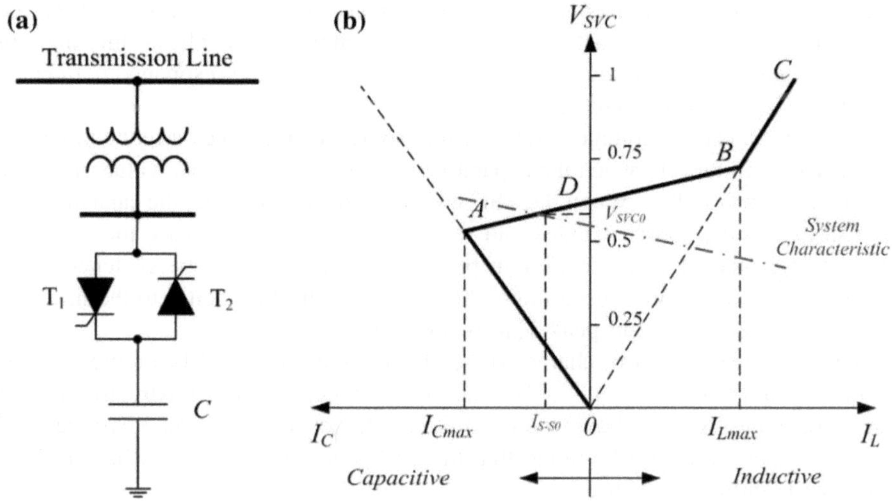

Fig. 8.14 Thyristor switched capacitor, **a** circuit diagram, **b** control characteristic of SVC

The thyristor valve is closed in case of steady-state conditions and TSC is connected to the transmission line where it is imposed to handle sinusoidal ac line voltage. The TSC current at this interval is

$$I_{S-S} = V\frac{n^2}{n^2 - 1}\omega C \cos \omega t \tag{8.14}$$

where V is the terminal voltage of TSC and n is given by

$$n = \frac{1}{\sqrt{\omega^2 LC}} = \sqrt{\frac{X_C}{X_L}} \tag{8.15}$$

The current flowing across the capacitor at a given interval t is determined by the following where it is assumed that the equivalent resistance of the TSC is nearly zero comparing to the transmission system;

$$\begin{aligned} i(t) = {} & \frac{V_m}{X_C - X_L}\cos(\omega t + \alpha) - \frac{V_m}{X_C - X_L}\cos(\alpha)\cos(\omega_r t) \\ & + \left[\frac{X_C V_m \sin(\alpha)}{\omega_r L(X_C - X_L)} - \frac{V_{CO}}{\omega_r L}\right]\sin(\omega_r t) \end{aligned} \tag{8.16}$$

$$\omega_r = \frac{1}{\sqrt{LC}} \tag{8.17}$$

where V_m is the instantaneous maximum voltage of the source, X_C and X_L are capacitive and inductive reactance of the compensator, α is the phase-shift angle of voltage when the capacitor is connected to the line, ω_r is the resonance frequency, and V_{C0} is the capacitor voltage at $t = 0^-$

Since the capacitor voltage V_{C0} is equal to source voltage, the current component of the TSC will be zero when the capacitor is connected at the peak value of source voltage V_m, ($\alpha = \pm 90°$). When the TSC is disconnected from line, the capacitor will be charged to the source voltage and remains at this value. In case the capacitor voltage is protected at the maximum peak value, the TSC can be switched again without any fluctuation. Furthermore, the reconnection of capacitor to the line may be done between zero and peak V_{C0} [6, 11, 15].

The steady-state control characteristic of SVC seen in Fig. 8.14b represents the control area on the line of ADB. The 0A line depicts the maximum limit of capacitor while BC line shows the maximum inductor limit in terms of voltage and current values. It should be noted that the SVC current is positive while the susceptance is inductive. Thus,

$$I_{S-S} = -B_{SVC}V_{SVC} \tag{8.18}$$

The slope of 0A line is determined by the susceptance of the capacitor B_C, while the slope of OBC line depends on susceptance of the reactor B_L. A positive slope in the range between 1–5% is presented to allow shunt operation of several SVCs connected to the same transmission line, and to prevent the operating at the lower and higher limits.

The steady-state degree of SVC line voltage is detected regarding to the system parameters and control features. The system characteristic seen in Fig. 8.14b with a negative slope is given by

$$V_{SVC} = V_{Th} - X_{Th}I_{S-S} \tag{8.19}$$

where V_{Th} and X_{Th} are the Thevenin voltage and reactance calculated from the SVC side to line as follows;

$$V_{Th} = \frac{V\cos(\delta/2)}{\cos(\delta/2)} \tag{8.20}$$

$$X_{Th} = \frac{Z_n}{2}\tan(\theta/2) \tag{8.21}$$

where Z_n is the surge impedance given by [26]

$$Z_n = \sqrt{\frac{L}{C}} \tag{8.22}$$

8.3.2 Thyristor Controlled Reactor (TCR)

The TCR is the most widely used compensator with thyristor that is constituted of a line transformer, a bi-directional thyristor valve, and an inductor as seen in Fig. 8.15a. In many cases, the compensator includes LC passive filter configured with fixed values to eliminate particular harmonic orders. The TCR provides frequently variable inductive reactance to the line owing to its shunt connection and controlled phase-angle. The voltage and current waveforms of the compensator are shown in Fig. 8.15b where it is controlled at switching angle α, which increases or decreases the fundamental magnitude of reactor current I_L. The switching angle α, can be adjusted to any degree from 90° to 180° starting from zero crossing interval of voltage. The reactor is completely connected to the line at 90° and is completely disconnected from line at 180° while the partial value connection is available between both cases. The equivalent admittance of a TCR as a function of the switching angle α is shown in Fig. 8.15c where it is depended on the partial connection and disconnection cases.

When the switching angle is increased, the reactor current is decreased and provides to reduce the absorbed reactive power by increasing the equivalent

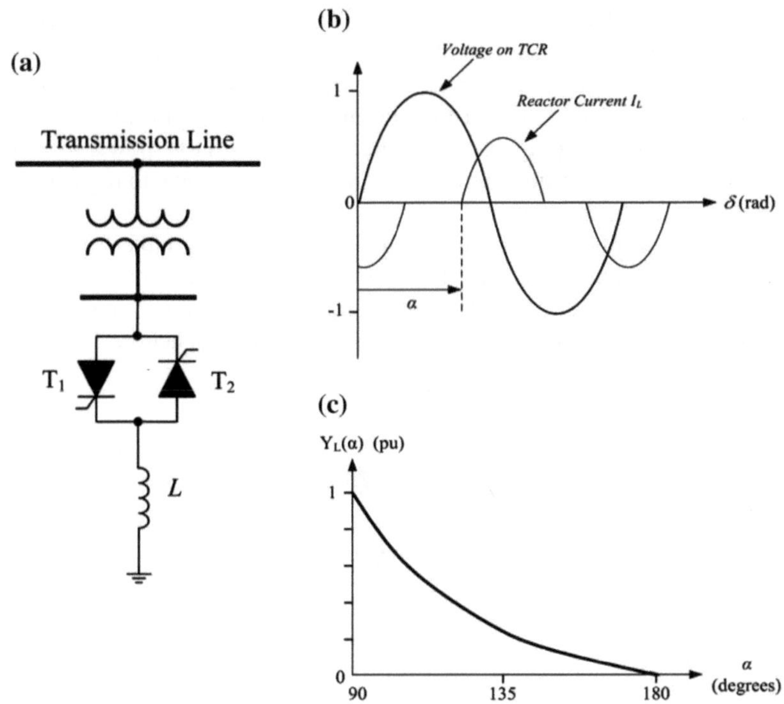

Fig. 8.15 Thyristor controlled reactor, **a** circuit diagram, **b** voltage and current waveforms, **c** equivalent admittance of a TCR as a function of the switching angle α

inductance. The reactor current is calculated depending on inductor and switching angle as given by

$$I_{TCR} = \frac{V_{rms}}{\pi \omega L}(2\pi - 2\alpha + \sin(2\alpha)) \tag{8.23}$$

The TCR compensators provide continuous control, maximum delay limited to one cycle at most, and very low transients during connection and disconnection. However, the significant drawback of the compensator is higher losses in inductive operation areas, and low ordered harmonics of current. The magnitude of current harmonics are defined as shown in Eq. (8.24)

$$I_k = \frac{4V_{rms}}{\pi X_L}\left[\frac{\sin(k+1)\alpha}{2(k+1)} + \frac{\sin(k-1)\alpha}{2(k-1)} - \cos(\alpha)\frac{\sin(k\alpha)}{k}\right] \tag{8.24}$$

where k is harmonic orders; k = 3, 5, 7,... [6, 11, 15, 27, 28].

8.3.3 Thyristor Controlled Series Compensator (TCSC)

The thyristor-controlled series compensator (TCSC) is shown in Fig. 8.16. This compensator is implemented regarding to series connection of a TCR to the transmission line. In the series connection case, a capacitor should be shunt connected to TCR in order to control the current. The most significant advantage of TCSC is balance the damping occurred while integrating large power systems, and to overcome SSR. The circuit diagram seen in figure is identical to regular SVC that is shunt connected to the line.

The TCSC presents two particular principles that are variable capacitive reactance and variable inductance. Both of these principles are managed by TCSC switching methods that are used to control the thyristor valve. Consequently, this configuration provides a continuously controlled series capacitor. The equivalent

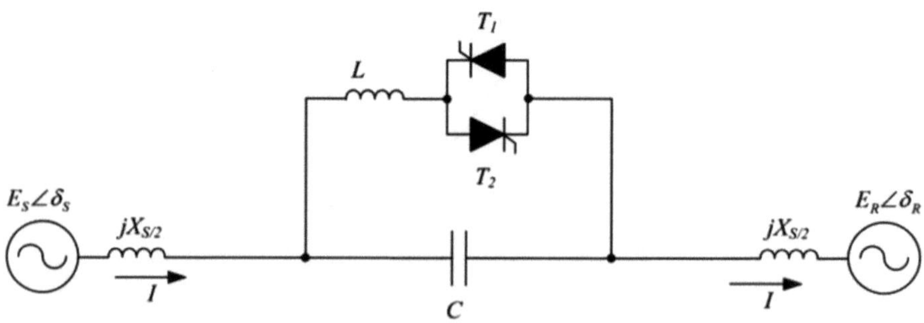

Fig. 8.16 Thyristor controlled series compensator

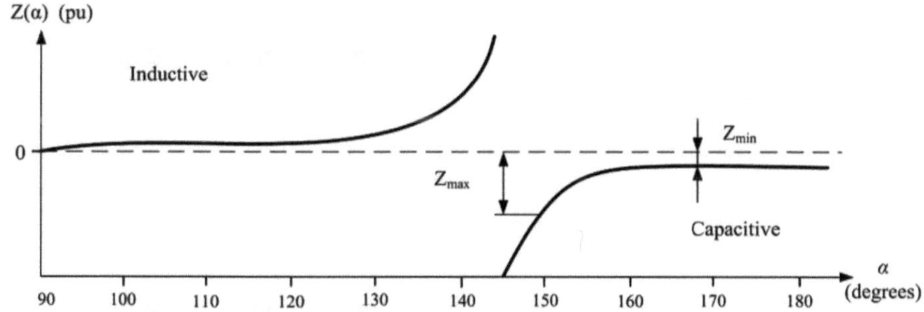

Fig. 8.17 Equivalent resistance of TCSC as a function of the switching angle α

resistance of TCSC is depicted in Fig. 8.17 as a function of the switching angle α where it is operated in two different areas as inductive and capacitive. The transition between operation ranges appears at around 145° in this case. The regular operation of TCSC is performed in capacitive area up to 180° that the minimum impedance is obtained around this angle. The maximum impedance $Z_{\max}$ is presented at 150° switching angle where the resonance around this area is not considered safe for operating.

The TCSC can also be operated in inductive region that is performed by limiting the switching angle between 90° and 140°, where the α lower than 90° causes to decrease the amount of the delivered power [6, 11].

8.4 Flexible AC Transmission System (FACTS)

The force-commutated or self-commutated compensators are implemented by using VSC in order to provide more flexible and robust compensation comparing to conventional line-commutated compensators. The FACTS are integrated to symmetrical systems in shunt, series, or shunt-series connections as conventional compensators and SVCs. The most widely used FACTS are STATCOM and SSSCs where the each compensator is based on VSC to generate compensating voltage and increase the system stability. The most significant contribution of FACTS is harmonic elimination capability that is provided by the sophisticated device configurations and control algorithms.

The improvement of high-power VSCs is promoted by the development of force-commutated semiconductors such as GTOs, IGBTs, and IGCTs. The primary applications of VSCs are shown in Fig. 8.18 where two-level six-pulse VSC is illustrated in Fig. 8.18a and three-level diode clamped topology is shown in Fig. 8.18b. The VSCs are configured with GTOs and antiparallel diodes in order to provide bi-directional current flow where the switches could be IGBT or IGCT. The voltage blocking capability of the VSCs given in the figure is unidirectional. In case of CSCs used, the voltage blocking capability is shifted to bi-directional while the

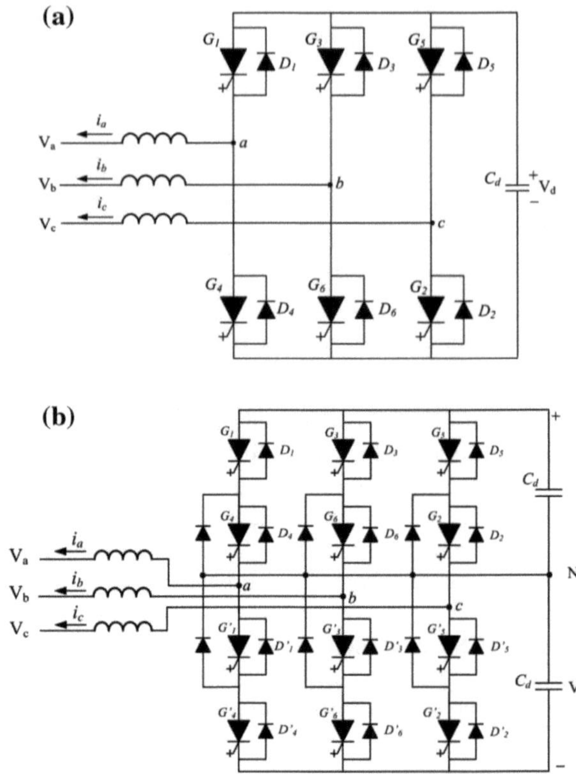

Fig. 8.18 VSC topologies used in FACTS, **a** basic six-pulse two-level VAR compensator, **b** basic three-level VAR compensator

current flow is performed in unidirectional mode. The regular operation of VSC is performed with a voltage source connected to the dc side that allows only reactive power compensation in a compensator application by replacing the dc source with a capacitor [6, 9, 11, 14].

However, in case that active power management is also required, then the circuit involves a small reactor on the ac side of compensator to limit the inrush currents and current spikes during the transients. The first installation of high-power VSC configured in six-pulse topology with 10 MW rated power is done in Japan, in 1991 [6, 11, 29]. The switching frequency was in the line frequency to ensure the stability of the compensator, and eight sets of six VSCs were building a 48-pulse VSC compensator with 80 MVA compensation capability. The ensuing applications of force-commutated VSCs have presented effective compensations and the researches on VSC accelerated their integration to industrial plants.

The emerging power systems and FACTS require more effective controllers such as STATCOMs, (SSSC, UPFC), the inter-phase power controller (IPC), the thyristor-controlled braking resistor (TCBR), the thyristor-controlled voltage limiter (TCVL), the battery energy storage system (BESS), and the superconducting magnetic energy storage (SMES) systems [30].

8.4.1 Static Synchronous Compensator (STATCOM)

A STATCOM is a VSC based compensating device that is shunt connected to the transmission line. A STATCOM integrated to the line and shunt to generators and load operates by generating or absorbing reactive load to compensate the transmission line. The operation principle of STATCOM makes it either a source or a load for the transmission line by replacing widely known series or shunt FACTS. The STATCOM increases the power quality by performing several compensations such as dynamic voltage control, oscillation damping of power line, pursuing the stability during transients, flicker and sag-swell controls, and active and reactive power (*PQ*) control in transmission systems. Furthermore, STATCOMs substitute SVC devices in the distribution systems [4, 6, 30].

Figure 8.19a shows the basic connection of STATCOM to transmission line where the reactive power control is performed by controlling the terminal voltages.

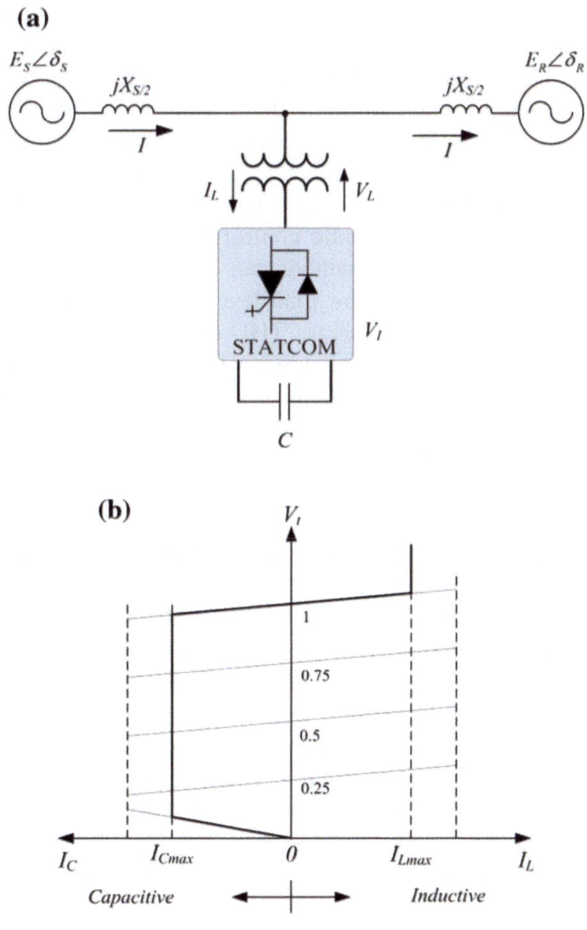

Fig. 8.19 STATCOM, **a** circuit diagram of line integration, **b** control characteristic of STATCOM

The transmission line is configured in symmetrical system that E_S denotes the source voltage, while the line impedance X_S, line voltage V_L, line current I_L, and receptor voltage E_R are indicated. The reactive power is controlled depending on the compensator voltage V_I control by assuming that is in phase with the source voltage E_S. There are three options to determine the reactive power of STATCOM according to correspondence among the source and STATCOM voltages. In case of equality, there is not any active or reactive power generated or absorbed in STATCOM. When the source voltage exceeds, than the reactive power of STATCOM is shifted to inductive. In the case of the source voltage is lower than STATCOM voltage, this causes to capacitive reactive power.

A typical *V-I* characteristic of STATCOM as shown in Fig. 8.19b, illustrates this situation where the STATCOM can perform inductive or capacitive compensation regarding to its line current. Besides, the V-I characteristics show that STATCOM provides fully capacitive or inductive power starting from lower voltages even from 0.15 p.u.

Thus, the main case that should be handled in reactive power compensation by using STATCOM is to control the magnitude of compensator voltage V_I. This operation is performed based on two basic principles where one is used in multi-pulse converters and the other one for PWM controlled multilevel converters. The capacitor voltage building the dc line can be fixed to a stable magnitude in PWM control of multipulse converter based STATCOM where the compensator voltage is controlled by the modulator. This is managed by adjusting the switching angle owing to PWM modulator. The second STATCOM configuration based on multilevel converters should control the capacitor voltage only to sustain its constant value. The control methods can be either scalar or vector based in this configuration. Furthermore, the multilevel configuration allows switching with high frequency PWM to decrease THD of reactive components [6, 9, 14]. The following sections are dedicated VSC topologies used in STATCOMs where multipulse configurations are indicated by 12-pulse and 24-pulse converters. Afterwards, multilevel topologies and controller structures are analysed to compare to multi-pulse STATCOMs at a glance.

8.4.1.1 Multi-pulse Converter Based STATCOM

The first VSC based STATCOM applications are developed with multi-pulse converters owing to their lower losses and lower harmonic contents comparing to conventional compensators. The multi-pulse converter topologies consist of several six-pulse VSC circuits where a basic VSC STATCOM in the configuration of six-pulse two-level was illustrated in Fig. 8.18a. The switching devices of the system were GTOs that the converter could generate balanced three-phase ac output voltages by using a dc capacitor as the voltage source. The frequency of the output voltage is adjusted by the modulating frequency of GTO switches and the phase voltages are coupled to ac grid over an interconnection reactor [9, 31, 32].

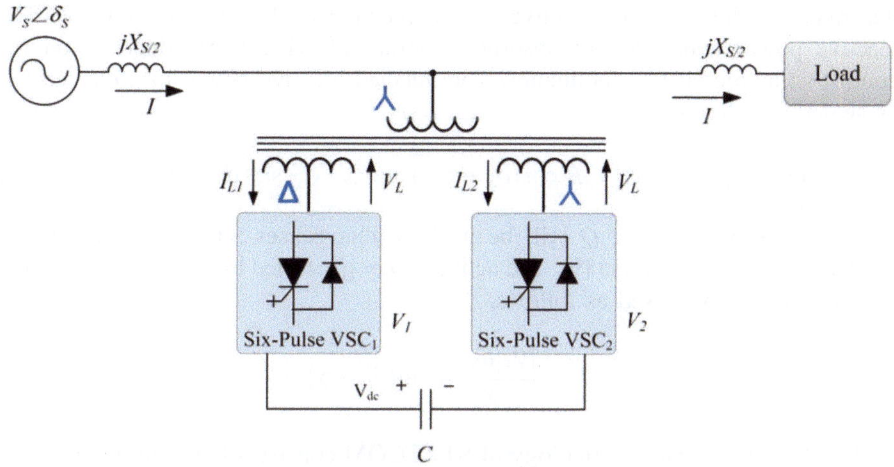

Fig. 8.20 Circuit diagram of two-level 12-pulse STATCOM

The 12-pulse converters are constituted of six-pulse converters connected in parallel as shown in Fig. 8.20. The expressed configuration is improved regarding to two-level topology shown in Fig. 8.18a while another configuration is in three-level structure as illustrated in Fig. 8.18b. The two-level topology requires non-identical winding types for each six pulse VSC that one is Δ-Y connected while other is Y-Y connected to ac grid. The Δ connected secondary of the transformer with three times higher turns according to Y connected secondary provides 30° phase shift comparing to other. Therefore, the lower ordered harmonics are easily eliminated.

The output phase voltage of STATCOM can be calculated by using Eq. (7.25) where n_1 and n_2 are the voltage ratios of the corresponding VSC transformers while v_{a-Y} and $v_{a-\Delta}$ are the output voltages of the Y-Y and Δ-Y connected converters, respectively [9].

$$v_a = n_1 v_{a,Y} + n_2 \frac{v_{a,\Delta}}{\sqrt{3}} \tag{8.25}$$

The voltage magnitude of VSC output depends on modulation index m_i of control signal that allows obtaining the output voltage at $m_i V_{dc}$ where V_{dc} is the voltage across the capacitor. The control signal that is based on some variation of PWM generates switching angle α that generates the STATCOM bus voltage at $V_S\angle\delta_S$ form. The Eq. (8.26) expresses the admittance of transformer

$$Y = \frac{1}{R + jX_S} = G + jB \tag{8.26}$$

that manages the active or reactive power injection to the transmission line. The reactive power generation or absorption ability of STATCOM depends on the control voltage of PWM modulator. The injected reactive power to utility grid is presented as follows

$$Q = V_S^2 B - m_i V_{dc} V_S B \cos(\delta_S - \alpha) + m_i V_{dc} V_S G \sin(\delta_S - \alpha) \tag{8.27}$$

In case of $V_S > m_i V_{dc}$, Q will be positive that causes STATCOM to absorb reactive power or vice versa [9]. The active power generated by the source to charge the capacitor is expressed as follows,

$$P = \frac{|V_S| m_i V_{dc}}{X_S} \sin(\delta_S - \alpha) \tag{8.28}$$

The 24-pulse converter topology of STATCOM consists of six-pulse converters where it eliminates the $6n \pm 1$ ordered harmonics. The 23rd, 25th, 47th, and 49th harmonics in the ac output waveform are naturally eliminated since the $n = 4$ for 24-pulse converter. The 24-pulse STATCOM constituting two-level VSCs is illustrated in Fig. 8.21 where this topology is also known as quasi 24-pulse [9, 33]. Each VSC and transformer connection configures a stage that each one is connected to the transmission line by Y-Y and Δ-Y connections of transformers. The transformers may connected in zigzag that generates 15° phase shift according to each other with −7.5°, 7.5°, 22.5°, and 37.5° angles [9, 33, 34].

Another topology of quasi 24-pulse STATCOM is presented in Fig. 8.22 [35] where the VSCs are constituted according to neutral point clamping (NPC) topology that is previously shown in Fig. 8.18b.

The conduction angle α defines the negative and positive cycles of intervals where $\alpha = 180 - 2\beta$, where 2β stands for the dead-band angle. The on-off spectral transitions are ensured by the dead-band angle where the output voltages of a six-pulse three-level VSC is given by Eq. (8.29).

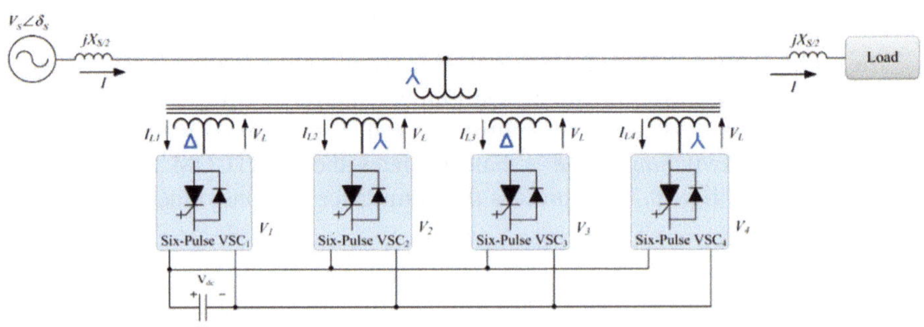

Fig. 8.21 Circuit diagram of a quasi two-level 24-pulse STATCOM

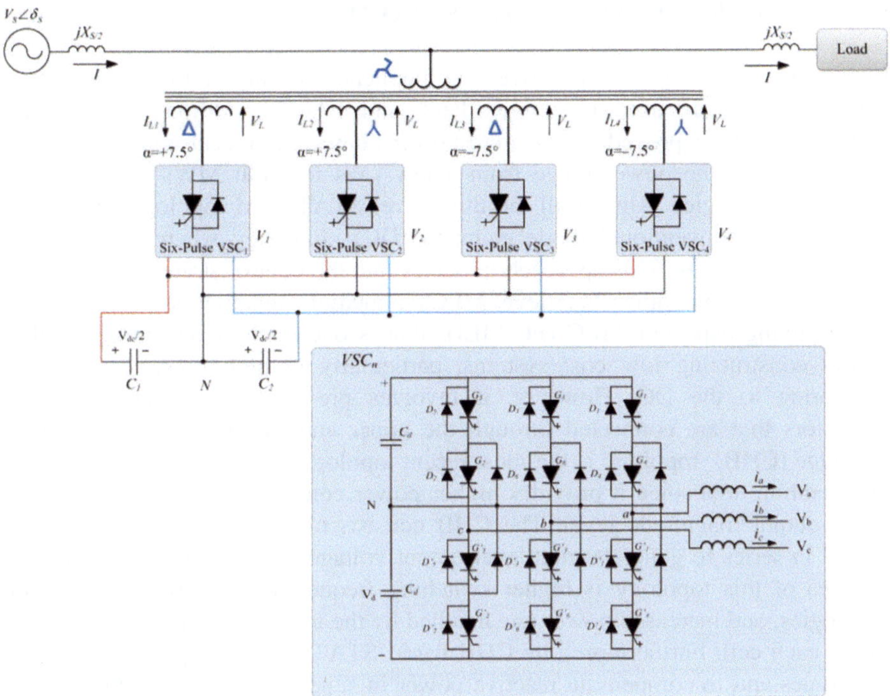

Fig. 8.22 Circuit diagram of a three-level 24-pulse STATCOM

$$V_1 = \frac{2\sqrt{6}}{\pi}\frac{v_{dc}}{2}\sin\left(\frac{180 - 2\beta}{2}\right) \tag{8.29}$$

On the other hand, output voltages of a 24-pulse three-level VSC is given as follows while the active and reactive powers are calculated as given below by considering X_S as the reactance along with STATCOM and the transmission line by using Eqs. (8.31) and (8.32) [9, 35]

$$V_{c1} = np\frac{2\sqrt{6}}{6\pi}\frac{v_{dc}}{2}\sin\left(\frac{180 - 2\beta}{2}\right)\cos\frac{\pi}{24} \tag{8.30}$$

$$P = \frac{V_L V_S \sin\delta_S}{X_S} \tag{8.31}$$

$$Q = V_S\left(\frac{V_S - V_L \sin\delta_S}{X_S}\right) \tag{8.32}$$

where the phase angle δ between the ac bus voltage V_S and STATCOM output voltage V_L sustains the dc line voltage V_{dc} at a constant value.

8.4.1.2 Multilevel Converter Based STATCOM

In addition to their power conversion applications, the multilevel converters (MLC) are widely used in active filters, and FACTS. Although the primary applications are improved for medium power ranges, the recent applications have been increased the power range from a few kWs to multi MWs owing to their cascaded topologies. Almost all the three-level STATCOM topologies are implemented with conventional diode clamped (DC) topology where the switching devices were GTOs. This approach indicates that the control strategy of DC is the simplest one among other three-level MLCs in medium and high power application.

The flying capacitor MLC (FC-MLI) that is one of the outstanding topology allows constructing low cost systems, particularly in high power applications, comparing to the DC. However, it involves pre-charge arrangement for the capacitors that are connected through the upper and lower arms. The cascaded H-bridge (CHB) topology is the most recent topology among others that extended its operation area since it provides higher power conversion and lower cost for a given output magnitude level. The CHB consists of several H-bridge cells connected in series to generate multilevel output voltage. One of the most significant features of this topology is higher switching frequency comparing the previous topologies, and increased power rate handled by the total device and equally shared to the each cell. Furthermore, the CHB based STATCOM is capable to eliminate harmonics and to compensate reactive power in superior comparing others.

The main requirement of this configuration is its separated dc sources. However, CHB can be easily integrated to various type of energy sources to obtain higher output levels and energy storage applications [9, 36–39].

Figure 8.23 shows the circuit diagram of STACOM configured by using three-phase five-level CHB including series connection of three-level H-bridge cells at each phase leg. The output voltage synthesis of H-bridges generate a combined phase voltage in staircase waveform oscillating in the ranges of $+V_{dc}$, 0, $-V_{dc}$.

Where V_{a1} and V_{a2} voltages of series connected H-bridge cells synthesize the phase voltage V_{an} in a stepped waveform as seen in Fig. 8.24. The positive levels of output voltage are indicated with P_1 and P_2 while the negatives are P_1^1 and P_2^1. Equation (8.33) expresses the Fourier series expansion of the general multilevel stepped output voltage while the transform applied to the STATCOM in Fig. 8.23 is shown in Eq. (8.34), where n represents harmonic order of the output voltage.

$$V_{an}(t) = \frac{4V_{dc}}{\pi} \sum_{n=1,3,5,\ldots}^{\infty} [\cos(n\alpha_1) + \cos(n\alpha_2) + \ldots + \cos(n\alpha_5)] \frac{\sin(n\omega t)}{n} \quad (8.33)$$

The switching angles α_1, α_3,…, α_5 in Eq. (7.33) are selected for the minimum voltage harmonics [36, 40].

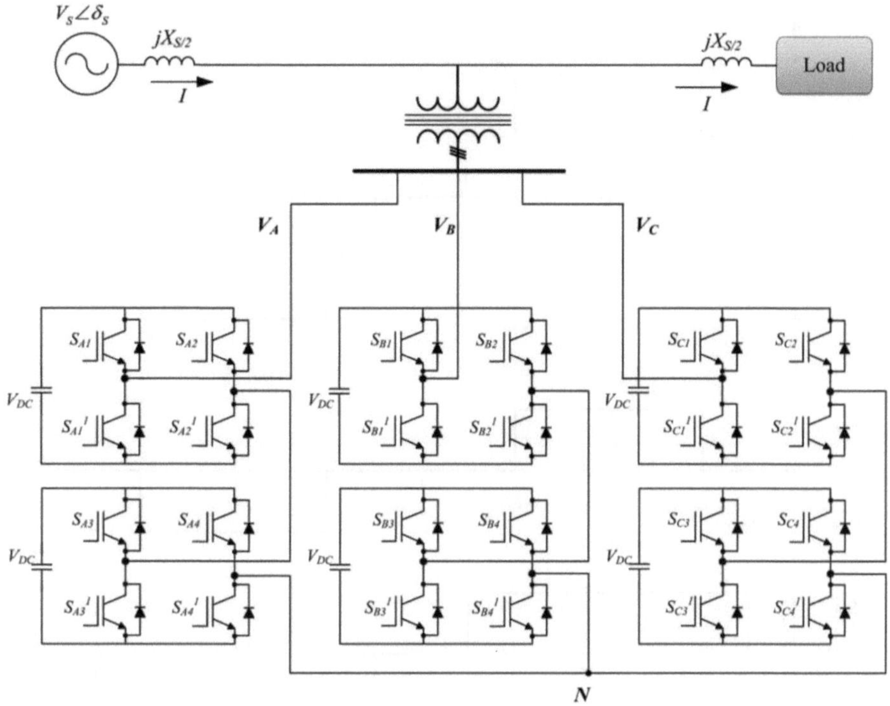

Fig. 8.23 Circuit diagram of a five-level CHB STATCOM

$$V_{an5}(t) = \frac{4V_{dc}}{\pi} \sum_{n=1,3,5,\ldots}^{\infty} [\cos(n\alpha_1) + \cos(n\alpha_2)] \frac{\sin(n\omega t)}{n} \tag{8.34}$$

Researchers have proposed novel MLCs based on hybrid topologies that are based on the existing devices [36, 40].

8.4.2 Static Synchronous Series Compensator (SSSC)

The series capacitor based compensation that brings some capabilities such as increasing the transient stability, reactance control, and load sharing is a conventional technique that was introduced in Sect. 8.3.3. Even though the series compensation is assumed to be used to decrease the reactive impedance of the transmission line, it is actually operated by increasing the line voltage, current, and thereby the delivered power from source to load. Although the mentioned advantages, series compensation has several drawbacks on SSR issues that large energy exchange is done by several sources under synchronous system frequency.

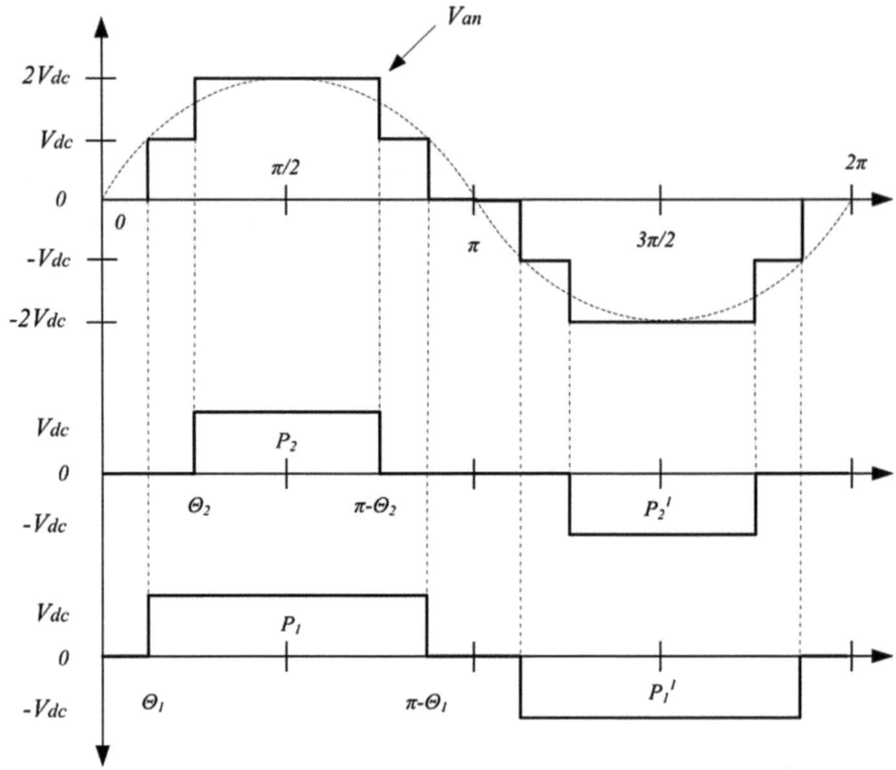

Fig. 8.24 Phase voltage generation of a five-level CHB STATCOM

There are several studies are performed to overcome SSR problem. The robust control methods and configurations of FACTS stand out of other methods proposed by researchers. The SSSC has been developed as an alternative to the TCSC devices where the SSR immunity is inherited from TCSC. In addition to regular advantages of series compensators, SSSC presents numerous capabilities to increase the reactive power control and stability issues. Furthermore, there are several studies are noted that SSSCs effectively damps the SSR comparing regular series compensators [6, 11, 14, 41–46].

Figure 8.25 illustrates basic circuit diagram of a SSSC based on VSC similar to STATCOM that is serially connected to the transmission line, and a capacitor on the dc side. The SSSC controller considers the line current and line voltage to compensate the transmission line by adjusting the reactance in steady-state operation. It should be noted that in case the current is flowing in the line, then SSSC generates compensation voltage V_C that is quadrature to the current and regulates the reactive power demand. The compensation characteristic of the device is same with the one shown in Fig. 8.11 for $0 < \delta < 180°$ during series compensation that yields positive magnitude power delivered.

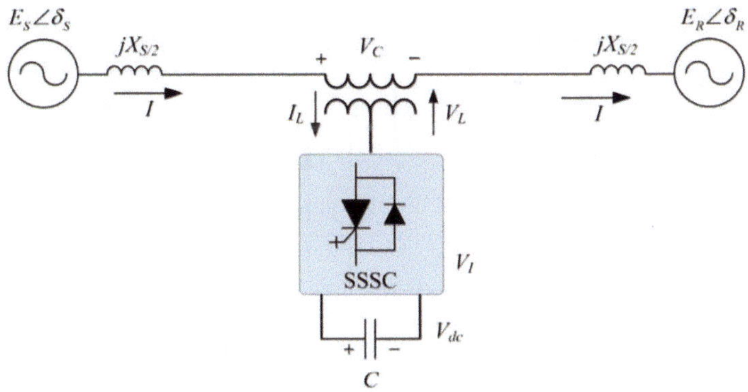

Fig. 8.25 Static synchronous series compensator (SSSC)

The fundamental operating principles of SSSC is compared to conventional series capacitive compensator as shown in Fig. 8.26 where the circuit diagrams for only series compensation with only capacitive compensator (Fig. 8.26a) and with SSSC (Fig. 8.26c), and respective phasor diagrams are analyzed. Despite it is assumed to decrease the line impedance, a series capacitor increases the line voltage, line current, and consequently the power through the line impedance. This approach may be considered irrelevant. However, the voltage along the transmission line is the significant parameter while the capacitor is cared to increase the voltage in order to provide the required current level. The steady-state power transmission can be realized when a synchronous voltage source follows the series compensation as shown in Fig. 8.26c that is analytically presented as follows

$$\overline{V_q} = \overline{V_C} = -jX_C\overline{I} \tag{8.35}$$

It should be noted that there are particular differences between series capacitive compensator and synchronous voltage source type compensator. The operational characteristic of a series capacitor can be illustrated with an ideal voltage source. However, SSSC is significantly different from the series capacitor in terms of delivered power and angle characteristics, active power exchange, and immunity to resonance and SSR, and control ranges.

Capacitor provides reactive impedance that causes proportional voltage to the line current when it is series connected to the line. The compensation voltage is changed regarding to the transmission angle δ and line current. The delivered power P_S is a function of the series compensation degree s where it is given by

$$P_q = \frac{V^2}{X_L(1-s)}\sin\delta \tag{8.36}$$

where $V = V_1 = V_2$ and $\delta = \delta_1 - \delta_2$. The delivered power P_S regarding to transmission angle δ is shown in Fig. 8.27a as a function of s ($s = 0$, $s = 1/3$, and

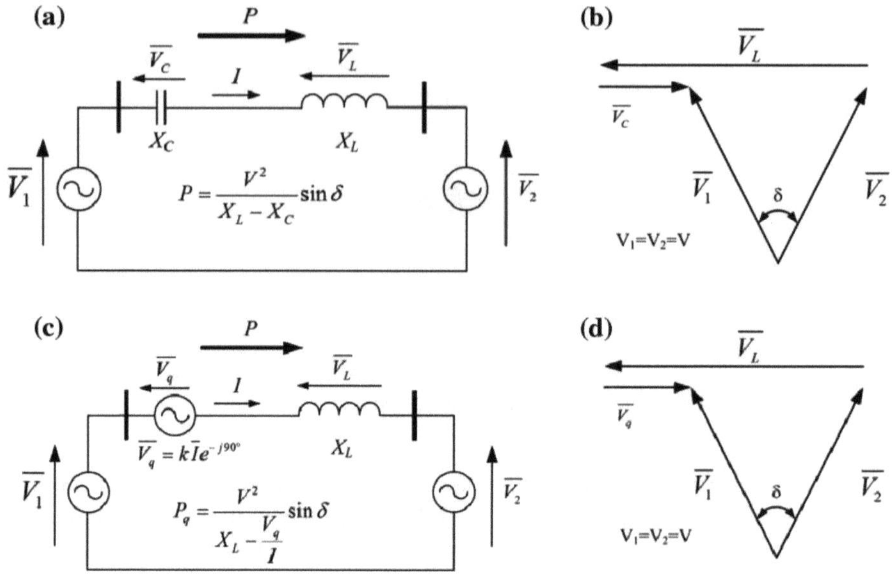

Fig. 8.26 Symmetrical system under series compensating, **a** circuit diagram of series compensating capacitor, **b** phasor diagram of capacitor, **c** circuit diagram of SSSC, **d** phasor diagram of SSSC

s = 1/5). Additionally, SSSC is independent from line current while compensating the voltage. The delivered power P_q is a parametric function of injected voltage $V_q = |\overline{V_q}|$ series to line where $\overline{V}_q = V_q[\overline{I}/|\overline{I}|]e^{\mp j90°}$ that is also given by

$$P_q = \frac{V^2}{X_L}\sin\delta + \frac{V}{X_L}V_q\cos\frac{\delta}{2} \tag{8.37}$$

The delivered power P_q is shown in Fig. 8.27b as a parametric function of injected voltage V_q against transmission angle δ. The normalized V_q voltage is selected to provide same maximum power with series capacitor configuration. The comparison of Fig. 8.27a, b expresses that the series capacitor increases the delivered power by a fixed rate of the delivered by the uncompensated line at a given transmission angle δ. On the other hand, the SSSC increases the delivered power by a fixed fraction of the maximum power available by uncompensated line independent from δ in the angle range of $0 \leq \delta \leq 90°$ [41, 45–47].

8.4.3 *Unified Power Flow Controller (UPFC)*

The UPFC is a VSC based shunt-series compensator that is constituted of STATCOM and SSSC combination with a common dc link as shown in Fig. 8.28.

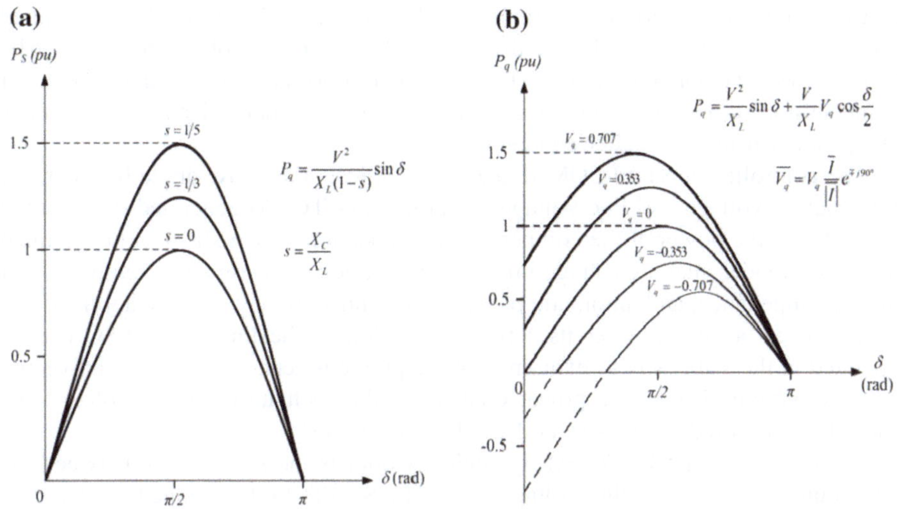

Fig. 8.27 Delivered power *P* and transmission angle *δ* relation, **a** series capacitive compensation, **b** SSSC

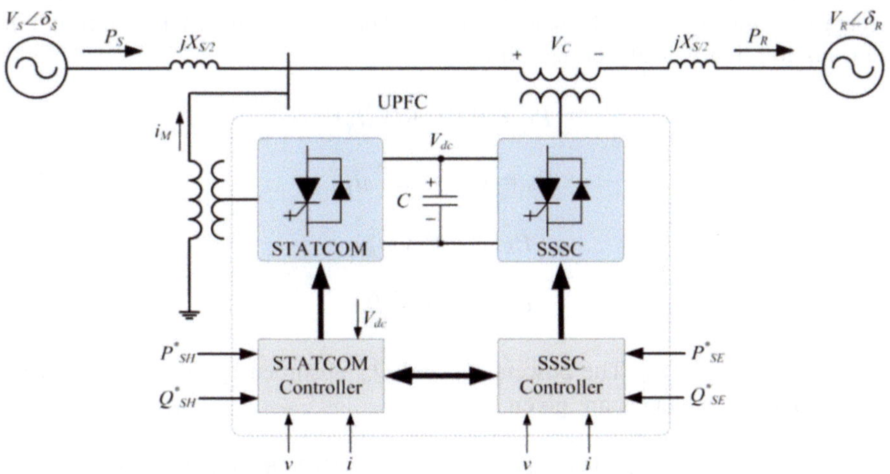

Fig. 8.28 The block diagram of UPFC

Separate observers of each compensator perform the control of UPFC. The shunt and series compensators of UPFC exchange reactive power at the terminals while they are operating as independent FACTS compensators. In this case, SSSC injects a voltage quadrature to the line current, and imitates a series inductive or capacitive reactance to the transmission line. On the other hand, STATCOM injects reactive current in order to imitate shunt reactance to the line while operating stand-alone. In case of the combination as a UPFC), the series compensator voltage can be at any phase corresponding to the predominant line current. Thereby, the exchanged power can be reactive as well as active power at the terminals of compensators. The active power exchanged by any of the compensators is transmitted to the other compensator across the common dc-link. However, exchanged reactive power is independent from each other compensator.

The controller of STATCOM dynamically adjusts the phase angle between the VSC output voltage and line voltage to get the STATCOM generated or absorbed desired reactive power at the compensation connection. The SSSC can be operated in several modes such as voltage injection, regulation of phase angle, emulation of the line impedance, and automatic power flow control. In any of these applications, SSSC injects a voltage over its series connection to the line. The SSSC can be operated in the same way as examined in the previous section to control active and reactive current. The active power control would fluctuate capacitor voltage that should be observed by STATCOM [6, 11, 14, 26, 47].

The controller of STATCOM dynamically adjusts the phase angle between the VSC output voltage and line voltage to get the STATCOM generated or absorbed desired reactive power at the compensation connection. The maximum value of series voltage V_C is limited by SSSC that is the most significant advantage comparing to thyristor-based compensators. In the case reactive power exchange, the UPFC increases the source voltage V_S while generating reactive power, and decreases while absorbing the reactive power. The power flow of UPFC through the transmission line is analyzed as follows;

$$P_S = P_{sh} + \Re(\overline{V_S I_R^*}) \tag{8.38}$$

$$Q_S = Q_{sh} + \Im(\overline{V_R I_R^*}) \tag{8.39}$$

$$P_R = -\Re(\overline{V_R I_R^*}) \tag{8.40}$$

$$Q_R = -\Im(\overline{V_R I_R^*}) \tag{8.41}$$

where $s_{sh} = \sqrt{3/8} \cdot r_{sh}$. The powers P_{sh} and Q_{sh} absorbed by shunt compensator, and common dc voltages are given by

$$P_{sh} = V_S^2 G_{sh} - s_{sh} V_{dc} V_S G_{sh} \cos(\theta_S - \alpha) - s_{sh} V_{dc} V_S B_{sh} \sin(\theta_S - \alpha) \tag{8.42}$$

$$Q_{sh} = V_S^2 B_{sh} - s_{sh} V_{dc} V_S B_{sh} \cos(\theta_S - \alpha) - s_{sh} V_{dc} V_S G_{sh} \sin(\theta_S - \alpha) \tag{8.43}$$

$$V_{dc} = \frac{P_{sh}}{C V_{dc}} + \frac{\Re(\overline{V_S I_R^*})}{C V_{dc}} - \frac{V_{dc}}{R_C C} - \frac{R_{sh}(P_{sh}^2 + Q_{sh}^2)}{C V_{dc} V_S^2} - \frac{R_{se} I_R^2}{C V_{dc}} \tag{8.44}$$

where G_{sh} is the conductance, and B_{sh} is the susceptance of STATCOM while θ is phase angle of the transmission line and α is the phase angle of the compensator [47].

8.5 State-of-the-Art Equipment for Reactive Power Compensation

8.5.1 Dynamic Power Flow Controller (DPFC)

The DPFC is one of the emerging compensators configured with a standard PST for tap-changing, and several series TSC and TCRs as shown in Fig. 8.29. All the compensators connected in the DPFC are introduced in the previous sections of this chapter. The resulting stepped reactance of DPFC is obtained owing to binary arrangement of reactor and capacitor reactance. The PST controls the phase of transmission line voltage in order to control the power flow across the line. Although the system response of PST is slower since its mechanical configuration, this issue is regulated by rapid structure of TSC and TCR. The series compensators are operated at zero current in order to eliminate harmonic contents. The most significant advantage of DPFC is its lower cost comparing other FACTS.

Furthermore, DPFC has several significant advantages according to PST that are related to the speed of power flow control, and reactive power compensation. The dynamic response of DPFC is quite effective on power oscillation damping (POD) and transient stability in addition to steady-state stability. Furthermore, DPFC is more appropriate in corrective compensation, thereby the system stability and voltage stability is improved owing to its TSC and TCR configuration. This is particularly efficient during overload situations by supporting the transmission line with rapid voltage response. The reactive power consumption of PST is a serious problem in weak grids. DPFC overcomes this issue by compensating the tap-dependency of series inductance, and improves voltage stability [48–50].

The reactance, impedance, and admittance of the transmission line starting from shunt connection of PST a to end of the TCR b are given by;

$$X_{ab} = k_L X_L + k^2 X_E + X_B + X_{Line} - k_C X_C \tag{8.45}$$

$$Z_{ab} = R_{ab} + jX_{ab} \tag{8.46}$$

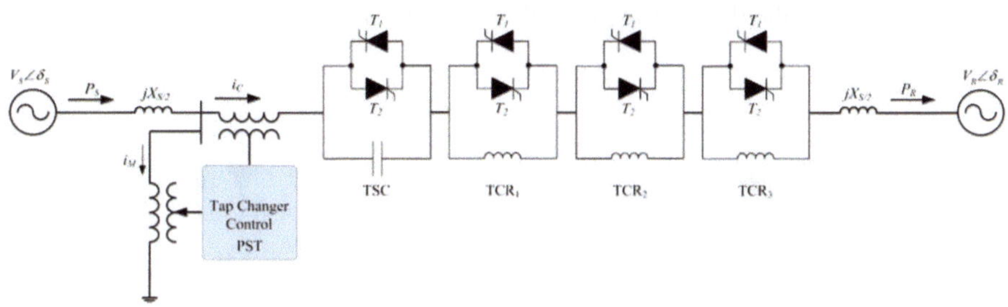

Fig. 8.29 The block diagram of DPFC

$$Y_s = Y_{ab} = \frac{1}{Z_{ab}} = G_S + jB_S \tag{8.47}$$

where $k_L X_L$ and $k_C X_C$ represent the reactive and capacitive modules by ignoring ohmic losses, and k is the PST voltage ratio which is between -1 and 1, X_E, X_B, X_P and R_v are PST internal parameters [51].

8.5.2 VSC-Based HVDC Transmission

The HVDC is an appropriate method used in dc transmission systems. The block diagram of HVDC system is illustrated in Fig. 8.30 that is constituted with VSCs connected B2B. The dc side of converters are coupled with capacitors to regulate the dc line voltage. One of the most significant advantages of VSC based HVDC system is its capability to control active and reactive power separately at each terminal sides that make this system convenient for weak power systems. Furthermore, VSC based HVDC is more effective than CSC on the improvement of system stability and power flow control, and isolation under transients owing to its rapid dynamic response.

The VSC based HVDC can also be used in the interline power flow controller IPFC that is utilized to transfer active power among dc link buses. HVDC system can be seen as an extension of UPFCas well as IPFC in some cases. The main difference between both systems is segmentation of transmission line while HVDC system does, UPFC does not divide.

The HVDC systems with thyristor-based converters control the dc-line current while the most widely used VSC-based configuration controls the dc-line voltage. This difference in the device topology allows using HVDC system in the integration of asynchronous sources. The increased power demand and high rate integration of large distributed sources are handled owing to modular multilevel converter (MMC) based HVDC that is installed by series and parallel connection of VSC cells as shown in Fig. 8.31 [6, 40, 52].

The HVDC systems with thyristor-based converters control the dc-line current while the most widely used VSC-based configuration controls the dc-line voltage. This difference in the device topology allows using HVDC system in the integration

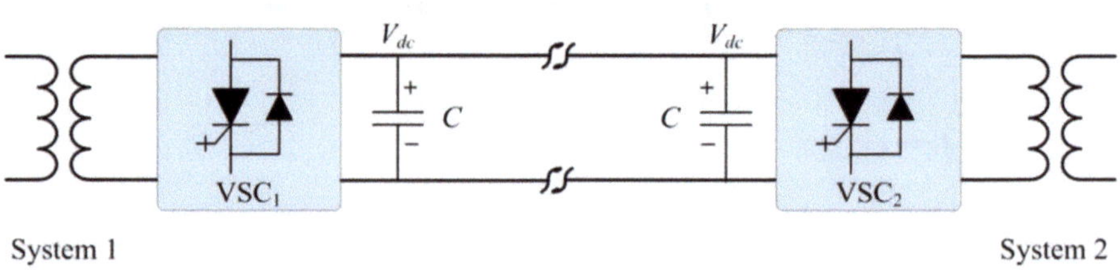

Fig. 8.30 The block diagram of HVDC transmission system

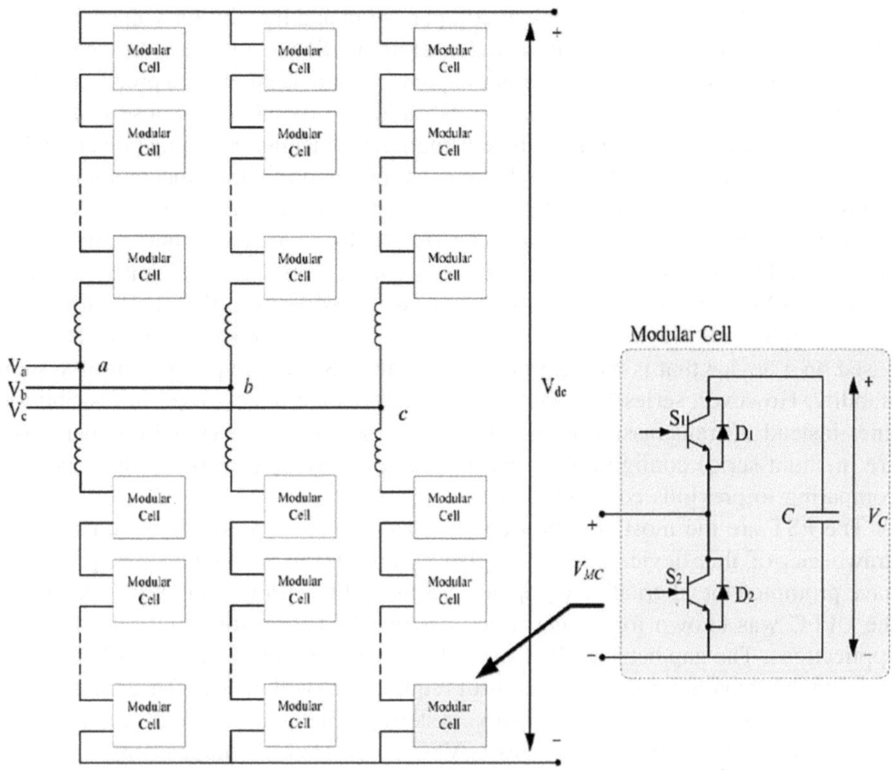

Fig. 8.31 The block diagram of MMC based HVDC system

of asynchronous sources. The increased power demand and high rate integration of large distributed sources are handled owing to MMC based HVDC that is installed by series and parallel connection of VSC cells. The most recent applications are performed by MMC topology since its footprint requirements are lower, and capable to control active and reactive power independently. Furthermore, it brings the advantages of lower switching losses, lower dv/dt ratio, modular structure, reconfigurable topology, and direct connection to the power systems without power transformer requirement that decreases the substation area. Besides, MMC configuration provides to use several types of energy storage systems [6, 40, 52].

8.6 Conclusion

The most widely used reactive power compensators in transmission line compensation are introduced in this chapter. The theoretical backgrounds, applications, and compensation methods are discussed in a comprehensive approach by listing the compensator according to their improvements. The shunt devices among others are being widely used for reactive power compensation and thereby for voltage control. The first improvement seen in compensators is SVCs that are replaced with mechanically switched devices in order to provide more accurate and precise control. The primary contribution of SVC was its rapid and dynamic response to the line fluctuations. However, it was not efficient enough against fast changing power quality problems such as voltage dips and flickers that required the implementation of more robust compensators. The FACTS are improved to overcome this kind of stability problems where the first one was STATCOM listed in the shunt compensator group. In addition to shunt devices, the series compensation devices that are capable to improve line impedance and stability are also considered in FACTS family.

Although the SSSC is a mature technology in series compensation, the TCSC is based on a device that is efficient on damping the SSR and improvement of system stability. However, series VSC based compensators are mostly used in distribution lines instead of transmission lines. The most extensively researched compensators are in shunt-series configurations that are more convenient in power flow control comparing to previous configurations.

The PST are the most common compensators used in power flow control. The drawbacks of this device such as higher footprint and maintenance requirement have promoted the alternative compensators as UPFC), IPFC, and DPFC. Although the UPFC was known for a long time, it is installed for testing and improvement applications. The gap between PST and UPFC) has been fulfilled by DPFC that was meeting the dynamic power flow control requirement with its simpler configuration by overcoming drawbacks of both systems. The most recent compensator responding the industry requirements is VSC-based HVDC compensator. The VSC and MMC configurations brought numerous advantages such as fully independent power control, modular structure, improved voltage and stability control and dynamic power flow control comparing the previous devices.

Although the FACTS are generally recognized as a new technology, there are lots of FACTS are being operated in the industry. The most widely used device is SVC first installed in 1970s and it is known that the total installed SVC compensator capacity is over than 100.000 MVA. It is also noted by literature that over than 700 conventional series compensation plants are installed worldwide that are capable to compensate around 350.000 MVA total power. The successor technologies as VSC-based and MMC HVDC B2B compensators, STATCOM, TSCS, and DPFC present quick improvements in their application areas.

References

1. W. Hofmann, J. Schlabbach, W. Just, Reactive Power Compensation: A Practical Guide, First Edition, John Wiley & Sons Ltd, 2012.
2. P.R. Sharma, A. Kumar, N. Kumar, Optimal Location for Shunt Connected FACTS Devices in a Series Compensated Long Transmission Line, Turk J. Elec. Engineering, vol. 15, no. 3, pp. 321–328, 2007.

3. M. Beza, Control of Energy Storage Equipped Shunt-Connected Converter for Electric Power System Stability Enhancement, Thesis, Department of Energy and Environment Chalmers University of Technology, Gothenburg, Sweden, 2012.
4. E. Acha, V. Agelidis, O. Anaya, T.J.E. Miller, Power Electronic Control in Electrical Systems, Newnes Power Engineering Series, ISBN-13: 978-0750651264, 2002.
5. H. Amaris, M.A. Carlos, A. Ortega, Reactive Power Management of Power Networks with Wind Generation, Lecture Notes in Energy, Springer-Verlag, London, 2013.
6. E.H. Watanabe, M. Aredes, P.G. Barbosa, F.K. de Araujo Lima, R.F. da Silva Dias, G. Santos Jr., 32 - Power Electronic Control in Electrical Systems, Power Electronics Handbook, Third Edition, Edited by Muhammad H. Rashid, Butterworth-Heinemann, Boston, pp. 851–877, 2011.
7. X.P. Zhang, C. Rehtanz, B. Pal, Flexible AC Transmission Systems: Modelling and Control, Springer-Verlag, Berlin, Heidelberg, 2012.
8. T. Petersson, Reactive Power Compensation, ABB Technical Report, no. 500-028E, pp. 1–25, Dec. 1993.
9. F. Shahnia, et al. (Eds.), Static Compensators (STATCOMs) in Power Systems, Springer Science+Business Media, Singapore, 2015.
10. H.K. Tyll, F. Schettle, Historical Overview on Dynamic Reactive Power Compensation Solutions from the Begin of AC Power Transmission Towards Present Applications, Power Systems Conference and Exposition, pp. 1–7, 15–18 March 2009.
11. J. Dixon, L. Moran, J. Rodriguez, R. Domke, Reactive Power Compensation Technologies: State-of-the-Art Review, Proceedings of the IEEE, vol. 93, no. 12, pp. 2144–2164, Dec. 2005.
12. A.A. Edris, et al., Proposed Terms and Definitions for Flexible AC Transmission Systems (FACTS), IEEE Trans. Power Delivery, vol. 12, no. 4, pp. 1848–1853, October 1997.
13. N. Flourentzou, V.G. Agelidis, G.D. Demetriades, VSC-Based HVDC Power Transmission Systems: An Overview, IEEE Transactions on Power Electronics, vol. 24, no. 3, pp. 592–602, March 2009.
14. K. Kalyan, M.L. Sen, Introduction to FACTS Controllers, The Institute of Electrical and Electronics Engineers, Inc., 2009.
15. L. Gyugyi, Power Electronics in Electric Utilities: Static VAR Compensators, Proceedings of the IEEE, vol. 76, no. 4, pp. 483–494, Apr. 1988.
16. L. Liu, H. Li, Y. Xue, W. Liu, Reactive Power Compensation and Optimization Strategy for Grid-Interactive Cascaded Photovoltaic Systems, IEEE Transactions on Power Electronics, vol. 30, no. 1, pp. 188–202, Jan. 2015.
17. H. Akagi, E.H. Watanebe, M. Aredes, Instantaneous Power Theory and Applications to Power Conditioning, IEEE Press, 2007.
18. F.L. Luo, H. Ye, M. Rashid, Digital Power Electronics and Applications, ISBN: 0-1208-8757-6, Elsevier, USA, 2005.
19. J. Brochu, F. Beauregard, J. Lemay, P. Pelletier, R.J. Marceau, Steady-State Analysis of Power Flow Controllers Using the Power Controller Plane, IEEE Transactions on Power Delivery, vol. 14, no. 3, pp. 1024–1031, Jul. 1999.
20. U.N. Khan, T.S. Sidhu, A Phase-Shifting Transformer Protection Technique Based on Directional Comparison Approach, IEEE Transactions on Power Delivery, vol. 29, no. 5, pp. 2315–2323, Oct. 2014.
21. R. Baker, G. Guth, W. Egli, P. Eglin, Control Algorithm for a Static Phase Shifting Transformer to Enhance Transient and Dynamic Stability of Large Power Systems, IEEE Transactions on Power Apparatus and System, vol. PAS-101, no. 9, pp. 3532–3542, Sept. 1982.
22. R. Mihalic, P. Zunko, Phase-Shifting Transformer with Fixed Phase between Terminal Voltage and Voltage Boost: Tool for Transient Stability Margin Enhancement, IEE Proceedings - Generation, Transmission and Distribution, vol. 142, no. 3, pp. 257–262, May 1995.
23. J. Verboomen, D. Van Hertem, P.H. Schavemaker, W.L. Kling, R. Belmans, Analytical Approach to Grid Operation with Phase Shifting Transformers, IEEE Transactions on Power Systems, vol. 23, no. 1, pp. 41–46, Feb. 2008.

24. J. Mescua, A Decoupled Method for Systematic Adjustments of Phase-Shifting and Tap-Changing Transformers, IEEE Transactions on Power Apparatus and Systems, vol. PAS-104, no. 9, pp. 2315, 2321, Sept. 1985
25. Ch.N. Huang, Feature Analysis of Power Flows Based on the Allocations of Phase-Shifting Transformers, IEEE Transactions on Power Systems, vol. 18, no. 1, pp. 266–272, Feb. 2003.
26. K.R. Padiyar, FACTS: Controllers in Power Transmission and Distribution, Anshan Publishers, 1st Edition (August 15, 2009), 978-1848290105.
27. A.K. Chakravorti, A.E. Emanuel, A Current Regulated Switched Capacitor Static Volt Ampere Reactive Compensator, IEEE Transactions on Industry Applications, vol. 30, no. 4, pp. 986–997, Jul./Aug. 1994.
28. H. Jin, G. Goos, L. Lopes, An Efficient Switched-Reactor-Based Static VAr Compensator, IEEE Transactions on Industry Applications, vol. 30, no. 4, pp. 998–1005, Jul./Aug. 1994.
29. S. Mori, K. Matsuno, M. Takeda, M. Seto, Development of a Large Static Var Generator Using Self-Commutated Inverters for Improving Power System Stability, IEEE Transactions Power Delivery, vol. 8, no. 1, pp. 371–377, February 1993.
30. R. Mohan Mathur, R.K. Varma, Thyristor Based FACTS Controllers for Electrical Transmission Systems, ISBN: 978-0-471-20643-9, Wiley-IEEE Press, 2002.
31. B. Singh, K.V. Srinivas, Three-Level 12-Pulse STATCOM with Constant DC Link Voltage, 2009 Annual IEEE India Conference (INDICON), pp. 1–4, 18–20 Dec. 2009.
32. S.R. Barik, B. Nayak, S. Dash, A Comparative Analysis of Three Level VSC Based Multi-Pulse STATCOM, International Journal of Engineering and Technology (IJET), vol. 6, no. 3, pp. 1550–1563, Jun.–Jul. 2014.
33. N. Voraphonpiput, S. Chatratana, Analysis of Quasi 24-Pulse STATCOM Operation and Control Using ATP-EMTP, IEEE Region 10 Conference TENCON 2004, vol. C, pp. 359–362, 21–24 Nov. 2004.
34. B. Singh, R. Saha, A New 24-Pulse STATCOM for Voltage Regulation, Int. Conf. on Power Electronics, Drives and Energy Systems, pp. 1–5, 12–15 Dec. 2006.
35. K.V. Srinivas, B. Singh, Three-Level 24-Pulse STATCOM with Pulse Width Control at Fundamental Frequency Switching, IEEE Industry Applications Society Annual Meeting (IAS), pp. 1–6, 3–7 Oct. 2010.
36. I. Colak, E. Kabalci, R. Bayindir, Review of Multilevel Voltage Source Inverter Topologies and Control Schemes, Energy Conversion and Management, vol. 52, no. 2, pp. 1114–1128, February 2011.
37. I. Colak, E. Kabalci, Practical Implementation of a Digital Signal Processor Controlled Multilevel Inverter with Low Total Harmonic Distortion for Motor Drive Applications, Journal of Power Sources, vol. 196, no. 18, pp. 7585–7593, 2011.
38. I. Colak, E. Kabalci, Developing a Novel Sinusoidal Pulse Width Modulation (SPWM) Technique to Eliminate Side Band Harmonics, International Journal of Electrical Power and Energy Systems, vol. 44, pp. 861–871, 2013.
39. I. Colak, E. Kabalci, Implementation of Energy Efficient Inverter for Renewable Energy Sources, Electric Power Components and Systems, vol. 41, no. 1, pp. 31–46, 2013.
40. D. Soto, T.C. Green, A Comparison of High-Power Converter Topologies for the Implementation of FACTS Controllers, IEEE Transactions on Industrial Electronics, vol. 49, no. 5, pp. 1072–1080, Oct. 2012.
41. L. Gyugyi, C.D. Schauder, K.K. Sen, Static Synchronous Series Compensator: A Solid-State Approach to the Series Compensation of Transmission Lines, IEEE Transactions on Power Delivery, vol. 12, no. 1, pp. 406–417, Jan. 1997.
42. A.C. Pradhan, P.W. Lehn, Frequency-Domain Analysis of the Static Synchronous Series Compensator, IEEE Transactions on Power Delivery, vol. 21, no. 1, pp. 440–449, Jan. 2006.
43. M. Farahani, "Damping of Subsynchronous Oscillations in Power System Using Static Synchronous Series Compensator, IET Generation, Transmission & Distribution, vol. 6, no. 6, pp. 539–544, Jun. 2012.

44. M. Saradarzadeh, S. Farhangi, J.L. Schanen, P.O. Jeannin, D. Frey, Application of Cascaded H-Bridge Distribution-Static Synchronous Series Compensator in Electrical Distribution System Power Flow Control, IET Power Electronics, vol. 5, no. 9, pp. 1660–1675, Nov. 2012.
45. X.P. Zhang, Advanced Modeling of the Multi-Control Functional Static Synchronous Series Compensator (SSSC) in Newton Power Flow, IEEE Transactions on Power Systems, vol. 18, no. 4, pp. 1410–1416, Nov. 2003.
46. R. Benabid, M. Boudour, M.A. Abido, Development of a New Power Injection Model with Embedded Multi-Control Functions for Static Synchronous Series Compensator, IET Generation, Transmission & Distribution, vol. 6, no. 7, pp. 680–692, Jul. 2012.
47. R. Natesan, G. Radman, Effects of STATCOM, SSSC and UPFC on Voltage Stability, Proceedings of the Thirty-Sixth Southeastern Symposium on System Theory, pp. 546–550, 2004.
48. C. Rehtanz, Dynamic Power Flow Controllers for Transmission Corridors, IREP Symposium Bulk Power System Dynamics and Control - VII, Revitalizing Operational Reliability, pp. 1–9, 19–24 Aug. 2007.
49. N. Johansson, Aspects on Dynamic Power Flow Controllers and Related Devices for Increased Flexibility in Electric Power Systems, Royal Institute of Technology School of Electrical Engineering Division of Electrical Machines and Power Electronics, PhD. Dissertation, Stockholm, 2011.
50. U. Hager, K. Gorner, C. Rehtanz, Hardware Model of a Dynamic Power Flow Controller, IEEE Trondheim PowerTech, pp. 1–6, 19–23 Jun. 2011.
51. R. Ahmadi, A. Sheykholeslami, A.N. Niaki, H. Ghaffari, Power Flow Control and Solutions with Dynamic Flow Controller, IEEE Canada Electric Power Conference, EPEC 2008, pp. 1–6, 6–7 Oct. 2008.
52. J. Dorn, H. Huang, D. Retzmann H., Multilevel Voltage-Sourced Converters for HVDC and FACTS Applications, Cigre SC B4 Colloquium, Bergen and Ullensvang, pp. 1–8, Norway, 2009.

Chapter 9
Optimal Placement of Reactive Power Compensators in AC Power Network

Hossein Shayeghi and Yashar Hashemi

Abstract A framework and versatile approach is presented in this chapter to extend a multi-objective reactive power planning (RPP) method for concurrently study of reactive power from flexible alternating current transmission system (FACTS) devices and capacitor banks. The proposed plan will enable system operators to define the optimal location of FACTS devices and capacitor banks that should be connected in the network to improve voltage stability, active power losses and the cost of VAR injection. A formulation and solution method are presented for the RPP problem, including FACTS devices and capacitor banks.

Nomenclature

FDM	Fuzzy decision making	NTVE	Nonlinear time-varying evolution
RPP	Reactive power planning	Z_{shi}, Z_{sein}	Shunt and series transformer impedances
FACTS	Flexible ac transmission system	V_{shi}, V_{sein}	Injected voltage of M-FACT
MO	Multi-objective	$\Delta P_{g,j}$	The change in the plan of the *j*th generator
SLP	Sequential linear programming	$x_i(k)$	Situation of *i*th particle at iteration k
MILP	Mixed integer linear programming	$x_i^l(k)$	Local best of *i*th particle at iteration k
OPF	Optimal power flow	x^g	Global best of all particles
SVC	Static VAR compensator	$v_i(k)$	Velocity of *i*th particle at iteration *k*

(continued)

H. Shayeghi (✉) · Y. Hashemi
Technical Engineering Department, University of Mohaghegh Ardabili, Ardabil, Iran
e-mail: hshayeghi@gmail.com

Y. Hashemi
e-mail: yashar_hshm@yahoo.com

N. Mahdavi Tabatabaei et al. (eds.), *Reactive Power Control in AC Power Systems*, Power Systems, DOI 10.1007/978-3-319-51118-4_9

(continued)

STATCOM	Static synchronous compensator	c_1	Cognitive parameter (acceleration coefficient)
TCSC	Thyristor controlled series compensator	c_2	Social parameter (acceleration coefficient)
UPFC	Unified power flow controller	ϕ_1, ϕ_2	Random numbers between 0 and 1
PST	Phase-shifting transformer	S_{iSC}, S_{iHFC}	Size related to the *i*th HFC and slow VAR device
MSC	Mechanically-switched shunt capacitor	k_S, k_R	Stator and rotor leakage factor
TSSC	Thyristor switched series capacitor	Q_{GSC}	Reactive power injected from GSR
TSSR	Thyristor switched series reactor	S_{GSC}	Apparent power of GSC
DM	Decision makers	k_L, k_C	The amount of X_C and X_L in service
HFC	Hybrid flow controller	C_{iHFC}, C_{iSC}	Cost function of the *i*th HFC and slow VAR device
M-FACTS	Multi-converter FACTS	k	PST voltage ratio
MOP	Multi-objective problem	r_T	Discount rate
MOPSO-NTVE	Multi-objective particle swarm optimization		

9.1 Introduction

Reactive power planning includes all management actions to improve the voltage profile, voltage stability and other objectives. Recent researches have represented optimization-based techniques to verify RPP issue. RPP problem is one of the most challenging issues in power system studies. Current costs such as variable and fixed VAR set up cost, power loss cost, and fuel cost may be used for target function of RPP [1]. Other targets may be deviation from a given plan of a control parameter (such as voltage) or voltage stability margin that is used as a multi-objective (MO) model [2]. Several network blackouts related to deficiency of reactive power in stiffly stressed situations have shown that voltage collapse issue is closely related to reactive power handling [3]. Thus, the main purpose of the RPP problem is to obtain minimum investment cost in new reactive plan which is necessary to preserve acceptable voltage profiles [4].

Power grids are stressed due to various causes; high cost of extending transmission lines, difficulty of getting right of way, and change in generating schemes related to pecuniary and environmental concerns, and enlarged loading of transmission grid. These issues have infused power utilities to discuss maintaining against voltage instability as an essential option in reactive power planning. Hence, the coordinated VAR management plan can be verified as specifying reactive power expansion plan that convinces requirements for voltage deviation and voltage permanency. This issue makes a great non-linear optimization subject.

Different methods are used to solve RPP problem. Sequential linear programming (SLP) was reported for VAR management and pricing in [5]. Other procedures have considered integer parameters to shape discrete scales and fixed costs, mainly based on Benders decomposition [6]. This technique mixes use of mixed integer linear programming (MILP) for finance sub-problem and classical optimal power flow (OPF) for operational sub-problem.

Reactive power management has gained more importance due to the inclusion of FACTS devices in transmission networks in the system [7]. Optimal allocation of fast reactive power tools, as for example static VAR compensator (SVC) [8], static synchronous compensator (STATCOM), static inter-phase power controller (SIPC) [9], thyristor controlled series compensator (TCSC) and unified power flow controller (UPFC), is a critical component in RPP problem solution or VAR management. Hybrid flow controller (HFC) is a novel FACTS device that is constructed from a conventional phase-shifting transformer (PST), a mechanically-switched shunt capacitor (MSC), a multi-module thyristor switched series capacitor (TSSC) and a multi-module thyristor switched series reactor (TSSR) [10].

With regard to property of series and shunt compensation in HFC structure, its function is similar to UPFC. Since HFC has some preference than the UPFC, it is expected that application of this device would be expanded. Cost efficiency of HFC is better than UPFC [11]. In [12], it was indicated that HFC may be used for the most effective satisfaction of dispatcher demand supported technical and economic characteristics. Simplicity of idea, management and operational ways, lower loss and higher efficiency are the advantages of HFC. Furthermore, HFC produces low harmonics and has no incompatible result on power quality index.

There are many structures by combining two or more converter blocks. Interline power flow controller (IPFC), the generalized unified power flow controller (GUPFC) and generalized interline power flow controller (GIPFC) are developed to regulate power flows of sub-grid rather than regulate power flow of single line by a UPFC or static synchronous series compensator (SSSC). The suitable control ability of the multi-converter FACTS (M-FACTS) presents an excellent potential in solving several problems in electric utilities [13]. The optimal power flow with the M-FACTS devices would be an effective tool to control, plan and manage power network [14].

9.1.1 Objectives

A framework for deriving FACTS steady state studies based on injection models is presented. The presented framework has the following specifications:

- Active and reactive power is applied as management parameters. It's a suitable technique which models nearly any kind of FATCS tools. In this model, FACTS control parameters are the injected powers.
- In this model, the injected power is an independent index that does not change with the amplitudes and phases of linked node voltage. In this case the Jacobian matrix does not vary within power flow iterations.
- Since in the model built, power injections are treated as values, which do not vary with the connected node voltage, thereby the need not to be modified during the. Thus, it is easy and efficient to implement.

In this chapter the injection model of FACTS devices is used to investigate its effect on power flow of the power system. At first, the injection pattern of FACTS tools is extracted. Then, a steady state study is defined by FACTS devices injection pattern. To find size and site of FACTS tools, a multi-objective optimization procedure is developed for decreasing the overall costs of RPP and improving voltage behavior of the power network. Multiple objectives are considered at the same time which addresses the multi-objective design technique. In a case that FACTS devices are connected in a power grid, its cost should be considered that it is related to the converter capacity. The purpose of RPP is to prepare adequate VAR sources for the system to be acted economically. The aims of VAR design issues contain decreasing investment cost and voltage deviation, and voltage stability criterion maximization.

In multi-target issues the objectives are in contrast with each other. Thus, a set of solutions is obtained instead of one. In this work, multi-objective optimization algorithms are used to get the non-dominated set. This survey implements an external archive to retain all non-dominated sets within the evolutionary procedure and a fuzzy decision-making technique is used to group these solutions based on their significance. Decision makers can choose the suitable solution among them by utilizing the fuzzy decision-making manner. Applying a multi-target optimization method is an appropriate method to plan and to placement reactive power tools by considering simultaneously an extensive range of target functions such as: correcting the voltage stability, power losses reduction or decreasing the cost of VAR sources.

The presented procedure is applied on IEEE 57 bus power grid to illustrate the optimal operation of power network and the achieved outcomes are compared. Two new FACTS devices, HFC and M-FACTS, are discussed in this chapter.

The chapter is organized into 5 sections. The second section verifies FACTS devices and their use in the intent of RPP and operational subjects with power flow issues. The overall aims and objectives of the FACTS technology are achieved in Sect. 9.2. This section reviews the steady condition behaviors of the transmission lines. Also, it contains the analysis of FACTS controller on power flow and voltage profile. Section 9.2 covers the modeling of the various FACTS devises, HFC and GUPFC. The second section presents the modeling and steady state characteristics of HFC and GUPFC. In each case of FACTS devices, the action of the FACTS tools is clarified with the explanation of the power circuit, associated controllers and operating modes. The idea of the application of multi-objective optimization method in achieving the desired level of techno-economic share of FACTS devices in power grid action and RPP issue is given in two Sects. 9.3 and 9.4. In these sections to extract size and site of FACTS tools and capacitor banks, multi-objective optimization algorithm based on multi-attribute decision making method is developed by formulating the overall costs of power generation and maximizing of profit. The method of investment analysis and the allocation process of FACTS devices in power network are explained in Sect. 9.5. Section 9.5 provides the test network. Simulation outcomes of the presented technique and conclusion remarks are presented in Sect. 9.5.

9.2 Steady State Characteristics of the FACTS Devices

9.2.1 Hybrid Flow Controller

The power injection model of [12] is used to model HFC. This model of HFC is suitable for conventional power flow analysis. The accuracy and the conformability of this model on any power system leads to accurate steady state analysis. The structure of the HFC has been given in Fig. 9.1. V_i and V_j are voltage phasors of buses *i*, *j*, respectively. The PST induces voltage V_P and V_i' is the voltage of the HFC internal node. Magnitude of V_{XC} and V_{XL} depend on value of line flow I_{ij} and the number of TSSC and TSSR units. The series injected voltage of PST and the limitation proposing power interchange between transformers of the PST, can be explained as follows equations

$$V_P = jkV_E \tag{9.1}$$

$$V_E = V_i - jX_E I_E \tag{9.2}$$

$$V_P I_{ij}^* = V_E I_E^* \tag{9.3}$$

where, *k* is the PST voltage proportion and X_E is the leakage reactance related to excitation transformer. Substituting (9.1) in (9.3), we obtain

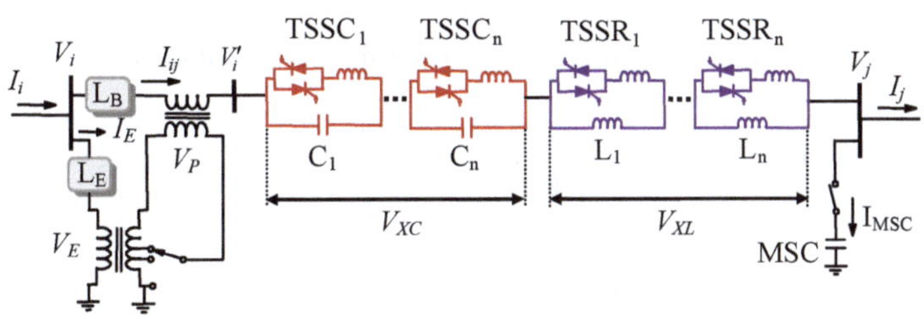

Fig. 9.1 Schematic diagram of HFC

$$I_E = -jkI_{ij} \tag{9.4}$$

The line current I_{ij} is as follows

$$I_{ij} = \frac{(1+jk)V_i - V_j}{j(k_L X_L + k^2 X_E + X_B - k_C X_C)} \tag{9.5}$$

where, k_L and k_C are the amount of X_C and X_L in service and X_B is the leakage reactance of series transformer. The currents I_i and I_j based on V_i and V_j is developed as follows:

$$I_i = \frac{((1+k^2)V_i - (1-jk)V_j)}{j(k_L X_L + k^2 X_E + X_B - k_C X_C + X_S)} \tag{9.6}$$

$$I_j = \frac{((1+jk)V_i - V_j)}{j(k_L X_L + k^2 X_E + X_B - k_C X_C + X_S)} + \frac{V_j}{jX_{MSC}} \tag{9.7}$$

If the current I_B be expressed as

$$I_B = \frac{V_i - V_j}{jX_B} \tag{9.8}$$

The injected current into bus i is as follows

$$I_{SS} = I_B - I_i = \frac{((1+k^2)V_i - (1-jk)V_j)}{j(k_L X_L + k^2 X_E + X_B - k_C X_C + X_S)} + \frac{(-V_i + V_j)}{jX_B} \tag{9.9}$$

The active and reactive power P_{SS} and Q_{SS} injected by I_{SS} into bus i are written as

$$P_{SS} = -\frac{|V_i||V_j|(k\cos(\theta_i - \theta_j) + \sin(\theta_i - \theta_j))}{(k_L X_L + k^2 X_E + X_B - k_C X_C + X_S)} + \frac{|V_i||V_j|}{X_B}\sin(\theta_i - \theta_j) \tag{9.10}$$

$$\begin{aligned} Q_{SS} = &-\frac{|V_i|(|V_i|(1+k^2) + k|V_j|\sin(\theta_i - \theta_j) - |V_j|\cos(\theta_i - \theta_j))}{(k_L X_L + k^2 X_E + X_B - k_C X_C + X_S)} \\ &+ \frac{|V_i|}{X_B}(|V_i| - |V_j|\cos(\theta_i - \theta_j)) \end{aligned} \tag{9.11}$$

The injected current into bus j can be determined as follows

$$\begin{aligned} I_{SR} = I_j - I_B = &\frac{((1+jk)V_i - V_j)}{j(k_L X_L + k^2 X_E + X_B - k_C X_C + X_S)} \\ &+ \frac{(-V_i + V_j)}{jX_B} + \frac{V_j}{jX_{MSC}} \end{aligned} \tag{9.12}$$

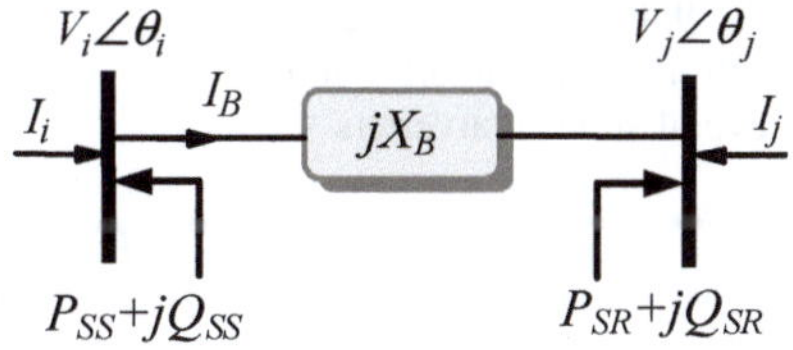

Fig. 9.2 Injection model for HFC

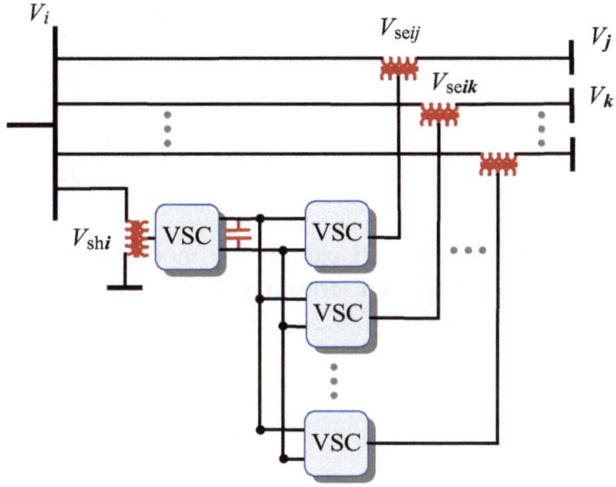

Fig. 9.3 M-FACTS structure with a shunt converter and several series converter

According to Fig. 9.2 the active and reactive power injected by I_{SR} into bus j are as

$$P_{SR} = -P_{SS} \tag{9.13}$$

$$\begin{aligned} Q_{SR} = & -\frac{|V_i|\left(|V_i|\cos(\theta_i-\theta_j) - k|V_j|\sin(\theta_i-\theta_j) - |V_j|\right)}{(k_L X_L + k^2 X_E + X_B - k_C X_C + X_S)} \\ & + \frac{|V_j|\left(|V_j| - |V_i|\cos(\theta_i-\theta_j)\right)}{X_B} + \frac{k_m|V_j|^2}{X_{MSC}} \end{aligned} \tag{9.14}$$

9.2.2 *Multi-converter FACTS Devices*

The principle aim of the M-FACTS is to adjust the voltage and power flow [15]. The circuit of the M-FACTS including of a shunt injected voltage source and two series injected voltage sources is given in Figs. 9.3 and 9.4. Active power can be interchanged among these converters based on the DC branch. In M-FACTS with one shunt converter and two series converters, it can adjust five power grid parameters such as node voltage and active and reactive power of two lines.

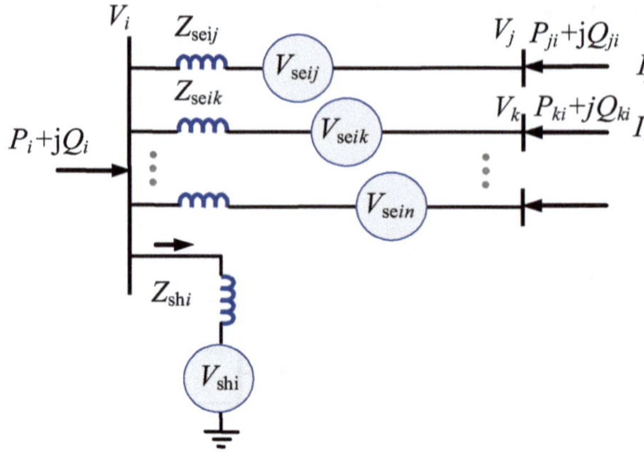

Fig. 9.4 M-FACTS diagram with a shunt converter and several series converters

More series converters can present more degrees of control. The Z_{shi} and Z_{sein} in Fig. 9.4 are shunt and series transformer impedances. The injected voltage sources that are shown in Fig. 9.2 are explained as follows:

$$V_{shi} = V_{shi}\angle\theta_{shi} \tag{9.15}$$

$$V_{sein} = V_{sein}\angle\theta_{sein} \tag{9.16}$$

where V_{shi}, V_{sein} are injected voltage of M-FACTS and $n = j, k, \ldots$.

9.2.2.1 Power Flow Equations of M-FACTS

As shown in Fig. 9.4, the power flow relations of M-FACTS can be given as

$$\begin{aligned} P_i = {} & V_i^2 g_{ii} - V_i V_{shi}(g_{shi}\cos(\theta_i - \theta_{shi}) + b_{shi}\sin(\theta_i - \theta_{shi})) \\ & - \sum_n V_i V_n (g_{in}\cos\theta_{in} + b_{in}\sin\theta_{in}) \\ & - \sum_n V_i V_{sein}(g_{in}\cos(\theta_i - \theta_{sein}) + b_{in}\sin(\theta_i - \theta_{sein})) \end{aligned} \tag{9.17}$$

$$\begin{aligned} Q_i = {} & -V_i^2 b_{ii} - V_i V_{shi}(g_{shi}\sin(\theta_i - \theta_{shi}) + b_{shi}\sin(\theta_i - \theta_{shi})) \\ & - \sum_n V_i V_n (g_{in}\sin\theta_{in} - b_{in}\cos\theta_{in}) \\ & - \sum_n V_i V_{sein}(g_{in}\sin(\theta_i - \theta_{sein}) - b_{in}\cos(\theta_i - \theta_{sein})) \end{aligned} \tag{9.18}$$

$$\begin{aligned} P_{ni} = {} & V_n^2 g_{nn} - V_i V_n (g_{in}\cos(\theta_n - \theta_i) + b_{in}\sin(\theta_n - \theta_i)) \\ & V_n V_{sein}(g_{in}\cos(\theta_n - \theta_{sein}) + b_{in}\sin(\theta_n - \theta_{sein})) \end{aligned} \tag{9.19}$$

$$\begin{aligned} Q_{ni} &= -V_n^2 b_{nn} - V_i V_n \left(g_{in} \sin(\theta_n - \theta_i) - b_{ij} \cos(\theta_n - \theta_i) \right) \\ & V_n V_{sein} \left(g_{in} \cos(\theta_n - \theta_{sein}) - b_{in} \cos(\theta_n - \theta_{sein}) \right) \end{aligned} \tag{9.20}$$

where $g_{shi} + jb_{shi} = 1/Z_{shi}$, $g_{in} + jb_{in} = 1/Z_{sein}$, $g_{nn} + jb_{nn} = 1/Z_n$, $g_{ii} = g_{shi} + \sum_n g_{in}$, $b_{ii} = b_{shi} + \sum_n b_{in}$, and $n = j, k, \ldots$

9.2.2.2 Operating Limitations of M-FACTS

Active power change in DC branch based on operating limitations among converters is as follows:

$$PE = \mathrm{Re}\left(V_{shi} I_{shi}^* - \sum_n V_{sein} I_{ni}^* \right) = 0 \tag{9.21}$$

or

$$\begin{aligned} PE = {} & V_{shi}^2 g_{shi} - V_i V_{shi} (g_{shi} \cos(\theta_i - \theta_{shi}) - b_{shi} \sin(\theta_i - \theta_{shi})) \\ & + \sum_n \left(V_{sein}^2 g_{in} - V_i V_{sein} (g_{in} \cos(\theta_i - \theta_{sein}) - b_{in} \sin(\theta_i - \theta_{sein})) \right) \\ & + \sum V_n V_{sein} (g_{in} \cos(\theta_n - \theta_{sein}) - b_{in} \sin(\theta_n - \theta_{sein})) = 0 \end{aligned} \tag{9.22}$$

where $n = j, k, \ldots$. The related injected voltage source bound limitations

$$\begin{aligned} V_{shi}^{\min} &\le V_{shi} \le V_{shi}^{\max} \\ \theta_{shi}^{\min} &\le \theta_{shi} \le \theta_{shi}^{\max} \\ V_{sei}^{\min} &\le V_{sei} \le V_{sei}^{\max} \\ \theta_{sei}^{\min} &\le \theta_{sei} \le \theta_{sei}^{\max} \end{aligned}$$

where $n = j, k, \ldots$

9.2.2.3 Injection Pattern of GUPFC

In general, GUPFC consist three voltage source converters and using this basic configuration, it can adjust simultaneously power flow in two transmission lines by varying device control parameters. For the sake of explanation, the complete voltage source-based mathematical modeling of GUPFC is presented in this section. The principle configuration of GUPFC linked between nodes *i*, *j* and *k* is illustrated in Fig. 9.5.

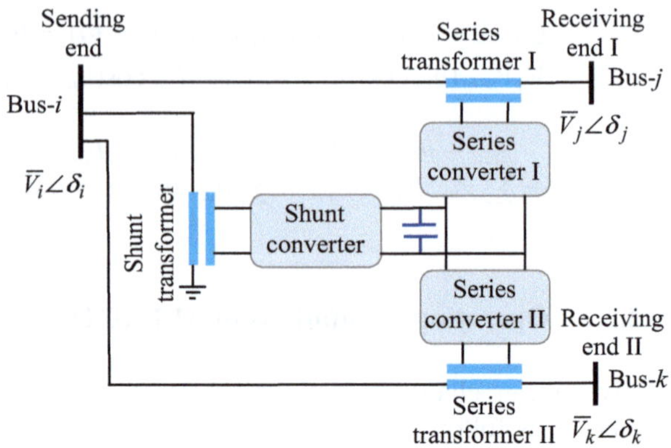

Fig. 9.5 Schematic diagram of GUPFC

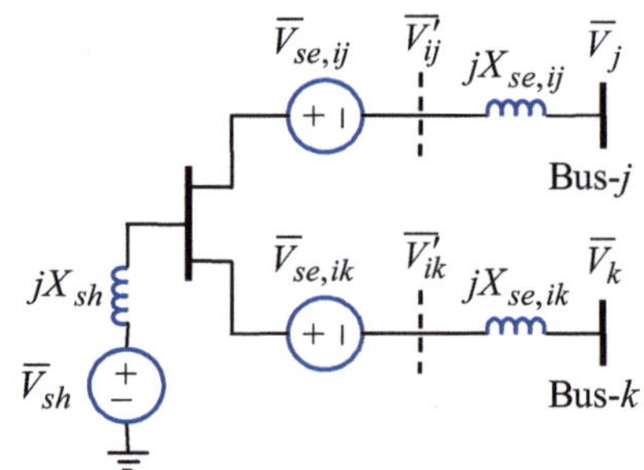

Fig. 9.6 Voltage source model of GUPFC

In this configuration, two voltage source converters are connected in two different transmission lines having a common bus. The third converter is connected at this common bus and acts as a shunt connected voltage source converter. This shunt converter supplies the power that is supplied by the series converters. All these converters are linked through a common DC branch to exchange the power flow. For the sake of simplification, it is assumed that, the voltage injected by the series converters is sinusoidal and the reactance of the coupling transformer is neglected. With these assumptions, the final voltage source plan of GUPFC is exhibited in Fig. 9.6. The voltages at GUPFC connected buses can be expressed as

$$V_m = |V_m|\angle\delta_m \quad \forall\ m = i, j, k \tag{9.23}$$

The applied voltage of the series converters can be represented as:

$$V_m = |V_m|\angle\delta_m \quad \forall\ m = i, j, k \tag{9.24}$$

In Fig. 9.2, the voltage behind the series voltage source can be expressed for both converters as:

$$\overline{V}'_{im} = \overline{V}_i + \overline{V}_{se,im} \quad \forall\ m = j, k \tag{9.25}$$

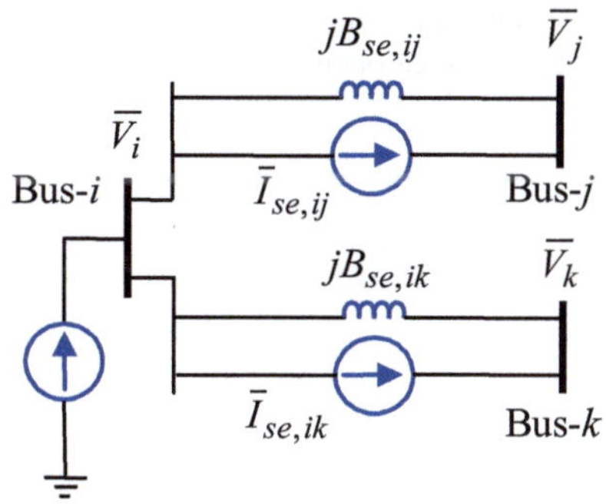

Fig. 9.7 Equivalent current source model GUPFC

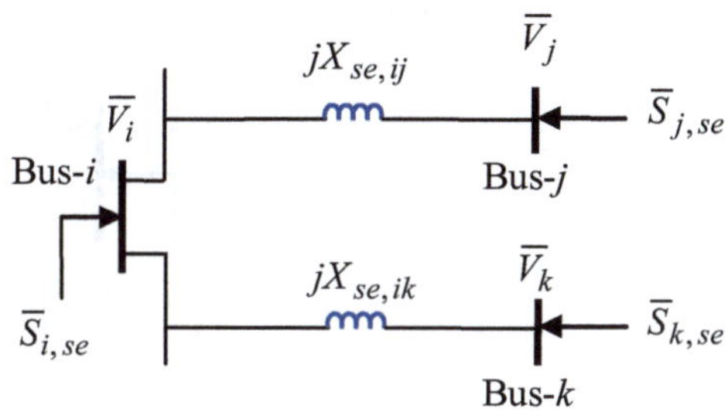

Fig. 9.8 Voltage source model of GUPFC

To develop the injection pattern, the voltage source exhibition is converted into an equivalent current source model using Norton's theorem and is shown in Fig. 9.7.

$$I_{se,im} = -jB_{se,im}\overline{V}_{se,im} \tag{9.26}$$

where $B_{se,im} = 1/jX_{se,im}$ is the admittance of the coupling transformer.

Using this, the power injected by these sources at the device connected buses can be expressed as

$$S_{i,se} = \overline{V}_i(-\overline{I}_{se,ij} - \overline{I}_{se,ik})^* \tag{9.27}$$

$$S_{m,se} = \overline{V}_m\overline{I}^*_{se,im} \tag{9.28}$$

Using Eqs. (9.4), (9.5) and (9.6) can be simplified as:

$$\overline{S}_{i,se} = \sum_{m=j,k} (-jV_iV_{se,im}B_{se,im}\angle(\delta_i - \theta_{se,im})) \tag{9.29}$$

$$\overline{S}_{m,se} = jV_mV_{se,im}B_{se,im}\angle(\delta_m - \theta_{se,im}) \quad \forall \ m = j,k \tag{9.30}$$

The final series voltage source with the related power injections is illustrated in Fig. 9.8.

Similarly, the shunt VSC can be modeled as a power injection pattern at the respective bus. In this modeling, it is assumed that, the reactive power applied by

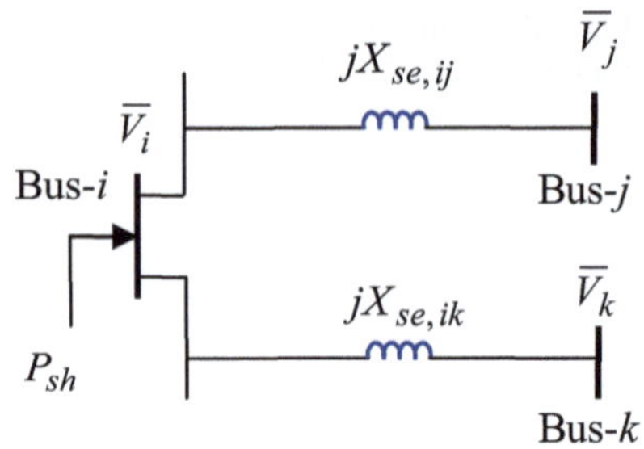

Fig. 9.9 Equivalent shunt voltage source pattern

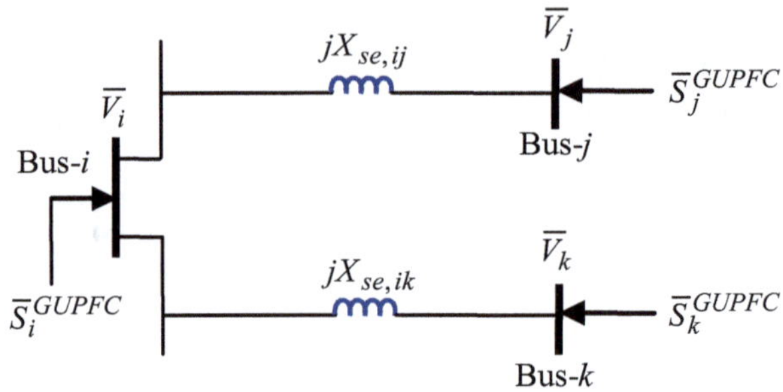

Fig. 9.10 Equivalent power injection model of GUPFC

the shunt VSC is zero, because the purpose of this reactive power is to preserve the voltage value at the converter connected node. The equivalent shunt voltage source pattern of GUPFC is illustrated in Fig. 9.9. The net active power applied at shunt converter connected bus can be expressed as:

$$P_{sh} = -P_{series,ij} - P_{series,ik} \tag{9.31}$$

The magnitude of the apparent power of series converters can be calculated as:

$$\overline{S}_{series,im} = \overline{V}_{se,im} I_{ij}^{*} = j\overline{V}_{se,im} B_{se,im} (\overline{V'}_{ij} - \overline{V}_m)^{*} \quad \forall \; m = j, k \tag{9.32}$$

Using Eq. (9.32), after simplifying, the expressions for active and reactive powers supplied by the series converters derived are

$$\begin{aligned} P_{series,im} &= V_i V_{se,im} B_{se,im} \sin(\theta_{se,im} - \delta_i) \\ &\quad - V_m V_{se,im} B_{se,im} \sin(\theta_{se,im} - \delta_j) \quad \forall m = j, k \\ Q_{series,im} &= -V_i V_{se,im} B_{se,im} \cos(\theta_{se,im} - \delta_i) \\ &\quad + V_m V_{se,im} B_{se,im} \cos(\theta_{se,im} - \delta_j) - V_{se,ij}^2 B_{se,ij} \quad \forall \; m = j, k \end{aligned} \tag{9.33}$$

The final power injection pattern is obtained by combining series voltage source pattern and shunt voltage source model. The combined model is displayed in Fig. 9.10. The respective power injections at GUPFC connected buses can be obtained as

$$
\begin{aligned}
P_i^{GUPFC} &= 2V_iV_{se,ij}B_{se,ij}\sin(\delta_i - \theta_{se,ij}) + 2V_iV_{se,ik}B_{se,ik}\sin(\delta_i - \theta_{se,ik}) \\
&\quad - V_iV_{se,ij}B_{se,ij}\sin(\delta_i - \theta_{se,ij}) - V_iV_{se,ik}B_{se,ik}\sin(\delta_i - \theta_{se,ik}) \\
Q_i^{GUPFC} &= -V_iV_{se,ij}B_{se,ij}\sin(\delta_i - \theta_{se,ij}) - V_iV_{se,ik}B_{se,ik}\sin(\delta_i - \theta_{se,ik}) \\
P_j^{GUPFC} &= -V_jV_{se,ij}B_{se,ij}\sin(\delta_j - \theta_{se,ij}) \\
Q_j^{GUPFC} &= V_jV_{se,ij}B_{se,ij}\cos(\delta_j - \theta_{se,ij}) \\
P_k^{GUPFC} &= -V_kV_{se,ik}B_{se,ik}\sin(\delta_k - \theta_{se,ik}) \\
Q_k^{GUPFC} &= V_kV_{se,ik}B_{se,ik}\cos(\delta_k - \theta_{se,ik})
\end{aligned} \tag{9.34}
$$

9.3 RPP by VAR Resources

The VAR management problem is the determination of an action and provision of new reactive tools that minimize RPP cost (*INV*), voltage deviation (V_D) and STC-criterion (STC).

RPP cost (INV): The first target function includes two sections. The first section determines the overall cost of energy loss (EL_C) as bellow:

$$EL_C = 24 \times 365 \times hP_{loss} \tag{9.35}$$

where h is per-unit energy cost (0.06 \$/KWh). The power loss in transmission branches can be given as follows

$$P_{loss} = \sum_{k=1}^{n_l} g_k\left[V_i^2 + V_j^2 - 2V_iV_j\cos\left(\delta_i - \delta_j\right)\right] \tag{9.36}$$

The cost of the new prepared VAR tools is systematized as the total investment cost of VAR device, as OC_{var}

$$OC_{var} = \frac{r_T(r_T+1)^{n_T}}{(1+r_T)^{n_T}-1} \times \sum_{i\in\Omega}(C_{iFACTS}S_{iFACTS} + C_{iSC}S_{iSC}) \tag{9.37}$$

where, C_{iFACTS} and C_{iSC} are the cost function of the *i*th FACTS and slow VAR device. S_{iFACTS} and S_{iSC} refer to size related to them and Ω is the set of all candidate locations. For long-term cost analysis, it is necessary to consider the lifetime and discount rate as n_T and r_T, respectively. The cost function of the HFC, M-FACTS and slow VAR device can be represented in a quadratic form [16, 17]

$$C_{i,HFC} = 0.00012\,S_{i,HFC}^2 - 0.10764\,S_{i,HFC} + 75.288 \tag{9.38}$$

$$C_{i,SC} = 0.000000014\,S_{i,SC}^3 - 0.0000014\,S_{i,SC}^2 + 0.0052\,S_{i,SC} + 0.91 \tag{9.39}$$

$$C_{M-FACTS} = 0.00045\,S_{M-FACTS}^2 - 0.40365\,S_{M-FACTS} + 282.33 \tag{9.40}$$

where S_{iSC}, S_{iHFC} are size related to the *i*th HFC. Thus, the overall cost can be presented as the sum of two costs as OC_{var} and EL_C. The annual cost is calculated as follows

$$INV = \min(OC_{var} + EL_C) \tag{9.41}$$

Voltage deviation (V_D): This target is to optimize the voltage magnitude variations at load nodes that can be presented by

$$V_D = \min\left\{\sum_{k=1}^{N_L} |V_K - 1|\right\} \tag{9.42}$$

where, N_L is the number of load nodes.

STC criterion (STC): This aim is to optimize the index for defining the voltage stability of the system and is given by [18]

$$STC = \min\left\{\max_{j\in N_L}\left(STC_j\right)\right\} \tag{9.43}$$

Equation (9.43) requires finding the magnitude for STC_j in total load buses. After determining all magnitudes of STC_j, the load node which has the maximum STC_j is found. The definition of STC_j among total load buses has been presented by Eq. (9.44). One method of determining STC is

$$STC = \min\left\{\max_{j\in N_L}\left|\frac{V_j - \sum_{i\in N_G} H_{ji}V_i}{V_j}\right|\right\} \tag{9.44}$$

The magnitudes of H_{ji} are received from the H_{LG} matrix as follows

$$H_{LG} = -[Y_{LL}]^{-1}[Y_{LG}] \tag{9.45}$$

where, Y_{LL} and Y_{LG} are separated sections of grid Y-bus matrix. The elements of H_{LG} matrix are complex and its columns related to the generator node numbers and rows related to the load node numbers. This matrix presents the data for each load node about the value of power that should be given from each generator in the normal and grid contingencies as far as the system efficiency is assumed based on voltage profiles. To guarantee the stability situation, the state of $STC_j \leq 1$ must not be deviated for any of the *j*th node.

9.4 MOPSO-NTVE Algorithm Implementation to Solve RPP Problem

Multi-objective optimization: To determine the collection of solutions related to the multi-target problem, Pareto optimality idea is applied [19]. The total multi-target optimization issue, without limitations, can be presented as bellow [20]

$$\min f(x) = (f_1(x), f_2(x), \ldots, f_m(x)) \tag{9.46}$$

where $x \in \Omega$ is a possible solution set, Ω is the possible area of the problem, m is the number of targets and $f_i(x)$ is the ith target function of the matter. The aim is to optimize m target functions concurrently, to determine a suitable trade-off of solutions that present the best agreement among the objectives. So, assuming $f(x) = (f_1(x), f_2(x), \ldots, f_m(x))$ and $f(y) = (f_1(y), f_2(y), \ldots, f_m(y))$, $f(x)$ dominates $f(y)$, denoted by $f(x) < f(y)$, if and only if (minimization) [21]:

$$\begin{aligned} &\forall i \in \{1, 2, \ldots, m\} : f_i(x) \leq f_i(y) \\ &\exists i \in \{1, 2, \ldots, m\} : f_i(x) < f_i(y) \end{aligned} \tag{9.47}$$

If there is no $f(y)$ that dominates $f(x)$, $f(x)$ is non-dominated. Moreover, if there is no solution y that dominates x, x is defined Pareto optimal and $f(x)$ is a non-dominated purpose vector. The Pareto optimal set is signified by P^*, and the collection of total non-dominated target vector is defined Pareto front, signified by PF^*.

Multi-objective particle swarm optimization method with nonlinear time-varying evolution (MOPSO-NTVE): The position and the velocity of the ith individual in the n-dimensional search region is verified as $x_i = [x_{i,1} \quad x_{i,2} \quad \ldots \quad x_{i,n}]$ and $v_i = [v_{i,1} \quad v_{i,2} \quad \ldots \quad v_{i,n}]$, respectively where $v_i(k)$ is velocity of ith particle at iteration k and $x_i(k)$ is situation of ith particle at iteration k. The local best of the ith individual is defined as $x_i^l = [x_{i,1}^l \quad x_{i,2}^l \quad \cdots \quad x_{i,n}^l]$ and the global best determined so far explained as $x^g = [x_1^g \quad x_2^g \quad \cdots \quad x_n^g]$. At each iteration, the new velocities of the individuals are updated by employing the given equation:

$$\begin{aligned} v_i(k+1) = C(\phi)\{&\omega(k)v_i(k) + c_1(k)\varphi_1(x_i^l(k) - x_i(k)) \\ &+ c_2(k)\varphi_2(x^g(k) - x_i(k))\} \quad \text{for } i = 1, 2, \ldots, m \end{aligned} \tag{9.48}$$

The first section illustrates the current velocity of the individual, second section represents the cognitive term of MOPSO-NTVE and the third section related to the social term of MOPSO-NTVE. Each individual moves from the current condition to the next one by the changed velocity as bellow

$$x_i(k+1) = x_i(k) + v_i\,(k+1) \quad \text{for } i = 1, 2, \ldots, m \tag{9.49}$$

where m and k are number of individuals and current iteration, $x_i(k)$ is the situation of ith individual at iteration k, $x_i^l(k)$ and x^g are local and global best, $v_i(k)$ is velocity of ith individual, c_1 and c_2 are cognitive and social index, and φ_1 and φ_2 are random numbers between 0 and 1.

The cognitive index c_1 starts with a high magnitude $c_{1\max}$ and non-linearity reduces to $c_{1\min}$. Moreover, the social index c_2 starts with a low magnitude $c_{2\min}$ and non-linearity enhances to $c_{2\max}$ according to the following functions [22]

$$\omega(k) = \omega_{\min} + \left(\frac{iter_{\max} - k}{iter_{\max}}\right)^{\alpha}(\omega_{\max} - \omega_{\min}) \tag{9.50}$$

$$c_1(k) = c_{1\min} + \left(\frac{iter_{\max} - k}{iter_{\max}}\right)^{\beta}(c_{1\max} - c_{1\min}) \tag{9.51}$$

$$c_2(k) = c_{2\max} + \left(\frac{iter_{\max} - k}{iter_{\max}}\right)^{\gamma}(c_{2\min} - c_{2\max}) \tag{9.52}$$

$$C(\phi) = \frac{2}{\left|2 - \phi - \sqrt{\phi^2 - 4\phi}\right|} \quad \text{where } 4.1 \le \phi \le 4.2 \tag{9.53}$$

where $iter_{\max}$ is the maximum iteration and α, β and γ are constant rates.

To determine the optimal mixture of α, β and γ, overall combinations must be analyzed ad c_1 is cognitive parameter (acceleration coefficient) and c_2 is Social parameter (acceleration coefficient). It is considered that:

$$\alpha, \beta, \gamma \in \{0, 0.5, 1, 1.5, 2\} \tag{9.54}$$

There are 5^3 feasible combinations for the indexes of α, β and γ. These three indexes have many feasible magnitudes, but it may not be possible to implement the experiments of all combinations. Thus, to sample a small subset of this large number of tests, an orthogonal design method will be employed. Details of the orthogonal way and its utilization have been explained in [23]. The following is an $L_{25}(5^6)$ orthogonal array that can deal with at most six variables in five feasible magnitudes with 25 tests. Instead of 5^3 feasible combinations, one only requires to implement 25 tests to find optimal combination of α, β and γ.

$$L_{25}(5^6) = \begin{bmatrix} 1 & 1 & 1 & 1 & 1 & 1 \\ 1 & 2 & 2 & 2 & 2 & 2 \\ 1 & 3 & 3 & 3 & 3 & 3 \\ 1 & 4 & 4 & 4 & 4 & 4 \\ 1 & 5 & 5 & 5 & 5 & 5 \\ 2 & 1 & 2 & 3 & 4 & 5 \\ 2 & 2 & 3 & 4 & 5 & 1 \\ 2 & 3 & 4 & 5 & 1 & 2 \\ 2 & 4 & 5 & 1 & 2 & 3 \\ 2 & 5 & 1 & 2 & 3 & 4 \\ 3 & 1 & 3 & 5 & 2 & 4 \\ 3 & 2 & 4 & 1 & 3 & 5 \\ 3 & 3 & 5 & 2 & 4 & 1 \\ 3 & 4 & 1 & 3 & 5 & 2 \\ 3 & 5 & 2 & 4 & 1 & 3 \\ 4 & 1 & 4 & 2 & 5 & 3 \\ 4 & 2 & 5 & 3 & 1 & 4 \\ 4 & 3 & 1 & 4 & 2 & 5 \\ 4 & 4 & 2 & 5 & 3 & 1 \\ 4 & 5 & 3 & 1 & 4 & 2 \\ 5 & 1 & 5 & 4 & 3 & 2 \\ 5 & 2 & 1 & 5 & 4 & 3 \\ 5 & 3 & 2 & 1 & 5 & 4 \\ 5 & 4 & 3 & 2 & 1 & 5 \\ 5 & 5 & 4 & 3 & 2 & 1 \end{bmatrix} \tag{9.55}$$

9.4.1 Fuzzy Decision Making

When solutions according to the extracted Pareto-optimal set are determined by MOPSO-NTVE method, it needs to select one of them for applying [24]. A linear membership equation is assumed for each of the target functions [12]. The membership equation is verified as bellow [25]

$$\chi_i = \begin{cases} 1, & OF_i \le OF_i^{\min} \\ \frac{OF_i^{\max} - OF_i}{OF_i^{\max} - OF_i^{\min}} & OF_i^{\min} < OF_i < OF_i^{\max} \\ 0 & OF_i \ge OF_i^{\max} \end{cases} \tag{9.56}$$

for minimized target functions and

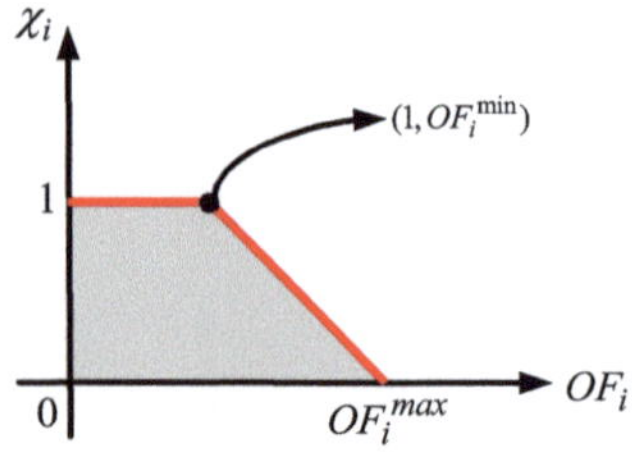

Fig. 9.11 Linear membership equation

$$\chi_i = \begin{cases} 0, & OF_i \leq OF_i^{\min} \\ \frac{OF_i - OF_i^{\max}}{OF_i^{\max} - OF_i^{\min}} & OF_i^{\min} < OF_i < OF_i^{\max} \\ 1 & OF_i \geq OF_i^{\max} \end{cases} \tag{9.57}$$

for maximized target functions. Where $OF_i^{\min}$ and $OF_i^{\max}$ are the minimum and the maximum magnitude of ith target function among total non-dominated solutions, respectively. The membership equation χ is changed between 0, 1. Where $\chi = 0$ demonstrates the incompatibility of the solution with the set, while $\chi = 1$ presents full compatibility. Figure 9.11 shows a typical structure of the membership equation.

For each non-dominated solution k, the normalized membership equation χ^k is extracted as:

$$\chi^k = \frac{\sum_{i=1}^{N_{ob}} \chi_i^k}{\sum_{k=1}^{M} \sum_{i=1}^{N_{ob}} \chi_i^k} \tag{9.58}$$

where M and N_{ob} are the number of non-dominated solutions and target functions. The function χ^k can be stated as a membership function of non-dominated solutions in a fuzzy set, where the solution having the maximum membership in the fuzzy collection is assumed as the best compromise solution.

9.5 Implementation

MOPSO-NTVE algorithm is applied to determine optimal reactive plan. A decision making method according to FDM algorithm is followed to determine the best solution from the collection of Pareto-solutions created by MOPSO-NTVE technique. Details of the solution synthesis are given in Figs. 9.12, 9.13 and 9.14. For this purpose, an initial population of MPSO-NTVE is randomly produced. For each particle, AC-OPF is directed and STC and V_D are calculated. Then, the investment cost is gained. The calculation is repeated until the stopping condition is obtained. As shown in Fig. 9.15, in the RPM problem codification, each chromosome is

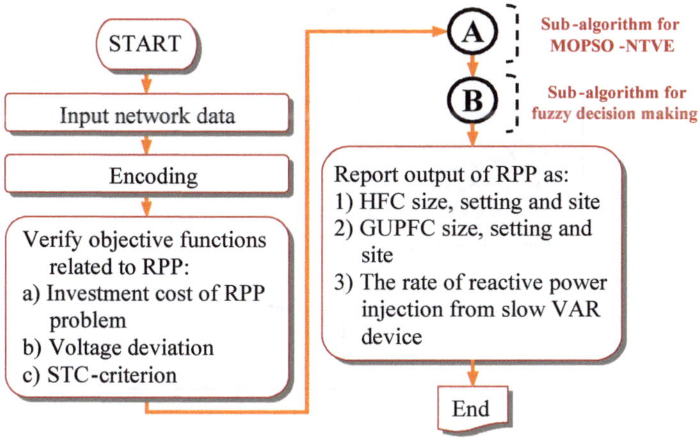

Fig. 9.12 Flowchart of the proposed RPP

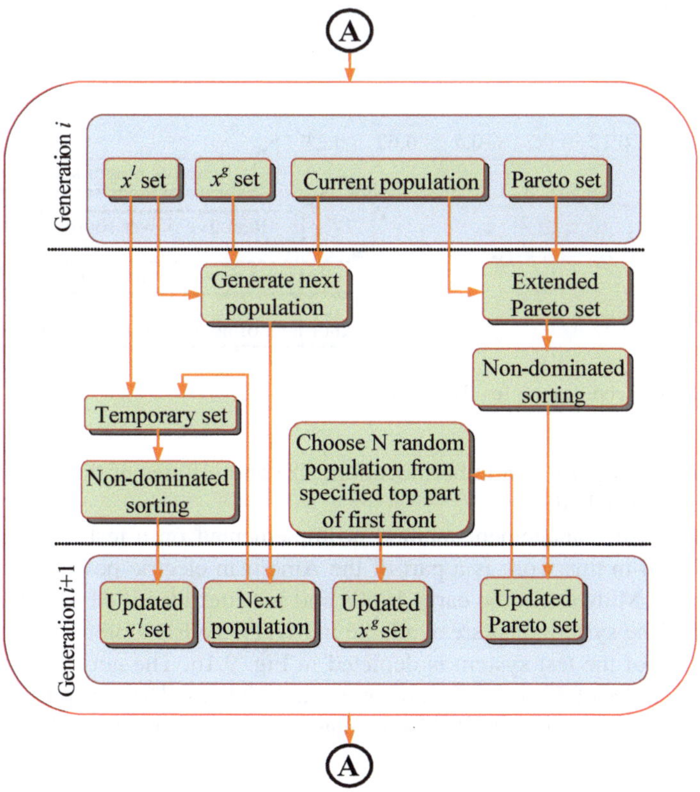

Fig. 9.13 Sub-algorithm of *A* for design procedure delineated in Fig. 9.12

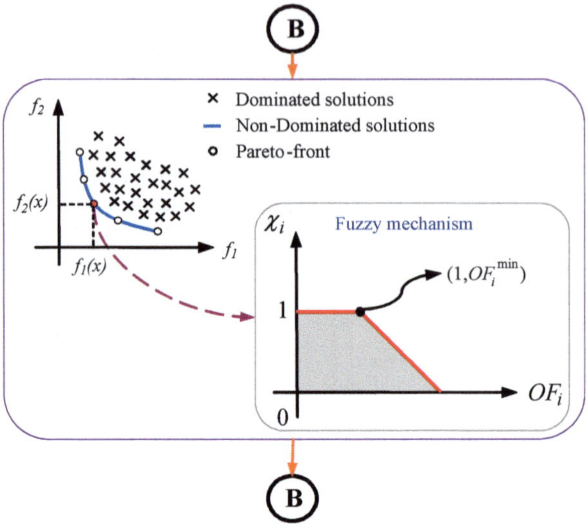

Fig. 9.14 Sub-algorithm of *B* for design procedure delineated in Fig. 9.12

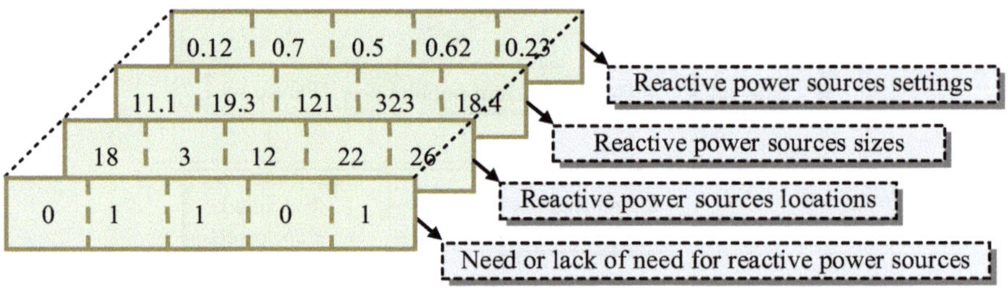

Fig. 9.15 Simple codification for RPP problem

formed from four parts of number, location, size and settings of the candidate VAR devices to be installed.

The proposed management method is implemented on a test system. The test grid employed in this work is a part of the American electric power system, AEP, applied in the Midwest in the early 1960s and is better identified as IEEE 57-node test system. The system data are available in MATPOWER toolbox [22]. The basic configuration of the test system is depicted in Fig. 9.16. The network as shown in Fig. 9.16 includes of 57 buses, 7 generators, and 80 lines. The generators are placed at nodes 1, 2, 3, 6, 8, 9, and 12. The voltage constraints are adjusted between 0.94 and 1.06 p.u.

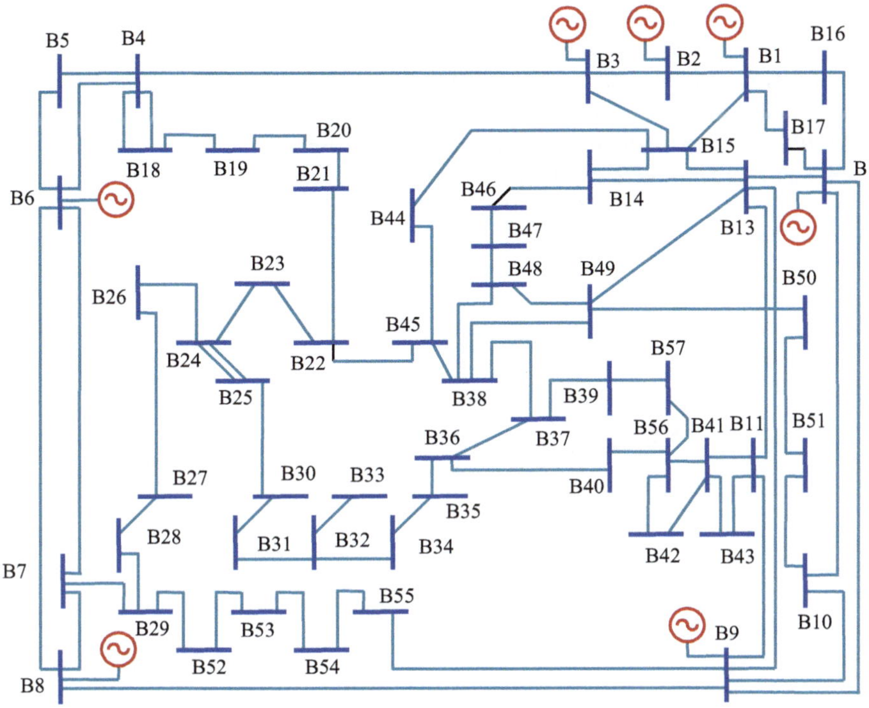

Fig. 9.16 IEEE 57-bus system

The constraints have been enforced based on Eq. (9.59):

$$\text{Evaluation Function} = \text{Objective Function} + \alpha(\text{Constraints Violations}) \quad (9.59)$$

where α is a large value and constraint violations are analyzed as the aggregate of the absolute magnitudes of all deviations. The solution will end up with the least cost selection without violating limitations. Also, discrete variables have been encoded based on "*ceil*" operators.

To analyze the effect of different FACTS tools on the power system operation, the following scenarios are studied:

- Framework 1: Slow VAR device allocation.
- Framework 2: Slow VAR device with HFC allocation.
- Framework 3: Slow VAR device with GUPFC allocation.

The performance of two frameworks is evaluated on the test system. For this purpose, the steps of the RPP problem (as given in Fig. 9.12) are performed. The Pareto-optimal archive obtained by MOPSO-NTVE algorithm in two-dimensional and three-dimensional target functions is depicted in Figs. 9.17, 9.18 and 9.19. It is observed that the gained solutions are distributed in the area, except some discontinuity, created by the discrete decision parameters. The trade-offs represented in

Figs. 9.17, 9.18 and 9.19 can help the decision maker to choose appropriate reference membership magnitudes.

A decision making approach by employing FDM theory is followed to determine the best solution as new lines from the collection of Pareto-solutions gained by MOPSO-NTVE method. FDM is used to choose size and site of the reactive power sources. Complete results related to two frameworks and four optimization methods, MOPSO, MOPSO-TVIW, MOPSO-TVAC and MOPSO-NTVE are tabled in Table 9.1.

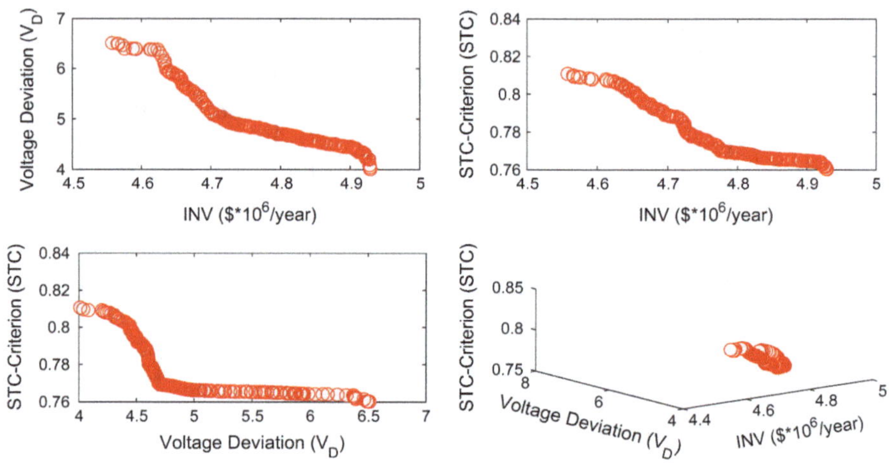

Fig. 9.17 The Pareto archive in two-dimensional and three-dimensional objective area based on framework 1

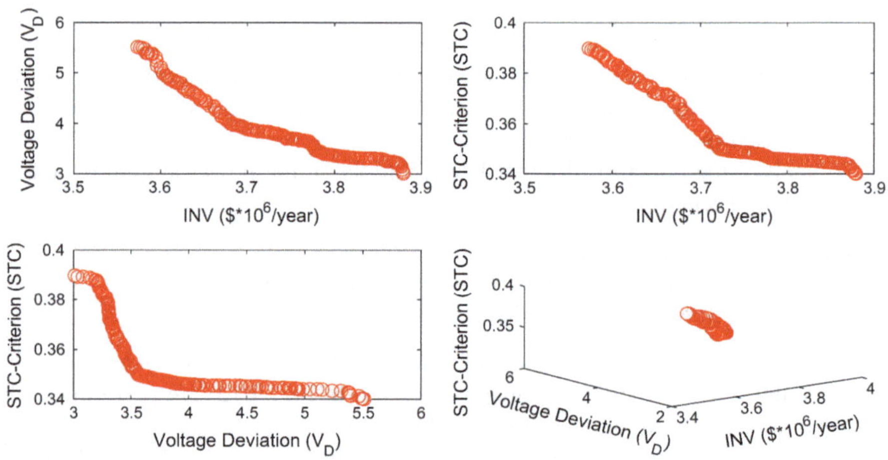

Fig. 9.18 The Pareto archive in two-dimensional and three-dimensional objective area based on framework 2

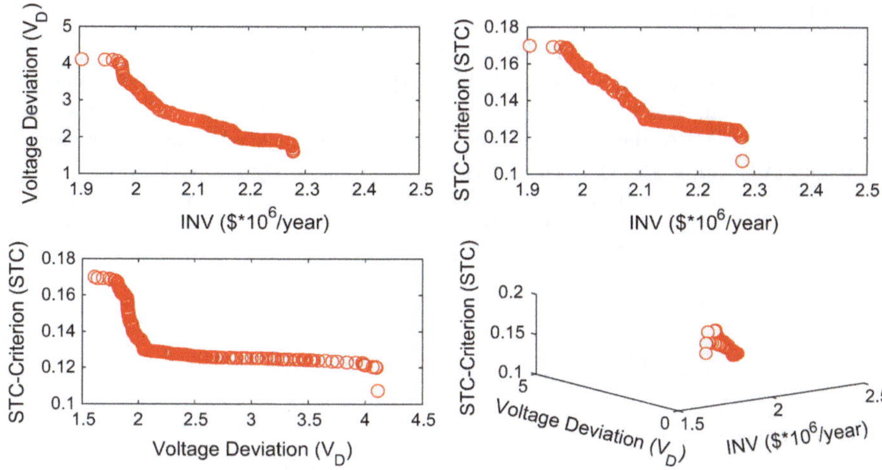

Fig. 9.19 The Pareto archive in two-dimensional and three-dimensional objective area based on framework 3

Table 9.1 Location, size and setting of reactive power sources added to network in frameworks under different solution methods

	Framework 1	Framework 2	Framework 3		
	Slow VAR devices MVAr (bus)	Slow VAR devices MVAr (bus)	HFC MVAr (location)	Slow VAR devices MVAr (bus)	GUPFC MVAr (location)
MOPSO	Q_{16} = 423.57 Q_4 = 355.56 Q_{46} = 436.56 Q_{21} = 365.56 Q_{49} = 367.97 Q_{39} = 314.56 Q_{22} = 431.56 Q_{11} = 478.03 Q_{32} = 434.67	Q_{14} = 476.56 Q_{19} = 523.56 Q_{35} = 437.67 Q_{55} = 456.73	33.34 (13-9)	Q_{11} = 475.23 Q_{53} = 478.25	13.26 (24-25, 24-23)
MOPSO-TVIW	Q_{41} = 414.76 Q_{57} = 432.45 Q_{39} = 421.45 Q_{43} = 431.56 Q_{44} = 476.67 Q_{14} = 365.76 Q_{50} = 432.67 Q_{12} = 391.67	Q_{48} = 394.67 Q_{56} = 512.56 Q_{43} = 476.56 Q_{30} = 456.65	32.9788 (22-21)	Q_{27} = 487.67 Q_{10} = 421.78	12.6786 (45-44, 45-38)
MOPSO-TVAC	Q_{18} = 409.65 Q_{46} = 434.74 Q_{50} = 446.76 Q_{49} = 346.76 Q_{39} = 476.65 Q_{41} = 421.76 Q_{56} = 365.74 Q_{31} = 418.76	Q_{11} = 535.87 Q_{35} = 465.76 Q_{32} = 421.85 Q_7 = 365.87	32.6457 (23-22)	Q_{57} = 353.64 Q_{20} = 365.87	11.6543 (29-28, 29-52)

(continued)

Table 9.1 (continued)

	Framework 1	Framework 2	Framework 3		
	Slow VAR devices MVAr (bus)	Slow VAR devices MVAr (bus)	HFC MVAr (location)	Slow VAR devices MVAr (bus)	GUPFC MVAr (location)
MOPSO-NTVE	$Q_4 = 529.32$ $Q_{20} = 440.84$ $Q_{46} = 377.67$ $Q_{50} = 457.34$ $Q_{38} = 401.83$ $Q_{24} = 277.65$ $Q_{43} = 357.67$ $Q_{53} = 401.86$	$Q_{21} = 163.75$ $Q_{14} = 189.78$ $Q_{11} = 163.75$ $Q_{33} = 189.78$	27 (49-38)	$Q = 229.32$ $Q_{40} = 340.84$	10 (15-13, 15-14)

Table 9.2 Frameworks results

	Solution methods	RPM cost (INV) ($ × 10^6/year)	Voltage deviation (V_D)	STC-criterion (STC)
Framework 1	MOPSO	4.9560	5.3219	0.7832
	MOPSO-TVIW	4.8598	5.0032	0.7755
	MOPSO-TVAC	4.7767	4.9821	0.7743
	MOPSO-NTVE	4.7257	4.9532	0.76432
Framework 2	MOPSO	4.0932	4.3217	0.3689
	MOPSO-TVIW	3.9517	4.2345	0.3578
	MOPSO-TVAC	3.8544	4.1234	0.3521
	MOPSO-NTVE	3.7713	3.9865	0.3456
Framework 3	MOPSO	2.9886	2.6422	0.1789
	MOPSO-TVIW	2.8361	2.5432	0.1675
	MOPSO-TVAC	2.3361	2.4321	0.1298
	MOPSO-NTVE	2.0085	2.3456	0.1234

To identify the advantage of the presented method, results of frameworks 1 and 2 in four solution methods are compared and tabled in Table 9.2. As shown in this table, RPP cost (INV) for framework 1 is: 4.9560, 4.8598, 4.7767 and 4.7257 and, for framework 2 it is: 4.0932, 3.9517, 3.8544 and 3.7713 and, for framework 3 it is: 2.9886, 2.8361, 2.3361 and 2.0085. RPP cost in framework 3 has 39.69, 41.64, 51.09 and 57.49% decrease compared to framework 1 and, 26.98, 28.23, 39.39 and 46.74% decrease compared to framework 2. This economic perspective of the presented way reveals priority of the obtained optimal configuration in framework 3 and the advantages of GUPFC than HFC and capacitor banks. From the viewpoint of voltage deviation (V_D) and STC-criterion, these two indices in framework 2 are significantly improved than the framework 1 in each four scenarios as displayed in Table 9.2.

Simulation test is applied for different discount rates to evaluate robustness of the proposed RPP problem. As shown in Table 9.3 twelve cases, C_1 to C_{12}, are studied. All the parameters for three discount rates (r_T = 5, 10 and 20%) and three

Table 9.3 Considered cases

Case	Discount rate (%)	Solution methods
C.1	$r_T = 5$	MOPSO
C.2	$r_T = 5$	MOPSO-TVIW
C.3	$r_T = 5$	MOPSO-TVAC
C.4	$r_T = 5$	MOPSO-NTVE
C.5	$r_T = 10$	MOPSO
C.6	$r_T = 10$	MOPSO-TVIW
C.7	$r_T = 10$	MOPSO-TVAC
C.8	$r_T = 10$	MOPSO-NTVE
C.9	$r_T = 20$	MOPSO
C.10	$r_T = 20$	MOPSO-TVIW
C.11	$r_T = 20$	MOPSO-TVAC
C.12	$r_T = 20$	MOPSO-NTVE

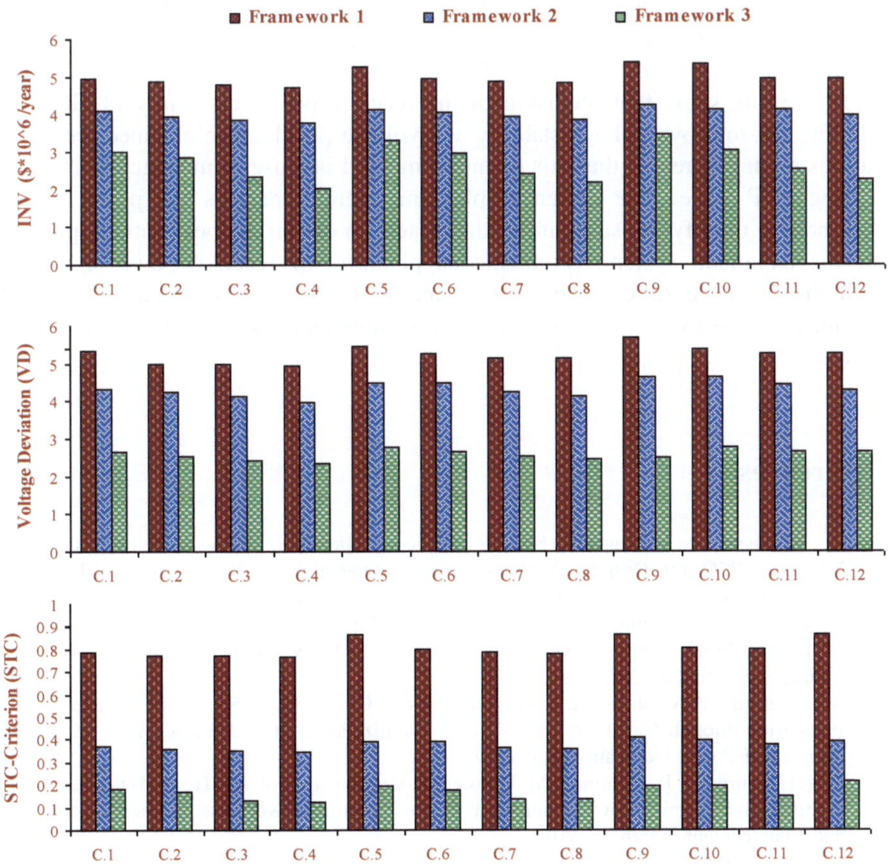

Fig. 9.20 Comparison of the performances

scenarios have been extracted, as depicted in Fig. 9.20. Results of both frameworks will change by increasing or decreasing the discount rate, but framework 3 has the best results and low changes than the nominal value of discount rate and achieves good robust performance.

9.6 Conclusion

An attempt has been made in this chapter to study solving RPP problem while HFC and GUPFC are involved in power system. A suitable model to compute reactive power that has to be injected by HFC and GUPFC is imported in the solution procedure. Multi-objective PSO-NTVE algorithm is employed and, from a decision maker perspective; the FDM method is used to define solutions by considering all attributes from the set of Pareto-solutions. The presented solution procedure is implemented on the IEEE 57-bus system as the first attempt for RPP. A comparative survey confirms that the presented management algorithm significantly improves the cost related to RPP.

The results show that inclusion of the reactive power capability of HFC and GUPFC can improve voltage stability and voltage profile. The obtained results for various scenarios reveal that this planning method is a useful management tool for solving RPP issue. The presented planning method permits the power system designers to modify the structure of the system to obtain the best optimal program for the expanded system. Simultaneous planning of reactive power expansion with the presence of capacitor banks and FACTS devices instead of separate planning can be led to more economic, applicable and optimal scheme in a power network.

References

1. H. Shayeghi, M. Ghasemi, FACTS Devices Allocation Using a Novel Dedicated Improved PSO for Optimal Techno-Economic Operation of Power System, Journal of Operation and Automation in Power Engineering, 1 (2), 2013.
2. A. Rabiee, M. Vanouni, M. Parniani, Optimal Reactive Power Dispatch for Improving Voltage Stability Margin Using a Local Voltage Stability Index, Energy Conversion and Management, 59:66-73, 2012.
3. M. Mahfouz, M.A. El-Sayed, Static Synchronous Compensator Sizing for Enhancement of Fault Ride-through Capability and Voltage Stabilization of Fixed Speed Wind Farms, IET Renewable Power Generation, 8 (1):1-9, 2014.
4. R.A. Hooshmand, R. Hemmati, M. Parastegari, Combination of AC Transmission Expansion Planning and Reactive Power Planning in the Restructured Power System, Energy Conversion and Management, 55:26-35, 2012.
5. I.P. Abril, J.A.G. Quintero, VAR Compensation by Sequential Quadratic Programming. IEEE Transactions on Power Systems, 18 (1):36-41, 2003.

6. T. Akbari, A. Rahimikian, A. Kazemi, A Multi-Stage Stochastic Transmission Expansion Planning Method, Energy Conversion and Management, 52 (8):2844-2853, 2011.
7. E. Ghahremani, I. Kamwa, Optimal Placement of Multiple-Type FACTS Devices to Maximize Power System Loadability Using a Generic Graphical User Interface, IEEE Transactions on Power Systems, 28 (2):764-778, 2013.
8. H. Shayeghi, M. Ghasemi, FACTS Devices Allocation Using a Novel Dedicated Improved PSO for Optimal Techno-Economic Operation of Power System, Journal of Operation and Automation in Power Engineering, 1 (2):124-135, 2013.
9. J. Pourhossein, G. Gharehpetian, S. Fathi, Static Inter-Phase Power Controller (SIPC) Modeling for Load Flow and Short Circuit Studies, Energy Conversion and Management, 64:145-151, 2012.
10. S.A.N. Niaki, R. Iravani, M. Noroozian, Power Flow Model and Steady-State Analysis of the Hybrid Flow Controller, IEEE Transactions on Power Delivery, 23 (4):2330-2338, 2008.
11. H. Shayeghi, Y. Hashemi, Technical-Economic Analysis of Including Wind Farms and HFC to Solve Hybrid TNEM-RPM Problem in the Deregulated Environment, Energy Conversion and Management, 80:477-490, 2014.
12. A. Lashkar Ara, A. Kazemi, S. Nabavi Niaki, Multiobjective Optimal Location of FACTS Shunt-Series Controllers for Power System Operation Planning, IEEE Transactions on Power Delivery, 27 (2):481-490, 2012.
13. Y. Hashemi, H. Shayeghi, B. Hashemi, Attuned Design of Demand Response Program and M-FACTS for Relieving Congestion in a Restructured Market Environment, Frontiers in Energy, 9:1-15, 2015.
14. B.S. Rao, K. Vaisakh, Multi-Objective Adaptive Clonal Selection Algorithm for Solving Optimal Power Flow Considering Multi-Type FACTS Devices and Load Uncertainty, Applied Soft Computing, 23:286-297, 2014.
15. Fardanesh B, Optimal utilization, sizing, and steady-state performance comparison of multiconverter VSC-based FACTS controllers. IEEE Transactions on Power Delivery, 19 (3):1321-1327. doi:10.1109/TPWRD.2004.829154, 2004.
16. K. Habur, D. OLeary, FACTS-Flexible Alternating Current Transmission Systems-for Cost Effective and Reliable Transmission of Electrical Energy, Siemens-World Bank Document-Final Draft Report, Erlangen, 2004.
17. M. Eghbal, N. Yorino, E. El-Araby, Y. Zoka, Multi-Load Level Reactive Power Planning Considering Slow and Fast VAR Devices by Means of Particle Swarm Optimization, IET Generation, Transmission & Distribution, 2 (5):743-751, 2008.
18. G. Yesuratnam, D. Thukaram, Congestion Management in Open Access Based on Relative Electrical Distances Using Voltage Stability Criteria, Electric Power Systems Research, 77 (12):1608-1618, 2007.
19. C. Coello, G. Lamont, D. Veldhuizen, Evolutionary Algorithms For solving Multi-Objective Problems, Springer, 2007.
20. F. Gunes, U. Ozkaya, Multiobjective FET Modeling Using Particle Swarm Optimization Based on Scattering Parameters with Pareto Optimal Analysis, Turk J. Elec Eng. & Comp. Sci., 20 (3):353-365, 2012.
21. A. Kavousi Fard, T. Niknam, Multi-Objective Probabilistic Distribution Feeder Reconfiguration Considering Wind Power Plants, International Journal of Electrical Power & Energy Systems, 55:680-691, 2014.
22. C. Ko, Y. Chang, C. Wu, A PSO Method with Nonlinear Time Varying Evolution for Optimal Design of Harmonic Filters, IEEE Transactions on Power Systems, 24 (1):437-444, 2009.
23. Y. Leung, Y. Wang, An Orthogonal Genetic Algorithm with Quantization for Global Numerical Optimization, IEEE Transactions on Evolutionary Computation, 5 (1):41-53, 2001.

24. A. Mazza, G. Chicco, A. Russo, Optimal Multi-Objective Distribution System Reconfiguration with Multi Criteria Decision Making-Based Solution Ranking and Enhanced Genetic Operators, International Journal of Electrical Power & Energy Systems, 54:255-267, 2014.
25. S. Christa, P. Venkatesh, Multi-Objective Optimization Problem for the Thyristor Controlled Series Compensators Placement with Multiple Decision-Making Approaches, European Transactions on Electrical Power, 23 (2):249–269, doi:10.1002/etep.658, 2011.

Chapter 10
Reactive Power Optimization in AC Power Systems

Ali Jafari Aghbolaghi, Naser Mahdavi Tabatabaei, Narges Sadat Boushehri and Farid Hojjati Parast

10.1 Introduction

Reactive Power is one of the most important features in power networks so that its appropriate production and distribution among consumers can affect performance, efficiency, and reliability of the power networks positively. Creating competitive mechanism via establishing a market to present different services and changing the current rules necessitate that in the new condition programming and controlling of reactive power as well as the voltage to be considered more accurately. The purpose of reactive power optimization in AC power systems is to recognize the best value for control variables in order to optimize the target function considering the possible constraints. With current developments of power grids, it has been growing in popularity among researchers to probe into how to use existent reactive power compensators in order to reduce active power losses and improve voltage profile.

The reactive power optimization is a complicated problem with a broad solution space, nonlinear and non-convex, in which there both continuous are and discrete variables. In general, reactive power optimization problem entails two separate branches as optimal placement of reactive power compensators and optimal oper-

A. Jafari Aghbolaghi (✉)
Zanjan Electric Energy Distribution Company, Zanjan, Iran
e-mail: ali.jafari.860@gmail.com

N. Mahdavi Tabatabaei
Electrical Engineering Department, Seraj Higher Education Institute, Tabriz, Iran
e-mail: n.m.tabatabaei@gmail.com

N.S. Boushehri
Department of Management, Taba Elm International Institute, Tabriz, Iran
e-mail: nargesboush@yahoo.com

F. Hojjati Parast
Andishmand Shomal Gharb Engineering Consultancy, Zanjan, Iran
e-mail: hojjatifarid@gmail.com

N. Mahdavi Tabatabaei et al. (eds.), *Reactive Power Control in AC Power Systems*,
Power Systems, DOI 10.1007/978-3-319-51118-4_10

ation of existent reactive power compensators. Optimal placement of reactive power compensators problem tries to determine three parameters as the type of the compensator, the rate of the output power and installation location. However, optimal operation of existing reactive power compensators is about determining optimal reactive power output for the compensators that have already been installed.

Although producing reactive power involves no fee in the operation step, it influences the final cost via affecting the active power losses in power transmission system significantly. Optimal distribution of reactive power is a sub-problem of Optimal Power Flow (OPF) and the parameters that should be controlled are actually the control variables such as output reactive power or the terminal voltage of generators, output reactive power of all reactive power compensators, and tap-changing transformers' tap settings. Since flowing reactive power through power transmission system results in active power losses, the main goal of reactive power optimization is to reduce the active power losses via optimal controlling of the above mentioned parameters considering the security aspects, which must be fulfilled in order to have a reliable and stable power network with which consumers can be fed continuously.

In addition, some other goals are pursuit beside the main goal, like improving voltage profile which involves the security measures of power systems. Therefore, the target function of reactive power optimization problem is active power losses equation, in which some constraints should be included as voltage rate boundaries, equipment's power output limits, and transmission system margins for carrying power etc. In the reactive power optimization problem, there are both continuous and discrete types of variables, of which should be determined by an optimization algorithm that leads the power system to contain least possible active power losses. In one hand, discrete variables are transformers' tap settings and capacitors' output powers. In the other hand, continuous variables are reactive power outputs of generators as well as synchronous compensators'.

In the recent years, wide variety of optimization algorithms have been recommended to solve the reactive power optimization problem, containing traditional algorithms such as quadratic programing (QP) and sequential quadratic programming (SQP) etc., and heuristic ones which are inspired by nature such as particle swarm optimization (PSO), genetic algorithm (GA), evolutionary programming (EP) and their derivatives etc. Traditional algorithms have too many defects encountering problems that have nonlinear essences, the ones which consist of discrete variables and have got many local optimum solutions. Eventually, traditional gradient based algorithms lose their capability to be efficiently solve such challenging optimization problems. Overall, the traditional optimization algorithms become nonfunctional confronting practical reactive power optimization in large power systems having huge dimensions. On the other hand, optimization algorithms inspired by nature have proved their high performance optimizing problems with so many variables covering discrete and continuous ones while having many local optimum points.

This chapter is focusing on the reactive power optimization using artificial optimization algorithms and trying to give a comprehensive perspective of all formulations and constraints that are needed in order to implement reactive power optimization problem and use its practical implications effectively. First, brief fundamentals of reactive power optimization containing some relevancies which are of crucial importance is presented. Second, to enlighten how reactive power optimization works the classic method of reactive power optimization is presented, thanks to [1]. Third, basic principles and problem formulation of reactive power optimization using artificial intelligent algorithms is elaborated. Fourth, particle swarm optimization algorithm and pattern search method and how to use them in reactive power optimization problem have been expounded defining PSO's parameters way back into reactive power optimization involving ones. Finally, the offered algorithms simulation result on two case studies have been presented.

10.2 Fundamentals of Reactive Power Optimization

There are two types of reactive power flow in power transmission systems; one for fulfilling transmission system needs and the other for feeding loads that naturally consume reactive power in order to work properly, that cause according to Eq. (10.1) active power losses and following that, the fuel consumption of the power plant raises [2]. Taking the Eq. (10.1) into consideration, it is obvious that the imaginary part of the current equation simply affects the absolute value of that which influences total active power losses in the next place.

In addition, there is no any control possible on I_R as it is corresponding current to the active power which should be supplied to the loads via transmission systems. The idealistic condition recommends compensating all the reactive power needs of the loads at the same place, just beside the loads, in order not to impose reactive current into transmission systems. The closer reactive power compensator to loads, the less reactive power flowing in power transmission systems. It is not feasible however, because of technical and economical restrictions upon that, such as the relevancy of voltage stability to the reactive power in AC power systems, of which was broadly investigated in [3, 4].

Reactive power is absorbed by different equipment as synchronous condensers being operated with lagging power factor, shunt reactors, transmission lines and transformers' inductances, static reactive power compensators and loads etc., while being injected to transmission systems by generators, synchronous condensers operating with leading power factor, static capacitors, static compensators, and transmission lines' capacitances etc. [2]. There is another type of equipment that avails power system operators to have a control on reactive power flow through transmission system and it is transformers with tap changing facility under the load condition. There is a straight connection between two neighboring buses' voltages and reactive power flow in the corresponding transmission system which connects them together. In simple words, reactive power flows from the bus having higher voltage magnitude to the bus which has lower voltage magnitude in order to naturally raise the second bus's voltage close to the same level. Take a simple network shown in Fig. 10.1 into consideration, reactive power will flow from the slack bus

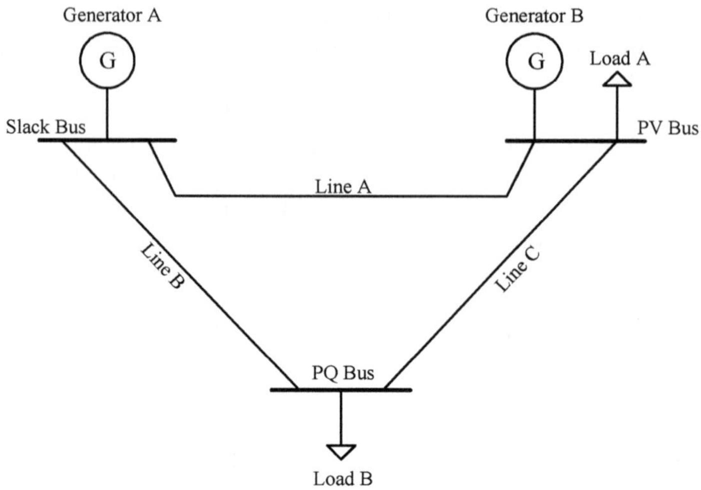

Fig. 10.1 Three-bus power system

to the PV bus if slack bus's voltage magnitude is higher than *PV* bus's. On the whole, there would not be any reactive power flow in transmission systems if the voltage magnitudes of all buses are equal.

$$\begin{cases} P_{loss} = R|I|^2 \\ I = I_R + jI_X \end{cases} \tag{10.1}$$

$$I_{ij} = \frac{V_i - V_j}{Z_{ij}} \tag{10.2}$$

$$P_{ij} + jQ_{ij} = V_i I_{ij}^* \tag{10.3}$$

(10.2) and (10.3)

$$\Rightarrow P_{ij} + jQ_{ij} = V_i \left[\frac{V_i - V_j}{Z_{ij}}\right]^* \tag{10.4}$$

$$\begin{cases} V_i^* = |V_i|(\cos\,\delta_i - j\,\sin\,\delta_i) \\ V_j^* = |V_j|(\cos\,\delta_j - j\,\sin\,\delta_j) \\ Z_{ij}^* = |Z_{ij}|(\cos\,\theta_{ij} - j\,\sin\,\theta_{ij}) \end{cases}. \tag{10.5}$$

where, $|V_i|$ and $|V_j|$ are voltage magnitudes of *i*th and *j*th busbars respectively. $|Z_{ij}|$ is absolute value of the impedance of transmission line between *i*th and *j*th busbars. δ_i and δ_i are the voltage angles of *i*th and *j*th busbars respectively, while θ_{ij} is angle of impedance between *i*th and *j*th busbars.

Expanding Eq. (10.4) using the assumptions of Eq. (10.5) will be as following

$$P_{ij}+jQ_{ij} = |V_i|(\cos\,\delta_i+j\,\sin\,\delta_i)\left[\frac{|V_i|(\cos\,\delta_i-j\,\sin\,\delta_i)-|V_j|(\cos\,\delta_j-j\,\sin\,\delta_j)}{|Z_{ij}|(\cos\,\theta_{ij}-j\,\sin\,\theta_{ij})}\right] \tag{10.6}$$

$$=|V_i|(\cos\,\delta_i+j\,\sin\,\delta_i)\left[\frac{|V_i|\cos\,\delta_i-j|V_i|\sin\,\delta_i-|V_j|\cos\,\delta_j+j|V_j|\sin\,\delta_j}{|Z_{ij}|(\cos\,\theta_{ij}-j\,\sin\,\theta_{ij})}\right]$$

$$=\frac{\left(\begin{array}{c}|V_i|^2\cos^2\delta_i-j|V_i|^2\sin\,\delta_i\cos\,\delta_i-|V_i||V_j|\cos\,\delta_i\cos\,\delta_j+j|V_i||V_j|\sin\,\delta_j\cos\,\delta_i\\+j|V_i|^2 sin\,\delta_i\cos\,\delta_i+|V_i|^2\sin^2\delta_i-j|V_i||V_j|sin\,\delta_i\cos\,\delta_j-|V_i||V_j|\sin\,\delta_j\sin\,\delta_i\end{array}\right)}{|Z_{ij}|(\cos\,\theta_{ij}-j\,\sin\,\theta_{ij})}$$

$$=\frac{\left(\begin{array}{c}\left(|V_i|^2\cos^2\delta_i-|V_i||V_j|\cos\,\delta_i\cos\,\delta_j+|V_i|^2\sin^2\delta_i-|V_i||V_j|\sin\,\delta_j\sin\,\delta_i\right)\\+j\left(-|V_i|^2\sin\,\delta_i\cos\,\delta_i+|V_i||V_j|\sin\,\delta_j\cos\,\delta_i+|V_i|^2 sin\,\delta_i\cos\,\delta_i-|V_i||V_j|sin\,\delta_i\cos\,\delta_j\right)\end{array}\right)}{|Z_{ij}|(\cos\,\theta_{ij}-j\,\sin\,\theta_{ij})}$$

$$=\frac{\left(|V_i|^2-|V_i||V_j|(\cos\,\delta_i\cos\,\delta_j+\sin\,\delta_j\sin\,\delta_i)\right)+j(|V_i||V_j|(\sin\,\delta_j\cos\,\delta_i-sin\,\delta_i\cos\,\delta_j))}{|Z_{ij}|(\cos\,\theta_{ij}-j\,\sin\,\theta_{ij})}$$

$$=\frac{\left(\begin{array}{c}|Z_{ij}||V_i|^2\cos\,\theta_{ij}-|Z_{ij}||V_i||V_j|\cos(\delta_j-\delta_i)\cos\,\theta_{ij}+j|Z_{ij}||V_i||V_j|\sin(\delta_j-\delta_i)\cos\,\theta_{ij}\\+j|Z_{ij}||V_i|^2\sin\,\theta_{ij}-j|Z_{ij}||V_i||V_j|\cos(\delta_j-\delta_i)\sin\,\theta_{ij}-|Z_{ij}||V_i||V_j|\sin(\delta_j-\delta_i)\sin\,\theta_{ij}\end{array}\right)}{|Z_{ij}|^2}$$

$$P_{ij}+jQ_{ij} =\frac{\left(\begin{array}{c}|Z_{ij}||V_i|^2\cos\theta_{ij}-|Z_{ij}||V_i||V_j|\cos(\delta_j-\delta_i)\cos\,\theta_{ij}-|Z_{ij}||V_i||V_j|\sin(\delta_j-\delta_i)\sin\,\theta_{ij}\\+j\left(|Z_{ij}||V_i||V_j|\sin(\delta_j-\delta_i)\cos\,\theta_{ij}+|Z_{ij}||V_i|^2\sin\,\theta_{ij}-|Z_{ij}||V_i||V_j|\cos(\delta_j-\delta_i)\sin\,\theta_{ij}\right)\end{array}\right)}{|Z_{ij}|^2} \tag{10.7}$$

$$Q_{ij}=\frac{|V_i|^2\sin\,\theta_{ij}+|V_i||V_j|sin(\delta_j-\delta_i-\theta_{ij})}{|Z_{ij}|} \tag{10.8}$$

The term $(\delta_j-\delta_i)$ has a minute value and can be neglected in order to simplify Eq. (10.8), then

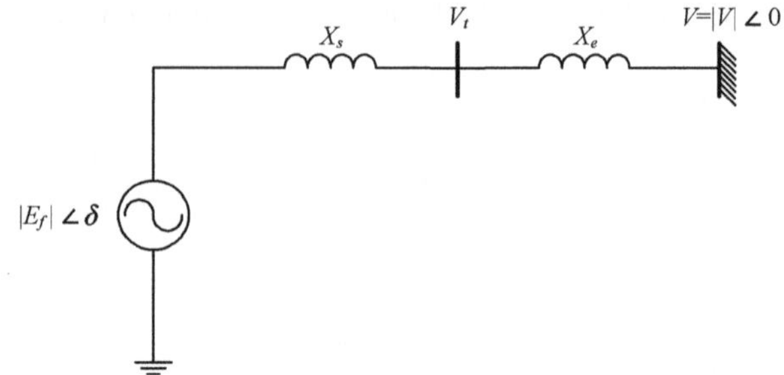

Fig. 10.2 The simplified electrical circuit of synchronous generator

$$Q_{ij} = \frac{|V_i|^2 \sin\theta_{ij} - |V_i||V_j| sin\,\theta_{ij}}{|Z_{ij}|} \tag{10.9}$$

$$Q_{ij} = \underbrace{\left(\frac{\sin\theta_{ij}}{|Z_{ij}|}\right)}_{\text{Constant}} |V_i| \underbrace{(|V_i| - |V_j|)}_{\text{Voltage Difference}} \tag{10.10}$$

Therefore, Eq. (10.10) shows that the value of the corresponding Q_{ij} flowing in transmission systems is directly dependent on the difference of voltage magnitudes between ith and jth busbars.

According to the electrical rules which synchronous machines work based on, the reactive power injected or absorbed by a synchronous generator can be controlled by its excitation system that affects the generator's terminal voltage straightforwardly [5]. Take Fig. 10.2 into consideration as a simplified electrical circuit of synchronous generators and power systems, then reactive power absorbing or producing by the generator will be as the following:

$$\begin{cases} V = |V|\angle 0 \\ E_f = |E_f|\angle\delta \\ X = X_e + X_s \end{cases} \tag{10.11}$$

$$I = \frac{|E_f|\angle\delta - |V|\angle 0}{X\angle 90} \tag{10.12}$$

The apparent power transmitted from the generator to the infinite bus is:

$$S = P + jQ = E_f I^* \tag{10.13}$$

Expanding Eq. (10.13) using Eqs. (10.11) and (10.12) will be as following

$$S = |E_f|\angle\delta \frac{|E_f|\angle-\delta - |V|}{X\angle-90} = \frac{|E_f|^2\angle 90 - |E_f||V|\angle 90+\delta}{X} \tag{10.14}$$

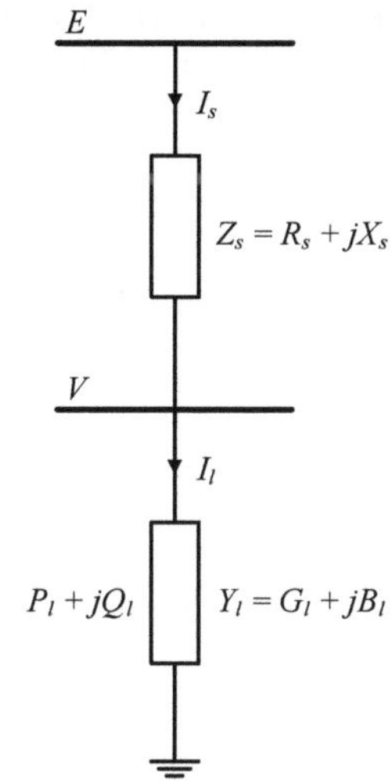

Fig. 10.3 The single-phase Thevenin equivalent circuit

Then by segregating Eq. (10.14), the reactive power absorbed or injected to the infinite busbar by the synchronous generator will be as following

$$Q = \frac{|E_f|}{X}\left(|E_f| - |V|\cos\delta\right) \tag{10.15}$$

where V_t is the voltage of infinite busbar and δ is the angle between infinite busbar and the generator's terminal under no-load condition.

If the voltage magnitude of the infinite busbar to be considered as constant, it is obvious that consuming or feeding reactive power by the generator is entirely correlated with the generator's voltage magnitude which is controlled by the excitation system, while the active power being transmitted assumed to be constant over the handling period [5]. There is another issue as voltage regulation which is associated with the reactive power control in AC power systems. The voltage regulation is defined as a proportional (per-unit) change in the voltage magnitude of supply terminal in relation to defined change to the load current (e.g., from no-load to full-load).

Taking Fig. 10.3 as Thevenin equivalent of supply system into consideration, the voltage drop in the transmission system in the absence of compensator is shown in Fig. 10.4 as ΔV, which is as the following

$$\Delta V = E - V = Z_s I_l \tag{10.16}$$

$$\begin{cases} Z_s = R_s + jX_s \\ I_l = \dfrac{P_l - jQ_l}{V} \end{cases} \tag{10.17}$$

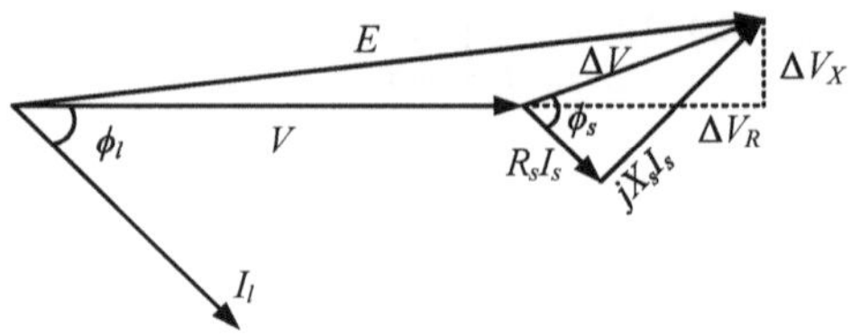

Fig. 10.4 The corresponding phasor diagram to Fig. 10.3, without compensation

So that

$$\Delta V = (R_s + jX_s)\left(\frac{P_l - jQ_l}{V}\right) \tag{10.18}$$

$$= \left(\frac{R_sP_l + X_sQ_l}{V}\right) + j\left(\frac{X_sP_l - R_sQ_l}{V}\right) \tag{10.19}$$

$$= \Delta V_R + j\Delta V_X \tag{10.20}$$

The voltage drop has two components as ΔV_R in the same phase with V and ΔV_X in quadrature with V which is illustrated in Fig. 10.4 elaborately. It is obvious that the magnitude and the phase of V are functions of the magnitude and the phase of load current so that the amount of voltage regulation depends straightly on the amount of real power as well as reactive power consuming of the load [2].

According to Eq. (10.10) reactive power flowing in transmission systems is a function of the difference of voltage magnitude between two neighboring busbars. Hence, the bigger the difference, the higher amount of reactive power flows toward busbar having lower voltage magnitude in order to raise it, and it causes additional active power losses in transmission systems. Therefore, it would be such a great asset if reactive power could be compensated at the same place as loads lie, which would help the power factor to reach near the unique value. However, a voltage drop that is relevant to active current flow in transmission systems would yet be consistent. Considering Eq. (10.18) while having loads' reactive power consumption compensated on-site, the equation will be as the following

$$\Delta V = (R_s + jX_s)\left(\frac{P_l}{V}\right) \tag{10.21}$$

$$= \left(\frac{R_sP_l}{V}\right) + j\left(\frac{X_sP_l}{V}\right) \tag{10.22}$$

$$= \Delta V_R + j\Delta V_X \tag{10.23}$$

In order to have a voltage regulation by the value of zero, the reactive power consumption by the load and the voltage drop being caused by active current flowing in the transmission system should be compensated. Referring to Fig. 10.5,

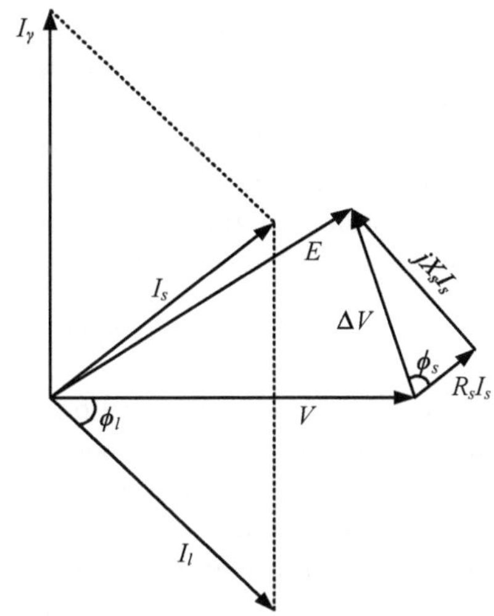

Fig. 10.5 The corresponding phasor diagram to Fig. 10.3, with compensation

it is possible to have $|E| = V$ via producing reactive power and injecting to the load busbar in presence of the load which will be greater than the load's absorption [2].

For sum up, all the components of power systems work together continuously as a united structure so that any single change in any part will influence all the other parameters directly, some of which are explained in the current section regarding that they are associated with reactive power's role in AC power systems. Therefore, all the relations should be evaluated carefully whenever a change happens in a specific parameter, which necessitates having a comprehensive control system to fulfill all the needs of power systems in order to have a reliable, stable, and a cost-effective energy produced and dispatched.

10.3 Reactive Power Optimization Using Classic Methods

The main aim of reactive power optimization is to minimize reactive power flowing in transmission systems including transmission lines, transformers etc., which leads to less real power losses. This will be approachable via specifying the best reactive power output value of the reactive power compensators and the other controllable parameters under a number of constraints which should be met in order to have a reliable, stable and cost-effective power system [1].

At first glance, it might come into mind that if loads' reactive power consumption need to be met just by the nearest reactive power source (in terms of electrical distance), the active power losses would be minimized, although it is not such a true interpretation. On one hand, while it is being talked about reactive power optimization, it should be considered that reactive power in AC power systems has plenty of associations with the other parameters of the system, most of which are crucial in terms of power system security and reliability. On the other hand, reactive power optimization is not a straight optimization, however, the power flow plays a key role in the reactive power optimization problem. For instance, the voltage magnitude is one of the most important parameters which influences and also is influenced by reactive power of the system.

According to Eq. (10.1), reactive power loss in AC power network can be considered as a function of load active and reactive power consumption as

$$P_L = P_L(P_1, P_2, \ldots, P_n, Q_1, Q_2, \ldots, Q_n) \tag{10.24}$$

Reactive power optimization is a constrained optimization problem that may be attacked formally using advanced calculus methods involving Lagrange function. In classic reactive power economic dispatch, the active power of all generators are already known and fixed during optimization procedure, except slack bus's, so that any change in active power losses associated to reactive power optimization could be distinguished easily. One of the well-known techniques to implement optimization in power network, of which consists of some constraints to be fulfilled, is the Lagrange method. In reactive power optimization, one of the main constraints is reactive power balance as Eq. (10.25), which is shown in the following [1]

$$\sum_{i=1}^{M} Q_{Gi} = Q_D + Q_L \tag{10.25}$$

where Q_G includes all reactive power sources, Q_D is reactive power demands of the loads, and Q_L is reactive power losses in the transmission system.

In order to establish the necessary conditions for an extreme value of the objective function, the constraint functions should be added to the objective function after the constraint functions has been multiplied by an undetermined multiplier. This is known as Lagrange function and is shown in Eq. (10.26).

$$L = F_T + \lambda\phi \tag{10.26}$$

where, F_T is target function which is aimed to be optimized, λ is an unspecified multiplier, and ϕ contains all the probable constraint functions that must be met during the optimization procedure.

Therefore, the Lagrange function formed to decrease active power losses, subjected to equality constraint of producing and consuming reactive power, constructed from Eqs. (10.24) and (10.25) will be as

$$L = P_L - \lambda\left(\sum_{i=1}^{M} Q_{Gi} - Q_D - Q_L\right) \tag{10.27}$$

Then, the necessary condition to have an extreme value for the Lagrange function is to set the first derivative of the function with respect to each independent variables (Q_G and λ) equal to zero [1].

$$\frac{\partial L}{\partial Q_{Gi}} = \frac{\partial P_L}{\partial Q_{Gi}} - \lambda\left(1 - \frac{\partial Q_L}{\partial Q_{Gi}}\right) = 0 \quad i = 1, 2, \ldots, M \tag{10.28}$$

$$\frac{\partial L}{\partial \lambda} = -\left(\sum_{i=1}^{M} Q_{Gi} - Q_D - Q_L\right) = 0 \tag{10.29}$$

From Eq. (10.28)

$$\frac{\partial P_L}{\partial Q_{Gi}} \times \frac{1}{\left(1 - \frac{\partial Q_L}{\partial Q_{Gi}}\right)} = \lambda \quad i = 1, 2, \ldots, N \tag{10.30}$$

Equation (10.30) is the formula of reactive power economic dispatch, where, $\frac{\partial P_L}{\partial Q_{Gi}}$ is incremental rate of active power losses with respect to ith reactive power source, $\frac{\partial Q_L}{\partial Q_{Gi}}$ is incremental reactive power losses with respect to ith reactive power source.

The terms $\frac{\partial P_L}{\partial Q_{Gi}}$ and $\frac{\partial Q_L}{\partial Q_{Gi}}$ can be calculated with impedance matrix method which is depicted below.

The real power losses in power transmission systems can be represented as [1]

$$P_L + jQ_L = V^T I^* = (ZI)^T \hat{I} = I^T Z^T I^* \tag{10.31}$$

$$I = I_P + jI_Q \tag{10.32}$$

$$Z = R + jX \tag{10.33}$$

where, I is the current vector in transmission lines, I^* is the conjugate current vector in transmission lines, Z is the impedance matrix of transmission lines, and V^T is the voltage vector of all busbars.

Substituting Eqs. (10.32) and (10.33) into Eq. (10.31), P_L and Q_L will be obtained as [1]

$$P_L = \sum_{j=1}^{n}\sum_{k=1}^{n} R_{jk}\left(I_{Pj}I_{Pk} + I_{Qj}I_{Qk}\right) \tag{10.34}$$

$$Q_L = \sum_{j=1}^{n}\sum_{k=1}^{n} X_{jk}\left(I_{Pj}I_{Pk} + I_{Qj}I_{Qk}\right) \tag{10.35}$$

The relation between injected power and current to the system is

$$P_i + jQ_i = (V_i \cos\theta_i + jV_i \sin\theta_i)(I_{Pi} - jI_{Qi}) \tag{10.36}$$

Then I_{Pi} and I_{Qi} will be as the following

$$I_{Pi} = \frac{P_i \cos\,\theta_i + jQ_i \sin\,\theta_i}{V_i} \tag{10.37}$$

$$I_{Qi} = \frac{P_i \sin\,\theta_i + jQ_i \cos\,\theta_i}{V_i} \tag{10.38}$$

Substituting Eqs. (10.37) and (10.38) into Eqs. (10.34) and (10.35) P_L and Q_L will be as [1]

$$P_L = \sum_{j=1}^{n}\sum_{k=1}^{n}\left[\alpha_{jk}\left(P_jP_k + Q_jQ_k\right) + \beta_{jk}\left(Q_jP_k - P_jQ_k\right)\right] \tag{10.39}$$

$$Q_L = \sum_{j=1}^{n}\sum_{k=1}^{n}\left[\delta_{jk}\left(P_jP_k + Q_jQ_k\right) + \gamma_{jk}\left(Q_jP_k - P_jQ_k\right)\right] \tag{10.40}$$

where

$$\alpha_{jk} = \frac{R_{jk}}{V_jV_k}\cos\left(\theta_j - \theta_k\right) \tag{10.41}$$

$$\beta_{jk} = \frac{R_{jk}}{V_jV_k}\sin\left(\theta_j - \theta_k\right) \tag{10.42}$$

$$\delta_{jk} = \frac{X_{jk}}{V_jV_k}\cos\left(\theta_j - \theta_k\right) \tag{10.43}$$

$$\alpha_{jk} = \frac{X_{jk}}{V_jV_k}\sin\left(\theta_j - \theta_k\right) \tag{10.44}$$

Then, taking the first derivative of Eq. (10.39) into consideration with respect to independent variables, the consequence will be as following [1]

$$\begin{aligned}\frac{\partial P_L}{\partial P_i} &= \sum_{j=1}^{n}\sum_{k=1}^{n}\frac{\partial}{\partial P_i}\left[\alpha_{jk}\left(P_jP_k + Q_jQ_k\right) + \beta_{jk}\left(Q_jP_k - P_jQ_k\right)\right] \\ &= 2\sum_{k=1}^{n}\left(P_k\alpha_{ik} - Q_k\beta_{ik}\right) + \underbrace{\sum_{i=1}^{n}\sum_{k=1}^{n}\left[\left(P_jP_k + Q_jQ_k\right)\frac{\partial\alpha_{jk}}{\partial P_i} + \beta_{jk}\left(Q_jP_k - P_jQ_k\right)\frac{\beta_{jk}}{\partial P_i}\right]}_{\approx 0}\end{aligned} \tag{10.45}$$

The second term of the Eq. (10.45) is negligible, and then it will be simplified as [1]

$$\frac{\partial P_L}{\partial P_i} \approx 2\sum_{k=1}^{n} (P_k\alpha_{ik} - Q_k\beta_{ik}) \tag{10.46}$$

In a high-voltage power network, $(\theta_j - \theta_k)$ is infinitesimal, then sin $(\theta_j - \theta_k)$ 0. Therefore, β_{jk} can be ignored as well.

$$\frac{\partial P_L}{\partial P_i} \approx 2\sum_{k=1}^{n} P_k\alpha_{ik} \tag{10.47}$$

Similarly

$$\frac{\partial P_L}{\partial Q_i} \approx 2\sum_{k=1}^{n} Q_k\alpha_{ik} \tag{10.48}$$

$$\frac{\partial Q_L}{\partial P_i} \approx 2\sum_{k=1}^{n} P_k\delta_{ik} \tag{10.49}$$

$$\frac{\partial Q_L}{\partial Q_i} \approx 2\sum_{k=1}^{n} Q_k\delta_{ik} \tag{10.50}$$

Considering real and reactive power consumption of loads constant during the optimization procedure, two assumptions as the following can be made [1]

$$dP_i = d(P_{Gi} - P_{Di}) = dP_{Gi} \tag{10.51}$$

$$dQ_i = d(Q_{Gi} - Q_{Di}) = dQ_{Gi} \tag{10.52}$$

Then, Eq. (10.47) to Eq. (10.50) can be written as

$$\frac{\partial P_L}{\partial P_{Gi}} \approx 2\sum_{k=1}^{n} P_k\alpha_{ik} \tag{10.53}$$

$$\frac{\partial P_L}{\partial Q_{Gi}} \approx 2\sum_{k=1}^{n} Q_k\alpha_{ik} \tag{10.54}$$

$$\frac{\partial Q_L}{\partial P_{Gi}} \approx 2\sum_{k=1}^{n} P_k\delta_{ik} \tag{10.55}$$

$$\frac{\partial Q_L}{\partial Q_{Gi}} \approx 2 \sum_{k=1}^{n} Q_k \delta_{ik} \tag{10.56}$$

Therefore, if the power system which is being investigated has enough amount of reactive power sources, the steps of classic reactive power optimization using Lagrange function would be as following [1]:

I. Power flow calculations should be carried out in order to have all generators' active power output, then fix them all in the current value as active power consumption of loads has been considered constant, the only exception is the slack generator. The output power of slack generators would not remain constant during reactive power optimization.
II. The value of λ should be computed for each reactive power source using Eqs. (10.54) and (10.56). For the sources having $\lambda < 0$, it means the active power losses of the system can be reduced by increasing the output amount of reactive power of the source. For the sources having $\lambda > 0$, it will be vice versa. Therefore, in order to decrease active power losses of the system, the amount of reactive power output should be increased for the sources having $\lambda < 0$ and also decreased for the sources having $\lambda > 0$. Each time, the source with minimum value of λ will be chosen to increase its output if $\lambda < 0$, and the source with maximum value of λ to decrease its output if $\lambda > 0$. Eventually, power flow calculations should be computed to have the result of optimization.
III. Using power flow results, the active power losses can be computed. Since the active power output of the reference unit was not fixed, whatever happening to the active power losses of the system can be sensed in the active power output of the slack generator. The reactive power process will be continued until active power losses cannot be reduced anymore.

It should be noted that limitations of reactive power sources were not considered in the procedure above. There is a limitation for each reactive power source as

$$Q_{Gi\,\min} \leq Q_{Gi} \leq Q_{Gi\,\max} \tag{10.57}$$

If they are supposed to be considered, the constraint in Eq. (10.57) should be checked in every iteration for each source. When it comes to choose an output amount for the power of the sources according to their λ value, if λ suggests to increase the output power of the *i*th source while it is exceeding either its above or its below margins, the amount of output reactive power of the corresponding source should be set to its margins accordingly. Thus, the source which its output power has been adjusted to either its maximum or its minimum values will not be considered any longer in the reactive power optimization procedure [1].

The above-mentioned structure of minimizing active power losses using reactive power economic dispatch was a simplified method depicted in order to represent the concept, and it is not much of a practical method to be used in a real system having large dimensions and plenty of constraints, as voltage profile, transmission lines bounds etc. There are a lot of conventional optimization methods which can be found in [6]. In the next section, reactive power optimization using artificial intelligence algorithms and the used model will be described elaborately, which is more applicable when it comes to a large system in which all the security constraints should be considered.

The reason why heuristic methods are highly concerned in reactive power optimization is a few complexities in such problems' nature, some of which known as non-convexity, having continuous and discrete variables, and having plenty of local and global optimums. Conventional methods and algorithms which were used to optimize such problems proved themselves rather unable to be applied on the problem satisfactorily, because conventional methods mostly use the gradient of the objective function, leading to a local minimum rather than a global one. Another problem is that conventional methods need the derivatives of objective functions which are not accessible in the ones consisting of discrete variables. Thus, there is a compelling need to some sort of algorithms, of which are able to redeem all defections of the conventional ones when applying to optimization problems in order to have as much performance as possible. Although pure mathematical methods like Lagrange function are very precise, the performance of intelligent algorithms has been proved in plenty of cases, and they happen to be more efficient than conventional methods in many of aspects.

10.4 Reactive Power Optimization Using Artificial Intelligent Algorithms

As it has been mentioned in the last part of previous section, the accuracy of conventional methods is their most redeeming feature, although they cause problems in terms of mathematical fulfilments. The conventional methods are not capable of finding the global optimum of the target function and they will get into trouble if the objective function consists of discrete variables. In addition, classic pure mathematical methods need the derivatives of target function, which will impose so many difficulties to the optimization procedure, such as the complexities that calculating derivatives of some functions possess. Heuristic algorithms use the target function itself during the optimization procedure instead of its derivatives and they are readily able to find the global optimum point. Fortunately, heuristic algorithms have solved many drawbacks that conventional algorithms had, and also they have proven their capabilities in the optimization respect. Therefore, it is worth to put some effort on using heuristic methods to optimize engineering problems like the reactive power optimization.

10.4.1 Basic Principles

There are two types of variables concerning reactive power optimization as control and state variables. The terminal voltage of generators, tap setting of transformers and reactive power output of reactive power sources are control variables that are changeable within their bounds. Control variables are the tools that heuristic algorithms use in order to optimize the target function. The voltage magnitudes and voltage angels of *PQ* buses are state variables, of which the active power losses can be computed using Eq. (10.58) and the corresponding values. The general procedure of optimization by heuristic algorithms is shown in Fig. 10.6, whereas the flowchart of reactive power optimization using intelligent algorithms is shown in Fig. 10.7.

$$F_{loss} = \sum_{k=1}^{N_L} g_k \left[V_{1,k}^2 + V_{2,k}^2 - 2V_{1,k}V_{2,k}\cos\left(\theta_{1,k} - \theta_{2,k}\right) \right] \tag{10.58}$$

where, g_k is conductance value of the transmission line between starting and ending buses, $V_{1,k}$ and $V_{2,k}$ are voltage magnitudes of starting and ending buses, and, $\theta_{1,k}$ and $\theta_{2,k}$ are voltage angles of starting and ending buses, respectively.

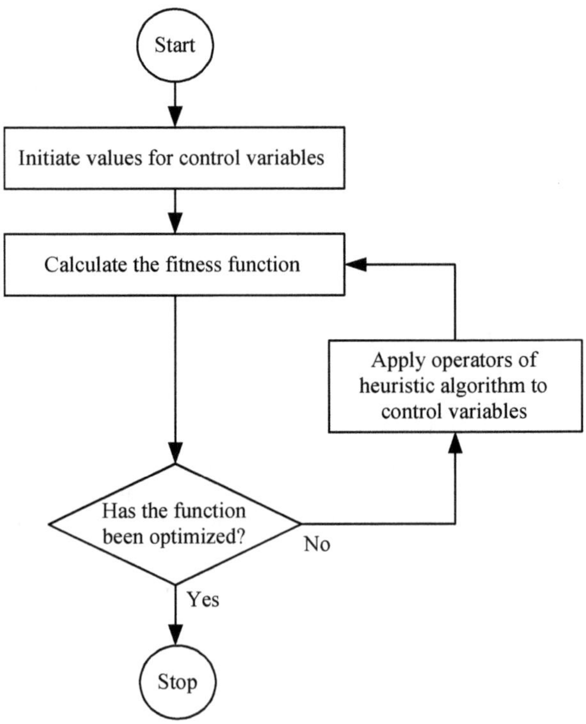

Fig. 10.6 General procedure of optimization by heuristic algorithms

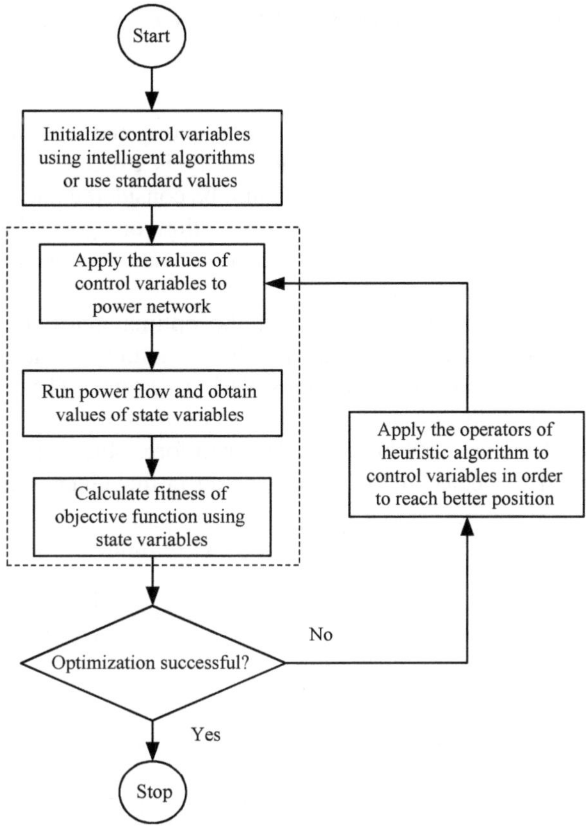

Fig. 10.7 General reactive power optimization trend using intelligent algorithms

As it can be inferred from Figs. 10.6 and 10.7, there is an important point in reactive power optimization problem to pay attention to, which is the calculation of objective function in order to examine whether it has improved or not. There is one extra step in reactive power optimization as the power flow calculation, shown with dashed rectangle in Fig. 10.7. In order to evaluate the objective function represented by Eq. (10.58) in which both state and control variables are involved, after initializing and/or determining control variables by heuristic algorithms, power flow should be run to make calculation of fitness function available. In other words, we have to run power flow calculations in order to achieve the values for the state variables which are mentioned before.

10.4.2 Problem Formulation of Reactive Power Optimization

There are several methods of formulating and modeling reactive power optimization problem according to what exactly is expected as its practical implications, for instance, what constraints should be considered and what objectives are aspired will determine the approach. In this section, one general model will be presented which is useful when it turns to be used in heuristic algorithms to optimize reactive power in AC power systems.

The objective function of the optimal reactive power flow (ORPF) includes technical and financial goals. The economic goal has mainly considered minimizing active power losses in transmission systems. The technical goals are to diminish voltage deviation of *PQ* buses (VD) from the ideal voltage setting and to increase voltage stability index (VSI). Therefore, objective functions for both the technical and monetary goals are considered in this chapter as following [7]

$$f(X) = \min(P_L)$$

$$\text{subject to:} \quad \begin{cases} (VD) \\ (VSI) \end{cases}$$

10.4.2.1 Active Power Losses Objective Function

Reducing active power losses is the most crucial aim of reactive power optimization problem, which influences final cost of dispatched energy. The active power losses can be computed by Eq. (10.59) as follows

$$F_{loss} = \sum_{k=1}^{N} g_k \left[V_{1,k}^2 + V_{2,k}^2 - 2V_{1,k}V_{2,k}\cos\left(\theta_{1,k} - \theta_{2,k}\right) \right] \tag{10.59}$$

where, g_k is conductance of the transmission line between starting and ending buses, $V_{1,k}$ and $V_{2,k}$ are voltage magnitudes of starting and ending buses, and $\theta_{1,k}$ and $\theta_{2,k}$ are voltage angles of starting and ending buses, respectively, and N is the number of transmission lines.

10.4.2.2 Voltage Deviation Constraint

Another significant goal of reactive power optimization in AC power systems is to shrink voltage deviation. Electrical equipment is designed for optimal operation at its nominal voltage. Any deviance from the nominal voltage can result in reducing the general effectiveness and decreasing longevity of electrical apparatus. Voltage deviance constraint is to enhance the voltage profile of power systems by minimizing the summation of voltage deviations at load buses. The voltage deviance constraint can be considered as the least possible amount of voltage deviation summation at each load bus. This function is defined as follows:

$$V_D = \sum_{j=1}^{M} \left| V_j - V_j^{ref} \right| \tag{10.60}$$

where, V_j is the actual voltage of jth load bus, V_j^{ref} is the ideal voltage of jth load bus, and M is number of load buses [8].

10.4.2.3 System Voltage Stability Index

A rather simple voltage stability index to define is V/V_0 ratio, where V is the voltage magnitude of all the PQ buses under load condition and V_0 is the voltage magnitude of all the PQ buses under no-load condition, both identified from load flow calculations or state estimation studies of the system. The ratio V/V_0 at each node offers a voltage stability diagram for the corresponding bus, providing power system operators with weak spots to be taken care of. There are wide range of indices for this purpose, while the current being chosen for the sake of simplicity.

$$VSI = \sum_{i=1}^{T} \left| 1 - \frac{V_i}{V_{i0}} \right| \tag{10.61}$$

where, V_i is the voltage magnitude of ith PQ bus under load condition, V_{i0} is the voltage magnitude of ith PQ bus under no-load condition, and T is number of load buses, respectively.

10.4.2.4 Constraints of Control and State Variables

The control variable constraints embrace tap changer settings of all transformers T, the output capacity of reactive power compensators C, and terminal voltage of all the generators V. The state variables consist of voltage magnitude of all PQ buses U, and reactive power output of all generators Q. Thus, the restriction expressions of control and state variables can be written as

$$V_{Gk,\min} < V_{Gk} < V_{Gk,\max} \tag{10.62}$$

$$T_{i,\min} < T_i < T_{i,\max} \tag{10.63}$$

$$C_{j,\min} < C_j < C_{j,\max} \tag{10.64}$$

$$Q_{Gk,\min} < Q_{Gk} < Q_{Gk,\max} \tag{10.65}$$

$$V_{l,\min} < U_l < V_{l,\max} \tag{10.66}$$

where, $V_{Gk,\min}$ ($V_{Gk,\max}$), $T_{i,\min}$ ($T_{i,\max}$), $C_{j,\min}$ ($C_{j,\max}$), $Q_{Gk,\min}$ ($Q_{Gk,\max}$) and $V_{l,\min}$ ($V_{l,\max}$) are lower (upper) boundary values of PV bus voltages, tap ratio of transformers, reactive power output of compensators, reactive power output of PV buses and voltage magnitude of load buses, respectively.

10.4.2.5 System Power Flow Constraint Equations

The reactive power optimization must fulfill the power flow balances, which are written as

$$P_{Gi} - P_{Li} - V_i \sum_{j=1}^{n} V_j \left(G_{ij} \cos \delta_{ij} + B_{ij} \sin \delta_{ij} \right) = 0 \tag{10.67}$$

$$Q_{Gi} - Q_{Li} - V_i \sum_{j=1}^{n} V_j \left(G_{ij} \sin \delta_{ij} - B_{ij} \cos \delta_{ij} \right) = 0 \tag{10.68}$$

where, n is number of buses, P_{Gi} and Q_{Gi} are generator active and reactive power of the ith bus. P_{Li} and Q_{Li} are load active and reactive power of ith bus. V_i and V_j are voltage magnitudes of ith and jth buses (two neighboring busbars). G_{ij}, B_{ij}, and δ_{ij} are conductance parameters and voltage angle between ith and jth buses, respectively.

It should be noted that the optimization algorithm that determines the control variables in relation to the optimization procedure, satisfies the corresponding constraints as well. The constraints related to state variables will be met by the standard power flow procedure and the algorithm being used, such as Newton or Gauss Sidle. In addition, violations happened in the state variables due to adjusting control variables, can be controlled using a penalty function added to the final objective function of reactive power optimization.

10.4.2.6 General Form of Objective Functions Used in Intelligent Algorithms

Most of the intelligent optimization algorithms tend to have an unconstrained objective function rather than a constrained one. However, different sorts of constraints such as linear and non-linear are typically inseparable part of optimization procedure. Therefore, there is a compelling need to be able to convert a constrained fitness function to an unconstrained one in order to use it in intelligent algorithms. The solution is the penalty function which exactly transforms both objective function and its constraints to a unique unconstrained function.

Considering the optimization problem generally as Eq. (10.69) [6]

$$\text{minimize } f(X)$$
$$\text{subject to: } \begin{cases} g_j(X) \leq 0 & j = 1, \ldots, m \\ h_i(X) = 0 & i = 1; \ldots, n \end{cases} \tag{10.69}$$

It has been converted into an unconstrained optimization fitness function by constructing a function of the form:

$$\phi(X) = f(X) + \sum_{i=1}^{m} r_i \langle g_i(X) \rangle^2 + \sum_{j=1}^{p} R_j \big(h_j(X)\big)^2 \tag{10.70}$$

where X is the vector of control variables, r_i and R_j are penalty multipliers which are constant for all the constraints during the optimization procedure, $g_i(X)$ and $H_j(X)$ are inequality and equality constraints, respectively. $\langle g_i(X) \rangle$ is the bracket function which is defined as Eq. (10.71).

$$\langle g_i(X) \rangle = \begin{cases} g_i(X) & g_i(X) > 0 \\ 0 & g_i(X) \leq 0 \end{cases} \tag{10.71}$$

Equation (10.70) can be considered for maximization or minimization problems appropriate to which optimization algorithm is being used. If a specific algorithm is naturally good at minimizing functions and it is used to minimize a function, then it will be better to use $F(X) = \phi(X)$ as the fitness function, Otherwise it will be better to use the form shown in Eq. (10.72) to define the maximization problem back to minimization one. For instance, Genetic Algorithm is naturally good at maximizing objective functions, thus the fitness function shown in (10.72) will be used if the considered problem aims to minimize the objective function.

$$F(X) = \frac{1}{\phi(X)} \tag{10.72}$$

As it can be seen in the Eq. (10.70), $\phi(X)$ is a new fitness function consisting of constraints and the objective function itself. It is forced by an additional value, multiplying r_i and R_j, which are allocated big values in case the optimization procedure violates the minimum and maximum bounds of constraints. Therefore, the new objective function will have a huge value when the limits are violated, having been influenced by the multipliers. It is the exact method optimization algorithms use to perceive whether the bounds are violated or not. The same method is highly appreciated when it comes to using intelligent algorithms in reactive power optimization problem.

Thus, the penalty function to use in intelligent algorithms for reactive power optimization will be as follows

$$f(V,\theta) = \underbrace{\sum_{k=1}^{N} g_k\left[V_{1,k}^2 + V_{2,k}^2 - 2V_{1,k}V_{2,k}\cos\left(\theta_{1,k} - \theta_{2,k}\right)\right]}_{P_{loss}} + \underbrace{k_v\sum_{j=1}^{M}\left|V_j - V_j^{ref}\right|}_{\text{Voltage Deviation}} + \underbrace{k_s\sum_{i=1}^{T}\left|1 - \frac{V_i}{V_{i0}}\right|}_{\text{Voltage Stability}} \quad (10.73)$$

where, f is an unconstrained objective function, k_v and k_s are penalty multipliers which are used to obligate optimization procedure not to violate the corresponding bounds.

It should be noted that the constraints related to control variables are usually managed by optimization algorithms, while the constraints pertinent to power flow calculations represented by Eqs. (10.67) and (10.68) will be met by power flow. Moreover, the constraints containing state variables should be included in the penalty function to be fulfilled. There is another way to consider constraints of state variables and that is multi-objective optimization methods, in which each constraint is defined as an objective function separately and the procedure drives the main objective function as well as all the constraints simultaneously to be optimized and fulfilled. There is a major difference between using multi-objective optimization methods and single-objective ones, and that is, all the objectives including main fitness function are driven to be minimized while the multi-objective approaches are used. Using penalty function method (single-objective), however, tries to keep the constraints within their bounds.

10.5 Particle Swarm Optimization Algorithm

Particle swarm optimization (PSO) is one of the most successful optimization approaches, of which was inspired by nature. It has broadly been used in all sorts of optimization problems that possess large search spaces, without needing gradients of objective functions. Particle swarm optimization algorithm is based on the performance of a colony or swarm, a flock of birds, a school of fish and/or any kind of creatures that live in groups. The particle swarm optimization algorithm tries to simulate the behavior of these collective organisms which have evolved along centuries in order to enhance their performance of finding food, encountering danger and keeping themselves more competitive in a world that adaptability is a prerequisite feature to survive. The word ‘particle’ notes to an individual in a swarm which acts in a distributed way using its own intellect and also the cooperative intelligence of the crowd.

Fig. 10.8 A fish swarm using their collective intelligence

Taking Fig. 10.8 into consideration, since there is sight limitation under the water, each individual can just see a near distance radially so that can put all of them into trouble all the time facing sharks or any offensive actions. Nevertheless, they have been capable of keeping their species alive in harsh nature of the underwater. As it can be seen in the photo, individuals try to move close to each other keeping precautionary distance so that whenever there is a danger the nearest individual will sense it and act, and since all of them follow the same rules, the farthest one in the group will sense the danger and act appropriately. That was just one of the advantages taken by swarms in the nature, using individualistic intelligence as well as the collective intelligence. Optimization methods founded on swarm intelligence are called behaviorally enthused processes which are called evolution-based procedures. The PSO algorithm was originally proposed by Kennedy and Eberhart in 1995 [1].

On the whole, each particle tries to observe three rules instinctively in swarms and these rules are the bases of the optimization algorithm as well. Rules are as follows [1]:

1. It tries not to come too near and not to go too far from other individuals simultaneously.
2. It directs toward the middling track of other individuals.
3. It tries to fit the "average position" among other individuals with no extensive gaps in the flock.

Three rules mentioned above lead to the behavior of the swarm which is based on a mixture of three simple features as follows [1]:

1. Cohesion, which tries to keep the swarm altogether.
2. Separation, which causes the individuals not to came too close to each other.
3. Alignment, which causes the swarm to keep an eye on the general heading point of the flock.

In general, size of the swarm is assumed to be fixed, whereas each particle situated primarily at accidental positions in the multidimensional space of the optimization problem. Each individual has two sorts of data as position and velocity, both of which are stored and compared to each other continuously during the optimization procedure in order to work out the best position discovered by the particles. The velocity and position in the previous iteration are used to determine the new values of them in the next iteration. This process runs consecutively until it has found the best position possible or has reached one of the stopping criteria. Therefore, in simple words the procedure of particle swarm optimization algorithm is generally as follows [6]:

1. All the particles exchange their latest information with each other simultaneously to figure out which particle has found the best location so far.
2. All the particles considering the location and velocity of each one incline to the best point found in relation to their current parameters.
3. The past memory of each particle as well as its current position affects the next position where it will be.

Overall, the particle swarm optimization algorithm quests for the optimum position via a group of individuals similar to other AI-based exploratory optimization methods. The presented model simulates a partly random search method that is armed with individualistic and collective artificial intelligence, leading the process to a global optimum point of the objective function.

10.5.1 Computational Implementation of PSO for Reactive Power Optimization

Since the concept of particle swarm optimization algorithm has been elaborated in many articles and book chapters like [6, 9], this section focuses on the computational implementation of particle swarm algorithm in reactive power optimization problem instead of the algorithm itself. A reactive power optimization problem including the constraints using penalty function can be considered as:

$$
\begin{aligned}
f(V', \theta') = & \sum_{k=1}^{N} g_k \left[V'^2_{1,k} + V'^2_{2,k} - 2V'_{1,k}V'_{2,k}\cos\left(\theta'_{1,k} - \theta'_{2,k}\right)\right] \\
& + k_v \sum_{j=1}^{M} \left| V'_j - V'^{ref}_j \right| + k_s \sum_{i=1}^{T} \left| 1 - \frac{V'_i}{V'_{i0}} \right|
\end{aligned}
\tag{10.74}
$$

As particle swarm optimization algorithm is naturally good at maximizing objective functions, then the appropriate fitness function in order to minimize the target function in Eq. (10.74) will be as follows

$$\text{maximize}\, F(V', \theta') = \frac{1}{f(V', \theta')} \tag{10.75}$$

Subject to control variables, such as terminal voltages of generator buses, tap settings of transformers, and reactive power output of compensators

$$X = [V \quad T \quad C] \tag{10.76}$$

$$V = [V_1 \quad V_2 \quad \ldots \quad V_n] \tag{10.77}$$

$$T = [T_1 \quad T_2 \quad \ldots \quad T_m] \tag{10.78}$$

$$C = [C_1 \quad C_2 \quad \ldots \quad C_d] \tag{10.79}$$

$$X^{(l)} = [V_{\min} \quad T_{\min} \quad C_{\min}] \tag{10.80}$$

$$X^{(u)} = [V_{\max} \quad T_{\max} \quad C_{\max}] \tag{10.81}$$

where, X is the vector of control variables, while $X^{(l)}$ and $X^{(u)}$ are lower and upper bounds of them, V is a vector containing terminal voltages of generator buses, T is a vector consisting of tap settings of transformers, and C is the vector of reactive power compensators. n, m and d are the number of generator buses, transforms having tap changing facility, and compensators, respectively.

The PSO procedure can be implemented through the following steps [1]:

1. Consider N as the size of the swarm which is mostly between 20 and 30. It is obvious that taking big numbers will raise the time of evaluation of the objective function and will influence the total calculation time. However, it should not be too small either as it can affect the performance of the PSO algorithm. In general, the size of the population is obtained using trial and error method for each optimization problem, although there as some approximate methods to apply in order to get appropriate numbers for them [6].
2. Produce the primary population of X in between $X^{(l)}$ and $X^{(u)}$ randomly as X_1, X_2, …, X_N. Henceforth, for the sake of convenience, the position of jth individual and its speed in ith iteration are signified as $X_j^{(i)}$ and $V_j^{(i)}$, respectively. Accordingly, the particles created initially are indicated as $X_1(0)$, $X_2(0)$, …, $X_N(0)$. The vectors $X_j(0)$ ($j = 1, 2, \ldots, N$) are called particles or coordinate vectors of particles [6].
3. Apply the generated control variables to the power network and run power flow in order to obtain the values of state variables, because the objective function in Eq. (10.75) needs the values of both state (voltage magnitudes of PQ buses) and control variables (voltage values of PV buses). After doing power flow, the bus voltages and voltage angles will be considered as follows [6]:

$$Y_j(i) = \begin{bmatrix} V' & \theta' \end{bmatrix} \tag{10.82}$$

where, V' is the vector of voltage magnitudes and θ' is the vector of voltage angles of all the buses.

4. Work out the values of the objective function for each particle as $F[(Y_1(0)), F(Y_2(0)), \ldots, F(Y_N(0)]$.
5. Determine the velocity values for all the particles. The velocity will help leading the particles through reaching the optimum point. Primarily, the speed value for all the particles are presumed zero. Then, set the iteration number to one ($i = 1$).
6. In the ith iteration, the two following significant parameters should be calculated using the data of jth particle:

   (a) Work out the best value of the objective function for X_j among all the iterations and allocate it to $P_{best,j}$ (personal best) which is the best value of the $F[Y_j(i)]$ found by jth individual so far. Find the best value of the objective function for all the particles (X) among all the iterations so far and assign it to G_{best} (global best), which is the best value found for the objective function $F[Y_j(i)]$ $j = 1, 2, 3, \ldots, N$ [1].

   (b) Find the velocity of jth particle in ith iteration by means of the following equation:

$$V_j(i) = \theta V_j(i-1) + c_1 r_1 \left[P_{best,j} - X_j(i-1)\right] + c_2 r_2 \left[G_{best} - X_j(i-1)\right] \\ j = 1, 2, \ldots, N \tag{10.83}$$

   where, c_1 and c_2 are the perceptive (individual) and collective (group) learning coefficients, respectively. r_1 and r_2 are uniformly distributed random numbers in the range of 0 and 1. The factors c_1 and c_2 signify the comparative rank of the memory (location) of the particles to the memory (location) of the swarm. c_1 and c_2 are commonly considered to have the amount of 2 in a lot of implementations. The inertia weight θ is a constant which is used in order to decrease the velocities as time goes by (or iterations), facilitating the swarm to congregate more precisely and proficiently compared to the original PSO algorithm. Equation (10.83) represents a formulation for adjusting the velocity, which helps the accuracy increase. Equation (10.83) demonstrates that a greater amount of θ supports global exploration, while a smaller value encourages a local exploration better. Hence, a great value of θ marks the algorithm continually discover new areas deprived of much local examination, failing to find the true optimum point. To strike a balance between global and local search in order to speed up converging to the exact optimal location, an inertia coefficient whose value declines linearly in relation to the iteration number has been used as [6]

$$\theta(i) = \theta_{\max} - \left(\frac{\theta_{\max} - \theta_{\min}}{i_{\max}}\right) i \tag{10.84}$$

where, $\theta_{\max}$ and $\theta_{\min}$ are primary and final values of the inertia weight, respectively. $i_{\max}$ is the maximum quantity of iterations used in PSO. Values of $\theta_{\max} = 0.9$ and $\theta_{\min} = 0.4$ are commonly used [6].

(c) Find the coordinate of the jth particle in ith iteration using [6]

$$X_j(i) = X_j(i-1) + V_j(i); \quad j = 1, 2, \ldots, N \tag{10.85}$$

Evaluate the values of objective function for all the particles as $F[Y_1(i)]$, $F[Y_2(i)]$, …, $F[Y_N(i)]$, in which the matrix Y is acquired from the power flow calculations after applying new control variables to the power network [6].

7. Check if the algorithm has reached the optimal point or not, which is applicable via creating and checking a few stopping criteria for the optimization problem. If the points of all particles congregate to the similar set of values, the technique is considered to be converged. If the convergence criterion is not met, step 6 will be reiterated by updating the iteration number to $i = i + 1$, and calculating the new values for $P_{best,j}$ and G_{best}. The reiterative procedure will continue until almost all of the particles congregate to the same optimal point. The flow chart of PSO algorithm founded on above declared process is shown in Fig. 10.9 [6].

There are plenty of stopping criteria to use, of which can be chosen in relation to what the algorithm is expected to do or reach. The first one can be the iteration number so that the process will stop after a specified number of iterations whether the algorithm has reached the optimum point or not. The second one can be considered as the time duration of the process running, which completely depends on how heavy the process might be. The third one can be to use a pre-specified fitness value and compare the last overcome of the optimization procedure to it, and then if the appropriateness of the solution is met, the process can be stopped. The fourth one can be taken as whether there is any substantial progression in the result or not, and stop the process if not. Usually a combination of several criteria is used in practice appropriate to what the expectation is from the algorithm and the procedure. Mostly, the criteria that consider just the calculation cost such as the time duration and the number of iterations can be used with some other criteria that consider the fitness of the result.

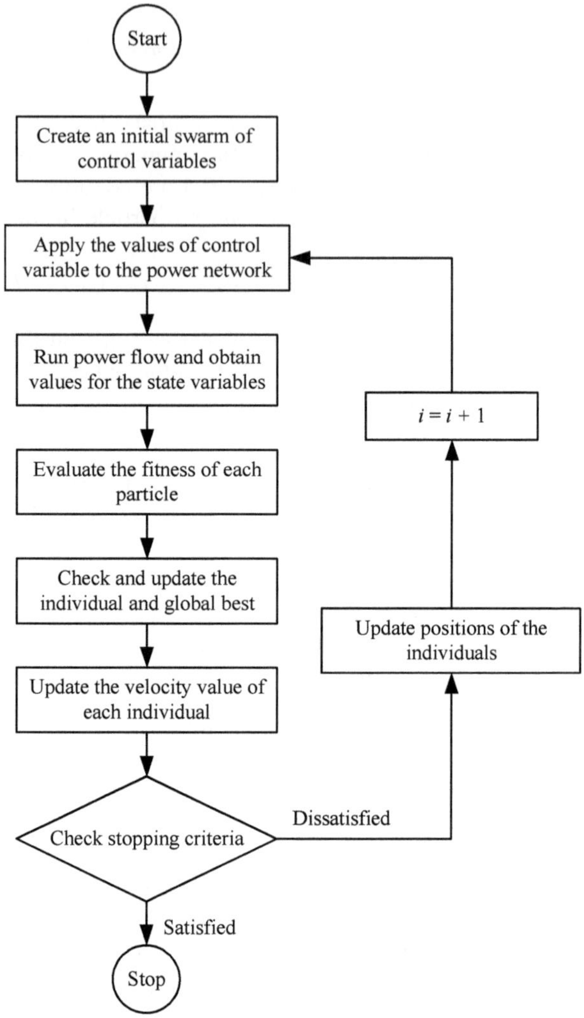

Fig. 10.9 The flowchart of reactive power optimization using PSO

10.6 Pattern Search Optimization Algorithm

Pattern search (PS) algorithm is one of the arithmetical optimization approaches that do not need the gradient of the objective function which is meant to be optimized. Therefore, PS can be used on discrete functions that are not differentiable as well. Such optimization means are also recognized as direct-search, derivative-free, or black-box methods. They are founded on producing search patterns which positively sweep the search space in order to detect the minimum points. The procedure of optimizing starts from a random starting point, though being highly dependent on what location they start from (initial point) is their biggest defect. This drawback can cause the optimization procedure to get stuck in one of the local minimums which is not favorable. However, because of their flexibility, they can be integrated with heuristic algorithms (global optimizers) for global search, which is a mode pattern search technique receives some of the properties of the imported global optimization technique (global optimization), without risking the convergence and being trapped in one of the local optimum points [10–12].

10.6.1 Mathematical Description of Pattern Search Optimization Algorithm

Pattern search algorithm can be counted as a direct search optimization algorithm in which the central notion is on the positive spanning sets. This section will present the PS algorithm in search/poll framework which is the best choice when it comes to cooperate with heuristic algorithms. One of the simplest positive spanning sets is formed by the vectors of the canonical foundation and also their negatives [10, 12]

$$D_{\oplus} = \{e_1, \ldots, e_n, -e_1, \ldots, -e_n\} \tag{10.86}$$

The set $D_{\oplus}$ is also a (highest) positive basis. The basic straight search technique based on this progressive spanning set is recognized as coordinate or compass search. Bearing in mind a progressive spanning set as D and the recent iteration $y(t)$, two groups of points are defined as the net M_t and the election set P_t. The net M_t is known as [10]

$$M_t = \left\{y(t) + \alpha(t)D_z,\ z \in Z_+^{|D|}\right\} \tag{10.87}$$

where $\alpha(t) > 0$ is the mesh dimension factor and also identified as the step-length controller, and Z_+ is the set containing nonnegative integer numbers. The mesh has to fulfil some integrality necessities for the technique to attain global convergence to static points from random initial points. The matrix D has to be in the formula of $G\hat{Z}$, where $G \in R^{n\times n}$ is a nonsingular producing matrix and $G \in R^{n\times n}$. The progressive foundation $D_{\oplus}$ meets the prerequisites when G is entitled the identity matrix. The pursuit step conducts a limited exploration in the net M_t. The poll step is performed only if the examination step fails to discover a position for which f is lesser than $f(y(t))$. The poll step assesses the function at the positions in the poll set P_t in order to discover a location where f is minor than $f(y(t))$ [10, 12]

$$P_t = \{y(t) + \alpha(t)d,\ d \in D\} \tag{10.88}$$

It should be noted that P_t is a subdivision of M_t. If f is constantly differentiable at $y(t)$, the poll step is assured to succeed if $\alpha(t)$ is appropriately small, since the progressive spanning set D comprises at least one pattern of descent which sorts an acute angle with $-\nabla f(y(t))$. Consequently, if the poll step flops to discover a coordination better than the former one, the net size factor must be made smaller. The poll step which is the key tool of pattern search approach to explore the optimal point guarantees the global convergence.

In order to extrapolate pattern search optimization procedure for bound constrained problems, it is indispensable to use a practicable primary guess $y(0)\in\Omega$ and to keep feasibility of the iterates safe by declining any trial position that is out of the acceptable region. Rejecting unviable test locations can be achieved by applying a pattern search algorithm to the subsequent penalty function [10, 12].

$$\hat{f}(z) = \begin{cases} f(z) & \text{if} \quad z \in \Omega \\ +\infty & \text{otherwise} \end{cases} \tag{10.89}$$

There is no big dissimilarity between constrained and unconstrained pattern search optimization algorithm excluding it is applied to the minimization of $f(z)$ subject to simple bounds and to the refusal of impractical test points. It is also essential to embrace in the exploration directions D those patterns that warranty the existence of a practicable descent track at any nonstationary location of the bound constrained problem [10].

In order to have an elaborate depiction of the basic pattern search algorithm, it is necessary to state in what way to expand and contract the net size or step-length control factor $\alpha(t)$. The growths and reductions use the parameter $\phi(t)$ and $\theta(t)$, respectively, which must observe the subsequent rules:

$$\begin{array}{lll} \phi(t) = \bar{T}^{l_t}, & \text{for some } l_t \in \{0, \ldots, l_{\max}\} & \text{if} \quad t \text{ is successful} \\ \theta(t) = \bar{T}^{m_t}, & \text{for some } m_t \in \{m_{\min}, \ldots, -1\} & \text{if} \quad t \text{ is unsuccessful} \end{array} \tag{10.90}$$

where, $\bar{T} > 1$ is a positive rational, $l_{\max}$ is a nonnegative integer, and $m_{\min}$ is a negative integer, selected at the commencement of the procedure and unaffected with t. For instance, it can be considered $\theta(t) = 0.5$ for unproductive iterations and $\phi(t) = 1$ or $\phi(t) = 2$ for up-and-coming iterations [10, 12].

The process of basic pattern search method has been offered in follows [10]:

1. Select a positive rational $\bar{T}$ and the tolerance $\alpha_{tol} > 0$ as the stopping criterion. Pick the positive spanning set $D = D_{\oplus}$.
2. Set $t = 0$. Choose an primary practical guess $y(0)$. Pick $\alpha(0) > 0$.
3. [Search Step], Assess f at a limited number of points in M_t. If a position $z(t) \in M_t$ is discovered for which $\hat{f}(z(t)) < \hat{f}(y(t))$ then set $y(t + 1) = z(t)$, $\alpha(t + 1) = \phi(t)\alpha(t)$ (optionally increasing the net size factor), and tag both the exploration step and the present iteration as successful.
4. [Poll Step], Avoid the poll step if the examination step was successful.
   - If there exists $d(t) \in D$ so that $\hat{f}(y(t) + \alpha(t)d(t)) < \hat{f}(y(t))$, then:

     Set $y(t + 1) = y(t) + \alpha(t)d(t)$ (poll step and iteration successful).
     Set $\alpha(t + 1) = \phi(t)\alpha(t)$ (optionally increase the net size factor).

   - Otherwise, $\hat{f}(y(t) + \alpha(t)d(t)) \geq \hat{f}(y(t))$ for all $d(t) \in D$, and:

     Set $y(t + 1) = y(t)$ (iteration and poll step unsuccessful).
     Set $\alpha(t + 1) = \theta(t)\alpha(t)$ (contract the net size factor).
5. If $\alpha(t + 1) < \alpha_{tol}$ then break, where α_{tol} is the least value which is defined as the mesh dimension factor. Otherwise, increase t by one and go to Step 3.

10.6.2 Pattern Search Algorithm in Simple Words

The flowchart of pattern search algorithm is shown in Fig. 10.10 in the simplest way possible. At first an initial point as well as initial step size are produced randomly as $x_0 \in R^n$ (for one dimensional problem) and $\Delta_0 > 0$, respectively. Then, the fitness function will be calculated in neighboring points x_0, $x_0 + \Delta$, $x_0 - \Delta$. If $f(x_0 + \Delta) < f(x_0)$ then the point $x_0 + \Delta$ will be considered as the center point and Δ will be added to it, then new fitness values will be computed. This procedure will be continued until the fitness value does not get better, then the value of Δ will be decreased and added to center point, afterwards new fitness values will be calculated again. This procedure will go on until the stopping criteria stop the algorithm. A simple graphical example has shown in the Fig. 10.11 in order to illustrate how the algorithm works.

10.7 Particle Swarm Pattern Search Algorithm

The most considerable feature that distinguishes heuristic optimization techniques from traditional ones is their capability to find the global optimum point, not needing the gradient of the objective function as a great asset, so that the cost of the calculation (considering time as a resource) will be decreased. Nevertheless, their local minimization is not as efficient as their global optimization, or in another words, is more time consuming in comparison with to the extent the function gets minimized. Moreover, pattern search algorithm is such a great direct local minimizer, although in some cases it has proved its power finding even global optimum point. The more problem space gets non-convex, however, the more pattern search algorithm seems to fail finding the global optimum point, because of the so many local optimum points which exist. Therefore, the pattern search algorithm will be more practical finding local optimum points rather than global ones. In addition, pattern search algorithm is highly dependent on the starting point which is normally chosen randomly so that different starting points can lead the algorithm to different solutions not probably being a global optimum point.

The idea is to take the advantages of both types using heuristic algorithms as the global optimizers and direct search methods as local optimizers, in this case pattern search algorithm. Since pattern search algorithm needs a starting point and the final answer is directly dependent on it, considering the global minimum point found by intelligent algorithms like PSO that is mostly near the exact solution as a starting point to pattern search algorithm will fasten the optimization procedure finding the more accurate point. Likewise, if there is a time limitation to calculation, using the particle swarm pattern search algorithm will not be empty of favor [10].

There are some different strategies combining an intelligent optimization algorithm and a direct search method, each of them has its own advantages and disadvantages. For instance, one can be considered as a series one where a heuristic algorithm is chosen to implement a global optimization, after reaching near the global optimum point the direct search method is used in order to pinpoint the local one of the neighboring search space in which the best optimum point lies. The first

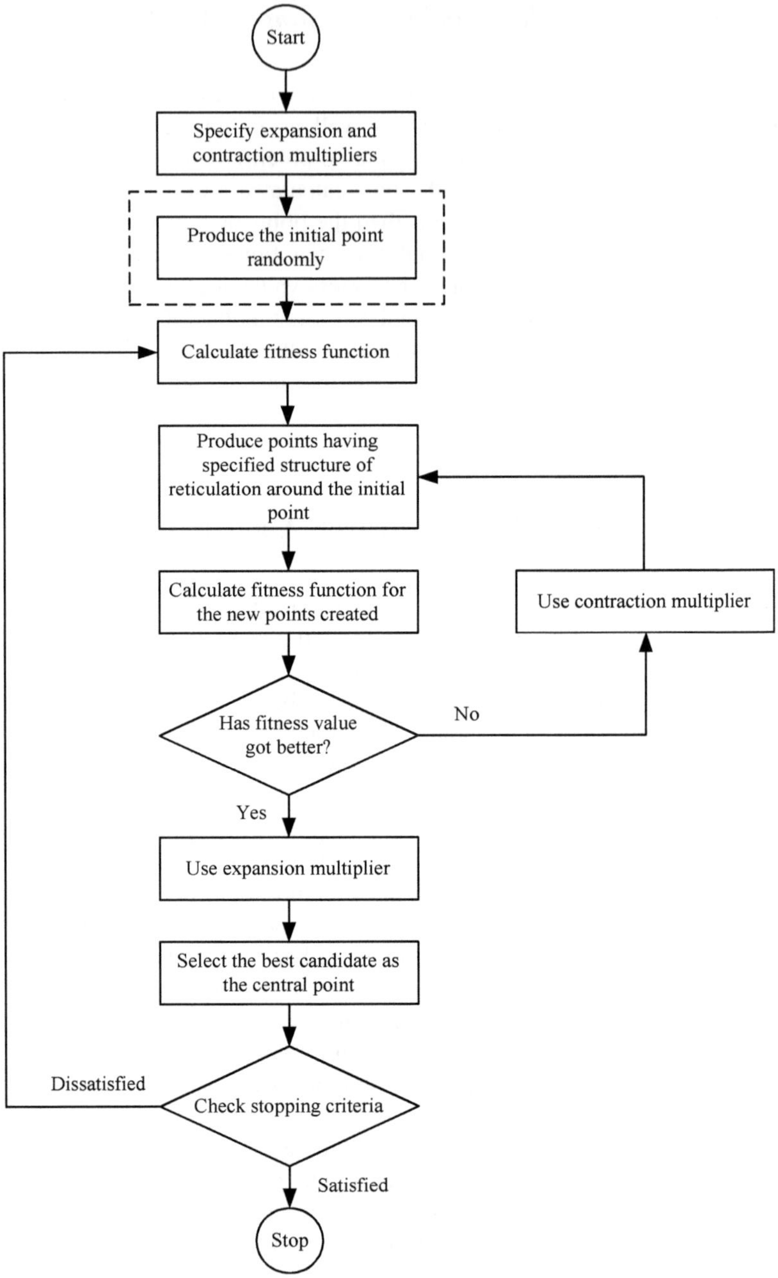

Fig. 10.10 Flowchart of pattern search optimization algorithm

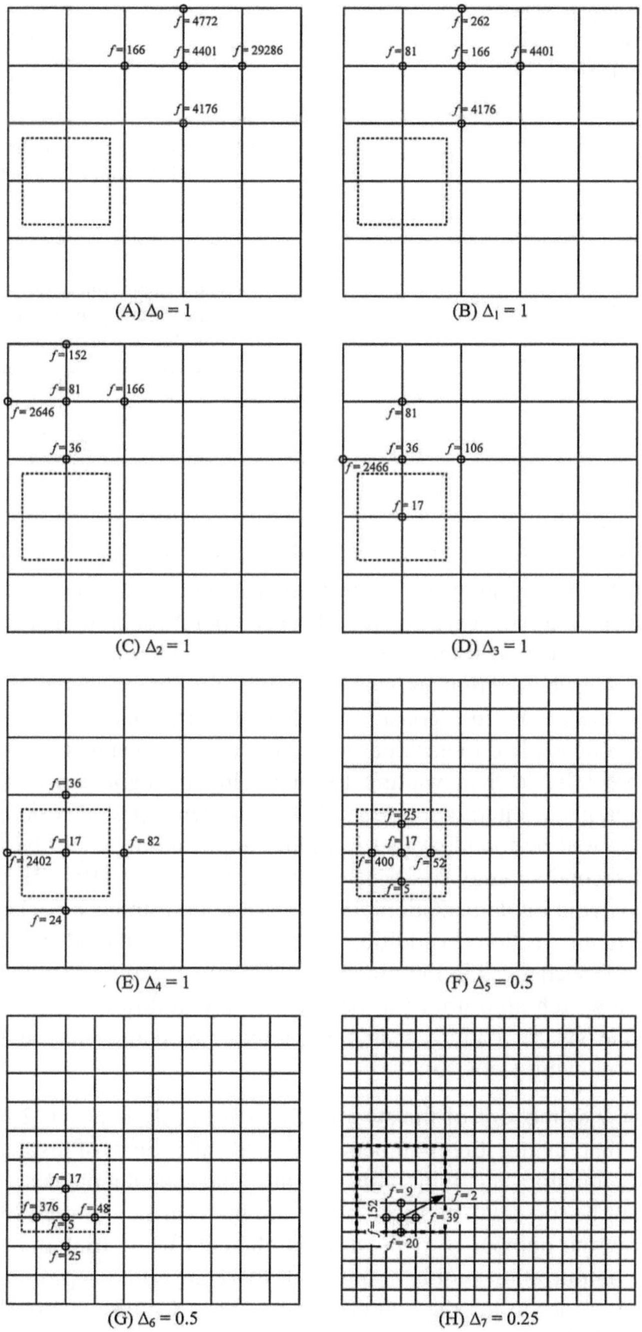

Fig. 10.11 Graphical show of how pattern search optimization algorithm works [12]

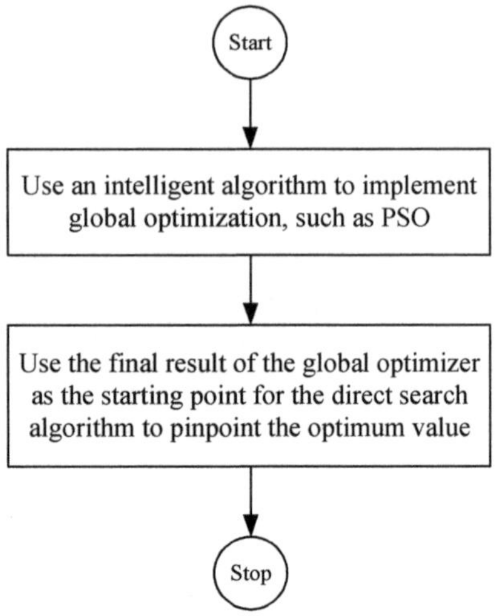

Fig. 10.12 Flowchart of the first strategy using heuristic algorithms incorporated with direct search method

strategy is the simplest and fastest one which can be taken into practice. The flowchart of the first strategy has been shown in Fig. 10.12. The second strategy can be like using a direct search method for each point found by particles (PSO) or genes (genetic algorithm) in order to find the exact optimum point using positions of particles of each iteration as starting points for direct search methods. This approach can be more time consuming than the first one, as the direct search process should be applied to each particle. Although, It sounds to be more accurate and also more reliable than waiting to an intelligent algorithm to find a global optimum point, in each iteration an exact optimum point (probably not global one at the beginning) will be found so that in online implications would be at work. The flowchart of second strategy has been shown in Fig. 10.13 [10, 12].

In order to avoid implementation difficulties, the first strategy has been chosen to be presented in this section. In addition, it can be turned to the other strategy readily considering a few changes in the procedure. Therefore, the first strategy as particle swarm pattern search algorithm can be coincided as follows:

The particle swarm pattern search algorithm procedure can be implemented through the following steps [6]:

1. Consider N as the size of the swarm which is mostly between 20 and 30. It is obvious that taking big numbers will raise the time of evaluation of the objective function and will influence the total calculation cost. However, it should not be too small as it can affect the performance of the PSO algorithm. In general, the number of population is obtained using trial and error method for each optimization problem, although there are some approximate methods to apply in order to get the appropriate numbers. Choose appropriate stopping criteria for PSO algorithm, considering the fact that optimization procedure will go on after getting to the neighborhood of global optimum point with pattern search algorithm which has its own stopping criteria [6].

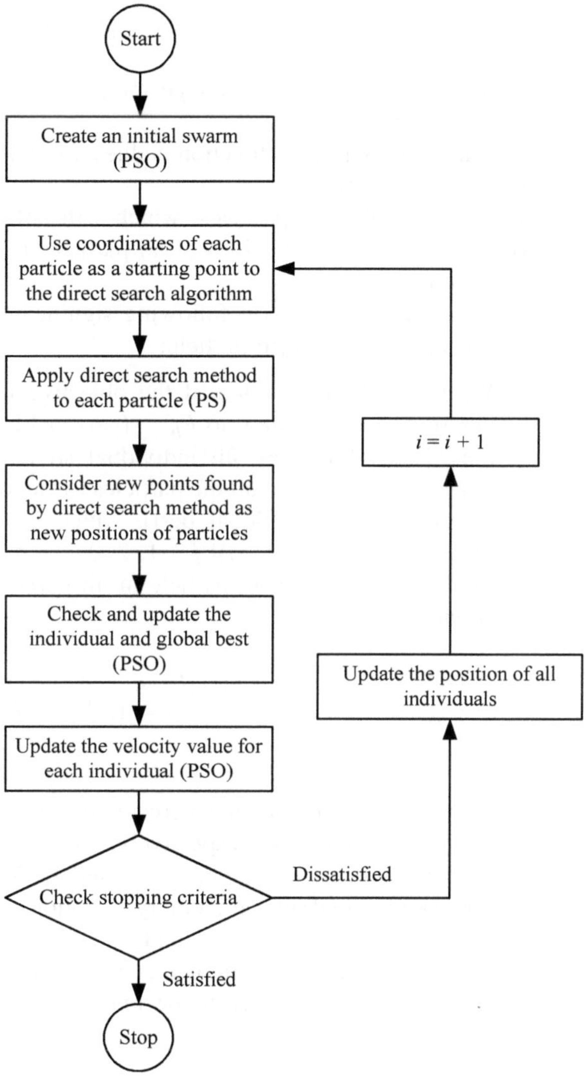

Fig. 10.13 The flowchart of the second strategy using heuristic algorithms incorporated with direct search methods

2. Produce the primary population of X in the assortments $X^{(l)}$ and $X^{(u)}$ randomly as $X_1, X_2, \ldots, X_N$. Henceforth, for the sake of convenience, the position of and velocity of jth individual in ith iteration are signified as $X_j^{(i)}$ and $V_j^{(i)}$, respectively. Accordingly, the particles created at the beginning are indicated as $X_1(0)$, $X_2(0)$, …, $X_N(0)$. The vectors $X_j(0)$ ($j = 1, 2, \ldots, N$) are called particles or vectors of coordinates of particles [6].
3. Apply the generated control variables to the power network and run power flow in order to obtain state variables as they are needed in the objective function represented by Eq. (10.75). Therefore, after doing power flow the bus voltages and voltage angels will be considered as follows [6]

$$Y_j(i) = \begin{bmatrix} V' & \theta' \end{bmatrix} \tag{10.91}$$

where, V' is the vector of voltage magnitudes of all buses and θ' is the bus voltage angles.

4. Calculate the objective function values for all the particles as $F(Y_1(0))$, $F(Y_2(0))$, …, $F(Y_N(0))$].
5. Find the velocities of particles, which will help particles reach the optimum point. Primarily, the velocities of all particles are presumed zero. Set the iteration number to $i = 1$.
6. In the *i*th iteration, the two following significant parameters should be calculated using the data of *j*th particle:
   (a) Work out the best value of the objective function for X_j among all the iterations and allocate it to $P_{best,j}$ (personal best) which is the best value of the $F[Y_j(i)]$ found by *j*th individual so far. Find the best value of the objective function for all the particles (X) among all the iterations so far and assign it to G_{best} (global best), which is the best value found for the objective function $F[Y_j(i)]$ $j = 1, 2, 3, \ldots, N$ [1].
   (b) Find the velocity of *j*th particle in *i*th iteration by means of the following equation

$$V_j(i) = \theta V_j(i-1) + c_1 r_1 \left[P_{best,j} - X_j(i-1)\right] + c_2 r_2 \left[G_{best} - X_j(i-1)\right] \quad j = 1, 2, \ldots, N \tag{10.92}$$

where, c_1 and c_2 are the perceptive (individual) and collective (group) learning coefficients, respectively. r_1 and r_2 are uniformly distributed random numbers in the range of 0 and 1. The factors c_1 and c_2 signify the comparative rank of the memory (location) of the particles to the memory (location) of the swarm. c_1 and c_2 are commonly considered to have the amount of 2 in a lot of implementations. The inertia weight θ is a constant which is used in order to decrease the velocities as time goes by (or iterations).

$$\theta(i) = \theta_{\max} - \left(\frac{\theta_{\max} - \theta_{\min}}{i_{\max}}\right) i \tag{10.93}$$

where, $\theta_{\max}$ and $\theta_{\min}$ are primary and final values of the inertia weight, respectively. $i_{\max}$ is the maximum quantity of iterations used in PSO. Values of $\theta_{\max} = 0.9$ and $\theta_{\min} = 0.4$ are commonly used [1].

(c) Calculate the location or coordinate of the jth particle in ith iteration as [6]

$$X_j(i) = X_j(i-1) + V_j(i); \quad j = 1, 2, \ldots, N \tag{10.94}$$

Evaluate the values of objective function for all the particles as $F[Y_1(i)]$, $F[Y_2(i)]$, …, $F[Y_N(i)]$, in which the matrix Y is acquired from the power flow calculations after applying new control variables to the power network [1].

7. Check if the algorithm has reached the optimal point or not, which is applicable via creating and checking a few stopping criteria for the optimization problem. If the points of all particles congregate to the similar set of values, the technique is considered to be converged. If the convergence criterion is not met, step 6 will be reiterated by updating the iteration number to $i = i + 1$, and calculating the new values for $P_{best,j}$ and G_{best}. The reiterative procedure will continue until almost all of the particles congregate to the same optimal point. The flow chart of PSO algorithm founded on above declared process is shown in Fig. 10.9 [6].
8. Select a positive rational $\bar{T}$ and the tolerance $\alpha_{tol} > 0$ as the stopping criterion. Pick the positive spanning set $D = D_{\oplus}$.
9. Set $t = 0$. Choose a primary practical guess $y(0)$ and pick $\alpha(0) > 0$. The best position found by PSO algorithm will be considered as the initial point in this step.
10. Apply control variables to the power network and implement power flow to calculate the state values as

$$Y = \begin{bmatrix} V' & \theta' \end{bmatrix} \tag{10.95}$$

11. [Search Step], Assess f using Eq. (10.74) at a limited number of positions in M_t. If a coordinate $Z'(t) \in M_t$ is discovered for which $f(Y_{Z'(t)}) < f(Y_{Z(t)})$, [where, $(Y_{Z'(t)})$ is a matrix containing the values for the state variables after doing power flow for $Z'(t)$], then consider $Z(t) = Z'(t)$ and $\alpha(t+1) = \phi(t)\alpha(t)$ (optionally increasing the net size factor), and state successful both the exploration step and the present iteration.

$$Z = \begin{bmatrix} V & T & C \end{bmatrix} \tag{10.96}$$

$$V = \begin{bmatrix} V_1 & V_2 & \ldots & V_n \end{bmatrix} \tag{10.97}$$

$$T = \begin{bmatrix} T_1 & T_2 & \ldots & T_m \end{bmatrix} \tag{10.98}$$

$$C = \begin{bmatrix} C_1 & C_2 & \ldots & C_d \end{bmatrix} \tag{10.99}$$

$$\alpha = \begin{bmatrix} \alpha_1 & \alpha_2 & \ldots & \alpha_n \end{bmatrix} \tag{10.100}$$

12. [Poll Step], Avoid the poll step if the exploration step was successful.

- If there exists $[Z(t) + \alpha(t)d(t)] \in D$ so that $f(Y_{Z(t)+\alpha(t)d(t)}) < f(Y_{Z(t)})$, then

Consider $Z(t+1) = Z(t) + \alpha(t)d(t)$ (poll stage and iteration successful).
Adjust $\alpha(t+1) = \phi(t)\alpha(t)$ (optionally increase the net size factor).

- Otherwise, $f(Y_{Z(t)+\alpha(t)d(t)}) \geq f(Y_{Z(t)})$ for all $d(t) \in D$, and

Consider $Z(t+1) = Z(t)$ (iteration and poll stage unsuccessful).
Adjust $\alpha(t+1) = \theta(t)\alpha(t)$ (contract the net size factor).

13. If $\alpha(t+1) < \alpha_{tol}$ then stop, where α_{tol} is the least value which is defined as the mesh dimension factor. Otherwise, increase t by one and go to step 10.
The flowchart of particle swarm pattern search algorithm which was mentioned above is shown in Fig. 10.14. In the next section practical implementation of reactive power optimization using the proposed particle swarm pattern search algorithm will be presented.

10.8 Simulation Results of Reactive Power Optimization

Two algorithms as particle swarm pattern search and genetic pattern search algorithms have been implemented on two standard systems as IEEE 6-bus and IEEE 14-bus and the results have been compared. In addition, the simulation results for IEEE 39-bus New England power network using particle swarm pattern search algorithm has been presented in this section. Besides, how to implement such optimization procedures using MATLAB and DIgSILENT has been presented in Chap. 11 step by step.

10.8.1 Case Study 1—IEEE 6-Bus Power Network

The corresponding data for IEEE 6-bus power grid has been presented in Appendix 1. The results for reactive power optimization on the related network will be offered in this section.

10.8.1.1 Reactive Power Optimization Using Particle Swarm Pattern Search Algorithm

The initial conditions and power flow results for 6-bus power network are presented in Table 10.19. In addition, the corresponding data after the optimization procedure on the same power grid are presented as below Figs. 10.15, 10.16 and 10.17.

Referring to Tables 10.1, 10.2 and 10.3, it is obvious that operational parameters have been improved considerably. Active power losses have decreased from

Fig. 10.14 Flowchart of particle swarm pattern search optimization algorithm

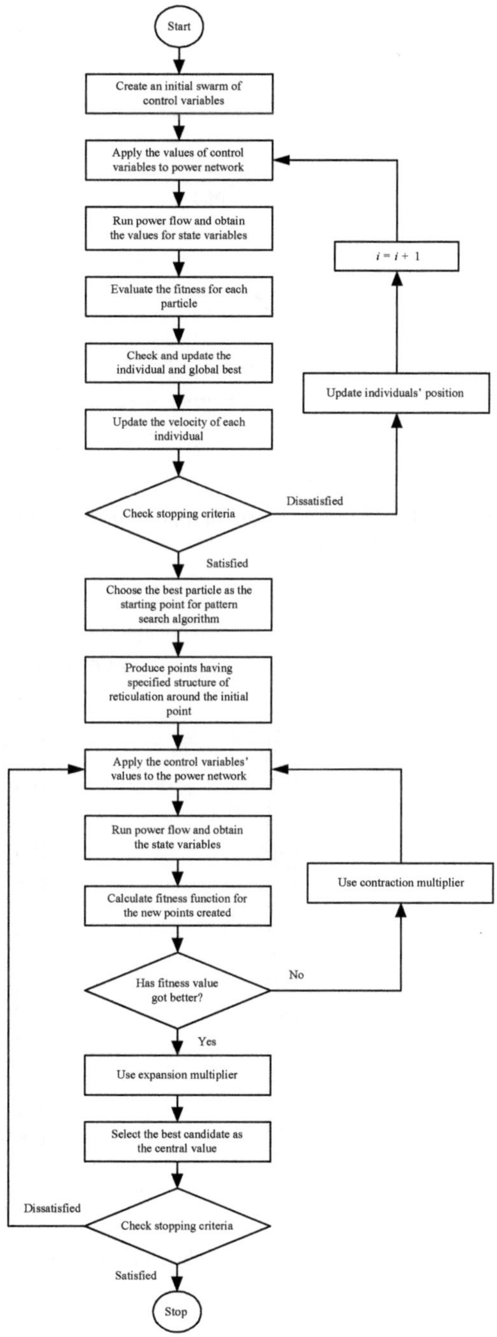

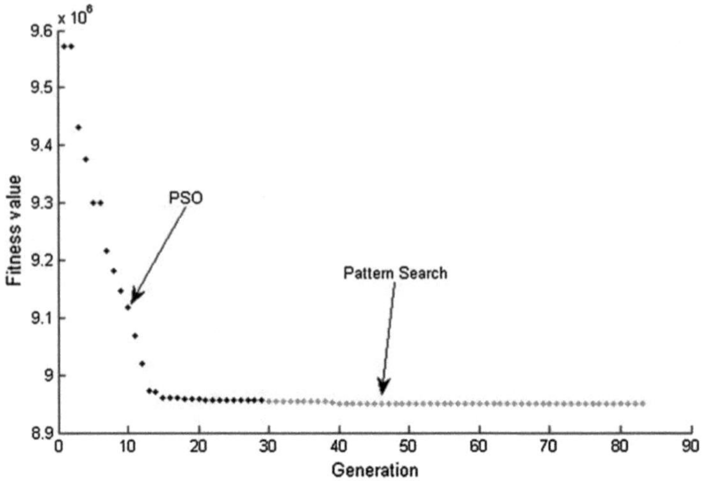

Fig. 10.15 Reactive power optimization trend for 6-bus power system using particle swarm pattern search algorithm

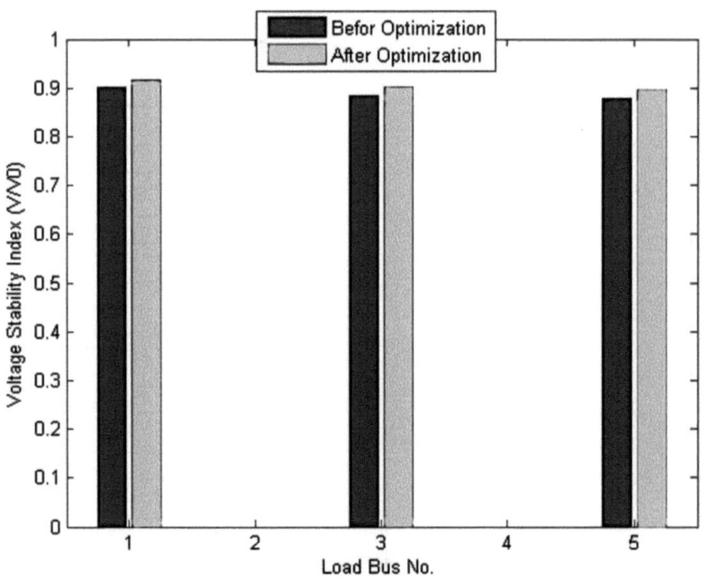

Fig. 10.16 Voltage stability index for 6-bus power system using particle swarm pattern search algorithm

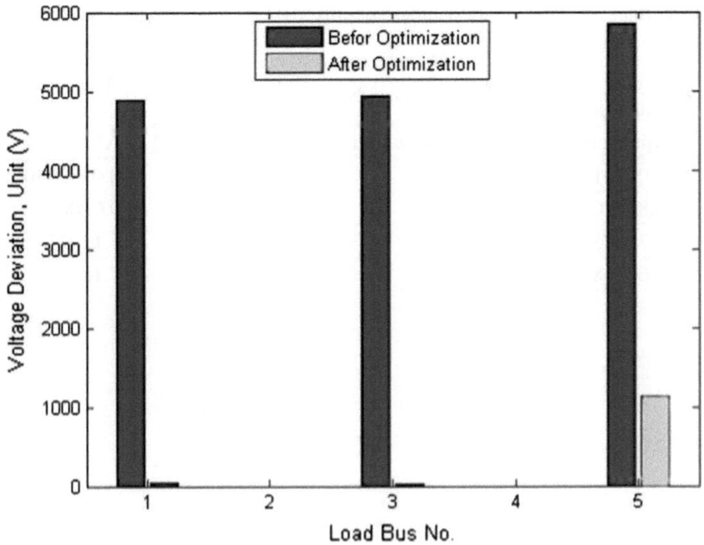

Fig. 10.17 Voltage deviation for 6-bus power system using particle swarm pattern search algorithm

Table 10.1 Power flow results after optimization for 6-bus power system using particle swarm pattern search algorithm

Bus no.	Voltage magnitudes and angles		Load consumption		Injection power	
	V (kV)	θ (degree)	P_l (MW)	Q_l (MVar)	P_G (MW)	Q_G (MVar)
1	62.97	−11.69	55	13	0	0
2	72.45	−3.59	0	0	50	22.65
3	62.98	−10.97	50	5	0	5
4	63.86	−8.79	0	0	0	0
5	61.82	−11.23	0.3	18	0	5
6	69.30	0	0	0	93.96	33.13

Table 10.2 Active power losses for 6-bus power system using particle swarm pattern search algorithm

Active power losses (initial condition)	10,778,370 (W)
Active power losses (after optimization)	8,950,232 (W)
Reduction percentage	16.96121%

10,778,370 to 8,950,232 (W). Likewise, the total value of voltage stability index has increased from 2.665148 to 2.715127, and the total value of voltage deviation of load buses has been reduced considerably from 15.65904 to 1.185288 kV.

Table 10.3 Voltage deviation and voltage stability data for 6-bus power system using particle swarm pattern search algorithm

Bus no.	Voltage stability index value V/V_0 ideal value = 1		Voltage deviation of load buses	
	Initial status	After optimization	Initial status kV	After optimization kV
1	0.9009016	0.9161407	4.881641	0.03538300
3	0.8845708	0.9021702	4.934923	0.02064700
5	0.8796755	0.8968166	5.842471	1.129258
Total	2.665148	2.715127	15.65904	1.185288

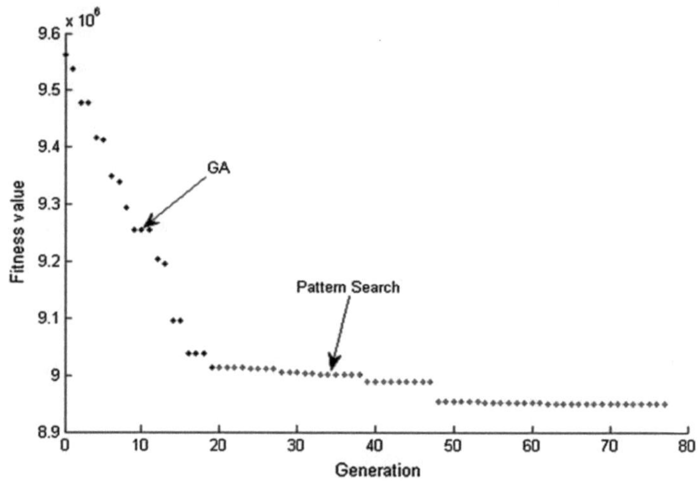

Fig. 10.18 Reactive power optimization trend for 6-bus power system using genetic pattern search algorithm

10.8.1.2 Reactive Power Optimization Using Genetic Pattern Search Algorithm

The initial conditions and power flow results for 6-bus power network are presented in Table 10.18. In addition, the corresponding data after the optimization procedure on the same power grid are presented as below Figs. 10.18, 10.19 and 10.20.

Considering Tables 10.4, 10.5 and 10.6, a marked improvement is obvious in the operational parameters. Active power losses have decreased from 10,778,370 to 8,950,550 (W). Similarly, the total value of voltage stability index has increased from 2.665148 to 2.712409, and the total value voltage deviation of load buses has reduced from 15.65904 to 2.127879 kV.

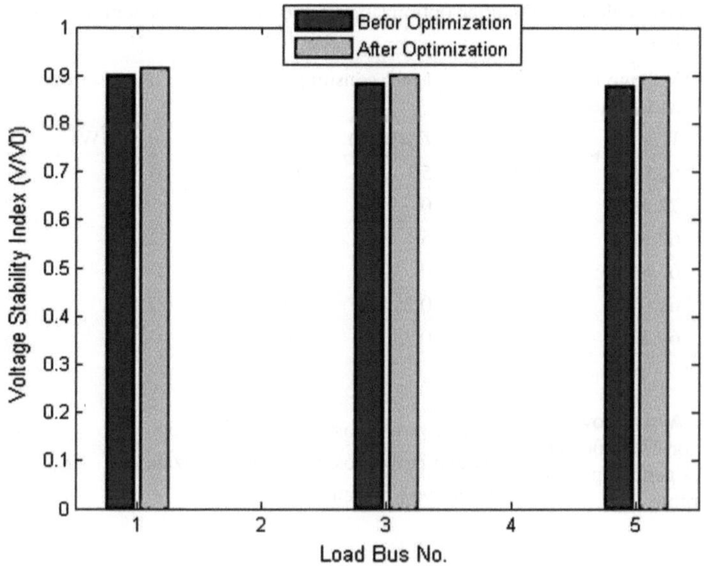

Fig. 10.19 Voltage stability index for 6-bus power system using genetic pattern search algorithm

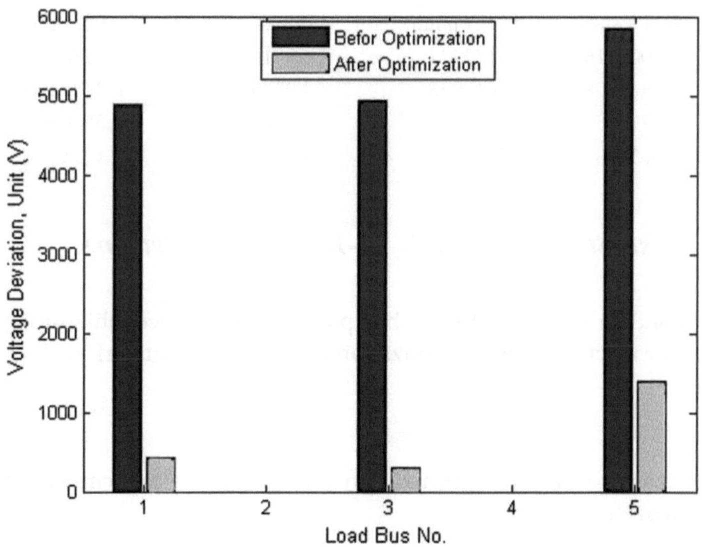

Fig. 10.20 Voltage deviation for 6-bus power system using genetic pattern search algorithm

Table 10.4 Power flow results after optimization for 6-bus power system using genetic pattern search algorithm

Bus no.	Voltage magnitudes and angles		Load consumption		Injection power	
	V (kV)	θ (degree)	P_l (MW)	Q_l (MVar)	P_G (MW)	Q_G (MVar)
1	62.96	−11.69	55	13	0	0
2	72.45	−3.59	0	0	50	22.65
3	62.98	−10.97	50	5	0	5
4	63.86	−8.79	0	0	0	0
5	61.87	−11.23	0.3	18	0	5
6	69.30	0	0	0	93.96	33.13

Table 10.5 Active power losses data for 6-bus power system using genetic pattern search algorithm

Active power losses (initial condition)	10,778,370 (W)
Active power losses (after optimization)	8,950,550 (W)
Reduction percentage	16.95826%

Table 10.6 Voltage deviation and voltage stability data for 6-bus power system using genetic pattern search algorithm

Bus no.	Voltage stability index value V/V_0 Ideal value = 1		Voltage deviation of load buses	
	Initial status	After optimization	Initial status kV	After optimization kV
1	0.9009016	0.9152744	4.881641	0.4280500
3	0.8845708	0.9011948	4.934923	0.3137400
5	0.8796755	0.8959399	5.842471	1.386089
Total	2.665148	2.712409	15.65904	2.127879

10.8.2 Case Study 2—IEEE 14-Bus Power Network

The corresponding data to IEEE 14-bus power grid are accessible in Appendix 2. The results for reactive power optimization on the related network will be presented in this section.

10.8.2.1 Reactive Power Optimization Using Particle Swarm Pattern Search Algorithm

The initial circumstances and power flow results for 14-bus power network are presented in Table 10.22. In addition, the corresponding data after the optimization procedure on the same power grid are presented as below Figs. 10.21, 10.22 and 10.23.

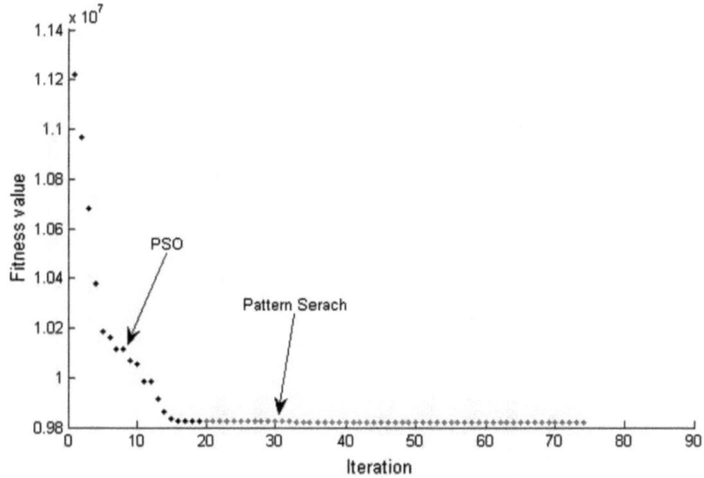

Fig. 10.21 Reactive power optimization trend for 14-bus power system using particle swarm pattern search algorithm

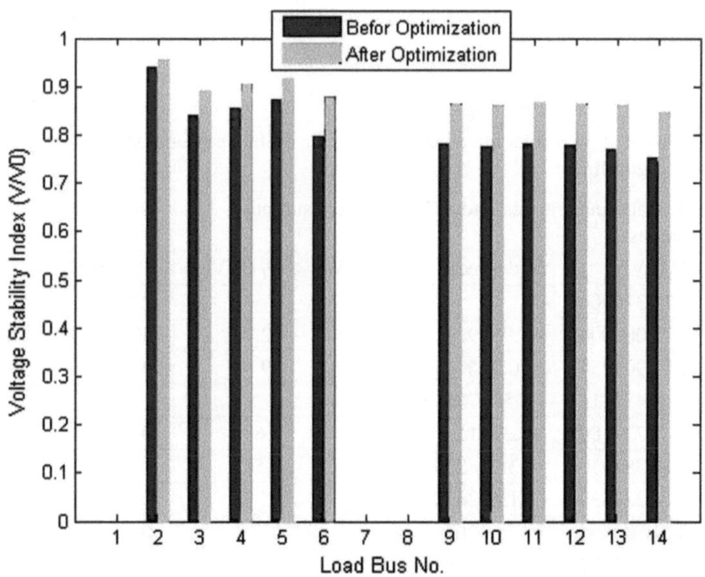

Fig. 10.22 Voltage stability index for 14-bus power system using particle swarm pattern search algorithm

Regarding Tables 10.7, 10.8 and 10.9, active power losses have decreased from 16,710,700 to 9,823,200 (W). Likewise, the total value of voltage stability index has increased from 8.953467 to 9.708002 and total value of voltage deviation of load buses has reduced considerably from 219.7092 to 56.17259 kV.

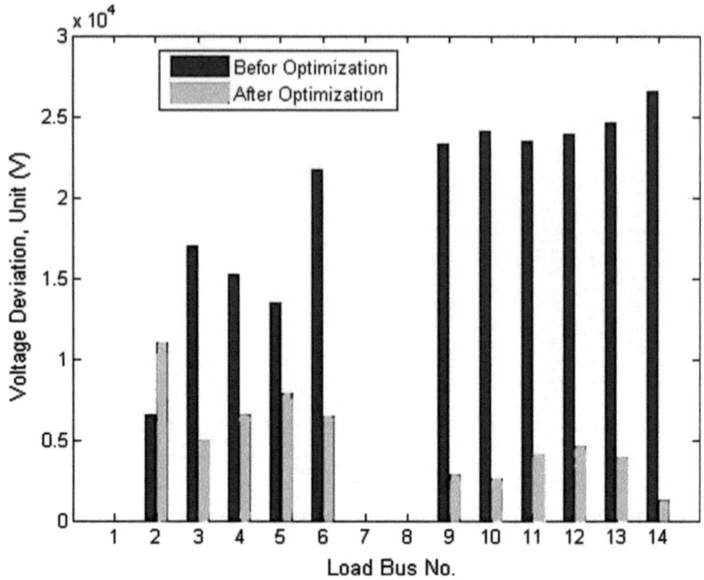

Fig. 10.23 Voltage deviation for 14-bus power system using particle swarm pattern search algorithm

Table 10.7 Power flow results after optimization for 14-bus power system using particle swarm pattern search algorithm

Bus no.	Voltage magnitudes and angles		Load consumption		Injection power	
	V (kV)	θ (degree)	P_l (MW)	Q_l (MVar)	P_G (MW)	Q_G (MVar)
1	125.376906	0	0	0	231.36	45.67
2	121.000004	−3.957964	21.7	12.7	40	16.73
3	114.908254	−10.725941	94.2	19	0	7.436791
4	116.456311	−8.576776	47.8	0	0	0
5	117.737189	−7.335156	7.6	1.6	0	0
6	116.381916	−13.008229	11.2	7.5	0	2.399997
7	114.224630	−11.632678	0	0	0	0
8	114.805564	−11.632674	0	0	0	3.129134
9	112.753677	−13.290581	29.5	16.6	0	0
10	112.559232	−13.530266	9	5.8	0	0
11	114.056146	−13.391523	3.5	1.8	0	0
12	114.573162	−13.866247	6.1	1.6	0	0
13	113.897577	−13.900069	13.5	5.8	0	0
14	111.226337	−14.602206	14.9	5	0	0

Table 10.8 Active power losses data for 14-bus power system using particle swarm pattern search algorithm

Active power losses (initial condition)	16710700 (W)
Active power losses (after optimization)	9823200 (W)
Reduction percentage	41.21612%

Table 10.9 Voltage deviation and voltage stability data for 14-bus power system using particle swarm pattern search algorithm

Bus no.	Voltage stability index value V/V_0 ideal value = 1		Voltage deviation of load buses	
	Initial status	After optimization	Initial status kV	After optimization kV
2	9.409176	0.9552817	6.499059	11
3	8.398882	0.8916201	16.96615	4.975350
4	8.563136	0.9047917	15.24372	6.534758
5	8.728775	0.9160301	13.45913	7.815445
6	7.977896	0.8778006	21.72454	6.421239
9	7.835140	0.8636542	23.25914	2.831756
10	7.761442	0.8598163	24.08241	2.630989
11	7.822556	0.8658360	23.42380	4.112191
12	7.780087	0.8651542	23.91052	4.615991
13	7.718829	0.8609605	24.58441	3.943326
14	7.538748	0.8470567	26.55627	1.291542
Total	8.953467	9.708002	219.7092	56.17259

10.8.2.2 Reactive Power Optimization Using Genetic Pattern Search Algorithm

The initial circumstances and power flow results for 14-bus power network are presented in Table 10.22. In addition, the corresponding data after the optimization procedure on the same power grid are presented as below Figs. 10.24, 10.25 and 10.26.

Referring to Tables 10.10, 10.11 and 10.12, active power losses have decreased remarkably from 16,710,700 to 9,862,619 (W). Similarly, the total value of voltage stability index increased from 8.953467 to 9.655891 and the total value of voltage deviation of load buses has reduced considerably from 219.7092 to 57.82206 kV.

10.8.3 Case Study 3—IEEE 39-Bus New England Power Network

The corresponding data to IEEE 39-bus New England power grid are accessible in Appendix 3. The results for reactive power optimization on the related network will be presented in this section.

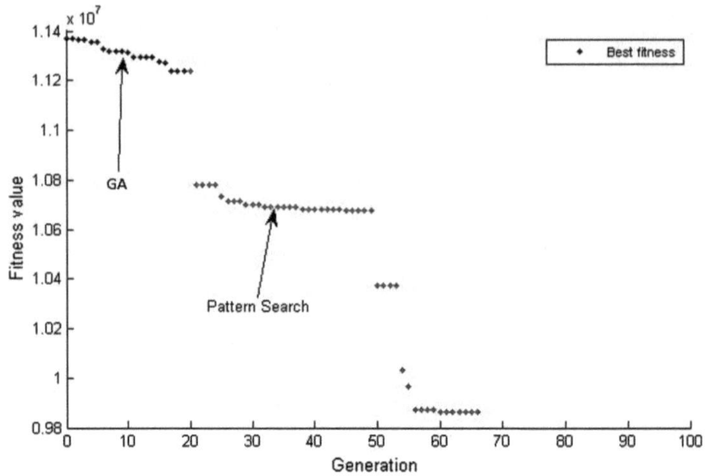

Fig. 10.24 Reactive power optimization trend for 14-bus power system using genetic pattern search algorithm

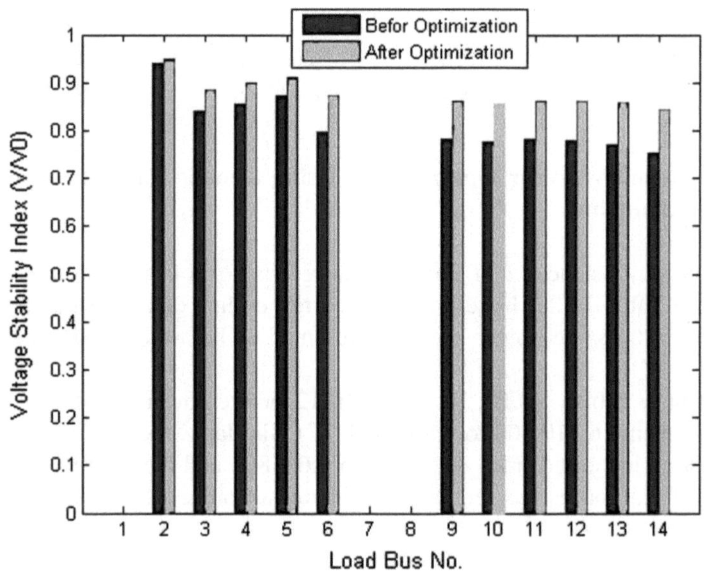

Fig. 10.25 Voltage stability index for 14-bus power system using genetic pattern search algorithm

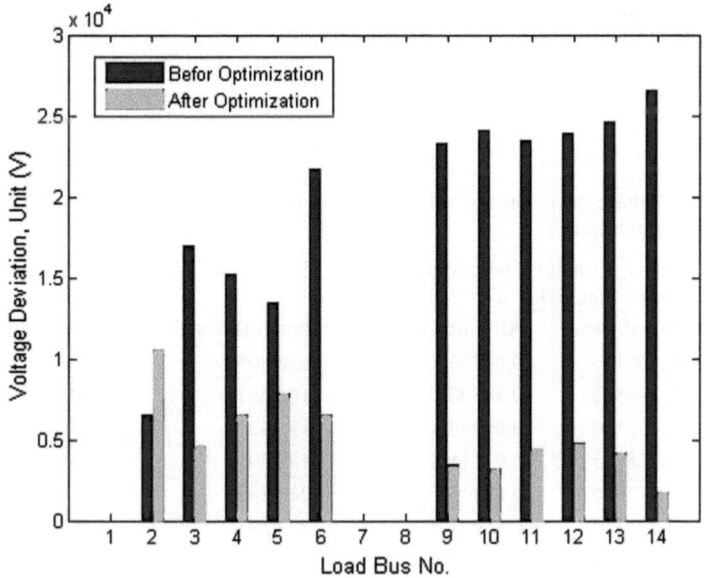

Fig. 10.26 Voltage deviation for 14-bus power system using genetic pattern search algorithm

Table 10.10 Power flow results after optimization for 14-bus power system using genetic pattern search algorithm

Bus no.	Voltage magnitudes and angles		Load consumption		Injection power	
	V (kV)	θ (degree)	P_l (MW)	Q_l (MVar)	P_G (MW)	Q_G (MVar)
1	125.803591	0	0	0	231.42	58.56
2	120.793094	−3.853065	21.7	12.7	40	−0.15
3	115.351591	−10.701274	94.2	19	0	12.74691
4	116.493973	−8.503816	47.8	0	0	0
5	117.785497	−7.265805	7.6	1.6	0	0
6	116.251125	−12.931138	11.2	7.5	0	23.72837
7	114.143428	−11.564184	0	0	0	0
8	114.561700	−11.564181	0	0	0	2.373186
9	112.686674	−13.225312	29.5	16.6	0	0
10	112.480325	−13.463648	9	5.8	0	0
11	113.951417	−13.320288	3.5	1.8	0	0
12	114.445424	−13.791402	6.1	1.6	0	0
13	113.773793	−13.826640	13.5	5.8	0	0
14	111.132858	−14.535142	14.9	5	0	0

Table 10.11 Active power losses data for 14-bus power system using genetic pattern search algorithm

Active power losses (Initial condition)	16710700 (W)
Active power losses (After optimization)	9862619 (W)
Reduction percentage	40.98023%

Table 10.12 Voltage deviation and voltage stability data for 14-bus power system using genetic pattern search algorithm

Bus no.	Voltage stability index value V/V_0 ideal value = 1		Voltage deviation of load buses	
	Initial status	After optimization	Initial status kV	After optimization kV
2	9.409176	0.9476370	6.499059	10.59916
3	8.398882	0.8850283	16.96615	4.580411
4	8.563136	0.8994890	15.24372	6.543957
5	8.728775	0.9111127	13.45913	7.844678
6	7.977896	0.8734883	21.72454	6.543719
9	7.835140	0.8599726	23.25914	3.448204
10	7.761442	0.8561004	24.08241	3.164167
11	7.822556	0.8618414	23.42380	4.445315
12	7.780087	0.8609751	23.91052	4.777684
13	7.718829	0.8568609	24.58441	4.141013
14	7.538748	0.8433849	26.55627	1.733747
Total	8.953467	9.655891	219.7092	57.82206

10.8.3.1 Reactive Power Optimization Using Particle Swarm Pattern Search Algorithm

The initial circumstances and power flow results for 39-bus New England power network are presented in Table 10.13. In addition, the corresponding data after the optimization procedure on the same power grid are presented as below Fig. 10.27.

One of the best criteria that can be taken into consideration as the calculation cost is Number of Function Evaluation (NFE) when it comes to comparing the performance of two or more intelligent algorithms, especially the ones which do not use the gradient of objective functions but the function itself. As it can be inferred from the name, each time the algorithm refers to the objective function to work its value out the amount of NFE increases by one. Eventually, the best optimum point found by every algorithm can be compared to one another in relation to their NFE. In addition, NFE itself can be considered as a stopping criterion as well based on the requirements of the corresponding study.

As mentioned before, one of the benefits of using MATLAB and DIgSILENT together is to take the advantages of built-in toolboxes and functions in both. After the optimization procedure small signal analysis (Eigenvalue or Modal Analysis) has been carried out for the grid in PowerFactory without experiencing

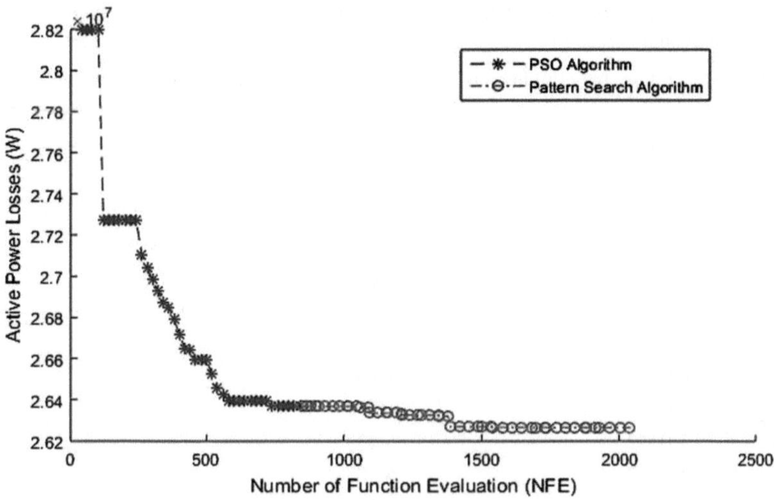

Fig. 10.27 Reactive power optimization trend for 39-bus New England power system using particle swarm pattern search algorithm

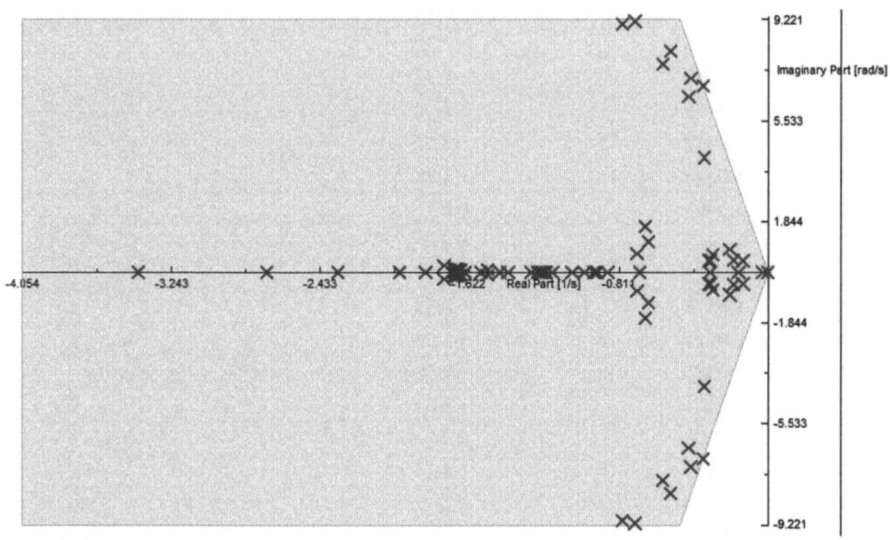

Fig. 10.28 Eigenvalues of 39-bus New England power system without power system stabilizers after optimization

programming difficulties and the result has been shown in Fig. 10.28. The steps how to calculate eigenvalues has been presented in Chap. 11.

According to Tables 10.14, 10.15 and 10.16, active power losses have decreased from 28,194,720 to 26,266,740 (W) via adjusting the terminal voltages of

Table 10.13 Active power losses data of IEEE 39-bus New England power system using particle swarm pattern search algorithm

Bus name	Voltage magnitudes p.u.	Voltage angles deg.	Generators	
			P (MW)	Q (Mvar)
Bus01	1.0474	−8.44	–	–
Bus02	1.0487	−5.75	–	–
Bus03	1.0302	−8.60	–	–
Bus04	1.0039	−9.61	–	–
Bus05	1.0053	−8.61	–	–
Bus06	1.0077	−7.95	–	–
Bus07	0.9970	−10.12	–	–
Bus08	0.9960	−10.62	–	–
Bus09	1.0282	−10.32	–	–
Bus10	1.0172	−5.43	–	–
Bus11	1.0127	−6.28	–	–
Bus12	1.0002	−6.24	–	–
Bus13	1.0143	−6.10	–	–
Bus14	1.0117	−7.66	–	–
Bus15	1.0154	−7.74	–	–
Bus16	1.0318	−6.19	–	–
Bus17	1.0336	−7.30	–	–
Bus18	1.0309	−8.22	–	–
Bus19	1.0499	−1.02	–	–
Bus20	0.9912	−2.01	–	–
Bus21	1.0318	−3.78	–	–
Bus22	1.0498	0.67	–	–
Bus23	1.0448	0.47	–	–
Bus24	1.0373	−6.07	–	–
Bus25	1.0576	−4.36	–	–
Bus26	1.0521	−5.53	–	–
Bus27	1.0377	−7.50	–	–
Bus28	1.0501	−2.01	–	–
Bus29	1.0499	0.74	–	–
Bus30	1.0475	−3.33	250.00	146.16
Bus31	0.9820	0.00	520.81	198.25
Bus32	0.9831	2.57	650.00	205.14
Bus33	0.9972	4.19	632.00	109.91
Bus34	1.0123	3.17	508.00	165.76
Bus35	1.0493	5.63	650.00	212.41
Bus36	1.0635	8.32	560.00	101.18
Bus37	1.0278	2.42	540.00	0.44
Bus38	1.0265	7.81	830.00	22.84
Bus39	1.0300	−10.05	1000.00	88.28

Table 10.14 Power flow results after optimization for 39-bus New England power system using particle swarm pattern search algorithm

Active power losses (Initial condition)	28194720 (W)
Active power losses (After optimization)	26266740 (W)
Reduction percentage	6.84%

Table 10.15 The tap setting of transformers of IEEE 39-bus New England power system after optimization

Bus no.	Voltage magnitudes and angles		Load consumption		Injection power	
	V (kV)	θ (degree)	P_l (MW)	Q_l (Mvar)	P_G (MW)	Q_G (Mvar)
1	364.810659	−6.452421	–	–	–	–
2	370.798179	−3.981883	–	–	–	–
3	372.902266	−6.776672	322.0	2.4	–	–
4	374.141517	−7.781993	500.0	184.0	–	–
5	374.772457	−6.949538	–	–	–	–
6	375.577162	−6.385371	–	–	–	–
7	372.536129	−8.253005	233.8	84.0	–	–
8	371.586660	−8.667435	522.0	176.0	–	–
9	366.270135	−8.296282	–	–	–	–
10	377.511546	−4.188488	–	–	–	–
11	376.869300	−4.942417	–	–	–	–
12	150.734523	−4.933962	7.5	88.0	–	–
13	377.168446	−4.770793	–	–	–	–
14	376.168387	−6.099718	–	–	–	–
15	375.663420	−6.121481	320.0	153.0	–	–
16	376.505858	−4.680447	329.0	32.3	–	–
17	375.344286	−5.652905	–	–	–	–
18	374.016041	−6.467971	158.0	30.0	–	–
19	379.190499	0.018998	–	–	–	–
20	237.505552	−0.769076	628.0	103.0	–	–
21	375.538973	−2.509637	274.0	115.0	–	–
22	378.827835	1.548230	–	–	–	–
23	377.526406	1.366011	247.5	84.6	–	–
24	377.641053	−4.568311	308.6	−92.2	–	–
25	375.556902	−2.870753	224.0	47.2	–	–
26	376.153057	−3.965808	139.0	17.0	–	–
27	374.094746	−5.811787	281.0	75.5	–	–
28	371.978374	−0.616475	206.0	27.6	–	–
29	370.455789	2.031981	283.5	26.9	–	–
30	17.130497	−1.599642	–	–	250.0	−54.17687004

(continued)

Table 10.15 (continued)

Bus no.	Voltage magnitudes and angles		Load consumption		Injection power	
	V (kV)	θ (degree)	P_l (MW)	Q_l (Mvar)	P_G (MW)	Q_G (Mvar)
31	17.926045	0.000000	9.2	4.6	517.55180622	179.62643204
32	17.584661	2.436932	–	–	650.0	82.93753618
33	17.681268	4.575524	–	–	632.0	328.35318619
34	16.557073	4.375941	–	–	508.0	−164.8434324
35	17.899414	6.135797	–	–	650.0	128.77506507
36	18.144109	8.626973	–	–	560.0	46.86633642
37	17.536826	3.495544	–	–	540.0	19.78359724
38	17.563459	8.560209	–	–	830.0	−58.73820034
39	355.349990	−7.996433	1104.0	250.0	1000	−332.33380789

Table 10.16 The data for transmission lines and transformers of IEEE 6-bus standard power system

From bus	To bus	Transformers tap magnitude (p.u.)	Transformers tap tap position
12	11	1.0000	0
12	13	1.0000	0
6	31	1.0350	1
10	32	1.0350	1
19	33	1.0700	2
20	34	1.0000	0
22	35	1.0250	1
23	36	1.0000	–
25	37	1.0250	1
2	30	1.0250	1
29	38	1.0000	0
19	20	1.0300	1

generators and the tap settings of transformers considering the voltage limitations for all the busbars.

10.9 Summary

Investigating active power losses caused in transmission system is of crucial importance in designing power systems as well as their operation and development, of which so many methods have been carried out practically on. Plenty of approaches have been recommended in order to decrease active power losses in power grid in which reactive power optimization is one of the most influential ones. Reactive power optimization investigates the operational condition of reactive power sources in AC power systems in order to have a minimum reactive power current flowing in transmission systems to lessen active power losses associated with it. There are three main parameters on which power system operators have control, as the voltage magnitude of *PV* busbars, reactive power output of compensators and tap setting of transformers which have under-load tap changer facility. It seems quite obvious that the number of controlling parameters can be so many in a real power network so as there are eight dimensions in a 14-bus power grid. In addition, since there is a tough association between plenty of parameters in electric power systems, reactive power optimization is a very nonlinear and non-convex problem comprising discrete and continuous variables simultaneously, which has a lot of local optimum points so that traditional optimization algorithms, most of which are based on the gradient of objective function, lose their performance.

Heuristic optimization methods have proven their performance in such complex optimization problems and it will be efficient to be able to use them in reactive power optimization problem as well. Two heuristic algorithms in combination with direct search optimization methods have been applied to three standard power grids and the results have been presented. Genetic and particle swarm optimization algorithms are great methods in terms of global optimization, whereas pattern search optimization method is remarkable in the respect of local optimization. Therefore, using a combination of a global optimizer and a local one will have so many benefits in its favor. There are plenty of articles, book chapters, books etc. in this respect so that each ones' advantages can be taken.

This chapter tries to focus on the fundamentals and implementation of reactive power optimization problem using MATLAB and DIgSILENT and creating an effectual link between them. It will avail engineers of using the professional tools of PowerFactory DIgSILENT in the respect of electrical engineering as stability analysis, power flow calculations etc. Besides, MATLAB also has got a lot of toolboxes and flexibility while optimizing problems, using artificial intelligence. The method has been presented for a simple standard power grid step by step using particle swarm pattern search algorithm in Chap. 11, while both MATLAB and DIgSILENT files for both approaches have been presented in book attachment as particle swarm pattern search and genetic pattern search for IEEE 6- and 14-bus power grids. In addition, reactive power optimization using built-in particle swarm and pattern search algorithm on IEEE 39-bus New England power system has been depicted in Chap. 11.

Appendices

Appendix 1: IEEE 6-Bus Standard Power System

The single-line diagram of IEEE 6-bus power network has been presented in Fig. 10.29 and the corresponding data are given in Tables 10.17, 10.18 and 10.19.

Appendix 2: IEEE 14-Bus Standard Power System

The single-line diagram of IEEE 14-bus power network has been presented in Fig. 10.30 and the corresponding data are given in Tables 10.20, 10.21 and 10.22.

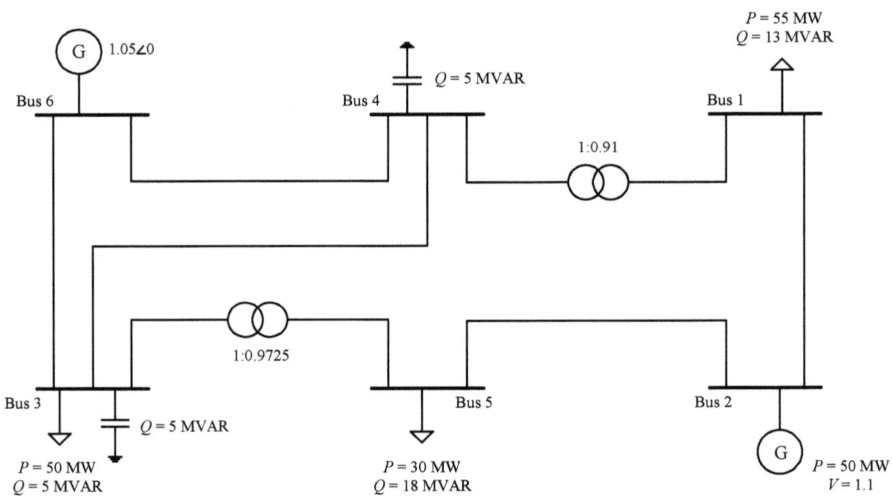

Fig. 10.29 Single-line diagram of IEEE 6-bus standard power system

Table 10.17 The corresponding limitations for control and state variables of IEEE 6-bus standard power system

Line no.	Starting busbar	Ending busbar	Line impedances		Line admittances		Transformer tap settings
			R (Ω)	X (Ω)	G (S)	B (S)	
1	6	3	4.88187	20.55942	0.0109	−0.0460	–
2	6	4	3.1752	14.6853	0.0141	−0.0651	–
3	4	3	3.84993	16.15383	0.0140	−0.0586	–
4	5	2	11.19258	25.4016	0.0145	−0.0330	–
5	2	1	28.69587	41.6745	0.0112	−0.0163	–
6	3	5	0	11.907	0	−0.0840	0.9725
7	4	1	0	5.27877	0	−0.1894	0.9100

Table 10.18 Initial statues and power flow results for IEEE 6-bus standard power system

	Transformer tap settings	Voltage magnitudes of PV busbars		Output capacity of reactive power sources		Load bus voltage magnitudes	Reactive power output of PV busbars
	T_{35}, T_{41}	V_6	V_2	Q_3	Q_4	V_1, V_3, V_5	Q_2, Q_6
Min	0.910	63	69.3	0.0	0.0	56.7	−20
Max	1.110	69.3	72.45	5.0	5.0	69.3	100

Table 10.19 The data for transmission lines and transformers of IEEE 14-bus standard power system

Bus no.	Voltage magnitudes and angles		Load consumption		Injection power	
	V (kV)	θ (degree)	P_l (MW)	Q_l (MVar)	P_G (MW)	Q_G (MVar)
1	58.12	−13.24	55	13	0	0
2	69.3	−4.78	0	0	50	27.63
3	58.07	−12.44	50	5	0	0
4	59.11	−9.85	0	0	0	0
5	57.16	−12.87	0.3	18	0	0
6	66.15	0	0	0	95.79	44.37

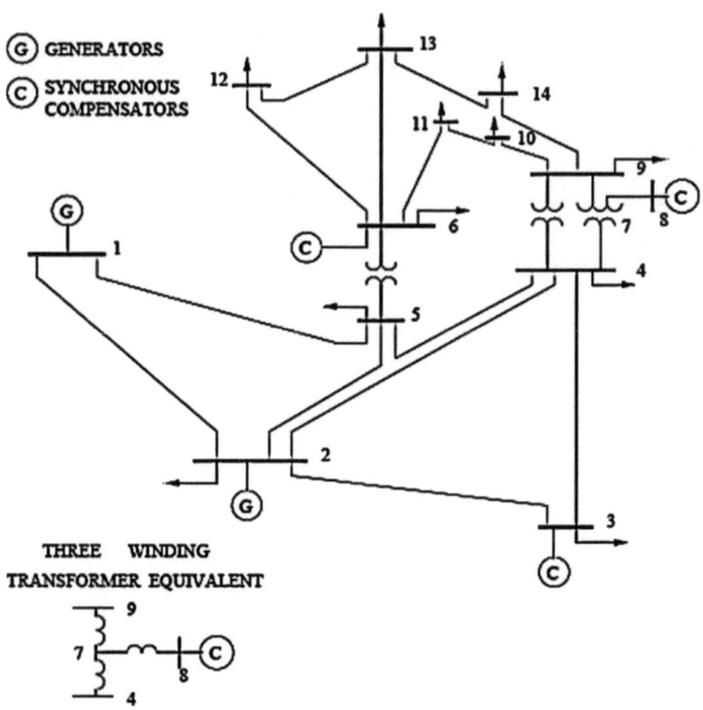

Fig. 10.30 Single-line diagram of IEEE 14-bus standard power system

Table 10.20 The corresponding limitations to control variables of IEEE 14-bus standard power system

Line no.	Starting busbar	Ending busbar	Line impedances		Line admittances		Line capacitances	Transformer tap settings
			R (Ω)	*X* (Ω)	*G* (S)	*B* (S)	*C* (F)	
1	1	2	2.34498	7.15957	0.041315	−0.12614	4.3636e–4	–
2	1	5	6.53763	26.98784	0.0085	−0.0350	4.066e–4	–
3	2	3	5.68579	23.95437	0.0094	−0.0395	3.61983e–4	–
4	2	4	7.03131	21.33472	0.0139	−0.0423	3.0909e–4	–
5	2	5	6.89095	21.03948	0.0141	−0.0429	2.80991e–4	–
6	3	4	8.10821	20.69463	0.0164	−0.0419	2.8595e–4	–
7	4	5	1.61535	5.09531	0.0565	−0.1783	1.05785e–4	–
8	4	7	0	25.30352	0	−0.0395	0	0.978
9	4	9	0	67.29778	0	−0.0149	0	0.969
10	5	6	0	30.49442	0	−0.0328	0	0.932
11	6	11	11.49258	24.0669	0.0162	−0.0338	0	–
12	6	12	14.87211	30.95301	0.0126	−0.0262	0	–
13	6	13	8.00415	15.76267	0.0256	−0.0504	0	–
14	7	8	0	21.31415	0	−0.0469	0	–
15	7	9	0	13.31121	0	−0.0751	0	–
16	9	10	3.84901	10.2245	0.0322	−0.0857	0	–
17	9	14	15.38031	32.71598	0.0118	−0.0250	0	–
18	10	11	9.92805	23.24047	0.0155	−0.0364	0	–
19	12	13	26.73132	24.18548	0.0206	−0.0186	0	–
20	13	14	20.68253	42.11042	0.0094	−0.0191	0	–

Table 10.21 Initial statues and power flow results for IEEE 14-bus standard power system

	Transformer tap settings	Voltage magnitudes of PV busbars		Output capacity of reactive power sources		
	T_{47}, T_{49}, T_{56}	V_1	V_2	Q_3	Q_6	Q_8
Min	0.910	110	110	0.0	−6	−6
Max	1.110	126.5	121	40	24	24

Table 10.22 Data of lines of IEEE 39-bus New England power system (100 MVA, 60 Hz) [13]

Bus no.	Voltage magnitudes and angles		Load consumption		Injection power	
	V (kV)	θ (°)	P_l (MW)	Q_l (MVar)	P_G (MW)	Q_G (MVar)
1	110	0	0	0	242.05	102.32
2	102.39	−6.54	21.7	12.7	40	50
3	91.83	−16.11	94.2	19	0	0
4	93.67	−12.60	47.8	0	0	0
5	95.53	−10.60	7.6	1.6	0	0
6	87.10	−19.66	11.2	7.5	0	0
7	88.14	−17.72	0	0	0	0
8	88.14	−17.72	0	0	0	0
9	85.51	−20.66	29.5	16.6	0	0
10	84.68	−21.02	9	5.8	0	0
11	85.36	−20.58	3.5	1.8	0	0
12	84.88	−21.21	6.1	1.6	0	0
13	84.19	−21.35	13.5	5.8	0	0
14	82.17	−22.85	14.9	5	0	0

Appendix 3: IEEE 39-Bus New England Power System

The Single-line diagram of IEEE 39-bus New England power system has been shown in Fig. 10.31 and the corresponding data are given in Tables 10.23, 10.24, 10.25, 10.26, 10.27 and 10.28. The nominal frequency of the New England transmission system is 60 Hz and the main voltage level is 345 kV (nominal voltage). For nodes at a different voltage level, following nominal voltages have been assumed for the PowerFactory model: Bus 12–138 kV, Bus 20–230 kV, Bus 30 and Bus 38–16.5 kV [13].

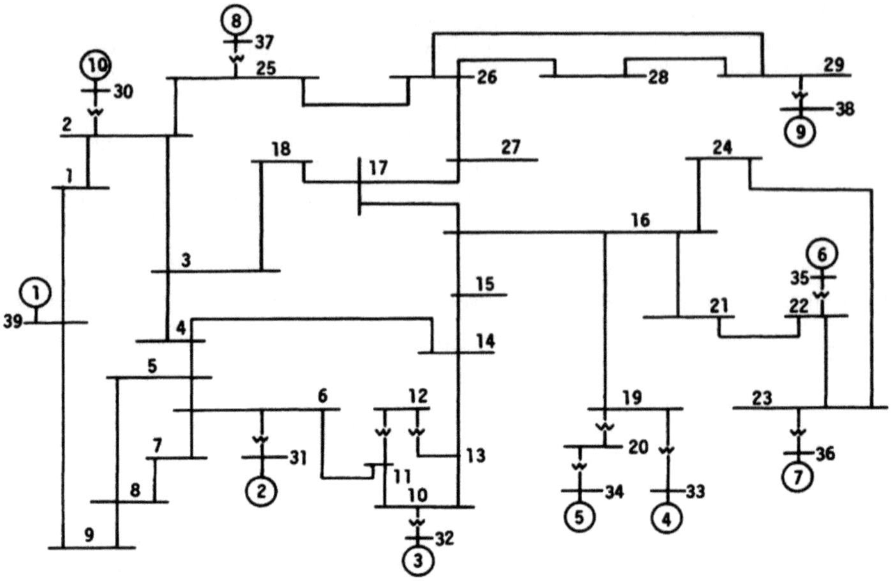

Fig. 10.31 Single-line diagram of IEEE 39-bus New England power system [13]

Table 10.23 Load demands of IEEE 39-bus New England power system [13]

Line	From bus	To bus	*R* in p.u.	*X* in p.u.	*B* in p.u.
Line 01–02	Bus 01	Bus 02	0.0035	0.0411	0.6987
Line 01–39	Bus 01	Bus 39	0.0010	0.0250	0.7500
Line 02–03	Bus 02	Bus 03	0.0013	0.0151	0.2572
Line 02–25	Bus 02	Bus 25	0.0070	0.0086	0.1460
Line 03–04	Bus 03	Bus 04	0.0013	0.0213	0.2214
Line 03–18	Bus 03	Bus 18	0.0011	0.0133	0.2138
Line 04–05	Bus 04	Bus 05	0.0008	0.0128	0.1342
Line 04–14	Bus 04	Bus 14	0.0008	0.0129	0.1382
Line 05–06	Bus 05	Bus 06	0.0002	0.0026	0.0434
Line 05–08	Bus 05	Bus 08	0.0008	0.0112	0.1476
Line 06–07	Bus 06	Bus 07	0.0006	0.0092	0.1130
Line 06–11	Bus 06	Bus 11	0.0007	0.0082	0.1389
Line 07–08	Bus 07	Bus 08	0.0004	0.0046	0.0780
Line 08–09	Bus 08	Bus 09	0.0023	0.0363	0.3804
Line 09–39	Bus 09	Bus 39	0.0010	0.0250	1.2000
Line 10–11	Bus 10	Bus 11	0.0004	0.0043	0.0729
Line 10–13	Bus 10	Bus 13	0.0004	0.0043	0.0729
Line 13–14	Bus 13	Bus 14	0.0009	0.0101	0.1723

(continued)

Table 10.23 (continued)

Line	From bus	To bus	*R* in p.u.	*X* in p.u.	*B* in p.u.
Line 14–15	Bus 14	Bus 15	0.0018	0.0217	0.3660
Line 15–16	Bus 15	Bus 16	0.0009	0.0094	0.1710
Line 16–17	Bus 16	Bus 17	0.0007	0.0089	0.1342
Line 16–19	Bus 16	Bus 19	0.0016	0.0195	0.3040
Line 16–21	Bus 16	Bus 21	0.0008	0.0135	0.2548
Line 16–24	Bus 16	Bus 24	0.0003	0.0059	0.0680
Line 17–18	Bus 17	Bus 18	0.0007	0.0082	0.1319
Line 17–27	Bus 17	Bus 27	0.0013	0.0173	0.3216
Line 21–22	Bus 21	Bus 22	0.0008	0.0140	0.2565
Line 22–23	Bus 22	Bus 23	0.0006	0.0096	0.1846
Line 23–24	Bus 23	Bus 24	0.0022	0.0350	0.3610
Line 25–26	Bus 25	Bus 26	0.0032	0.0323	0.5130
Line 26–27	Bus 26	Bus 27	0.0014	0.0147	0.2396
Line 26–28	Bus 26	Bus 28	0.0043	0.0474	0.7802
Line 26–29	Bus 26	Bus 29	0.0057	0.0625	1.0290
Line 28–29	Bus 28	Bus 29	0.0014	0.0151	0.2490

Table 10.24 Generator dispatch of IEEE 39-bus New England power system [13]

No.	Load	Bus	*P* (MW)	*Q* (Mvar)
1	Load 03	Bus 03	322.0	2.4
2	Load 04	Bus 04	500.0	184.0
3	Load 07	Bus 07	233.8	84.0
4	Load 08	Bus 08	522.0	176.0
5	Load 12	Bus 12	7.5	88.0
6	Load 15	Bus 15	320.0	153.0
7	Load 16	Bus 16	329.0	32.3
8	Load 18	Bus 18	158.0	30.0
9	Load 20	Bus 20	628.0	103.0
10	Load 21	Bus 21	274.0	115.0
11	Load 23	Bus 23	247.5	84.6
12	Load 24	Bus 24	308.6	−92.2
13	Load 25	Bus 25	224.0	47.2
14	Load 26	Bus 26	139.0	17.0
15	Load 27	Bus 27	281.0	75.5
16	Load 28	Bus 28	206.0	27.6
17	Load 29	Bus 29	283.5	26.9
18	Load 31	Bus 31	9.2	4.6
19	Load 39	Bus 39	1104.0	250.0

Table 10.25 Data of transformers (100 MVA) of IEEE 39-bus New England power system [13]

Generator	Bus	Bus type	*P* in MW	*V* in p.u.
G 01	Bus 39	PV	1000.0	1.0300
G 02	Bus 31	Slack	N.A.	0.9820
G 03	Bus 32	PV	650.0	0.9831
G 04	Bus 33	PV	632.0	0.9972
G 05	Bus 34	PV	508.0	1.0123
G 06	Bus 35	PV	650.0	1.0493
G 07	Bus 36	PV	560.0	1.0635
G 08	Bus 37	PV	540.0	1.0278
G 09	Bus 38	PV	830.0	1.0265
G 10	Bus 30	PV	250.0	1.0475

Table 10.26 Data of generators (100 MVA) of IEEE 39-bus New England power system [13]

From bus	To bus	*R* (p.u.)	*X* (p.u.)	Transformers tap magnitude (p.u.)	Transformers tap angle (degree)	Lower and upper limits of taps
12	11	0.0016	0.0435	1.0060	0.00	$1 \pm (1 \times 0.006)$
12	13	0.0016	0.0435	1.0060	0.00	$1 \pm (1 \times 0.006)$
6	31	0.0000	0.0250	1.0700	0.00	$1 \pm (2 \times 0.035)$
10	32	0.0000	0.0200	1.0700	0.00	$1 \pm (2 \times 0.035)$
19	33	0.0007	0.0142	1.0700	0.00	$1 \pm (2 \times 0.035)$
20	34	0.0009	0.0180	1.0090	0.00	$1 \pm (1 \times 0.009)$
22	35	0.0000	0.0143	1.0250	0.00	$1 \pm (1 \times 0.025)$
23	36	0.0005	0.0272	1.0000	0.00	–
25	37	0.0006	0.0232	1.0250	0.00	$1 \pm (1 \times 0.025)$
2	30	0.0000	0.0181	1.0250	0.00	$1 \pm (1 \times 0.025)$
29	38	0.0008	0.0156	1.0250	0.00	$1 \pm (1 \times 0.025)$
19	20	0.0007	0.0138	1.0600	0.00	$1 \pm (2 \times 0.03)$

Table 10.27 Data of AVRs of IEEE 39-bus New England power system [13]

Unit no.	H (s)	R_a (p.u.)	x'_d (p.u.)	x'_q (p.u.)	x_d (p.u.)	x_q (p.u.)	T'_{d0} (s)	T'_{q0} (s)	x_l (p.u.)	x'' (p.u.)	T''_{d0} (s)	T''_{q0} (s)
1	500.0	0.0000	0.0060	0.0080	0.0200	0.0190	7.000	0.700	0.0030	0.0040	0.050	0.035
2	30.3	0.0000	0.0697	0.1700	0.2950	0.2820	6.560	1.5000	0.0350	0.0500	0.050	0.035
3	35.8	0.0000	0.0531	0.0876	0.2495	0.2370	5.700	1.5000	0.0304	0.0450	0.050	0.035
4	28.6	0.0000	0.0436	0.1660	0.2620	0.2580	5.690	1.5000	0.0295	0.0350	0.050	0.035
5	26.0	0.0000	0.1320	0.1660	0.6700	0.6200	5.400	0.4400	0.0540	0.0890	0.050	0.035
6	34.8	0.0000	0.0500	0.0814	0.2540	0.2410	7.300	0.4000	0.0224	0.0400	0.050	0.035
7	26.4	0.0000	0.0490	0.1860	0.2950	0.2920	5.660	1.5000	0.0322	0.0440	0.050	0.035
8	24.3	0.0000	0.0570	0.0911	0.2900	0.2800	6.700	0.4100	0.0280	0.0450	0.050	0.035
9	34.5	0.0000	0.0570	0.0587	0.2106	0.2050	4.790	1.9600	0.0298	0.0450	0.050	0.035
10	42.0	0.0000	0.0310	0.0500	0.1000	0.0690	10.200	0.0000	0.0125	0.0250	0.050	0.035

Table 10.28 Initial statues and power flow results for IEEE 39-bus New England power system [13]

Unit no.	$K_a = K_A$	$T_a = T_A$	$V_{r\min} = V_{R\min}$	$V_{r\max} = V_{R\max}$	$K_e = K_E$	$T_e = T_E$	$K_f = K_F$	$T_f = T_F$	$S_{e1} = C_1$	$S_{e2} = C_2$	$E_1 = E_{X1}$	$E_2 = E_{X2}$
2	6.2	0.05	−1.0	1.0	−0.6330	0.405	0.0570	0.500	0.660	0.880	3.036437	4.048583
3	5.0	0.06	−1.0	1.0	−0.0198	0.500	0.0800	1.000	0.130	0.340	2.342286	3.123048
4	5.0	0.06	−1.0	1.0	−0.0525	0.500	0.0800	1.000	0.080	0.314	2.868069	3.824092
5	40.0	0.02	−10.0	10.0	1.0000	0.785	0.0300	1.000	0.070	0.910	3.926702	5.235602
6	5.0	0.02	−1.0	1.0	−0.0419	0.471	0.0754	1.246	0.064	0.251	3.586801	4.782401
7	40.0	0.02	−6.5	6.5	1.0000	0.730	0.0300	1.000	0.530	0.740	2.801724	3.735632
8	5.0	0.02	−1.0	1.0	−0.0470	0.528	0.0854	1.260	0.072	0.282	3.191489	4.255319
9	40.0	0.02	−10.5	10.5	1.0000	1.400	0.0300	1.000	0.620	0.850	4.256757	5.675676
10	5.0	0.06	−1.0	1.0	−0.0485	0.250	0.0400	1.000	0.080	0.260	3.546099	4.728132

References

1. S.R. Singiresu, Engineering Optimization - Theory and Practice, Hoboken, New Jersey: John Wiley & Sons, Inc., 2009.
2. J. Zhu, Reactive Power Optimization, in Optimization of Power System Operation, New Jersey, John Wiley & Sons, pp. 409–454, 2009.
3. T.J. Miller, Reactive Power Coordination, in Reactive Power Control in Electric Systems, New York, John Wiley & Sons, pp. 353–363, 1982.
4. P. Kundur, Power System Stability and Control, New York: McGraw-Hill, 1994.
5. J. Machowski, J.W. Bialek, J.R. Bumby, Power System Dynamics, John Wiley & Sons, 2008.
6. H. Saadat, Power System Analysis, New York: McGraw-Hill, 1999.
7. X. Hugang, C. Haozhong, L. Haiyu, Optimal Reactive Power Flow Incorporating Static Voltage Stability Based on Multi-Objective Adaptive Immune Algorithm, Energy Conversion and Management, vol. 49, pp. 1175–1181, 2008.
8. T. Niknam, M.R. Narimani, R. Azizipanah Abarghooee, B. Bahmani Firouzi, Multi-Objective Optimal Reactive Power Dispatch and Voltage Control: A New Opposition-Based Self-Adaptive Modified Gravitational Search Algorithm, IEEE Systems Journal, vol. 7, pp. 742–753, 2013.
9. S.A.H. Soliman, H.A.A. Mantawy, Modern Optimization Techniques with Applications in Electric Power Systems, New York Heidelberg Dordrecht London: Springer, 2012.
10. A.I.F. Vaz, L. Vicente, A Particle Swarm Pattern Search Method for Bound Constrained Global Optimization, Department of Production and Systems, School of Engineering, University of Minho, Campus of Gualtar, Braga, Portugal, 2006.
11. Wikipedia, the Free Encyclopedia, https://en.wikipedia.org/wiki/Pattern_search_(optimization).
12. M.A. Abramson, C. Audet, J. Dennis, Generalized Pattern Search Algorithms - Unconstrained and Constrained Cases, IMA Workshop - Optimization in Simulation Based Models.
13. DIgSILENT PowerFactory Version 15 User Manual, Gomaringen, Germany: DIgSILENT GmbH, 2014.

Chapter 11
Reactive Power Optimization Using MATLAB and DIgSILENT

Naser Mahdavi Tabatabaei, Ali Jafari Aghbolaghi, Narges Sadat Boushehri and Farid Hojjati Parast

Abstract There are plenty of methods to implement reactive power optimization, such as coding in MATLAB on its own or DIgSILENT, both of which are powerful pieces of software in the electrical engineering respect. However, on one hand MATLAB has a lot of toolboxes, functions, flexibility etc. when it comes to be used in implementation of artificial intelligence and heuristic algorithms. On the other hand, DIgSILENT is such a powerful tool while studying power network stability, power flow calculations, ready-to-use equipment blocks etc., which provides operators with the simplicity of carrying out their studies. Therefore, it would be very efficient to be available of using both pieces of software in parallel in order to take each one's advantages. In this method, calculations related to power network are executed in DIgSILENT, the ones related to heuristic algorithms are carried out in MATLAB. Although both of them are able to put the whole procedure into practice solely, the point is to take the redeeming features of both to do the process as fast, simple, accurate as possible.

N. Mahdavi Tabatabaei (✉)
Electrical Engineering Department, Seraj Higher Education Institute, Tabriz, Iran
e-mail: n.m.tabatabaei@gmail.com

A. Jafari Aghbolaghi · F. Hojjati Parast
Andishmand Shomal Gharb Engineering Consultancy, Zanjan, Iran
e-mail: ali.jafari.860@gmail.com

F. Hojjati Parast
e-mail: hojjatifarid@gmail.com

N.S. Boushehri
Department of Management, Taba Elm International Institute, Tabriz, Iran
e-mail: nargesboush@yahoo.com

N. Mahdavi Tabatabaei et al. (eds.), *Reactive Power Control in AC Power Systems*,
Power Systems, DOI 10.1007/978-3-319-51118-4_11

11.1 Introduction

There are plenty of methods to implement reactive power optimization, such as coding in MATLAB on its own or DIgSILENT, both of which are powerful pieces of software in the electrical engineering respect. However, on one hand MATLAB has a lot of toolboxes, functions, flexibility etc. when it comes to be used in implementation of artificial intelligence and heuristic algorithms. On the other hand, DIgSILENT is such a powerful tool while studying power network stability, power flow calculations, ready-to-use equipment blocks etc., which provides operators with the simplicity of carrying out their studies. Therefore, it would be very efficient to be available of using both pieces of software in parallel in order to take each one's advantages. In this method, calculations related to power network are executed in DIgSILENT, the ones related to heuristic algorithms are carried out in MATLAB. Although both of them are able to put the whole procedure into practice solely, the point is to take the redeeming features of both to do the process as fast, simple, accurate as possible.

The reason why it is beneficial to employ MATLAB and DIgSILENT both together to optimize reactive power in AC power systems is to simplify some basic approaches which should be done, although they are not the very case on which need to be concentrated. Therefore, using some ready-to-use tools and techniques will avail researchers of skipping the affairs that do not need to be spent much time on. For instance, in the reactive power optimization problem power flow calculations, stability analysis, modelling equipment etc. are not inevitable to use, whereas they are readily accessible without programming difficulties in DIgSILENT software. Thus, avoiding some routine and basic affairs, researchers could focus their energy on the further studies. This section will present an efficient approach using both MATLAB and DIgSILENT software in order to put reactive power optimization into practice so that both pieces of software will work in parallel until the optimum point has been found.

11.2 How to Implement Reactive Power Optimization Using MATLAB and DIgSILENT

The theoretical aspect of reactive power optimization has been presented elaborately in Chap. 10. In this section, DIgSILENT will do power flow calculations and other processes related to optimization procedure will be carried out by MATLAB. One of the best ways to couple MATLAB and DIgSILENT is to use text files to exporting data, importing data and also to issue orders to the both applications. It should be noted that it is assumed the reader has an intermediate level of knowledge of using DIgSILENT and MATLAB.

Having traced the power network in DIgSILENT and coded the optimization algorithm in MATLAB, the general procedure of reactive power optimization coupling DIgSILENT and MATLAB will be as follows:

1. At the beginning, control variables are initiated by optimization algorithm within feasible bounds and then are written in a text file named "ToDIg.txt", while DIgSILENT is already waiting to a command from MATLAB to do the power flow calculations.
2. MATLAB sends a command to run the power flow to DIgSILENT via a text file named "Couple.txt", which contains either "0" or "1" as stop or start order and/or "2" noting end of the procedure (completely stop DIgSILENT). Since the commands will be commuted between DIgSILENT and MATLAB steadily in a text file as "Couple.txt", if "0" is stop command to DIgSILENT, it is start command to MATLAB and/or vice versa. In this chapter, 1 is used to start DIgSILENT and 0 to stop it.
3. After getting run command, DIgSILENT reads the new control variables created and written by MATLAB into "ToDIg.txt" and applies them to the power network and implements power flow calculations. After finishing power flow, DIgSILENT writes new state values as voltage magnitudes and angles of all busbars into another text file named "LoadFlow.txt". Then it writes "0" in "Couple.txt" to send an order as to start MATLAB to continue the procedure.
4. MATLAB starts to calculate and investigate the fitness function. If the stopping criteria have been reached, the optimization procedure gets over and the best coordinates found so far will be chosen as the best values for the control variables, if not, the optimization algorithm produces new set of control variables and writes in "ToDIg.txt" to send to DIgSILENT. Then the procedure continues from step 2 again.

The simple flowchart of above-presented approach has been shown in Fig. 11.1 to elaborate the trend. It will be presented step by step in the next section as elaborate as possible considering the fact that reader has intermediate level of knowledge using MATLAB and DIgSILENT.

Hence, optimizing reactive power in order to reduce active power losses in AC power systems based on the approach presented in this chapter has two different stages. One is preparing a power network in DIgSILENT software and the other is coding the optimization algorithm in MATLAB software, both of which will be explained step by step in the next phases of the chapter.

11.2.1 First Step—Tracing a 6-Bus IEEE Standard Power Network in DIgSILENT

The first step toward putting reactive power optimization into practice using MATLAB and DIgSILENT is to have a network ready in DIgSILENT to apply the

Fig. 11.1 The flowchart of reactive power optimization using DIgSILENT and MATLAB

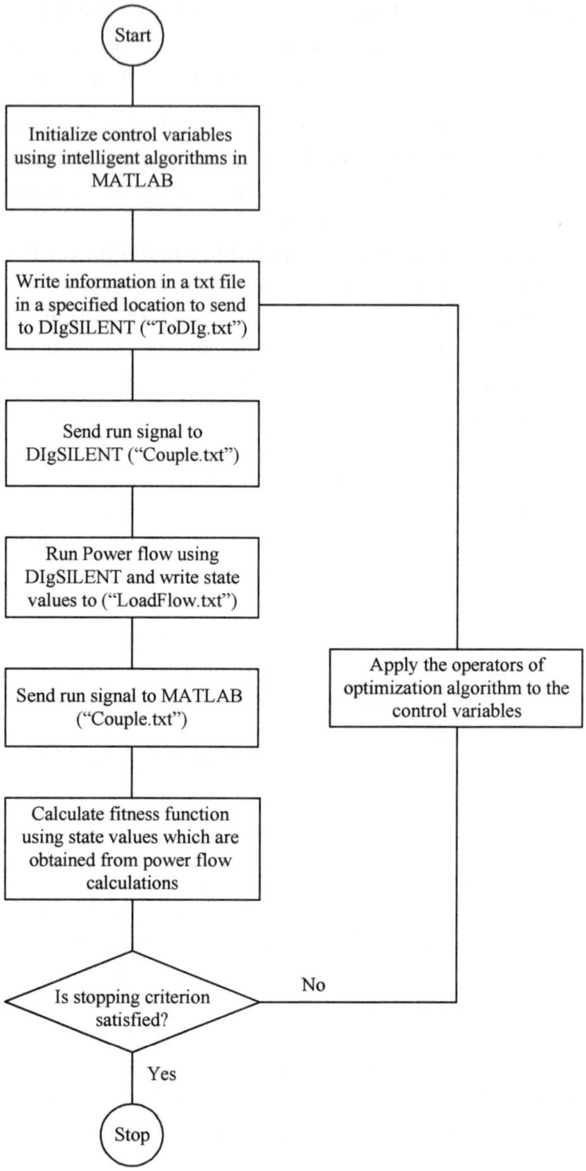

optimization algorithm on. The IEEE 6-bus standard system given in the appendices section of Chap. 10 is used in order to elaborate the approach. The power grid has already traced for you and the corresponding file has been presented in the book attachments, although a brief explanation how to create a project is presented. The base values are considered as follows:

$$V_{base} = 63 \text{ kV} \tag{11.1}$$

$$S_{base} = 100 \text{ MVA} \tag{11.2}$$

11.2.1.1 Creating a New Project in DIgSILENT

To build a new project, the instruction below should be followed:

Open DIgSILENT software, go to File Menu → New → Project (Fig. 11.2), then enter the project name in the matching field (Fig. 11.3), then leave all other settings as default and click on OK. Start tracing the 6-bus power grid (Fig. 11.4) with the given data in appendices section of Chap. 10 (use the software's user manual for more information).

Figure 11.5 shows the 6-bus power grid after being traced in DIgSILENT, its file has been presented in the book attachments as well and it can be imported to the software in order to avoid drawing it again. After tracing the basic network, there are some extra steps which should be done.

11.2.1.2 How to Enter Operational Limitations of Generators

To enter operational limitations of generators, double click on bus-6 generator (Fig. 11.6) → go to "Load Flow" tab and tick "Use Limits Specified in Type" in Reactive Power Operational Limits section (Fig. 11.7). Then go to "Basic Data" tab → "Right Arrow" in front of Type (Fig. 11.8), then enter the values of Nominal

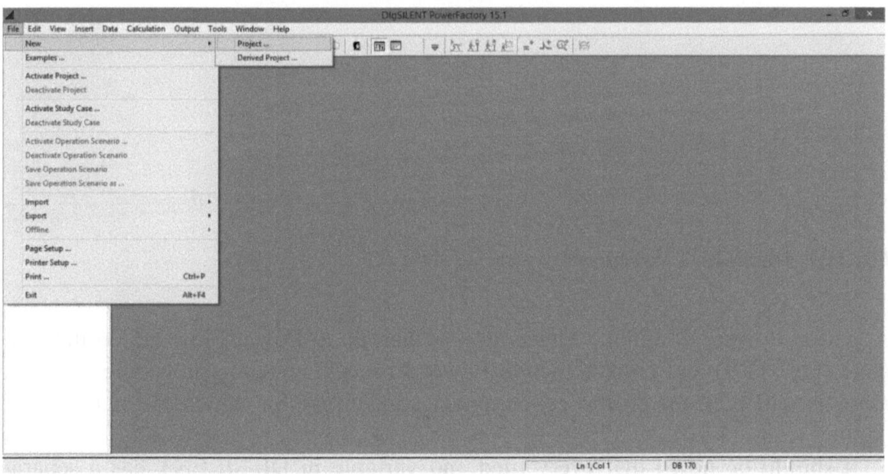

Fig. 11.2 DIgSILENT PowerFactory—creating a new project [1]

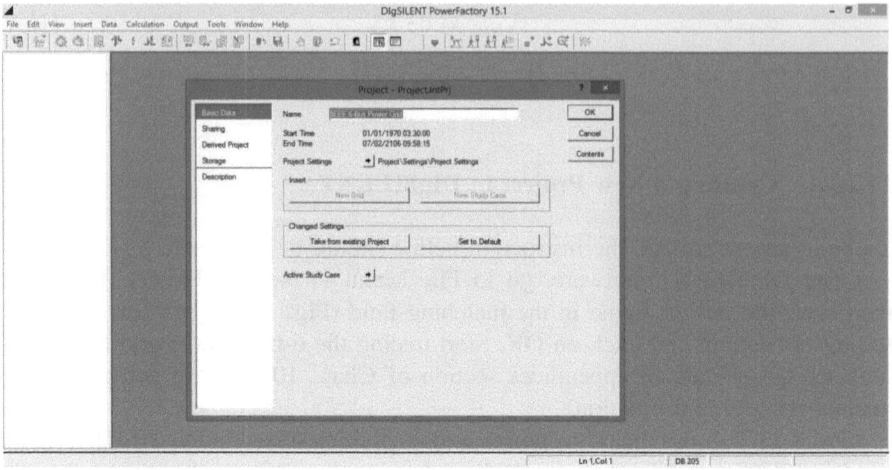

Fig. 11.3 DIgSILENT PowerFactory—assigning a name to the project [1]

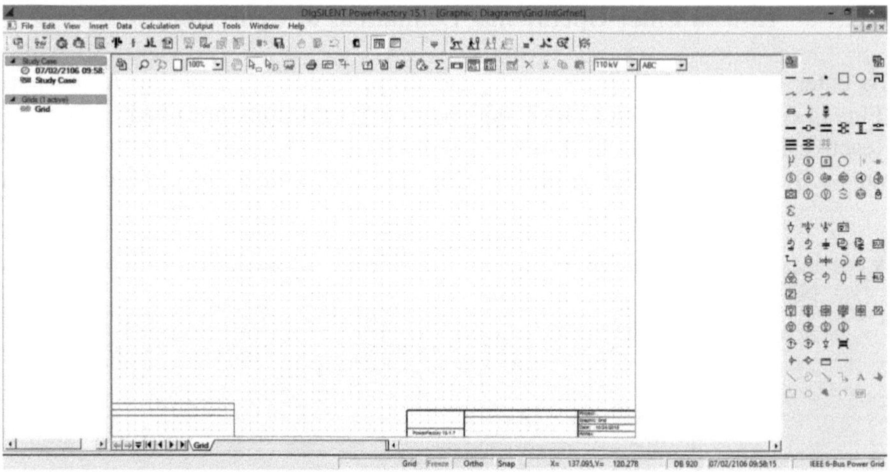

Fig. 11.4 DIgSILENT PowerFactory—tracing sheet [1]

Apparent Power, Nominal Voltage etc. → then go to Power Flow tab on the same page (Fig. 11.9) and enter Reactive Power Limits. The same procedure as generators should be done to the synchronous condensers for which the parameter is "qgini" (Fig. 11.10).

It should be noted that every filed and variable in DIgSILENT has a separate parameter which will be at work when it comes to be used in DIgSILENT

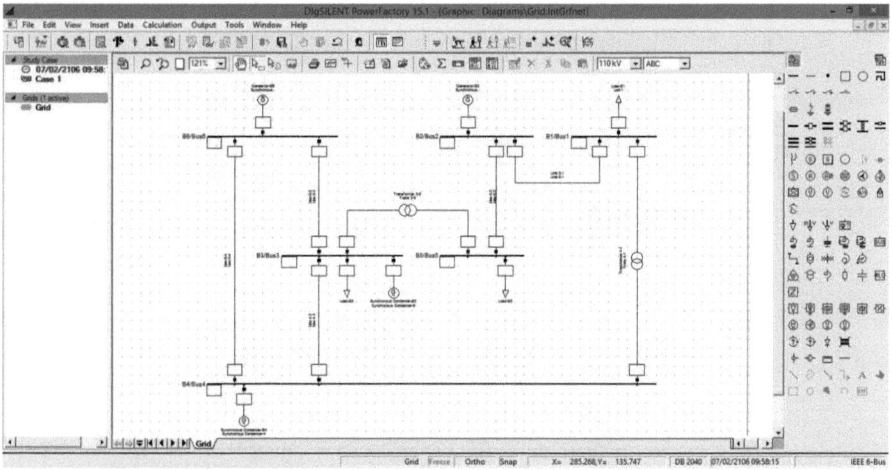

Fig. 11.5 DIgSILENT PowerFactory—IEEE 6-bus power grid [1]

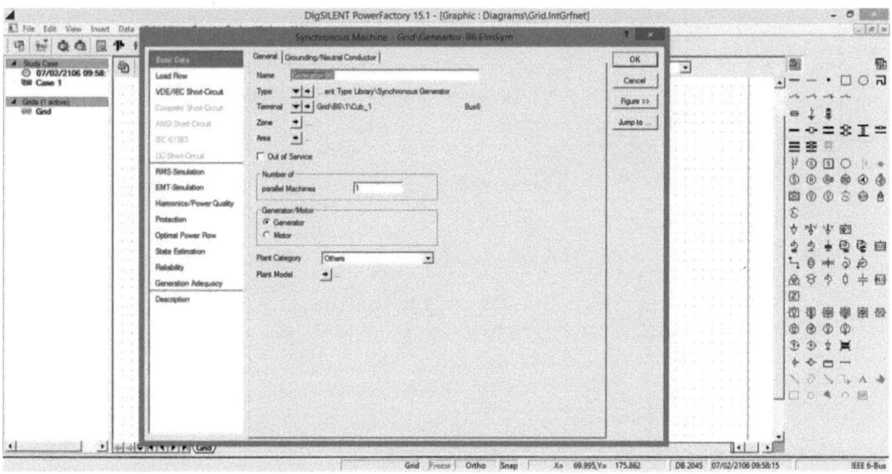

Fig. 11.6 DIgSILENT PowerFactory—the main page of synchronous generators [1]

Programming Language (DPL) to assign and/or read the values of different parameters. To work out what the parameter for a specific field is, just hold mouse pointer on the field then the parameter will be shown, for example, the voltage magnitude parameter for generators is "usetp" (Fig. 11.11).

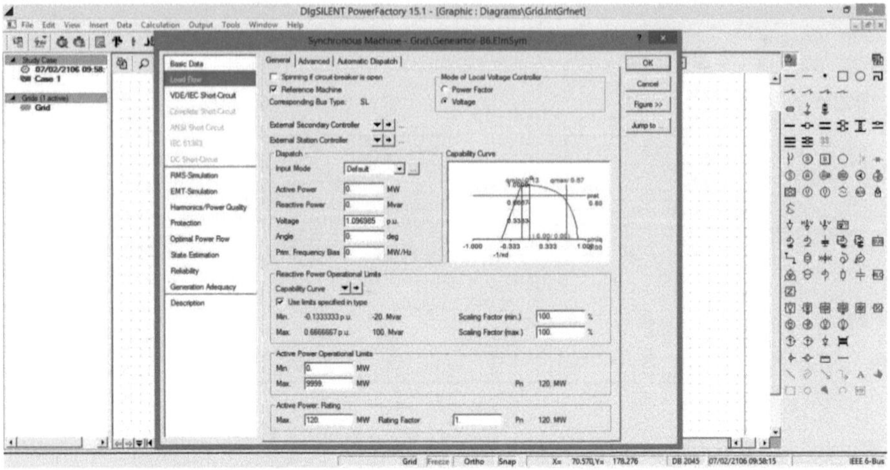

Fig. 11.7 DIgSILENT PowerFactory—the main page of synchronous generators, "Load Flow" tab [1]

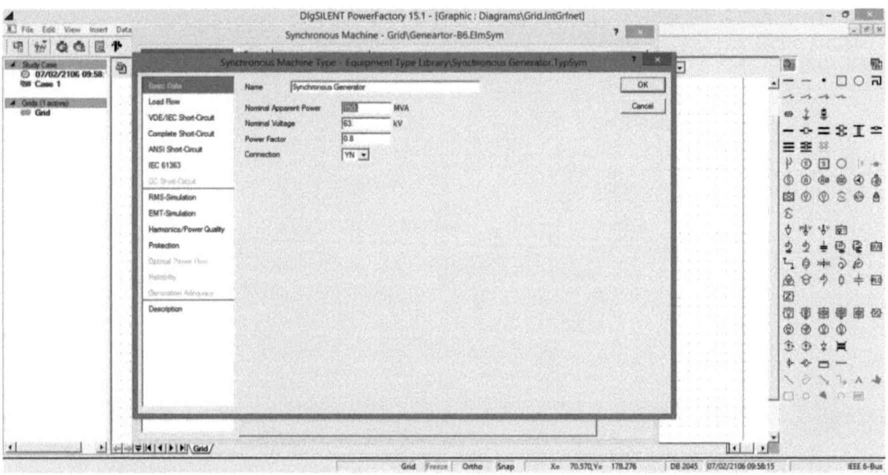

Fig. 11.8 DIgSILENT PowerFactory—adjusting nominal apparent power of synchronous generators [1]

11.2.1.3 How to Adjust Tap Settings for Transformers

In order to set initial tap settings for the transformers to be in an appropriate form for controlling by DPL, double click on a transformer → click on "Right Arrow" in front of "Type" (Fig. 11.12) → enter the values for "Rated Power", "Rated Voltage" etc. (Fig. 11.13) → then go to VDE/IEC "Short-Circuit" tab (Fig. 11.14)

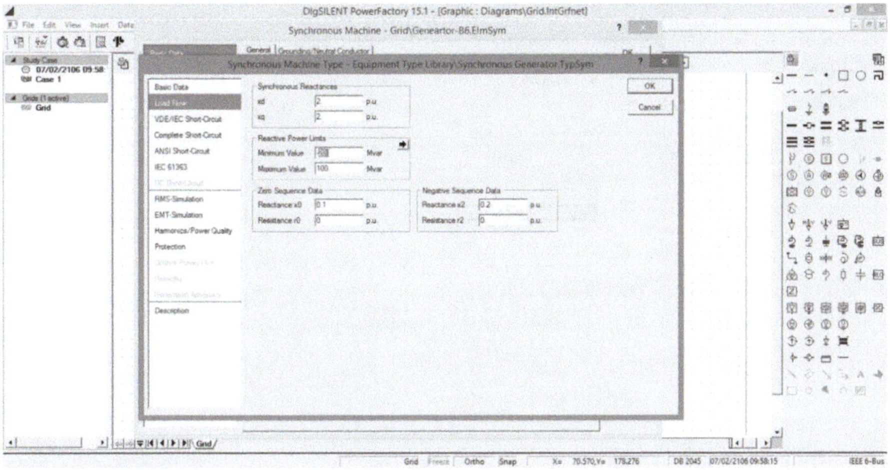

Fig. 11.9 DIgSILENT PowerFactory—adjusting reactive power limitations of synchronous generators [1]

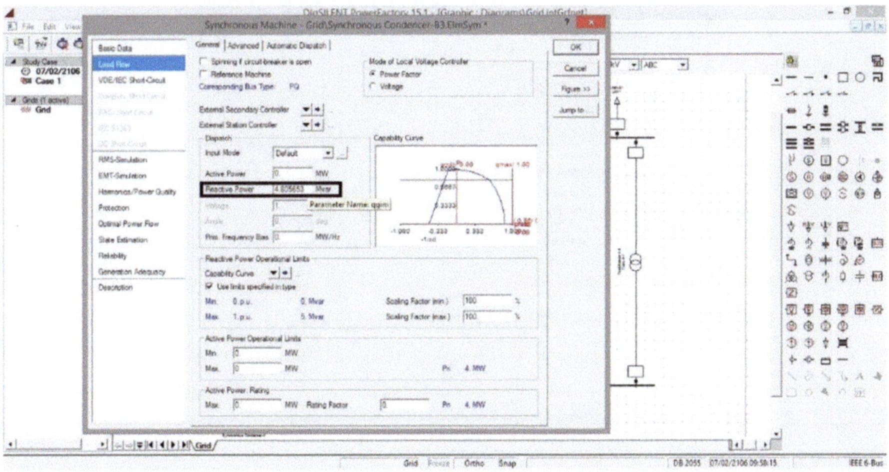

Fig. 11.10 DIgSILENT PowerFactory—adjusting voltage magnitudes of synchronous generators [1]

and enter the values for "Additional Voltage Per Tap" = 0.0001, "Neutral Position" = 10,000, "Minimum Position" = 9100 and "Maximum Position" = 11,100 according to 6-bus IEEE power grid then click OK → go to "Power Flow" tab (Fig. 11.15), you will see the tap changer data that you have already assigned. The parameter for tap setting is "nntap" which is shown when holding mouse pointer on "Tap Position" field.

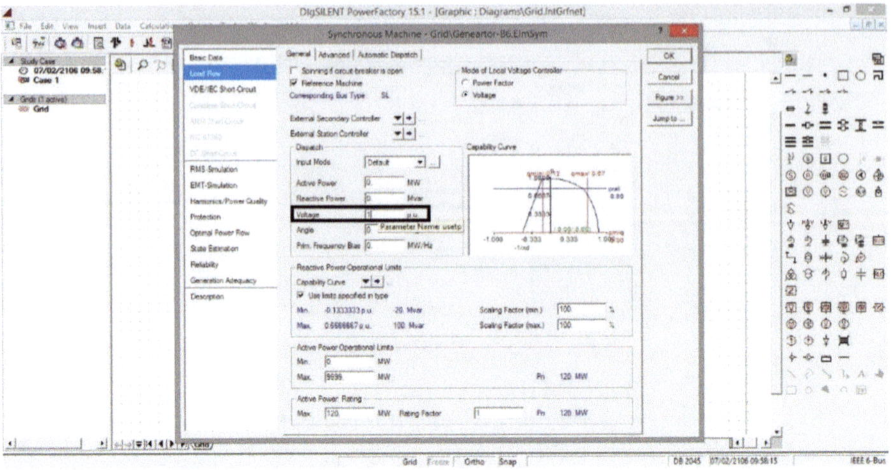

Fig. 11.11 DIgSILENT PowerFactory—the main page for transformers [1]

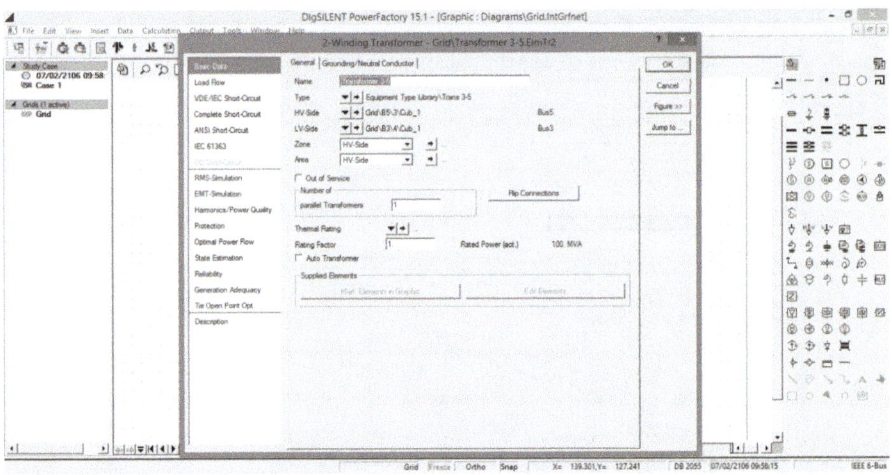

Fig. 11.12 DIgSILENT PowerFactory—the main page for transformers, “Basic Data” tab [1]

11.2.2 Second Step—How to Use DIgSILENT Scripting Facility

There are some prerequisite steps while using DIgSILENT scripting facility, such as creating a DPL command set, creating a general set and introducing external variables to the DPL command set, each of which are representatives of parameters that are needed in order to control or extract values of parameters from equipment

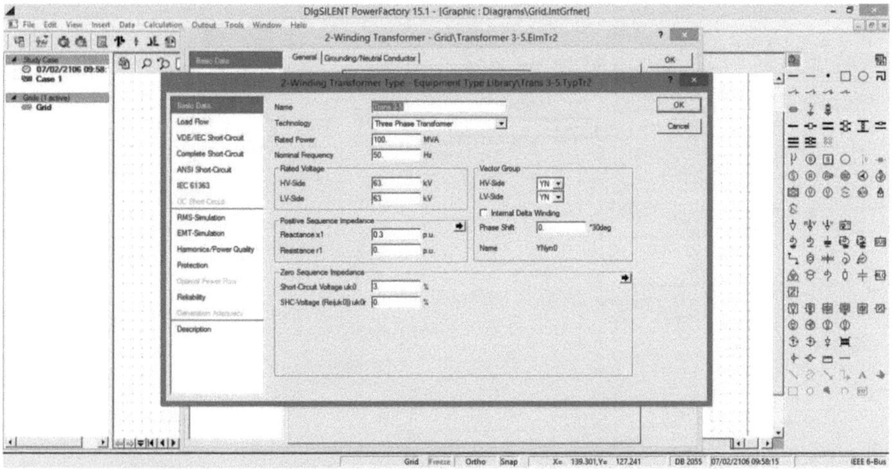

Fig. 11.13 DIgSILENT PowerFactory—adjusting tap settings of transformers [1]

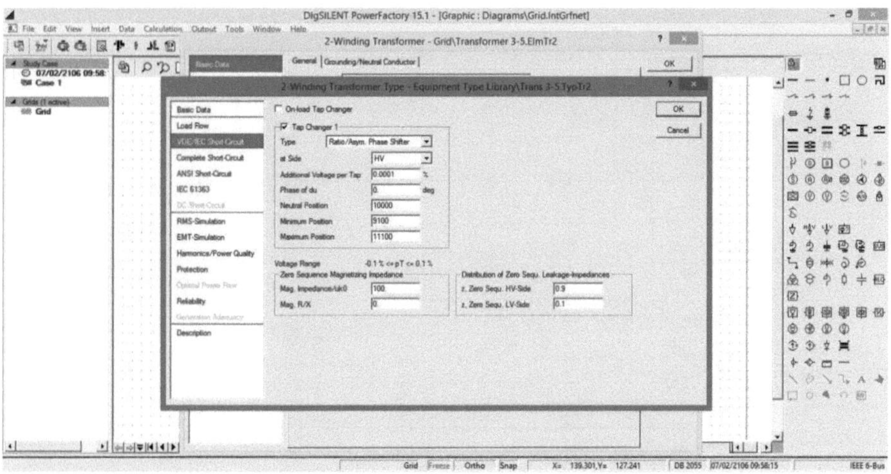

Fig. 11.14 DIgSILENT PowerFactory—the main page for transformers, “Load Flow” tab [1]

during the optimization procedure. All the steps have been represented in the following sections.

11.2.2.1 How to Create a DPL Command Set

In order to have an efficient communication between MATLAB and DIgSILENT scripting is inevitable. To write a script in DIgSILENT click on “Open Data

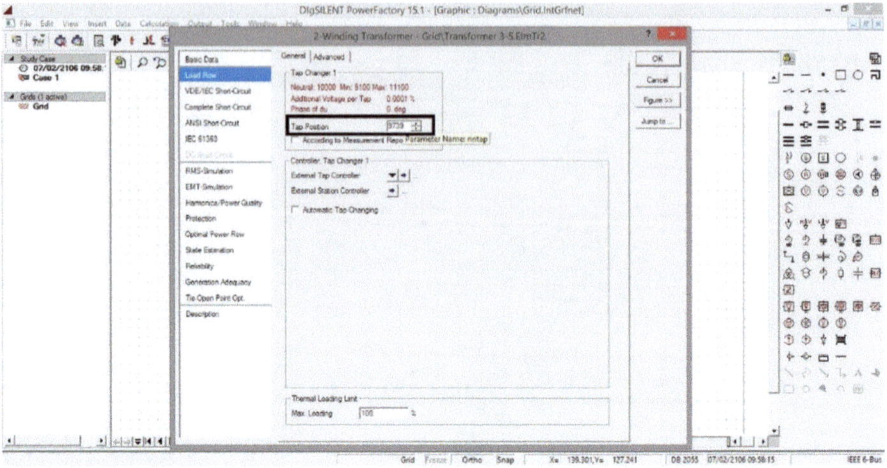

Fig. 11.15 DIgSILENT PowerFactory—setting up operational limitations for synchronous condensers [1]

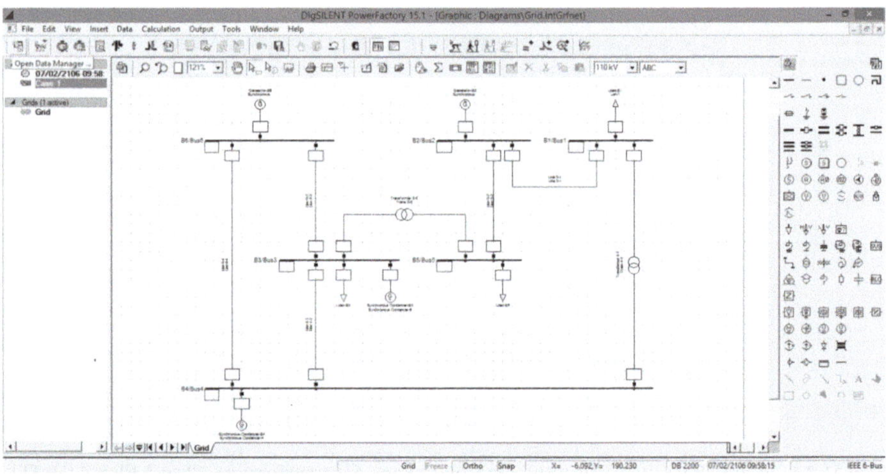

Fig. 11.16 DIgSILENT PowerFactory—"data manager" [1]

Manager" on the main toolbar (Fig. 11.16) → click on "New Object" on the "Data Manager" window and choose "DPL Command and more" (Fig. 11.17), then click on OK → write the script in the "Scrip" tab of "DPL Command" window (Fig. 11.18). The corresponding script to reactive power optimization for IEEE 6-bus power network has been given in the attachments.

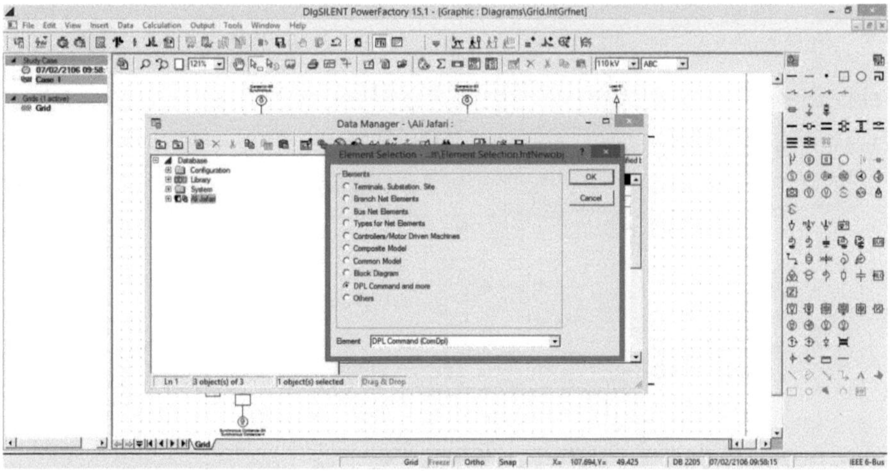

Fig. 11.17 DIgSILENT PowerFactory—creating DPL command [1]

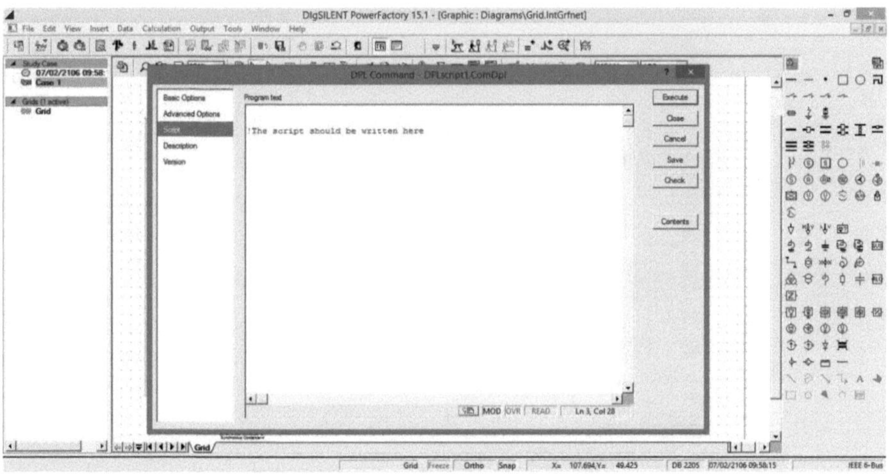

Fig. 11.18 DIgSILENT PowerFactory—"DPL command" window [1]

11.2.2.2 How to Create a "General Set" and Introduce It to DPL

There are a few other steps toward being able to use DPL dynamically in order to assign and export data, for instance, defining control and state variables to the written script in DPL. First, select all the equipment and right click on → "Define" → "General Set" (Fig. 11.19). To introduce the control variables of the network to the corresponding DPL, click on "Open Data Manager" on the main toolbar (Fig. 11.16) → right click on the script you have already created and

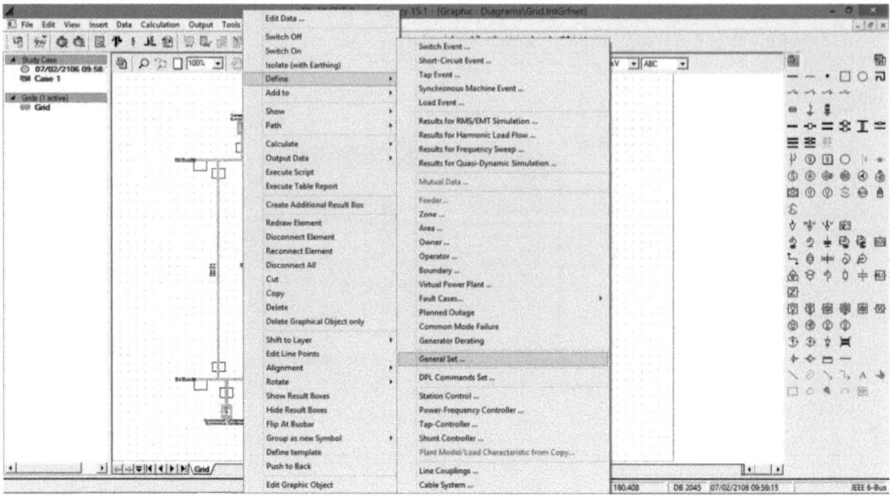

Fig. 11.19 DIgSILENT PowerFactory—creating general set [1]

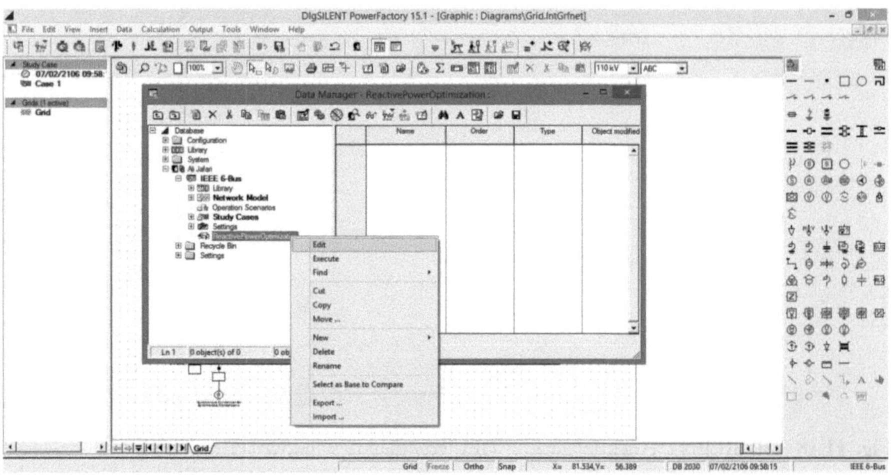

Fig. 11.20 DIgSILENT PowerFactory—introducing "General Set" to DPL command [1]

then click on "Edit" (Fig. 11.20) → in the "Basic Options" tab, select the general set to your script (Figs. 11.21 and 11.22). Please note that the settings related to the "General Set" section may be necessary to be done again after exporting the network in order to use it elsewhere, otherwise the DPL would not be able to access the equipment.

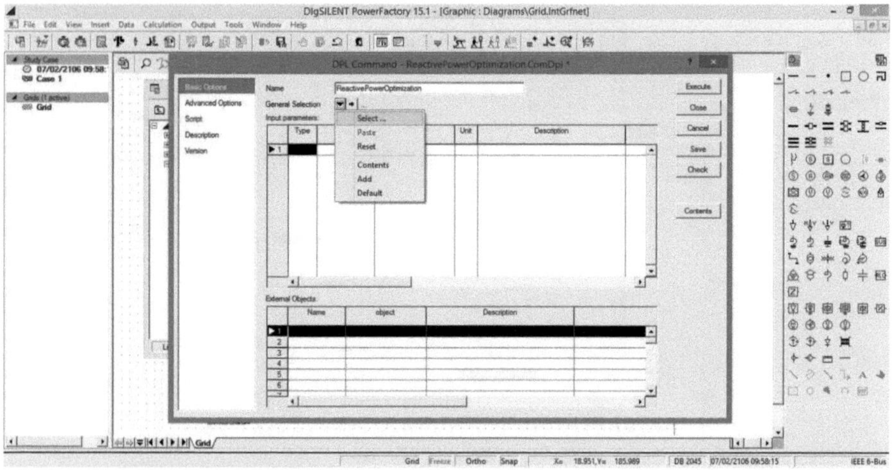

Fig. 11.21 DIgSILENT PowerFactory—introducing "General Set" to DPL command [1]

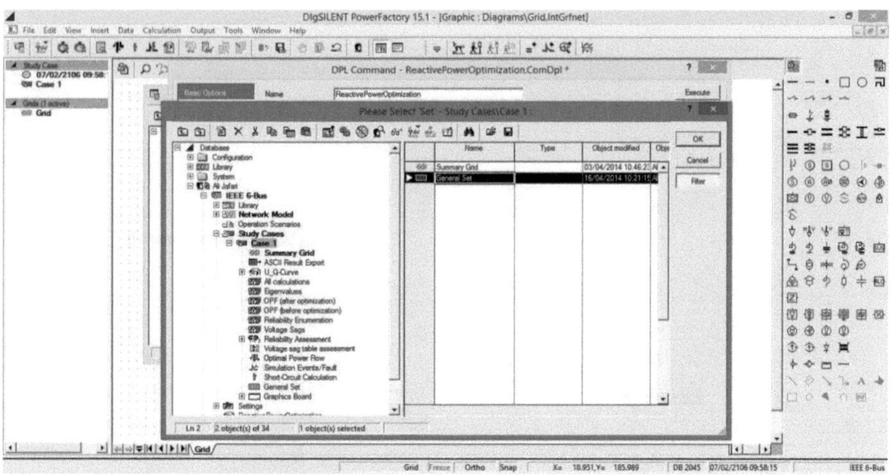

Fig. 11.22 DIgSILENT PowerFactory—introducing "General Set" to DPL command [1]

11.2.2.3 How to Introduce External Variables to DPL

In the "DPL Command" window and in the "External Objects" section the state variables should be defined. To do this, click on "Name" filed and assign a variable name like "G2" as the voltage magnitude variable of the generator of bus-2 → double click in the "Object" filed and select Generator-2 from the list (Fig. 11.23). In order to be available of using power flow calculations in the script, it should be defined just like control variables (Fig. 11.24). The final phase is to write the script in the scripting window. The corresponding script has written and explained step by step in the next stages.

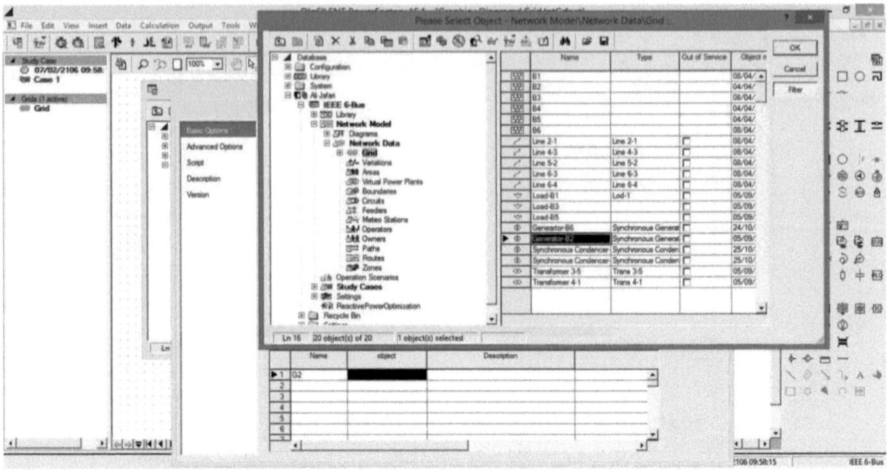

Fig. 11.23 DIgSILENT PowerFactory—defining external variables to the DPL [1]

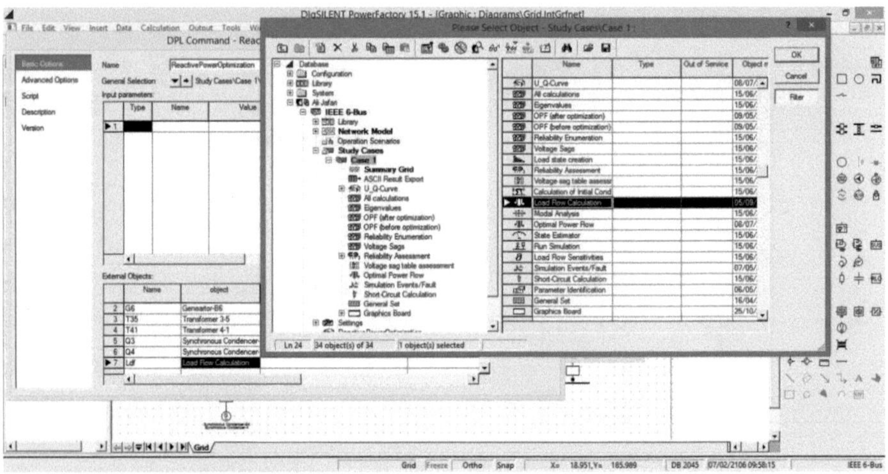

Fig. 11.24 DIgSILENT PowerFactory—introducing “Load Flow Calculation” function to DPL [1]

11.2.2.4 How to Write Scripts for DIgSILENT and MATLAB Step by Step

This section will present an expressive explanation of the written script for DIgSILENT and MATLAB step by step. It would be better to describe the written script for reactive power optimization using presented method dynamically switching from one piece of software to the other, so the procedure of the written program in both pieces of software will be easy to comprehend. The procedure has been explained in following just as it will run and switch between two pieces of software practically.

Initializing MATLAB Software

Since the suggested procedure of optimizing reactive power using DIgSILENT and MATLAB has programmed the way to consider MATLAB as the main calculator and DIgSILENT as the implicit one, the procedure turns out to be started by MATLAB [2].

Initializing	MATLAB

```
clear
clc
close all;
hold on;

CostFunction=@(x) OF_GA_PSO_PS(x);        % Defining Cost Function for PSO Algorithm

tic

%% Global Parameters
nVar=6;
VarMin=[1.1 1 9100 9100 0 0];              % Lower Bounds of Control Variables
VarMax=[1.15 1.1 11100 11100 5 5];         % Upper Bounds of Control Variables

%Creating 'ToDIg.txt' file to export the control variables values to be imported in
DIgSILENT
t=fopen('ToDIg.txt','w');
fprintf(t,'%d %d %d %d %d %d',1.1,1.05,9725,9100,0,0);
fclose(t);

% Making a blank "Couple.txt" file to order the switching process between MATLAB and
DIgSILENT
t=fopen('Couple.txt','w');

fprintf(t,'%d',0);
fclose(t);

% Waiting for DIgSILENT to Carry Out Power Flow Calculation
a=0;
while a==0
    pause(0.01);
    a=load('Couple.txt');
end
```

In the codes above, 'clear' was used to clear all possible existent variables in workplace, 'clc' to clear the workplace itself, 'close all' to close all possible open figures, and 'hold on' to retain the current graph and to add another graph to it. MATLAB adjusts the axes ranges to display full range of the added graphs. 'CostFunction = @(x) OF_GA_PSO_PS(x);' is used to define 'OF_GA_PSO_PS (x)' as the 'CostFunction' which is used within the PSO codes that will be presented in the next steps. 'tic' in the beginning and 'toc' at the end of the entire code are used to measure calculation time. In addition, 'VarMin' and 'VarMax' consist of lower and upper bounds of the control variables which were presented in appendices section of Chap. 10, and 'nVar' is the number of control variables, respectively [2].

In order to enter the values for control variables into DIgSILENT a '.txt' file as 'ToDIg.txt' is created using the codes presented above. In the same way, it is needed to have a '.txt' file as 'Couple.txt' to command both pieces of software one to proceed and the other to stop. It is noted that, in 'Couple.txt', 0 is written to mandate DIgSILENT to carry on and 1 to MATLAB, whenever one works the other stops. Thus, in general when 'Couple.txt' in the scripting section emerges, it means that process is being switched from one piece of software to the other. Therefore, at the end of codes above 'Couple.txt' switches the rest of the process to be continued in DIgSILENT.

Initializing DIgSILENT Software

Like any other programming language, there are some primary steps to take in DPL, such as defining and initiating the variables that are used in the script [1].

Initializing	**DIgSILENT**

```
int a;
object o;
double V,U,Phi,VL;
double VG2,VG6,TT35,TT41,QC3,QC4;
set s1,s2;
string str1;
a=0;
```

where, 'double' and 'int' are used to define real and integer variables respectively, 'object' is a reference to apparatus in DIgSILENT, 'set' is also used to address an array of components, and 'string' is used to define string class variables. DPL uses the following internal parameter types [1]:

- double, a 15 digits real number
- int, an integer number
- string, a string
- object, a reference to a PowerFactory objects
- set, a container of objects

Vectors and Matrices are available as external objects. The syntax for defining variables is as follows:

- *[VARDEF] = [TYPE] varname, varname, ..., varname;*
- *[TYPE] = double | int | object | set*

All the parameters should be defined in the first lines of the DPL script. The semicolon is obligatory. Examples [1]:

- *double Losses, Length, Pgen;*
- *int NrOfBreakers, i, j;*
- *string txt1, nm1, nm2;*
- *object O1, O2, BestSwitchToOpen;*
- *set AllSwitches, AllBars;*

Running Power Flow in DIgSILENT Based on Initial Condition

Equations (10.59) and (10.61) will be used to calculate active power losses and voltage stability index in 6-bus power network, of which contain state variables that should be taken from power flow calculations. Thus, in this step DIgSILENT will run power flow on the given power grid [1].

Running Power Flow in DIgSILENT Based on Initial Condition	DIgSILENT

```
! IEEE Standard Circumstances
fopen('d:\Project\ToDIg.txt','r',0);
fscanf(0,'%d %d %d %d %d %d',VG2,VG6,TT35,TT41,QC3,QC4);
fclose(0);
G2:usetp=VG2;
G6:usetp=VG6;
T35:nntap=TT35;
T41:nntap=TT41;
Q3:qgini=QC3;
Q4:qgini=QC4;

! Load Flow Calculation
ClearOutput();
ResetCalculation();
Ldf.Execute();

! Exporting Load Flow Calculation's Result
s1=SEL.AllBars();
s1.SortToName(0);
fopen('d:\Project\LoadFlow.txt','w',0);
o=s1.First();
while (o) {
V=o:m:Ul;
U=o:m:u1;
Phi=o:m:phiu;
str1=o:loc_name;
fprintf(0,'%f\t %f\t %f\t',V,U,Phi);
o=s1.Next();
}
fclose(0);
```

In the codes above, "! IEEE Standard Circumstances" section is used to import and apply control variables written by MATLAB in "ToDIg.txt" to the power grid in DIgSILENT. "! Load Flow Calculation" sector is used to carry load flow out on the power grid. Finally, "! Exporting Load Flow Calculation's Result" step is used to write the corresponding state variables in a file as "LoadFlow.txt" to be imported in MATLAB for further calculations.

In order to find a parameter name for each kind of equipment follow the instruction below:

Do right click on the "Result Box" of the equipment → "Edit Format for Nodes" (Fig. 11.25) → Right click on "Variable" column → "Insert Row(s)" (Fig. 11.26) → Double click on the newly created row → a new window entitled

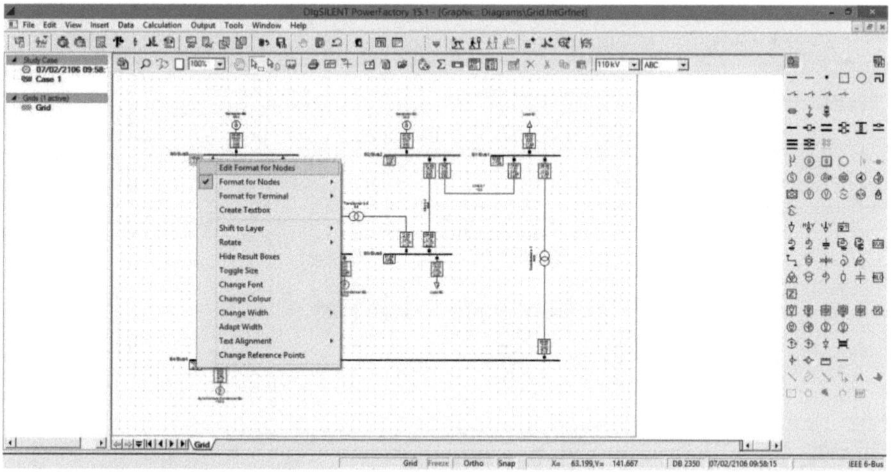

Fig. 11.25 DIgSILENT PowerFactory—“Edit Format for Nodes” [1]

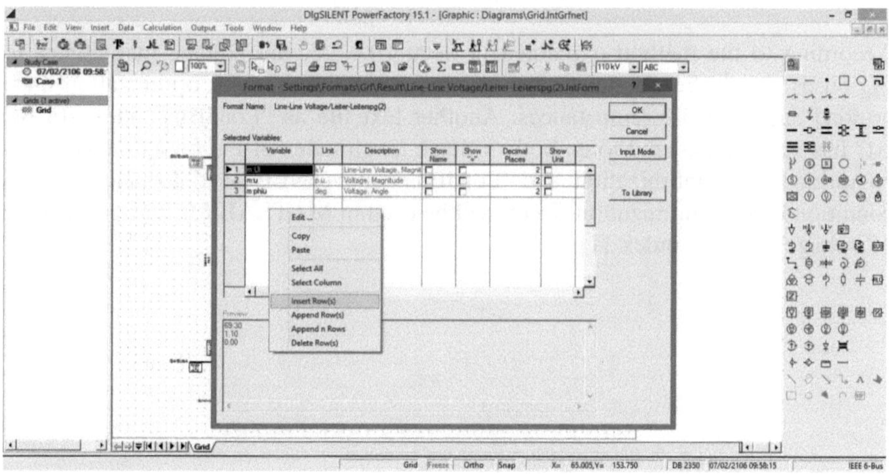

Fig. 11.26 DIgSILENT PowerFactory—“Edit Format for Nodes” “Insert Row(s)” [1]

“Variable Selections” will be opened (Fig. 11.27) → Select “Load Flow” tab in the left side of the window → In the “Available Variables” subsection choose the set of variables which is needed → In the “Selected Variables” subsection see the parameters which can be directly used in the DPL.

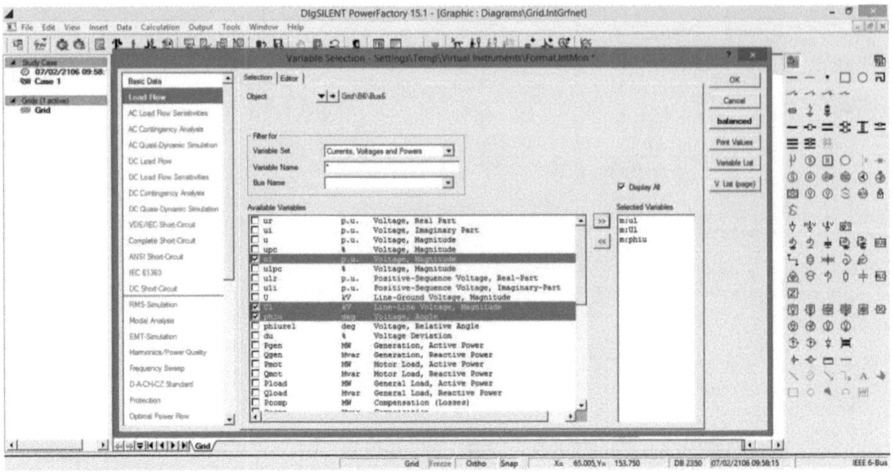

Fig. 11.27 DIgSILENT PowerFactory—"Variable Selection" window [1]

Voltage Stability Index Process Before Reactive Power Optimization

According to the method explained in Sect. 10.4.2.3, the voltage magnitudes for PQ buses should be obtained after and before reactive power optimization under no-load and loaded circumstances. Another text file as "LoadBusVoltagesFLBO.txt" has been considered to embody voltage magnitudes of PQ buses under loaded condition before optimization, and "LoadBusVoltagesNLBO.txt" for under no-load condition before optimization which will be used in MATLAB for further processes of voltage stability index [1].

The Needed Calculations for Voltage Stability Index	**DIgSILENT**

```
!! Start of Voltage Stability Index Process Before Optimization
fopen('d:\Project\LoadBusVoltagesFLBO.txt','w',0);
o=s1.First();
while (o) {
VL=o:m:Ul;
str1=o:loc_name;
fprintf(0,'%f\t',VL);
o=s1.Next();
}
fclose(0);

! Making All Loads OutofService
s2=SEL.AllLoads();
o=s2.First();
while(o){
o:outserv=1;
o=s2.Next();
}

! Load Flow calculation without Load
ClearOutput();
ResetCalculation();
s1=SEL.AllBars();
s1.SortToName(0);
Ldf.Execute();
fopen('d:\Project\LoadBusVoltagesNLBO.txt','w',0);
o=s1.First();
while (o) {
VL=o:m:Ul;
str1=o:loc_name;
fprintf(0,'%f\t',VL);
o=s1.Next();
}
fclose(0);

! Making All Loads Back to Service
s2=SEL.AllLoads();
o=s2.First();
while(o){
o:outserv=0;
o=s2.Next();
 }
!! End of Voltage Stability Index Process Before Optimization
```

```
! Send Run Message to MATLAB
fopen('d:\Project\Couple.txt','w',1);
fprintf(1,'%d',1);
fclose(1);

! Waiting to Get Run Signal from MATLAB
a=1;
while(a=1){
fopen('d:\Project\Couple.txt','r',0);
fscanf(0,'%d',a);
fclose(0);
}
```

There is a parameter to make a component of the network out of or back to service by the name "outserve". As it has shown in the "! Making All Loads Out of Service" section, at first all the loads should be selected as a set, then allocating "outserv = 1" for them one by one to make them out of service. After having all the data gained which is needed to calculate voltage stability index before optimization for further comparisons with the data after optimization procedure, DIgSILENT will send a run signal to MATLAB using "Couple.txt" to continue the procedure.

Active Power Losses Calculation Based on Initial Circumstances

In order to calculate active power losses based on initial conditions to figure out how much active power losses has decreased in the reactive power optimization procedure the following code will be proceeded in MATLAB [2].

Initial Active Power Losses Calculation	**MATLAB**

```
% Waiting for DIgSILENT to Carry Out Power Flow Calculation
a=0;
while a==0
    pause(0.01);
    a=load('Couple.txt');
end

% Loading Power Flow Results From DIgSILENT
BUSData=load('LoadFlow.txt');
BUSData(:,1)=BUSData(:,1)*1000;                  %Transferring Voltage from KV to
V
G=[0.0109 0.0141 0.0140 0.0145 0.0112 0 0];      %The line's Conductance

PL1=abs(G(1)*(BUSData(6,1)^2+BUSData(3,1)^2-
2*BUSData(6,1)*BUSData(3,1)*cos(deg2rad(BUSData(6,3)-BUSData(3,3)))));
PL2=abs(G(2)*(BUSData(6,1)^2+BUSData(4,1)^2-
2*BUSData(6,1)*BUSData(4,1)*cos(deg2rad(BUSData(6,3)-BUSData(4,3)))));
PL3=abs(G(3)*(BUSData(4,1)^2+BUSData(3,1)^2-
2*BUSData(4,1)*BUSData(3,1)*cos(deg2rad(BUSData(4,3)-BUSData(3,3)))));
PL4=abs(G(4)*(BUSData(5,1)^2+BUSData(2,1)^2-
2*BUSData(5,1)*BUSData(2,1)*cos(deg2rad(BUSData(5,3)-BUSData(2,3)))));
PL5=abs(G(5)*(BUSData(2,1)^2+BUSData(1,1)^2-
2*BUSData(2,1)*BUSData(1,1)*cos(deg2rad(BUSData(2,3)-BUSData(1,3)))));

OF_IEEE=PL1+PL2+PL3+PL4+PL5;

GlobalBest.Cost=OF_IEEE;
```

Scripting Particle Swarm Optimization

Scripting PSO algorithm has done based on the PSO theory which was presented in Sect. 10.5.1. Some further explanations have been given in the script to enlighten what is going on. The "Particle Swarm Optimization Algorithm" is as follows [2].

Particle Swarm Optimization Algorithm	MATLAB

```
%% Using PSO as Optimization Algorithm
VarSize=[1 nVar];            % Size of Decision Variables Matrix

%PSO Parameters
MaxIt=30;                    % Maximum Number of Iterations
nPop=20;                     % Population Size (Swarm Size)

% Constriction Coefficients
phi1=2.05;
phi2=2.05;
phi=phi1+phi2;
chi=2/(phi-2+sqrt(phi^2-4*phi));
w=chi;                       % Inertia Weight
wdamp=1;                     % Inertia Weight Damping Ratio
c1=chi*phi1;                 % Personal Learning Coefficient
c2=chi*phi2;                 % Global Learning Coefficient

% Velocity Limits
VelMax=0.1*(VarMax-VarMin);
VelMin=-VelMax;

% Initialization
empty_particle.Position=[];
empty_particle.Cost=[];
empty_particle.Velocity=[];
empty_particle.Best.Position=[];
empty_particle.Best.Cost=[];

particle=repmat(empty_particle,nPop,1);

for i=1:nPop

   % Initialize Position
   particle(i).Position=unifrnd(VarMin,VarMax,VarSize);
   % Initialize Velocity
   particle(i).Velocity=zeros(VarSize);

   % Evaluation
   [particle(i).Cost particle(i).Sol]=CostFunction(particle(i).Position);

   % Update Personal Best
   particle(i).Best.Position=particle(i).Position;
   particle(i).Best.Cost=particle(i).Cost;
```

```
    particle(i).Best.Sol=particle(i).Sol;

    % Update Global Best
    if particle(i).Best.Cost<GlobalBest.Cost
            GlobalBest=particle(i).Best;
    end
end

BestCost=zeros(MaxIt,1);

% PSO Main Loop
for it=1:MaxIt
     for i=1:nPop

     % Update Velocity
     particle(i).Velocity = w*particle(i).Velocity ...
        +c1*rand(VarSize).*(particle(i).Best.Position-particle(i).Position) ...
        +c2*rand(VarSize).*(GlobalBest.Position-particle(i).Position);

     % Apply Velocity Limits
     particle(i).Velocity = max(particle(i).Velocity,VelMin);
     particle(i).Velocity = min(particle(i).Velocity,VelMax);

     % Update Position
     particle(i).Position = particle(i).Position + particle(i).Velocity;

     % Velocity Mirror Effect
     IsOutside=(particle(i).Position<VarMin | particle(i).Position>VarMax);
     particle(i).Velocity(IsOutside)=-particle(i).Velocity(IsOutside);

     % Apply Position Limits
     particle(i).Position = max(particle(i).Position,VarMin);
     particle(i).Position = min(particle(i).Position,VarMax);

     % Evaluation
     [particle(i).Cost particle(i).Sol] = CostFunction(particle(i).Position);

     % Update Personal Best
     if particle(i).Cost<particle(i).Best.Cost

        particle(i).Best.Position=particle(i).Position;

        particle(i).Best.Cost=particle(i).Cost;
        particle(i).Best.Sol=particle(i).Sol;
```

```
        % Update Global Best
        if particle(i).Best.Cost<GlobalBest.Cost
            GlobalBest=particle(i).Best;
          end
        end
      end

  clc
  BestCost(it)=GlobalBest.Cost;
  per=round(it*100/MaxIt);
  fprintf('Processing : %d\n %',per);
  disp(['Best Cost = ' num2str(BestCost(it))]);

  w=w*wdamp;

  %plot(BestCost,'LineWidth',2);
  % set(gca,'xlim',[0,MaxIt]);
  %xlabel('Iteration');
  %ylabel('Fitness value');
  plot(it,GlobalBest.Cost,'.k');
  %semilogy(BestCost,'LineWidth',2);
  %ylabel('Best Cost');

end

OF_PSO=GlobalBest.Cost;
Var_PSO=GlobalBest.Position;
```

In the "% Evaluation" sections in the codes above, PSO algorithm uses "CostFunction" to evaluate how much the fitness function has gotten optimized, which was defined as "CostFunction = @(x) OF_GA_PSO_PS(x);" at the beginning of the codes in another m.file as a function "OF_GA_PSO_PS(x)" [2].

Cost Function Evaluation of Reactive Power Optimization

In order to have the coherency aspect in the coding section, fitness function of the optimization procedure has been programmed in a separate function m.file entitled "OF_GA_PSO_PS.m" which will be called within the main code wherever it is needed. Evaluation process applies on every particle in each iteration in order to get their appropriateness amount. The function file is as follows [2].

Cost Function of Reactive Power Optimization - A	**MATLAB**

```
function [F ContVar]=OF_GA_PSO_PS(x)

ContVar=x;
Vref=1;
%Kv=10;

G=[0.0109 0.0141 0.0140 0.0145 0.0112 0 0];                   %The lines' Conductance
%B=[-0.0460 -0.0586 -0.0330 -0.0163 -0.0840 -0.1894];          %The lines' Susceptance

%% Calculation of Objective Function (Ploss)
  t=fopen('ToDIg.txt','w');
  fprintf(t,'%d %d %d %d %d
%d',ContVar(1),ContVar(2),ContVar(3),ContVar(4),ContVar(5),ContVar(6));
  fclose(t);

  % Send a Signal to Run Power Flow in DIgSILENT
  t=fopen('Couple.txt','w');
  fprintf(t,'%d',0);
  fclose(t);

  % Wait to DIgSILENT Runs Power Flow Calculation
  a=0;
  while a==0
      pause(0.01);
      a=load('Couple.txt');
  end
```

It should be mentioned that in the cost function above, just the active power losses have been considered as an objective function for the sake of simplicity. The other constraints can be added easily to the main fitness function using penalty function. In the "%% Calculation of Objective Function (Ploss)" section, at first the gained positions (values for control variables) by particles are written in the "ToDIg.txt" in order to be imported in DIgSILENT. Then, another command is used to run power flow in DIgSILENT via "Couple.txt". Whilst the codes above in MATLAB is being proceeded, DIgSILENT is waiting to receive an order from MATLAB to run power flow calculations for the new values of control variables as the codes below [1, 2].

RUNNING POWER FLOW TO EVALUATE FITNESS VALUE OF EACH PARTICLE	DIgSILENT

```
a=1;
while(a=1){
fopen('d:\Project\Couple.txt','r',0);
fscanf(0,'%d',a);
fclose(0);
}

! The Main Loop of Fitness Function Evaluation of Particles
while(a<>2)
{
fopen('d:\Project\ToDIg.txt','r',0);
fscanf(0,'%d %d %d %d %d %d',VG2,VG6,TT35,TT41,QC3,QC4);
fclose(0);
G2:usetp=VG2;
G6:usetp=VG6;
T35:nntap=TT35;
T41:nntap=TT41;
Q3:qgini=QC3;
Q4:qgini=QC4;

! Load Flow Calculation
ClearOutput();
ResetCalculation();
s1=SEL.AllBars();
s1.SortToName(0);
Ldf.Execute();
fopen('d:\Project\LoadFlow.txt','w',0);
o=s1.First();
while (o) {
V=o:m:Ul;
U=o:m:u1;
Phi=o:m:phiu;
str1=o:loc_name;
fprintf(0,'%f\t %f\t %f\t',V,U,Phi);
o=s1.Next();
}
fclose(0);

!Send Run Message to MATLAB
fopen('d:\Project\Couple.txt','w',1);
```

```
fprintf(1,'%d',1);
fclose(1);

! Wait for MATLAB Run
a=1;
while(a=1){
fopen('d:\Project\Couple.txt','r',0);
fscanf(0,'%d',a);
fclose(0);
}
} ! End of the Main Loop
```

In the codes above, DIgSILENT gets new values for control variables and applies them to the network, then runs power flow and writes the state values in the “LoadFlow.txt” to be exported to MATLAB. In the end, DIgSILENT sends a signal to MATLAB to declare its work has finished and results are ready to use. MATLAB gets the signal and starts processing upon the state variables. It should be noted that, in this section the number of power flow calculations equals to the number of particles which is determined at the beginning of MATLAB codes multiplied by the number of iterations. Therefore, if there are 10 particle and 10 iteration, the number of power flow calculations will be 100.

Cost Function of Reactive Power Optimization - B	MATLAB

```
% Wait to DIgSILENT Runs Power Flow Calculation
  a=0;
  while a==0
      pause(0.01);
      a=load('Couple.txt');
  end

  BUSData=load('LoadFlow.txt');              % Bus Data Loading
  BUSData(:,1)=BUSData(:,1)*1000;            % Changing Voltage from KV to V

  PL1=abs(G(1)*(BUSData(6,1)^2+BUSData(3,1)^2-
2*BUSData(6,1)*BUSData(3,1)*cos(deg2rad(BUSData(6,3)-BUSData(3,3)))));
  PL2=abs(G(2)*(BUSData(6,1)^2+BUSData(4,1)^2-
2*BUSData(6,1)*BUSData(4,1)*cos(deg2rad(BUSData(6,3)-BUSData(4,3)))));
  PL3=abs(G(3)*(BUSData(4,1)^2+BUSData(3,1)^2-
2*BUSData(4,1)*BUSData(3,1)*cos(deg2rad(BUSData(4,3)-BUSData(3,3)))));
  PL4=abs(G(4)*(BUSData(5,1)^2+BUSData(2,1)^2-
2*BUSData(5,1)*BUSData(2,1)*cos(deg2rad(BUSData(5,3)-BUSData(2,3)))));
  PL5=abs(G(5)*(BUSData(2,1)^2+BUSData(1,1)^2-
2*BUSData(2,1)*BUSData(1,1)*cos(deg2rad(BUSData(2,3)-BUSData(1,3)))));
  PL=PL1+PL2+PL3+PL4+PL5;
  %VD1=Kv*((abs(BUSData(1,2)-Vref)>0.05)*(abs(BUSData(1,2)-Vref)));
  %VD3=Kv*((abs(BUSData(3,2)-Vref)>0.05)*(abs(BUSData(3,2)-Vref)));
  % VD5=Kv*((abs(BUSData(5,2)-Vref)>0.05)*(abs(BUSData(5,2)-Vref)));
  %VD=VD1+VD3+VD5;
  %F=PL+VD;
  F=PL;

end
```

The PSO algorithm will be proceeded until the given stopping criteria have been exceeded. After having the optimization procedure by PSO finished, it is time to use Patter Search algorithm to pinpoint the exact optimum location [1, 2].

Using Pattern Search Algorithm as Supplementary Optimization Algorithm

As immediately as PSO algorithm finishes its procedure, the Pattern Search algorithm starts its local optimizing. The best result discovered by PSO is considered as the starting point for pattern search algorithm to pinpoint the optimum point. There is a very complete toolbox for pattern search algorithm as a built-in function in MATLAB, of which is going to be used in series with PSO algorithm. In order to use pattern search combined with PSO algorithm as a collective one, it is needed to have the codes of pattern search. Fortunately, MATLAB has given this availability

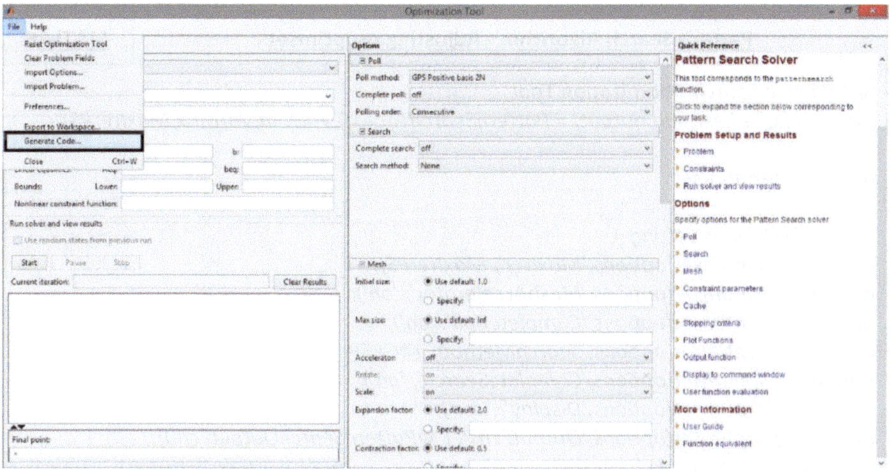

Fig. 11.28 MATLAB—Pattern Search toolbox [2]

to its users to extract the codes form the toolbox. It is readily accessible clicking on "File" menu then "Generate Code" after adjusting all the corresponding parameters in the toolbox window (Fig. 11.28). Referring to help documentation of MATLAB and following the written codes will reveal what has been done.

After generating the script of pattern search via the related toolbox, a general structure of the algorithm will be given which has been formed to optimize reactive power as follows [2].

Pattern Search Algorithm - Main Code	MATLAB

```
%% Using Pattern Search Algorithm as Second Optimization Algorithm
x0=Var_PSO;    % Using Best Control Variable Obtained from PSO to Start Point of
Pattern Search Algorithm

MaxIterPS=60;   % Maximum Number of Iteration for Pattern Search Algorithm

[Var_PS,OF_PS,exitflag,output] = PatternSearchTool(x0,VarMin,VarMax,MaxIterPS);

LossReductionPercent=((OF_IEEE-OF_PS)/OF_IEEE)*100;
```

In order to adjust corresponding parameters for Pattern Search algorithm in the scripts, another function as "psoptimset" is needed. The related function has been implemented as follows for reactive power optimization. MATLAB software itself has an elaborate guide on different adjustments of the Pattern Search algorithm which is the best resource to pursue for more information [2].

Pattern Search Algorithm - Adjusting psoptimset	**MATLAB**

```
%% Pattern Search Optimization Tool
function [x,fval,exitflag,output] = PatternSearchTool(x0,VarMin,VarMax,MaxIterPS)

options = psoptimset;

% Modify Options Setting
options = psoptimset(options,'MaxIter', MaxIterPS);
options = psoptimset(options,'MeshAccelerator', 'on');
options = psoptimset(options,'CompletePoll', 'on');
options = psoptimset(options,'SearchMethod', @GPSPositiveBasis2N);
options = psoptimset(options,'CompleteSearch', 'on');
options = psoptimset(options,'Display', 'iter');
options = psoptimset(options,'OutputFcns', { @PatternSearchOutputFcn });
[x,fval,exitflag,output] =
patternsearch(@OF_GA_PSO_PS,x0,[],[],[],[],VarMin,VarMax,[],options);
```

Finishing the Optimization Procedure

After having the optimization procedure finished and the optimum point found by pattern search algorithm, MATLAB sends a stop signal to the main power flow calculation loop of DIgSILENT. In addition, "toc" is used to stop measuring of how much time has been spent during optimization procedure. The codes are as follows [2].

Finishing the Optimization Process	**MATLAB**

```
%% Finishing Process
% Send a Signal to Stop DIgSILENT
t=fopen('Couple.txt','w');
fprintf(t,'%d',2);
fclose(t);
hold off;

toc

% Wait to DIgSILENT Runs Power Flow Calculation
  a=2;
  while a==0
      pause(0.01);
      a=load('Couple.txt');
  end
```

Voltage Stability Index Process After Reactive Power Optimization

After having the optimization procedure finished, DIgSILENT runs a power flow considering the best optimum point found under loaded and no-load circumstances, then writes the results in "LoadBusVoltagesFLAO.txt" and "LoadBusVoltagesNLAO.txt", respectively. Afterwards, all the loads should be set back to service in order to have the power network in normal condition. DIgSILENT has finished its duty in this section and the procedure is switched to MATLAB to further calculations [1].

The Needed Calculations for Voltage Stability Index	DIgSILENT

```
!!! Start Voltage Stability Index Process After Optimization
ClearOutput();
ResetCalculation();
s1=SEL.AllBars();
s1.SortToName(0);
Ldf.Execute();
fopen('d:\Project\LoadBusVoltagesFLAO.txt','w',0);
o=s1.First();
while (o) {
VL=o:m:Ul;
str1=o:loc_name;
fprintf(0,'%f\t',VL);
o=s1.Next();
}
fclose(0);

! Making All Loads Out of Service
s2=SEL.AllLoads();
o=s2.First();
while(o){
o:outserv=1;
o=s2.Next();
}

! Load Flow calculation without Load
ClearOutput();
ResetCalculation();
s1=SEL.AllBars();
s1.SortToName(0);
Ldf.Execute();
fopen('d:\Project\LoadBusVoltagesNLAO.txt','w',0);
o=s1.First();
while (o) {
VL=o:m:Ul;
str1=o:loc_name;
fprintf(0,'%f\t',VL);
o=s1.Next();
}
fclose(0);

! Make All Loads Back to Service
s2=SEL.AllLoads();
o=s2.First();
```

```
while(o){
o:outserv=0;
o=s2.Next();
}
!! End Voltage Stability Index Process After Optimization

! Send Run Message to MATLAB
fopen('d:\Project\Couple.txt','w',1);
fprintf(1,'%d',1);
fclose(1);
```

Final Calculations of Voltage Stability Index

After having all the data obtained via power flow calculations in DIgSILENT that is initially needed thorough processing the voltage stability index before and after optimization procedure, MATLAB gets the data and carries further calculations on them to have kind of tangible information of how much the index has been improved. The related codes are as follows [2].

Voltage Stability Index Final Calculations	**MATLAB**

```
%% Voltage Stability Index Calculation
VLNLBF=1000*load('LoadBusVoltagesNLBO.txt'); % Voltage-Load Bus-No load-Before-
Optimization
VLFLBF=1000*load('LoadBusVoltagesFLBO.txt'); % Voltage-Load Bus-Full load-Before-
Optimization

VLNLAF=1000*load('LoadBusVoltagesNLAO.txt'); % Voltage-Load Bus-No load-Before-
Optimization
VLFLAF=1000*load('LoadBusVoltagesFLAO.txt'); % Voltage-Load Bus-Full load-Before-
Optimization

VSI.BO(1)=VLFLBF(1)/VLNLBF(1);
VSI.BO(2)=VLFLBF(3)/VLNLBF(3);
VSI.BO(3)=VLFLBF(5)/VLNLBF(5);

VSI.AO(1)=VLFLAF(1)/VLNLAF(1);
VSI.AO(2)=VLFLAF(3)/VLNLAF(3);
VSI.AO(3)=VLFLAF(5)/VLNLAF(5);

hold on;
figure;
BarSort=[VSI.BO(1),VSI.AO(1);0,0;VSI.BO(2),VSI.AO(2);0,0;VSI.BO(3),VSI.AO(3)];
bar(BarSort);
xlabel('Load Bus No.');
ylabel('Voltage Stability Index (V/V0)');
legend('Befor Optimization','After Optimization');
hold off;
```

Voltage Deviation Calculations Before and After Optimization

Another index as voltage deviation before and after reactive power optimization has been considered in order to work out the impact of the optimization on the voltage deviation of PQ busbars. The related data have been put in the bar figures to have an easy comparison between them. The codes are as following [2].

Voltage Deviation Calculations	MATLAB

```
%% Deviation of LoadBus's Voltages Before and After Optimization
Vref=63000;
VD.BO(1)=abs(VLFLBF(1)-Vref);
VD.BO(2)=abs(VLFLBF(3)-Vref);
VD.BO(3)=abs(VLFLBF(5)-Vref);

VD.AO(1)=abs(VLFLAF(1)-Vref);
VD.AO(2)=abs(VLFLAF(3)-Vref);
VD.AO(3)=abs(VLFLAF(5)-Vref);

figure;
BarSort=[VD.BO(1),VD.AO(1);0,0;VD.BO(2),VD.AO(2);0,0;VD.BO(3),VD.AO(3)];
bar(BarSort);
xlabel('Load Bus No.');
ylabel('Voltage Deviation, Unit (V)');
legend('Befor Optimization','After Optimization');
hold off;
```

Reduction Figures of Active Power Losses

A bar type figure has been considered to indicated how much active power losses have reduced due to reactive power optimization using particle swarm algorithm as the main and global optimizer with the association of Pattern Search as the supplementary one, which is rather a local optimizer [2].

Active Power Losses Reduction	MATLAB

```
%% Plot Bars of Loss Reduction
figure;
BarSort=[OF_IEEE,OF_PS];
bar(BarSort,0.1);
xlabel('1-IEEE-Standard    2-PSO+PS Algorithm');
ylabel('Ploss, Unit (W)');
hold off;
```

Representing All Data in MATLAB Command Window

In addition to the figures which are planned to indicate the data that provide us with having an easy comparison of the power network state before and after optimization, the additional and accurate data will be presented in the command window of MATLAB which are produced using the codes below [2].

Specifying All the Data Related to Reactive Power Optimization	MATLAB

```
%% Indicating Results
disp('=========================================================');

str = sprintf('Total Real Power Losses (W) on IEEE-Standard Circumstance = %d
.',OF_IEEE);
disp(str);

disp('=========================================================');

str = sprintf('Total Real Power Losses (W) After Optimization Using PSO+PS = %d
.',OF_PS);
disp(str);

disp('=========================================================');

str = sprintf('The Percentage of Power Loss Reduction = %d .',LossReductionPercent);
disp(str);

disp('=========================================================');

disp('The Voltage Stability Index for Load Buses Before Optimization:');
disp('----------');
str = sprintf('VSI_Bus_1 = %d .', VSI.BO(1));
disp(str);
str = sprintf('VSI_Bus_3 = %d .', VSI.BO(2));
disp(str);
str = sprintf('VSI_Bus_5 = %d .', VSI.BO(3));
disp(str);
disp('----------');
disp('The Voltage Stability Index for Load Buses After Optimization:');
disp('----------');
str = sprintf('VSI_Bus_1 = %d .', VSI.AO(1));
disp(str);
str = sprintf('VSI_Bus_3 = %d .', VSI.AO(2));
disp(str);
str = sprintf('VSI_Bus_5 = %d .', VSI.AO(3));
disp(str);
disp('----------');

disp('=========================================================');

disp('The Sum of Voltage Stability Index for all Load Buses Before and After Optimization
are as follow:');
```

```
disp('Considering the Ideal Value is 3');
disp('----------');

VSI_Total_BO=VSI.BO(1)+VSI.BO(2)+VSI.BO(3);
VSI_Total_AO=VSI.AO(1)+VSI.AO(2)+VSI.AO(3);

str = sprintf('Befor Optimization = %d .', VSI_Total_BO);
disp(str);
str = sprintf('After Optimization = %d .', VSI_Total_AO);
disp(str);

disp('=====================================================');

disp('The Voltage Deviation for Load Buses Before Optimization:');
disp('----------');
str = sprintf('VD_Bus_1 = %d .', VD.BO(1));
disp(str);
str = sprintf('VD_Bus_3 = %d .', VD.BO(2));
disp(str);
str = sprintf('VD_Bus_5 = %d .', VD.BO(3));
disp(str);
disp('----------');
disp('The Voltage Deviation for Load Buses After Optimization:');
disp('----------');
str = sprintf('VD_Bus_1 = %d .', VD.AO(1));
disp(str);
str = sprintf('VD_Bus_3 = %d .', VD.AO(2));
disp(str);
str = sprintf('VD_Bus_5 = %d .', VD.AO(3));
disp(str);
disp('----------');

disp('=====================================================');

disp('The Sum of Voltage Deviations for all Load Buses Before and After Optimization are
as follow:');
disp('----------');

VD_Total_BO=VD.BO(1)+VD.BO(2)+VD.BO(3);
VD_Total_AO=VD.AO(1)+VD.AO(2)+VD.AO(3);

str = sprintf('Before Optimization = %d .', VD_Total_BO);
disp(str);
str = sprintf('After Optimization = %d .', VD_Total_AO);
disp(str);
```

```
disp('========================================================');

disp('The Control Variables After Optimization are as follow:');
disp('---------');

VG_Bus2 = Var_PS(1,1);
VG_Bus6 = Var_PS(1,2);
TT_Bus35=Var_PS(1,3);
TT_Bus14=Var_PS(1,4);
QC_Bus3=Var_PS(1,5);
QC_Bus4=Var_PS(1,6);

str = sprintf('Generator Voltage in Bus_2 Before and After Optimization are %d and %d,
respectively.',1.1,VG_Bus2);
disp(str);
str = sprintf('Generator Voltage in Bus_6 Before and After Optimization are %d and %d,
respectively.',1.05,VG_Bus6);
disp(str);
str = sprintf('Tap Position of transformer 3 to 5 Before and After Optimization are %d
and %d, respectively.',9725,TT_Bus35);
disp(str);
str = sprintf('Tap Position of Transformer 1 to 4 Before and After Optimization are %d
and %d, respectively.',9100,TT_Bus14);
disp(str);
str = sprintf('Reactive Power Output of Synchronize Condenser on Bus_3 Before and
After Optimization are %d and %d, respectively.',0,QC_Bus3);
disp(str);
str = sprintf('Reactive Power Output of Synchronize Condenser on Bus_4 Before and
After Optimization are %d and %d, respectively.',0,QC_Bus4);
disp(str);

disp('========================================================');
```

11.2.3 Third Step—How to Run MATLAB and DIgSILENT to Start Optimizing Reactive Power

After having the mfiles of MATLAB and scripts of DIgSILENT ready, it is the final step to run the optimization process. It should be noted that, according to the structure used in the script files enclosed with this chapter, the mfile of MATLAB is the first one which should be run and then DIgSILENT. To run the optimization procedure via MATLAB by "RPObyPSOplusPS.m" simply click on "Run" icon in

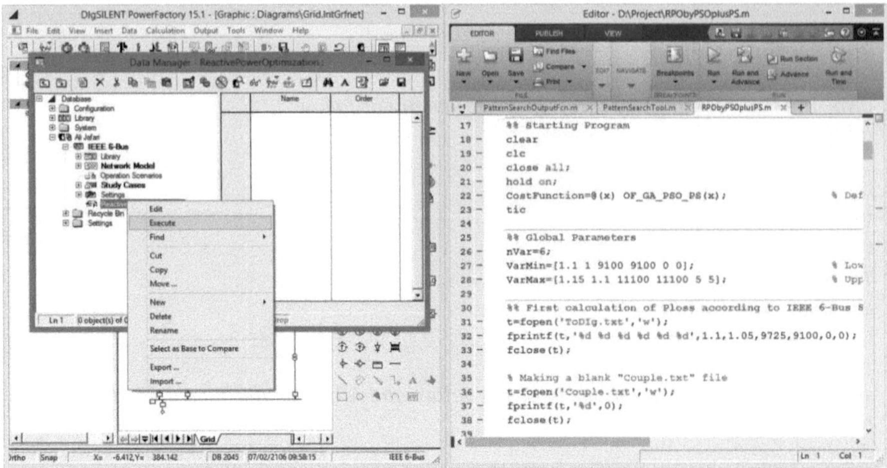

Fig. 11.29 DIgSILENT and MATLAB—running the optimization procedure [1], [2]

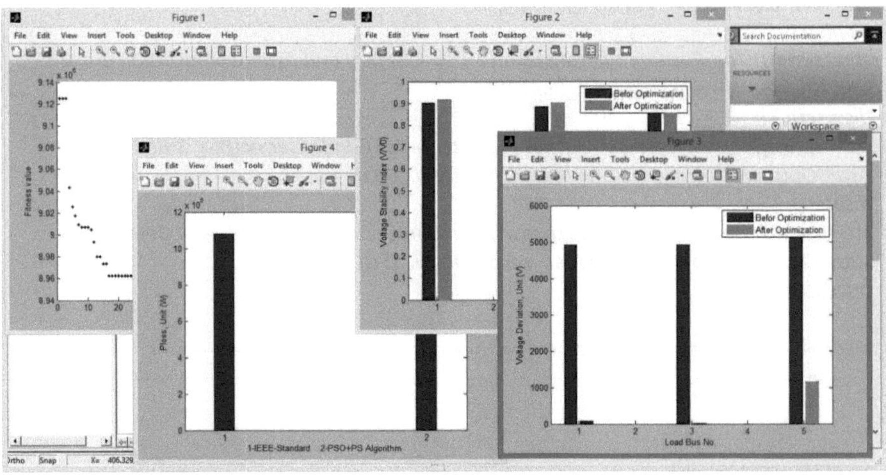

Fig. 11.30 MATLAB—the figures of results after reactive power optimization [2]

the mfile toolbar, and to run the script written in DIgSILENT right click on the script and click on "Execute" (Fig. 11.29) [1, 2].

Figures 11.30, 11.31 and 11.32 show the result of optimization after the process is done. The elaborate report of reactive power optimization for the same power grid has been presented in the Simulation Results (Sect. 10.8) in Chap. 10.

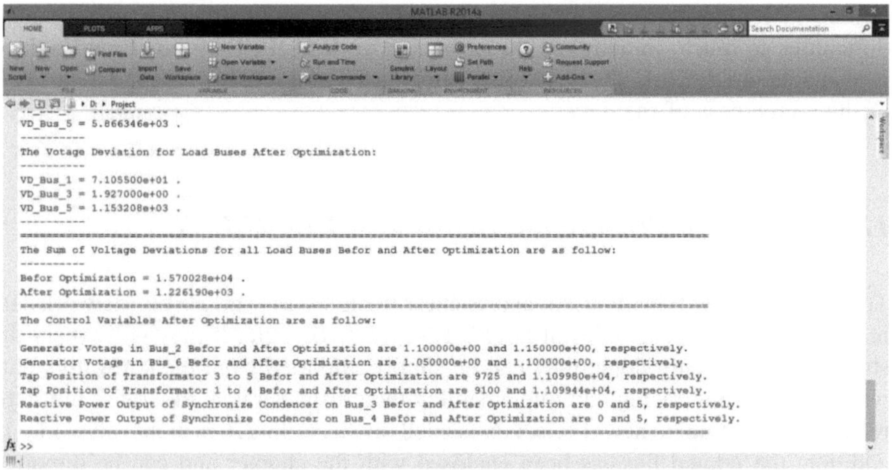

Fig. 11.31 MATLAB—the results which are represented in command window [2]

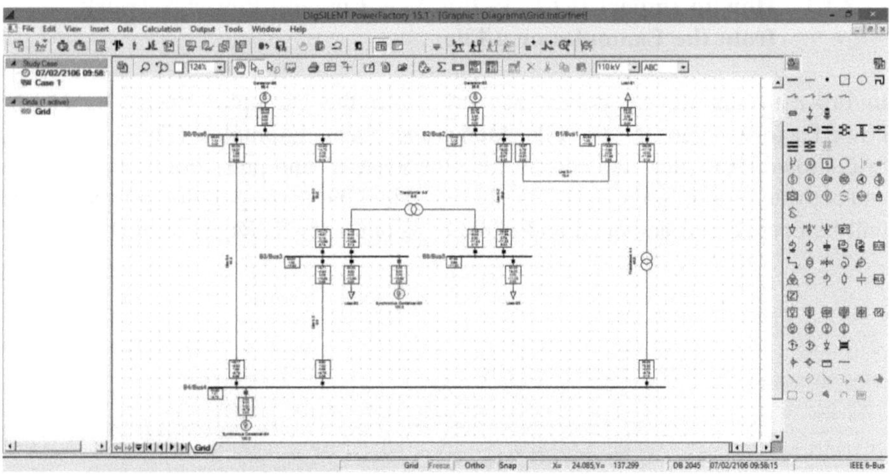

Fig. 11.32 DIgSILENT PowerFactory—the power network after reactive power optimization [1]

11.3 How to Use Built-in Optimization Functions to Implement Reactive Power Optimization

This section will present the simplest and more efficient way of carrying out reactive power optimization on 39-bus New England power network using built-in functions of MATLAB software and the corresponding ready-to-use network of DIgSILENT. The optimization algorithm will be particle swarm pattern search algorithm as its efficiency has been investigated before. Fortunately, both the optimization algorithms are available in MATLAB without suffering programming difficulties so much. In addition, the power network of IEEE 39-bus New England is available in DIgSILENT 15.1 which can be used as a reliable case study in terms of the correctness of the data.

11.3.1 How to Prepare the Network in DIgSILENT

To do this there will be different approaches. IEEE 9-, 14- and 39-bus networks are readily available as examples of DIgSILENT 15.1 software. If they will satisfy the studies that are being done, they can be simply imported from the software's data without bothering tracing and implementing them from scratch, otherwise the steps related to creating a new project presented in Sect. 11.2.1 will be at work. In addition, the related file has been enclosed with the chapter and can be imported easily in DIgSILENT.

11.3.1.1 How to Import IEEE 39-Bus New England Power Network from the Examples or a File

In order to import the power network from the examples open DIgSILENT PowerFactory → go to the File Menu and press Examples (Fig. 11.33) → in "Additional Examples" tab choose "39 Bus System" and press Folder Icon (Import and activate Example Project) (Fig. 11.34). Then the corresponding power network will be opened and activated and is ready to put a few changes on it (Fig. 11.35).

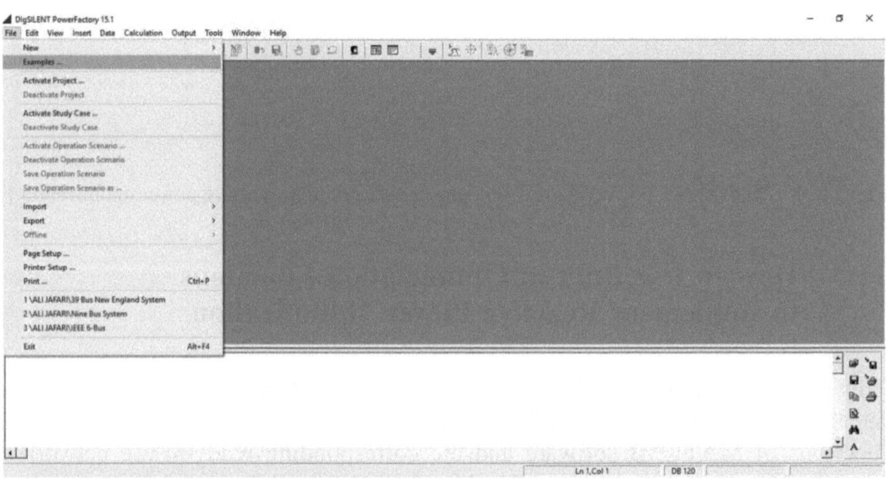

Fig. 11.33 DIgSILENT PowerFactory—File Menu → Examples

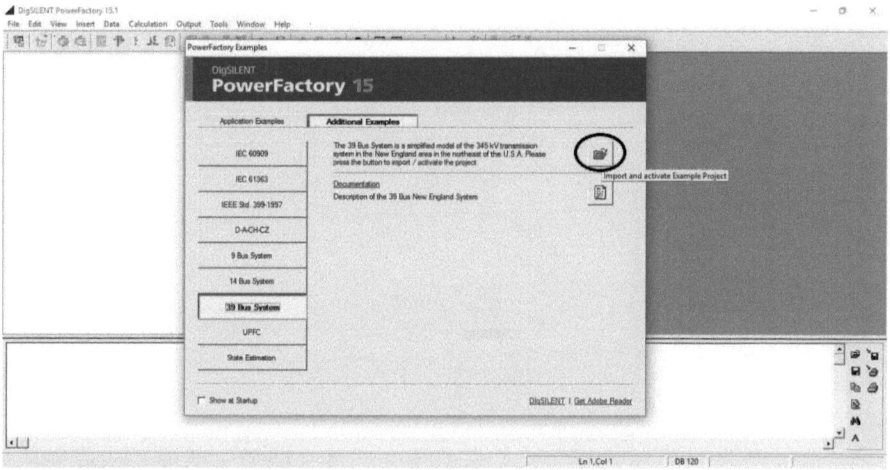

Fig. 11.34 DIgSILENT PowerFactory—Examples → 39 Bus System

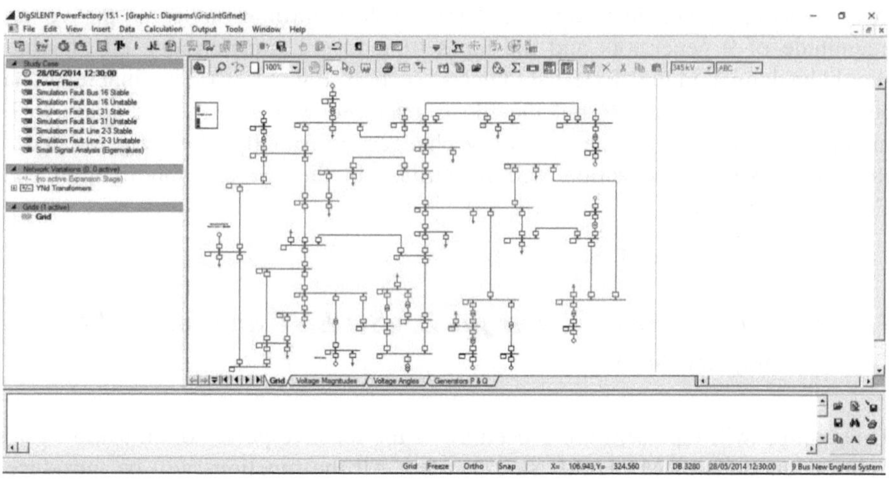

Fig. 11.35 DIgSILENT PowerFactory—39 Bus System ready to use

11.3.1.2 Putting Finishing Touches on the Power Network in DIgSILENT

The next stage will be preparing the prerequisite conditions toward using DIgSILENT scripting facility. The first step will be "Creating a DPL Command Set", refer to the Sect. 11.2.2.1 to find out how to do so. The next step will be creating a general set and introducing it to the DPL which can be easily found in Sect. 11.2.2.2. The third step is introducing external variables to DPL which is

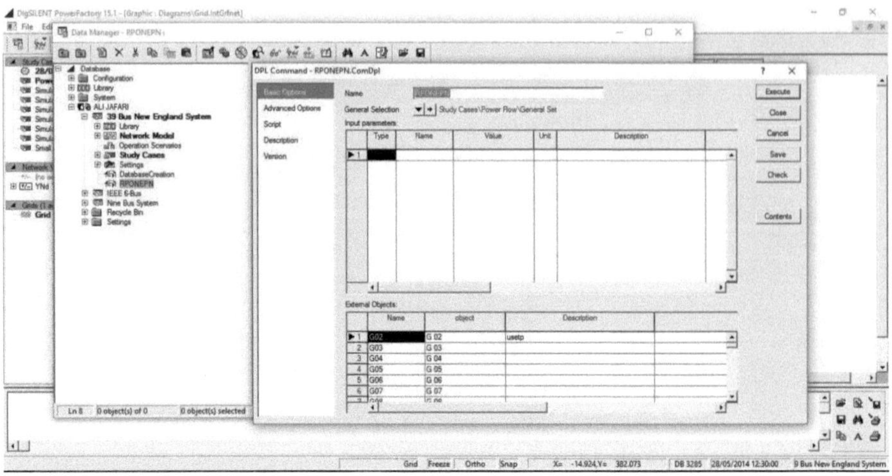

Fig. 11.36 DIgSILENT PowerFactory—DPL command set

already represented in Sect. 11.2.2.3. According to IEEE 39-bus New England Power network given in Chap. 10, there are 20 control variables as the voltage magnitude of 9 generators and the tap setting for 11 transformers, all of which should be defined in the “External Objects” section (Fig. 11.36). The final result has been shown in Fig. 11.36 and the file is enclosed with the chapter ready to be imported, containing all the necessary changes. Eventually, the script related to Reactive Power Optimization will be written in the “Script” tab in the “DPL Command” window.

11.3.2 How to Write Scripts for DIgSILENT and MATLAB Step by Step

This section will represent the procedure of the reactive power optimization using built-in functions of MATLAB step by step. All the functions will be introduced where necessary in order to make it easy to comprehend. The optimization algorithm has considered to be particle swarm optimization algorithm, both of which have got their own efficient functions in MATLAB.

11.3.2.1 Initializing and Defining the Problem in MATLAB

As presented in the codes below, global variables have been considered to be used. Global variables are at work when a variable comes to be used in several functions and each function makes its own changes on the variable, all of which are vital in the procedure. Their values should be set to zero right at the beginning in order to avoid any unexpected interference during several implementations of the process. Cost function has been written as a function itself and defined to the procedure using "Cost function" section in the scripts below. The rest of the codes represents the definition of the upper and lower boundaries of the control variables.

Initializing and Defining the Problem	**MATLAB** **RPONewEngland.m**

```
clc;
clear;
close all;

%% Problem Definition

global nfe NFE BVMin BVMax BestFunValuePSO IterationPSO BestFunValuePS
IterationPS;
NFE = 0;
nfe = 0;
BestFunValuePSO = 0;
IterationPSO = 0;
BestFunValuePS = 0;
IterationPS = 0;

CostFunction = @(OCV) RPO_Function(OCV);        % Cost Function

% Decision Variables %
% VG02,VG03,VG04,VG05,VG06,VG07,VG08,VG09,VG10
% TT1211,TT1213,TT0631,TT1032,TT1933,TT2034,TT2235,TT2537,TT0230
% TT2938,TT1920

% Busbars Voltage Limits P.U. %
% Acceptable Voltage Limitations, +-10% for 16.5, 138, and 230 kV Busbars,
% +-5% for 345 kV Busbars
% BVMin = [0.95 0.95 0.95 0.95 0.95 0.95 0.95 0.95 0.95 0.95 0.95 0.90...
%    0.95 0.95 0.95 0.95 0.95 0.95 0.95 0.90 0.95 0.95 0.95 0.95 0.95...
%    0.95 0.95 0.95 0.95 0.9 0.9 0.9 0.9 0.9 0.9 0.9 0.9 0.9 0.9];
% BVMax = [1.05 1.05 1.05 1.05 1.05 1.05 1.05 1.05 1.05 1.05 1.05 1.10...
%    1.05 1.05 1.05 1.05 1.05 1.05 1.05 1.10 1.05 1.05 1.05 1.05 1.05...
%    1.05 1.05 1.05 1.05 1.10 1.10 1.10 1.10 1.10 1.10 1.10 1.10 1.10 1.10];

% Busbars Voltage Limits P.U. %
% Acceptable Voltage Limitations, +-10% for 16.5, 138, 230 and 345 kV Busbars
BVMin = 0.9*ones(1,39);
BVMax = 1.1*ones(1,39);

% Generators Voltage Limits P.U. %
GVMin = [0.9 0.9 0.9 0.9 0.9 0.9 0.9 0.9 0.9];
GVMax = [1.1 1.1 1.1 1.1 1.1 1.1 1.1 1.1 1.1];

% Transformators Tap Limits %
```

```
TTMin = [-1 -1 -2 -2 -2 -1 -1 -1 -1 -1 -2];
TTMax = [1 1 2 2 2 1 1 1 1 1 2];

VarMin = [GVMin TTMin];        % Lower Bound of Variables
VarMax = [GVMax TTMax];        % Upper Bound of Variables
```

11.3.2.2 Initial Power Losses According to New England 39-Bus Network

In order to calculate initial value of active power losses according to IEEE 39-bus New England it is needed to apply initial condition to the network and after running power flow calculation of losses will be feasible. The matrix "SVCV" represents Standard Values for Control Variables which are taken from the data given in the appendix section of Chap. 10. Then they will be exported in a ".txt" file as "OCV. txt" to be imported in DIgSILENT. Finally, MATLAB will make a "Couple.txt" file as a means of exchanging commands between two pieces of software, waiting for DIgSILENT to implement its duty.

Initial Power Losses	MATLAB RPONewEngland.m

```
%% Initial Power Losses According to New England 39-Bus Network

% Standard Values for Control Variables %
SVCV = [0.982 0.9831 0.9972 1.0123 1.0493 1.0635 1.0278 1.0265 1.0475...
   1 1 2 2 2 1 1 1 1 1 2];

t = fopen('OCV.txt','w');     % Optimal Control Variables
fprintf(t,'%d %d %d %d %d %d %d %d %d %d %d %d %d %d %d %d %d %d %d %d',SVCV);
fclose(t);

% Making a Blank "Couple.txt" File, 0 => DIgSILENT, 1 => MATLAB %
t = fopen('Couple.txt','w');
fprintf(t,'%d',0);
fclose(t);

% Waiting for DIgSILENT to Complete the PowerFlow Process %
a = 0;
while a == 0
    pause(0.02);
    a = load('Couple.txt');
end
```

11.3.2.3 Initializing DIgSILENT Software

Initializing DIgSILENT consists in defining variables, which will be used throughout the procedure of optimization. Integer variables for storing and using integer values, object variables for storing equipment used in the power network, double variables for storing values for control variables, set variables for storing a group of objects, and string variables for storing string values.

Initializing DIgSILENT Software	DIgSILENT

```
int a;
object o;
double V,U,Phi,Ploss,Qloss;
double VG02,VG03,VG04,VG05,VG06,VG07,VG08,VG09,VG10,TT1211,TT1213,
TT0631,TT1032, TT1933,TT2034,TT2235,TT2537,TT0230,TT2938,TT1920;
set s1;
string BusName,LineName;
a=1;
```

11.3.2.4 How to Apply Control Variables to the Network

The main loop of reactive power optimization in DIgSILENT consist in running power flow and exporting the results. Each time MATLAB wants DIgSILENT run, the procedure in DIgSILENT starts from this session, and if MATLAB sends command signal as "2", the procedure in DIgSILENT will be over. Referring to the codes below, DIgSILENT first reads the "OCV.txt" file written by MATLAB and then allocated the values to the control variables "VG02" to "TT1920". Then applying the values using the external objects defined in Sect. 11.3.1.2 will be feasible as done below. It means that now the values considered to be control variables are set to the values created by MATLAB. The next step will be running the power flow according to new values.

Applying Control Variables to the Network	DIgSILENT

```
!! The Main Loop of Reactive Power Optimization
while(a<>2)    ! Checking to Finish the Process
{

fopen('C:\Users\ALI JAFARI\Documents\MATLAB\
IEEE 39-Bus New England Power System\Reactive Power Optimization\OCV.txt','r',0);
fscanf(0,'%d %d %d %d %d %d %d %d %d %d %d %d %d %d %d %d %d %d %d
%d',VG02,VG03,VG04,VG05,VG06,VG07,VG08,VG09,VG10,TT1211,TT1213,
TT0631,TT1032,TT1933,TT2034,TT2235,TT2537,TT0230,TT2938,TT1920);
fclose(0);

G02:usetp = VG02;
G03:usetp = VG03;
G04:usetp = VG04;
G05:usetp = VG05;
G06:usetp = VG06;
G07:usetp = VG07;
G08:usetp = VG08;
G09:usetp = VG09;
G10:usetp = VG10;

T1211:nntap = TT1211;
T1213:nntap = TT1213;
T0631:nntap = TT0631;
T1032:nntap = TT1032;
T1933:nntap = TT1933;
T2034:nntap = TT2034;
T2235:nntap = TT2235;
T2537:nntap = TT2537;
T0230:nntap = TT0230;
T2938:nntap = TT2938;
T1920:nntap = TT1920;
```

11.3.2.5 Running Load Flow in the Script Section in DIgSILENT

After applying values for control variables, it is time to carry the load flow calculations out. To do so, load flow should be introduced to the DPL command set as an external object, which is illustrated in Sect. 11.2.2.3. The procedure is as the following.

Running Load Flow	DIgSILENT

```
! Running Load Flow Calculation
ClearOutput();
ResetCalculation();
Ldf.Execute();
```

11.3.2.6 Exporting Load Flow Results from DIgSILENT

The results of power flow will be exported using "fprintf" function via "LoadFlow.txt" as a comma separated value (CSV) file. The data being exported completely depends on what type of information is exactly needed within the calculations in MATLAB pertaining to the study being done. In this section voltage magnitudes, p.u. values for the voltages and voltages degrees have been exported.

Exporting Load Flow Results	DIgSILENT

```
! Exporting Load Flow Results
fopen('C:\Users\ALI JAFARI\Documents\MATLAB\IEEE 39-Bus New England Power
System\Reactive Power Optimization\LoadFlow.txt','w',0);
fprintf(0,'%s,%s,%s,%s','BusName','Voltage_V','Voltage_PU','VoltageAngle_Deg');
s1 = SEL.AllBars();
s1.SortToName(0);
o = s1.First();
while (o){
V = o:m:Ul;
U = o:m:u1;
Phi = o:m:phiu;
BusName = o:loc_name;
fprintf(0,'%s,%f,%f,%f',BusName,V,U,Phi);
o = s1.Next();
}
fclose(0);
```

11.3.2.7 Exporting Transmission Lines Power Losses from DIgSILENT

The transmission lines power losses can also be exported from DIgSILENT instead of calculating them in MATLAB using the active power losses of transmission lines equation expressed in Chap. 10. The only thing which should be done is to find the appropriate parameter name for active power losses to be able to export the corresponding value for all the transmission lines. DIgSILENT is able to calculate wide range of parameters related to power flow without suffering programming difficulties. How to find the parameter name which is used in the scripting session to export the related value has been presented in Section "Running Power Flow in DIgSILENT Based on Initial Condition". Following the same instruction, "Ploss_bus1" is representative of active power losses in transmission lines and can be used in the scripting segment to export the results. The following code exports power losses for transmission lines in "TLPowerLosses.txt" as a CSV file. Eventually, DIgSILENT sends a signal to MATLAB via "Couple.txt" to let it know that power flow is done, results are exported and ready to use, then it waits to get another set of values for control variables and another command from MATLAB to run power flow based on new condition again.

Exporting Transmission Lines Power Losses	DIgSILENT

```
! Exporting Transmission Lines Power Losses
fopen('C:\Users\ALI JAFARI\Documents\MATLAB\IEEE 39-Bus New England Power
System\Reactive Power Optimization\TLPowerLosses.txt','w',0);
fprintf(0,'%s,%s,%s','LineName','ActivePowerLoss_W','ReactivePowerLoss_VAR');
s1 = SEL.AllLines();
s1.SortToName(0);
o = s1.First();
while (o){
Ploss = o:m:Ploss:bus1;
Qloss = o:m:Qloss:bus1;
LineName = o:loc_name;
fprintf(0,'%s,%f,%f',LineName,Ploss,Qloss);
o = s1.Next();
}
fclose(0);

! Sending Run Signal to MATLAB
fopen('C:\Users\ALI JAFARI\Documents\MATLAB\IEEE 39-Bus New England Power
System\Reactive Power Optimization\Couple.txt','w',1);
fprintf(1,'%d',1);
fclose(1);

! Waitign for MATLAB to Finish the Corresponding Calculations
a = 1;
while(a = 1){
fopen('C:\Users\ALI JAFARI\Documents\MATLAB\IEEE 39-Bus New England Power
System\Reactive Power Optimization\Couple.txt','r',0);
fscanf(0,'%d',a);
fclose(0);
}
}
```

11.3.2.8 Importing Active Power Losses from Text File in MATLAB

In order to import power losses results into MATLAB from the "TLPowerLosses.txt" file, a built-in function called "readtable('TLPowerLosses')" can be used, which creates a table by reading column oriented data from a file. “readtable” creates one variable in “TLPL” for each column in the file and reads variable names from the first row of the file. In the following code MATLAB imports the data written by DIgSILENT in the “TLPowerLosses.txt” file and then sums up the values in the “ActivePowerLoss_W” column to work out the total value of active power losses in transmission lines and stores it in “PowerLosses.Initial.APL”. In addition, it saves the values for control variables in “PowerLosses.Initial.Position”.

Importing Active Power Losses from “TLPowerLosses.txt”	**MATLAB** **RPONewEngland.m**

```
% Loading Active Power Losses from DIgSILENT Calculation %
TLPL = readtable('TLPowerLosses'); % Transmission Lines Power Losses Data
PowerLosses.Initial.APL = sum(TLPL.ActivePowerLoss_W);
PowerLosses.Initial.Position = SVCV;
```

11.3.2.9 Reactive Power Optimization Using Built-in Particle Swarm Function

MATLAB provides a built-in function for particle swarm optimization algorithm in its recent versions. It can be effortlessly used in optimization algorithms adjusting some parameters without programming the algorithm itself. A user defined function has been created as “ParticleSwarmTool” in order to gather all the corresponding parameters and the Particle Swarm built-in function in a unique mfile. The inputs to this function is just lower and upper bounds for control variables and the initial position which is optional.

Reactive Power Optimization Using PSO Algorithm	**MATLAB** **RPONewEngland.m**

```
%% Reactive Power Optimization Using PSO Algorithm
[PowerLosses.PSO.Position,PowerLosses.PSO.APL,exitflagPSO] = ...
   ParticleSwarmTool(VarMin,VarMax);
LNFE = length(nfe);
```

The main function which carries out the optimization procedure is “particleswarm”, while there are a few prerequisite adjustments to be made before using the function itself. In order to access the details of the particle swarm optimization function MATLAB provides another function as “optimoptions”. The detailed information about how to use it is available either online or MATLAB software local help. In this section just the needed adjustments have been presented. “options = optimoptions (options,‘InitialSwarm’, [0.982 0.9831 0.9972, 1.0123 1.0493 1.0635 1.0278 1.0265 1.0475 1 1 2 2 2 1 1 1 1 1 2]);” has been used to define initial point to PSO algorithm which is optional. Self-adjustment and social-adjustment rates have been specified to the function using "options = optimoptions(options,'SelfAdjustment',1.49);" and "options = optimoptions(options,'SocialAdjustment',1.49);". In addition, the member of members in the swarm has been defined using "options = optimoptions (options,'SwarmSize',20);" and so forth, to see more information please refer to MATLAB software help.

Reactive Power Optimization Using PSO Algorithm	**MATLAB** **ParticleSwarmTool.m**

```
function [x,fval,exitflagPSO] = ParticleSwarmTool(VarMin,VarMax)

nVar = 20;            % Number of Decision Variables

options = optimoptions('particleswarm');
options = optimoptions(options,'CreationFcn',@pswcreationuniform);
options = optimoptions(options,'Display','Iter');
options = optimoptions(options,'DisplayInterval',1);
options = optimoptions(options,'FunValCheck','off');
options = optimoptions(options,'HybridFcn',[]);
options = optimoptions(options,'InertiaRange',[0.1 1.1]);
options = optimoptions(options,'InitialSwarm',[0.982 0.9831 0.9972...
   1.0123 1.0493 1.0635 1.0278 1.0265 1.0475 1 1 2 2 2 1 1 1 1 1 2]);
options = optimoptions(options,'InitialSwarmSpan',2000);
options = optimoptions(options,'MaxIter',Inf);
options = optimoptions(options,'MaxTime',Inf);
options = optimoptions(options,'MinFractionNeighbors',0.25);
options = optimoptions(options,'ObjectiveLimit',-Inf);
options = optimoptions(options,'OutputFcns',@ParticleSwarmOutputFcn);
options = optimoptions(options,'PlotFcns',[]);
options = optimoptions(options,'SelfAdjustment',1.49);
options = optimoptions(options,'SocialAdjustment',1.49);
options = optimoptions(options,'StallIterLimit',20);
options = optimoptions(options,'StallTimeLimit',Inf);
options = optimoptions(options,'SwarmSize',20);
options = optimoptions(options,'TolFun',1e-2);
options = optimoptions(options,'UseParallel',0);
options = optimoptions(options,'Vectorized','off');
% The Main Function of Particle Swarm Optimization Algorithm %
[x,fval,exitflagPSO] = ...
   particleswarm(@RPO_Function,nVar,VarMin,VarMax,options);
```

The general form to use the built-in function of particle swarm is like "[x,fval,exitflag, output] = particleswarm(fun,nvars,lb,ub,options)" in which "x" returns the position of control variables, "fval" represents the best cost, "exitflag" describes the exit condition, "output" containing information about the optimization process, "fun" is a function containing objective function, "nvars" refers to number of control variables, "lb" and "lu" embodies lower and upper bounds for control variables and "options" is a variables consisting of all the setting parameters of PSO function. The objective function will be defined as the following codes.

Reactive Power Optimization Objective Function	MATLAB RPO_Function.m

```
function [OF] = RPO_Function(OCV)

  global NFE BVMin BVMax;
  NFE = NFE+1;
  k = 1e15;           % Penalty Factor

  % Calculation of Objective Function (Ploss) %
  t = fopen('OCV.txt','w');   % Consisting of Optimal Control Variables
  fprintf(t,'%d %d %d %d %d %d %d %d %d %d %d %d %d %d %d %d %d %d %d
%d',OCV);
  fclose(t);

  % Sending Run Signal to DIgSILENT; 0 => DIgSILENT, 1 => MATLAB %
  t = fopen('Couple.txt','w');
  fprintf(t,'%d',0);
  fclose(t);

  % Waiting for DIgSILENT to Complete the PowerFlow Process %
  a = 0;
  while a == 0
   pause(0.03);
   a = load('Couple.txt');
  end

  % Loading Power Losses from DIgSILENT Calculation %
  TLPL = readtable('TLPowerLosses');              % Transmission Lines Power Losses Data
  FAPL = sum(TLPL.ActivePowerLoss_W);              % Transmission Systems Active
Power Losses
%   FRPL = sum(TLPL.ReactivePowerLoss_VAR);      % Transmission Systems Reactive
Power Losses

  % Applying Penalty to the Cost Function for Keeping Voltages of PQ Busbars Within
the Limitations %
  PF = readtable('LoadFlow');      % Load Flow Results
  PQBV = (PF{1:29,3})';          % Extracting PQ Busbars Voltatges

  PV = k*sum(BVMax(1:29)-PQBV<0|BVMin(1:29)-PQBV>0);   % To Apply a Penalty to
the Cost Function
                                             % if There is a Voltage Violation
  OF = FAPL + PV;

end
```

In the reactive power optimization function above "NFE" has been considered to work the number of function evaluation out. "% Calculation of Objective Function (Ploss) %" section tries to apply the new set of control variables to the power network in DIgSILENT and carries out power flow and the calculates active power losses accordingly. Section "% Applying Penalty to the Cost Function for Keeping Voltages of PQ Busbars Within the Limitations %" has been considered to apply a penalty to the cost value if there is a violation in the voltages of PQ busbars. The particle swarm built-in function refers to the objective function several times to figure out if the objective function has been optimized or not.

In order to have access to the detailed data throughout the procedure a user defined function "ParticleSwarmOutputFcn" has been used, which is a facility that built-in PSO function provides. It also can be used in order to stop the function using a specific criterion, to see more information refer to MATLAB software help documentation. The code is as following.

Particle Swarm Output Function	**MATLAB ParticleSwarmOutputFcn.m**

```
function stop = ParticleSwarmOutputFcn(optimValues,state)

global nfe NFE BestFunValuePSO IterationPSO;

stop = false;          % This function does not stop the solver

switch state

  case 'init'

  case 'iter'

    BestFunValuePSO(optimValues.iteration) = optimValues.bestfval;
    nfe(optimValues.iteration) = NFE;

  case 'done'

    IterationPSO = optimValues.iteration;

end
```

After finishing the optimization procedure of the PSO function, the optimum positions for control variables are stored in "PowerLosses.PSO.Position" and the cost value for the values of control variables in "PowerLosses.PSO.APL". The next step will be using Pattern Search as the supplementary optimization algorithm to pinpoint the exact optimum control variables.

11.3.2.10 Reactive Power Optimization Using Built-in Pattern Search Function

Pattern search optimization algorithm has been used as a supplementary optimization algorithm in case PSO is not able to pinpoint the optimum control variables. User defined "PatternSearchTool" has been considered to gather all the parameters and settings which are needed to adjust and run pattern search algorithm all together. The codes below call the function.

Reactive Power Optimization Using Pattern Search	MATLAB RPONewEngland.m

```
%% Reactive Power Optimization Using PatternSearch Algorithm
[PowerLosses.PS.Position,PowerLosses.PS.APL,exitflagPS] = ...
  PatternSearchTool(PowerLosses.PSO.Position,VarMin,VarMax);
```

The main function which carries out the optimization procedure is "patternsearch", while there are a few requirements needed to be met before using the function itself. In order to access the details of the pattern search function, MATLAB provides another function as "psoptimset". The detailed information about how to use it is available either online or MATLAB software local help. In this section just the needed adjustments have been presented. "options = psoptimset (options,'TolFun',1e-4);" is the tolerance on the objective function which stops the optimization procedure if both the change in function value and the mesh size are less than "TolFun".

The general form to use the built-in function of pattern search is like "[x,fval, exitflag,output] = patternsearch(fun,x0,A,b,Aeq,beq,LB,UB)" in which "x" returns the position of control variables, "fval" represents the best cost, "exitflag" describes the exit condition, "output" containing information about the optimization process, "fun" is a function containing objective function, "x0" is the starting point to the pattern search algorithm, "A", "b", "Aeq" and "beq" are parameters of inequality and equality constraints, "LB" and "UB" lower bounds and upper bounds of the control variables, and "options" is a variables consisting of all the setting parameters of pattern search function.

Reactive Power Optimization Using Pattern Search	MATLAB PatternSearchTool.m

```
function [x,fval,exitflagPS] = PatternSearchTool(x0,VarMin,VarMax)

options = psoptimset;
options = psoptimset(options,'TolMesh',[]);
options = psoptimset(options,'TolCon',[]);
options = psoptimset(options,'TolX',[]);
options = psoptimset(options,'TolFun',1e-4);
options = psoptimset(options,'TolBind',[]);
options = psoptimset(options,'MaxIter',Inf);
options = psoptimset(options,'MaxFunEvals',[]);
options = psoptimset(options,'TimeLimit',[]);
options = psoptimset(options,'MeshContraction',[]);
options = psoptimset(options,'MeshExpansion',[]);
options = psoptimset(options,'MeshAccelerator','on');
options = psoptimset(options,'MeshRotate',[]);
options = psoptimset(options,'InitialMeshSize',[]);
options = psoptimset(options,'ScaleMesh',[]);
options = psoptimset(options,'MaxMeshSize',[]);
options = psoptimset(options,'InitialPenalty',[]);
options = psoptimset(options,'PenaltyFactor',[]);
options = psoptimset(options,'PollMethod',[]);
options = psoptimset(options,'CompletePoll',[]);
options = psoptimset(options,'PollingOrder','Success');
options = psoptimset(options,'SearchMethod',@GPSPositiveBasis2N);
options = psoptimset(options,'CompleteSearch',[]);
options = psoptimset(options,'Display','iter');
options = psoptimset(options,'OutputFcns',@PatternSearchOutputFcn);
options = psoptimset(options,'PlotFcns',[]);
options = psoptimset(options,'PlotInterval',[]);
options = psoptimset(options,'Cache',[]);
options = psoptimset(options,'CacheSize',[]);
options = psoptimset(options,'CacheTol',[]);
options = psoptimset(options,'Vectorized',[]);
options = psoptimset(options,'UseParallel',[]);

% The Main Function of Pattern Search Optimization Algorithm %
[x,fval,exitflagPS] = ...
  patternsearch(@RPO_Function,x0,[],[],[],[],VarMin,VarMax,[],options);
```

"options = psoptimset(options,'OutputFcns',@PatternSearchOutputFcn);" is used to access the detailed information during the optimization procedure in every iteration as presented for PSO function in Sect. 11.3.2.9. See the help documentation of MATLAB for more information how to use it.

Pattern Search Output Function	MATLAB PatternSearchOutputFcn.m

```
function [stop,options,optchanged] =
PatternSearchOutputFcn(optimValues,options,state)

global nfe NFE BestFunValuePS IterationPS IterationPSO;

stop = false;         % This function does not stop the solver
optchanged = false;

switch state

  case 'init'

  case 'iter'

    BestFunValuePS(optimValues.iteration) = optimValues.fval;
    nfe(optimValues.iteration + IterationPSO) = NFE;

  case 'done'

    IterationPS = optimValues.iteration;

end
```

After finishing the optimization procedure by Pattern Search function, the optimum positions for control variables are stored in "PowerLosses.PS.Position" which will be best point found, and the best cost value will be saved in "PowerLosses.PS.APL" which is the best minimum cost.

11.3.2.11 Finishing the Optimization Procedure and Exporting Results

After finishing optimization procedure, the final step will be sending a stop signal to DIgSILENT and printing the data in relation to what information was meant to be exported. The following codes are an example of such processes.

Finishing the Optimization Procedure	MATLAB RPONewEngland.m

```
%% Sending Stop Signal to DIgSILENT

t = fopen('Couple.txt','w');
fprintf(t,'%d',2);
fclose(t);

%% Results

clc
hold on
plot(nfe(1:LNFE),BestFunValuePSO,'-o','LineWidth',1);
plot(nfe(LNFE+1:end),BestFunValuePS,'-.*','LineWidth',1);
xlabel('Number of Function Evaluation (NFE)');
ylabel('Active Power Losses (W)');

LR = PowerLosses.Initial.APL - PowerLosses.PS.APL;  % Losses Reduction

LRP = 100*((PowerLosses.Initial.APL-PowerLosses.PS.APL)...
   /PowerLosses.Initial.APL);         % Losses Reduction Percentage, Total

disp('------------------------------------------------------------');
str = sprintf('Initial Active Power Losses (W) = %d',...
   PowerLosses.Initial.APL);
disp(str);

disp('------------------------------------------------------------');
str = sprintf('Active Power Losses After Optimization (W) = %d',...
   PowerLosses.PS.APL);
disp(str);

disp('------------------------------------------------------------');
str = sprintf('Losses Reduction (W), Total = %d',LR);
disp(str);

disp('------------------------------------------------------------');
str = sprintf('Losses Reduction Percentage , Total = %d',LRP);
disp(str);
```

11.3.3 How to Run the Optimization Procedure

After having the mfiles of MATLAB and scripts of DIgSILENT ready, it is the final step to run the optimization process. It should be noted that, according to the structure used in the script files enclosed with this chapter, the mfile of MATLAB is the first one which should be run and then DIgSILENT. To run the optimization procedure via MATLAB by "RPO_NewEngland.m" simply click on "Run" icon in the mfile toolbar, and to run the script written in DIgSILENT right click on the script and click on "Execute" (Fig. 11.37).

Figure 11.38 shows the result of optimization after the process is done for one run. The elaborate report of reactive power optimization for the same power grid has been presented in the Simulation Results in Chap. 10.

11.3.4 How to Run Small Signal Analysis in DIgSILENT

There are a few steps through running small signal analysis in DIgSILENT. The IEEE 39-bus New England power system is already prepared for carrying out such analysis. To run the process, activate Small Signal Analysis in the Study Case section (Fig. 11.39), then choose Modal Analysis toolbar (Fig. 11.40). The equation system has to be initialized (Calculate Initial Conditions) (Fig. 11.41), then

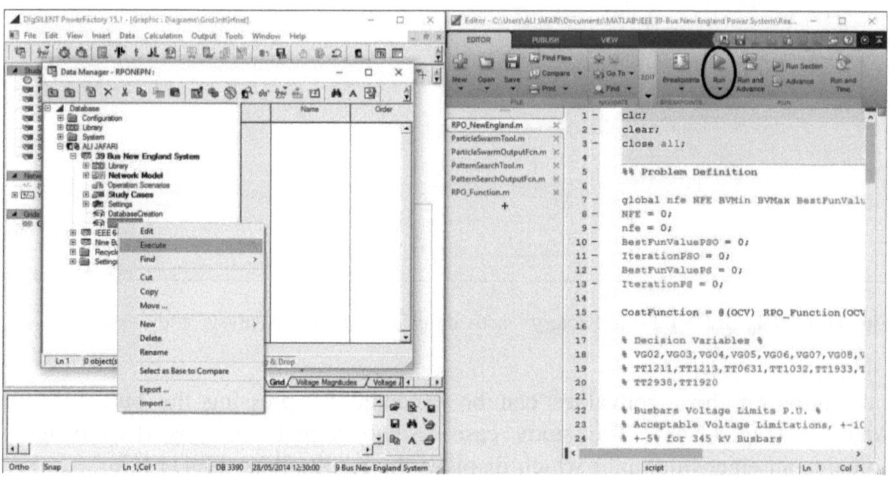

Fig. 11.37 DIgSILENT and MATLAB—how to run the optimization procedure

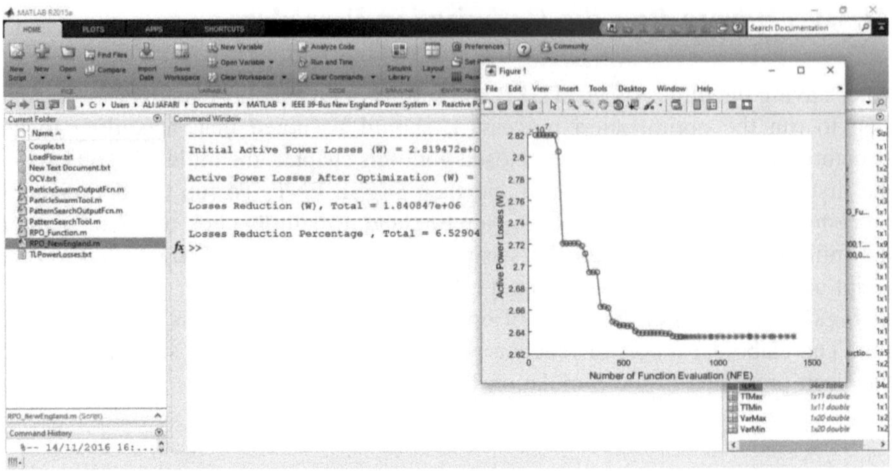

Fig. 11.38 MATLAB—the result after the optimization

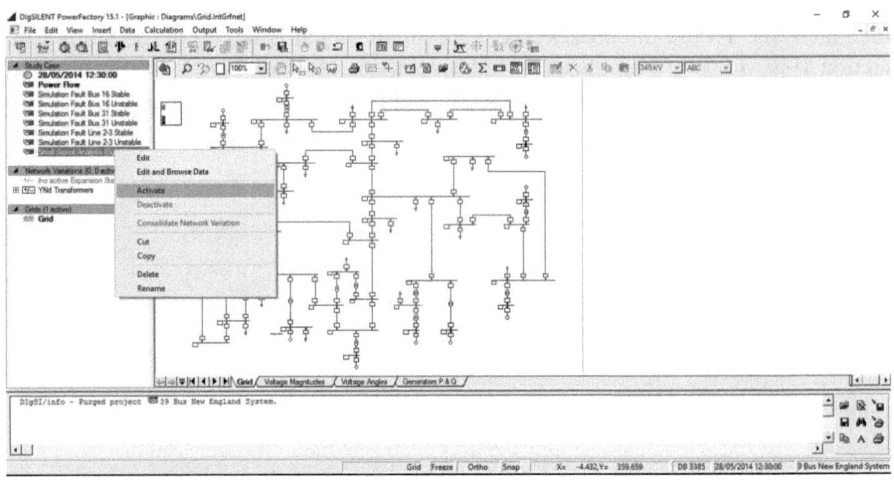

Fig. 11.39 DIgSILENT PowerFactory—activating Small Signal Analysis study case

calculation of the eigenvalues can be executed by pressing the button Modal Analysis (Fig. 11.41). The study case "Small Signal Analysis (Eigenvalues)" contains an eigenvalue plot which displays the results (Fig. 11.41) [1].

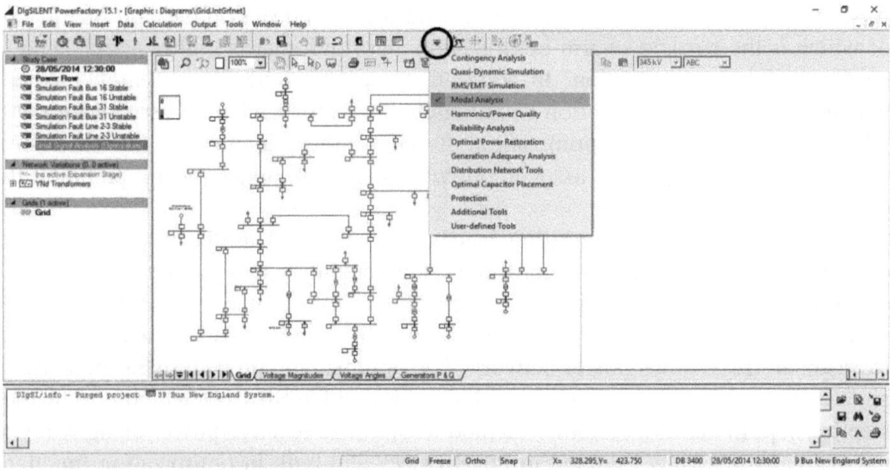

Fig. 11.40 DIgSILENT PowerFactory—Modal Analysis tool

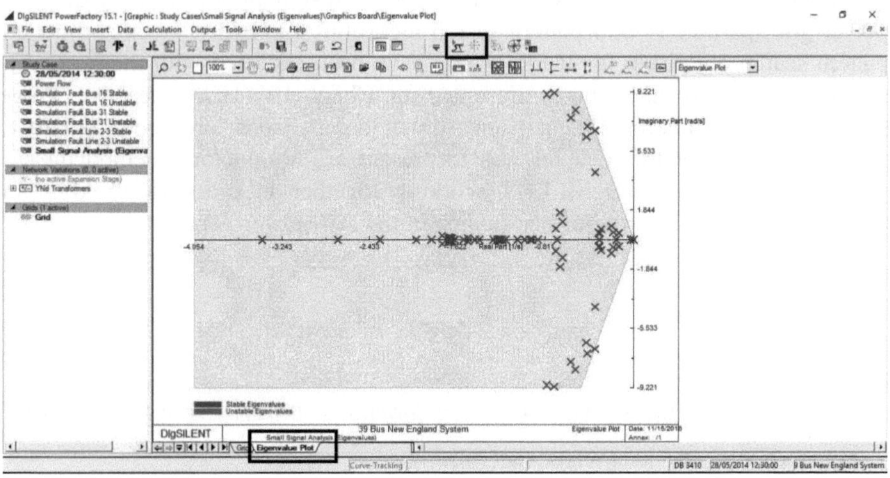

Fig. 11.41 DIgSILENT PowerFactory—Eigenvalue Plot

11.4 Summary

The main objective of reactive power optimization is to reduce active power losses in the power network and improve voltage profile. Reactive power optimization problem is a multidimensional problem which has plenty of local optimum points that causes the traditional optimization methods, which mostly are based on gradient method, to have difficulty finding the global point. Heuristic algorithms have proved their high performance in optimizing these kind of complex problems, one of which is the particle swarm optimization algorithm that is highly appreciated in terms of global optimization. In addition, there are some direct search methods which do not use target function's gradient but itself, of which are very performable when it comes to local optimization. It seems, then, applicable to use a local optimization algorithm in association with a global optimizer to pinpoint the optimum point.

In this chapter, PSO algorithm has been considered as the main and global optimizer and pattern search algorithm as supplementary and the local one through optimizing reactive power in the power grid. This chapter represents a particle swarm pattern search algorithm to optimize reactive power to decrease active power losses in an IEEE 6-bus power network. The optimization procedure has done by an approach consisting of two software MATLAB and DIgSILENT which is explained step by step. In addition, another method to optimize reactive power using ready to use IEEE 39-bus New England in DIgSILENT and built-in optimization functions in MATLAB has been presented while considering the fact that the reader has intermediate knowledge how to use both pieces of software. The corresponding files have been attached with the book to use. Additional approaches containing reactive power optimization using genetic pattern search algorithm for IEEE 6-bus power grid, particle swarm pattern search algorithm for IEEE 14-bus and genetic pattern search algorithm for IEEE 14-bus have been enclosed with the book. The simulations results have been presented in Chap. 10 (10.8—Reactive Power Optimization Simulation Results) and further comparisons have been put there. Overall, the aim of this chapter was to present an introductory method through using MATLAB and DIgSILENT working together in order to avoid a few unimportant steps towards further studies in the AC power network.

References

1. DIgSILENT PowerFactory Version 15 User Manual, Gomaringen, Germany: DIgSILENT GmbH, 2014.
2. MATLAB Help, The MathWorks, Inc., 2014.

Chapter 12
Multi-objective Optimal Reactive Power Dispatch Considering Uncertainties in the Wind Integrated Power Systems

Seyed Masoud Mohseni-Bonab, Abbas Rabiee and Behnam Mohammadi-Ivatloo

Abstract One of the most principle optimization problems which gained the attention of power system operators around the world is optimal power flow (OPF). The OPF basically performs an intelligent power flow and optimizes the system operation condition by optimally determination of control variables. It also considers a specific set of operational constraints and technical limits for this aim, which guaranties both feasibility and optimality of the scheduled operation condition. Generally, this problem can be categorized into two main sub-problems, i.e. optimal reactive power dispatch (ORPD) and optimal real power dispatch, which are differ in their aims and control variables. This chapter deals with the first one, ORPD, which has significant impact on power system security. ORPD is modeled as an optimization problem with nonlinear functions and mixed continuous/discrete variables. Thus, it is a complicated optimization problem. The multi-objective ORPD (MO-ORPD) problem is studied, taking into account different operational constraints such as bus voltage limits, power flow limits of branches, limits of generators voltages, transformers tap ratios and the amount of available reactive power compensation at the weak buses. Three different objective functions are considered in the proposed MO-ORPD framework, which are minimizing total active power losses, minimizing voltage variations and minimizing voltage stability index (*L*-index). These conflicting objectives are optimized via ε-constraint method. In order to model the stochastic behavior of demand and wind power generation, it is necessary to modify the MO-ORPD problem, and develop a probabilistic approach to handle the uncertainties in MO-ORPD problem. Hence, a two-stage stochastic MO-ORPD (SMO-ORPD) is suggested to handle the load and wind power uncertainties in the MO-ORPD problem. In the proposed two-stage

---

S.M. Mohseni-Bonab · A. Rabiee (✉)
Department of Electrical Engineering, University of Zanjan, Zanjan, Iran
e-mail: rabiee@znu.ac.ir

S.M. Mohseni-Bonab
e-mail: s.m.mohsenibonab@ieee.org; masoud.mohseni-bonab.1@ulaval.ca

B. Mohammadi-Ivatloo
Faculty of Electrical and Computer Engineering, University of Tabriz, Tabriz, Iran
e-mail: mohammadi@ieee.org

N. Mahdavi Tabatabaei et al. (eds.), *Reactive Power Control in AC Power Systems*, Power Systems, DOI 10.1007/978-3-319-51118-4_12

stochastic optimization model, the decision variables are classified into two categories, namely, "*here and now*" and "*wait and see*" variables. The optimal values of "*here and now*" variables should be known before realization of scenarios, and therefore, their values are the same for all scenarios while the optimal values of "*wait and see*" variables are based on the realized scenario, and hence their values are scenario dependent. Moreover, in order to examine performance of the proposed SMO-ORPD and the impact of wind power generation on the results of SMO-ORPD, deterministic ORPD (DMO-ORPD) has also been solved in two modes: DMO-ORPD without wind farms (WFs) and any uncertainty, for the sake of comparison with the available methods in recent literature, and DMO-ORPD with WFs. In this chapter the reactive power compensation devices are modeled as discrete/continuous control variables. DMO-ORPD and SMO-ORPD are formulated as mixed integer non-linear program (MINLP) problems, and solved by General Algebraic Modeling System (GAMS). Also, the IEEE 30-bus standard system is utilized for evaluation of the proposed MO-ORPD models.

12.1 Introduction and Problem Statement

Optimal reactive power dispatch (ORPD) is an important problem for power utilities from the viewpoints of system security and energy losses. ORPD is a specific form of optimal power flow (OPF) problem, in which various objective functions such as transmission losses or voltage stability enhancement indices are optimized by adjusting the generator voltages set-points, reactive power compensation in weak buses and optimal setting of transformers tap ratios.

12.1.1 Background and Review of the Recent Literature

Active power losses are the most important objective function used in the classical OPRD problem. Many heuristic algorithms such as multi-agent-based particle swarm optimization approach (MA-PSO) [1], seeker optimization algorithm (SOA) [2], shuffled frog leaping algorithm (SFLA) and Nelder–Mead SFLA (NM-SFLA) [3], combination of hybrid modified imperialist competitive algorithm (MICA) and invasive weed optimization (IWO) named as (MICA-IWO) [4], combination of modified teaching learning algorithm (MTLA) and double differential evolution (DDE) algorithm [5] have been developed to obtain optimal solutions of ORPD while satisfying technical constraints. In [6], active power losses are minimized in wind farms (WFs) by PSO algorithm taking into consideration the reactive power requirement at the point of common coupling (PCC).

In some papers, active power losses are considered with two other common objective functions of ORPD. At the first category active power losses are considered with voltage deviation, while in the second category active power losses are optimized along with a voltage stability enhancement index. In the third category active power losses are considered with both of the other objectives. In order to handle these categories, various intelligent algorithms are employed in the existing literature, such as feasible solutions differential evolution (SF-DE), self-adaptive penalty DE (SP-DE), ε-constraint DE (EC-DE) and stochastic ranking DE (SR-DE) in [7], evolutionary programming (EP), PSO, DE and the hybrid differential evolution (HDE) in [8], combination of NM and Firefly Algorithm (FA) in [9], PSO with an aging leader and challengers (ALC-PSO) in [10] and gray wolf optimizer (GWO) in [11], linear programming (LP) based method using FACTS devices in [12], nonlinear programming (NLP) based method for optimizing local voltage stability index in [13], SOA in [14], DE in [15], PSO, simple genetic algorithm (SGA) and harmony search algorithm (HSA) in [16], gravitational search algorithm (GSA) in [17].

A different research carried out on the single objective ORPD problem, in which steady state voltage stability analysis has been carried out in [18] considering initial conditions for transient stability (TS), small disturbance (SD), and continuation power flow (CPF) for minimizing real power losses. Also, penalty function based method has been presented in [19] to convert discrete ORPD model to a continuous and differentiable model.

The multi-objective ORPD (MO-ORPD) has attracted the attention researchers in recent years. Similar to single objective OPRD, many algorithms utilized to deal with MO-ORPD by considering two or three objective conflicting functions, simultaneously. Most popular objective function of MO-ORPD problem is real power losses which is considered with voltage deviation (as the conflicting objective function) in [20], and solved by strength Pareto evolutionary algorithm (SPEA). Also, it is solved in [21] by non-dominated sorting genetic algorithm-II (NSGA-II).

MO-ORPD problem is modeled as a mixed integer nonlinear program (MINLP) and solved in some literature with considering real power losses and $L_{\max}$ index. It is worth to note that, $L_{\max}$ is a voltage stability index, where it varies from 0 at no-load condition to 1 at voltage collapse point. In [22], NSGA-II and modified NSGA-II (MNSGA-II), In [23] hybrid fuzzy multi-objective evolutionary algorithm (HFMOEA) and in [24], chaotic improved PSO based multi-objective optimization (MOCIPSO) and improved PSO-based multi-objective optimization (MOIPSO) are used for this aim. In more advanced MO-ORPD problem, three objective functions are considered simultaneously. For instance, teaching learning based optimization (TLBO) and quasi-oppositional TLBO (QOTLBO) in [25], chaotic parallel vector evaluated interactive honey bee mating optimization (CPVEIHBMO) in [26], and strength Pareto multi-group search optimizer (SPMGSO) [27] are used to solve MO-ORPD problem.

Moreover, in recent literature ORPD or MO-ORPD problems are solved considering the effects of technical uncertainties raised from the power system restructuring, probable disturbances, or the integration of renewable energies. In [28], chance constrained programming technique is proposed to solve ORPD problem with the aim of minimizing active power losses. Nodal power injections and random branch outages are considered as uncertainty sources. In [29, 30] the load uncertainty is included in MO-ORPD problem, considering different objective functions. Monte Carlo simulations (MCS) are used for handling the load uncertainties in the nonlinear constrained multi objective ORPD problem.

12.1.2 Chapter Contributions

The main focus of this chapter is to solve the MO-ORPD problem in a wind integrated power system considering the uncertainties of load demand and wind power generation. Even these two sources of uncertainty are considered here, but the proposed method is generic and other uncertainty sources could be included via the proposed scenario based approach. The normal probability distribution function (PDF) and Rayleigh (PDF) are used for modeling the load and wind speed uncertainties, respectively.

Three different objective functions, namely active power losses, voltage stability index ($L_{\max}$), and voltage deviations at load buses are considered. The multi-objective problem is handled using ε-constraint technique and optimal Pareto sets are obtained for each pair of the above objective functions. In this chapter, for the sake of comparison with existing methods, the reactive power compensation by shunt VAR compensators is modeled as continuous variable in deterministic MO-ORPD (DMO-ORPD), while in real world problems discrete model is employed for these devices in the proposed stochastic MO-ORPD (SMO-ORPD). Thus, the DMO-ORPD is a NLP problem, while the SMO-ORPD is a MINLP problem. The proposed optimization problems are implemented in GAMS [31], and solved by SNOPT [32] and SBB [33] solvers, for NLP and MINLP problems, respectively.

Hence, the main contributions of this chapter are outlined as follows:

- Modeling and including stochastic nature of loads and wind power generations in the MO-ORPD problem (i.e. development of SMO-ORPD problem).
- Investigation of the impact of renewable power generation on the results obtained by ORPD problem.
- Utilizing discrete model for shunt VAR compensation devices in the proposed SMO-ORPD problem, since most of the pervious literature used continuous modeling for capacitor banks.
- Implementation of ε-constraint technique and fuzzy satisfying criteria to solve the MO-ORPD problem, and for selection of the best compromise solution, respectively.
- Providing comprehensive illustrative studies for different types of ORPD problem, such as DMO-ORPD with and without wind integration and SMO-ORPD.
- Comparison of the obtained results for DMO-ORPD case with previously published methods in literature, which confirms the efficiency of the proposed method.

12.2 Uncertainty Modeling

Uncertain parameters in power systems are classified to technical and economical parameters. Technical parameters consist of operational (like demand and generation) and topological parameters, whereas economical parameters include macroeconomic and microeconomic parameters [34]. There are different methods for modeling these uncertain parameters which are summarized in Fig. 12.1. In this chapter, scenario based probabilistic method is utilized to handle to uncertainties. At the following, the detailed description of considered scenarios is given.

12.2.1 Demand Uncertainty Characterization via Scenario Based Modeling

On account of stochastic nature of the load demand in electric power systems, it is required to model the load uncertainty in operation and planning of power systems. In the general manner load uncertainty can be modeled using the normal or Gaussian PDF [35]. In this chapter, it is presumed that the mean and standard deviation of the load PDF, μ_D and σ_D are known. Probability of *d*-th load scenario is represented by π_d (probability of demand scenario *d*) and calculated using (12.1). Figure 12.2 shows the load levels. It is worth to note that $P_{D_d}^{\min}$ and $P_{D_d}^{\max}$ (minimum/maximum value of real power demand at *d*-th load scenario) are the boundaries of *d*-th interval (or *d*-th load scenario), as shown in Fig. 12.2.

$$\pi_d = \int_{P_{D_d}^{\min}}^{P_{D_d}^{\max}} \frac{1}{\sqrt{2\pi\sigma^2}} \exp\left[-\frac{(P_D - \mu_D)^2}{2\sigma^2}\right] dP_D \tag{12.1}$$

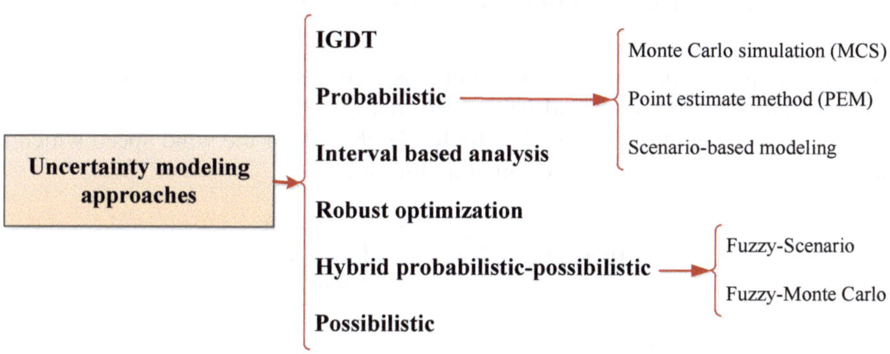

Fig. 12.1 Uncertainty modeling approaches [34]

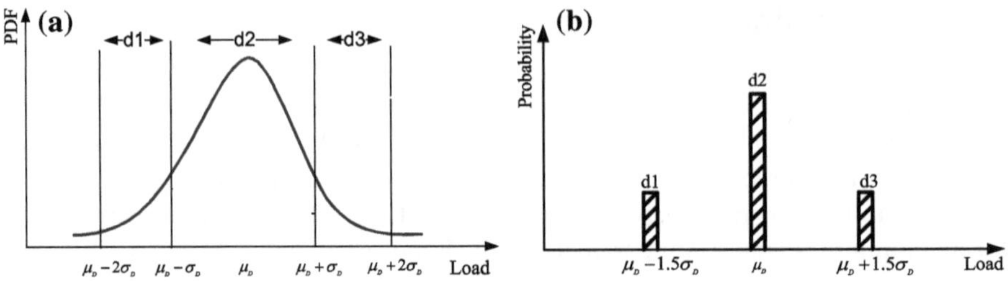

Fig. 12.2 The load PDF and load scenarios, **a** Normal PDF, **b** considered scenarios

Fig. 12.3 Rayleigh PDF for wind speed characterization

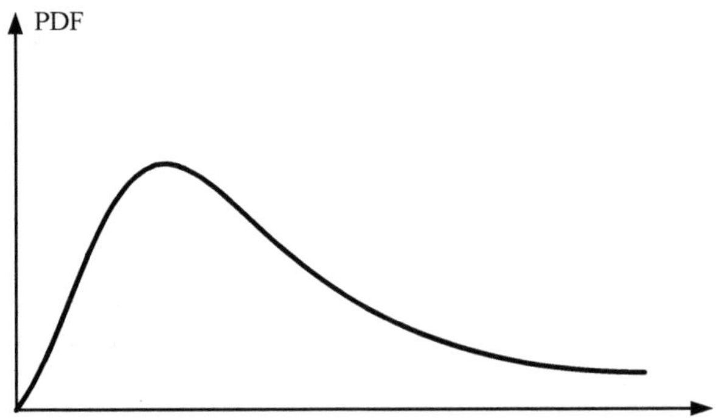

$$P_{D_d} = \frac{1}{\pi_d} \int_{P_{D,d}^{\min}}^{P_{D,d}^{\max}} \left(P_D \times \frac{1}{\sqrt{2\pi\sigma^2}} \exp\left[-\frac{(P_D - \mu_D)^2}{2\sigma^2} \right] \right) dP_D \tag{12.2}$$

12.2.2 Wind Power Generation Uncertainty Modeling

Principally the wind speed uncertainty is modeled using the Rayleigh or Weibul PDF [36, 37]. It should be noticed that the Weibull distribution is a generalized form of the Rayleigh PDF. The Rayleigh PDF of the wind speed which is depicted in Fig. 12.3, can be exhibited as follows

$$\mathrm{PDF}(v) = \left(\frac{v}{c^2}\right) \exp\left[-\left(\frac{v}{\sqrt{2}c}\right)^2 \right] \tag{12.3}$$

where v is wind speed in m/s.

The wind speed variation range is classified into intervals, which is named wind scenarios. The probability of each scenario can be calculated from the following equation. The probability of scenario s and the corresponding wind speed v_s is calculated using the following equations.

$$\pi_w = \int_{v_{i,w}}^{v_{f,w}} \left(\frac{v}{c^2}\right) \exp\left[-\left(\frac{v}{\sqrt{2}c}\right)^2\right] dv \tag{12.4}$$

$$v_w = \frac{1}{\pi_w} \int_{v_{i,w}}^{v_{f,w}} \left(v \times \left(\frac{v}{c^2}\right) \exp\left[-\left(\frac{v}{\sqrt{2}c}\right)^2\right]\right) dv \tag{12.5}$$

where, v_w is the wind speed at w-th wind scenario, and $v_{i,w}, v_{f,w}$ are the starting and the last points of wind speed's interval at w-th scenario, respectively. Also, c is scaling parameter which is acquired by historical wind data.

The characteristics curve of a wind turbine determines the correspondence between the available wind speed and generated wind power. A linearized characteristics curve is presented in Fig. 12.4 [38]. Using this curve, the predicted production power of the wind turbine for various wind speeds can be obtained using the following equation.

$$P_w^{avl} = \begin{cases} 0 & v_w \le v_{in}^c \text{ or } v_w \ge v_{out}^c \\ \frac{v_w - v_{in}^c}{v_{rated} - v_{in}^c} P_r^w & v_{in}^c \le v_w \le v_{rated} \\ P_r^w & v_{rated} \le v_w \le v_{out}^c \end{cases} \tag{12.6}$$

where, P_w^{avl} is available wind power generation, v_{in}^c, v_{out}^c are cut-in/out speed of wind turbine in m/s, and v_{rated} is rated speed of wind turbine in m/s.

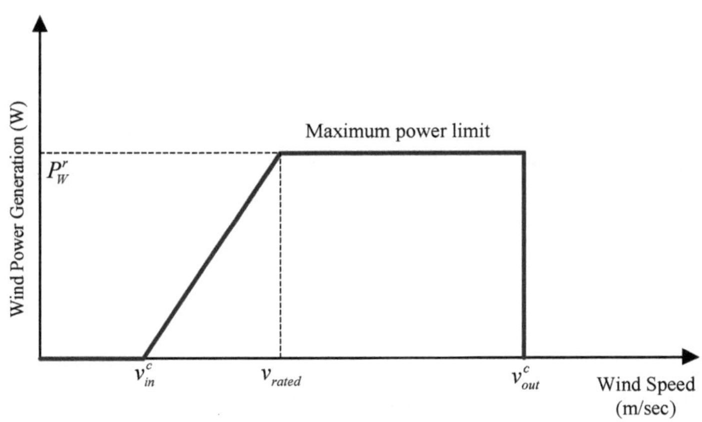

Fig. 12.4 The power curve of a wind turbine

By producing the appropriate number of scenarios for wind power and load demand, the whole number of compound wind-load scenarios is attained by multiplying the number of wind and load individual scenarios. The probability of scenario s, which is acquired considering w-th scenario of wind and d-th scenario of load demand, can be attained using the following equation.

$$\pi_s = \pi_w \times \pi_d \tag{12.7}$$

where, π_s is probability of scenario s, and π_w is probability of wind power generation scenario w.

Figure 12.5 illustrates the procedure used to generate the total number of 15 wind-load scenarios with related probabilities. As presented in Fig. 12.5, the normal probability distribution function (PDF) and Rayleigh (PDF) are used for modeling the load and wind speed uncertainties. The mean value of this PDF is the rated power of loads given in [39]. The standard deviation is supposed to be 2% of the mean load. Also, the entire PDF of loads is divided into three discrete areas and hence three scenarios are considered for loads. The parameters of wind speed PDF and the correlating wind power generation scenarios for this wind farm are adjusted from [40]. There are five wind power generation scenarios, which their features are summarized in Fig. 12.5.

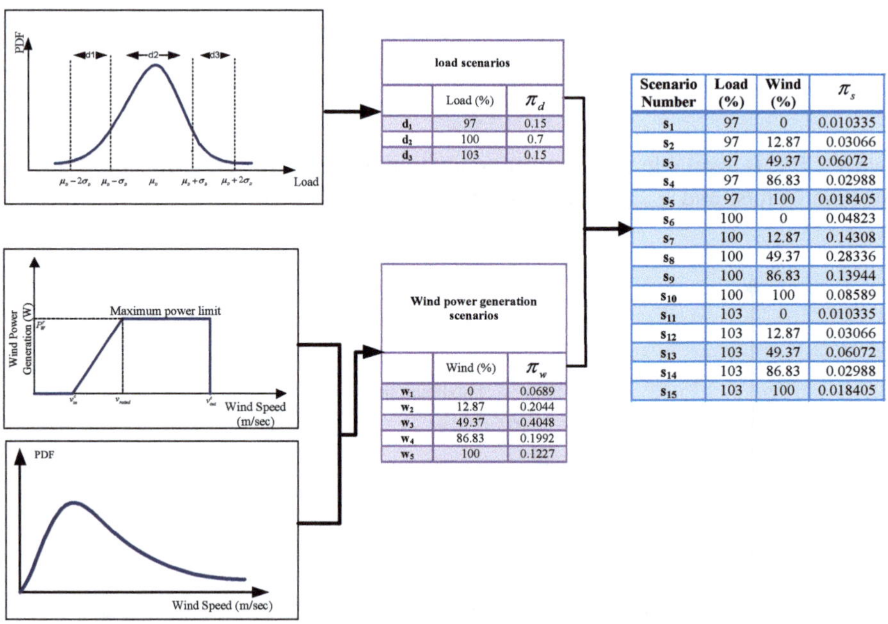

Fig. 12.5 Illustration of scenario generation procedure

12.3 The Problem Formulation

In this section, the studied objective functions, description of ε-constraint method for dealing with the MO-ORPD, fuzzy satisfaction method for selection of the best compromise solution, and the operational constraints such as load flow equations along with various technical limits, are represented.

12.3.1 Objective Functions

Voltage stability of the power system is greatly related to reactive power management. Hence, voltage stability improvement is also considered as another objective function along with the total active power losses and voltage deviations. These objective functions may be conflicted [2, 24]. The subsets of ORPD problem variables can be stated as follows.

$$
\begin{aligned}
\bar{\Theta} &= \begin{bmatrix} V_{i,} & \forall i \in \Omega_G & \\ t_m, & \forall m \in \Omega_T & \\ Q_{C_i,s}, & \forall i \in \Omega_C, & \forall s \in \Omega_S \\ P_{W_i,s}, & \forall i \in \Omega_W, & \forall s \in \Omega_S \\ Q_{W_i,s}, & \forall i \in \Omega_W, & \forall s \in \Omega_S \end{bmatrix} \\
\bar{X} &= \begin{bmatrix} V_{i,s}, & \forall i \in \Omega_{PQ}, & \forall s \in \Omega_S \\ \theta_{i,s}, & \forall i \in \Omega_B, & \forall s \in \Omega_S \\ S_{\ell,s}, & \forall \ell \in \Omega_L, & \forall s \in \Omega_S \\ Q_{G_i,s}, & \forall i \in \Omega_G, & \forall s \in \Omega_S \\ P_{G_i,s}, & \forall i = sb, & \forall s \in \Omega_S \end{bmatrix}
\end{aligned}
\tag{12.8}
$$

where, $\overline{\Theta}$ is vector of control variables, $\overline{X}$ is vector of state (or dependent) variables, Ω_B is set of all buses, Ω_L is set of all branches, Ω_G is set of all generating units, Ω_W is set of wind farms, Ω_S is set of all possible scenarios, Ω_T is set of all tap changing transformers, Ω_C is set of all VAR compensators, Ω_{PQ} is set of system *PQ* buses, and Ω_{B_i} is set of buses connected to bus *i*.

12.3.1.1 Minimization of Total Active Power Losses

Minimizing the total power losses in transmission system is critical objective in power systems for improvement of the total energy efficiency and economic reasons. The active power losses in scenario *s* is mathematically expressed as follows.

$$PL_s\left(\overline{\Theta}_s, \overline{X}_s\right) = \sum_{i \in \Omega_G} P_{G_i,s} + \sum_{i \in \Omega_W} P_{W_i,s} - \sum_{i \in \Omega_B} P_{D_i,s} \tag{12.9}$$

where, PL_s is active power losses in scenario s, $P_{G_i,s}$ is active power production of generator at bus i in scenario s, $P_{W_i,s}$ active power generation of wind farm at scenario s, and $P_{D_i,s}$ real power demand of i-th bus in scenario s.

Expected value of power losses (*EPL*), Ξ_{PL}, over the whole scenarios is considered as the first objective function, F_1. It is computed using the following equation.

$$F_1 = \Xi_{PL} = \sum_{s \in \Omega_S} (\pi_s \times PL_s) \tag{12.10}$$

where, $\Xi_{PL}^{\min}/\Xi_{PL}^{\max}$ are minimum/maximum value for expected real power loss.

12.3.1.2 Minimization of Voltage Deviations

The second objective of ORPD problem is to sustain a proper voltage level at load buses. Electrical equipment is constructed for optimal operation of nominal voltage. Any violation from this rated voltage can result in reduced efficiency and life reduction of electric devices. Thus, the voltage profile of the system could be optimized by minimization of the total voltage deviations (*VD*) from the corresponding rated values at load buses. This objective function is represented as follows.

$$VD_s\left(\overline{\Theta}_s, \overline{X}_s\right) = \sum_{i \in \Omega_{PV}} \left| V_{i,s} - V_{i,s}^{spc} \right| \tag{12.11}$$

where, VD_s is voltage deviations value in scenario s, $V_{i,s}$ is voltage magnitude of bus i in scenario s, and Ω_{PV} is set of system *PV* buses.

Similar to the F_1, here, the expected value of voltage deviations (*EVD*), Ξ_{VD}, is the second conflicting objective function (F_2). *EVD* is calculated as follows.

$$F_2 = \Xi_{VD} = \sum_{s \in \Omega_S} (\pi_s \times VD_s) \tag{12.12}$$

where, $\Xi_{VD}^{\min}/\Xi_{VD}^{\max}$ are minimum/maximum value of expected voltage deviation.

12.3.1.3 Minimization of Voltage Stability Index (L-Index)

Some methods can be used for associating static voltage stability enhancement in ORPD problem. For example, power-voltage curves is utilized as an index in [41] for static voltage stability modeling. Static voltage stability index based on the modal analysis was used in [42] for determination of voltage stability margin. Minimum singular value of the load flow Jacobin matrix [43] and minimum *L*-index [44] are other indexes used for determining the voltage stability margin of the system. In this chapter *L*-index is selected for quantifying voltage stability. This index illustrates the distance of the current state of power system from the voltage stability limit point, which is evaluated using power flow solution. It should be mentioned that the value of *L*-index varies between 0 and 1. *L*-index value less than 1 (voltage collapse point) and close to 0 (no load point) corresponds with more voltage stability margin. The voltage magnitude and phase angle of network buses are functions of system load and generation. By enlarging the transmitted power and for near maximum power transfer position, the voltage stability index values for load buses becomes closer to 1, which shows that the system is closer to voltage collapse. For any load node *j*, *L*-index can be expressed as [25]

$$L_j = \left| 1 - \sum_{i \in \Omega_G} \overline{\lambda}_{ji} \frac{\overline{V}_i}{\overline{V}_j} \right|, \quad \forall j \in \Omega_{PQ} \tag{12.13}$$

where, $\overline{V}_i = V_i \angle \delta_i$ and $\overline{V}_j = V_j \angle \delta_j$. In order to calculate $\overline{\lambda}_{ji}$, the system Y_{BUS} matrix is reorganized as follows

$$\begin{bmatrix} \overline{I}_L \\ \overline{I}_G \end{bmatrix} = \begin{bmatrix} Y_{GG} & Y_{GL} \\ Y_{LG} & Y_{LL} \end{bmatrix} \begin{bmatrix} \overline{V}_L \\ \overline{V}_G \end{bmatrix} \tag{12.14}$$

where $\overline{V}_L$ and $\overline{I}_L$ are the vectors of *PQ* buses voltage and injected currents phasors, whereas $\overline{V}_G$ and $\overline{I}_G$ are the ones for generator buses (including slack and *PV* buses). Since, $\overline{I}_L = 0$, the following expression can be inferred from (12.14).

$$\overline{\Gamma} = -[Y_{LL}]^{-1}[Y_{LG}] \tag{12.15}$$

Since $\overline{\Gamma}$ is a complex matrix, then $\overline{\lambda}_{ji}$ in (12.13) can be obtained as follows.

$$\overline{\Gamma} = \left[\overline{\lambda}_{ji}\right], (\forall j \in \Omega_{PQ}, \forall i \in \Omega_G) \tag{12.16}$$

where, *i*, *j* are indexes of bus numbers.

Therefore, for each scenario *s*, the maximum value of *L*-index among all load buses is considered as the voltage stability index as follows

$$LM_s\left(\overline{\Theta}_s, \overline{X}_s\right) = \max(L_j), \forall j \in \Omega_{PQ} \tag{12.17}$$

where, LM_s is $L_{\max}$ value in scenario *s*.

The third objective (F_3) is the expected value of $L_{\max}$, Ξ_{LM}, for all scenarios, which is obtained as follows

$$F_3 = \Xi_{LM} = \sum_{s \in \Omega_S} (\pi_s \times LM_s) \tag{12.18}$$

where, $\Xi_{LM}^{\min}/\Xi_{LM}^{\max}$ are minimum/maximum value of $EL_{\max}$.

12.3.2 ε-Constraint Method

ε-constraint method is an approach in which the multi-objective optimization problem is converted to a conventional single-objective problem [35]. In this method, all objective functions except one, considered as inequality constraints by assigning a proper value of control parameter named as ε parameter. In the proposed SMO-ORPD problem, two different multi-objective cases are studied.

In the first case (Case-I), F_1 and F_2 are minimized. In order to solve this multi-objective optimization problem by ε-constraint method, one of the objective functions (here we assumed F_2) is moved to the constraints, i.e. it is considered as a constraint and the other objective (here, F_1) is minimized subject to this new constraint and the constraints of the original multi-objective problem, as follows.

$$\begin{aligned} &\min\ (F_1) \\ &\text{s.t.: } F_2 \leq \varepsilon_{F_2} \end{aligned} \tag{12.19}$$

It can be observed from (12.19) that Ξ_{VD} is constrained by the parameter ε. This parameter varies from the minimum value to the maximum value of F_2 (from $F_2^{\min}$ to F_2^{max}) and F_3 (from $F_3^{\min}$ to $F_3^{\max}$) gradually, and for any value of ε_{F_2} the modified single objective optimization problem (12.19) is solved. The set of obtained solutions for the entire variations of ε_{F_2} are called Pareto optimal front of the multi-objective optimization problem.

Similarly, in the second case (Case-II), F_1 and F_3 are minimized. In the ε-constraint method, one of the objectives (here F_3) is moved to the constraints, i.e. it is considered as a constraint and the other objective (F_1) is minimized subject to this new constraint and the constraints of the original multi-objective problem, as follows.

$$\begin{aligned} &\min(F_1) \\ &\text{s.t. :} \quad F_3 \leq \varepsilon_{F_3} \end{aligned} \tag{12.20}$$

Similarly, in this case Ξ_{LM} is constrained by the parameter ε_{F_3}. This parameter varies from the minimum value to the maximum value of F_3 (from $F_3^{\min}$ to F_3^{max})

gradually, and for any value of ε_{F_3} the modified single objective optimization problem (12.20) is solved and the Pareto optimal front of the problem is obtained.

12.3.3 Fuzzy Decision Maker

By solving the MO-ORPD problem a Pareto front is derived. The solutions corresponding to Pareto front are non-dominated solution and another method is required to select the best compromising solution among the obtained non-dominated solutions. Fuzzy decision maker is utilized in this chapter for this purpose. In this method a fuzzy membership function is assigned to each solution in the Pareto front. The fuzzy membership is in the interval [0, 1]. The linear fuzzy membership functions can be expressed for k-th objective function using the following equation [35].

$$\mu_k = \begin{cases} 1 & F_k \leq F_k^{\min} \\ \frac{F_k - F_k^{\max}}{F_k^{\min} - F_k^{\max}} & F_k^{\min} \leq F_k \leq F_k^{\max} \\ 0 & F_k \geq F_k^{\max} \end{cases} \tag{12.21}$$

where, k is index of objective functions, F_k is individual value of k-th conflicting objective function and $\hat{F}_k$ is normalized value of k-th objective function.

For the obtained Pareto optimal set, the best compromise solution can be selected using the min-max method described in [45]. In this method, for r-th Pareto optimal solution the minimum membership number $(\hat{\mu}_r)$ is obtained as follows:

$$\hat{\mu}_r = \min(\mu_k), \forall k \in \Omega_{OF} \tag{12.22}$$

where, Ω_{OF} is set of conflicting objective functions.

Now, the best compromise solution is that which has the maximum value of minimum membership number, as follows:

$$\widehat{\mu} = \max(\hat{\mu}_r), \forall r \in \Omega_{POS} \tag{12.23}$$

where, Ω_{POS} is set of all optimal Pareto solutions of a multi-objective optimization problem.

12.3.4 Constraints

12.3.4.1 Equality Constraints (AC Power Balance Equations)

The feasible solution should assure the power flow equations in each scenario, which are represented mathematically in the following.

$$\begin{cases} P_{G_i,s}+P_{W_i,s}-P_{D_i,s}=\sum\limits_{j\in\Omega_{B_i}} V_{i,s}V_{j,s}\left(G_{ij}\cos\left(\theta_{i,s}-\theta_{j,s}\right)+B_{ij}\sin\left(\theta_{i,s}-\theta_{j,s}\right)\right) \\ Q_{G_i,s}+Q_{W_i,s}+Q_{C_i,s}-Q_{D_i,s}=\sum\limits_{j\in\Omega_{B_i}} V_{i,s}V_{j,s}\left(G_{ij}\sin\left(\theta_{i,s}-\theta_{j,s}\right)-B_{ij}\cos\left(\theta_{i,s}-\theta_{j,s}\right)\right) \end{cases} \tag{12.24}$$

where, $P_{G_i,s}, Q_{G_i,s}$ are active and reactive powers production of generator at bus i in scenario s, $P_{W_i,s}, Q_{W_i,s}$ active and reactive power generations of wind farm at scenario s, $P_{D_i,s}, Q_{D_i,s}$ real and reactive power demand of i-th bus in scenario s, and G_{ij}, B_{ij} are real and imaginary parts of ij-th element of Y_{BUS} matrix (pu/radian).

12.3.4.2 Inequality Constraints

The active and reactive power output of generators and voltage magnitudes of all buses should be kept in the predefined ranges as follows.

$$P_{G_i}^{\min} \le P_{G_{sl},s} \le P_{G_i}^{\max}, \forall i = sb, \forall s \in \Omega_S \tag{12.25}$$

$$Q_{G_i}^{\min} \le Q_{G_i,s} \le Q_{G_i}^{\max}, \forall i \in \Omega_G, \forall s \in \Omega_S \tag{12.26}$$

$$V_i^{\min} \le V_{i,s} \le V_i^{\max}, \forall i \in \Omega_B, \forall s \in \Omega_S \tag{12.27}$$

where, $P_{G_i}^{\min}, P_{G_i}^{\max}$ are minimum and maximum value for active power, $Q_{G_i}^{\min}, Q_{G_i}^{\max}$ are minimum and maximum value for reactive power of generator at bus i, and $V_i^{\min}/V_i^{\max}$ are minimum and maximum value for voltage magnitude of i-th bus.

The power flowing from the branches is constrained to its maximum value as follows.

$$\left|S_{\ell,s}\right| \le S_\ell^{\max}, \forall \ell \in \Omega_L, \forall s \in \Omega_S \tag{12.28}$$

where, ℓ is index of transmission lines, $S_{\ell,s}$ is power flow of ℓ - th branch in scenario s, and $S_\ell^{\max}$ is maximum transfer capacity of line ℓ.

The tap amounts of tap changers are also restricted as follows.

$$t_m^{\min} \le t_m \le t_m^{\max}, \forall m \in \Omega_T \tag{12.29}$$

where, m is index of tap changing transformers, and $t_m^{\min}, t_m^{\max}$ are minimum/maximum value for m-th tap changer.

It is noteworthy that the reactive power output of VAR compensation devices are modeled as a multi-step compensation, i.e. a discrete variable is utilized for each VAR compensation node as follows, which determines the required steps for VAR injections.

$$Q_{C_i,s} = Q_{C_i}^b \times u_{C_i,s}, \forall i \in \Omega_C, \forall s \in \Omega_S \tag{12.30}$$

where, $Q_{C_i,s}$ is reactive power compensation at bus i in scenario s, $Q_{C_i}^b$ is VAR compensation capacity in each step at bus i, and $u_{C_i,s}$ is reactive power compensation step at bus i in scenario s.

The reactive power compensation stages are restricted as follows

$$u_{C_i}^{\min} \leq u_{C_i,s} \leq u_{C_i}^{\max} \quad \forall i \in \Omega_C, \forall s \in \Omega_S \tag{12.31}$$

where, $u_{C_i}^{\min}, u_{C_i}^{\max}$ are minimum and maximum value for reactive power compensation at bus i.

Also, for the available active and reactive power outputs of wind farms, the following constraints should be satisfied

$$0 \leq P_{W_i,s} \leq \zeta_{W_{i,s}} \times P_{W_i}^r, \forall i \in \Omega_W, \forall s \in \Omega_S \tag{12.32}$$

$$Q_{W_i}^{\min} \leq Q_{W_i,s} \leq Q_{W_i}^{\max}, \forall i \in \Omega_W, \forall s \in \Omega_S \tag{12.33}$$

where, $P_{W_i}^r$ is wind farm rated capacity installed in bus i, $\zeta_{W_{i,s}}$ is percentage of wind power rated capacity realized at scenario s in bus i, and $Q_{W_i}^{\min}, Q_{W_i}^{\max}$ are minimum/maximum value of reactive power produced by wind farm.

In this chapter in line with references [35–37], the reactive power output of wind farms are related to the active power output as follows.

$$\begin{cases} Q_{W_i}^{\max} = \tan\left(\cos^{-1}(PF_{lg,i})\right) \times P_{W_i,s} \\ Q_{W_i}^{\min} = -\tan\left(\cos^{-1}(PF_{ld,i})\right) \times P_{W_i,s} \end{cases} \tag{12.34}$$

where, $PF_{lg,i}, PF_{ld,i}$ are lag/lead power factor limits of the wind farms located at node i.

12.4 Scenario Generation and Two-Stage Stochastic Programming

In this chapter two-stage stochastic programming method is utilized for decision making in an uncertain environment. In this method, the decision variables are classified as "*here and now*" and "*wait and see*" variables [46]. The optimal values of "*here and now*" or *"first stage"* variables should be recognized before realization of scenarios. In other words, their values are scenario independent and are similar for all scenarios. In other words, the optimal values of "*wait and see*" or *"second stage"* variables should be considered after realization of the scenarios. In other words, their values are scenario dependent and may be different for different scenarios. In the suggested SMO-ORPD problem the decision variables (DVs) are generator voltages, tap values of tap changing transformers and reactive power output of VAR compensators in the weak buses.

As it is mentioned before, the set of control variables is categorized into two separated subsets, i.e. *here and now* and *wait and see* control variables. The set of here and now decision variables (DV_{HN}) are as follows:

$$DV_{HN} = \left\{ \begin{array}{ll} V_i, & \forall i \in \Omega_G \\ t_m, & \forall m \in \Omega_T \end{array} \right\} \tag{12.35}$$

where, t_m is value of m-th tap changer setting.

Also, the set of wait and see decision variables (DV_{WS}) are as follows.

$$D_{WS} = \left\{ \begin{array}{lll} Q_{C_i,s}, & \forall i \in \Omega_C, & \forall s \in \Omega_S \\ P_{W_i,s}, & \forall i \in \Omega_W, & \forall s \in \Omega_S \\ Q_{W_i,s}, & \forall i \in \Omega_W, & \forall s \in \Omega_S \end{array} \right\} \tag{12.36}$$

where, $Q_{C_i,s}$ is reactive power compensation at bus i in scenario s.

12.5 Simulations on a Standard Test System

Simulations are carried out on the IEEE 30-bus test system. In order to show the effectiveness of the presented approach, several cases are studied as follows.

(A) Deterministic optimization without wind farms (by neglecting the uncertainties of load and wind farms)
(B) Deterministic optimization with the expected value of wind farms power output and mean value of load (by neglecting uncertainty)
(C) Stochastic optimization with load and wind farms power generation uncertainties (uncertainty representation using scenario based approach).

For the sake of comparison with available methods in literature, the VAR compensation devices are modeled as continuous control variables in case (A). While in cases (B) and (C) the VAR compensations are modeled with discrete steps as defined in previous section.

In cases (B) and (C), it is assumed that a wind farm with 56 MW rated capacity is installed at bus 20.

12.5.1 Test System

IEEE 30-bus system [47] includes 30 buses with 6 generator buses as represented in Fig. 12.6. Bus 1 is the slack bus and buses 2, 5, 8, 11 and 13 are *PV* buses whereas

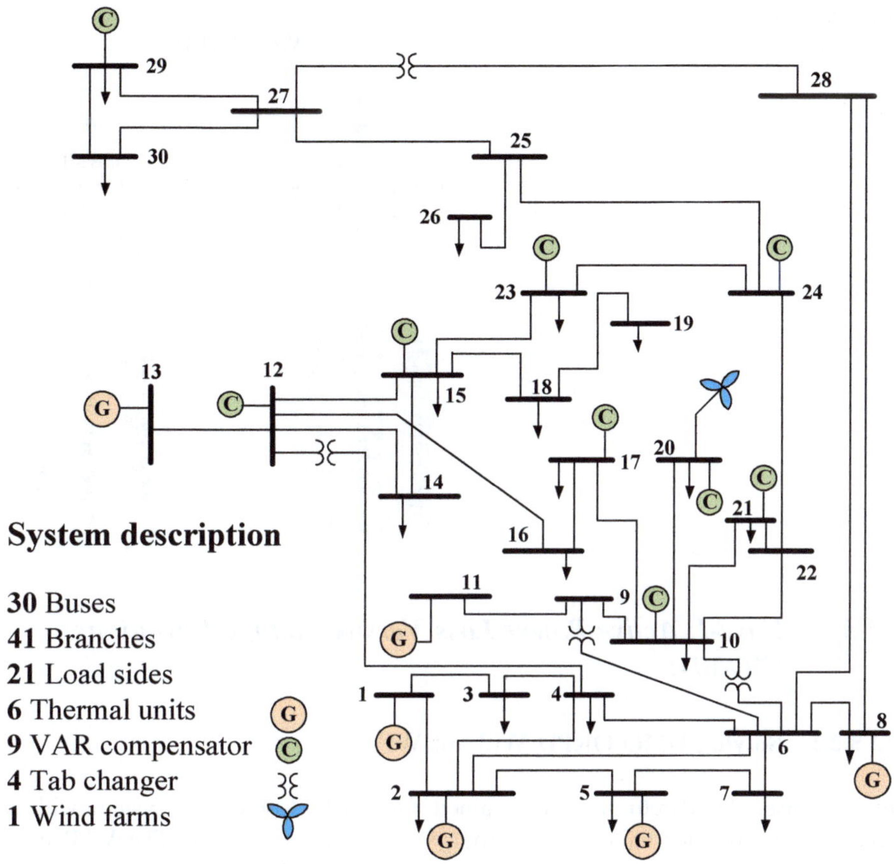

Fig. 12.6 One-line diagram of IEEE 30-bus test system

the remaining 24 buses are *PQ* buses. The network consists of 41 branches, 4 transformers and 9 capacitor banks. Four branches, 6–9, 6–10, 4–12 and 28–27 are under load tap changing transformers. The tap ratios are within the interval [0.9, 1.1]. Additionally, buses 10, 12, 15, 17, 20, 21, 23, 24 and 29 are selected as shunt VAR compensation buses.

In each study we have considered two cases regarding the objective functions. As mentioned in Sect. 12.2, in MO-ORPD problem F_1 with F_2 and F_1 with F_3 are conflicting. Minimization is classified in two part PL/Ξ_{PL} and VD/Ξ_{VD} (Case I) and *PL/* and LM/Ξ_{LM} (Case II). Figure 12.7 shows the case studies conducted in this chapter. For each case, at first the DMO-ORPD is solved for the sake of comparison with the existing methods, and then SMO-ORPD is solved to investigate the impact of uncertainties on the obtained results.

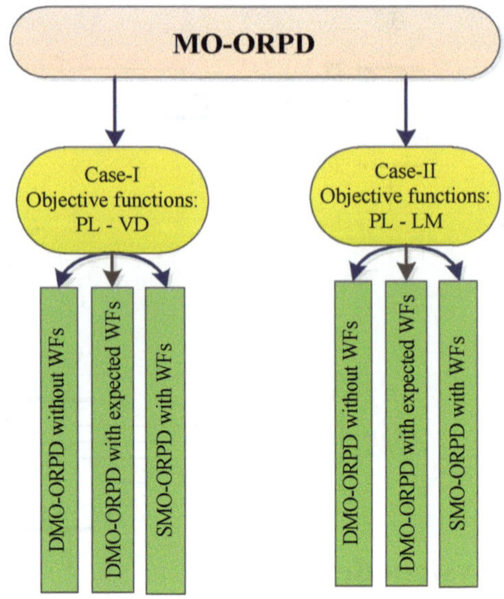

Fig. 12.7 Illustration of the studied cases

12.5.2 Case-I: Active Power Loss Versus Voltage Deviations (PL-VD)

12.5.2.1 Solving DMO-ORPD Without WFs

In this case, the Pareto front is attained for IEEE 30-bus test system without considering any uncertainty and without wind power integration. The VAR compensation devices are modeled by continuous variables for the sake of comparison with the previously published works in literature. The data of compensation limits are available in Appendix (Table 12.15). Table 12.1 summarizes the acquired Pareto solutions for this case. By using min-max fuzzy satisfying criterion, it is observed from Table 12.1 that the best compromise solution is *Solution#16*, with the maximum weakest membership function of 0.7734. The corresponding PL and *VD* are equal to 4.4438 MW and 0.0092, respectively. It is also noticeable that *Solution#1* corresponds to the loss minimization case in *Solution#1,* only *PL* is minimized, and the minimum value of *PL* is obtained 4.2875 MW. The Pareto optimal front of the two objective functions is depicted in Fig. 12.8. For this solution, the optimal values of control variables are given in Table 12.2.

12.5.2.2 Solving DMO-ORPD with Expected WFs

According to Fig. 12.6, IEEE-30 bus is modified and it is presupposed that wind turbine is located in bus 20. In this part, by using Fig. 12.5, the expected value of wind turbine generation is taken account in simulation. Wind power capacity

Table 12.1 Pareto optimal solutions for DMO-ORPD without WFs (Case-I)

#	*PL* (MW)	*VD*	μ_1	μ_2	$\hat{\mu}_r$
1	4.2875	0.0387	1	0	0
2	4.2882	0.0367	0.999	0.0521	0.0521
3	4.2904	0.0348	0.9958	0.1047	0.1047
4	4.2944	0.0328	0.990	0.1574	0.1574
5	4.3005	0.0308	0.9811	0.2101	0.2101
6	4.3092	0.0288	0.9686	0.2627	0.2627
7	4.3190	0.0269	0.9543	0.3154	0.3154
8	4.3293	0.0249	0.9395	0.3681	0.3681
9	4.3401	0.0229	0.9238	0.4207	0.4207
10	4.3515	0.0210	0.9072	0.4734	0.4734
11	4.3637	0.0190	0.8896	0.5260	0.5260
12	4.3767	0.0170	0.8706	0.5787	0.5787
13	4.3909	0.0151	0.8501	0.6314	0.6314
14	4.4064	0.0131	0.8276	0.6840	0.6840
15	4.4238	0.0111	0.8024	0.7367	0.7367
16	4.4438	0.0092	0.7734	0.7894	0.7734
17	4.4680	0.0072	0.7384	0.8420	0.7384
18	4.4999	0.0052	0.6921	0.8947	0.6921
19	4.5544	0.0033	0.6132	0.9473	0.6132
20	4.9774	0.0013	0	1	0

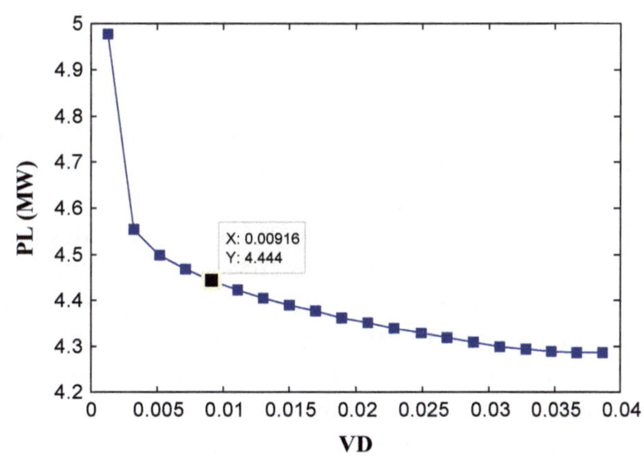

Fig. 12.8 Pareto optimal front for DMO-ORPD without WFs (Case-I)

amounted to twenty percent of the total load on the network is considered. Expected value of wind farm generation is calculated as:

$$P_{w,\exp} = \sum_{s \in \Omega_S} \left(\pi_s \times \zeta_{W_{i,s}} \times P^r_{W_i} \right) \tag{12.37}$$

Table 12.2 Optimal control variables for the best compromise solution (Solution#16) in (Case-I)

Control parameters	Parameter	DMO-ORPD (without WF)
Generator parameters	V_{g1} (pu)	1.0295
	V_{g2} (pu)	1.0204
	V_{g5} (pu)	1.0093
	V_{g8} (pu)	1.0030
	V_{g11} (pu)	1.0148
	V_{g13} (pu)	1.0258
	P_{g1} (MW)	62.8438
Compensation	Q_{c10} (MVar)	0
	Q_{c12} (MVar)	0
	Q_{c15} (MVar)	3.8495
	Q_{c17} (MVar)	3.6873
	Q_{c20} (MVar)	2.6983
	Q_{c21} (MVar)	6.6481
	Q_{c23} (MVar)	1.7635
	Q_{c24} (MVar)	3.0855
	Q_{c29} (MVar)	2.2464
Transformer tap changer	$t_{6\text{-}9}$	0.9706
	$t_{6\text{-}10}$	1.1000
	$t_{4\text{-}12}$	0.9812
	$t_{28\text{-}27}$	0.9970

It is explained that, discrete steps for compensation devices is used during this section based on the data provided in appendix (Table 12.15). Various Pareto solutions for this part are provided in Table 12.3. In order to solve the multi-objective ORPD problem by ε-constraint method, maximum and minimum values of the real power loss (F_1) and voltage deviation (F_2) are considered, which are equal to 3.8771 MW, 3.1651 MW, 0.0428 pu and 0.0012 pu, respectively. These extreme values are reached by maximizing and minimizing the objective functions of MO-ORPD individually. It means that in *Solution#1* and #20 the objectives are minimizing the *PL* and *VD*, respectively. Among these optimal solutions, *Solution#16* is minimizing both objectives, with the equal to 3.2983 MW PL and the *VD* of 0.0099. Pareto front of answers and control variables for BCS of this case are reported in Fig. 12.9 and Table 12.4, respectively.

12.5.2.3 Solving SMO-ORPD with WFs

In this case the load and wind power uncertainties are considered in the MO-ORPD using the previously described two stage stochastic programming approach.

Table 12.3 Pareto optimal solutions for DMO-ORPD with expected WFs (Case-I)

#	*PL* (MW)	*VD*	μ_1	μ_2	$\hat{\mu}_r$
1	3.1651	0.0428	1	0	0
2	3.166	0.0406	0.9987	0.0526	0.0526
3	3.1696	0.0384	0.9937	0.1052	0.1052
4	3.1759	0.0362	0.9849	0.1579	0.1579
5	3.183	0.0340	0.9749	0.2105	0.2105
6	3.1904	0.0318	0.9645	0.2631	0.2631
7	3.1982	0.0296	0.9535	0.3158	0.3158
8	3.2064	0.0275	0.9421	0.3684	0.3684
9	3.215	0.0253	0.9300	0.4210	0.4210
10	3.2241	0.0231	0.9172	0.4737	0.4737
11	3.234	0.0209	0.9032	0.5263	0.5263
12	3.2445	0.0187	0.8886	0.5789	0.5789
13	3.2556	0.0165	0.8729	0.6316	0.6316
14	3.2681	0.0143	0.8553	0.6842	0.6842
15	3.2823	0.0121	0.8355	0.7368	0.7368
16	3.2983	0.0099	0.8129	0.7895	0.7895
17	3.3182	0.0077	0.7850	0.8421	0.7850
18	3.3453	0.0055	0.7470	0.8947	0.7470
19	3.3928	0.0033	0.6801	0.9474	0.6801
20	3.8771	0.0012	0	1	0

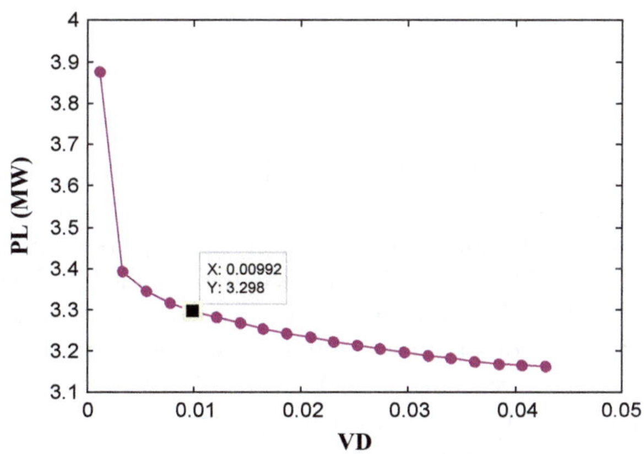

Fig. 12.9 Pareto front of DMO-ORPD with WFs (Case-I)

The attained Pareto optimal solutions in this case are presented in Table 12.5. It is observed from this table that Ξ_{PL} varies from 3.4003 to 4.1891 MW, while Ξ_{VD} varies from 0.0431 to 0.0013, respectively. The *Solution#1* corresponds to the Ξ_{PL} minimization case, where the minimum value of 3.4003 MW is obtained for Ξ_{PL},

Table 12.4 Optimal control variables for the best compromise solution (Solution#16) in DMO-ORPD with expected wind-(Case-I)

Control parameters	Parameter	DMO-ORPD (with WF)
Generator parameters	V_{g1} (pu)	1.0236
	V_{g2} (pu)	1.0161
	V_{g5} (pu)	1.0062
	V_{g8} (pu)	1.0015
	V_{g11} (pu)	1.0156
	V_{g13} (pu)	1.0248
	P_{g1} (MW)	48.2003
Compensation (switching steps)	u_{c10}	0
	u_{c12}	0
	u_{c15}	1
	u_{c17}	2
	u_{c20}	1
	u_{c21}	3
	u_{c23}	1
	u_{c24}	3
	u_{c29}	1
Wind farms parameters	P_{w19} (MW)	29.222
	Q_{w19} (MVar)	2.2632
Transformer tap changer	$t_{6\text{-}9}$	0.9711
	$t_{6\text{-}10}$	1.1000
	$t_{4\text{-}12}$	0.9840
	$t_{28\text{-}27}$	0.9936

whereas *Solution#20* deals with the case of Ξ_{VD} minimization, in which the minimum value of Ξ_{VD} is 0.0431. It is observed from Table 12.5 that *Solution#16* is the best compromise solution, with Ξ_{PL} equals to 3.5475 MW and Ξ_{VD} equals to 0.0101. Also, Fig. 12.10 depicts the obtained optimal Pareto front in this case.

Table 12.6 summarizes the obtained optimal *here and now* control variables for the best compromise solution. Also, the optimal values of *wait and see* control variables are depicted in Figs. 12.11, 12.12 and 12.13, in all possible scenarios. Figure 12.11 shows the active power generation at the slack bus (bus 1) in all 15 scenarios. Figure 12.12 represents the active/reactive power output of the wind farm in all scenarios. The optimal amount of reactive power compensation steps are also given in Fig. 12.13.

Table 12.5 Pareto optimal solutions for SMO-ORPD (Case-I)

#	Ξ_{PL} (MW)	Ξ_{LM}	μ_1	μ_2	$\hat{\mu}_r$
1	3.4003	0.0431	1	0	0
2	3.4188	0.0409	0.9765	0.0526	0.0526
3	3.4194	0.0387	0.9758	0.1053	0.1053
4	3.4213	0.0365	0.9734	0.1579	0.1579
5	3.4253	0.0343	0.9683	0.2105	0.2105
6	3.4322	0.0321	0.9595	0.2632	0.2632
7	3.4405	0.0299	0.9490	0.3158	0.3158
8	3.4492	0.0277	0.9380	0.3684	0.3684
9	3.4584	0.0255	0.9264	0.4211	0.4211
10	3.4681	0.0233	0.914	0.4737	0.4737
11	3.4785	0.0211	0.9008	0.5263	0.5263
12	3.4897	0.0189	0.8867	0.5790	0.5790
13	3.5018	0.0167	0.8713	0.6316	0.6316
14	3.5152	0.0145	0.8543	0.6842	0.6842
15	3.5302	0.0123	0.8353	0.7368	0.7368
16	3.5475	0.0101	0.8133	0.7895	0.7895
17	3.5687	0.0079	0.7865	0.8421	0.7865
18	3.5949	0.0057	0.7533	0.8947	0.7533
19	3.6396	0.0035	0.6967	0.9474	0.6967
20	4.1891	0.0013	0	1	0

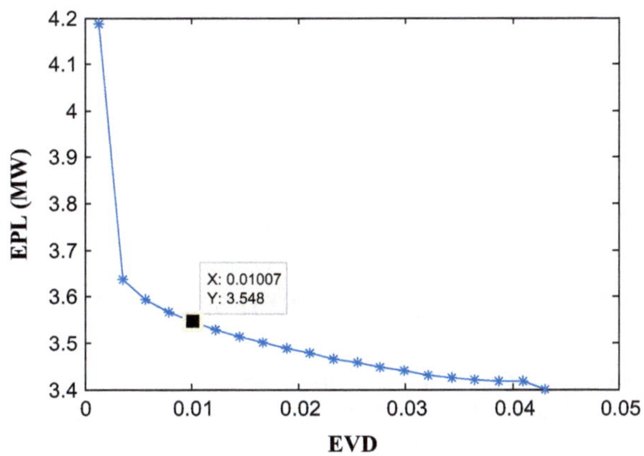

Fig. 12.10 Pareto front of SMO-ORPD (Case-I)

Table 12.6 Optimal values for here and now control variables at the best compromise solution (Solution#16) in Case-I

Control parameters	Parameter	DMO-ORPD(without WF)
Generator parameters	V_{g1} (pu)	1.0270
	V_{g2} (pu)	1.0193
	V_{g5} (pu)	1.0089
	V_{g8} (pu)	1.0043
	V_{g11} (pu)	1.0165
	V_{g13} (pu)	1.0267
Transformer tap changer	$t_{6\text{-}9}$	0.9694
	$t_{6\text{-}10}$	1.1000
	$t_{4\text{-}12}$	0.9841
	$t_{28\text{-}27}$	0.9933

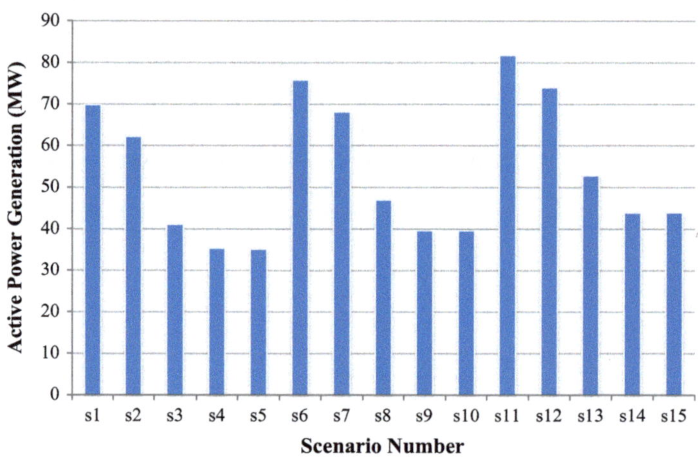

Fig. 12.11 Active power generation in slack bus (bus 1) in all scenarios (in MW)-(Case-I)

12.5.3 *Case-II: Active Power Loss Versus* L_{max}*, (PL-LM)*

12.5.3.1 Solving DMO-ORPD Without WFs

In this case, *PL* and *LM* are considered as the two considered conflicted objectives. The Pareto front is obtained without considering any uncertainty. Table 12.7 summarizes the obtained Pareto solutions for this case. By using min-max fuzzy

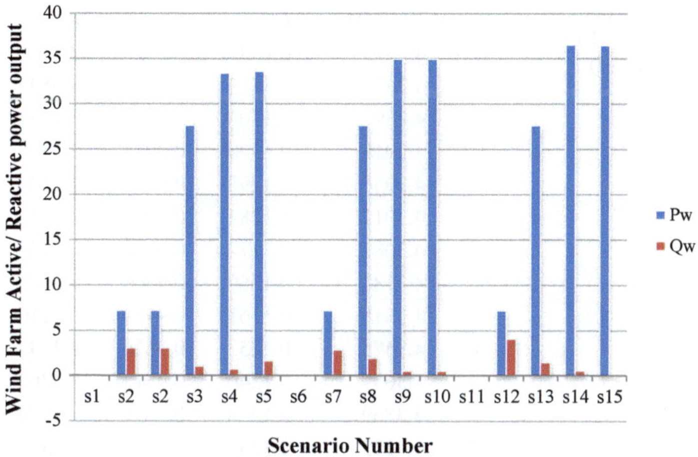

Fig. 12.12 Active/reactive power output of wind farm (located at bus 20) in all scenarios (in MW and MVAR) - (Case-I)

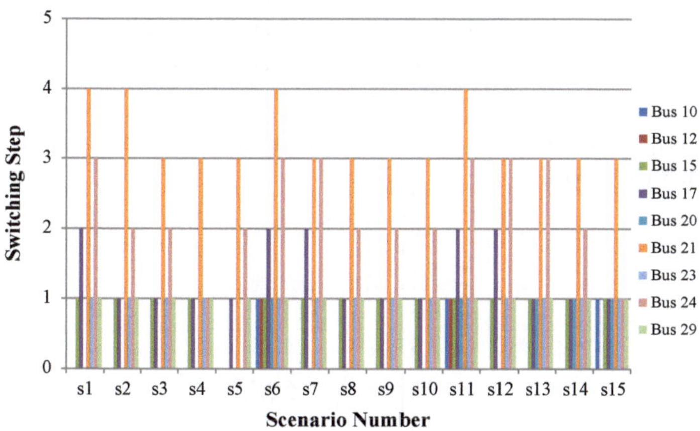

Fig. 12.13 Switching steps in VAR compensation buses at different scenarios (Case-I)

satisfying method, it is evident from Table 12.7 that the best compromise solution is *Solution#17*, with the maximum weakest membership function of 0.7988. The corresponding PL and *LM* are equal to 4.6106 MW and 0.1199 pu, respectively. Pareto front of this case is available in Fig. 12.14. Table 12.8 summarizes the obtained control variables for the best compromise solution, *Solution#17*, of this case.

Table 12.7 Pareto optimal solutions for DMO-ORPD without WFs (Case-II)

#	*PL* (MW)	*LM*	μ_1	μ_2	$\hat{\mu}_r$
1	4.2875	0.1323	1	0	0
2	4.2879	0.1315	0.9997	0.0526	0.0526
3	4.2893	0.1308	0.9989	0.1053	0.1053
4	4.2918	0.1300	0.9973	0.1579	0.1579
5	4.2956	0.1292	0.9949	0.2105	0.2105
6	4.3012	0.1284	0.9915	0.2632	0.2632
7	4.3087	0.1277	0.9868	0.3158	0.3158
8	4.3187	0.1269	0.9806	0.3684	0.3684
9	4.3318	0.1261	0.9724	0.4210	0.4210
10	4.3475	0.1253	0.9626	0.4737	0.4737
11	4.3659	0.1246	0.9512	0.5263	0.5263
12	4.3869	0.1238	0.9381	0.5789	0.5789
13	4.4105	0.1230	0.9234	0.6316	0.6316
14	4.4403	0.1222	0.9048	0.6842	0.6842
15	4.4835	0.1215	0.8779	0.7368	0.7368
16	4.5413	0.1207	0.8420	0.7895	0.7895
17	4.6106	0.1199	0.7988	0.8421	0.7988
18	4.7219	0.1191	0.7295	0.8947	0.7295
19	5.0212	0.1184	0.5431	0.9474	0.5431
20	5.8934	0.1176	0	1	0

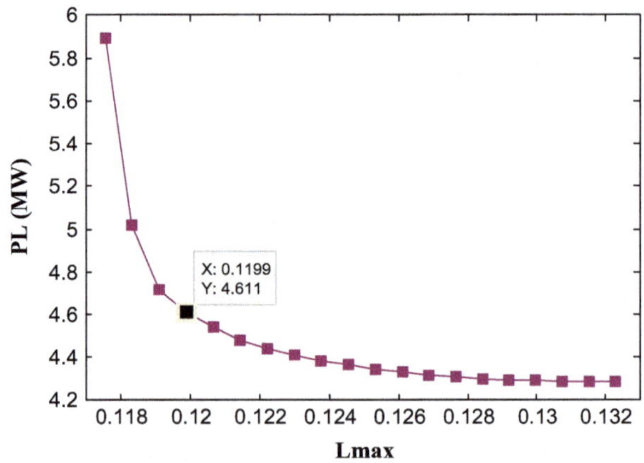

Fig. 12.14 Pareto front of DMO-ORPD without WFs (Case-II)

12.5.3.2 Solving DMO-ORPD with Expected WFs

Similar to Case-I, in this case, expected value of wind power as an input power is calculated by Eq. 12.37. Table 12.9 shows the obtained 20 selected Pareto solutions for this case. Among this answers, *Solution#17* has minimized both objective simultaneously with $PL = 3.2751$ and $LM = 0.1146$. Figure 12.15 depicts the

Table 12.8 Optimal control variables for the best compromise solution (Solution#17) in DMO-ORPD without wind-Case-II

Control parameters	Parameter	DMO-ORPD (without WF)
Generator parameters	V_{g1} (pu)	1.0366
	V_{g2} (pu)	1.0286
	V_{g5} (pu)	1.0182
	V_{g8} (pu)	1.0134
	V_{g11} (pu)	1.0110
	V_{g13} (pu)	1.0434
	P_{g1} (MW)	63.0106
Compensation	Q_{c10} (MVar)	8.2754
	Q_{c12} (MVar)	0
	Q_{c15} (MVar)	0
	Q_{c17} (MVar)	7.0185
	Q_{c20} (MVar)	4.0128
	Q_{c21} (MVar)	14.2677
	Q_{c23} (MVar)	2.9065
	Q_{c24} (MVar)	7.7563
	Q_{c29} (MVar)	0
Transformer tap changer	$t_{6\text{-}9}$	1.0706
	$t_{6\text{-}10}$	0.9000
	$t_{4\text{-}12}$	0.9956
	$t_{28\text{-}27}$	0.9719

Pareto front of this case. According to Fig. 12.15, it is clear that when wind power injected into the network, it substantially reduces network losses and increases network stability. Table 12.10 presents the obtained control variables for best solution of this case (*Solution#17*).

12.5.3.3 Solving SMO-ORPD with WFs

The uncertainty is modeled using scenario based approach for Case-II and results of Pareto solutions are presented in Table 12.11. Figure 12.16 shows the Pareto front for Case-II with considering load and wind uncertainties using scenario based approach.

Similar to Case-I, *here and now* control variables are shown in Table 12.12 and *wait and see* control variables are shown in Figs. 12.17, 12.18 and 12.19. It should be noticed that, these variables are presented for BCS (*Solution#16*).

Table 12.9 Pareto optimal solutions for DMO-ORPD with expected WFs (Case-II)

#	*PL* (MW)	*LM*	μ_1	μ_2	$\hat{\mu}_r$
1	3.1651	0.1247	1	0	0
2	3.1654	0.1241	0.9994	0.0526	0.0526
3	3.1662	0.1234	0.9979	0.1053	0.1053
4	3.1682	0.1228	0.9943	0.1579	0.1579
5	3.1707	0.1222	0.9895	0.2105	0.2105
6	3.1742	0.1215	0.9832	0.2632	0.2632
7	3.1786	0.1209	0.9749	0.3158	0.3158
8	3.1841	0.1203	0.9647	0.3684	0.3684
9	3.1905	0.1196	0.9528	0.4210	0.4210
10	3.1977	0.1190	0.9395	0.4737	0.4737
11	3.2056	0.1184	0.9247	0.5263	0.5263
12	3.2144	0.1178	0.9084	0.5789	0.5789
13	3.2242	0.1171	0.8901	0.6316	0.6316
14	3.235	0.1165	0.8700	0.6842	0.6842
15	3.2468	0.1159	0.8481	0.7368	0.7368
16	3.2597	0.1152	0.8241	0.7895	0.7895
17	3.2751	0.1146	0.7955	0.8421	0.7955
18	3.2944	0.1140	0.7594	0.8947	0.7594
19	3.3189	0.1133	0.7139	0.9473	0.7139
20	3.7027	0.1127	0	1	0

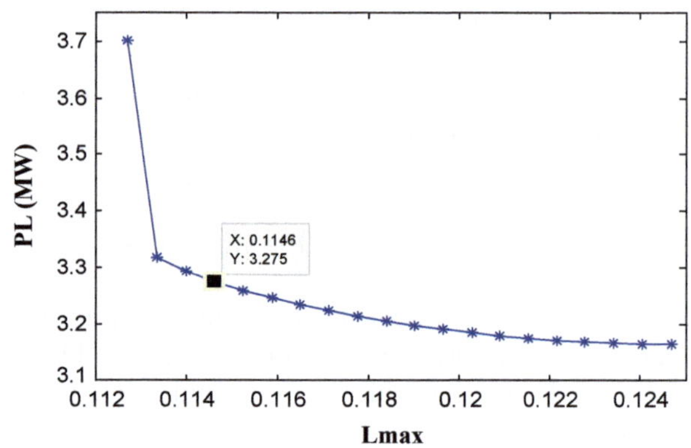

Fig. 12.15 Pareto front of DMO-ORPD with WFs (Case-II)

Table 12.10 Optimal control variables for the best compromise solution (Solution#17) in Case-II

Control parameters	Parameter	DMO-ORPD (with WF)
Generator parameters	V_{g1} (pu)	1.0371
	V_{g2} (pu)	1.0301
	V_{g5} (pu)	1.0204
	V_{g8} (pu)	1.0161
	V_{g11} (pu)	1.0136
	V_{g13} (pu)	1.0461
	P_{g1} (MW)	49.3914
Compensation (switching steps)	u_{c10}	0
	u_{c12}	0
	u_{c15}	1
	u_{c17}	2
	u_{c20}	1
	u_{c21}	4
	u_{c23}	1
	u_{c24}	3
	u_{c29}	1
Wind Farms parameters	P_{w20} (MW)	29.2220
	Q_{w20} (MVar)	2.4846
Transformer tap changer	$t_{6\text{-}9}$	1.0550
	$t_{6\text{-}10}$	0.9000
	$t_{4\text{-}12}$	0.9930
	$t_{28\text{-}27}$	0.9761

12.6 Discussions on the Results

12.6.1 Comparison of DMO-ORPD Performance with Pervious Literature

In Tables 12.13 and 12.14 the obtained *PL*, *VD* and *LM* are compared with the results reported by some recently published algorithms. Table 12.13 shows the obtained *PL* and *VD* for *Solution#1* (*PL* minimization), *Solution#20* (*VD* minimization), and *Solution#16* (compromise solution), whereas Table 12.14 give a comparison for *PL* and *LM*. *Solution#1* (*PL* minimization), *Solution#20* (*LM* minimization), and *Solution#17* (compromise solution) are compared with recently published works. According to these tables, it can be observed that the obtained

Table 12.11 Pareto optimal solutions for SMO-ORPD) (Case-II)

#	Ξ_{PL} (MW)	Ξ_{LM}	μ_1	μ_2	$\hat{\mu}_r$
1	3.4815	0.1258	1	0	0
2	3.4865	0.1251	0.9915	0.0635	0.0635
3	3.4881	0.1245	0.9887	0.1264	0.1264
4	3.4914	0.1239	0.9831	0.1890	0.1890
5	3.4918	0.1233	0.9824	0.2513	0.2513
6	3.4942	0.1227	0.9783	0.3131	0.3131
7	3.4989	0.1221	0.9702	0.3745	0.3745
8	3.5048	0.1215	0.9601	0.4352	0.4352
9	3.5118	0.1209	0.9481	0.4952	0.4952
10	3.5196	0.1204	0.9347	0.5541	0.5541
11	3.5279	0.1198	0.9204	0.6117	0.6117
12	3.5375	0.1193	0.9040	0.6675	0.6675
13	3.5485	0.1188	0.8851	0.7208	0.7208
14	3.5602	0.1183	0.8650	0.7703	0.7703
15	3.573	0.1179	0.8431	0.8143	0.8143
16	3.5873	0.1175	0.8187	0.8488	0.8187
17	3.6039	0.1174	0.7901	0.8658	0.7901
18	3.6214	0.1172	0.7600	0.8863	0.7600
19	3.6399	0.1171	0.7283	0.8942	0.7283
20	4.0646	0.1161	0	1	0

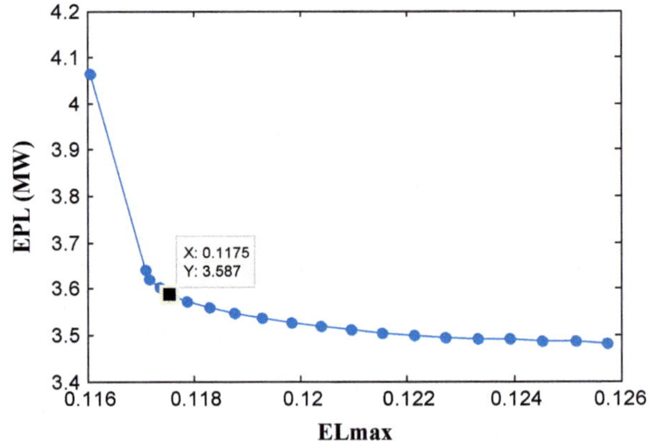

Fig. 12.16 Pareto front of SMO-ORPD (Case-II)

Table 12.12 Optimal values for here and now control variables at the best compromise solution (Solution#16) in Case-II

Control parameters	Parameter	DMO-ORPD (without WF)
Generator parameters	V_{g1} (pu)	1.0151
	V_{g2} (pu)	0.9976
	Vg_5 (pu)	1.0083
	V_{g8} (pu)	1.0092
	V_{g11} (pu)	1.0246
	V_{g13} (pu)	1.0398
Transformer tap changer	$t_{6\text{-}9}$	1.0455
	$t_{6\text{-}10}$	0.9000
	$t_{4\text{-}12}$	0.9821
	$t_{28\text{-}27}$	0.9698

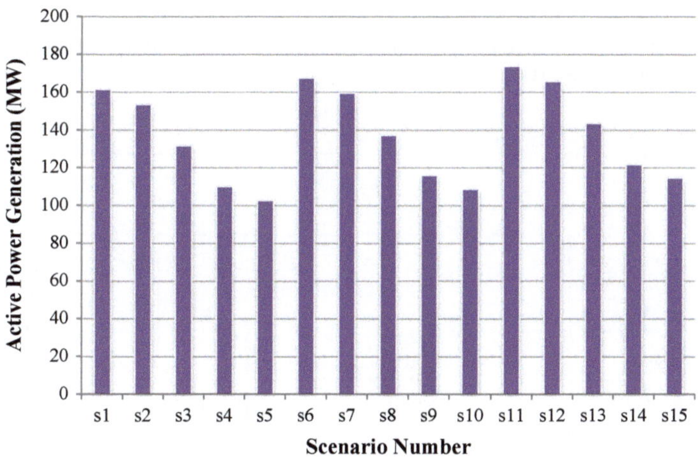

Fig. 12.17 Active power generation in slack bus (bus 1) in all scenarios (in MW)-(Case-II)

solutions are superior to the previously reported ones like as gravitational search algorithm (GSA) [17], differential evolutionary algorithm [15], quasi-oppositional teaching learning based optimization (TLBO) algorithm [25] and chaotic parallel vector evaluated interactive honey bee mating optimization (CPVEIHBMO) [26].

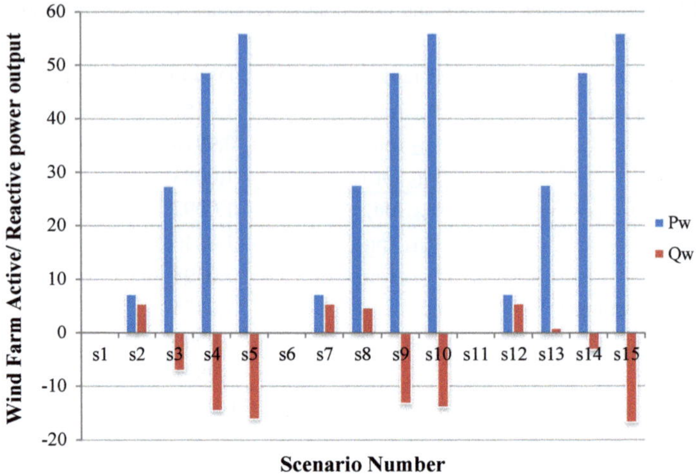

Fig. 12.18 Active and reactive power output of wind farm (located at bus 20) in all scenarios (in MW and MVAR)—(Case-II)

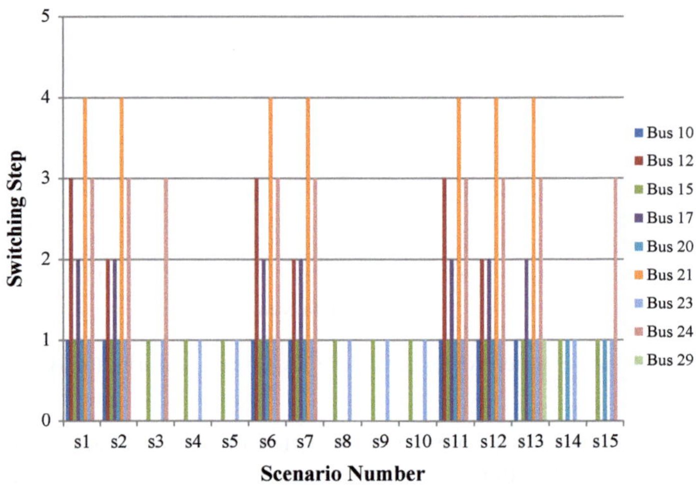

Fig. 12.19 Switching steps in VAR compensation buses at different scenarios (Case-II)

Table 12.13 Comparison of the obtained *PL* and *VD* in DMO-ORPD (without WFs), with the published methods

	PL Minimization	*VD* Minimization	*PL* (MW) Compromise solution
Proposed	4.2875	4.9774	4.4438
CPVEIHBMO [26]	4.37831	4.994831	5.3243
HBMO [26]	4.40867	5.2092	5.535
QOTLBO [25]	4.5594	6.4962	5.2594
NSGA-II [27]	5.137	5.686	–
DE [15]	4.555	6.4755	–
SPMGSO [27]	5.123	5.96	–
GSA [17]	4.5143	4.9752	–
	PL minimization	*VD* minimization	*VD* (pu) Compromise solution
Proposed	0.0387	0.0013	0.0092
CPVEIHBMO [26]	0.67352	0.198756	0.7397
HBMO [26]	0.87364	0.2106	0.87664
QOTLBO [25]	1.9057	0.0856	0.121
NSGA-II [27]	0.6443	0.1789	–
DE [15]	1.9989	0.0911	–
SPMGSO [27]	0.73986	0.1438	–
GSA [17]	0.8752	0.2157	–

12.6.2 *Impact of Wind Energy on MO-ORPD Problem*

Figures 12.20 and 12.21 summarize the obtained results of DMO-ORPD with and without WFs and SMO-ORPD for both Case-I and Case-II. According to Fig. 12.20a, it is clear that by installing wind farms on the system, real power losses (*PL* in deterministic and *EPL* in stochastic model) are reduced considerably. It is also observed from this figure that in the case of SMO-ORPD, the *EPL* is higher than the case of DMO-ORPD with expected wind. Besides, Fig. 12.20b depicts the obtained values for *VD* and *EVD* in both Case-I and II. It is evidently observed from this figure that, installation of wind farm leads to deterioration of voltage deviations, especially in the case of SMO-OPRD. This is mainly because of fluctuations of wind farm output power, in different scenarios.

Also, Fig. 12.21 shows the obtained values of *PL* (and *EPL*) along with L_{max} (and ELM) in Case-II. According to Fig. 12.21a active power losses are decreased when the wind farm is considered, both in DMO-ORPD (with wind farm) and SMO-ORPD. Besides, according to Fig. 12.21b the voltage stability of system increases in the presence of wind farm, since the L_{max} and *ELM* reduced in both DMO-ORPD (with wind farm) and SMO-ORPD cases.

Table 12.14 Comparison of the obtained PL and LM in DMO-ORPD (without WFs), with the published methods

	PL minimization	*LM* minimization	*PL* (MW) Compromise solution
Proposed	4.2875	5.8934	4.6106
CPVEIHBMO [26]	4.7831	6.6501	5.3243
HBMO [26]	4.40867	6.66	5.5352
QOTLBO [25]	4.5594	5.2554	5.2594
MOCIPSO [24]	5.174	5.419	5.232
MOPSO [24]	5.233	5.528	5.308
DE [15]	4.555	7.0733	–
RGA [22]	4.951	5.0912	–
CMAES [22]	4.945	5.129	–
GSA [17]	4.5143	6.6602	–
	PL minimization	*LM* minimization	*LM* (pu) Compromise solution
Proposed	0.1323	0.1176	0.1199
CPVEIHBMO [26]	0.141	0.116	-
HBMO [26]	0.12101	0.111	0.1163
QOTLBO [25]	0.1263	0.1147	0.1203
MOCIPSO [24]	0.1273	0.1242	0.1254
MOPSO [24]	0.12664	0.1141	11.821
DE [15]	0.1317	0.1192	0.12191
RGA [22]	0.5513	0.1246	–
CMAES [22]	0.13965	0.1386	–
GSA [17]	0.13944	0.1382	–

The obtained results imply the positive impact of wind power generation on the voltage stability enhancement and decreasing system real power losses. Also, it can be observed that the installed wind farm has little effects on the voltage deviations.

Fig. 12.20 Comparison of the obtained results of Case-I in different conditions, **a** *PL* and *EPL* (MW), **b** *VD* and *EVD* (pu)

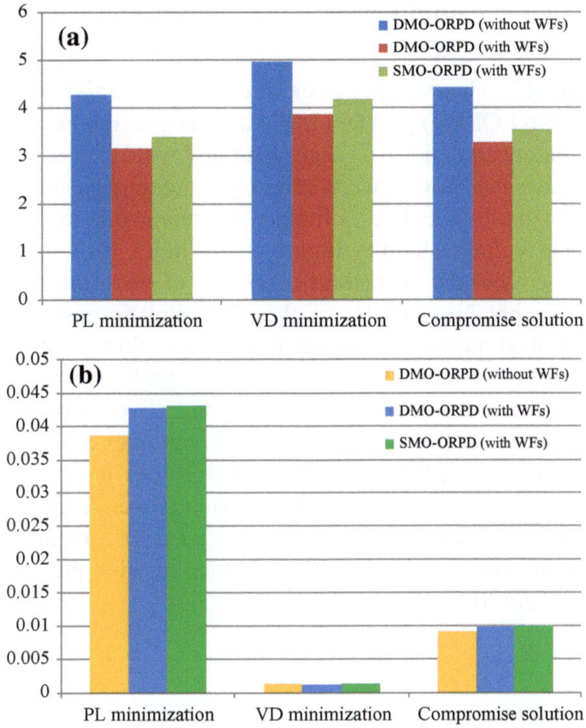

Fig. 12.21 Comparison of the obtained results of Case-II in different conditions, **a** *PL* and *EPL* (MW), **b** *LM* and $EL_{\max}$

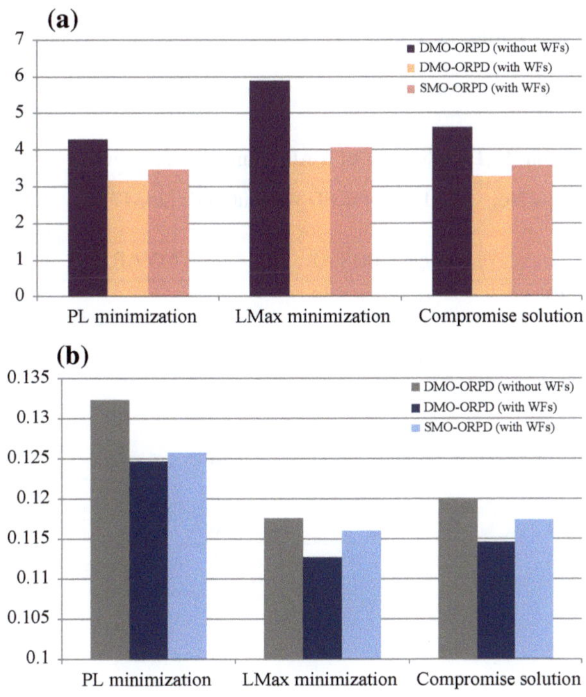

12.7 Conclusions

In this chapter the stochastic multi-objective optimal reactive power dispatch (SMO-ORPD) problem in a wind integrated power system is studied by considering the uncertainties of system load and wind power generations. For decision making under the above uncertainties two-stage stochastic optimization model is utilized. In the multi-objective optimization framework real power losses, voltage deviation and voltage stability improvement index (*L*-index) are optimized simultaneously. The ε-constraint method is utilized to solve multi-objective optimization problem. The obtained results in the deterministic case are compared with the recently applied intelligent search-based algorithms and it is found that the suggested method can find better solutions for both objective functions in this case.

In the stochastic case, a comprehensive set of decision variables including *here and now* and *wait and see* control variables are obtained. The offered SMO-ORPD model is verified using the IEEE 30-bus test system. The numerical results substantiate that in the presence of wind power generation, the expected value of active power losses and *L*-index are reduced in comparison with the deterministic case. This confirms the positive influence of wind power generation on the reduction of system losses.

Appendix

See Table 12.15.

Table 12.15 The data of VAR Compensation devices

Bus No.	DMO-ORPD without WFs [26, 47]		DMO-ORPDwith WFs SMO-ORPD with WFs		
	$Q_{C_i}^{\min}$ (MVAR)	$Q_{C_i}^{\max}$(MVAR)	$I_{C_i}^{\min}$	$I_{C_i}^{\max}$	Q_{C_i} (MVAR)
10	0	36	0	1	1
12	0	36	0	3	1.5
15	0	36	0	1	1.5
17	0	36	0	2	2
20	0	36	0	1	0.5
21	0	36	0	4	2
23	0	36	0	1	1
24	0	36	0	3	1.5
29	0	36	0	1	0.5

References

1. B. Zhao, C. Guo, Y. Cao, A Multiagent-Based Particle Swarm Optimization Approach for Optimal Reactive Power Dispatch, IEEE Transactions on Power Systems, vol. 20, pp. 1070–1078, 2005.
2. C. Dai, W. Chen, Y. Zhu, X. Zhang, Seeker Optimization Algorithm for Optimal Reactive Power Dispatch, IEEE Transactions on Power Systems, vol. 24, pp. 1218–1231, 2009.
3. A. Khorsandi, A. Alimardani, B. Vahidi, S. Hosseinian, Hybrid Shuffled Frog Leaping Algorithm and Nelder-Mead Simplex Search for Optimal Reactive Power Dispatch, IET Generation, Transmission & Distribution, vol. 5, pp. 249–256, 2011.
4. M. Ghasemi, S. Ghavidel, M.M. Ghanbarian, A. Habibi, A New Hybrid Algorithm for Optimal Reactive Power Dispatch Problem with Discrete and Continuous Control Variables, Applied Soft Computing vol. 22, pp. 126–40, 2014.
5. M. Ghasemi, M.M. Ghanbarian, S. Ghavidel, S. Rahmani, E. Mahboubi Moghaddam, Modified Teaching Learning Algorithm and Double Differential Evolution Algorithm for Optimal Reactive Power Dispatch Problem: A Comparative Study, Information Sciences, vol. 278, pp. 231–249, 2014.
6. M. Martinez Rojas, A. Sumper, O. Gomis Bellmunt, A. Sudria Andreu, Reactive Power Dispatch in Wind Farms Using Particle Swarm Optimization Technique and Feasible Solutions Search, Applied Energy, vol. 88, pp. 4678–4686, 2011.
7. R. Mallipeddi, S. Jeyadevi, P.N. Suganthan, S. Baskar, Efficient Constraint Handling for Optimal Reactive Power Dispatch Problems, Swarm and Evolutionary Computation, vol. 5, pp. 28–36, 2012.
8. C.M. Huang, S.J. Chen, Y.C. Huang, H.T. Yang, Comparative Study of Evolutionary Computation Methods for Active-Reactive Power Dispatch, IET Generation, Transmission & Distribution, vol. 6, pp. 636–645, 2012.
9. A. Rajan, T. Malakar, Optimal Reactive Power Dispatch Using Hybrid Nelder-Mead Simplex Based Firefly Algorithm, International Journal of Electrical Power & Energy Systems, vol. 66, pp. 9–24, 2015.
10. R.P. Singh, V. Mukherjee, S. Ghoshal, Optimal Reactive Power Dispatch by Particle Swarm Optimization with an Aging Leader and Challengers, Applied Soft Computing, vol. 29, pp. 298–309, 2015.
11. M.H. Sulaiman, Z. Mustaffa, M.R. Mohamed, O. Aliman, Using the Gray Wolf Optimizer for Solving Optimal Reactive Power Dispatch Problem, Applied Soft Computing, vol. 32, pp. 286–292, 2015.
12. D. Thukaram, G. Yesuratnam, Optimal Reactive Power Dispatch in a Large Power System with AC-DC and FACTS Controllers, Generation, Transmission & Distribution, IET, vol. 2, pp. 71–81, 2008.
13. A. Rabiee, M. Vanouni, M. Parniani, Optimal Reactive Power Dispatch for Improving Voltage Stability Margin Using a Local Voltage Stability Index, Energy Conversion and Management, vol. 59, pp. 66–73, 2012.
14. C. Dai, W. Chen, Y. Zhu, and X. Zhang, Reactive Power Dispatch Considering Voltage Stability with Seeker Optimization Algorithm, Electric Power Systems Research, vol. 79, pp. 1462–1471, 2009.
15. A. Ela, M. Abido, S. Spea, Differential Evolution Algorithm for Optimal Reactive Power Dispatch, Electric Power Systems Research, vol. 81, pp. 458–464, 2011.
16. A. Khazali, M. Kalantar, Optimal Reactive Power Dispatch Based on Harmony Search Algorithm, International Journal of Electrical Power & Energy Systems, vol. 33, pp. 684–692, 2011.
17. S. Duman, Y. Sonmez, U. Guvenc, N. Yorukeren, Optimal Reactive Power Dispatch Using a Gravitational Search Algorithm, IET Generation, Transmission & Distribution, vol. 6, pp. 563–576, 2012.

18. J.M. Ramirez, J.M. Gonzalez, T.O. Ruben, An Investigation about the Impact of the Optimal Reactive Power Dispatch Solved by DE, International Journal of Electrical Power & Energy Systems, vol. 33, pp. 236–244, 2011.
19. E.M. Soler, E.N. Asada, G.R. Da Costa, Penalty-Based Nonlinear Solver for Optimal Reactive Power Dispatch with Discrete Controls, IEEE Transactions on Power Systems, vol. 28, pp. 2174–2182, 2013.
20. M. Abido, J. Bakhashwain, Optimal VAR Dispatch Using a Multi-Objective Evolutionary Algorithm, International Journal of Electrical Power & Energy Systems, vol. 27, pp. 13–20, 2005.
21. L. Zhihuan, L. Yinhong, D. Xianzhong, Non-Dominated Sorting Genetic Algorithm-II for Robust Multi-Objective Optimal Reactive Power Dispatch, Generation, Transmission & Distribution, IET, vol. 4, pp. 1000–1008, 2010.
22. S. Jeyadevi, S. Baskar, C. Babulal, M. Willjuice Iruthayarajan, Solving Multi-Objective Optimal Reactive Power Dispatch Using Modified NSGA-II, International Journal of Electrical Power & Energy Systems, vol. 33, pp. 219–228, 2011.
23. A. Saraswat, A. Saini, Multi-Objective Optimal Reactive Power Dispatch Considering Voltage Stability in Power Systems Using HFMOEA, Engineering Applications of Artificial Intelligence, vol. 26, pp. 390–404, 2013.
24. G. Chen, L. Liu, P. Song, Y. Du, Chaotic Improved PSO-Based Multi-Objective Optimization for Minimization of Power Losses and L Index in Power Systems, Energy Conversion and Management, vol. 86, pp. 548–560, 2014.
25. B. Mandal, P.K. Roy, Optimal Reactive Power Dispatch Using Quasi-Oppositional Teaching Learning Based Optimization, International Journal of Electrical Power & Energy Systems, vol. 53, pp. 123–134, 2013.
26. A. Ghasemi, K. Valipour, A. Tohidi, Multi-Objective Optimal Reactive Power Dispatch Using a New Multi-Objective Strategy, International Journal of Electrical Power & Energy Systems, vol. 57, pp. 318–334, 2014.
27. B. Zhou, K. Chan, T. Yu, H. Wei, J. Tang, Strength Pareto Multi-Group Search Optimizer for Multi-Objective Optimal VAR Dispatch, IEEE Transactions on Industrial Informatics, vol. 10, pp. 1012–1022, 2014.
28. Z. Hu, X. Wang, G. Taylor, Stochastic Optimal Reactive Power Dispatch: Formulation and Solution Method, International Journal of Electrical Power & Energy Systems, vol. 32, pp. 615–621, 2010.
29. S. M. Mohseni Bonab, A. Rabiee, S. Jalilzadeh, B. Mohammadi Ivatloo, and S. Nojavan, "Probabilistic Multi Objective Optimal Reactive Power Dispatch Considering Load Uncertainties Using Monte Carlo Simulations, Journal of Operation and Automation in Power Engineering, vol. 3(1), pp. 83–93, 2015.
30. S. M. Mohseni Bonab, A. Rabiee, B. Mohammadi Ivatloo, Load Uncertainty Analysis in Multi Objective Optimal Reactive Power Dispatch Considering Voltage Stability, Majlesi Journal of Energy Management, vol. 4(2), pp. 23–30, 2015.
31. A. Brooke, D. Kendrick, A. Meeraus, GAMS Release 2.25: A User's Guide: GAMS Development Corporation Washington, DC, 1996.
32. P.E. Gill, W. Murray, M.A. Saunders, SNOPT: An SQP Algorithm for Large-Scale Constrained Optimization, SIAM Journal on Optimization, vol. 12, pp. 979–1006, 2002.
33. The GAMS Software Website, http://www.gams.com/dd/docs/solvers/sbb.pdf, 2013.
34. A. Soroudi, T. Amraee, Decision Making under Uncertainty in Energy Systems: State of the Art, Renewable and Sustainable Energy Reviews, vol. 28, pp. 376–384, 2013.
35. A. Rabiee, A. Soroudi, B. Mohammadi Ivatloo, M. Parniani, Corrective Voltage Control Scheme Considering Demand Response and Stochastic Wind Power, IEEE Transactions on Power Systems, vol. 29, pp. 2965–2973, 2014.
36. A. Soroudi, B. Mohammadi Ivatloo, A. Rabiee, Energy Hub Management with Intermittent Wind Power, in Large Scale Renewable Power Generation, Springer, pp. 413–438, 2014.

37. A. Rabiee, A. Soroudi, Stochastic Multiperiod OPF Model of Power Systems with HVDC-Connected Intermittent Wind Power Generation, IEEE Transactions on Power Delivery, vol. 29, pp. 336–344, 2014.
38. A. Soroudi, A. Rabiee, A. Keane, Stochastic Real-Time Scheduling of Wind-Thermal Generation Units in an Electric Utility, IEEE System Journal, 2014.
39. R.D. Zimmerman, C.E. Murillo Sanchez, D. Gan, A MATLAB Power System Simulation Package, 2005.
40. S. Wen, H. Lan, Q. Fu, D. Yu, L. Zhang, Economic Allocation for Energy Storage System Considering Wind Power Distribution, IEEE Transactions on power Systems, vol. 30(2), pp. 644–52, 2015.
41. C.A. Canizares, Voltage Stability Assessment: Concepts, Practices and Tools, Power System Stability Subcommittee Special Publication IEEE/PES, thunderbox.uwaterloo.ca/~claudio/claudio.html#VSWG, 2002.
42. F. Jabbari, B. Mohammadi Ivatloo, Static Voltage Stability Assessment Using Probabilistic Power Flow to Determine the Critical PQ Buses, Majlesi Journal of Electrical Engineering, vol. 8, pp. 17–25, 2014.
43. H. Xiong, H. Cheng, H. Li, Optimal Reactive Power Flow Incorporating Static Voltage Stability Based on Multi-Objective Adaptive Immune Algorithm, Energy Conversion and Management, vol. 49, pp. 1175–1181, 2008.
44. R. Raghunatha, R. Ramanujam, K. Parthasarathy, D. Thukaram, Optimal Static Voltage Stability Improvement Using a Numerically Stable SLP Algorithm, for Real Time Applications, International Journal of Electrical Power & Energy Systems, vol. 21, pp. 289–297, 1999.
45. X. Wang, Y. Gong, C. Jiang, Regional Carbon Emission Management Based on Probabilistic Power Flow with Correlated Stochastic Variables, IEEE Transactions on Power Systems, vol. 30, pp. 1094–1103, 2014.
46. M. Alipour, B. Mohammadi Ivatloo, K. Zare, Stochastic Risk-Constrained Short-Term Scheduling of Industrial Cogeneration Systems in the Presence of Demand Response Programs, Applied Energy, vol. 136, pp. 393–404, 2014.
47. R.D. Zimmerman, C.E. Murillo Sanchez, R.J. Thomas, MATPOWER: Steady-State Operations, Planning, and Analysis Tools for Power Systems Research and Education, IEEE Transactions on Power Systems, vol. 26, pp. 12–19, 2011.

Part III
Challenges, Solutions and Applications in AC Power Systems

Chapter 13
Self-excited Induction Generator in Remote Site

Ezzeddine Touti, Remus Pusca, J. Francois Brudny and Abdelkader Chaari

Abstract This chapter is devoted to the analysis of a self-excited asynchronous generator working in autonomous generation mode. It proposes a method of detection of the generator operating points and the control procedures which makes it possible to maintain quasi constant the frequency values. These strategies prevent the disengagement of the generator and use a Thyristor Controlled Reactor to regulate the reactive power. The control strategies present the study of the frequency in the steady case by the use of an appropriate single phase equivalent circuit. The space vector formalism is also used to study the transitory and steady output signals. At the end of the chapter, one will discuss the experimental validation of theoretical and numerical results.

13.1 Introduction

During the last decades, due to technological and industrial developments which impose the continuous demand of energy and when the cost of different types of energy increase, the efforts were continued to look for other solutions to diversify alternative energy sources. In this context, the wind energy seems to be a promising alternative source which can accompany the classical energy sources [1, 2]. The use

E. Touti (✉)
University of Northern Border, Arar, Saudi Arabia
e-mail: esseddine.touti@nbu.edu.sa

R. Pusca · J.F. Brudny
LSEE, University of Artois, Bethune, France
e-mail: remus.pusca@univ-artois.fr

J.F. Brudny
e-mail: jfrancois.brudny@univ-artois.fr

A. Chaari
Taha Hussein, University of Tunis, Tunis, Tunisia
e-mail: nabile.chaari@yahoo.fr

N. Mahdavi Tabatabaei et al. (eds.), *Reactive Power Control in AC Power Systems*,
Power Systems, DOI 10.1007/978-3-319-51118-4_13

of wind turbines is an attractive solution because they can provide energy at a competitive price and mainly it does not cause environmental contamination [3].

The work presented in this chapter treats the study of a three-phase Self-excited Induction Generator (SEIG) generating power in remote site. The strategy presented allows the regulation of the reactive power. Consequently it enables the maintaining of the frequency at quasi constant values.

Including Induction Machine (IM) in the wind generator, in the cases of limited electric power demand, becomes more and more expanded which explain their increased use in the electrical energy generation [3–6]. The IM is widely used because of its robustness, reliability, low cost and little maintenance [7–9]. Nevertheless, these types of generators require an external supply on reactive power needed for the machine magnetization and to establish the remnant rotating magnetic flux wave [10, 11]. In the case of Induction Generator (IG) connected at the grid, the IM use current drawn from the grid to creates the magnetic field.

In remote site, the IG can operate in autonomous mode only when capacitors are connected at its stator terminals to produce the necessary reactive power for machine but also for its load.

The Self-excited Induction Generator (SEIG) has been discovered for a few decades and was experienced at several remote sites, but the wave quality is not optimized. Also, one encounters difficulties such as frequency and voltage control [12, 13]. These difficulties are due to the inverse nature of the problem when no theory is available to describe the system. Indeed in this working mode the voltage, the frequency and the rotor sleep are unknown.

In order to overcome these difficulties and to increase robustness of regulation system this chapter presents a control law established for a resistive-inductive load supplied by a SEIG and which can be considered as a preliminary study for an electrical motors load. When a 'real' load is presented the difficulty consist in adapting the control low, because during the regulation an important transient state appears. In this regard, it is assumed that L will be maintained constant and that the voltage regulation will be realized acting only on R load.

This chapter presents the theoretical background of output voltage and frequency regulation for autonomous induction generators. It covers the analysis, modeling and simulation of an isolated self-excited induction generator working in remote site. Indeed, we are interested in the studying of a system able to vary the C capacitor when the SEIG supplies a resistive-inductive load (R–L). It suffices to parallel connect a fixed capacitor C_M and a regulating inductive load (L_R) connected to the isolated network via a dimmer. This system will control the reactive power consumption variation. The total reactive power is delivered by fixed capacitor C_M and can be variable and so adjustable continuously by means of the TCR and the regulating inductive load.

One treats, in the first section, the steady state analysis of IM. This study is based on a single phase equivalent circuit definition while considering the concept of voltage source. One will discuss subsequently the developed frequency control law. The operating points of the SEIG are fixed such that they give rated output at the rated conditions of voltage, current and speed. The objective of the various developments presented is to determine the exact operating points, and for that it is proposed to use as mathematical solution an iterative (Newton-Raphson) method or a graphical approach. The self-excited induction generator modeling based on space phasor formalism is treated in the second section. Subsequently, one presents the analysis of the control law effectiveness in the steady and transient state cases. In a standalone SEIG, the capacitors connected across the SEIG terminals can supply the reactive power requested by the SEIG as well as the load.

In the next part of this chapter, is presented the reactive power control strategy and the operating intervals imposed on the dimmer to stabilize the frequency. The dimmer allows the capacitors value variation across the machine terminals.

13.2 Proposed Autonomous Configuration

Figure 13.1 presents the global SEIG system configuration proposed for operating in remote site. Considering the wind as a prime mover, this asynchronous generator, which works in autonomous mode, will provide electrical power to supply a *R–L* load.

In this configuration the capacitor bank allows to provide the reactive power needed for generator magnetization [14–17] as well as the load. This section starts by studying the induction machine and its modeling. To establish the model of the asynchronous machine, we adopt the following simplifying hypotheses:

- The air gap has a uniform thickness by neglecting the slotting effects and saturation effects
- The winding resistances does not vary with temperature,

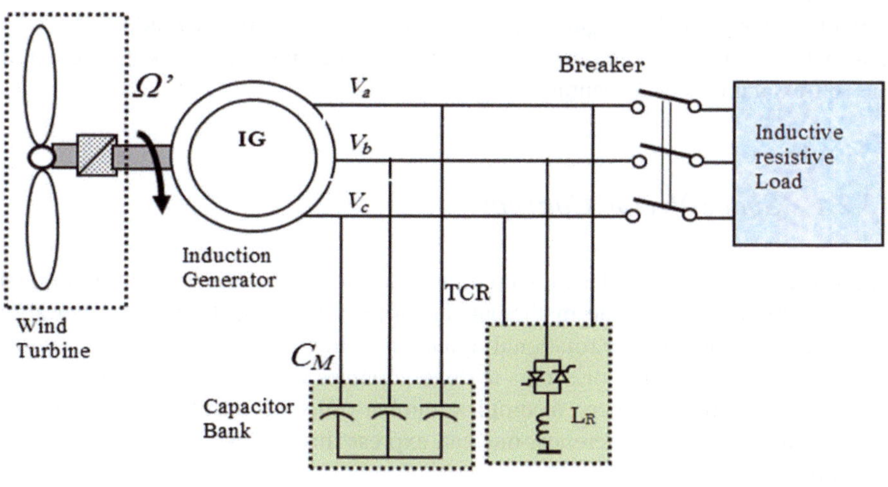

Fig. 13.1 Global diagram configuration of an isolated SEIG

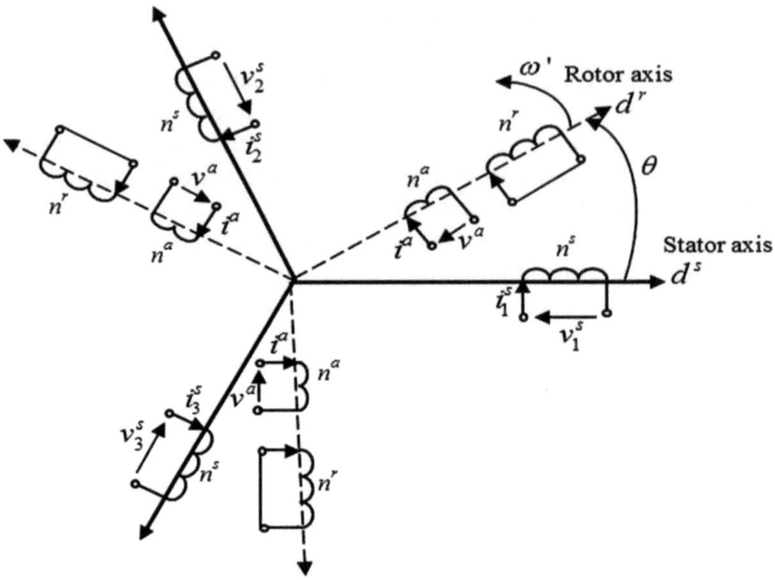

Fig. 13.2 Representation of the machine statoric and rotoric windings

– The iron losses are neglected.
– It is admitted that the magnetomotive force, which is created by each phase of the stator and rotor, has a sinusoidal spatial distribution along the air gap.

These assumptions lead to consider a constant inductance. However, the mutual inductances between the stator and rotor windings are variable according to a sinusoidal law and depend on the electrical angle of their magnetic axes.

Figure 13.2 represents the asynchronous machine in the electrical space (real variables reference). This figure shows a three phase n^a fictitious winding. They are distributed as the rotor windings of the reel coil in the objective of taking account of the self-excitation phenomenon.

13.2.1 Space Vector Concept

Space vector is used in order to overcome the difficulties of the working with three variables by simplifying the mathematical formulation through only two variables [18]. Hence, a view of the rotational dynamics is improved using the space vector formalism. Consequently, one joins at three variables x_1, x_2 and x_3 which belong to the assembly of real numbers, a complex number called vector of direct and inverse components. In a fixed reference, one can express this vector, denoted $\bar{x}$, using the given relationship

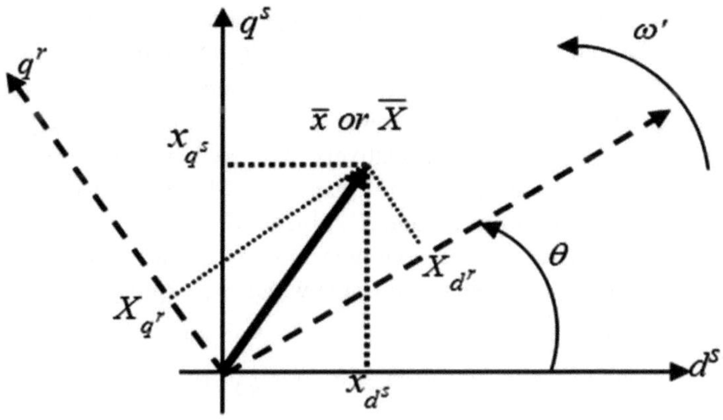

Fig. 13.3 Changing of reference frame

$$\bar{x} = x_d + jx_q = (2/3)\left[1 + \underline{a} + \underline{a}^2\right]\left[x_1 \;\; x_2 \;\; x_3\right]^t \tag{13.1}$$

where $\underline{a} = e^{j2\pi/3}$ is a rotational operator.

One will consider a moving reference with rotational angle θ. This vector denoted $\bar{X}$ presented in Fig. 13.3 is given using the following equation

$$\bar{X} = X_D + jX_Q = \bar{x}\, e^{-j\theta} \tag{13.2}$$

In the case of the study of a time derivative, following equation must be used

$$e^{-j\theta}\frac{d\bar{x}}{dt} = (j\omega')\bar{X} + \frac{d\bar{X}}{dt} \tag{13.3}$$

13.2.2 Induction Generator Model

The real machines, with their different kinds of windings and geometries, are too complex to lend itself to an analysis taking into account their exact configurations, hence the need to develop a model whose behavior is as close as possible to that of the original [19, 20]. In this context, it is essential to use a simple model to facilitate digital implementation and minimize the simulator computation time.

Through a numerical implementation of the SEIG model based on the space vector, one can study its steady and transient state. This study concerns a three-phase induction machine with p pole pair cage rotor. One will confuse the rotor and stator spatial reference (d^r and d^s) with the rotor and stator phase 1 axes. The asynchronous generator must have a rotor residual magnetic field [21, 22], to produce the stator voltage. Let us consider a fictitious n^a winding to create this magnetic field during the SEIG start-up step. These coils of n^a turns are considered supplied by i_q^a DC currents (q = 1, 2 or 3). These all currents are null after the start-up step. The stator space vector variables referred to d^s are denoted $\bar{x}^s$. Also, one will denote $\bar{x}^r$ the rotor space vector variables defined relatively to d^r. The stator space vector variables defined relatively to d^r are labelled $\bar{x}'^s$. One can deduce $\bar{x}'^s$ from $\bar{x}^s$ using variable change: $\bar{x}'^s = \bar{x}^s \exp(-j\theta)$ where: $\theta = \theta_0 + \omega' t$ represents the spatial angular gap between d^r and d^s, also $\omega' = p\Omega'$ where Ω' is the machine rotor angular speed. The asynchronous generator is characterized by the voltage equations written in the rotor reference frame which are expressed by the following system

$$\begin{cases} \bar{v}'^s = r^s \bar{i}'^s + \frac{d}{dt}\bar{\varphi}'^s + j\omega'\bar{\varphi}'^s \\ \bar{v}^r = r^r \bar{i}^r + \frac{d}{dt}\bar{\varphi}^r \\ \bar{v}^a = r^a \bar{i}^a + \frac{d}{dt}\bar{\varphi}^a \end{cases} \tag{13.4}$$

The per phase winding resistances, in system (13.4), are represented by r. Whereas the fluxes linked by the rotor, the stator and the fictitious windings are labeled $\bar{\varphi}'^s$, $\bar{\varphi}^r$ and $\bar{\varphi}^a$. They are given by

$$\begin{cases} \bar{\varphi}^r = L^r(1+\lambda^r)\bar{i}^r + M^{rs}\bar{i}'^s + M^{ra}\bar{i}^a \\ \bar{\varphi}^a = L^a(1+\lambda^a)\bar{i}^a + M^{as}\bar{i}'^s + M^{ar}\bar{i}^r \\ \bar{\varphi}'^s = L^s(1+\lambda^s)\bar{i}'^s + M^{sr}\bar{i}^r + M^{sa}\bar{i}^a \end{cases} \tag{13.5}$$

To take into account the remnant flux, one use $\lambda^a = l^a/L^a$ to characterize the fictitious windings leakage inductance. M^{sr}, M^{sa} and M^{ra} are the mutual inductances between the stator, the rotor and the fictitious windings (it is considered that $M^{sr} = M^{rs}$, $M^{sa} = M^{as}$ and $M^{ra} = M^{ar}$). The L quantities given in (13.5) represent the main winding cyclic inductances, whereas the cyclic mutual inductances between various windings are denoted M.λ^r and λ^s are expressed by: $\lambda^r = l^r/L^r$ and $\lambda^s = l^s/L^s$ where l^r and l^s represent respectively the rotoric and the statoric leakage inductances. Assuming that $n^s/n^r = \sqrt{2}$, the M inductance can be given by $M^{ra} = \sqrt{L^r L^a}$. Regarding the studied model, one considers that the saturation effect is neglected, which make constant all the main and mutual inductance coefficients. Through Eqs. (13.4) and (13.5), the statoric voltage defined in d^r reference will be expressed as

$$\bar{v}'^s = [r^s + j\omega' L'^s]\bar{i}'^s + L'^s\frac{d\bar{i}'^s}{dt} + j\omega' M^{sr}\bar{i}^r + M^{sr}\frac{d\bar{i}^r}{dt} + j\omega' M^{sa}\bar{i}^a \tag{13.6}$$

where $L'^s = L^s(1+\lambda^s)$ and $L'^r = L^r(1+\lambda^r)$.

From rotoric voltage equation we can deduce

$$r^r \bar{i}^r + L'^r \frac{d\bar{i}^r}{dt} = -M^{sr} \frac{d\bar{i}'^s}{dt} \tag{13.7}$$

For the R–L load, one can write the SEIG output current as

$$\bar{i}^s = -(\bar{i}_R + \bar{i}_C + \bar{i}_L) \tag{13.8}$$

where

$$\bar{i}_R = \frac{\bar{v}'^s}{R}, \bar{i}_C = C\frac{d\bar{v}'^s}{dt}, \bar{i}_L = \frac{1}{L}\int \bar{v}'^s dt \tag{13.9}$$

ω can be calculated considering the slip s given in: $\Omega' = (1-s)\Omega$ while taking into account the synchronous angular speed $\Omega' = \omega'/p$. Furthermore, to complete the SEIG modelling during transients and steady states, one has to consider the mechanical relationship given by

$$T_w = T_e + T_f + J\frac{d\Omega'}{dt} \tag{13.10}$$

where J represents the overall system inertia and T_f include the machine friction and windage torque. The T_e electromagnetic torque and the wind power P_w acting on the blades are related by

$$P_w = (T_e + T_f)\Omega' \tag{13.11}$$

The T_e machine electromagnetic torque is defined by the cross product

$$T_e = 3p(\bar{\psi}'^s \times \bar{i}'^s)/2 \tag{13.12}$$

As one has already said that all coefficients of the main and mutual inductances are considered constant. Also, the $\bar{i}^a$ time derivative is null as it is a DC current. For numerical implementation of the SEIG during the step of the self-exciting the $\bar{i}^a$ current in each phase is not null ($\left|\bar{i}^a_q\right| = 0.2$ A).

13.2.3 Startup of the SEIG

The variation of the voltage space phasor modulus is obtained using the record of each line to neutral voltage values v_1, v_2, v_3 and then one applies the transformation given by

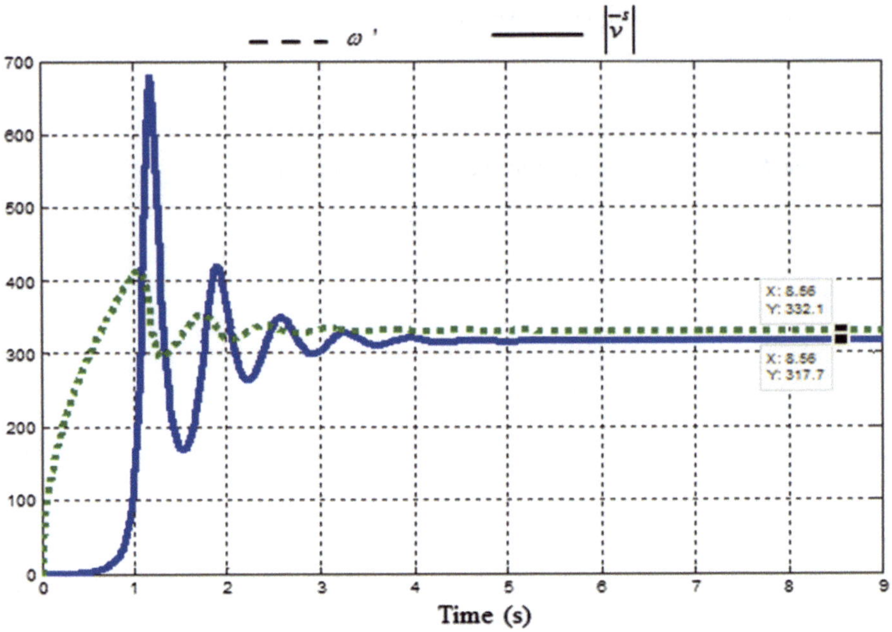

Fig. 13.4 Variations of ω' and $|\bar{v}^s|$ at SEIG startup

$$|\bar{v}^s| = \left|\frac{2}{3}(v_1 + av_2 + a^2v_3)\right| \tag{13.13}$$

Figure 13.4 presents the simulation results of $|\bar{v}'^s| = |\bar{v}^s|$ (solid line) and ω' (dashed line). The curves of these variables are represented versus time expressed in seconds (s).

Considering a constant power $P_w = 1884$ W which is applied to SEIG shaft, the variations of the last variables throughout startup are presented in Fig. 13.4 for an initial given load $R = R_0 = 111\,\Omega$, $C = C_0 = 87.5\,\mu\text{F}$, $L_0 = 170$ mH.

From a zero speed, the steady state is reached at $t = 6$ s, when $\bar{i}'^s$, $\bar{i}^r$ and $\bar{v}'^s$ have constant moduli while turning at the same $s\omega$ speed.

It is obtained: $|\bar{i}'^s| = 3.92$ A, $|\bar{i}^r| = 4.54$ A, $|\bar{v}^s| = |\bar{v}^s| = 317.7$ V, $s\omega = 18.9$ rad.s^{-1}, $\omega' = 332.1$ rad.s^{-1} and $\omega = 313.2$ rad.s^{-1} because $\omega' = \omega - s\omega$. One can deduce the slip value $s = -6.03\%$. The line to line RMS voltage obtained can be deduced from $|U^s| = |\bar{v}^s| \times \sqrt{3/2} = 389.1$ V. By against, the line to line voltage evolution versus time obtained by simulation is shown in Fig. 13.5a, b. In steady state, the line to line RMS voltage obtained is $550.1/\sqrt{2} = 389$ V. The simulation model shows that there is an important voltage peak which appears at startup if cares are not taken.

In order to avoid such voltage important variations in practical testing, it is recommended to perform the machine startup at no load and low power. It is also

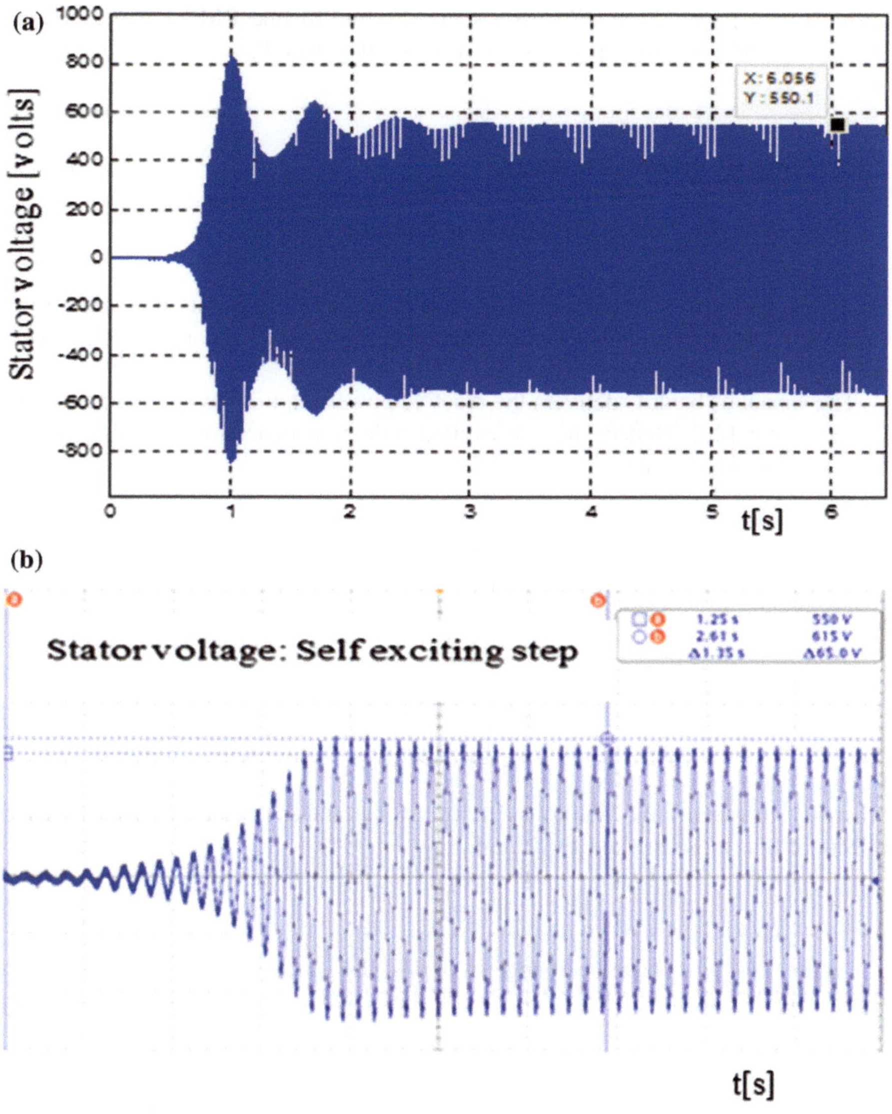

Fig. 13.5 Stator voltage at startup, **a** simulation result, **b** experimental result

necessary to gradually increase the SEIG shaft power (if possible). Also the starting procedure must follow specific steps for coupling C and R–L at the SEIG terminals. Such proceedings may be implemented in a control circuit or performed manually. This startup (at no load and low wind power) is presented in Fig. 13.5b. In practice, the duration of the transient state at startup depends on the procedure chosen according to the load and the wind turbine characteristics. Starting from $t = 0$ with null speed and voltage, the SEIG evolves to attain stable regime for $R = R_0 = 111$

Ω and $C = C_0 = 87.5$ μF. In steady state, the line to line RMS voltage obtained ($550.1/\sqrt{2} = 389$ V) corresponds to a SEIG shaft power P_w which is constant.

13.3 Steady State Analysis

13.3.1 Single Phase Equivalent Circuit Model

Using the concept of induced voltage source, Fig. 13.6 presents the single phase equivalent circuit of an induction machine with *p* pole pair. A parallel (*R*–*L*) load is connected across the stator terminals.

The capacitor bank *C* delivers the reactive power required by the load as well as the generator [23]. Hence, the self-excited induction generator operating is characterized by given system

$$\begin{cases} \underline{V}^s = r^s \underline{I}^s + jx^s \underline{I}^s + \underline{E}^s_R \\ \underline{E}^s_R = \underline{E}^s_{(s)} + \underline{E}^s_{(r)} \end{cases} \tag{13.14}$$

where

$$\begin{cases} \underline{E}^s_{(s)} = jX^s \underline{I}^s \\ \underline{E}^s_{(r)} = jX^s \underline{I}'^r \\ \underline{E}^s_R = jX^s \underline{I}^s_R \end{cases} \tag{13.15}$$

where $\underline{I}^s_R$ represents the fictitious current in the stator having to same magnetic effects of that those created by $\underline{I}^s$ and $\underline{I}'^r$

$$\underline{I}^s_R = \underline{I}^s + \underline{I}'^r \tag{13.16}$$

As the iron losses have been neglected, this SEIG model will be characterized by the given parameters of $X^s = L^s\omega$, with L^s the cyclic magnetizing inductance and ω

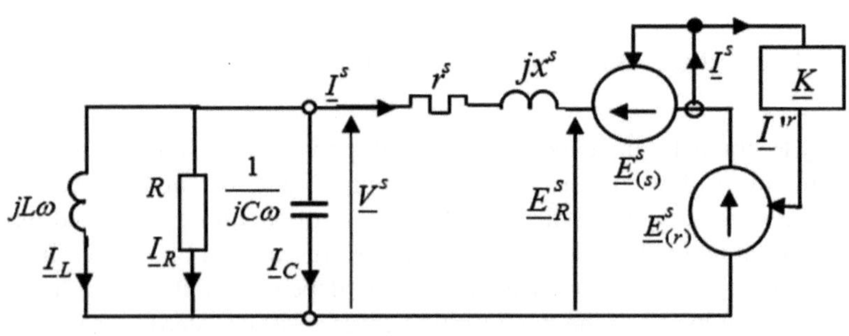

Fig. 13.6 Single phase equivalent circuit with parallel *R*–*L* load

stator angular frequency, $x^s = l^s\omega$, with l^s the statoric leakage inductance, and $x'^r = l'^r\omega$, with l'^r the rotoric leakage inductance related to the stator.

The $\underline{K}$ variable shown in Fig. 13.6 is expressed by

$$\begin{cases} \underline{K} = \underline{I}'^r/\underline{I}^s \\ \underline{K} = -jL^s s\omega/[r'^r + j(L^s + l'^r)s\omega] \end{cases} \quad (13.17)$$

Considering the equations from (13.14) to (13.16), one can deduce the time vector diagram. Taking into account the real and imaginary parts of the complex variables, one finds the following system

$$\begin{cases} s^2\omega^4 RLCA' + s^2\omega^2[LE'' - RA'] + s\omega^2 LL^{s2}r'^r - \omega^2 RLCD' \\ \qquad + RD' + (R + r^s)Lr'^{r2} = 0 \\ s^2\omega^4 LA'' - s\omega^4 LB + s^2\omega^2 r^s E' + s\omega^2 L^{s2}Rr'^r \\ \qquad -\omega^2 LD'' + Rr^s r'^{r2} = 0 \end{cases} \quad (13.18)$$

The constants A'', B, D'', E'' are expressed by

$$\begin{cases} A'' = L^{s2}(1 + \lambda^r)\{L^s[1 - (1 + \lambda^s)(1 + \lambda^r)] - r^s RC(1 + \lambda^r)\} \\ B = L^{s2}RCr'^r \\ D'' = r'^{r2}\{L^s(1 + \lambda^s) + r^s RC\} \\ E'' = L^{s2}(1 + \lambda^r)^2[R + r^s] \end{cases} \quad (13.19)$$

To obtain the values of the quantities A', D' and E', one will considers that $r^s = 0$ in the expression of A'', B and E''. The electromagnetic torque T_e of IM can be expressed from the cross product

$$T_e = k\overrightarrow{E}^s(s) \times \overrightarrow{E}^s(r) \quad (13.20)$$

where $k = 3p/\omega X^s$. The RMS voltage at the output of the generator is given by

$$V^s = \frac{RL\omega}{\sqrt{(L\omega)^2 + R^2(LC\omega^2 - 1)^2}} I^s \quad (13.21)$$

Hence the expression of the electromagnetic torque

$$\Gamma_e = \frac{3P}{\omega X^s} \frac{SX^{s3}r'^r[(L\omega)^2 + R^2(LC\omega^2 - 1)^2 V^{s2}]}{(r'^{r2} + S^2L^{s2}(1 + \lambda^r)^2\omega^2)RL\omega^2} \quad (13.22)$$

which allows to express V^s versus T_e

$$V^s = \frac{RL}{L^s}\sqrt{\frac{T_e\omega}{3p}\frac{r'^{r2} + s^2L^{s2}(1+\lambda^r)^2\omega^2}{s[L^2\omega^2 + R^2(LC\omega^2 - 1)^2]r'^r}} \tag{13.23}$$

13.3.2 *Identification of IM Operating Points*

The analysis of the system (13.18) can conclude that ω and s depend only on the load (R, L), the capacitor C and the machine parameters. Therefore, if one imposes these parameters, the self-excited induction generator will be locked on fixed ω and s independently of P_w values. In the objective of developments illustration, the studied generator is characterized by rated values given in appendix. When necessary, one uses the lower index "rat". For practical tests, one connects the stator windings in star, to prevent the machine from high destructive transients which occurs during P_w and load variations, but also to evade magnetic saturation. The R, C and L shall be chosen such a way that the electrical reference of the Operating Point (OP), indicated with the lower index "ref", may decreases from the rated values by a $\sqrt{3}$ ratio. To determine the ω_{ex} and s_{ex} exact values, one must take into account the parasitic elements. In order to solve system (13.18), two methods are used: an iterative (Newton-Raphson) and graphical methods.

13.3.2.1 Graphical Operating Point Determination

Starting with the first relationship of system (13.18), one can obtain

$$M_1\omega^4 + N_1\omega^2 + P_1 = 0 \tag{13.24}$$

where M_1, N_1 and P_1 are expressed by

$$\begin{cases} M_1 = LRCA's^2 \\ N_1 = (LE'' - RA')s^2 + LL^{s2}r'^r s - RLCD' \\ P_1 = RD' + L(R + r^s)r'^{r2} \end{cases} \tag{13.25}$$

Since M_1 is negative, one will consider only the positive ω value because the quantity under the radical is greater than unity. The root of (13.24) can be given by

$$\omega_1 = \left[-N_1\left\{1 + \sqrt{1 - 4M_1P_1/N_1^2}\right\}/2M_1\right]^{1/2} \tag{13.26}$$

Similarly, solutions of the second equation of system (13.18) $\omega^2_{2(1\text{or}2)}$ are given by

$$\omega_{2(1\text{or}2)} = \left[-N_2\left\{1 \pm \sqrt{1 - 4M_2P_2/N_2^2}\right\}/2M_2\right]^{1/2} \tag{13.27}$$

where

$$\begin{cases} M_2 = Ls(A''s - B) \\ N_2 = r^sE's^2 + L^{s2}Rr'^rs - LD'' \\ P_2 = Rr^sr'^{r2} \end{cases} \tag{13.28}$$

In Eq. (13.27) one will associate the sign + before the radical with the lower index 1 between brackets and sign − with the lower index 2. One indicates that M_2 has a positive value only for $0 > s > s^*$, given that $s^* = B/A$. Consequently, $\omega_{2(1)}$ exists only in this margin of s changes however $\omega_{2(2)}$ exists for s range between −1 and 0 as it is shown in Fig. 13.7 plotted for R_0,C_0 and L_0($s^* = -35.8\%$). As the OPs are given by the intersections of ω_1 with $\omega_{2(1)}$ and $\omega_{2(2)}$, Fig. 13.7 shows that $\omega_{2(2)}$ do not intervene in the OP determination.

There are only two OPs, tied to $\omega_{2(1)}$ which will be taken into account. The first OP 'A' with coordinates $s_{ex} = -6.03\,\%$ and $\omega_{ex} = 313.2$ rad.s^{-1}, is placed in the $\omega_{2(1)}$ characteristic flattened part situated around $\omega_{\min}$ where the changes of ω are limited [24]. When the slip values tend to $s_{\min}$ then ω corresponds to $\omega_{\min}$ given by

$$\omega^2_{\min} = -\frac{\left[s^2_{\min}r^sE' + s_{\min}L^{s2}Rr'^r - LD''\right]}{s_{\min}L[s_{\min}A'' - B]} \tag{13.29}$$

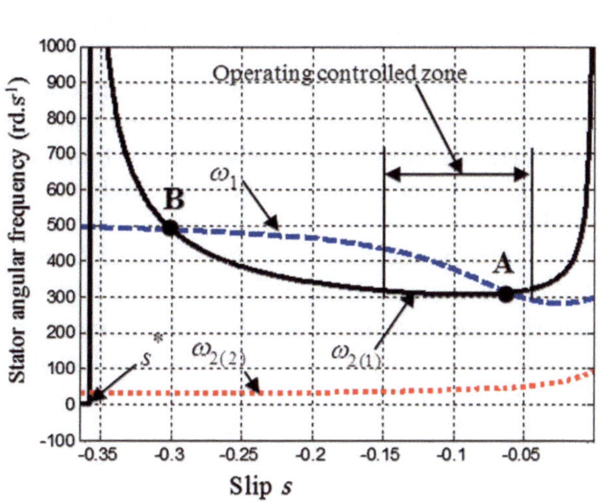

Fig. 13.7 Operating points for R_0,C_0 and L_0 load

Now we will consider the second equation with unknown s to find $s_{\min}$, we will have

$$s_{1,2} = \frac{\omega^4 LB - RL^{s2} r'^r \omega^2 \pm \sqrt{(RL^{s2} r'^r \omega^2 - \omega^4 LB)^2 - 4[LA'' \omega^4 + r^s E' \omega^2].[-\omega^2 LD'' + r^s r'^{r2} R]}}{2LA'' \omega^4 + 2\omega^2 E' r^s} \tag{13.30}$$

For the s existence, it is necessary that the quantity under the radical is positive or zero. So we obtain

$$s_{\min} = \frac{LB\omega_{\min}^2 - RL^{s2} r'^r}{2LA'' \omega_{\min}^2 + 2r^s E'} \tag{13.31}$$

which gives

$$s_{\min}^2 \left[L^{s2} R r'^r LA'' + r^s E' LB\right] + s_{\min}\left[-2A'' D'' L^2\right] + D'' BL^2 = 0 \tag{13.32}$$

where $\omega_{\min}$ is considered the minimum angular frequency. The roots of this equation are given by

$$s_{\min} = \frac{A'' D'' L}{L^{s2} R r'^r A'' + BE' r^s}\left[1 - \sqrt{1 - \frac{B[L^{s2} R r'^r A'' + r^s E' B]}{A''^2 D'' L}}\right] \tag{13.33}$$

The area, considered as stable operating zone, is limited on the left taking into account the machine energetic performances. Whereas the OP 'B' is placed far from the stable zone and possesses high ω and s values which can destroy the SEIG energetic performance.

13.3.2.2 Evolution of Operating Point for Variation of the L Load

To distinguish the L effect on the OPs, one has presented in Fig. 13.8 various cases for L values such as 0.17 H (L_0), 1, 10 and 100 H. The case of $L = 100$ H may be considered practically as a resistive load. From system (13.18), one deduces that $R = R_0$, however C should have respectively the values C_0, 32.52 μF, 22.4 μF and 21.37 μF.

As already indicated that, irrespective of L value, $\omega_{2(2)}$ never intervenes in the OP determination. Therefore, one does not take into account $\omega_{2(2)}$ in the next few analyses. Furthermore, it is noted that ω_{ex} increases with L however s_{ex} possesses quasi constant values. Also, one can note that the margin for ω nearly constant increases as L increases. This behavior can be benefit if there are no constraints on the energetic performance of the machine.

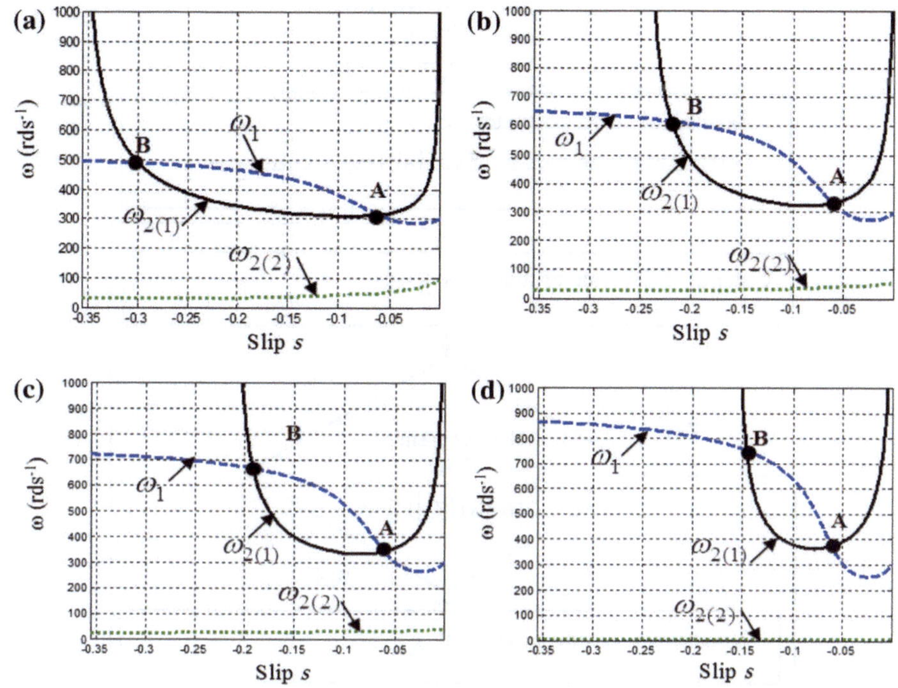

Fig. 13.8 Impact of L on the operating points positions, **a** $L = 0.17$ H, **b** $L = 0.55$ H, **c** $L = 1$ H, **d** $L = 100$ H

13.3.2.3 Newton-Raphson Numerical Solution

The numerical analysis is performed using the equations system (13.18) denoted by g_1 and g_2 respectively for the first and the second equation where the unknown are ω and s.

$$\begin{cases} g_1 = s^2\omega^4 RLCA' + s^2\omega^2[LE'' - RA'] + s\omega^2 LL^{s2} r'^{r} \\ \qquad -\omega^2 RLCD' + RD' + (R + r^s) L r'^{r2} \\ g_2 = s^2\omega^4 LA'' - s\omega^4 LB + s^2\omega^2 r^s E' \\ \qquad + s\omega^2 L^{s2} R r'^{r} - \omega^2 LD'' + R r^s r'^{r2} \end{cases} \quad (13.34)$$

Using the iterative Newton-Raphson method based on the successive use of the Jacobian for the regulation of the variables adjustment pitch, it is possible to find a numerical solution of the nonlinear static system (13.34) [25–27].

The approximation of the g_i (i = 1, 2) functions by a Taylor development stopped at the first order around the initial condition $\underline{x}_0(\omega_0 = 314, s_0 = -1)$ to a zero target for these functions, allows to deduce the necessary variation of the variables ω and s.

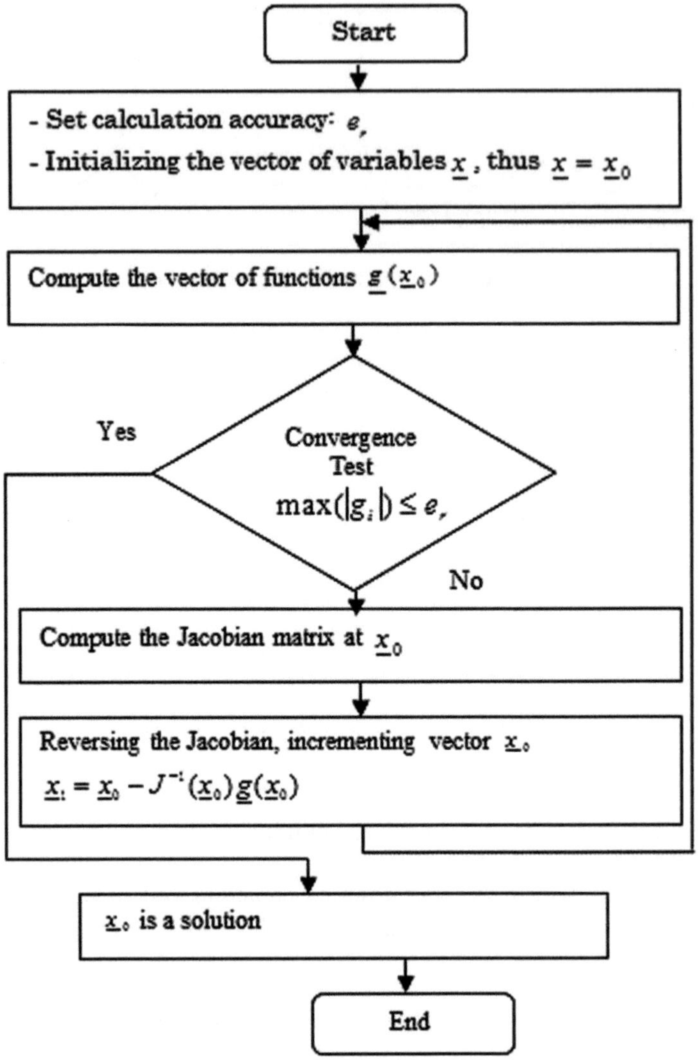

Fig. 13.9 Flowchart of the iterative method of Newton-Raphson

The calculation is redone as g_1 and g_2 continue to exceed calculation accuracy fixed in advance. The algorithm of the Newton-Raphson method is summarized by the flowchart presented by Fig. 13.9. To stop the algorithm, a convergence test which usually comprises a double control is established:

The first concerns a limited number of iterations and the second concerns the accuracy. This second test could be expressed as

$$\max(|g_i|) \le e_r \tag{13.35}$$

where e_r is a tolerance computation fixed by the operator ($e_r = 0.001$).This approach allowed to find $\omega_{ex} = 315.6$ rad/s and $s_{ex} = -6\%$.

13.4 Impact of the *C* Variation on the Frequency

Figure 13.7 revealed the possibility of stabilizing the SEIG output frequency for an operating point defined at the bottom of the curve. Therefore, it is necessary to choose the load *R–L* and the capacitor *C* required for the wind turbine operating at this point. Given that the curve is flattened around this point, limited variations of the *R* load around 19% will keep quasi constant the frequency by changing the capacitor *C*. Table 13.1 gives calculation and measurement example.

There are five cases presented in Table 13.1. The first was carried out for $R_0 = 111\ \Omega$ and $C_0 = 87.5\ \mu F$. This is the central SEIG operating point (Fig. 13.7). In the following, some variations in the *R* load will be tested around this central operating point « A » which is characterized by $s = -5.7\%$ and $\omega = 312.7$ rad/s ($f = 49.8$ Hz).

These changes move the operating point to right and left of the central operating point. Regarding the second case, we choose $R = 132\ \Omega$ (variation of 19%) that corresponds to the machine discharge characterized by the operating point whose cordinates are $s = -4.8\%$ and $\omega = 315.8$ rad/s (a displacement to the right, $f = 50.3$ Hz, $\Delta f = 0.5$ Hz). The capacitor value should be 83.12 μF, but considering our capacitor bank we used $C = 83\ \mu F$. For comparison, we have also tested (test 3) a variation of R value ($R = 132\ \Omega$) without changing *C* ($C = C_0 = 87.5\ \mu F$). We notice a slightly greater change in the stator angular frequency ($\omega = 307$ rad/s, $f = 48.8$ Hz, $\Delta f = 1$ Hz).

A fourth pair (*R*, *C*) was considered (a displacement to the left) such as R = 86 Ω (variation of 22.5%) and $C = 95.5\ \mu F$ which corresponds to an increase in the load. The change in C provides a smaller variation of ω ($\omega = 311.5$ rad/s, $f = 49.6$ Hz, $\Delta f = 0.2$ Hz). This case will be compared to another fifth case where C is kept constant when *R* varies ($R_2 = 86\ \Omega$ and $C_2 = C_0 = 87.5\ \mu F$). It is also noted a

Table 13.1 Changes in angular frequency, slip and voltage versus load at constant power

$L = 170$ mH	Experimental values				Theoretical values				
$P_w = 1884$ W	*R* Ω	*C* μF	ω rad/s	*f* Hz	*s* %	*V* volts	ω rad/s	*f* Hz	*s* %
C variable	111	87.5	312.7	49.8	−5.7	224.5	313.2	49.8	−6.03
	132	83	315.8	50.3	−4.8	245.4	317.4	50.5	−5.08
	86	95.5	311.5	49.6	−7.7	192.3	308.3	49.09	−7.8
C = 87.5 μF = constant	132	87.5	307	48.8	−4.9	244.2	308.7	49.1	−5.08
	86	87.5	322.3	51.3	−7.6	192.8	324.2	51.6	−7.8

greater variation of the stator angular frequency ω ($\omega = 322.3$ rad/s, $f = 51.3$ Hz, $\Delta f = 1.5$ Hz).

According to the results predicted by the developed model, Table 13.1 confirms that when considering a *R–L* load we can improve the stator frequency stability at the wind generator output by changing *C*. In fact, we obtain a frequency change near zero (1%) for R variation about 19%. This change is more important (3%) if *C* has not changed.

When the load is purely resistive, this variation may become much more important (11%). Other tests have been carried out for other operating points. They confirmed identically the test already presented. The theoretical values shown in Table 13.1 are obtained from graphical method (paragraph II) or numerically computed based on the analytical study already developed. It is noted that these values are close to those experimentally obtained. By maintaining constant $\tan \rho$, one can obtain an operating while keeping the frequency at quasi constant values without need of any control loop of this variable. Here ρ is the argument of the whole load and capacitor which are placed in parallel and connected to the SEIG stator.

13.4.1 Load Variation at Constant C

The angular frequencies variations for a load R change about 22% are shown in Fig. 13.10a, b.

The reduction of *R* values from reference point $R_0 = 111\ \Omega$ can displace the OPs to the curved area in the direction of increasing speed. This can cause a significant increase in the stator voltage frequency. This is due to the fact that *C* value is kept constant while varying *R*. In a second test we will increase the *R* value while keeping *C* constant. We can note a significant frequency change. By against, the

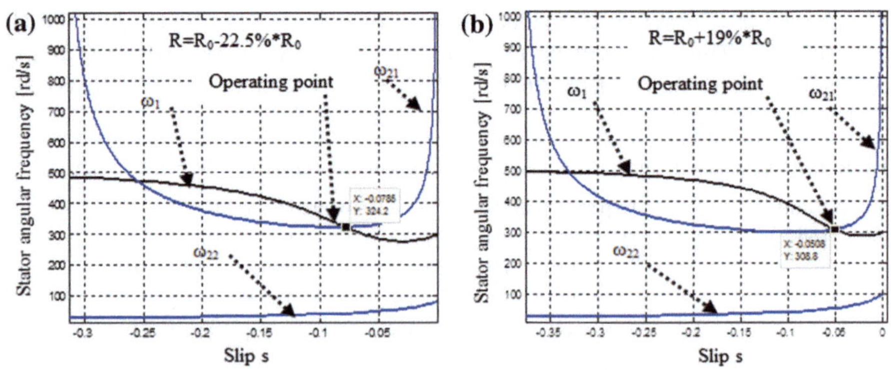

Fig. 13.10 **a** Decrease in *R* value at constant $C = 87.5$ μF, **b** Increase in *R* value at constant $C = 87.5$ μF

increases in R values are unlimited unless the machine energetic performance and voltage stability are lost.

13.4.2 R Load Variation at Variable C

In this case we will accompany to any R variation a change in the value of C. Figure 13.11 shows the considered study with two tests of load tilting around the initial operating point (R_0 and C_0). The strategy proposed for controlling the frequency needs the definition of pairs (R, C) allowing to reach the OP in the controlled zone. With the considered machine parameters, we lead to the point A (Fig. 13.7) corresponding to R_0 = 111 Ω and C_0 = 87.5 μF. The next stage is to act on R and therefore C while respecting the following law which is deduced from $\tan \rho = \text{constant}$.

$$C = \frac{(R + R_0(LC_0\omega_{rat}^2 - 1))}{LR\omega_{rat}^2} \tag{13.36}$$

In Fig. 13.11 is noted that the change of the pair (R, C) while respecting the Eq. (13.36), the frequency is kept quasi constant during the wind generator operating. We also note a small change in frequency ($\Delta f \approx 1.5\%$) if R varies about 20% from R_0. Through against, if C is kept constant while changing R close to 20%, the frequency variation can exceed 3.5%. Figure 13.10a, b show the changes of the OP

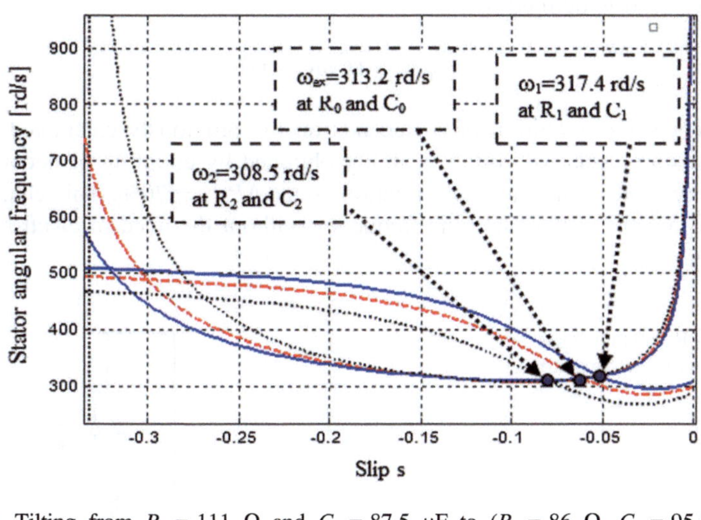

Fig. 13.11 Tilting from R_0 = 111 Ω and C_0 = 87.5 μF to (R_2 = 86 Ω, C_2 = 95.5 μF) and (R_1 = 132 Ω, C_1 = 83 μF)

positions for only R load variations (L and C kept constant). Figure 13.11 presents the results of the R variations with variable C. It is noted that the three characteristics are confused in the stable region shown in Fig. 13.7 when applying the f frequency control law.

13.5 Reactive Power Control

We consider the SEIG operation in remote site. By mean of the analytical method developed in Sect. 13.3, it is possible to find a couple R, C adapted for a given inductive load L. This study leads to some operating points around the point A within the stable zone (Fig. 13.7). When the power P_w varies, the voltage regulation may be ensured by changing a resistive R load. This change concerns only the consumers who are not priority and they can be disconnected or introduced in accord with the wind power changes. The R load changes must be accompanied by a modification of the reactive power provided by the capacitor C_M in order to stabilize the frequency. One can assure the reactive power changes using thyristor controlled reactor. This dimmer mounting is controlled by the firing angle ψ as in Fig. 13.12 [28].

Let us consider that regulating inductive load L_R have also an internal resistance R_{L_R}. We talk about a load $\underline{Z}_R$. By means of this system, a reactive power consumer can be disconnected or introduced in the isolated grid. According to Eq. (13.36), the R load variation is accompanied by a capacitor change $C = C_{eq}(\psi)$ through changing the firing angle ψ. In this case, the equivalent reactive power available on the isolated grid is defined by

$$Q_{eq} = -U^2\omega C_{eq}(\psi) \tag{13.37}$$

In practice, the evolution of ψ depends on the previously calculated values in order to ensure a relative voltage stability obtained by coupling or decoupling the non-priority loads with maximal variation about $\Delta R \approx \pm 20\%$. This change keeps quasi constant the voltage and the frequency without the need of electronic power converters.

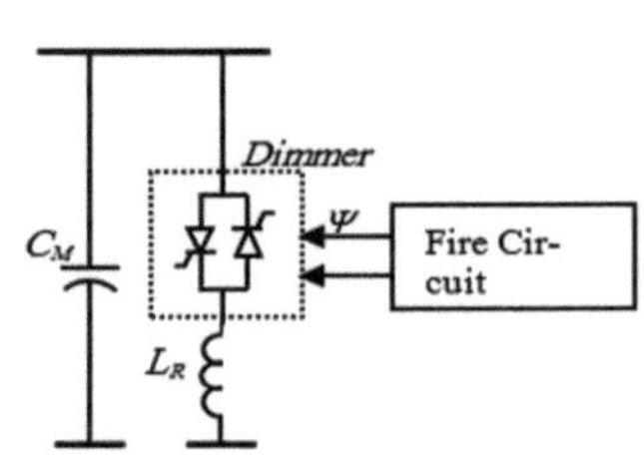

Fig. 13.12 Regulating load in parallel with the capacitor C_M

13.5.1 Variation of the Capacity

To regulate the reactive power, a three-phase system using a balanced load $\underline{Z}_R$ is considered. The C_M capacitor modification is realized using balanced regulating impedance $\underline{Z}_R$ which is constituted by a resistance R_{L_R} in series with an inductance L_R. This load is supplied through a three-phase dimmer as it is shown in Fig. 13.13a. The dimmer choice is due to its low cost and its ability to compensate reactive power. Let us denote ϕ the regulating impedance argument.

To simplify the analytical study, it has been presented the $v_1'(\theta)$ voltage curve and that of $i_1'(\theta)$ corresponding current in Fig. 13.13b. It is assumed that the $i_3'(\theta)$ third phase current will be null for an angle $\theta = \theta_1$. The thyristor firing angle will be noted ψ.

We will start the analytical study of the system, cited above, by considering a single phase dimmer model (Fig. 13.14a) which feed an inductive load $\underline{Z}_R$.

In order to characterize the voltage $V_q'(\theta)$ (q = 1, 2, 3) applied to the regulating load, a Fourier analysis is carried out. This development provides the RMS fundamental voltage value V_{q1}' given by [29]

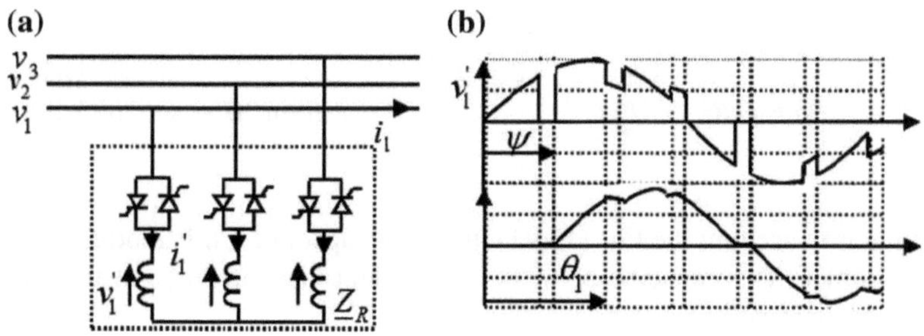

Fig. 13.13 **a** Dimmer connected to regulating load, **b** Voltage and current characteristics: 1st mode

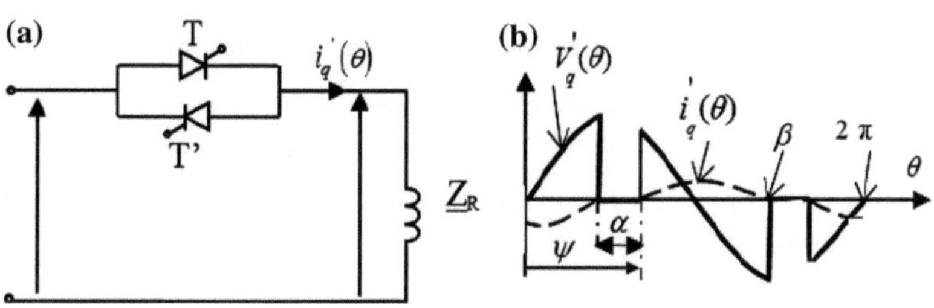

Fig. 13.14 **a** Single-phase dimmer, **b** current and voltage waveforms

$$V'_{q1} = \frac{V_q}{\pi}\begin{bmatrix} \frac{1}{2} - \frac{1}{2}\cos 2(\beta - \psi) + (\beta - \psi)^2 + \\ + (\beta - \psi)(\sin 2\psi - \sin 2\beta) \end{bmatrix}^{\frac{1}{2}} \tag{13.38}$$

where V_q is the RMS value of phase to neutral voltage applied to the dimmer input and β is the angle for which the current becomes zero in phase q. For $\omega L_R \gg R_{L_R}$ the reactive power absorbed by L_R is determined by

$$Q_{L_R} = V'_{q1} I'_{q1} \sin \phi_1 \tag{13.39}$$

with

$$I'_{q1} = V'_{q1}/|\underline{Z}_R| = V'_{q1}/\left[R^2_{L_R} + (L_R\omega)^2\right]^{\frac{1}{2}} \tag{13.40}$$

where I'_{q1} is the RMS value of the fundamental current and $\phi_n = \mathrm{a}\tan(n.L_R\omega/R_{L_R})$ with $n = 1$. Let us assume that L_R consumes a part of reactive power which is produced by the total capacitor C_M, the capacitance of capacitor removed $C(\psi)$ will be expressed as

$$C(\psi) = \frac{1}{\omega\pi^2\sqrt{R^2_{L_R} + (L_R\omega)^2}}\begin{bmatrix} \frac{1}{2} - \frac{1}{2}\cos 2(\beta - \psi) + (\beta - \psi)^2 + \\ + (\beta - \psi)(\sin 2\psi - \sin 2\beta) \end{bmatrix} \sin \phi_1 \tag{13.41}$$

Taking into account the Eq. (13.41), the total capacitor C_M is modified by means of the dimmer. The capacitance of the equivalent capacitor C_{eq} at the SEIG terminals is expressed as

$$C_{eq}(\psi) = C_M - C(\psi) \tag{13.42}$$

In the three phase operating case, the reactive power absorbed can be monitored by the dimmer control while $\phi < \psi < 5\pi/6$. Depending on the load parameters (L_R, R_{L_R}), two operating modes of the dimmer are possible [30].

- Mode 1: 2 or 3 thyristors conduct simultaneously ($\varphi < \psi < \psi_{lim}$)
- *Mode 3*: 2 or 0 thyristors conduct simultaneously ($\psi_{lim} < \psi < 5\pi/6$)

The transition from *mode 1* to *mode 3* occurs for a $\psi > \psi_{lim}$. The ψ_{lim} value can be obtained from the following relationship

$$\sin(\psi_{lim} - \phi - \frac{4\pi}{3}) = -\sin(\psi_{lim} - \phi)\frac{1 - 2e^{-\pi/3\rho}}{2 - e^{-\pi/3\rho}} \tag{13.43}$$

where $\rho = L_R\omega/R_{L_R}$. Therefore, it is possible that the dimmer can operates in one of two modes. Nevertheless, the mode 3 corresponds to a small variation in the current RMS value which causes a discontinuity for the current weak value. For a given reactive power variation at the SEIG output, it is preferred that the dimmer operates in the mode 1 because there is a possible larger variation range of $C(\psi)$ and more stability was noticed. The Eq. (13.41) enables to compute this capacitance if β is well known. Considering $q = 1$ (Fig. 13.14b) and taking into account the system symmetry for the mode 1, the current cancellation angle β can be expressed as

$$\beta = \pi + \psi - \alpha = \frac{2\pi}{3} + \theta_1 \tag{13.44}$$

where $\alpha = \psi + \pi/3 - \theta_1$ (Fig. 13.14b) and θ_1 represents the current cancellation angle in the third phase $i_3'(\theta)$ obtained from

$$\sin(\theta_1 - \phi - \frac{4\pi}{3})e^{(\theta_1-\psi)/\rho} = -\sin(\psi - \phi)\frac{1 - 2e^{-\pi/3\rho}}{2 - e^{-\pi/3\rho}} \tag{13.45}$$

However it is noted that for $\theta \in [\psi \div \beta]$ the Eq. (13.41) considers only V_{q1}' applied to the load. Nevertheless, during this interval, the signals are not sinusoidal at the dimmer output. In the case of three-phase system, the RMS fundamental voltage value V_1' is given by the following relationship

$$\begin{aligned} V_1' = \frac{3V}{2\pi}((\theta_1 - \psi + \frac{\pi}{3} - \frac{1}{2}\sin 2(\theta_1 + \frac{2\pi}{3}) + \frac{1}{2}\sin 2\psi)^2 \\ + \frac{1}{4}(\cos 2\psi - \cos 2(\theta_1 + \frac{2\pi}{3}))^2)^{1/2} \end{aligned} \tag{13.46}$$

where V is the phase to neutral voltage RMS value applied to the dimmer input.

The θ_1 angle is determined by the numerical resolution of the Eq. (13.45). The reactive power consumed by the regulating inductive load is defined by

$$Q_{L_R} = 3VI_1' \sin\phi_1 \tag{13.47}$$

where ϕ_1 is given by

$$\phi_1 = a\ \tan(L_R\omega/R_{L_R}) \tag{13.48}$$

and the I_1' current RMS value is deduced from (13.46) according to the relationship

$$I_1' = V_1'/\sqrt{\left(R_{L_R}^2 + (L_R\omega)^2\right)} \tag{13.49}$$

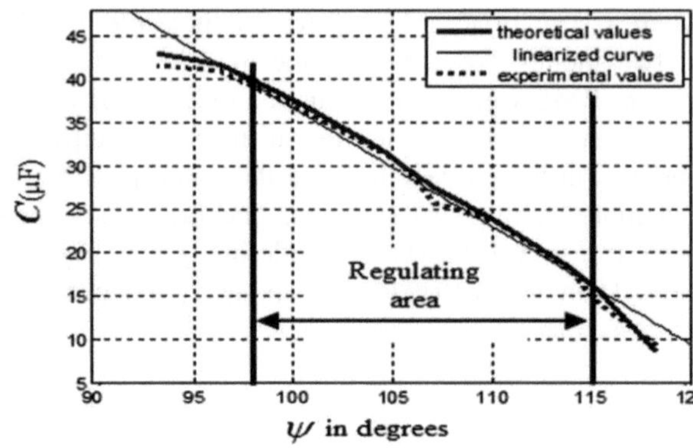

Fig. 13.15 Variation of C versus the firing angle ψ

This reactive power will be subtracted from that provided by the C_M capacitor which is computed by the following equation

$$Q_{C_M} = -3V^2\omega C_M \tag{13.50}$$

Consequently, one deduces the C capacitor (removed) expression depending on the thyristor firing angle

$$C(\psi) = \frac{3\sin\phi_1}{2\omega\pi\sqrt{R_{L_R}^2 + (L_R\omega)^2}}\sqrt{(\theta_1 - \psi + \frac{\pi}{3} - \frac{1}{2}\sin 2(\theta_1 + \frac{2\pi}{3}) + \frac{1}{2}\sin 2\psi)^2 + \frac{1}{4}(\cos 2\psi - \cos 2(\theta_1 + \frac{2\pi}{3}))^2} \tag{13.51}$$

Using Eq. (13.51), the C variation curve versus the angle ψ is plotted in Fig. 13.15. In this figure, the C theoretical values are plotted with solid line. They are compared with those experimental which are represented by dashed line. The curve plotted in continuous thin line represents the linearization of the characteristic $C = f(\psi)$. It is noted that there is a variation interval where the three characteristics are confused. That makes it possible a linear variation of C (regulating area) using a dimmer.

13.5.2 Dimmer Operating Interval

Considering a three-phase operation, the reactive power absorbed by $\underline{Z}_R$ can be controlled by the dimmer as soon as ψ is between ϕ_1 and $5\pi/6$. For a considered $\underline{Z}_R$ $(R_{L_R} \ll \omega L_R)$ to increase the system robustness, mode 1 was chosen because it allows a better operating stability and enables an important variation range of $C(\psi)$ given by Eq. (13.51). The voltage RMS value at the $\underline{Z}_R$ terminals is given by the following expression

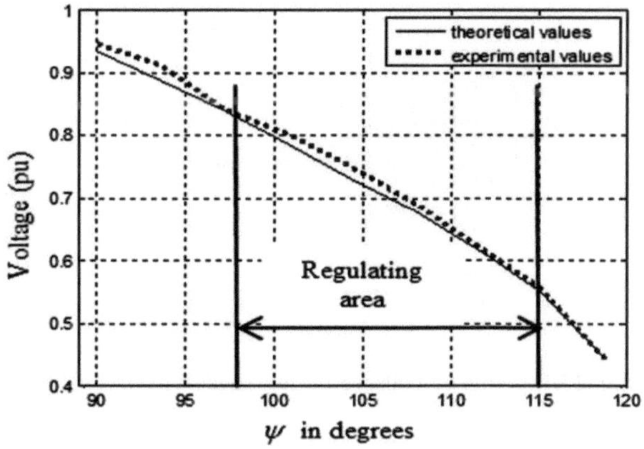

Fig. 13.16 Voltage at the dimmer output versus the ψ firing angle

$$V' = V\left\{\frac{1}{2} + \frac{3}{4\pi}\left[2(\theta_1 - \psi) + \sin 2\psi + \sin(2\theta_1 + \frac{\pi}{3})\right]\right\}^{1/2} \qquad (13.52)$$

Figure 13.16 shows the voltage evolution versus the firing angle ψ for the operation first mode. This evolution given by Eq. (13.52) (solid line) is experimentally validated (dashed line) for $V = 224.5$ V.

It shows a voltage linear variation for a clearly defined interval of ψ called "regulating area". The analytical study leads to $\psi_{lim} = 119.5°$ but to avoid a significant drop in voltage, it is necessary to limit the ψ variations although ψ_{lim} is not reached. Taking into account the Eqs. (13.51) and (13.52), constrained by the already mentioned regulating zone in Figs. 13.15 and 13.16, the implemented firing angle in the dSPACE system corresponds to the linear interval of the two characteristics. For this operating interval, it is possible to compensate the R variations respecting the low given by the Eq. (13.36).

13.6 Voltage Collapse and Self-excitation Practical Procedure

The self-excitation of IM is generally realized by increasing the rotor speed in presence of a remnant magnetization and it is easier to be obtained without R–L load connected. After self-excitation and obtained of the stable state of the generator output voltage, the connection of the load must be progressively introduced. However at instant of load connection or during a sudden load change a voltage collapse can appear. This case is shown in Fig. 13.17. Here the load requiring more active power is varied without regulation of the reactive power. We note in simulation case (Fig. 13.17a) a quickly decrease of the generator voltage.

In experimental test (Fig. 13.17b) the transient state is accompanied by more severe oscillations around of a new decreased voltage reference value before the

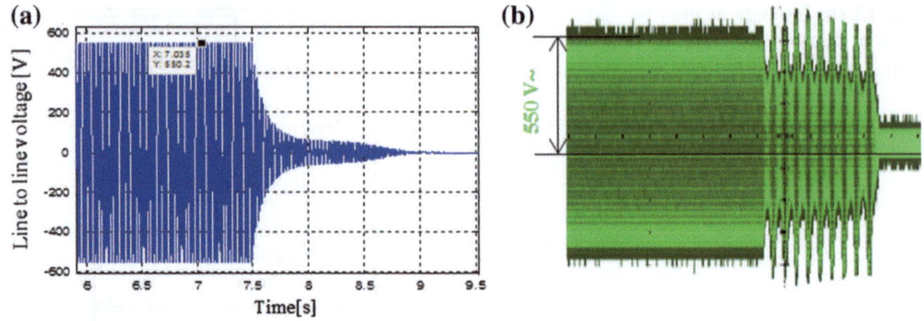

Fig. 13.17 Voltage collapse during sudden load variation, **a** simulation result, **b** experimental result

voltage collapse and the machine dropping out. These voltage oscillations are dangerousness for IM or for the load and their amplitude must be controlled. However, this relatively long transitory interval which is visible in experimental tests show that the system has a good stability and the SEIG voltage collapse appear more letter. So it can be easier to readjust the load in order to avoid the machine dropping out. Therefore in order to decrease the voltage oscillation amplitude, the load variation must be accompanied by capacitors with appropriate values.

Generally voltage collapse is accompanied by a dropping out of the machine and the self-excitation of IM can be assured only in the presence of a residual magnetic field. If IG work in remote site, a specific procedure is necessary for return it at the self-excitation step. A specific procedure is proposed in [31] where the reactive energy is given by a capacitor banc charged by a DC source. To make a success of the self-excited procedure the IG rotational speed must be increased at estimated calculated value taking in consideration the load.

Another procedure which we propose is shown in Fig. 13.18. Here the load is disconnected by breaker 1 and a DC supply is applied at the stator windings in two steps: the first by closing breakers 1–2, 3–4 and the second by changing 3–4 with

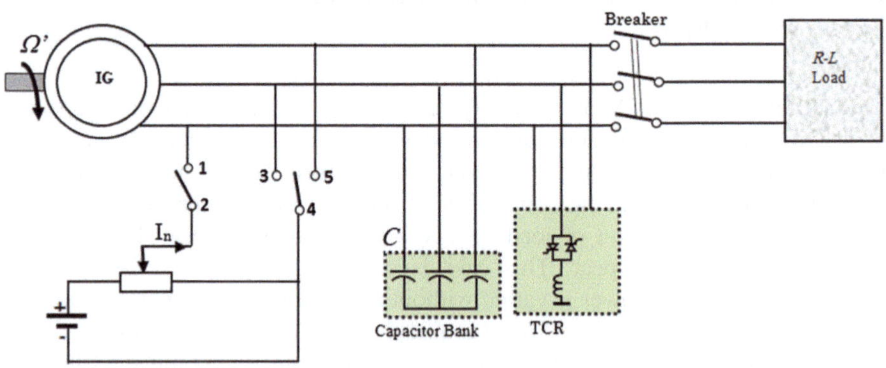

Fig. 13.18 Practical startup procedure

4–5. The current must be limited at the generator nominal value I_n. It is also possible after its disengagement to supply the machine with AC power system in order to operate as an asynchronous motor mode.

13.7 Conclusion

The work presented in this chapter treated a numerical model of a three phase Self Excited Induction Generator which operates in remote site. This wind generator supplies an inductive-resistive load connected in parallel with capacitor banks. This latter provides the reactive power required by the SEIG as well as the load. A control strategy, based on the reactive power change was studied. This study was conducted while respecting the low $\tan\rho =$ constant from which, any change of C is realized by adjusting the thyristors firing angle of the dimmer which feed a regulating load. The first part treats the use of space vector formalism in order to develop the SEIG model. The second part deals with the steady state studies by means of a single phase equivalent circuit. Another section of this chapter discusses the frequency control law. The operating points of the SEIG are fixed such that they give rated output at the rated conditions of voltage, current and speed. In the last part of this chapter, is presented the reactive power control strategy and the operating intervals imposed on the dimmer for stabilizing the frequency. The dimmer allows obtaining a variable capacitor across the machine terminals.

Appendix

The parameters of Induction Machine used in this study are: $380\,\mathrm{V}/660\,\mathrm{V}$, $7.3\,\mathrm{A}/4.2\,\mathrm{A}$, $\cos\varphi = 0.8$, $p = 2$, f=50 Hz, 1420 rpm, 3 kW, $n^s/n^r = \sqrt{2}$, $r^r = 3\,\Omega$, $L^r = 267\,\mathrm{mH}$, $M^{sr} = 377\,\mathrm{mH}$, $r^a = 10\,\Omega$, $M^{sa} = 100\,\mathrm{mH}$, $l^a = 5\,\mathrm{mH}$, $L^a = 18.7\,\mathrm{mH}$, $\lambda^a = 0.26$, $M^{ra} = \sqrt{L^a L^r} = 70.6\,\mathrm{mH}$, $L^s = 534\,\mathrm{mH}$, $r^s = 8.66\,\Omega$, $r'^r = 6\,\Omega$, $l^s = 24.24\,\mathrm{mH}$, $l'^r = 36.36\,\mathrm{mH}$, $\lambda^s = 0.0454$, $\lambda^r = 0.068$, $T_f = 1.3$ Nm, $L_R = 215\,\mathrm{mH}$, $C_M = 113\,\mu\mathrm{F}$, $380\,\mathrm{V}/660\,\mathrm{V}$, $7.3\,\mathrm{A}/4.2\,\mathrm{A}$, $V^s_{ref} = 223\,\mathrm{V}$, $s_{ref} = 5.4\%$, $I^s_{ref} = 2.42\,\mathrm{A}$, $\omega_{ref} = 314.2\,\mathrm{rad\,s^{-1}}$.

References

1. G.M. Joselin Herbert, S. Iniyan, E. Sreevalsan, S. Rajapandian, A Review of Wind Energy Technologies, Renewable and Sustainable Energy Reviews, 11, pp. 1117–1145, 2007.
2. A.H.M.A. Rahim, M. Ahsanul Alam, M.F. Kandlawala, Dynamic Performance Improvement of an Isolated Wind Turbine Induction Generator, Computers and Electrical Engineering, 35, pp. 594–607, 2009.

3. G.K. Kasal, B. Singh, Voltage and Frequency Controllers for an Asynchronous Generator-Based Isolated Wind Energy Conversion System, IEEE Trans. Energy Convers., vol. 26, no. 2, pp. 402–416, 2011.
4. Y.Y. Deng, K. Blok, K. van der Leun, Transition to a Fully Sustainable Global Energy System, Energy Strategy Reviews, 1, pp. 109–121, 2012.
5. M. Godoy Simoes, F.A. Farret, Renewable Energy Systems: Design and Analysis with Induction Generators, CRC PRESS, USA, 2004.
6. P. Gipe, Evaluating the Technology - What Works and What Doesn't, Chelsea Green Publishing Company, Wind Power for Home & Business, 1993.
7. S. Kumar, R. Narayan, Effect of Capacitive VAR on Performance of Three-Phase Self-Excited Induction Generator, International Journal of Emerging Technology and Advanced Engineering, vol. 2, issue 12, pp. 253–258, December 2012.
8. Sh. Vadhera, K.S. Sandhu, Constant Voltage Operation of Self-Excited Induction Generator Using Optimization Tools, International Journal of Energy and Environment, issue 4, vol. 2, pp. 191–198, 2008.
9. H. Kumar, N. Kamal, Steady State Analysis of Self-Excited Induction Generator, International Journal of Soft Computing and Engineering, vol. 1, issue 5, pp. 248–253, November 2011.
10. Sh. Boora, Analysis of Self-Excited Induction Generator under Balanced or Unbalanced Conditions, Int. J. on Electrical and Power Engineering, vol. 1, no. 3, pp. 59–63, 2010.
11. A.L. Alolah, M.A. Alkanthal, Optimization Based Steady State Analysis of Three Phase SEIG, IEEE Trans. Energy Convers., vol. 15, no. 1, pp. 61–65, 2000.
12. T. Ahmed, O.H. Noro, E.M. Nakaoka, Terminal Voltage Regulation Characteristics by Static Var Compensator for a Three Phase Self-Excited Induction Generator, IEEE Trans. Ind. Appl., vol. 40, no. 4, pp. 978–988, 2004.
13. A.L. Alolah, M.A. Alkanthal, Optimization Based Steady State Analysis of Three Phase SEIG, IEEE Trans. Energy Convers., vol. 15, no. 1, pp. 61–65, 2000.
14. A.E. Kalas, M.H. Elfar, S.M. Sharaf, Particle Swarm Algorithm-Based Self-Excited Induction Generator Steady State Analysis, The Online Journal on Electronics and Electrical Engineering (OJEEE), vol. 3, no. 1, pp. 369–373, 2013.
15. A. Nesba, R. Ibtiouen, O. Touhami, Dynamic Performances of Self-Excited Induction Generator Feeding Different Static Loads, Serbian Journal of Electrical Engineering, vol. 3, no. 1, pp. 63–76, June 2006.
16. S. Devabhaktuni, S.V. Jayaram Kumar, Performance Analysis of Wind Turbine Driven Self-Excited Induction Generator with External Rotor Capacitance, International Journal of Advanced Engineering Sciences and Technologies, vol. 10, no. 1, pp. 1–6, 2011.
17. L. Louze, A.L. Nemmour, A. Khezzar, M.E. Hacil, M. Boucherma, Cascade Sliding Mode Controller for Self-Excited Induction Generator, Renewable Energies Revue, vol. 12, no. 4, pp. 617–626, 2009.
18. R. Romary, J.F. Brudny, Chapter on: Harmonic Torques of Electrical AC Electrical Drives, Second Edition of the Industrial Electronics Handbook, CRC Press & IEEE Press, Part II, Chapter 11, Editors: B.M. Wimamowski, J.D. Irwin, ISBN: 9781439802854, ISBN 10: 1439802858, pp. 10.1–10.27, February 2011.
19. M. Radic, Z. Stajic, N. Floranovic, Performance Characteristics of a Three-Phase Self-Excited Induction Generator Driven by Regulated Constant Speed Turbine, Automatic Control and Robotics, vol. 11, no. 1, pp. 57–67, 2012.
20. R.J. Harrington, F.M.M. Bassiouny, New Approach to Determine the Critical Capacitance for Self-Excited Induction Generators, IEEE Transactions on Energy Conversion, vol. 13, no. 3, pp. 244–249, 1998.
21. I. Boldea, Variable Speed Generator, The Electric Generator Handbook, Edition Tailor & Francis Group, USA, pp. 1–30, 2006.
22. A. Kheldoun, L. Refoufi, Dj. Eddine Khodja, Analysis of the Self-Excited Induction Generator Steady State Performance Using a new Efficient Algorithm, Electric Power Syst. Res., 86, pp. 61–67, 2012.

23. J.F. Brudny, R. Pusca, H. Roisse, Wind Turbines Using Self-Excited Three-Phase Induction Generators: An Innovative Solution for Voltage-Frequency Control, Eur. Phys. J. Appl. Phys., vol. 43, pp. 173–187, 2008.
24. E. Touti, R. Pusca, J.P. Manata, J.F. Brudny, A. Chaari, Enhancement of the Voltage and Frequency at the Self-Excited Induction Generator Outputs by Adjusting Terminal Capacitor, 9th International Conference on Industrial Power Engineering, Bacau, Romania, pp. 199–208, 22–24 May 2014.
25. S.P. Singh, B. Singh, M.P. Jain, Comparative Study on the Performance of a Commercially Designed Induction Generator with Induction Motors Operating as Self-Excited Induction Generators, IEE Proceedings - C, vol. 140, no. 5, pp. 374–380, September 1993.
26. S.S. Murthy, O.P. Malik, A.K. Tandon, Analysis of Self-Excited Induction Generator, Proc. Inst. Elect. Eng. C, vol. 129, no. 6, pp. 260–265, November 1982.
27. J.J. Grainger, W.D. Stevenson, Power System Analysis, ISBN 0-07-061293-5, McGraw-Hill, Inc., USA, pp. 342–356, 1994.
28. E. Touti, R. Pusca, A. Chaari, Dimmer Control for Voltage Transient Performance Improvement in Isolated Wind Turbine, Wulfenia Journal, vol. 20, no. 4, Klagenfurt, Austria, pp. 108–117, April 2013.
29. E. Touti, R. Pusca, J.P. Manata, J.F. Brudny, A. Chaari, Active and Reactive Power Regulation in Remote Site Equipped with Self-Excited Induction Generator, 12th International Conference on Science and Techniques of Automatic Control and Computer Engineering, STA'2011, Tunisia, pp. 593–607, 2011.
30. Ch. Rombaut, G. Seguier, R. Bausiere, Power Converters - Alternative Conversion, vol. 2, TEC& DOC, ISBN 2-85206-316-6, 1986.
31. J.F. Brudny, H. Roisse, G. Peset, Practical Approaches for the Study of Wind Machine Self-Excited Induction Generators, International Conference Wind Energy and Remote Regions, Magdalen Islands, pp. 1–10, 19–21 October 2005.

Chapter 14
Communications for Electric Power System

Maaruf Ali and Nicu Bizon

Abstract This chapter is an overview on Communications applied for the Electric Power Systems. Thus, in the first section of this chapter, the Standards for Electric Power Systems Communications are briefly shown in order to understand the communication infrastructure requirements for the Smart Grids. The layers of the Smart Grid Network are (1) Power Grid, (2) Smart Grid and (3) Application. This chapter is focused on the Smart Grid layer, which has three primary functions to accomplish in real-time the requests of both consumers and suppliers based on Communications Technologies for Smart Grid Metering. So, Smart Grid Metering is an important interface to be implemented and integrated with Smart Grids based on technologies such as Fibre Optic Communication, x Digital Subscriber Line/Loop, Power Line Communications, and Wireless Technologies. The Wireless Technologies for the Smart Grid Architecture of Communication System used for Power System Control is approached here as well. Some examples of Communication Systems for the Electric Power System based on IEEE standard (such as IEEE 802.11 Mesh Networking, IEEE 802.15.4 Wireless Sensor Networks and so on) are presented at the end of this chapter. Last section concludes the chapter.

M. Ali
Department of Science and Technology,
University of Suffolk Neptune Quay, Ipswich, UK
e-mail: dr.maaruf.ali@gmail.com

N. Bizon (✉)
Faculty of Electronics, Communication and Computers,
University of Piteşti, Piteşti, Romania
e-mail: nicu.bizon@upit.ro

N. Mahdavi Tabatabaei et al. (eds.), *Reactive Power Control in AC Power Systems*,
Power Systems, DOI 10.1007/978-3-319-51118-4_14

14.1 Introduction

With the need to increase autonomous and automated monitoring of the electrical power system, various communication technologies are being considered for exploitation. This is an evolving process and the global market still has not reached a mature stable point in its adoption of any one particular technology. One major consideration is the maintenance of the power grid in a stable state, with minimal power outages.

The implementation of a smart grid requires the essential components of an effective communication technology coupled with an intelligent low maintenance and low energy drain device. The system must require the absolute minimal consumer interaction, preferably a set and forget device. Studies have focused on using both wireless and wired technologies. These are outlined in this chapter. However, they all have their limitations, especially for wireless coverage using frequencies higher than 2 GHz. This is due to propagation characteristics and signal attenuation through building materials.

A definition of a smart grid is provided by the European Commission: "Smart grids are energy networks that can automatically monitor energy flows and adjust to changes in energy supply and demand accordingly. When coupled with smart metering systems, smart grids reach consumers and suppliers by providing information on real-time consumption." [1].

It has to be stressed that smart grid communication is bidirectional, not necessarily at the same time, between the customer and the utility company. The smart grid is normally considered to consist of the parts of: generation, distribution and consumption of the energy [2]. The transmission network itself is not usually considered, as it can use any existing networking technology. This concept can be summarized in Fig. 14.1.

The use of a well-integrated smart grid is of immense benefit to both the consumer and the energy producer, these include [2]:

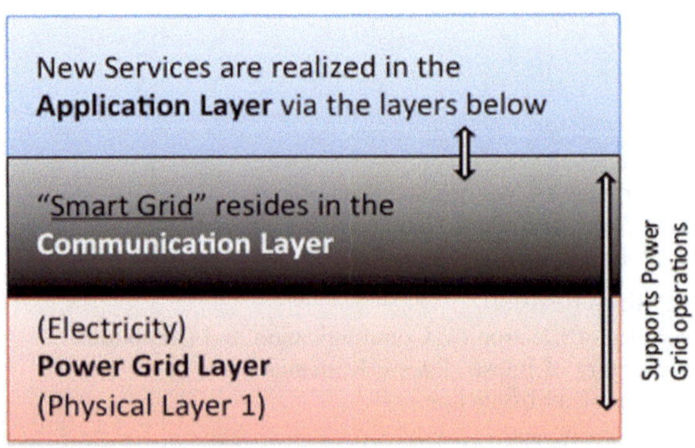

Fig. 14.1 The layer view of the smart grid network

- A better power distribution management policy to handle peak demands, minimizing load shedding;
- Efficient integration of renewable power sources into the grid;
- Consumers being able to utilize cheap power;
- Consumers being able to sell excess power during peak demand by the grid—for those generating their own, for example using a CHP [Combined Heat and Power] unit and
- Maintaining a stable power grid.

  The smart grid layer has three primary functions to fulfill, these being of providing [3, 4]:
- **Smart Network Management**—such as data acquisition, transmission, switching and monitoring of the energy flow between the customer and the supplier, ensuring an acceptable quality of service;
- **Smart Integrated Generation**—encompassing energy distribution, storage, fusion from diverse sources integration with provision of future grid scalability. The need to provide charging points to EV (electric vehicle) should also be provisioned as they become more commonly accepted by the general populace;
- **Smart Market**—two-way communication functionality must be implemented between the meter and the utility company in order to support advanced features such as dynamic pricing and load control. The meter in itself must offer advanced functionalities beyond just registering energy usage. Thus the smart meter must be coupled with inbuilt networking technologies linked to network gateways, much like how modern multimedia services are internetworked.

An EU Commission report also estimates that by 2020, about 245 million smart meters will be deployed just in the EU alone.

Whatever technology is employed, data protection, privacy and security are of the utmost importance to protect everyone in the revenue stream. Thus some form of encryption, preferably end-to-end from the consumer premises to the utility company should be implemented over the data channel.

14.2 Standards for Electric Power Systems Communications

For any implementation of a smart grid system, it must adhere to these communication infrastructure requirements [5]:

(a) **Scalability**

The number of consumers requiring connections to the energy grid is predicted to grow exponentially. Fortunately the reliance upon the network communication technology requirements in terms of protocols and signaling are expected to be the same. Thus scalability is an important factor to be dimensioned into a smart grid network for future growth. This will entail the use of a myriad of M2M (machine-to-machine) devices being linked to the low voltage electricity grid (11 kV level). The addressing task of each network device or devices will be helped by the utilization of IPv6 (Internetwork Protocol version 6) with its 128 bit addressing space.

(b) **Network Performance**

Under this term are specific network performance parameters such as the data rate, the quality of service, the network delay. Fortunately these parameters are not so stringent as those required for conveying multimedia real-time network traffic though the Internet. Smart grids require only data transfer rates of 10 Kb or less with latencies less than 100 ms, for network originating control and signaling information. Customer premises meter-originating data are even less demanding, being able to function with higher latencies of up to 72 h!

(c) **Availability**

Any power utility company must ensure they can deliver power ideally with 100% uptime or at best with the absolute minimum downtime or power outage of say at most a few hours. Even if an outage is unavoidable, then power must be maintained by rerouting to such critical facilities like hospitals. Load shedding must be avoided wherever possible. For smart grids, availability also means being in active communication link, this is less of a problem for power being provided by fixed network connections. However, radio links must be maintained for smart metering being provided by wireless radio transmission links. Radio interference must be combatted and error correction must be built into the transmission protocol to survive in a noisy radio transmission environment. To maintain network efficiency, FEC (Forward Error Correction) coding should be employed, avoiding the need for any ARQ (Automatic Repeat reQuest) of the data packets.

Some smart grid services, especially at the industrial level require an SLA (Service Level Agreement) of between 99.5 and 99.9%. This can be made possible by service reservation and prioritization of queues. Again, at the consumer premises level, the requirements are less stringent and can be relaxed by having the metering data being stored locally before being uploaded when the smart grid network becomes online or less congested again. Hence, an SLA of less than 90% may be tolerable.

(d) **Costs**

When a smart meter is to be installed, two cost factors need to be taken into account, these being the price of the smart meter itself and the actual installation cost of the meter into the consumer premise or the home. The smart meter needs to offer several functionalities and also need to be made appealing to the user whilst keeping the cost to a minimum for the power utility company. Most consumers expect their smart meters to be installed at zero cost to them, with the entire cost being borne by the power utility company themselves. The installation of the smart meter should be as simple as possible with automatic self-configuration and diagnostics. This will minimize retraining of the installers. A third factor that also needs to be considered is the actual transmission cost of each reading, in either direction. These exchanges of data need to be reliable and cost effective, without overwhelming the network capacity. This is of particular concern when expecting readings from a large city from millions of potential customers. Thus some form of data scheduling may be necessary.

(e) **Security**

Protection of the smart grid network is extremely important, not only to protect the personal billing information of the consumer, but also to protect the network from spoofing and infiltration by hackers attempting to defraud and bring down the entire network. Thus, the user data must be protected from: misuse, the network from hackers and the messages must be conveyed securely to arrive uncorrupted. The technology for secure transmission of sensitive data already exist in conveying data through the internet and this can be applied in the case for the future advanced fully integrated smart grid network.

(f) **Smart meter grid/system life-cycle**

The smart grid community would like to see a life-cycle of at least 15 years with standardization and interoperability as key features. This is to ensure longevity of the network and ease of maintenance with minimal cost. Implementing a new smart gird network is quite an expensive endeavor and these considerations of using standardized, COTS (Customised-of-The-Shelf) smart meter components should allow a cost effective implementation, deployment, maintenance and safe realization of the national and international smart grid network. The utility companies' goal is to minimize their maintenance of their smart meters, thus achieving considerable cost savings. They also do not like to be locked into purchasing from one specific vendor for the entire lifetime of their smart grid network.

(g) **Control**

To enable a smart grid to be under the effective control of the power generating utility, it may become necessary for the utility to use their own network - totally under their control. This is especially relevant when the network demands may exceed previously agreed SLAs with their third party data network capacity providers. The network must be always made available to the utility, this can only be guaranteed if the data network is under the total control of the utility. This will thus avoid complications of legal interpretations of contract laws and what to do when the network services beyond the initially agreed service levels are required. A solution may be is to use power line data transmission technology or to set-up a dedicated wireless network, for example like the GSM-R (Global System for Mobile Communications—Railway) network. This is the GSM mobile communications network, specifically employed by the railway operators, using their own frequencies.

14.3 Communications Technologies for Smart Grid Metering

Whatever technology is chosen, they must adhere to the above requirements. It may also be noted that some situations may require the use of more than one type of technology. The types of technologies that are currently being utilized in the smart grid can be categorized into two, these being wired or wireless [6–8]. For most practical implementation both are required together.

14.3.1 Wired Technologies for Smart Grid [3]

a. **Fibre Optic Communication**

Without doubt, the use of fibre optic cabling to network the smart grid offers the advantage of the availability of a high bandwidth, compared to using copper cables or twisted pair wires. However, the penetration of optical fibre to the home is still at a low level. The main reason for this is the high cost of deployment. This situation is slowly changing as the uptake of 'Fibre to the Home' (FTTH) continues to accelerate. New regional developments and new build towns, villages and cities are far easier to network cost effectively using fibre cabling technology to the local electricity generating substation, however.

b. **xDSL (x Digital Subscriber Line/Loop)**

For most of the industrialized nations, broadband services are provided by a form of DSL technology, hence the 'x' in the acronym. The cities are well served in this regard. It would appear that this form of technology is ideally suited to implement the smart grid. However, there are many challenges, these being the regulatory issues and the cost of the equipment, to actually interface into the DSL line itself, along with being dependent on the customer. It must also be noted that the smart grid is at the mercy of the customer paying for the DSL line and not terminating the contract. Despite these potential problems, smart meters are already deployed that use WiFi connections and the DSL line to carry the data back to the power generating companies.

c. **Power Line Communications** [7, 8]

Using the actual power line infrastructure itself to both carry electricity and data makes much commercial sense as the network is already there and the company has full control over their grid. Such technologies are already being utilized. However, the actual transmission is functional and possible between the transformers only and not through the transformers. The data is transmitted by carrier modulation. The use of multiple carriers, much like DSL technology, now enables broadband communication to also take place. For power line communications to be effective, the signal must operate in a very electrically noisy environment. The technology currently offers two modes of operation, this being narrow band and broadband. Wideband communication suffers from greater attenuation and restricted range. Depending on the scenario, a mixture of both narrow band and broadband power line communication is generally utilized. In the event of a break in the power line due to say a storm or from solar radiation, then complete breakdown in smart metering will occur. Thus some form of a redundant backup network must be in place. This will add a further cost to the power line smart grid network.

The implementation of power line communication is still in its infancy and no internationally agreed standard currently exists. Pilot schemes are in operation and performance data continues to be gathered.

14.3.2 Wireless Technologies for Smart Grid [3, 6]

There are two ways of implementing a smart grid using wireless technology. One is to use the currently available wireless technologies and mobile cellular networks. The other is to set up a private wireless localized network. Clearly the first one has the advantage of a lower cost as the infrastructure is already in place. The major disadvantage is the lack of control by the utility itself. This is not a problem, however, when a private wireless network is used. The major disadvantage here would be the high initial cost in setting up the private wireless network. Using any shared or unlicensed part of the electromagnetic spectrum would not be an advantage, as the utility company would have to compete with other users and the consequent radio noise, both affecting the effective data throughput. For in-building use, frequencies up to 1 GHz may be employed before attenuation becomes a severe problem. This would mean that the lower GSM and GPRS (General Packet Radio Service) frequencies of 450, 800 and 900 MHz should only be utilized.

a. **GSM and GPRS services at 900 MHz**

The bandwidth is adequate and the technology is mature to implement a smart grid solution using GSM/GPRS at 900 MHz currently. Indoor coverage also exists with the use of microcells and femtocells. However, in dense urban areas, network congestion and availability is a serious problem and alternative, more secure solution is needed. However, latency is a problem and the network is being upgraded to other services. Thus the long stability of the technology is of major concern to the utilities.

b. **800 MHz and LTE (Long Term Evolution)**

The use of this fourth generation mobile technology with its broadband data capacity offers clearly a superior alternative for implementing the smart grid. The technology is also available over different frequency bands. The lower band of 800 MHz is of particular interest for implementing the smart grid. As LTE continues to be rolled out and deployed, it will make implementing a smart grid using this solution an economic viability. Though this is not currently the state and not all regions offer LTE services at 800 MHz. Primarily it is designed for voice and internet traffic and not for M2M (Machine-to-Machine) data traffic. The utility also has to rely on the commercial third party provider, thus losing control of the network.

c. **450 MHz and CDMA (Code Division Multiple Access)**

With its wider cellular coverage requiring fewer base stations and greater in-building penetration depth and hence lower cost, this under-utilized technology can be effectively employed to form regional smart grids. The smart grids will be regional due to the regional availability of these cellular technologies, as they are not deployed globally. Growth in areas like Brazil is possible. Compared with the other cellular technologies, the use of CDMA over 450 MHz is technically the most viable solution, where it exists geographically.

14.4 Architecture of Communication System Used for Power System Control

The complexity of the information flow in a smart grid framework is shown in Fig. 14.2. The quantity of monitored data will vastly exceed that of command and control data, this is so, because of the myriad of devices that have to be monitored [9]. Clearly the control, command and billing channels need to be highly secure and the information conveyed in a timely manner to be acted upon when demand for

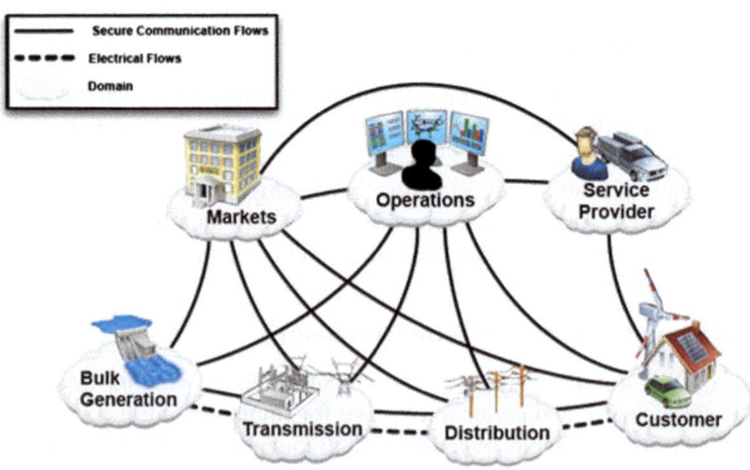

Fig. 14.2 NIST Smart Grid Framework 3.0 [9]

action requires it. In some parts of the world, effective monitoring of "stolen energy" is of paramount importance. Figure 14.2, also identifies five areas of a power generation system: the residence, the distribution system, the substation, the transmission system and the generating station.

14.5 Examples of Communication Systems for Electric Power System

A summary of the main technologies is given in Table 14.1 [3]. These technologies should not be seen as competitive, but rather as complementary depending on the scale, cost and geographical region of the intended smart grid implementation.

Table 14.1 Properties of the Communication Systems for the Smart Grid [3]

Parameters	GSM 900 MHz	LTE 800 MHz	CDMA 450 MHz	Fibre optic	DSL	Power line
Scalability—competition on resources	Yes	Yes	No	No	No	No
Low latency	No	Yes	Yes	Yes	Yes	Narrow band—no
Data rate sufficiency	Problematic	Yes	Yes	Yes	Yes	Narrow band—no
Enhanced resiliency	Not available	Not available	Available	Available	Only limited SLAs	Not available
Indoor penetration	Fair	Fair	Good	Good	Good	Good
System availability	Constrained	Constrained	Yes	Yes	Yes	Constrained
M2M optimized	No	No	Yes	No	No	No
Interference expectability	No	No	No	No	No	Expected
Economic nationwide coverage	Yes	Yes	Yes	Very limited	Partially limited	Limited
Installation/rollout	Simple	Simple	Simple	Difficult	Difficult	Simple
Security	Public grid	Public grid	Closed network	Public grid	Public grid	Closed network
Long-term availability	No	Yes	Yes	Yes	Yes	Yes
Customer behaviour dependent	No	No	No	Yes	Yes	No
Exposure to broadband market developments	No	Yes	No	No	Yes	No

Seven important parameters need to be carefully evaluated for the communication network, these being: latency, bandwidth, resilience, security, scalability, coverage and the life expectancy [10].

The latency is the round trip delay between the transmitter and the receiver, which includes the media access time, the propagation delay and queuing but not the processing time at the destination.

When considering the bandwidth of the communication link, the effect of channel coding and error coding in reducing the effective goodput of the information transfer capacity must be calculated. The ability of the communication link to recover from noise and natural disasters such as earthquakes is in the domain of the resilience of the link. Security is of ever increasing importance when deciding to adopt the appropriate wireless technology in our uncertain world. Unauthorized access must be prevented and confidentiality of data must be protected. Attacks from 'vampire' nodes can be prevented by using techniques such as mutual authentication. Clearly low powered wireless technologies are only suitable for small-scale applications, thus the coverage of the wireless technology has to be appropriate to the application. With a rapidly expanding market, scalability of the technology must be seamless and implementable with ease. For example it must be relatively easy to poll a rapidly increasing number of measuring devices at the correct rate. The life expectancy of the communication network must match that of the distribution generation system, otherwise it does not make any economic sense

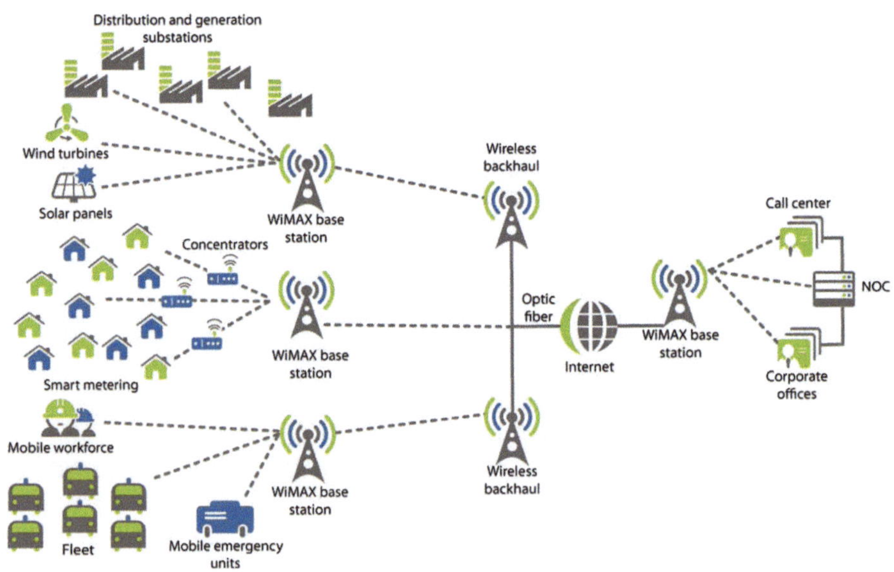

Fig. 14.3 The Integration of WiMAX for Smart Grid Applications [11]

to select the communication network in the first place. This is not an easy decision to make with technologies having a short life lifetime often considerably less than the electrical distributed generation system. When updating the communication network, fault tolerance will become more of a critical factor to prevent power outages in times of excessive or peak demands, especially when handling the signaling to implement an efficient load shedding algorithm.

A practical implementation of a smart grid network, as shown in Fig. 14.3, is to combine several technologies together, both wired and wireless including public and private networks.

14.6 Overview of IEEE 802.11 Mesh Networking

Whatever incarnation a smart grid takes, it will likely consist of a mesh network as show in Fig. 14.4, applied to a smart grid. The main difference with this type of network is that not all the nodes have access to the wired network, instead access is gained via anchor access points. Having key access points means that congestion management is a priority to free up bottlenecks. Latency and jitter throughout the mesh must also be taken into account. To enhance radio resource planning and security monitoring, a smart grid should employ a controller-based WLAN architecture.

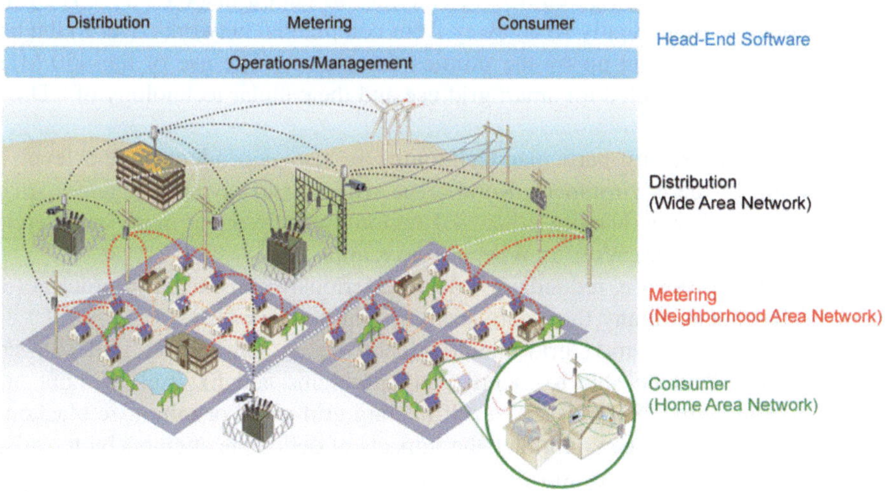

Fig. 14.4 Mesh Networking in Smart Grid Application [12]

14.7 IEEE 802.15.4 Wireless Sensor Networks

At the heart of the smart grid will be the sensors that are used to monitor the smart grid and to implement it. Clearly a wireless sensor network will afford the greatest flexibility in its placement. The use of an already existing standard, such as the IEEE 802.15.4 is judicious. This IEEE standard is mainly concerned with specifying the requirements for power efficient low data rate wireless networks. The ZigBee protocol is the main driver behind this protocol. Other competing standards such as the WirelessHART and ISA100.11a use the physical layer of this standard but have their own respective different layer two MAC (Media Access Control) definitions. Clearly using the lower frequency bands operation below 1 GHz, suffering less attenuation will give longer radio propagation paths than using the 2.4 GHz microwave band.

14.8 Conclusion

Those technologies that are resilient to network limitations should be considered for implementing the future smart grid. An overview of the various technologies, both wired and wireless, have shown that they have their respective advantages and disadvantages. The conclusion of this chapter author is that of a hybrid solution utilizing both wired and wireless technologies. The initial fast deployment should be to utilize the underlying cellular technologies to gain a market share. Then the future would be to gradually wean away from commercial networks over to the use of a private network set up by the utilities themselves. The use of the 450 MHz frequency is ideally suited for smart grid use and the cellular technology of CDMA in particular.

The deployment any particular technology is also dependent on the level of urbanization of the environment. Rural areas will tend to favour a wireless smart grid solution, whereas heavily urbanized regions will favour a more commercial cellular solution, using WiFi or WiMax for example.

For the interim phase, redundancy must be seriously considered in order not to be over dependent on any third party commercial network [13].

Finally the future smart grid should ideally be a "self-healing grid". A self healing grid would be able to identify any problems and fix itself without any human intervention. "Not only can a self-healing grid avoid or minimize blackouts and associated costs, it can minimize the impacts of deliberate attempts by terrorists or others to sabotage the power grid" [14].

References

1. https://ec.europa.eu/energy/en/topics/markets-and-consumers/smart-grids-and-meters.
2. Nicu Bizon, N.M. Tabatabaei, H. Shayeghi (Ed.), Analysis, Control and Optimal Operations in Hybrid Power Systems - Advanced Techniques and Applications for Linear and Nonlinear Systems, Springer Verlag London Limited, London, UK, 2013.
3. B. Sorries, https://www.cdg.org/resources/files/white_papers/CDG450SIG_Communication%20_Technologies_Networks_Smart_Grid_Smart_Metering_SEPT2013.Pdf.
4. M. Shahidehpour, Y. Wang, Communication and Control in Electric Power Systems: Applications of Parallel and Distributed Processing, Wiley-IEEE Press, London, UK, 2003.
5. M. Uslar, M. Specht, C. Danekas, J. Trefke, S. Rohjans, J.M. Gonzalez, C. Rosinger, R. Bleiker, Standardization in Smart Grids: Introduction to IT-Related Methodologies, Architectures and Standards, Verlag London Limited, London, UK, 2013.
6. S. Borlase, Smart Grids: Infrastructure, Technology, and Solutions, Wiley-IEEE Press, London, UK, 2014.
7. S.F. Bush, Smart Grid: Communication-Enabled Intelligence for the Electric Power Grid, Wiley-IEEE Press, London, UK, 2014.
8. N. Bizon, N.M. Tabatabaei (Ed.), Advances in Energy Research: Energy and Power Engineering, Nova Science Publishers Inc., USA, 2013.
9. http://www.nist.gov/smartgrid/images/FrameworkGraphic_1_1.jpg.
10. M.S. Thomas, John Douglas McDonald, Power System SCADA and Smart Grids, 2015, CRC Press.
11. M. Paolini, Empowering the smart grid with WiMAX, http://www.senzafiliconsulting.com/Blog/tabid/64/articleType/ArticleView/articleId/32/Empowering-the-smart-grid-with-WiMAX.aspx.
12. http://www.trilliantinc.com/solutions/multi-tier-architecture/.
13. M. Shahidehpour, H. Yamin, Z. Li, Market Operations in Electric Power Systems: Forecasting, Scheduling, and Risk Management, Wiley-IEEE Press, 2002.
14. M. Amin, The Self Healing Grid: A Concept Two Decades in the Making, IEEE Smart Grid Newsletter Compendium, 2015, http://www.qmags.com/qmct.asp?q=382&qt=3872236&u=/OLHit.asp?pub=SGNC&upid=19496&s=ML&url=/SGNC2015.

References

1. [illegible] in Power Systems: Advanced Techniques and Applications [illegible] Springer [illegible] London, U.K., 2013.
2. [illegible] MA, 2010.
3. [illegible]
4. [illegible]

Chapter 15
SCADA Applications for Electric Power System

Florentina Magda Enescu and Nicu Bizon

Abstract Main objective of this chapter is to present the Supervisory Control and Data Acquisition (SCADA) technology applied in the energy sector which requires distributed control and monitoring at different levels. If the process is distributed, then the advantages of SCADA system will be seen through low costs related to movements of the equipment to improve the performance tracking. Processes that need to be monitored on a large area, and request frequent and immediate interventions, can be solved more efficiently through a SCADA system. For this, the connection between the master station and remote units must be done via a communication system which can use different communication technologies such as cable, radio, mobile phones, and even satellites. The optimum data transfer may be obtained by using a specific communication protocol. It is worth to mention that such SCADA-based communications are developed using the optical fiber communications technology via the Internet. The information flow between remote and central SCADA units could be designed to be bidirectional for high performance and reliability of the distributed control system, but note that both digital and analog signals are involved in such systems. Besides, an important issue is the information security related to such systems. It is known that the automation and real-time control are used via the Internet and wireless technology, but these technologies have also brought some security problems, having a strong impact both in the business and to the users. The SCADA applications analyzed in this chapter is focused on Electrical Power Systems (EPS). The stepwise step design is shown using the programming environment named VIJEO CITECT SURVEILLANCE SOFTWARES —version 7.40®. The main objective is to show how can be optimized the real-time control to obtain affordable solutions for the EPS based on Renewable Energy

F.M. Enescu (✉)
Department of Electronics, Computers and Electrical Engineering, Faculty of Electronics, Communications and Computers, University of Piteşti, Piteşti, Romania
e-mail: florentina.enescu@upit.ro

N. Bizon
Faculty of Electronics, Communications and Computers,
University of Piteşti, Piteşti, Romania
e-mail: nicu.bizon@upit.ro

N. Mahdavi Tabatabaei et al. (eds.), *Reactive Power Control in AC Power Systems*,
Power Systems, DOI 10.1007/978-3-319-51118-4_15

Sources (RESs). The both current and optimized solutions are presented, and the role of the reactive power is highlighted in the comparative solutions shown, which are implemented in practice as well. This was achieved by optimizing the SCADA solution of the operation, transmission and overseeing the execution of programs for the operation of power plant trough: (1) the description of the existing solution; (2) the description of the proposed solution with redundant SCADA servers; (3) the disadvantages of the proposed solution with Remote Terminal Unit 32 (RTU32). The optimization of the applications has been carried out because the graphical interface is poor, as long as the only information displayed is from the distribution stations, the states of the switches in medium-voltage lines and of the equipment in the stations. Therefore the dispatcher has not displayed the graphical information about the status of the line equipment, and does not have a quick overview to state of the medium-voltage line as well. Thus, the dispatcher must consult the printed diagrams, which means time consumed and such complicated maneuvers will occupy the most time their activity during a day. Thus, here it will be shown the steps to optimize the graphical interface in order to quickly see the status of the EPS. Furthermore, by improving the graphic interface, the efficiency in carrying out the appropriate maneuvers will increase, as well as the degree of safety. The applications have been designed with high flexibility and can be used either for small applications, either for large size systems. Also, the graphic interface was designed to display online the page's status and the trends in state variables.

15.1 Introduction

In last decades the automation devices have evolved from simple electronics circuits used to monitor and control the parameters of the industrial processes [1], to smart devices that communicate with each other in automation networks [2]. The evolution of the automation devices has led to the development of technique for visualization of control parameters and state variables from the automation process [3]. The companies producing automation equipment are those that have achieved the first visualization systems. In time, they have developed the generic software named Viewer, which is adaptable to various situations possible in distributed processes [4, 5].

The Supervisory Control and Data Acquisition (SCADA) system collects data from the distributed processes through sensors. The central computer will store this big data after a preliminary processing. The system can receive commands automatically or manual commands based on the processed data by the human operator [6].

The first SCADA systems, which allowed acquisition and real-time data analysis, have appeared in the 1960s, but then were rarely used in installations. The term SCADA was referred to a comprehensive system of measurement and control. The Programmable Logic Controllers (PLCs), which were designed and built by Gould Modicon in 1971, have changed the philosophy to use the SCADA. The PLCs in EPS have been introduced in 1977 by Allen-Bradley, and this was easily accepted by electricians. The term of SCADA has been used for the first time in the late 1980s. This term was not widely used until the 1990s when the technology evolved. In 1998, most of the PLC manufacturers integrate the Human Machine Interface (HMI) in/SCADA systems using communication systems and open protocols, nonproprietary. The HMI/SCADA systems offer total compatibility with PLC-s by including the following components [7]: the HMI, the controllers, the input-output devices, the networks, the appropriate software etc.

A generic SCADA system must to implement a distributed database that contains elements called endpoints, which can be either hardware or software. Thus, the distributed database is composed by hardware and software items. The system SCADA controls and monitors the input or output endpoints. The record of each endpoint, including its dynamic, is stored in the database to have the history of all endpoints. This will help in predictive maintenance and the security of the EPS as well [8, 9].

So, the SCADA system has a multi-layered structure composed from basic functions and graphical user interfaces which are hardware and software supervised in real-time [10, 11]. From the point of view of the design environment, a SCADA system contains some specific instruments such as: text editing applications, graphics editors, tools to import/export from/to SCADA libraries to achieve configuration parameters, drivers & tools with advanced features for HMI interfaces, etc. [12].

The carrying-out of a project in the SCADA system, the following aspects must be taken into consideration [10, 11]:

- the needed components to be easily accessible from the point of view of visibility and how these are collaborating;
- timing sequence;
- scalability;
- flexibility to extend the system by the addition of new subsystems;
- redundancy;
- the initial statuses that should be determined;
- the operation diagram of the system.

The functions of the SCADA system are the following [1, 2, 8]:

- acquires data collected from the process;
- manages alarms;
- allows the needed actions for automation:
  - a technological preset action;
  - an initiating action of the events;
  - a monitoring of the pre-defined sequences.
- stores and archived data;

- generates and pursues reports:
  - a default graphics;
  - a multiple graphics scaling;
  - a history of graphs.
- allows the dispatcher to control the process via the HMI;
- allows the communication with user interface via the HMI using:
  - a libraries with symbols;
  - a connection between process and graphic elements;
  - a link between screens;
  - a collection of command's operators:
  - an animation based on multimedia features;
  - a possibility of connection with other platforms [5].

SCADA systems may be extended to Large Scale System by architecture, maintenance, post-processing, decision support systems, and economic planning [4, 7].

The architecture of hydro energy system is proposed in [12] and extended here to the architecture shown in Fig. 15.1.

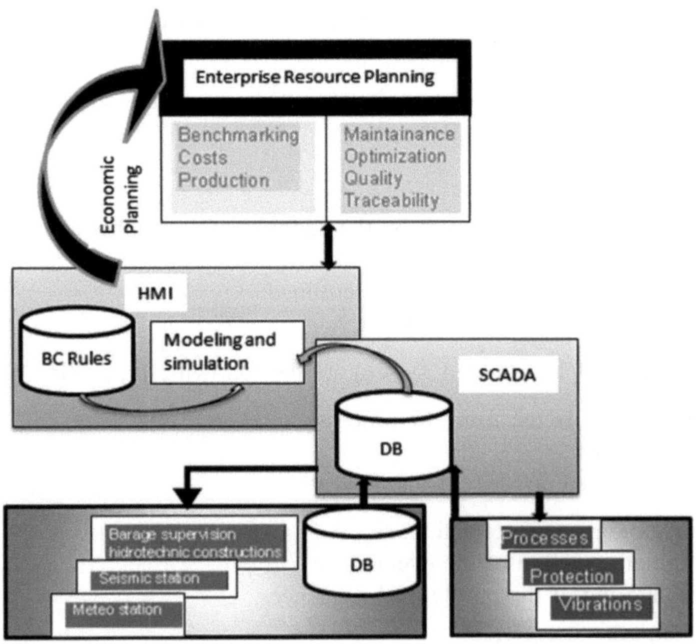

Fig. 15.1 Extended SCADA architecture with application in hydro-energetics

A number of recent studies in the scientific literature have highlighted the importance of the SCADA systems for RES hybrid EPS [1, 13–15], and in particular, their use for hydropower plants as below:

- the SCADA system applied for hydropower plants connected to national EPS;
- the use of metering systems for the electric power and water from hydropower plants;
- monitoring and control of the hydro power plants;
- monitoring and control of the hydropower plants and RESs used for hybrid EPS.

The SCADA architecture for a hydropower plant is presented in this chapter at level of hardware and software. Five levels of SCADA hierarchy are essentially in complex system [1, 9–11]:

- the devices level for instrumentation and control operations;
- the Remote Terminal Unit (RTU) terminals level;
- the communications system level;
- the master station level;
- the data processing level.

The evolution of SCADA systems is also shown as system generations for cyber security. The cyber security risk assessment for SCADA includes:

- Causes which can lead to catastrophic situations from common technical incidents;
- Cyber Security concepts;
- How incidents are treated;
- The tolerance which is acceptable;

A survey is conducted for a specific hydropower plant from Romania based on SCADA system. Also, an overview of the programming in VIJEO CITECT SURVEILLANCE SOFTWARES 7.40® is shown, and the projects to monitor the medium power subsystem of the EPS based on RESs is described step-by-step [10, 16–19].

SCADA system allows the command, control and operation of installations both from the control room of the station and from the center point. The main operative functions of the management system are as follows:

- acquisition, processing and exchange of data;
- instant recording of data;
- transmission by the higher command center (subsequently implemented);
- remote control (control room) and the local branch of the circuit breakers, traps, etc.;
- indication of the position of switching equipment;
- measurement of analog quantities;
- metering;
- sequential recording of data;
- the processing and management of alarms;

- voltage control circuits and plots;
- remote control and adjustment in plants;
- marking;
- long-term information archiving;
- registration of damage;
- interlock of primary equipment;
- switching sequences with checking of signals' timing;
- surveillance system.

All these features are available in new SCADA user interface with VIJEO CITECT SURVEILLANCE SOFTWARES.

Today, a SCADA system operates on large distances, up to the level of a country or continent, if we consider that the national EPSs will be integrated based on smart grids and these technologies. Examples of such processes are given below [20–24]:

- the groups of small hydro stations that are switched on and off in response to energy demand of customers and which are located, in general, in remote places. These systems can be controlled by opening or closing the turbine valves, but must be pursued continuously in order to quickly respond to the dispatcher's energy demands;
- the electrical transmission systems, which cover thousands of km^2, can be controlled by closing or opening the grid breakers, but these must respond almost instantly to the load demand on the line;
- the areas of oil and gas extraction, including sensors, collection systems, measurement equipment and pumps. These systems are generally distributed over wide areas, require relatively simple controls, such as starting or stopping motors, and metrological information on a regular basis, but must quickly respond to the exploitation conditions;
- the distribution network of gas, petroleum, chemicals, water, which have elements that are located at different distances from a central point of control. These can be controlled by closing or opening valves or starting and stopping the pumps, but must rapidly respond to market conditions and alarms due to the losses from toxic or dangerous materials;
- irrigation systems covering hundreds of km^2 can be controlled by closing or opening of simple valves, but also require a relative complex measurement of water supplied to consumers based on centralized system.

In conclusion, the SCADA system is useful because it helps to [7, 10]:

- optimize the process;
- take a decision objectively;
- increase the efficiency of use for the exploitation equipment;
- decrease energy costs;
- reduce the capital in the future;
- improve the services for the future;
- obtain competitive advantages on the market;

- decrease the number of staff;
- solve the problems of the environment;
- use of a new system instead of the old ones which are exceeded.

15.2 Research Extracts from Literature

The study shown in this section is linked to:

- The Electric Power Systems (EPS) as a critical infrastructure;
- Advanced SCADA architectures;
- SCADA system for optimal energy management;
- Cyber security risk assessment for SCADA;
- New SCADA applications.

A. SCADA system for driving the stock records from the hydropower plants developed in Valcea region

The SCADA system presented in this section is used in Valcea region, Romania, for driving the hydropower plants from the dispatcher station. The driving system from the Hydro-energetic Dispatcher (HD) contains (Fig. 15.2):

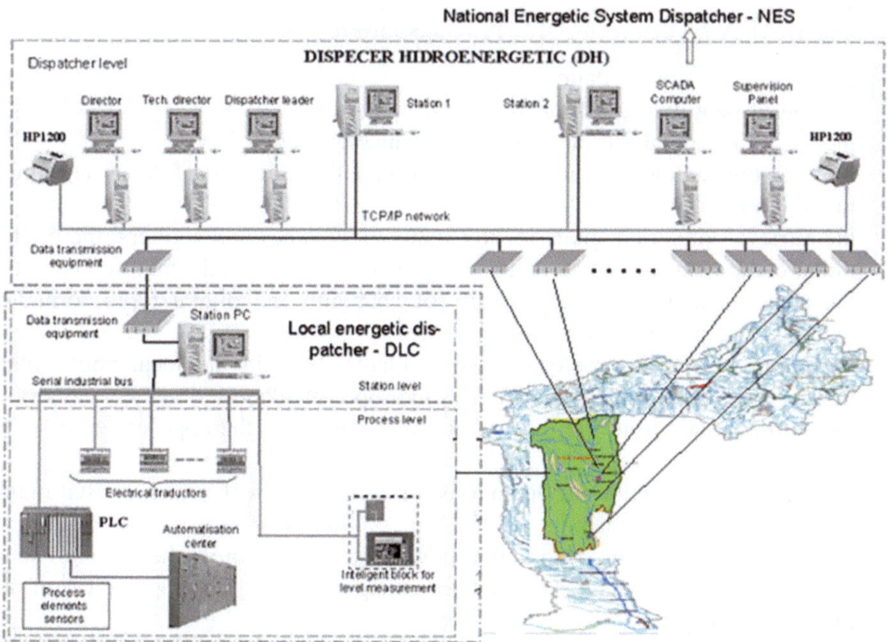

Fig. 15.2 Research extracts from literature

- SCADA systems in central dispatcher;
- SCADA systems in local dispatcher.

The HD is connected with the Dispatcher in Local Center (DLC) by modems on wire or using the Global System for Mobile communications (GSM), in last case a large communication network dedicated to this process is obtained [12, 25, 26]. This system allows the control of the operative functions, the automatic acquisition of data, and generation of the databases needed. All these actions and data records are necessary for rational utilization of RESs. Note that this SCADA architecture is of open system type, which ensures integration of the regional power plants into national system and flexibility to further develop this system.

The SCADA system from HD station provides information for water flow planning in relation with energy demand from dispatcher, which is of following type [12]:

- for monitoring and ensuring the service continuity for power stations and power substations subordinated to them.
- for implementation of the governmental program to monitor the operation of the main power stations;
- for energy management and operative management in each power stations;
- for operative exploitation of water accumulations and optimal distribution of the water flow to power stations;
- for operation in normal and hazard situations based on the remote control units, switching equipment and primary outlet installations (Fig. 15.2).

The SCADA system from DLC provides information for:

- primary processing of data;
- storage of data during a limited period;
- increasing the reliability and the maintenance capacity of system.
  Because this architecture is of distributed type, the functions of the system must to show in real time:
- the retrieval numeric signals for the status of switches and splitters, the operational equipment, the power cells, the protection loops, and timing cells;
- the acquired and processed signals (of analogue and digital type) form the hydroelectric power stations monitored;
- the commands to hydroelectric power plants, both in the local and central interfaces of the dispatcher from each hydropower plants;
- the alarm signals from the dispatchers in the event of the appearance of some defects;
- the record of the number of operations for each equipment, which is used in the management of revisions;
- the schematic diagrams of the hydropower plants monitored, the status of state variables from these diagrams, the wiring diagrams and the synoptic chart of the installations used [13];

- the stored records using the graphical or tabular configurations, features which are selectable by the operator [17].

B. Metering systems of electricity and water for a residential house

The metering of electricity and water flow requested by a residential house is also a usual application of SCADA system. The SCADA system are responsible for automatic reading of the meters installed and storing the data in SQL database as a record which contains the index of consumption, the location, the time, the state of protection circuits etc. These data generate some alarms and reports related to monthly consumptions, which are used for the allocation of costs facilities for each location and meter. Data are provided directly in Comma Separated Values (CSV) format to be imported directly into the billing software.

The network architecture is standard [10, 27, 28], and contains the following levels:

- monitoring software: it is composed from separate applications, which are running on different servers; it can be accessed directly on the web through monitoring portal, which displays graphical reports in real-time based on the data received;
- data concentrators, which are used to link the software that is run on the server and the meters installed in each home;
- metering level, which provides support for remote transmission of the index of consuming or other monitored parameters.
- Such mattering systems allow the following automatic operations [29–33]:
- reading the consumption and other parameters in the days and at the times programmed by the user; note that data supplied by the meters installed can also be read out manually, at the request of a user, in order to determine current consumption or to verify the proper operation of equipment;
- disconnection of the consumers can be remote controlled using the SCADA software and flow's distributors installed in the network;
- graphical analysis of the daily profile for the load demand or the carbon footprint reported to a day, week, or month, or other parameters available directly in web portal;
- detection and location of the faults arising in operation of the equipment (meters, concentrators, etc.) or the software modules, and generation of alarms via email to the authorized users;
- computing of daily consumption and monthly consumption (which is used for the forecast of the energy sources) by using the software management based on the difference between the automatic readings;
- exporting of the various query reports generated from the database to Excel or other formats.

C. Monitoring and control of the renewable energy sources

The RESs (solar energy and wind power especially) are used in hybrid power systems based on distributed generation because:

Table 15.1 The evolution of energy systems

Year	Energy based on	
The future	Hydrogen—combustion chef	
2010	Solar	
2003	Wind	
1969	Electric Power Plant	
1948	Petroleum	
1900	Steam	

- the actual consumption increased and the trend for the energy; price is to rise even higher in years to come;
- the classical resources begin to run out;
- the importance of RESs in climate change [13–15].

The evolution of energy systems can be monitored in the Table 15.1.

Two examples of RES hybrid power systems are given below:

1. Diesel-wind turbine hybrid EPS to supply the industrial consumers (Fig. 15.3). The application is submitted in under (Fig. 15.3) Sect. 15.7.2
2. Home applications based on solar power (Fig. 15.4)

Supervision and monitoring of the small photovoltaic parks

The dispatchers of small photovoltaic parks make the surveillance with dedicated appliances that enable measurements to be processed via Bluetooth from inverters. Supervision and monitoring in parks large photovoltaic

Supervision and monitoring of the large photovoltaic parks

The dispatchers of large photovoltaic parks use SCADA systems to monitoring the energy generated based on compact reconfigurable items which may avoid the fall

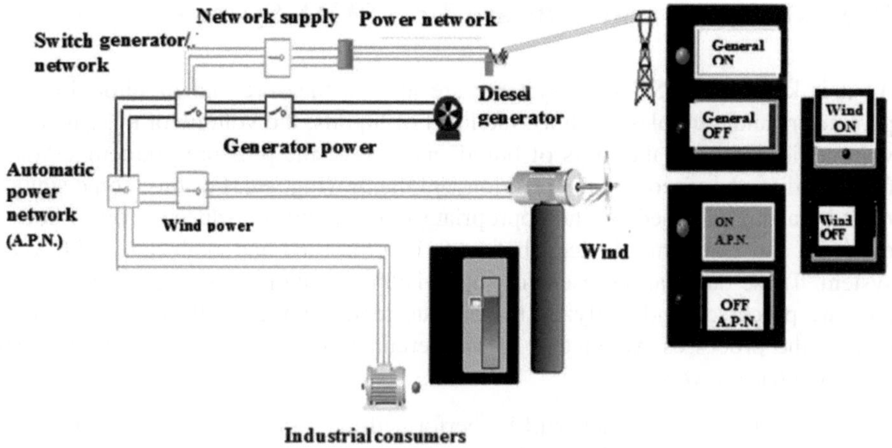

Fig. 15.3 SCADA system for hybrid EPS based on RESs

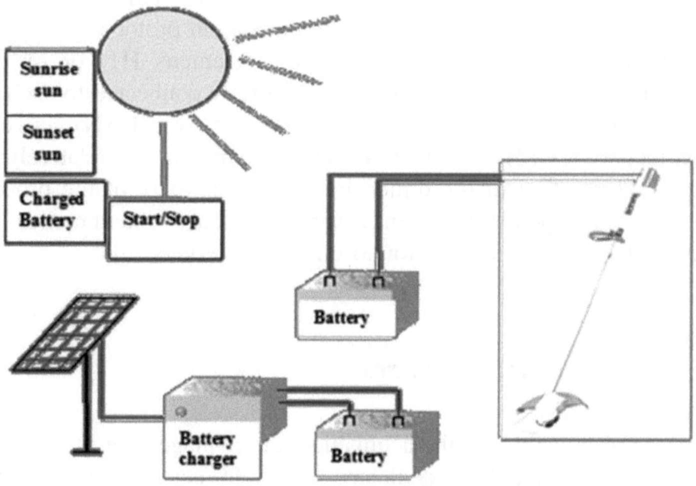

Fig. 15.4 Exploiting the RESs for usual home applications

of some subsystems. These items can be easily changed, reconfigured and maintained [10].

The optimal energy management strategies may be the following [20]:

- to control the balance between supply and demand with a lesser impact on carbon emissions and high profitability in maintaining the business;
- to distribute the energy generated for different energy sources (in particular from RESs);
- to offer clean energy at reasonable costs and under a control of the impact to the ambient.

15.3 Hardware and Software for SCADA Systems

As it is known, the SCADA systems are able to measure and monitor different parameters and variables, such as: the level of liquids, the volume of the gases, the volumetric flow rate, the mass of liquids and gases, the pressure, the temperature, the humidity, the viscosity, the position and the moving, etc. For this, some signals must be acquired based on the appropriate sensors and transducers from different electrical and electronic devices, before to be processed and analyzed by SCADA system. These data can be used for supervision of remote processes. For this, the data are processed and analyzed to generate reports which will be transmitted to control the processes parameters within certain limits. The SCADA applications have two basic levels:

- the customer level, which will be performed the human-machine interaction;
- the data server level which will control the entire process.

Thus, the data server sends these data to the customer level via remote equipment such as PLCs (connected to the database server either directly or via a communications network, using various communication protocols). The developed protocols can be property of a company (e.g. Siemens H1) or of open type (Modbus, Profibus etc.). Database servers can be fully connected to each other and also to the customer stations through Local Area Network (LAN) networks based on Ethernet technology [21–23]. The SCADA systems may be relatively simple or very complex, depending of the dimension and complexity of the process monitored. These systems operate in real-time using a database system named Real Time Data Base (RTDB), which can be found on one of servers [3].

15.3.1 Hardware Architectures

The servers are responsible for the acquisition and management of data for a set of parameters. Is it possible to have dedicated servers for certain tasks (for example, if the objective is to develop the process servers), servers dealing with the handling of alarms, or file servers [23].

A SCADA server will be connected directly to the PLC via an Open Platform Communications (OPC) or a RTU, which both collects and consolidates data from and from PLC-uri (Fig. 15.5). The SCADA takes on request the data from indicator reading appliances and equipment condition.

The OPC and RTU work as a data concentrator. The data is structured using a man-machine interface (HMI) in a convenient format for the operator in order to be able to take any decisions that would optimize the communication between the equipment.

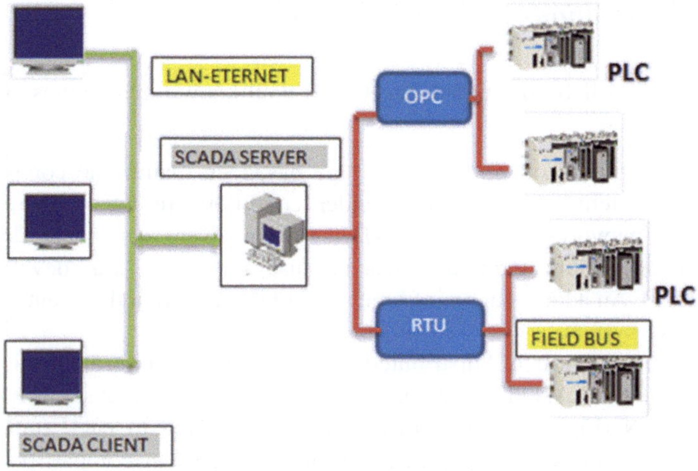

Fig. 15.5 SCADA hardware

RTU performs:

- the connection with supervised equipment;
- reading of equipment status (such as open/closed position of the valve or relay);
- acquisition of measured signals, such as the pressure, flow, voltage or current;
- the control of equipment by sending command signals, such as the closing of a valve or relay or setting the speed of a pump;
- reading the digital or analog signals, and sending the commands using both digital or analog signals.

The PLC contains a microprocessor used for the automatization of processes. A PLC is provided with input/output sensors and relays. PLC is less expensive alter to old systems which used tens or hundreds of relays and timers. The PLC is programmed using the structured programming languages and elementary logic operations.

The term "Supervisory Station" refers to the servers and software responsible for the communication with the equipment (RTUs, PLCs, etc.) and HMI software which run on master-station in the control room or elsewhere. The master station may be composed of one PC in small SCADA system. In large SCADA systems, the master station may include multiple servers, distributed software applications, and disaster recovery strategies. To increase the integrity of the system, the multiple servers will be often configured in a dual-redundant or hot-standby topology, providing control and monitoring even in the event of a server failure [9].

15.3.2 Generations of SCADA Systems

In this section will be presented briefly the well-known generations of SCADA systems.

- First generation is of "monolithic" type based on mainframe computers. The SCADA system was developed under conditions in which there were no computer networks. So, the SCADA systems were dependent designed to number of remote terminal units because the RTU was developed later (Fig. 15.6). An additional mainframe computer is added in the event of failure of the master system.
- Second generation is of "distributed" type because the information is shared in real time in the LAN. The data processing is done on several stations connected to the LAN (Fig. 15.7). Responsibilities are divided to each station, which enables high speed data processing at reduced cost in comparison with first generation.
- Third generation is of "network" type because the SCADA system uses the open system architecture, where can be easily connected peripherals such as printers, hard drives etc. The communication between the SCADA master and networked remote terminal unit is based on Wide Area Network (WAN) protocols such Internet Protocol (IP) address (Fig. 15.8). But the use of IP address standard via Internet makes the SCADA systems vulnerable to cyber-attacks. Consequently, new network protocols are developed by companies, but still remain important security issues to be solved.

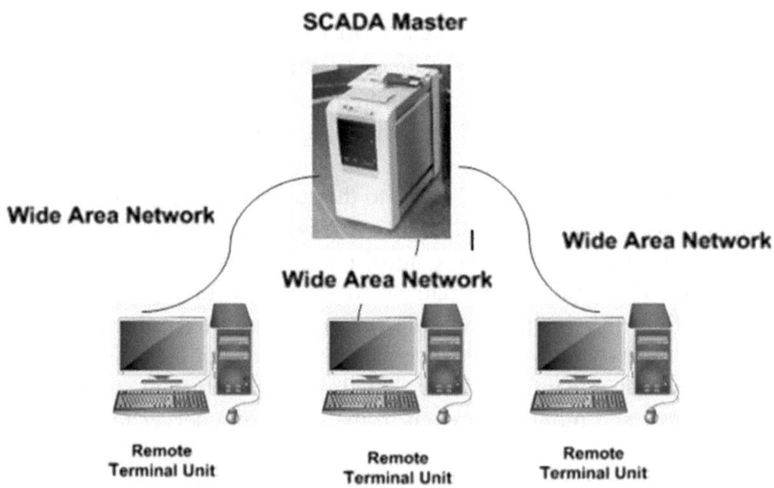

Fig. 15.6 First generation of SCADA systems

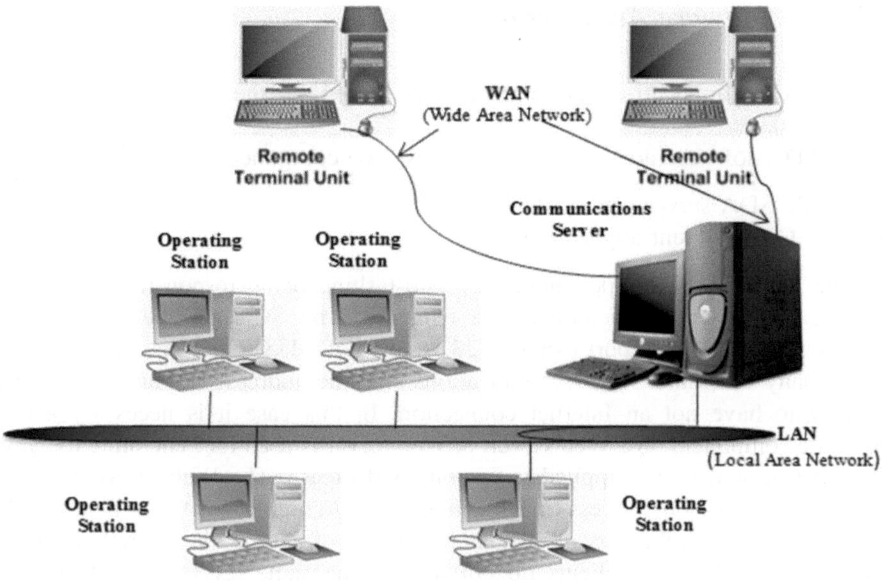

Fig. 15.7 Second generation of SCADA systems

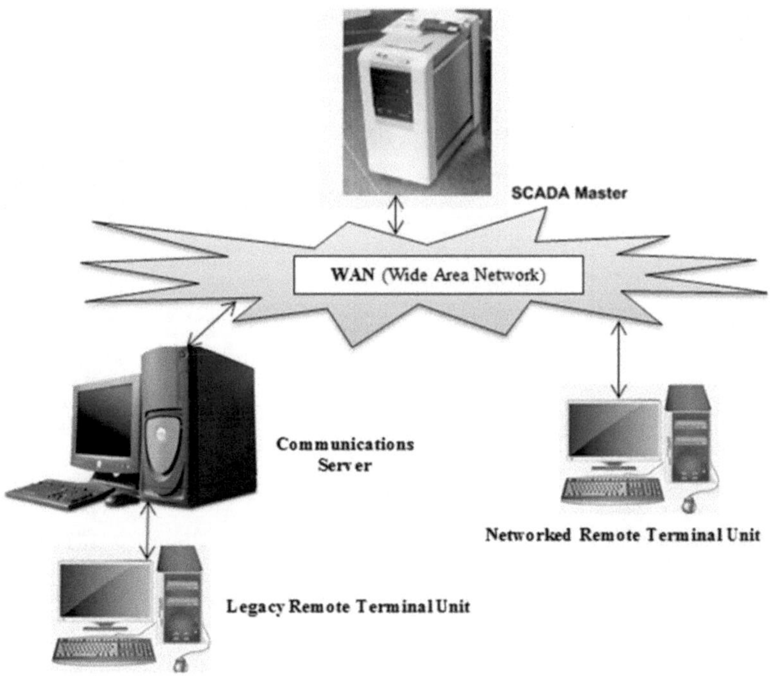

Fig. 15.8 Third Generation of SCADA systems

15.3.3 Software Architectures

The SCADA applications are implemented based on the software architecture (Fig. 15.9).

SCADA software architecture has at least two components:

- the SCADA server application;
- the SCADA client application;

The SCADA server application; is multi-tasking being responsible for storing the SCADA client applications in a database. The SCADA client application receives data via serial port using RS232 and MODBUS RTU protocols.

In many cases the SCADA applications shall be addressed to and remove the users who have not an Internet connection. In this case it is necessary a new software component on a Web server. A Web server is used for generating dynamic Web pages, which are supplied to customers decrease. The Web SCADA clients can benefit to real-time access (in a manner similar to regular SCADA clients) to the lists of parameters, the lists of events and the list of alarms via a simple browser. But, the Web SCADA clients do not receive the same level of graphical user interfaces and have the same HMI features as the regular SCADA clients. The regular SCADA clients are accustomed to running specific applications as opposed to The Web SCADA clients who view the Web pages provided by Web SCADA server. In order to reduce as much difference between the regular SCADA clients

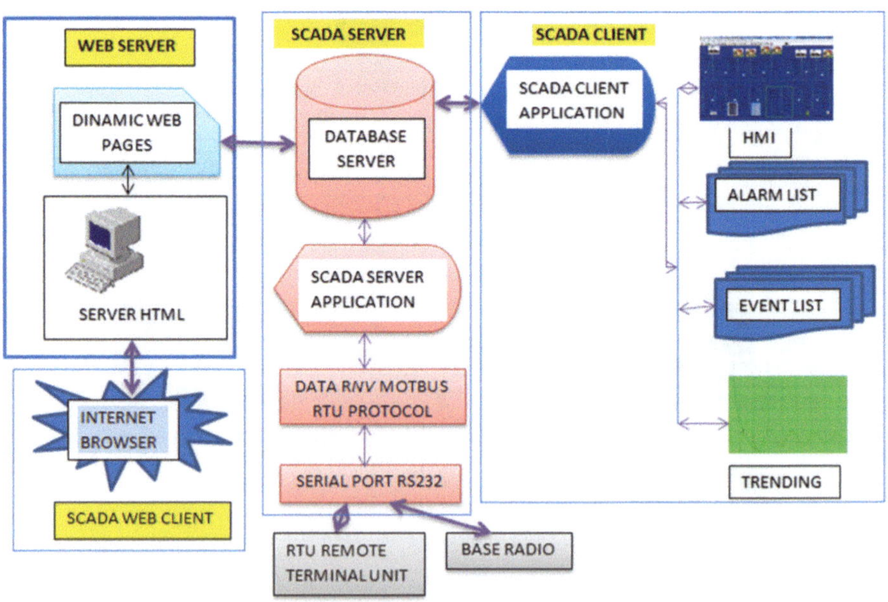

Fig. 15.9 SCADA software architecture

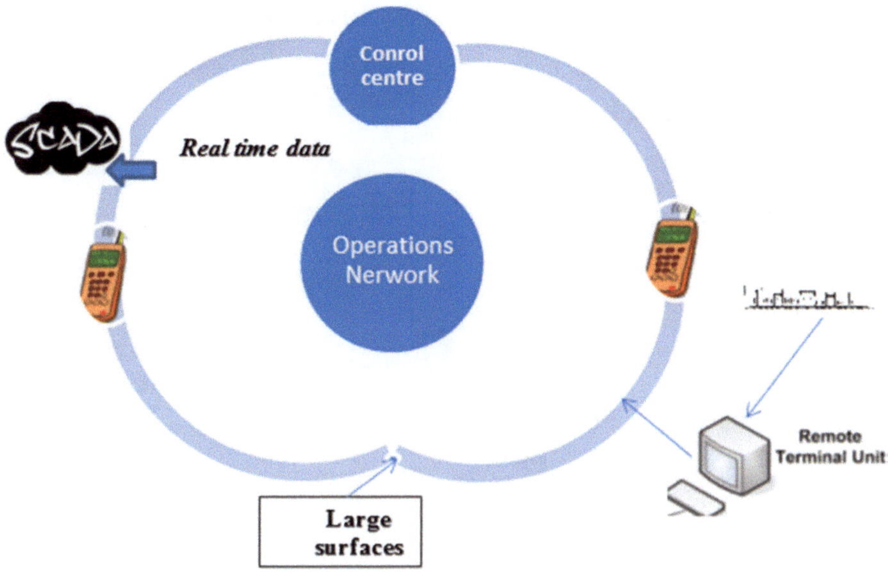

Fig. 15.10 SCADA systems

and Web SCADA clients some drives, services and virtual instruments are implemented in HMI. Visualization software used for monitoring and control of industrial process is called "software" HMI or "SCADA software". HMI as it is called generically, means any switch, interface through which you can control a device. The term "automation" refers to the graphical interfaces implemented through interaction with the mouse and/or the keyboard or by touch.

HMI/SCADA appeared as a terminal dedicated for users from the system implemented with PLC units. A PLC is programmed to automatically control a process. Because the PLC units are distributed in a large system, the data from PLC are stored and then transmitted. HMI/SCADA has the role to gather and combine the data from the PLC using a specific protocol. HMI software can be connected to a database in order:

- to fulfill the charts and diagrams in real time;
- to analyze the PLC data;
- to treat the planned maintenance;
- to drill the schemes for a particular sensor or machine;
- to apply the troubleshooting methods of the system.

The basic SCADA protocol is MODBUS, which is designed to send data to the master station even when the master interrogates the RTU station. MODBUS is based on a master-slave architecture known also as client-server architecture. The protocol is designed to be used by Modicon in own PLCs. Now, it became a standard of communication for the industry and is currently the most widely used to

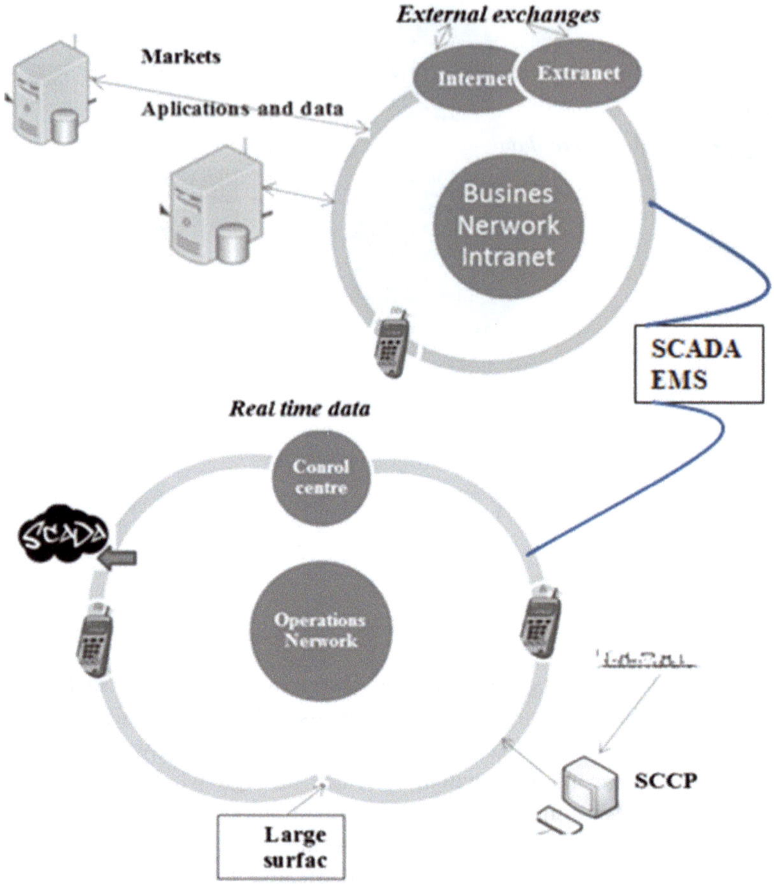

Fig. 15.11 SCADA energy management system

connect all the industrial devices. The protocols contain extensions to operate over Transmission Control Protocol (TCP/IP). In practice, it is suggested to avoid the security of Internet connection in order to reduce risks of cyber-attacks.

This protocol has the following features:

1. it is an open protocol with available documentation;
2. it can be quickly implemented (in few days, not months);
3. it works with bits and bytes, so no particular requirements on implementation.

MODBUS allows the management of the network devices, which gives new features to the SCADA system related to flexibility, reliability and safety in operation at high performance.

15.4 Assessment of Cyber Security Risk for SCADA

The developments in Information Technology (IT) systems regarding the safety of SCADA systems are shown in Table 15.2 [9–11].

Catastrophic situations which may occur due to incidental technical at level of the automation systems, Energy Management System (EMS), SCADA system, communications system, and support infrastructure. Besides these, other causes may be the incidents of cyber security and problems in the management of emergency situations.

New risks which occurred

"The Stuxnet incident has underlined what many ICS security experts had assumed for a long time: A sophisticated cyber-induced attack against an industrial facility would try to attack control systems rather than IT systems"—CIGRE WG B5-D2.46

- June 2010, Stuxnet was discovered;
- September 1, 2011, Duqu was discovered;
- May 2012, Flame Retardants was discovered;
- June 2012, Gauss was discovered;
- etc.

What is the concept of Cyber Security?

Security of those systems cannot be treated as a whole, and nobody can guarantee perfect security, because there is any system of defense perfect. Measures to protect and prevention are not enough, because emergencies and viruses could act in time using hidden forms or different rules. It is important as process to be defined from the point of view of management and responses to the incidents [8]. Incidents may be treated as shown in Fig. 15.12.

Safety in the field of energy must be monitored on the basis of production, transport and distribution, until to the end-consumer. In general, the current

Table 15.2 Systems developments in IT and SCADA security

Past	Present and future
SCADA (Fig. 15.10)	SCADA + Energy management system (EMS) + protection to electromagnetic pulse (EMP) (Fig. 15.11)
Closed systems	Partially open systems
Push-to-talk communications and cyber safe	Bi-directional communication and high cyber security (CIGRE—WG D2.34)
Few information	More information
Dedicated operating system (OS)	Standard OS + commercial OS
Few security requirements	Consistent data requirements

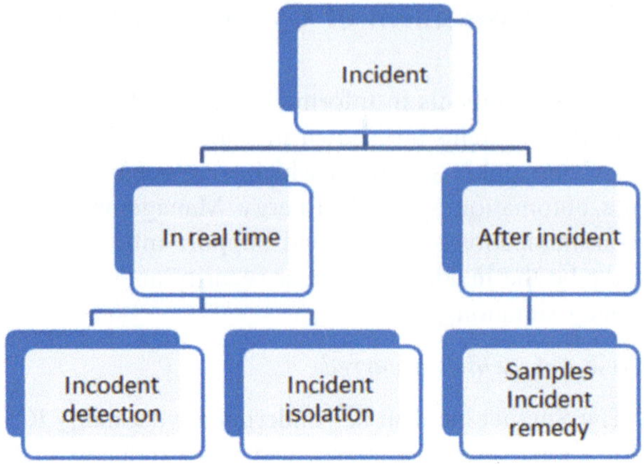

Fig. 15.12 Procedure for treating the incidents

problems are treated without taking into account new risks which can appear and could make huge damage, if will not be considered.

Risk assessment management involves the following stages:

- the analysis by which risks are identified in order to be assessed;
- the risks assessment from the point of view of the impact and likelihood of occurrence within tolerance imposed;
- the future decision by identifying and understanding risk [24].

Risk assessment can be done using several tools, one of the most popular being the probability-impact matrix which encompasses:

- the assessment of the probability that a risk to materialize and measured the degree of certainty:
- the assessment of the impact as the consequence or effect, where the risk would materialize. In the impact assessment, account shall be taken of the fact that the risks that have a strong impact on acute misconduct, while those which manifests itself frequently and have a low impact signals a chronic problem;
- the evaluation of risk exposure as a combination of probability and impact can be felt where the risk would be materialize. In case of the materialization, the risk exposure is, in fact, an impact [24].

As a result of completing the stages of identifying and assessing of possible risk, finally, a synthetic and clear representation of the risks must be drawn, mentioning their ranking in accordance with the level of exposure. The prioritization of risks is a very important tool in risk management. Tolerance is estimated by determining the level of risk accepted. The level of risk accepted is established after the adoption of the response. The following should be taken into account:

- inherent risks;
- residual risks.

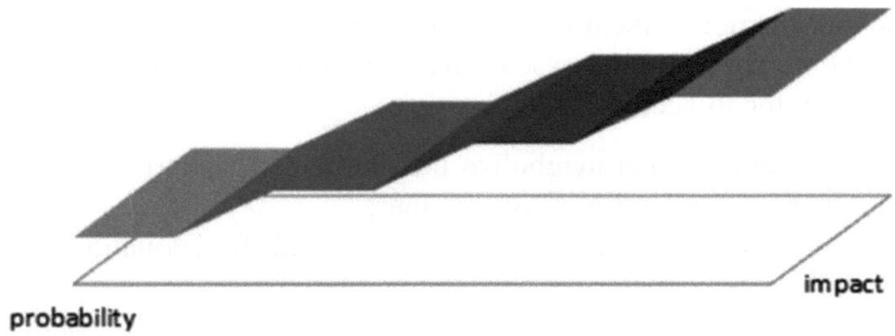

Fig. 15.13 Assessment of exposure to risks

Risk tolerance involves some subjectivity because it depends to a large extent by the way in which this is perceived as risk. Risks which have a high level of the exposure is located on top of the tolerance and should be dealt with by means of measures by which exposure to residual risks must be made under this tolerance [24] (Fig. 15.13).

15.5 SCADA Applications

The HMI/SCADA software can be installed on personal computers or embedded computers with touch-screen, distributed through the house or into the industrial process. It runs a real time operating systems because it must respond in the shortest time to changes in the parameters of the process monitored. Consequently, the HMI/SCADA software must meet the following requirements:

- to be easy to use;
- to have an intuitive interface;
- to provide increased security;
- to enable the remote access (if it is the case);
- to be easily to access and use the information;
- to be flexible in use;
- to have the possibility of communication through the various services (sms, e-mail, fax).

The HMI/SCADA software must be capable of performing certain functions, such as:

- control and supervision of process;
- alarming the user;
- acquisition of the data needed;
- storing of the events;
- keeping the databases updated;
- developing of graphical evolution of the process;
- coupling with Enterprise Resource Planning (ERP) software.

The most important element of an integrated HMI/SCADA is the interface that connects the command and the system that is monitored. This interface is used by the human operator to access:

- the graphic elements that symbolize the automated technological process in a systemic, appropriate and intuitive manner;
- the most important functions in order to enable all functionalities;
- the state of alarm variables and the alarms history;
- and many other information displayed in a form easy to be followed by the human operator to see the evolution over time of the process monitored.

Optimization through SCADA solutions

Currently, in the world and in our country, the energy industry is characterized by profound structural transformations. Besides these requests for structural changes, the function in a competitive market leads to the amplification of the current efforts made by the power companies to increase efficiency and quality of their service. One of the ways of achieving this aim is the orientation towards the smart grids based on a new strategy for energy management. Thus, information of the EP system is one of the basic prerequisites for increasing the effectiveness and safety of the operation of the national energy system. An example of optimization of the electric power stations along a river is shown in next section.

15.5.1 SCADA System Existing in a Hydropower Plant

The existing SCADA system (Fig. 15.14) consists of:

- the SCADA systems associated with each Hydro Electric central Computer (HEC) (HEC1, HEC2, HECi);
- the SCADA System for the Hydro Energy Dispatching (HED) from based Hydroelectric Station (HS).
- At the level of the HECs arrangement, the SCADA system is composed of:
- Distributed Antenna System (DAS) which represents the equipment for the data transmitted between the SCADA processes in HEDs from the Hydroelectric Station (HS).

The DAS has available four RS232 1 ports and an Ethernet port. The communication protocols supported are: (1) the DAS Serial, (2) the Modbus RTU, (3) the International Electrotechnical Commission (IEC) 1107 Mode C for serial ports, (4) the DAS TCP/IP, and (5) the Modbus TCP for the Ethernet port. Note that the communication with HEDs is made through the DAS serial ports on two redundant paths: (1) using the telephone line and the communication protocol; (2) using the radio communication protocol.

- Functional Assemblies (FAs 1–4) may represent Hydro-aggregate automation systems, automation systems, General Services (GS), and HEC barrage.

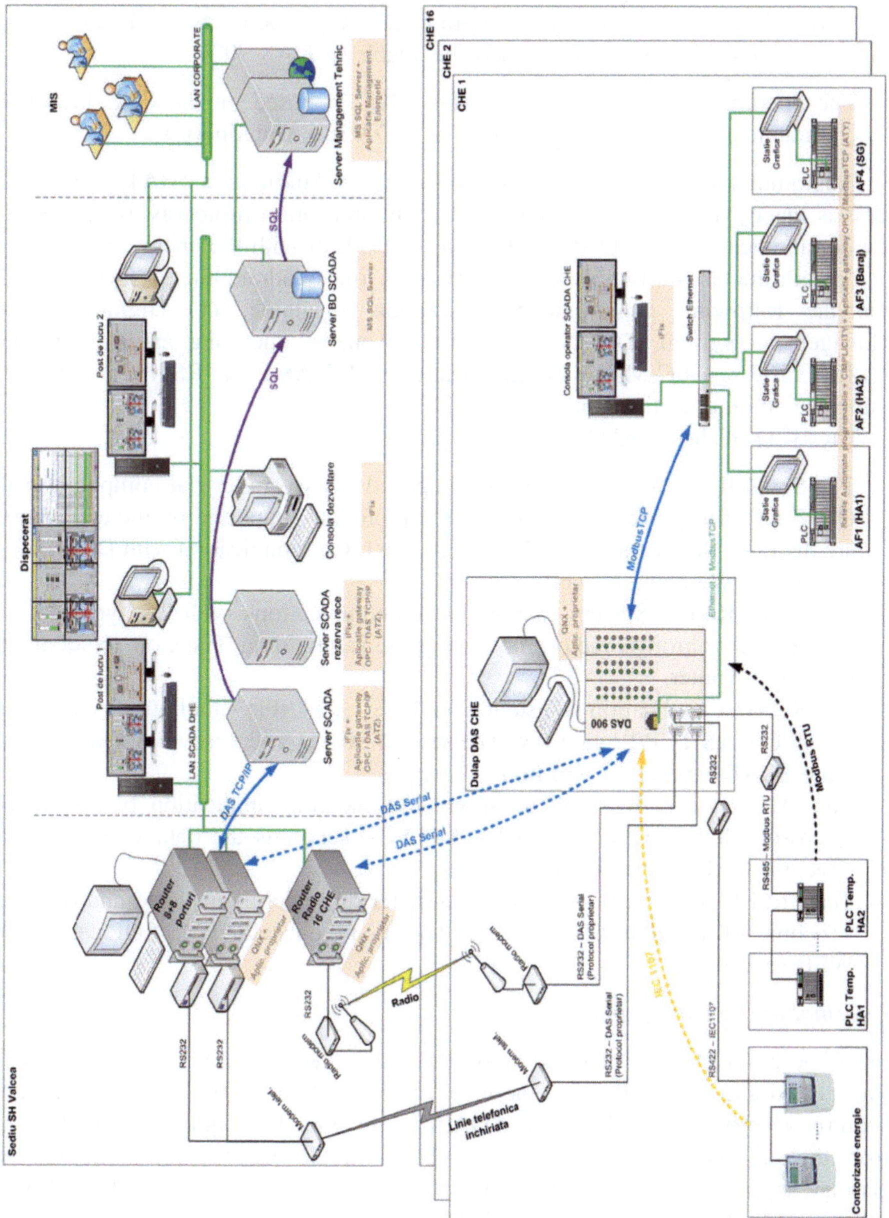

Fig. 15.14 Architecture of the existing SCADA System

The communication protocols implemented in the FA are the OPC industrial communication protocol (available in Wincc, Cimplicity). Communication with DAS is via Ethernet using the Modbus TCP protocol, which has the role of protocol gateway between OPC industrial communication protocol (available in Wincc, Cimplicity) and Modbus TCP protocol (available in DAS-900).

- The operating Console of the SCADA HEC is based on the OPC industrial communication protocol available in SCADA iFix (Intellution).

Programmable thermometry in Hydro-agregate Automation (HA1, HA2) systems is made based on (1) Modbus RTU communication protocols, which assure the communication with DAS via RS485 (or RS232) with the energy meters.

CewePrometer is a family of high-precision four-quadrant energy meters which has the following communication protocols available: (1) Digital Library Management System (DLMS) (used by the energy metering application) and (2) IEC 1107 serial C Mode. Communication with DAS is via RS-422 (or RS232) using the communication protocol IEC 1107.

The HED of the SCADA system is composed of:

- The Serial Router (8 + 8 ports) is composed by two industrial computers with QNX operating system. Communication protocols supported by the application are the DAS Serial and DAS TCP/IP Ethernet. Communication with DAS HECs is done in two paths as well.
- The radio Router System assures the communications with 16 hydroelectric power plants through radio-relay, using the same communication protocols as the Serial Router.
- The SCADA application is installed on the SCADA iFix (Intellution) server and the OPC/DAS TCP/IP gateway. The second SCADA server will operates as the main SCADA server if will be the case.
- The Data Base (DB) SCADA server use the iFix application based on the Microsoft SQL Server (MS SQL). MS SQL server is the data source for the "Technical Management" application.
- Technical Management server is a computer on which is installed and runs the "Technical Management" application based on the installed MS SQL server and WEB server.

Information flow diagram

From the point of view of information flow, the existing SCADA is presented in Figs. 15.15 and 15.16. Thus, at the level of DAS-900, the HEC data concentrator acquires specific data for HECs SCADA from the FAs. FA1 and FA2 read the data from the automation systems HA1 and HA2. The General Services (GS) and the Barrage application use the OPC protocol (existing graphic stations as base driver) via the Modbus-TCP from the DAS-900. This conversion from OPC to Modbus-TCP has the following disadvantages:

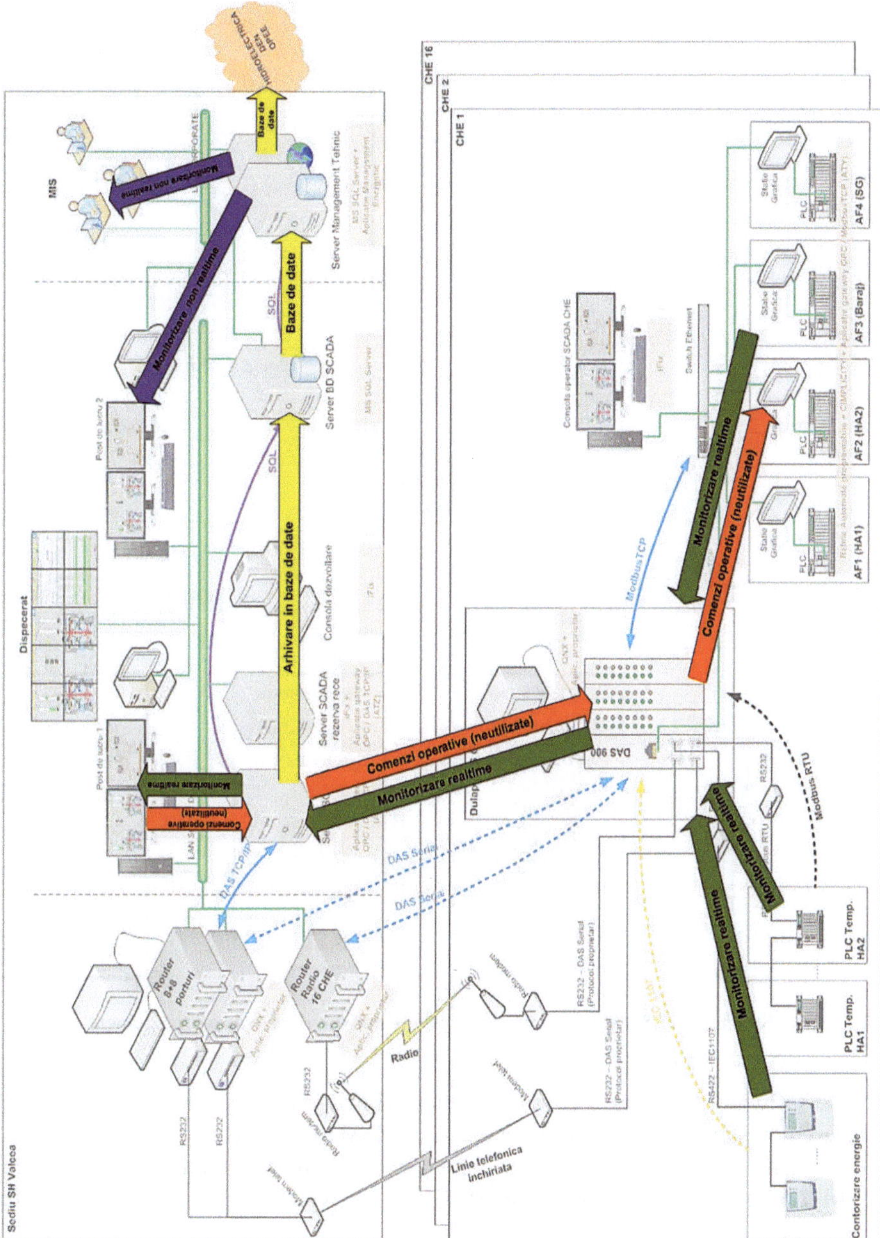

Fig. 15.15 SCADA system information feeds

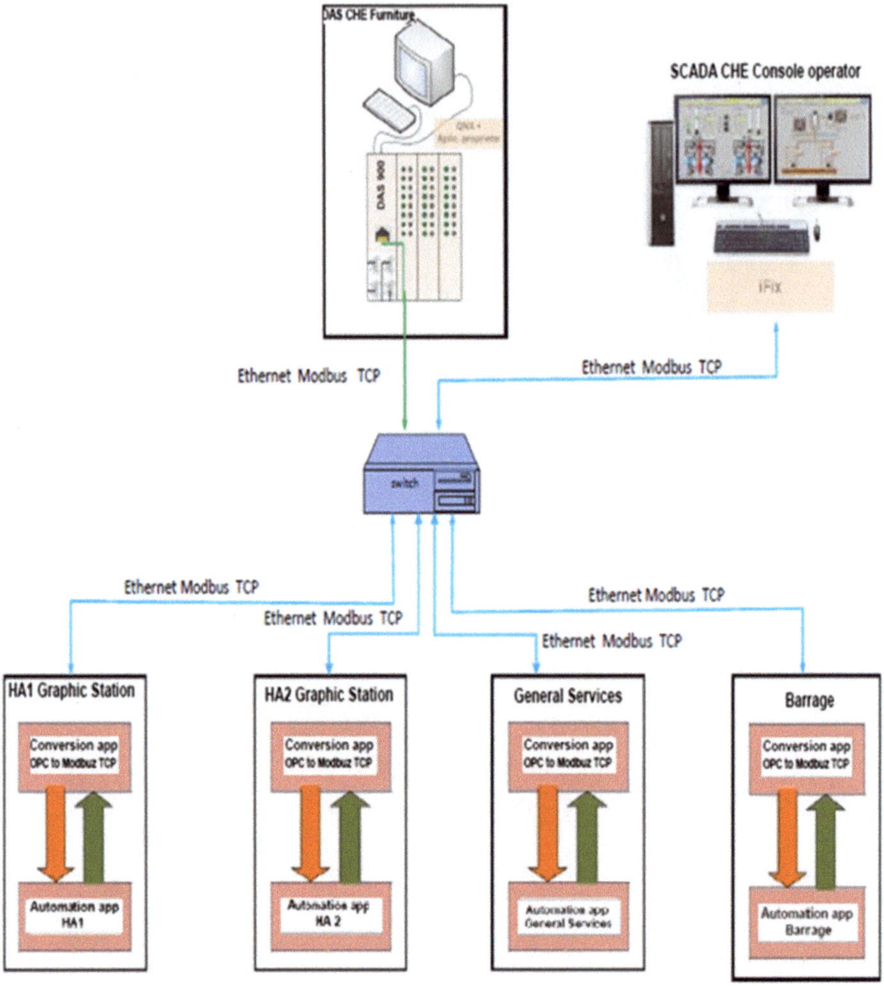

Fig. 15.16 Streams and SCADA system architecture for the river hydro-arrangement

- the software application is not documented and it is very hard to be interfaced based on standard protocols with existing SCADA architectures. It appeared as a necessity to interface DAS-900, which doesn't support for the OPC protocol, with the process charts from the HEC stations, which have locally implemented the OPC protocol;
- it requires automatic start and stop upon launching the application of monitoring and control on each processor of the workstation graphics or restarting the operating system. So, it must be individually monitored for any restart in the event of a communication blockage;

- as was seen in operation, there were frequent blockages of this application, which means that data and commands are not passed to and from the hierarchical level;
- the application for operation relies on two text files with strict rules of editing (spaces, TABs, bookmarks, etc.), which must be configured identically for each workstation graphics (HA1, HA2, General Services, and Barrage) and DAS-900. Note that the existing SCADA system has the possibility of transmission the commands, but at the present time these are not used.

Thermometry system for the HA1 and HA2 will read the temperatures monitored and the CewePrometer meters will give the main electric values: active power, reactive power, voltage, current etc. The data concentrated in DAS900 are transmitted by the SCADA system to the HED through the two redundant paths mentioned above. DAS900 system has the option of receiving SCADA commands and records from HED, but at the present time these are not used.

The SCADA data acquired and processed by the HED are stored in the MS SQL database in order to be used by the "Technical Management" application.

15.5.2 SCADA System Proposed at the River Hydro-Arrangement

The proposed architecture (Figs. 15.17 and 15.18) was chosen to obtain an increased security by physical separation of the LANs HEC, SCADA LAN, LAN DSZ, and LAN CORPORATE using Router/Firewall equipment.

The implementation of a fiber optic network (of redundant ring type) will facilitate the communication between the HEDs related to East River and HEC. The IEC 60870-5-104 communication protocol is proposed, which is an industrial protocol suitable for communication between servers and optimized SCADA network of HED and HEC WAN.

Note that the existing radio network will remain functional as a backup plan for fiber-optic (FO) network. Thus, the IEC 60870-5-101 radio protocol will be used as a hot backup (with automatic switching) for the IEC 60870-5-104 main protocol based on FO network. Consequently, the HED communication with other HECs will not be affected.

The managers of the Hydro-electrica (HE) Company aim to implement a central SCADA system data of "real time" type on all hydro-stations. For reasons of security and reliability, it is not recommended to send the data directly from HE SCADA LAN (which is a closed and secure network), but it is possible to transmit data from the isolated nodes. The Technical Management server has all the necessary facilities for data transmission to HE Company.

The proposal to modernize the HEC architecture by replacing the existing equipment with new DAS-900 equipment and two SCADA redundant servers (with

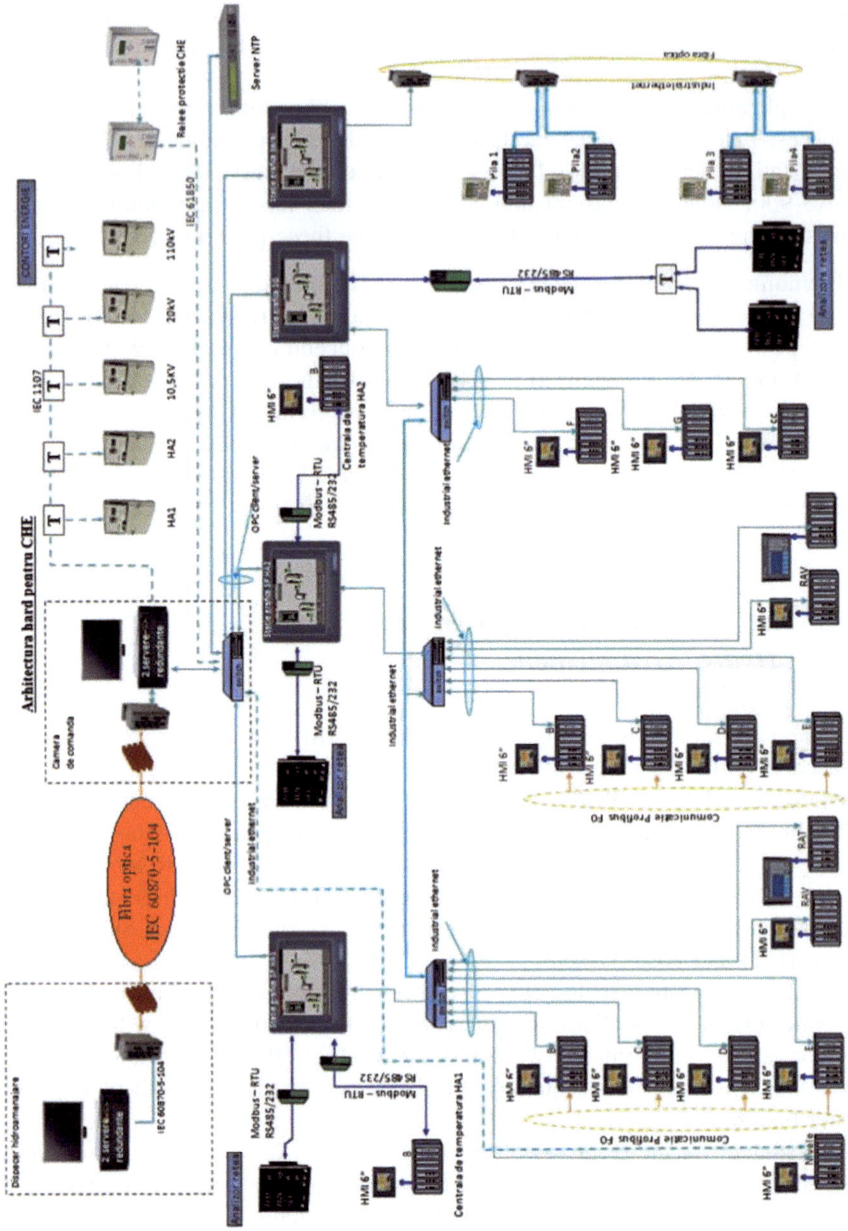

Fig. 15.17 Architecture of the proposed SCADA system

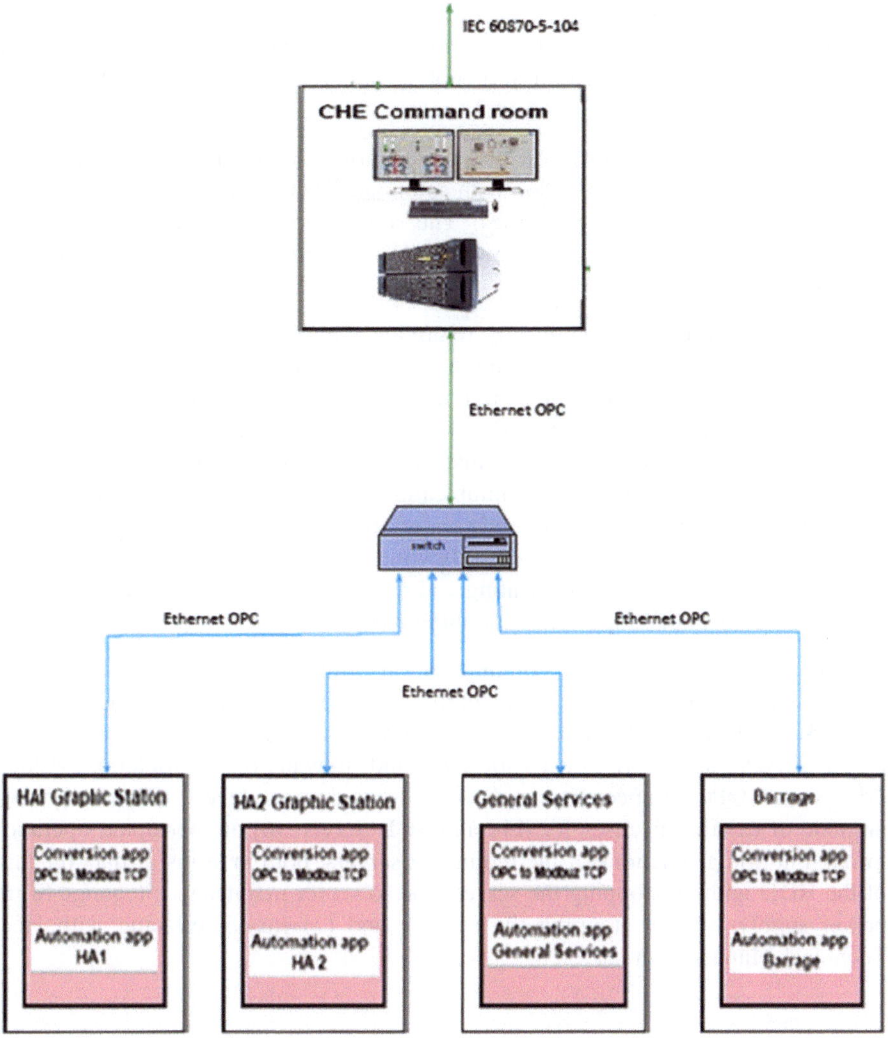

Fig. 15.18 Streams and SCADA system architecture proposed for river hydro-arrangement

hot redundancy) to eliminate the existing malfunctions and in addition to implement new communication functions needed for operation in HED such as

- Bidirectional communication between the automation systems (HA1, HA2, Barrage, GS) using the OPC industrial protocol, without the need of other applications for data conversion;
- Communication (in terms of realtime monitoring of the main electrical variables) with existing CewePrometer meters using the IEC1107 and Ethernet TCP-IP communication protocols;

- Bidirectional communication (in terms of data monitoring and transmission of commands) of HED SCADA system using two paths communication: IEC 60870-5-104 communication protocol (primary path) and IEC 60870-5-101 (backup path).

The SCADA applications dedicated servers hold a large number of standardized protocols for direct communication with automation devices used in monitoring and management of industrial processes. The advantages for data acquisition and remote management of the hydropower plants retrofitted with redundant SCADA servers are:

- Simplification of the structure by removing the SCADA DAS-900 equipment (or other similar equipment) and use of the redundant servers for sending commands and data from the central dispatcher and vice versa; this allows easy capacity expansion and further development of software, no constraints;
- The use of a SCADA software installed on the server, ensuring direct connection via OPC industrial communication protocol to the existing equipment for the management and supervision of the processes, without intermediate software; servers will provide the following functions: acquisition, operation of data, storage and archiving, printing and reporting to the higher and lower levels of communication, supervision of automated processes.

This will also assure: (1) communication with existing equipment installed (meters, protection relays etc.), (2) high speed communication through the FO network of the each HEC and HED, (3) improved security provided by the two levels SCADA servers (two network cards and firewalls on the routers), (4) local drivers and OPC connectivity with inferior equipment levels, (5) storage and upgrade of existing licenses for iFix and SQL Server, (6) uninterrupted operation through hot-plug functions of the redundant servers, both in terms of data (change of the HDD without stopping the server) and to server maintenance (change of the power supply without stopping the server), and (7) easy interfacing with HED Technical Management server.

15.5.3 Issues Related to Data Acquisition and Remote Management

Issues related to data acquisition and remote management applications of the power plants, which running on hydropower facilities with RTU, are the following:

- RTU32 in the proposed SCADA architecture is not hot redundant at equipment level;
- Unable to communicate with graphic stations of the RTU process and SCADA servers via the OPC industry standard protocol;

- The need to create a new application (named "Date Gateway") at the level of SCADA servers, as well as the "Technical Management" server, for bidirectional communication between the servers;
- Managing under Windows XP of the Setup software in RTU with Straton;
- Bringing of electrical signals directly through wires to the level 1 (RTU 32 level) is not indicated, but the solution to bring them to level 0 (PLC and RTU) it is recommended.

15.6 Overview of the Programming Environment VIJEO CITECT 7.40

CitectSCADA system is considered one of the most complex and complete products for the SCADA systems [34, 35], being part of the VIJEO CITECT SURVEILLANCE SOFTWARES 7.40® of to the Schneider company. This software product includes over 150 drivers for communication with the input-output devices and over 350 models for PLCs, RTUs, controllers and regulators, barcode readers etc.

For any type of SCADA application, this system offers the needed flexibility in implementation and execution of the project, in terms of speed, efficiency and accessibility. More detailed, the CitectSCADA system offers the following opportunities to user:

- Use of the HMI graphic pages for central or local control stations;
- Realization of operation stages by creating the graphical control buttons;
- Display the status of the actuators via graphical charts;
- Display the alarm messages and their evolution;
- Language selectable to display the messages;
- Achieving of universal commands through the keystrokes;
- Controlling, storing, monitoring and displaying of the alarms;
- View the graphs in real time or in "history record";
- Efficient and effective monitoring of the system;
- Creating of reports;
- Supervision of product quality by Statistical Process Control (SPC);
- Acceptance of transferring data with other applications;
- Security system can be configured on the priority levels.

VIJEO CITECT SURVEILLANCE SOFTWARES has the following modules:

- Citect Explorer;
- Project Editor;
- Graphics Builder;
- Cicode Editor.

Citect Explorer

The Citect Explorer module is used to create and manage a CitectSCADA project. This controls the configuration permitting enforcement of project in the rest of modules. Citect Explorer screen is shown in Fig. 15.19a:

The launch of the Citect Explorer module will automatically run the Project Editor and Graphics Builder modules, but these are minimized in Fig. 15.19a. The close of Citect Explorer module will close all CitectSCADA applications.

Project Editor

The Project Editor module is used to create and manage the configuration information of CitectSCADA project. The Project Editor window can be viewed by all projects. Special commands are accessible from the buttons or submenus (Fig. 15.19b).

Citect Graphics Builder

The Graphics Builder module is used to achieve pages and editing graphical objects used on other new graphics (Fig. 15.19c). Graphics Builder editor is activated by a double click on a graphic object in Citect Explorer.

Cicode Editor

The programs are edited in the window of Cicode Editor. The Cicode Editor menu contains a list of functions. The support information for functions is obtained by a right click on the name of the function (Fig. 15.19d).

Configuring CitectSCADA Projects

The schematic of a CitectSCADA project looks like in Fig. 15.19e.

As can be seen in Fig. 15.19e, the CitectSCADA project includes the following items:

- Graphics;
- Databases;
- Cicode programs.

Graphics mean a graphic page which allows the monitor to display the graphical interface with control buttons.

Databases allow the storage of process information collected for monitoring and controlling the system. These may be linked to the graphics page if desired.

Cicode programs allow functionality and contain a number of useful functions stored.

In all projects carried out, the user can:

- Create;
- Select;
- Open;
- Shut down;
- Delete;
- Etc.

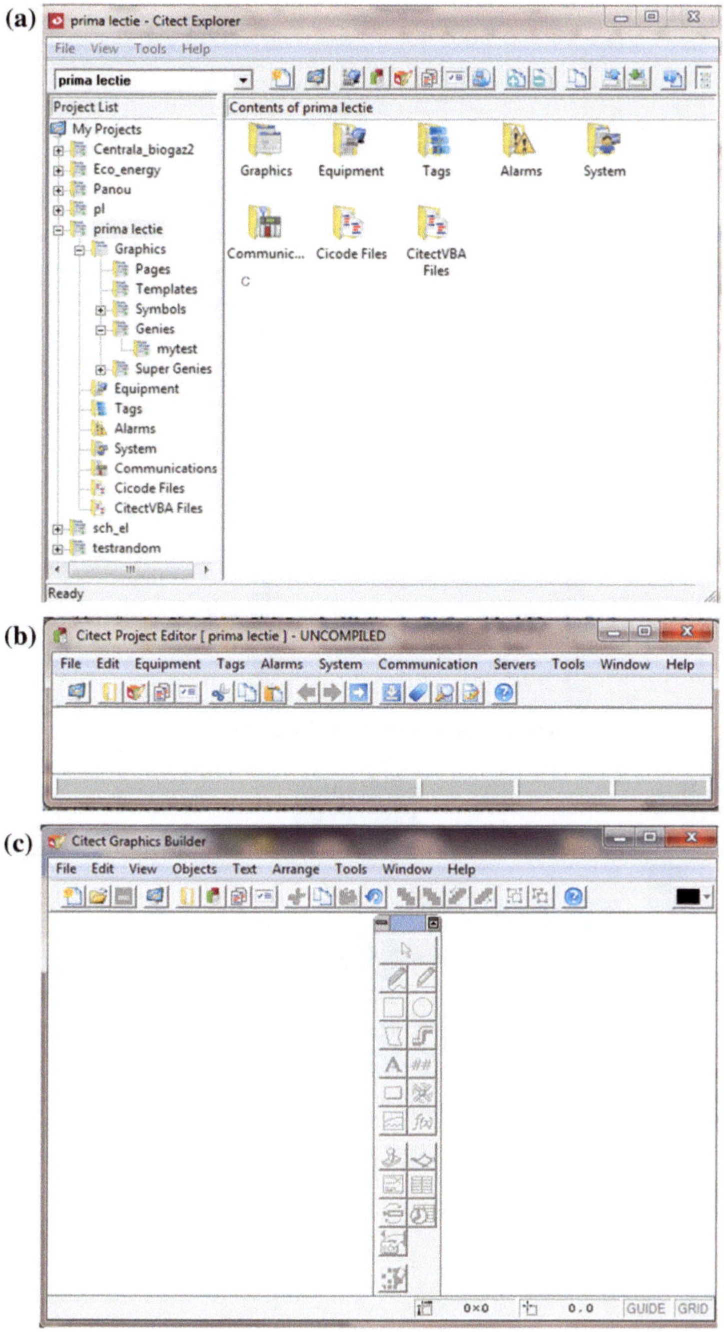

Fig. 15.19 **a** Screen of citect explorer. **b** Screen of project editor. **c** Screen of graphics builder **d** Screen of cicode editor. **e** Scheme of the CitectSCADA project

(d)

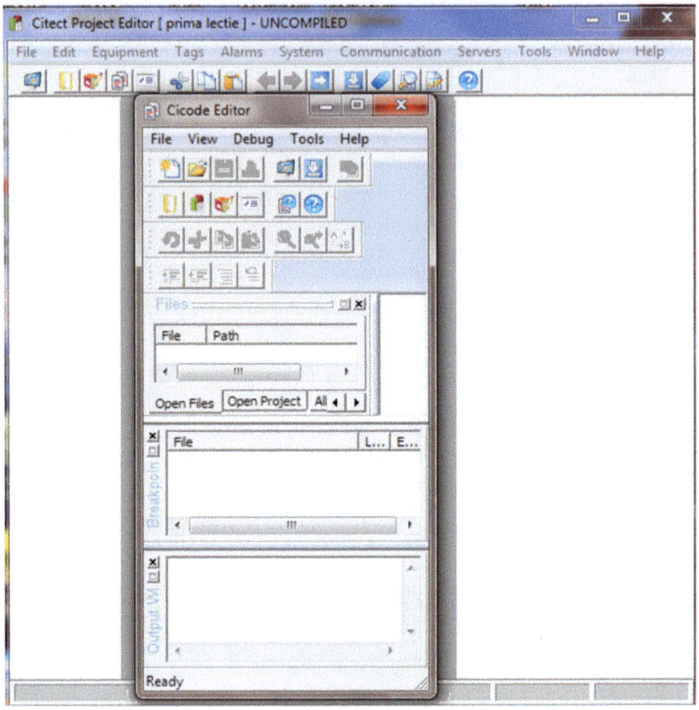

(e)

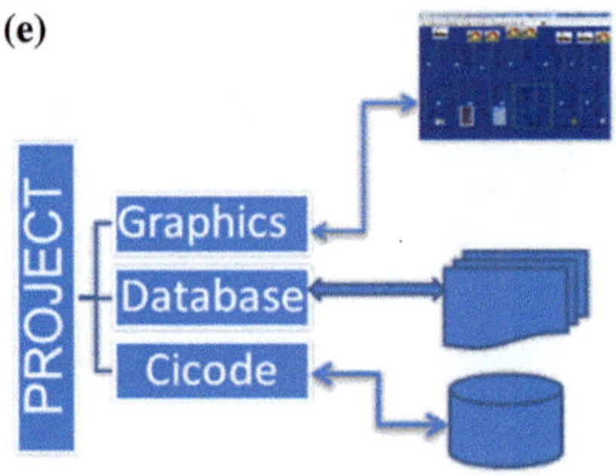

Fig. 15.19 (continued)

15.7 Developing Step by Step a SCADA Application for Electric Power System

Application 1

15.7.1 Implementing a Medium Voltage EPS Based on RESs

The CitectSCADA environment is used here to develop step by step a SCADA application for the Electric Power System based on RESs. The both interfaces will be shown in this section in order to shown the advantages of the new SCADA interface, which is a friendly and intuitive interface, with a high degree of integration schemes, facilitating for more dispatchers to work simultaneously.

The old interface

As can be seen in Fig. 15.20, the old interface used is poor. In case of maneuvers in power installations, the dispatcher has a limited overview on the system.

The new interface

A SCADA application is developed step-by-step starting with the identification of needed elements and objects, which then will be set for data acquisition, command and control equipment etc. The software structure will be identified correspondingly in the hardware structure.

A new project must be opened and created, and finally must be saved. The project will define the appropriate hardware to achieve a simulation structure using the graphical page. For the simulation process, the structure must be monitored to be identified and set properly in the graphic pages as associations of hardware items

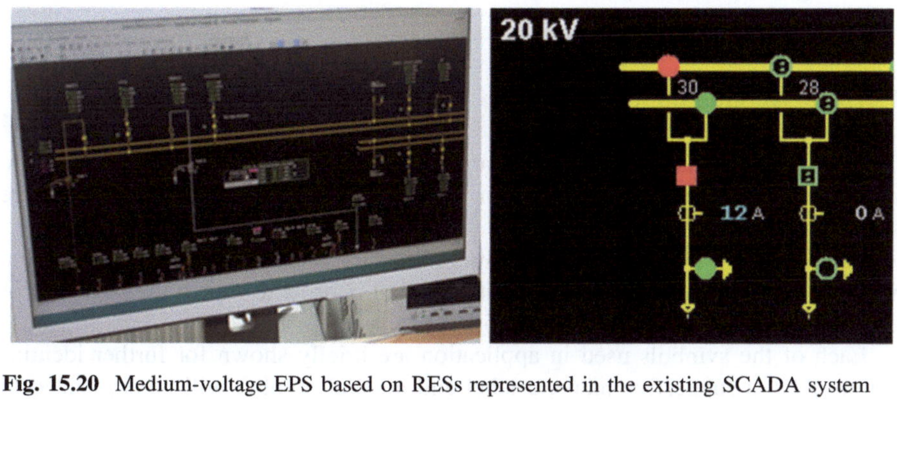

Fig. 15.20 Medium-voltage EPS based on RESs represented in the existing SCADA system

and their appropriate functions. An important step will be dedicated to security, namely:

- setting application users;
- setting their rights.

After implementation of the simulation structure, each item must be check using the software to test all functional features (Fig. 15.21a).

The steps for creating a new project are:

- From the Citect Explorer → File → New Project, a new project called "Micro_energy" is open (see Fig. 15.21b);
- From the Citect Explorer → gen → Communications → Clusters, define a cluster of communication called "sch_el_cluster" (Fig. 15.21c);
- From the Citect Project Editor → Servers → Network Address, define the TCP/IP address 127.0.0.1 called "sch_el_adr";

Address "sch_el_adr" is the TCP/IT data acquisition system which are allocated to the system (Fig. 15.21d).

- Then, the further facilities will be configured as below:

  1. "Alarm": From the Citect Project Editor → Servers → Alarm Server (Fig. 15.21e).
  2. "Trending": From the Citect Project Editor → Servers → Server Report (Fig. 15.21f).
  3. "Report": From the Citect Project Editor → Servers → Server Trend (Fig. 15.21g).
  4. "I/O Server": From the Citect Project Editor → Servers → I/O Server (Fig. 15.21h).
  5. "Express Communications Wizard": From the Citect Explorer → gen → Communications → Express I/O Device Setup → Next → Use an existing I/O → Disk I/O Device → Citect generic Protocol → Automatic refresh of tags → Finish (Fig. 15.21i).

15.7.1.1 Making the Graphic Page

To this end, in this section the libraries of the associated elements, which are required or which must be designed or imported if there are no such elements, are shown below (Fig. 15.21j).

The following symbols and window is used for the transformer and to set the voltage ratio, and to obtain the animation symbol in order to switch the power on or off the power for transformer (Fig. 15.21k).

Each of the symbols used in application are briefly shown for further identification. Also a suggestive name, a label and a serial number will be associated to

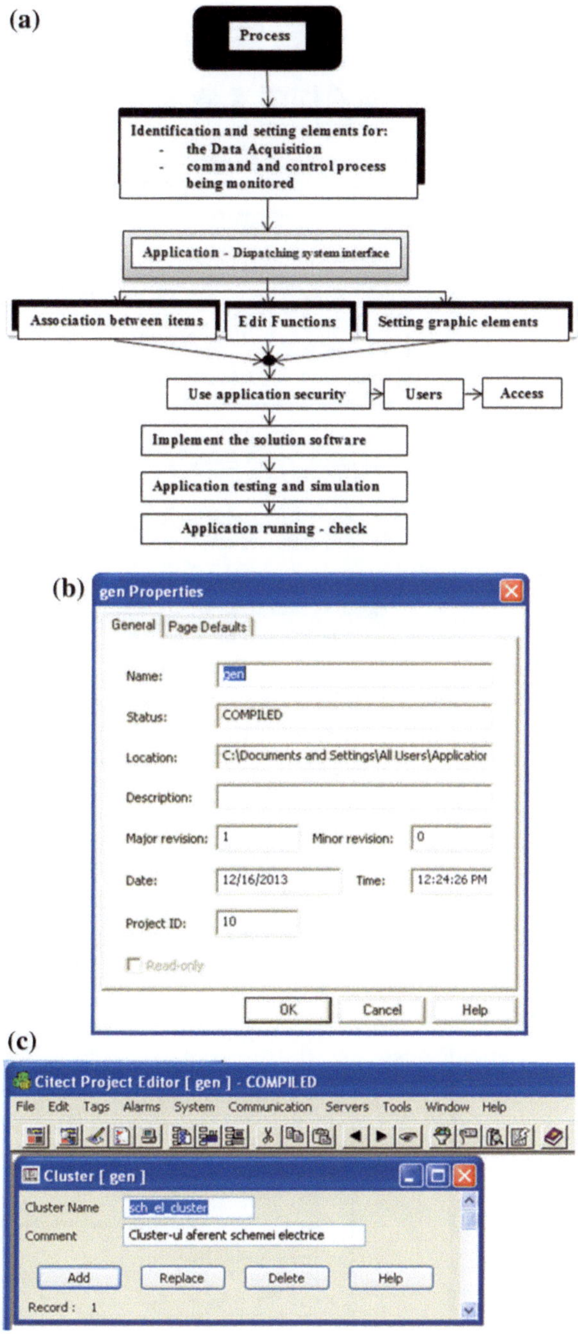

Fig. 15.21 **a** Concept diagram of the SCADA application. **b** Window "New Project". **c** Window "Cluster". **d** Window "Network Address". **e** Window "Alarm Server". **f** Window "Report Server". **g** Window "Trend Server" **h** Window "I/O Server". **i** Window "Express communications wizard". **j** Imported images. **k** "Transformer" symbol and window "Symbol Set Properties". **l** "Electrical splitter" symbol and window "Symbol Set Properties" **m** "Switch" symbol and window "Symbol Set Properties". **n** Animated symbols. **o** Window "Variable Tags". **p** Window "Symbol Set Properties". **q** Window for proposed process. **r** Graphic window of the active process. **s** The sequence of code in the graphical user interface (GUI)

(d)

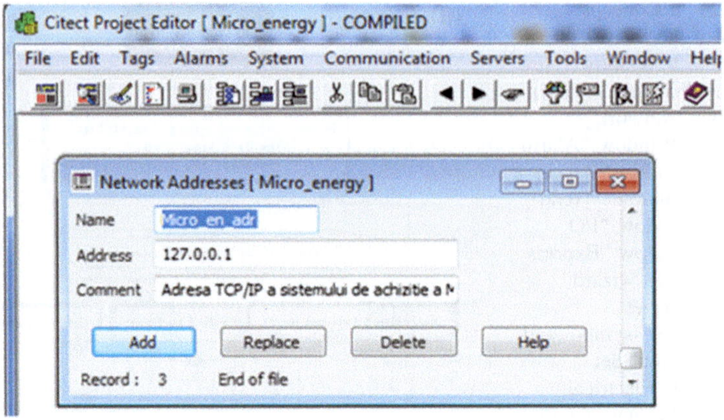

(e)

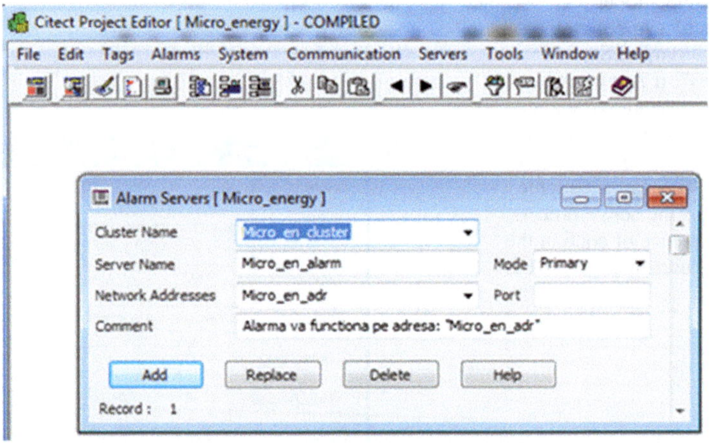

(f)

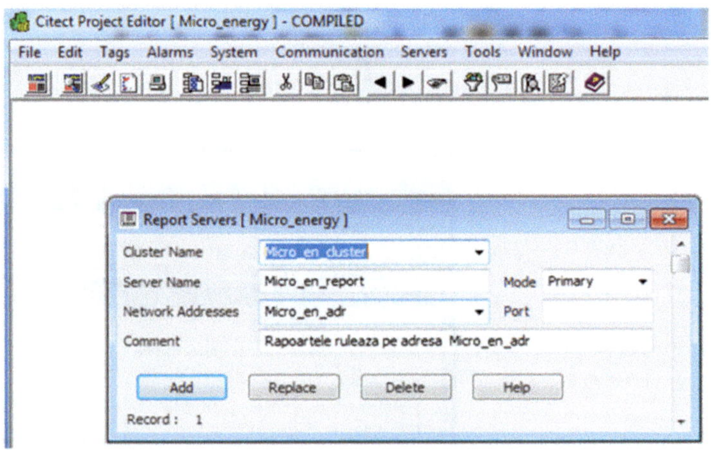

Fig. 15.21 (continued)

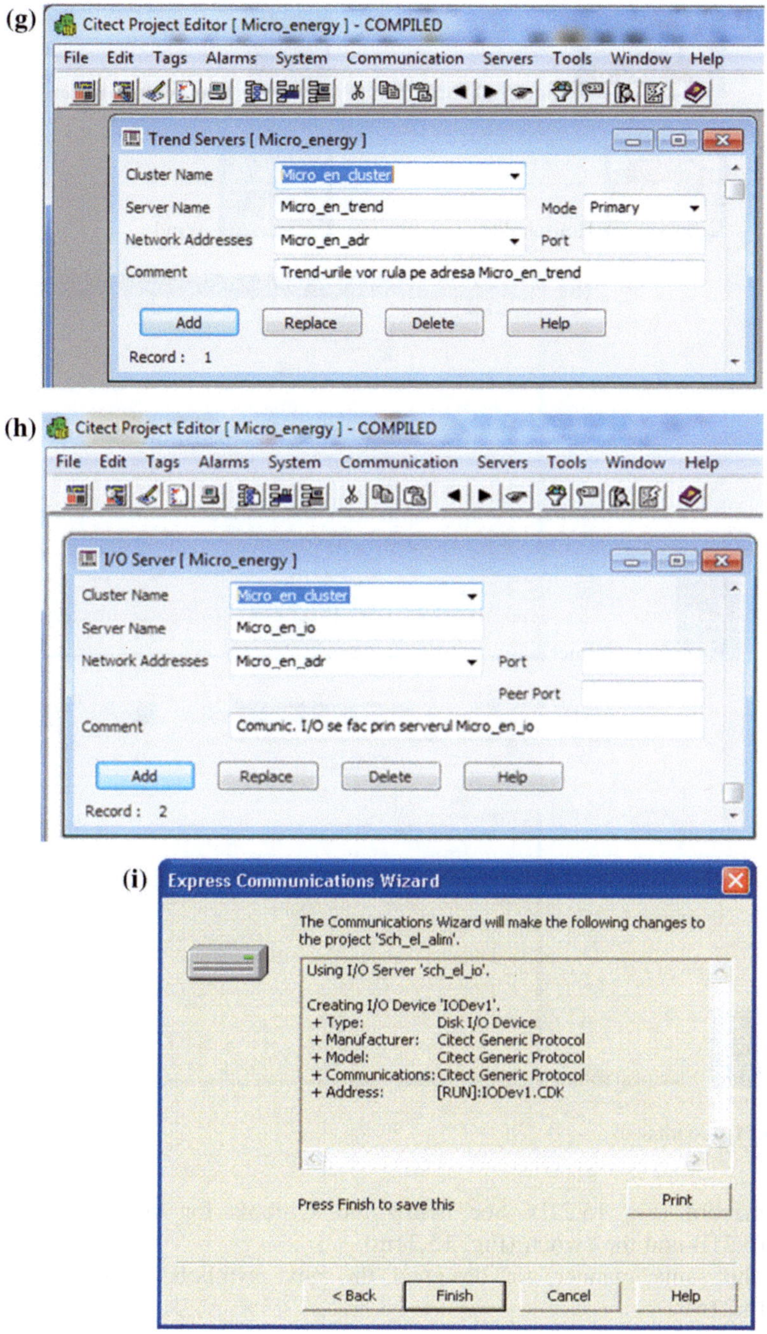

Fig. 15.21 (continued)

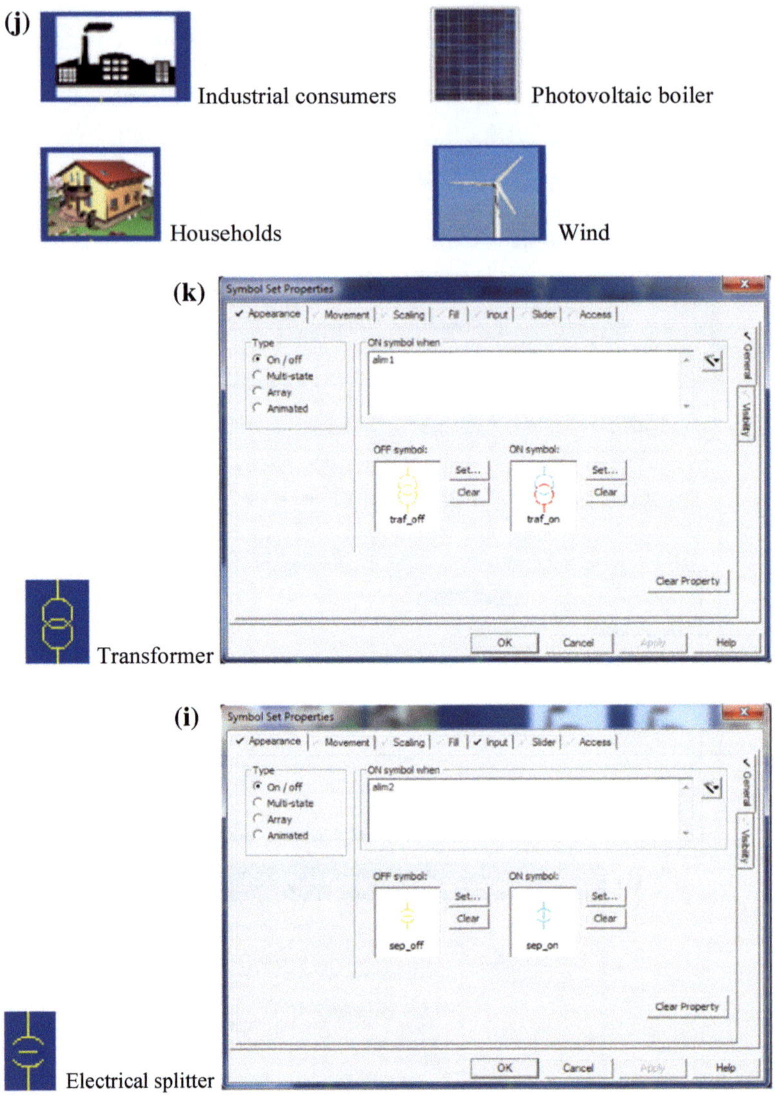

Fig. 15.21 (continued)

each symbol (Fig. 15.21l). See below the symbols for the electrical splitter (Fig. 15.21l) and the switch (Fig. 15.21m).

In the same manner are designed the next symbols used in application (Fig. 15.21n).

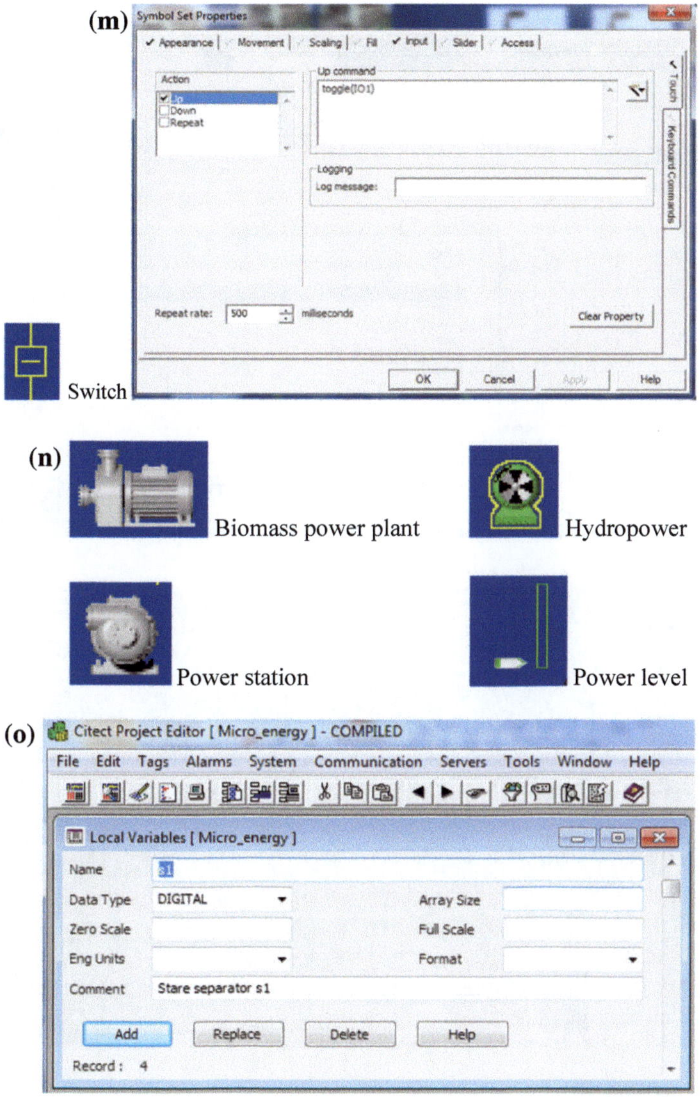

Fig. 15.21 (continued)

15.7.1.2 Configuring the Variables

All SCADA application are based on the setting and using of the tag-type variables. Tags are the interface between the user and the monitor. There are several types of tag-type variables:

(p)

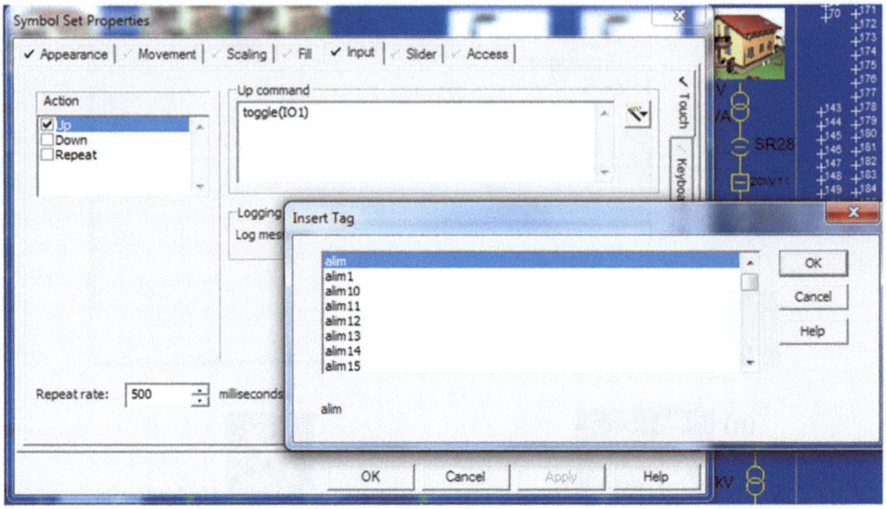

(q)

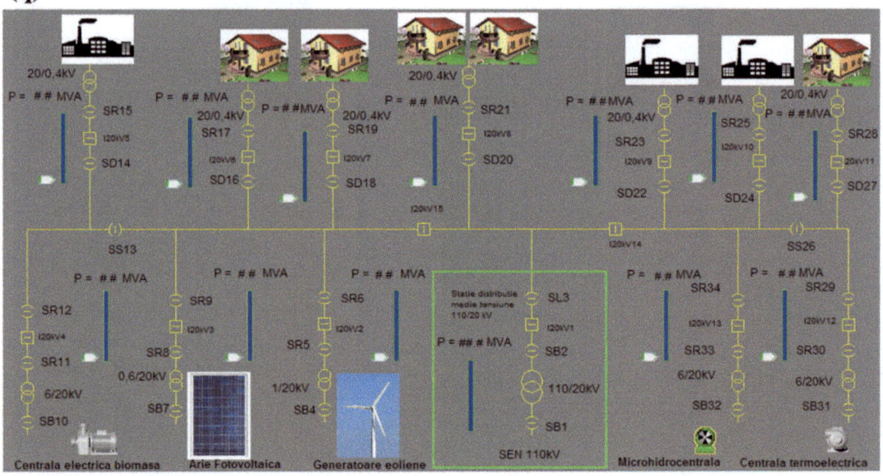

Fig. 15.21 (continued)

- Tag Variables are variables that are directly related to the process, being modified directly by the acquisition and control system, and similarly, they also have direct access to the SCADA application;
- Local Variables—required in the SCADA application development;
- Trends Variables—required for plotting the variables acquired from the process monitored.

To add a tag variable for a switch or a splitter remote controlled, proceed as follows, using Citect Project Editor → Tags → Variable Tags (Fig. 15.21o).

(r)

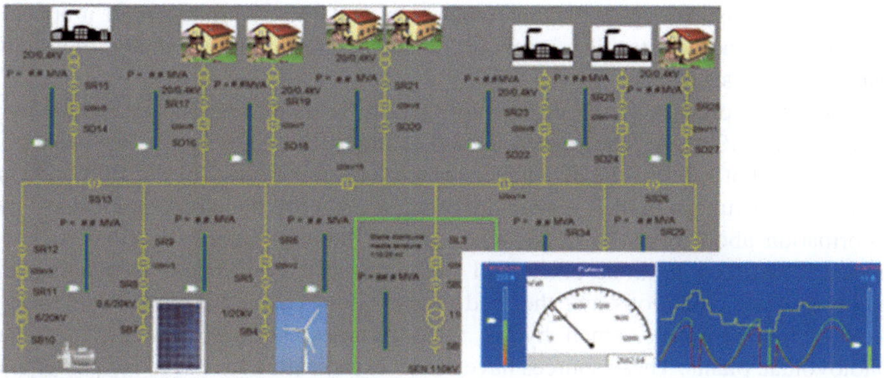

(s)

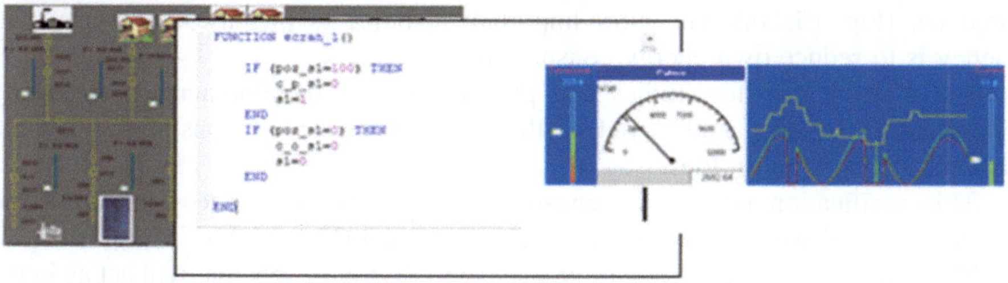

Fig. 15.21 (continued)

As shown, the variable called "s1" will be associated with the separator s1. Similarly, the rest of the variables were introduced and associated with each switching element or existing splitter shown in the diagram from Fig. 15.20

To associate an element to a symbol variable, the Properties window must be open for that element, as shown in the following figure for "Up command". Here, access the list of variables using the right button. From the list that appears, select the setting required for an ease of identification which containing a serial number associated with its name (Fig. 15.21p).

15.7.1.3 Method of Implementation

A new graphic page with the name "L 20 kV SCADA" will be created from the Citect Explorer → gen → Graphics → Pages → Next → Create new page → (Fig. 15.21q). Place the elements in the graphic page using the library and then configure the related functions as indicated above. One can start the graphical page in accordance with the operating mode of each one, and then the variables will be set; or vice versa, the design program being very flexible.

The simulation diagram is tested if running proper. Operation for each item is tested with the left button of the mouse, in different combinations of the tabs. Also, the operation of the power level indicators is checked. Adjustment is possible through the existing cursor for both consumers and energy producers. If it appears that there are inconsistencies, errors and changes are required. Finally, the application will be saved.

The whole process of the distribution system is monitored and controlled permanently by SCADA. The interface used is improved, showing the new features implemented at level of the dispatcher, which has an overview of the EPS and can perform maneuvers in the system in order to control the process. Besides the information about the state of switches at distribution stations, distribution lines appear illustrated to both household and industrial consumers. Depending on environmental conditions and the load demands are enabled the available energy sources: power plants, small hydro, wind turbines, biomass power plants and photovoltaic plants. These sources have the level indicators for the active power and the reactive power that can be modified and tracked at the consumer and the producer (Fig. 15.21s). The most important resource to increase the energy efficiency is to reduce own energy consumption.

An effective tool for studies and planning of the development of electricity distribution networks, and to reduce the domestic electricity consumption, is the analysis of the load profiles.

This verification process is repeated as many times as necessary, until the application will work correctly and as it is specified (Fig. 15.21r).

The application contains circuit breakers and disconnectors that will act as in the "General Regulation" of the maneuvers in the electrical installations and EPS (Fig. 15.21s).

Application 2

15.7.2 Diesel—Wind Turbine Hybrid EPS to Supply the Industrial Consumers

15.7.2.1 Concept Diagram of the Application

The concept diagram of this application is shown in Fig. 15.22.

15.7.2.2 Description of the Operation Diagram

The scheme shown in Fig. 15.23 considers the case of an industrial consumer powered from three different energy sources: wind, diesel, national EPS. Energy management strategy implemented is simple: the industrial consumer will be feed from the RES available, but not from two or more sources.

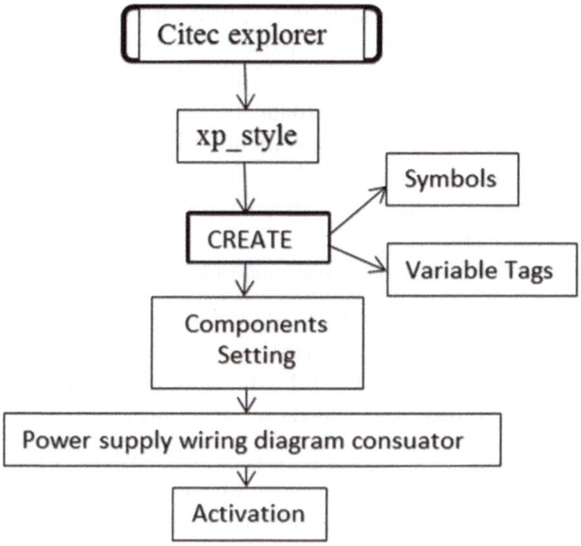

Fig. 15.22 Concept diagram of the application

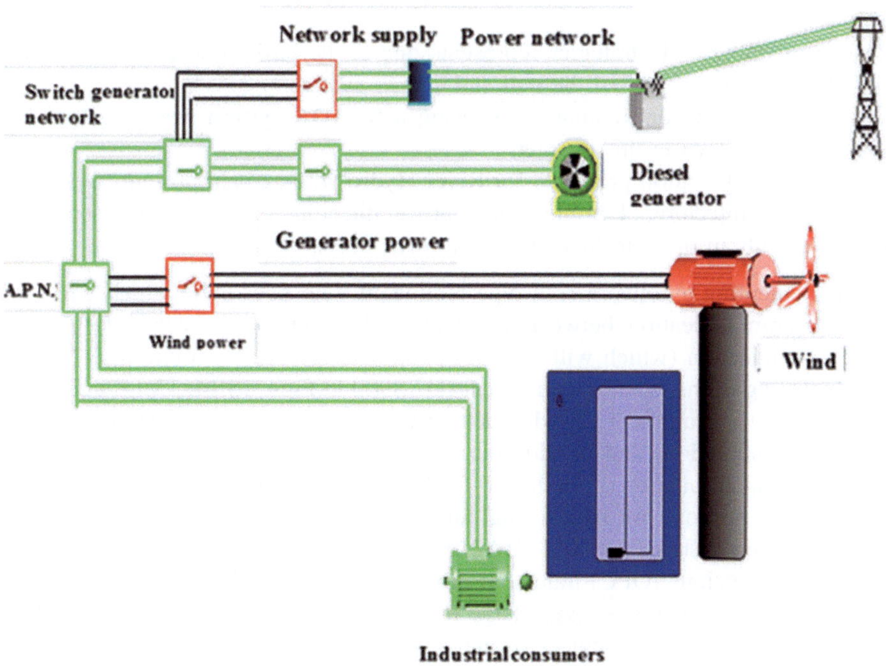

Fig. 15.23 EPS based on RESs

In this conception, the wind system is a priority, following on same level diesel generator and EPS. If the wind starts and can rotate the wind turbine, then the electric consumers will switch to energy produced by the wind turbine. If the wind stops, then the consumer will be automatically feed from the other two sources, whichever is convenient based on the specification: if the chosen second source will be interrupted, then it will automatically switch to the third variant. The energy management strategy will switch the consumer to energy line of the wind turbine if the wind will appear again.

In the scheme shown in Fig. 15.3 were also introduced three buttons as follows:

- "WIND ON" AND "WIND OFF" to start/stop voluntary the wind farm;
- "APN ON" and "APN OFF" to start/stop the industrial consumers;
- "GENERAL_ON" "GENERAL_OFF" to start/stop automatically the power flow between the national grid and diesel generator.

If wind turbine works, then the network between "SWITCH GEN/NETWORK" and "APN" switch is not active for reasons related to consumer protection.

The panel with cursor simulates wind speed which is established in a certain margin and which can turn on/off the wind turbine.

15.8 Conclusion

The focus on new IT strategies represents one of the basic premises of increasing the efficiency and safety of the national energy system.

Advantages and disadvantages of using a SCADA system were illustrated by applications shown in this chapter.

The first application proposed for River Hydro arrangement presents a complex SCADA architecture. For the existing architecture, it was proposed some improvements in order to increase its performance:

- Implementing a network ring-type redundant with fiber-optic which will facilitate communication between HED and HEC network, instead of the radio communication (which will remain operational as standby solution);
- Replacing of the existing equipment (DAS900 in Figs. 15.14, 15.15, and 15.16) with a new equipment (2 redundant servers), which will eliminate the existing DAS dysfunctions and in addition will implement new functions that are necessary in exploitation of the hydro facilities: (1) a bi-directional communication (for the purpose of monitoring and transmission of commands) with the automation systems (as HA1, HA2, SG, and Barrage) using the Open Platform Communications (OPC) industrial protocol; (2) a dedicated server for SCADA applications which has a very large number of standardized protocols for direct communication with devices of automation used in monitoring and driving of the industrial processes.

- Simplification of the SCADA architecture (Fig. 15.17) by (1) using the DROP software on the servers, (2) ensuring the direct connection between equipment without intermediate software, (3) ensuring a high speed communication with existing equipment based on the standard protocols, and implementing of high security levels for servers;
- Assuring of local drivers and OPC connectivity (Fig. 15.18) will permit (1) the functioning without interruption using the hot-plug redundant servers, (2) a future expansion and easy software development, without constraints, and (3) a simple interfacing of the equipment starting from the top-level to the bottom-level.

The second application proposed in this chapter has shown the stepwise implementation of a medium-voltage EPS based on RESs. The application is implemented using the HMI interfaces of the programming environment VIJEO CITECT®. This application provides an improved and friendly interface, with a high degree of integration of the schemes that facilitates easy operation of the system. Compared to the initial application, this new application enables:

- Monitoring and handling of the active and reactive power level indicators; the power factor is possible to be evaluated and improved in real-time;
- Monitoring is done both for consumers and producers using visualization in both value and graphic representations.
- The application will allow a quick view of the irregularities occurring in the system, so it will be possible to eliminate them (Fig. 15.21q, r, s).

Another application shows a diesel—wind turbine hybrid EPS used to supply the industrial consumers. This system has the advantage that the wind speed is permanently monitored and adjusted if is the case. The advantages of this supply system are the low costs and the flexibility in choosing of best variants to use the RESs in relation with the environment conditions.

Even if the initial costs related to SCADA implementation are relatively high, these will be recovered in short time due to the advantages offered by the SCADA:

- improved operation of the installation and of the process, whence a number of savings will appear as a result of the optimization operation;
- increased productivity;
- full operational control of the process being monitored;
- safety, robustness, minimum expenditure in exploitation;
- improved system safety due to a better information of the dispatchers and improved supervision;
- energy savings due to optimization of the manufacturing processes;
- improving the access to information, and their quality and visualization in real-time;
- extremely easy to extend the system;
- the possibility to be connected with different systems.

15.9 Trends in the Evolution of SCADA Systems

The trends in the evolution of SCADA systems refer to communications, RTU and MTU as follow:

- The development of new communication technology to optimize the equipment used as size and energy consumption, making it possible to integrate them even in the RTU;
- The upgrade of the radio equipment with numerical subsystems to perform the auto-calibration, which reduce the starting time of the transmitters up to seconds;
- The use of the geostationary satellites for SCADA systems using portable broadcasting stations which have the price comparable with that of mobile phones;
- The use of communications on fiber-optic, which offers the advantages of high speed transmission, increased safety and confidentiality; this technology is well adapted to the needs of communications in field of energy as well;
- The development in the RTU units based on embedded computers allowed more flexibility related to its functionality as controllers, meters, energy management units etc.

The later developments of the MTU units are focused along three levels: (1) improved operator interface (graphical user interfaces, windows, objects, graphics-oriented etc.), (2) increased autonomy of the intelligent self-trained systems, and (3) improved car-to-car communication based on LAN networks.

References

1. J. Figueiredo, J. Sa da Costa, A SCADA System for Energy Management in Intelligent Buildings, Energy and Buildings, vol. 49, pp. 85–98, 2012.
2. J. Tomic, et al., Smart SCADA System for Urban Air Pollution Monitoring, Measurement, vol. 58, pp. 138–146, 2014.
3. O. Barana, et al., Comparison between Commercial and Open Source SCADA Packages - A Case Study, Fusion Engineering and Design, vol. 85, pp. 491–495, 2010.
4. A. Gligor, T. Turc, Development of a Service Oriented SCADA System, Procedia Economics and Finance, vol. 3, pp. 256–261, 2012.
5. Z. Vale, et al., Distribution System Operation Supported by Contextual Energy Resource Management Based on Intelligent SCADA, Renewable Energy, vol. 52, pp. 143–153, 2013.
6. P. Novak, et al., Integration Framework for Simulations and SCADA Systems, Simulation Modelling Practice and Theory, vol. 47, pp. 121–140, 2014.
7. A. Daneels, What is SCADA, International Conference on Accelerator and Large Experimental Physics Control Systems, Trieste, Italy, pp. 339–343, 1999.
8. A. Rezai, et al., Secure SCADA Communication by Using a Modified Key Management Scheme, ISA Transactions, vol. 52, pp. 517–524, 2013.
9. B. Genge, C. Siaterlis, Physical Process Resilience-Aware Network Design for SCADA Systems, Computers and Electrical Engineering, vol. 40, pp. 142–157, 2014.

10. F.M. Enescu, Start to Design the HMI/SCADA - Applications (in Romanian), ISBN 978-606-560-425-4, Ed. Univ. Pitesti, 2015.
11. A. Fahad, et al., PPFSCADA: Privacy Preserving Framework for SCADA Data Publishing, Future Generation Computer Systems, vol. 37, pp. 496–511, 2014.
12. E. Ozdemir, M. Karacor, Mobile Phone Based SCADA for Industrial Automation, ISA Transactions, vol. 45, pp. 67–75, 2006.
13. M. Schlechtingen, et al., Wind Turbine Condition Monitoring Based on SCADA Data Using Normal Behavior Models, Part 1: System Description, Applied Soft Computing, vol. 13, pp. 259–270, 2013.
14. M. Schlechtingen, I.F. Santos, Wind Turbine Condition Monitoring Based on SCADA Data Using Normal Behavior Models, Part 2: Application Examples, Applied Soft Computing, vol. 14, pp. 447–460, 2014.
15. C.D. Dumitru, A. Gligor, A Management Application for the Small Distributed Generation Systems of Electric Power Based on Renewable Energy, Procedia Economics and Finance, vol. 15, pp. 1428–1437, 2014.
16. R.H. McClanahan, The Benefits of Networked SCADA Systems Utilizing IP Enabled Networks, Rural Electric Power Conference, 2002 IEEE, pp. C5- C5_7, 5–7 May 2002.
17. http://www.7t.dk/igss/default.asp - IGSS SCADA System - 2009.
18. http://www.7t.dk/igss/default.asp?showid=374 - IGSS Online SCADA Training - 2009.
19. http://www.7t.dk/free-scada-software/index.html - IGSS Free SCADA Software - 2009.
20. C.P. Nicolaou, et al., Measurements and Predictions of Electric and Magnetic Fields from Power Lines, Electric Power Systems Research, vol. 81, pp. 1107–1116, 2011.
21. I. Morsi, L.M. El-Din, SCADA System for Oil Refinery Control, Measurement, vol. 47, pp. 5–13, 2014.
22. A. Salihbegovic, et al., Web Based Multilayered Distributed SCADA/HMI System in Refinery Application, Computer Standards & Interfaces, vol. 31, pp. 599–612, 2009.
23. R. Ahiska, H. Mamur, A Test System and Supervisory Control and Data Acquisition Application with Programmable Logic Controller for Thermoelectric Generators, Energy Conversion and Management, vol. 64, pp. 15–22, 2012.
24. C. Damian, et al., Analysis of an Event in an Electric Transformer Station by Means of the SCADA System, Procedia Technology, vol. 12, pp. 740–746, 2014.
25. http://www.automation.ro/pdf/fisa_prez_SCADA_somes.pdf.
26. G. Clarke, D. Reynders, Practical Modern SCADA Protocols: DNP3, 60870.5 and Related Systems, Elsevier Ltd, 2003.
27. http://www.scada.ro/index.php%3Foption%3Dcom_content%26id%3D116%26lang%3Dro.
28. D. Bordea, Smart Grid or Secure Grid - New Opportunities and New Risks (in Romanian), http://cnrcme.ro/foren2014/presentations/RTF%202/pdf/02%20Dan%20Bordea%20foren%202014.pdf.
29. K. Jamuna, K.S. Swarup, Optimal Placement of PMU and SCADA Measurements for Security Constrained State Estimation, Electrical Power and Energy Systems, vol. 33, pp. 1658–1665, 2011.
30. G.N. Korres, N.M. Manousakis, State Estimation and Bad Data Processing for Systems Including PMU and SCADA Measurements, Electric Power Systems Research, vol. 81, pp. 1514–1524, 2011.
31. W. Yang, et al., Wind Turbine Condition Monitoring by the Approach of SCADA Data Analysis, Renewable Energy, vol. 53, pp. 365–376, 2013.
32. P. Cross, X. Ma, Nonlinear System Identification for Model-Based Condition Monitoring of Wind Turbines, Renewable Energy, vol. 71, pp. 166–175, 2014.
33. E Luiijf, et al., Assessing and Improving SCADA Security in the Dutch Drinking Water Sector, International Journal of Critical Infrastructure Protection, vol. 4, no. 3–4, pp. 124–134, 2011.
34. http://www.citect.com/ - CITECTSCADA -2009.
35. http://www.free-scada.org/ - Free SCADA - 2009.

[illegible]

Chapter 16
Effect of Geomagnetic Storms on Electric Networks

Daniel Mayer and Milan Stork

Abstract The beginning of this chapter summarizes the findings of the physical nature of Earth's magnetic field and its measurements in geomagnetic observatories. Experience shows that the variation of the geomagnetic field may affect the operation of various distracting electronic devices, such as systems for the transmission of electrical energy. It is derived an algorithm for calculating the currents that can be induced in the power lines and can lead to the violation of stability of the system. The algorithm is demonstrated in the illustrative example.

16.1 Introduction

Solar activity causes changes (variation) of the magnetic field on the surface of the Earth. If these changes are insignificant, affect life on Earth, or the technical equipment. Exceptionally, however, a large and rapid variations of the geomagnetic field, the so-called magnetic storms, which have the character of shock waves, may adversely affect living organisms (including humans), and the technical systems that people use. There are also pessimistic forecast of extremely strong magnetic storms, according to which may be threatened the modern civilization. The sophisticated technology you use, the more this technique is vulnerable and thus the threat of strong magnetic storms becomes actual. Geomagnetic variations, one

D. Mayer (✉)
Department of Theory of Electrical Engineering, University of West Bohemia, Plzen, Czech Republic
e-mail: mayer@kte.zcu.cz

M. Stork
Department of Applied Electronics and Telecommunications, University of West Bohemia, Plzen, Czech Republic
e-mail: stork@kae.zcu.cz

N. Mahdavi Tabatabaei et al. (eds.), *Reactive Power Control in AC Power Systems*, Power Systems, DOI 10.1007/978-3-319-51118-4_16

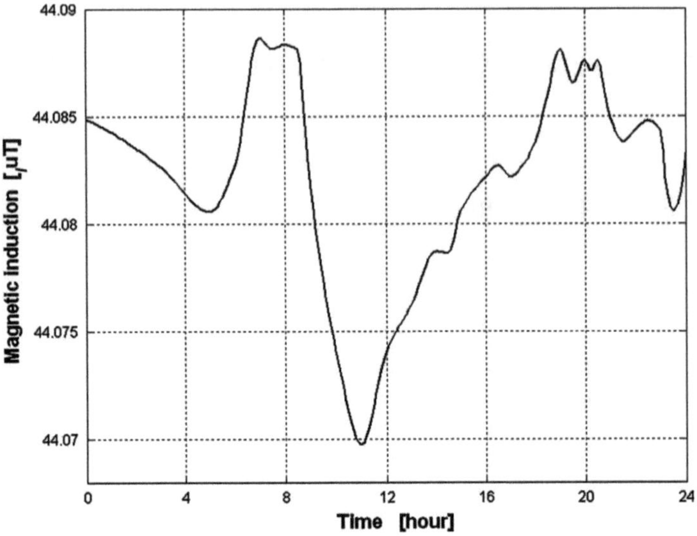

Fig. 16.1 The course of the vertical component of the geomagnetic field on 22 March 2013, as measured by certified geomagnetic observatory at Budkov, Czech [10]

cannot avoid, but they can be a threat to some approximation of the short-term predictor. Magnetic storms can cause serious disorders in technical systems. Reduce the quality of the radio connections, may damage the cable telecommunication networks and satellite communications systems and thus block the data transmission, can interfere with navigational systems and thus jeopardizing the safety of air transport.

Induced currents accelerate corrosion of pipelines transporting oil and gas, on railways may disrupting signaling devices and, in particular, threatens the stability of large electrification systems which can cause widespread blackouts and in limit case can cause a total collapse of the system and global chaos, with disastrous consequences [1–8].

Geomagnetic activity is monitored by the global network of regional centers [9], one of which is in Prague, Czech, with geomagnetic observatories in the Budkov near Prachatic. Geomagnetic measurements carried out here continuously, at intervals of one minute, or every second, with an accuracy of 0, 1 [nT] and the results are forwarded to the World Data Center in Boulder (USA) and to the International Centre in Edinburgh (Scotland). Geophysical Institute issues from 1994 forecasts of geomagnetic activity for Central Europe, Fig. 16.1, [10]. The issue of the geomagnetic field is also successfully engaged in work teams Institute of Atmospheric Physics of AS CR and the Astronomical Institute of Academy of Science, Czech [11].

This chapter presents a conceptual design that allows prediction of electrical power system emergency conditions by magnetic storms.

16.2 The Emergence of Geomagnetism and Geomagnetic Storms

16.2.1 Earth Magnetic Field

Internal geomagnetic field: W. Gilbert is more than 400 years ago believed that at the center of the Earth is a permanent magnet—a magnetic dipole which generates a magnetic field on the surface of the Earth. In the northern hemisphere is the south magnetic pole and the axis of magnetic dipole is inclined by 11.8° of the Earth's axis (Fig. 16.2). For dipole configuration performed in 1828, C.F. Gauss mathematical solution of the geomagnetic field using the spherical harmonic analysis. C. F. Gauss also developed a method of measuring absolute geomagnetic field and together with A. Humboldt and E. Weber established a worldwide network of observatories for continuous measurement of the geomagnetic field. The first geomagnetic observatory in Central and Eastern Europe was founded in 1839, K. Kreil in Prague. Measurement of the geomagnetic field and plotting global magnetic maps in different epochs showed that the geomagnetic field is not stationary, but with time changes—we are talking about *secular variations*. These are changes that are caused by physical processes in the Earth's core. These changes are very slow, are detected on the scale of decades, and our problem geomagnetic field effect on the electricity system are completely irrelevant [12, 13].

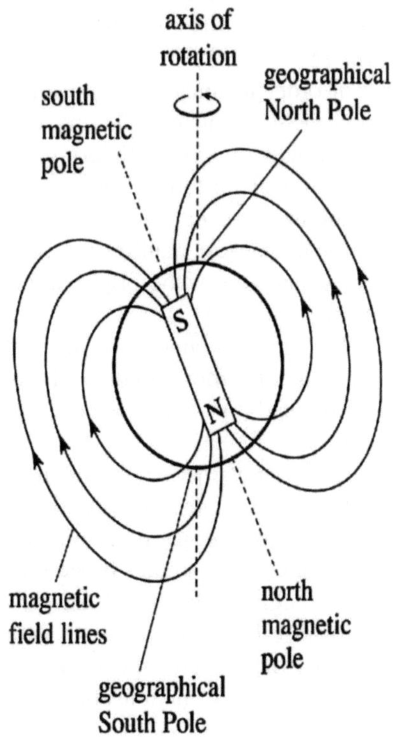

Fig. 16.2 Geomagnetic field undisturbed with the solar wind

The idea of geomagnetism with the permanent dipole magnet was abandoned after finding that the Earth's core there is a temperature above the Curie temperature; the ferromagnetic magnetic domains are randomly oriented. For example, at a depth of 100 km below the surface temperature is of the order of 1000 °C, which is significantly higher than the Curie temperature of the ferromagnetic material. In the first period at least century model was designed to explain geomagnetism—called geodynamo [12]. Its physical nature is the rotational flow liquid outer core of the Earth, which is composed mainly of iron and nickel. Since it is an electrically conductive environment that in a magnetic field (maybe a weak magnetic field such as the Sun), induce in it a very high electric current (about 10^9 A), which generates the geomagnetic field.

Model of geodynamo and his subsequent mathematical description underwent a complex evolution, in which participated many leading astrophysicists. Soon he had shown that simply cannot be represented geodynamo, with an axially symmetrical, stationary flow, as initially anticipated. The most accurate model, developed by Glatzmaier and Roberts [14], has a much more complex configuration: there are the three-dimensional, turbulent, unsteady magneto-hydrodynamic process, in which the applied laws of electromagnetic induction, the laws of fluid flow, electromagnetic and gravitational forces, Coriolis force, temperature convection and electric currents in the ionosphere. Although contemporary computers allow only an approximate solution of this complex dynamic process, computer simulation can be explained by secular variation and even inversion (i.e. change of polarity) geomagnetic field in the distant geological past of the Earth. The development of the theory of geodynamo deals e.g. the work of Chicago astrophysicist professor Parker [15, 16].

In order to understand the mechanism of the Earth's geomagnetic field generation including its chaotic reversals several attempts to create an adequate enough, but sufficiently simple theory of mathematical models have been undertaken [14]. Some models in form of physically motivated sets of nonlinear ordinary differential equations were derived. Also a variety of different dynamo models based on a physical analogy, manifesting strongly nonlinear phenomena, has often been used for explanation of the origin and chaotic-like behavior related to stellar and planet magnetic fields [17].

The irregular changes in the Earth's magnetic field polarity occurring in irregular intervals of order hundred thousands to millions of years have been widely investigated by means a concept of geodynamo in the second half of the last century.

Results obtained by computer simulation are compared with measured values obtained by geomagnetic observatories, and also using satellites. About secular variations of the geomagnetic field in the past centuries, resp. millennia inform archaeomagnetic studies based on the determination of remanent magnetism burned archaeological objects of known age. For example, the samples derived from prehistoric kilns or hearths to determine direction and magnitude of the magnetic intensity of the geomagnetic field, in which there was examined object at the time of burning. Information about the geomagnetic field of the Earth before tens of

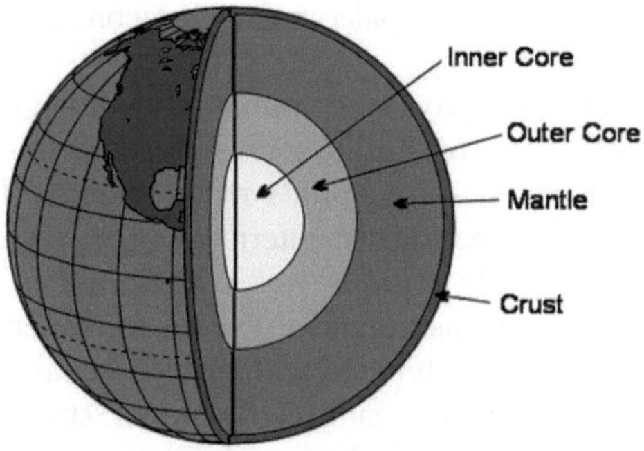

Fig. 16.3 Simplified view of the Earth's interior

thousands to millions of years of research provide paleomagnetic remanent magnetization Quaternary of volcanic and sedimentary rocks, geological and radiocarbon dating methods, based on the decay of atmospheric carbon ^{14}C. This will also find information on the geomagnetic poles wandering around the geographic poles in various geological epochs and then on multiple geomagnetic polarity inversion [18–24].

Intuitively, any real-world system which is able to convert mechanical power into its electric or electromagnetic form can be considered as a dynamo. When stable equilibrium between the mechanical input and the generated output electrical current is reached we get a self-sustaining magnetic dynamo. In theory of the Earth's geomagnetic field generation the physical concept of geodynamo has been introduced. The simplified view of the Earth's interior, which structure is known to play a crucial role in this context, is displayed in the Fig. 16.3.

The key idea of [25], the most popular Larmor's explanation for planetary magnetism is very simple. Any real-world dynamo generates a current. Every current produces magnetic field accumulating an energy. Therefore input power, both mechanical and magnetic, must be put into the system in order to maintain its motion. This is the reason why an input torque has to be applied, because otherwise the dynamo would run down due to Ohmic heating.

Now comes a new idea concerning internal power-informational interactions and well known destabilizing effects of the positive feedback. Let's assume existence of a current-carrying wire around the rotation axle of a rotating electromechanical system. It is easy to imagine that under certain conditions the resulting magnetic field reinforces the external field already present. The stronger field increases the Lorenz force and thus increase of the current follows. The growing current in turn creates an even stronger field, which increases the current again, and so forth. It can be demonstrated that under proper conditions the induced magnetic field will completely supplant the external field and as a result only mechanical energy needs be supplied. At about the same time as the Larmor's theory of planetary magnetism was published, evidence for chaotic-like reversals in the Earth's magnetic field appeared and began to accrue. This evidence came from paleomagnetic record, particularly from lava flows. This record indicates the presence of magnetic field over some billions of years [26], necessitating a self-regenerative energy source like a dynamo. The paleomagnetic record also indicates numerous reversals over geological history.

The geophysics community has long conjected that a geodynamo in the outer core (Fig. 16.1) is the source of the Earth's geomagnetic field. Now, it became clear that the cause of Earth's main magnetic field is the turbulent flow of conducting material deep in the interior, which can only mean the outer core. It means that Larmor's hypothesis was correct. About the question of the energy source that stimulates this turbulent flow, or equivalently about the physical nature of energy to drive the geodynamo with chaotic behavior, still remains considerable uncertainty. The problem is subject of further structurally oriented experimental as well as theoretical physical research.

In order to understand the mechanism of the magnetic field generation in the Earth's outer core the so called "dynamo effect" plays a crucial role and requires further research efforts. An alternative fundamentally different research direction, totally ignoring any information about the actual physical system structure seem to be possible [25, 27–31]. One such an approach is based on a very simple structure of the electromechanical system illustrated by the Fig. 16.4.

The physical analogy of the geodynamo with different modifications of the one-disc, as well as the two-disc unipolar dynamos including the Rikitake system depicted in the Fig. 16.4 is obvious [29]. The structure of the double-disc dynamo, in which the field of one dynamo is applied to a second dynamo and vice versa in such a way that a destabilizing nonlinear positive feedback arises can exhibit sudden reversals and chaotic-like field changes. However, it is fair to notice that in the Earth's interior context such a model is structurally highly unrealistic. It follows

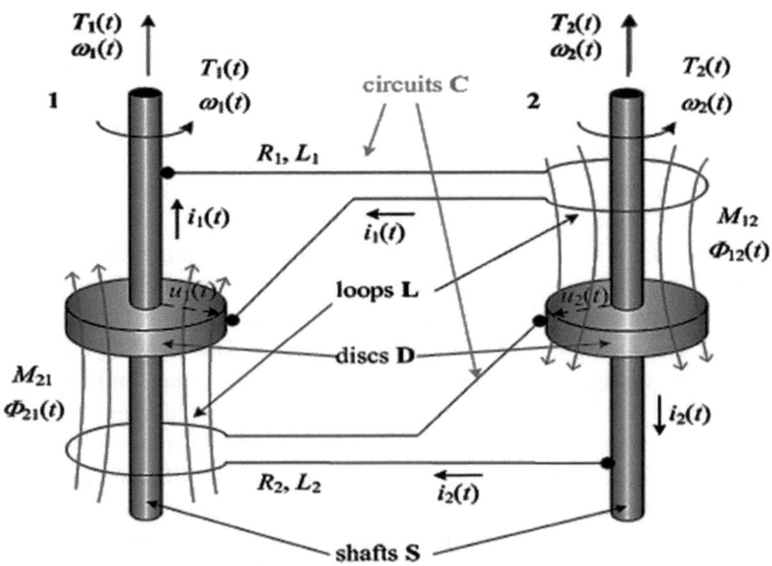

Fig. 16.4 Physical structure of the Rikitake dynamo

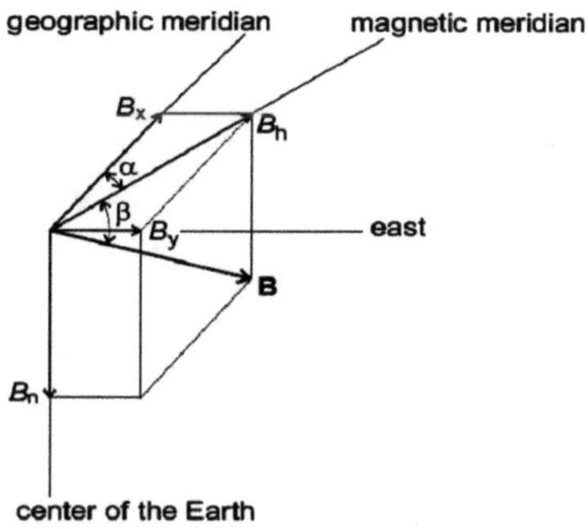

Fig. 16.5 Vector of the geomagnetic field and its components: $B = iB_x + jB_y + kB_n$

that the paleomagnetic record, particularly from lava flow, as well as observed data produced by complicated magneto-hydro-dynamical effects including geomagnetic inversion, cannot be expected to be completely explained by such simple models.

For instance, Hide showed in [28] that neglecting effects of mechanical friction is unwarranted for the two-disc dynamo. Moreover, in case of a symmetric mechanical friction the conventional model of the two-disc dynamo was due effects of a positive feedback proven to be structurally unstable and hence incapable to produce any chaotic oscillations.

In [30] a further natural question was discussed, whether the last statements concerning instability and chaotic oscillations remain valid when the azimuthal current distribution will be considered. In such a way interesting positive answers for modified two-disc Rikitake as well as one-disc dynamo have been gained.

Geomagnetic field generates the Earth's *magnetosphere*. Magnetic induction vector B has components B_x and B_y lying on the surface of the Earth (the resultant of the horizontal component B_h) and the vertical component B_n, which has a direction normal to the surface of the earth (Fig. 16.5). In geophysics usually determines the vector B from angles α (magnetic declination) and β (magnetic inclination) and from the normal component B_n. The vector magnetic field on the surface of the Earth is at the poles $B \sim 60$ μT, the magnetic equator $B \sim 30$ μT and in Czech is $B \sim 44$ μT [10].

16.2.2 Solar Activity

External geomagnetic field: The magnetic field effects contribute their electric currents in the ionosphere and magnetic field of the Sun and other planets. Over a hundred years ago it was found that there is a fairly rapid geomagnetic field variations that correlate with solar activity and show some periodicity—we are talking about variations of daily, annual, eleven year and lunar. For our problem are crucial

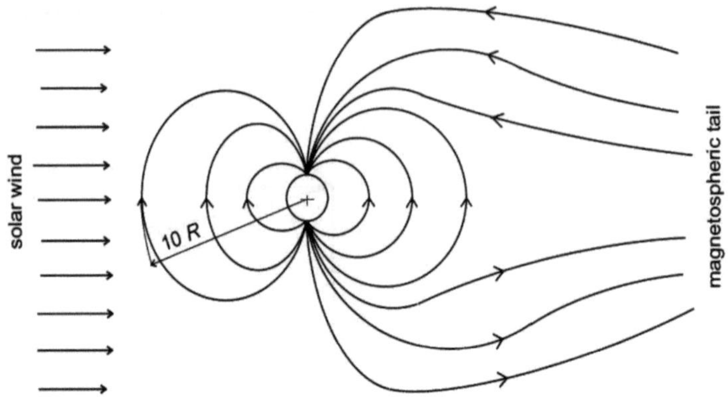

Fig. 16.6 Magnetosphere deformed with onslaught of the solar wind

variations, which have their origin in solar flares, when there is a surge of coronal mass containing electrons and high-energy protons and heavier ions, which then spread with great speed in interplanetary space. This product solar activity is known as a CME (Coronal Mass Ejection) and its flow is called *the solar wind*. The delay between the occurrence of solar flares and geomagnetic field variations usually lasts one to two days. The process flow of the solar wind in the interplanetary space undertook the marking *space weather*. Input electrically charged particles of the solar wind around the Earth to prevent the geomagnetic field [12, 31–34].

Solar wind interaction with the magnetosphere is reflected by the fact that the action of the Lorentz force is electrically charged particles moving in the direction of the magnetic field lines and flow around the magnetosphere. Thus the solar wind distorts the shape of the magnetosphere (Fig. 16.6): the front (daily, windward) side of the magnetosphere is compressed to about ten Earth radii (about 60,000 km) and dark (night, leeward) side is extended into a long tail of magnetosphere which extends far beyond the orbit of the moon [12]. Geomagnetism thus shields Earth from solar wind and thus protects the Earth's biosphere. Without geomagnetism would not exist on our planet, the existing forms of life. Just around the geomagnetic poles, where the magnetic induction vector approximately perpendicular to the surface of the Earth, created channels through which the solar wind in particles can partially penetrate into the deeper layers of the magnetosphere and endanger passengers and crew transoceanic years.

If CME has a sufficiently high energy of solar wind hits the Earth's magnetosphere, the geomagnetic activity reflected not only in the polar regions, but some of this energy is transferred to the magnetosphere, resulting in rapid variations of the geomagnetic field. Their intensity is different: from minor defects, which occur several times a day, to strong variations, caused by the solar wind in shock wave—called *magnetic storms* [19, 35, 36]. There are accompanied by polar lights.

Aurora is luminous phenomena in Earth's atmosphere. They are caused by the excitation of atoms of the atmosphere in heights between 70 and 1000 km with particles of the solar wind. They can watch them as green glowing walls and columns that changes in seconds to minutes. During increased solar activity constitutes the entire auroral ring around the northern and southern magnetic field.

Minor variations in the time of day quiet geomagnetic field are 20–30 nT and during the day, resp. year show some periodicity. On the other hand, for big magnetic storms geomagnetic field reaching up to hundreds nT.

In contrast, the changes of the internal magnetic field is changing in units nT per year. In order to perform these precise measurements are supplemented by continuous variation measurements with accurate absolute measurements. Note that the link between solar and geomagnetic activity is not as a close as previously assumed. The dynamics of the magnetosphere CME activity is complicated and can be described mathematically by a system of partial differential equations of Navier-Stokes, Maxwell's equations and the equations for heat conduction. This is called a *coupled nonlinear problem* that has so far been found only an approximate solution.

Any contact with the Earth's magnetosphere significant object CME with consequent development of geomagnetic storms beyond our control, but we try to construct mathematical models whose simulations of this natural phenomenon can predict the time for us to perform technical measures to prevent greater damage.

16.3 Definition of the Problem

In this work we calculate parasitic currents during geomagnetic storms can induce electricity wires in the network. The question of the influence of variations of the geomagnetic field on the electricity supply system, caused by solar activity, became current at the end of the last century, in connection with several big accidents. Examples include massive power supply outage on the east coast of North America in October 1989. The largest solar flare observed followed by a strong magnetic storm ($B \sim 3000\ \mu T$) registered British astronomer R. Carrington in August 1859. Since the introduction of electricity into practice at that time was at the very beginning, although magnetic storm knocked out telegraph network in the entire northern hemisphere, but did not cause more damage. At a time when the entire human civilization dependent electrical and electronics might causes incalculable similar storm damage.

In recent times there has been a strong magnetic storm, coupled with the rapid changes in the magnetic field 13 March 1989, in the Canadian province of Quebec. Total mains power failures hit Ontario, British Columbia and parts of North America. After a period of nine hours or more were 6 million people without electricity. After 24 h have been taken out of operation US navigation satellites. That event began systematically monitor space weather. List of magnetic storms, which damaged electrical equipment in different parts of the country, is given in [1].

Research into this phenomenon is discussed by many authors, among whom a significant proportion by Finnish authors [1, 2, 37–41]. In addition to electrical systems were also considered telecommunications networks, network for transport oil and gas [16] and the rail track. Despite great efforts has been devoted to resolving this issue, no matter reliable prediction of the effects of magnetic storms closed.

Physical causes harm of electricity networks are parasitic currents induced by geomagnetic variations in the branches of the transmission network—the abbreviation indicate the GIC (Geomagnetically Induced Currents) [42]. Geomagnetic variations have the character of shock waves, which are changes in seconds. GIC are superimposed together with the operating currents in the lines electricity system that transmits active and reactive power. In terms of electrical networks, these changes are very slow and therefore GIC denoted as quasi-DC currents. It follows that it is sufficient to count only with the resistance of the transmission lines, while the effect of inductance, capacity coupling and skin effect are negligible. GIC causing more load of conductor by current, increase the transmission losses, increase the voltage drop in the lines and cause oversaturation magnetic circuits of transformers and thus may cause a violation of stability and eventually black-out of system.

16.4 Geomagnetic Field on the Surface of the Earth and Its Rapid Variations

For electromagnetic induction GIC in the branches of the network is important the rate of time of normal (vertical) components of $B_n(t)$ of vector magnetic induction B. The magnetic storm on Earth is delayed eruption on the Sun for about 25–60 h. Drawing on many years of observations of solar and geomagnetic variations subsequent measurements can thus be predicted with a certain probability what will be $B_n(t)$. Estimated process $B_n(t)$ will be for us starting point for a further specified size calculation algorithm GIC.

16.5 Calculation of Geomagnetically Induced Currents

Investigated power system will be replaced by network N distributed in the plane (x, y), which consists from n irregular polygons having m branches and p nodes etc. (Fig. 16.7). We know: The topological structure of the network N, i.e. the graph G, the metric structure of network N, i.e. the geometric dimensions of the network N which are given by the coordinates of the vertices of each polygon and its physical structure, which are given by the resistance R_i $(i = 1, \ldots, m)$ of each branch of network. On the network N operates time-varying geomagnetic field; we know its axial (vertical) component $B_n(t)$ and we assume that in the field (x, y) of networks

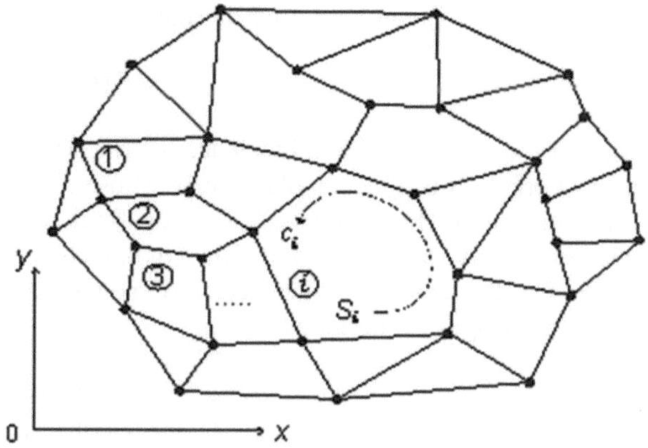

Fig. 16.7 Polygonal network *N*—general model of the transmission system

N, the value of $B_n(t)$ does not change. Using knowledge of matrix analysis of electrical circuits [43, 44], let the *i*th polygon containing a loop c_i is irregular and his vertices have coordinates (x_1, y_1), ..., (x_p, y_p). Polygons graph *G* is a set of *n* independent loops $\{c_1, \ldots, c_n\}$. According to the second Maxwell's equations apply to the *i*th loop c_i.

$$\oint_{c_i} Edl = -\frac{d\Phi_i}{dt}, \; i = 1, \ldots, n \tag{16.1}$$

where *E* is the intensity of electric field and

$$\Phi_i = \int_{S_i} B_n dS = S_i B_n \tag{16.2}$$

is the magnetic flux coupled with loop c_i. Equation (16.1) can be written as

$$\sum_k \pm u_k = S_i \frac{dB_{ni}}{dt} \tag{16.3}$$

where u_k is the voltage of the *k*th branch, which is incident with *i*th loop c_i and S_i is the content of the *i*th polygon. It is calculated from the formula

$$\begin{aligned} S_i &= \frac{1}{2}\left|\{x_1(y_2 - y_p) + x_2(y_3 - y_1) + \ldots + x_1(y_1 - y_{p-1})\}\right| \\ &= \frac{1}{2}\left|\{y_1(x_2 - x_{p-1}) + y_2(x_3 - x_1) + \ldots + y_n(x_1 - x_{p-1})\}\right| \end{aligned} \tag{16.4}$$

The Eq. (16.3) can be written in matrix form

$$Cu = F(t) \tag{16.5}$$

where $C(n, m)$ is the second incidence matrix, which is determined by the topology of network N. It is a rectangular matrix whose rows (columns) correspond to loops (branches) of G. Her elements are: +1 (or −1) if kth branch is incident with ith loop consistently (or vice versa), or 0 if the kth branch is not incident with ith loop (i = 1, …, n, k = 1, …, m), u is the vector of voltage branch and $F(t)$ is the vector of induced voltages in the loops.

$$u = \begin{bmatrix} u_1 \\ \ldots\ldots \\ u_m \end{bmatrix}, \quad F(t) = \begin{bmatrix} S_1 \\ \ldots\ldots \\ S_n \end{bmatrix} \frac{dB_n}{dt} \tag{16.6}$$

We introduce the vector of currents in the branches i and the vector of current in the loop i'

$$i(t) = \begin{bmatrix} i_1 \\ \ldots\ldots \\ i_m \end{bmatrix}, \quad i'(t) = \begin{bmatrix} i'_1 \\ \ldots\ldots \\ i'_n \end{bmatrix} \tag{16.7}$$

Both matrices i and i' are linked together with a second incidence matrix

$$i = C^T i' \tag{16.8}$$

where the upper index "T" denotes the matrix transpose.

Between vector of branches voltage u and a vector of branch currents i is true the Ohm's Law, which written by matrix is

$$u = Ri \tag{16.9}$$

where R is a diagonal matrix of branch resistances: R = diag $[R_1, \ldots.. R_m]$.

We multiplying Eq. (16.9) by the left matrix C and substituting the vector i from Eq. (16.8)

$$Cu = CRi = CRC^T i' \tag{16.10}$$

with help of Eq. (16.5), we determine

$$F(t) = CRC^T i' \tag{16.11}$$

Thence calculate the vector of loop currents i' and using Eqs. (16.8), (16.11) and (16.5) we find vector of instantaneous values of GIC in the branches of the network N

$$i(t) = C^T\left[CRC^T\right]S\frac{dB_n}{dt} \tag{16.12}$$

Using the vector $i(t)$ can be easily determined the vector of rms current in the branches I_{ef}.

$$i(t) = C^T\left[CRC^T\right]S\frac{dB_n}{dt} = Af(t) \tag{16.13}$$

16.6 Illustrative Example

The network N in the plane (x, y) has the topological structure characterized by graph G in Fig. 16.8, his metric structure is determined by the coordinates of nodes of the graph G: (0, 0), (220, 40), (180, 180), (30, 140) and (80, 70) km, and the physical structure of the network N is determined by the resistance of branches: $R_1 = 32.1\ \Omega$, $R_2 = 50.2\ \Omega$, $R_3 = 32.1\ \Omega$, $R_4 = 34.6\ \Omega$, $R_5 = 19\ \Omega$, $R_6 = 24.1\ \Omega$, $R_7 = 31.7\ \Omega$, $R_8 = 33\ \Omega$. Measurements were found the normal component of the magnetic induction of the geomagnetic field with a time step of 1 s; graph of $B_n(t)$ is the upper Fig. 16.7, determine the immediate and effective values GIC in branches of network N.

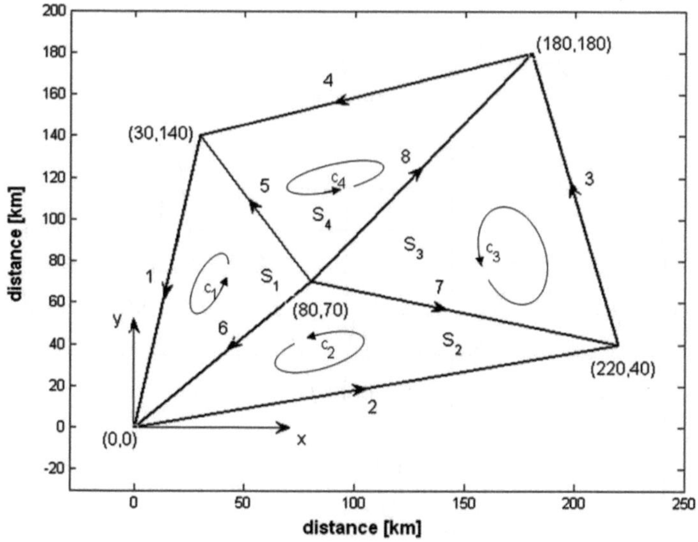

Fig. 16.8 Network N, which is solved

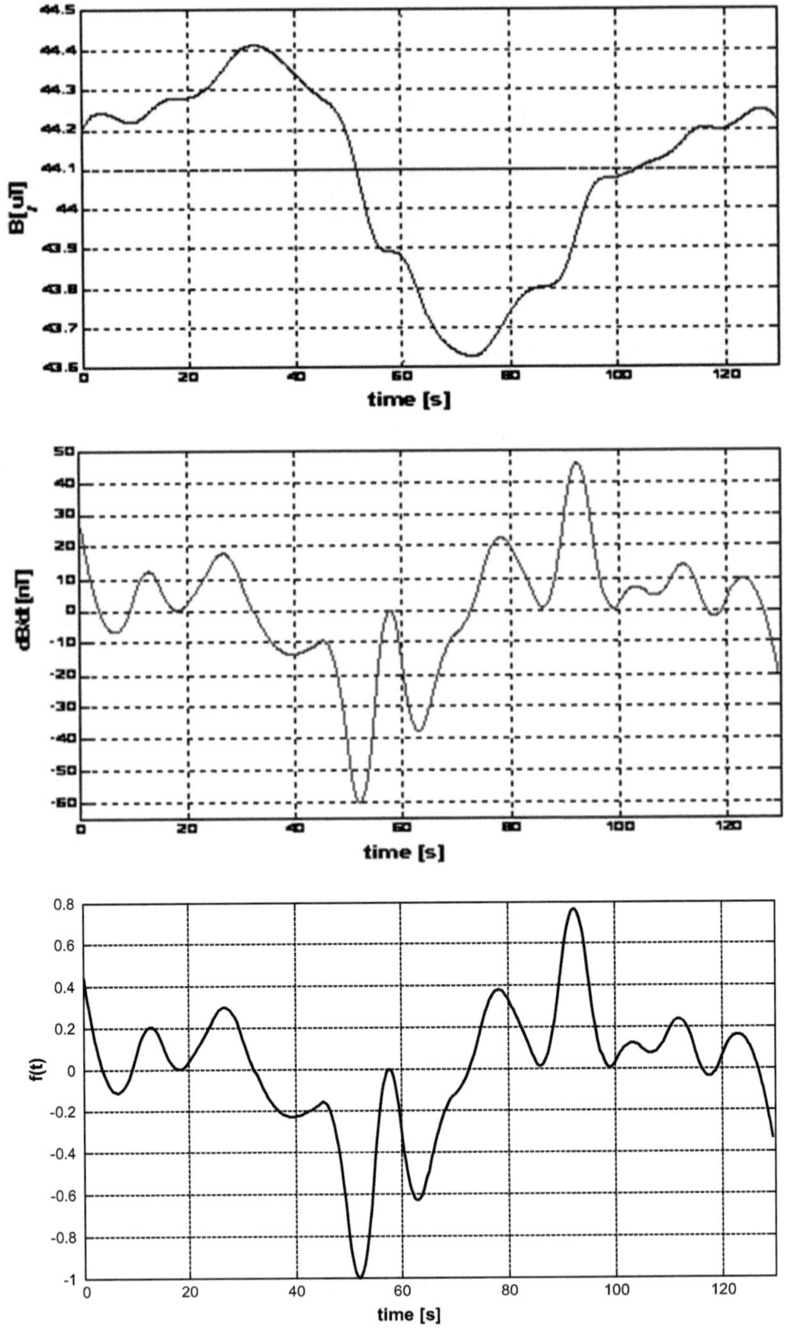

Fig. 16.9 Course of the normal component of the magnetic induction $B_n(t)$ depending on time (*top*), derivative of $B_n(t)$ (*middle*) and $f(t)$ (*bottom*)

Using the spline interpolation and subsequent numerical derivative of the function $B_n(t)$ can be found dB_n/dt. This function was normalized by its maximum size, $\max[|dB_n/dt|] = 60$ [nT].

Graph of standardized function $f(t) = \frac{1}{\max|dB_n/dt|} \cdot \frac{dB_n}{dt} = \frac{1}{60} \cdot \frac{dB_n}{dt}$ is in the bottom part of Fig. 16.9.

Oriented graph G (arbitrarily). The second incidence matrix, matrix of resistances of branches and vector of loops of polygon surfaces of network are

$$C = \begin{bmatrix} 1 & 0 & 0 & 0 & 1 & -1 & 0 & 0 \\ 0 & 1 & 0 & 0 & 0 & 1 & -1 & 0 \\ 0 & 0 & 1 & 0 & 0 & 0 & 1 & -1 \\ 0 & 0 & 0 & 1 & -1 & 0 & 0 & 1 \end{bmatrix}$$

$$R = \mathrm{diag}[\,32.150.232.134.61924.131.733\,] \;\; [\Omega] \tag{16.14}$$

$$S = \begin{bmatrix} 4.5 \\ 6.1 \\ 9.2 \\ 6.25 \end{bmatrix} \times 10^9 \;\; [\mathrm{m}^2]$$

Substituting of matrix (16.14) into Eq. (16.13) get the vector of instantaneous values of currents in the branches of the network $i(t)$. Can easily calculate the rms currents in the branches of network I_{ef}, in the time interval $t \in \langle 0,\, 130 \rangle$ [s].

$$i(t) = Af(t) = \begin{bmatrix} 9.5 \\ 9.4 \\ 12.6 \\ 11.2 \\ -1.7 \\ -0.09 \\ 3.2 \\ -1.4 \end{bmatrix} f(t)\,,\; I_{ef} = \begin{bmatrix} 2.9 \\ 2.8 \\ 3.8 \\ 3.4 \\ 0.53 \\ 0.03 \\ 0.98 \\ 0.42 \end{bmatrix} [\mathrm{A}] \tag{16.15}$$

The time evolution of currents $i(t)$ [from $i_1(t)$ to $i_8(t)$] are shown in Fig. 16.10.

16.7 Effect of Changes in the Earth's Magnetic Field on the Reliability of the Distribution Power Network

In the previous section we explain the physical causes of time-variable Earth's magnetic field and subsequent we found the solution of relatively slow variable phenomenon in the electricity network. We formulate the algorithm of calculation of the values of the quasi-direct current that induce in the branches of the electricity

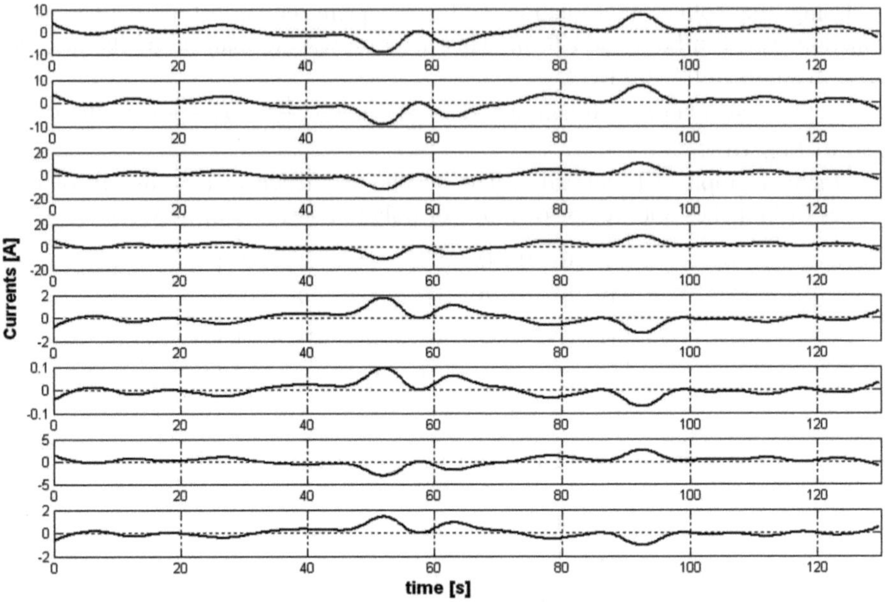

Fig. 16.10 Currents i_1 (*top*) to i_8 (*bottom*) for network according Fig. 16.8

network. These currents generate electric power which is completely converted to electric heat in conductors of transmission lines and in the winding of the transformers in the electricity system. These currents are purely parasitic and do not participate on the transmission of power from generators to appliances, as well as reactive power during normal operation of network. During magnetic storms (i.e. in the relatively quick changes in the Earth's magnetic field), this can cause the dissipation of heat damage to wires of transmission lines and windings of the transformers. This will avoid using the protection network, but at the cost of the interruption of power supply, which leads to a collapse of the electricity system.

Let in the *i*th transmission line is the current

$$i_{ik}(t) = I_{oi} + \sqrt{2}\{I_{ik}\sin(k\omega t + \varphi_{ik}\}, \quad k = 1, 2, \ldots, \infty \tag{16.16}$$

where I_{0i} is the value of quasi-direct current components induced by changing the magnetic field of the Earth, and the I_{ik} and φ_{ik} are the effective value of the *k*th harmonic currents and their phase shift, both in the *i*th wire of transmission line. Then the dissipated power in the *i*th wire, according to the Parseval identity, may be expressed as a Joule's power

$$P_i = R_i \sum_{k=0}^{\infty} I_{ik}^2 \tag{16.17}$$

A more complicated case occurs in the transformer winding. A quasi-direct current cause not only increases Joule losses in the windings, but also causes reduction of magnetization of magnetic circuit of the transformer.

Decreases permeability of the magnetic circuit and also reduction of the inductance of the windings can leads to a decrease the stability of the electricity system and cause also reactive power increasing.

16.8 Conclusion

The input value when calculating the GIC is the time course of the vertical component of the magnetic induction $B_n(t)$, which can only be estimated based on a larger number of long-term measurements. It is therefore a typical task with uncertain input data. These tasks are resolved in principle, two different approaches: stochastic (probabilistic), or the method of reliable solutions (i.e. using worst-case scenario), we assume the "worst" possible over $B_n(t)$, even if the probability of its occurrence is small. The second of the two approaches is therefore pessimistic, but it is "safer".

The size of the GIC is not crucial size of the normal component of the magnetic intensity B_n, but the size of its derivative, i.e. dB_n/dt. Based on the observation of the geomagnetic field it can be assumed that strong geomagnetic shock waves, which entitle to predict the state of emergency power system, events are very rare.

On presented example it was shown, that for electrical networks are strong magnetic storms dangerous, because of induction of quasi-direct current. This leads to current overload of lines. Most dangerous is it this quasi-dc part for transformers especially for magnetic saturation [45, 46]. These phenomena can be avoided by means of protections which excludes some parts of networks from work.

Should such a quite exceptional magnetic storms occurred, can the system operator to prevent destructive situations only intervention in the topological structure of the transmission system. Example, it can transform the transmission system into smaller units, or may interrupt galvanic connection some exposed line of system included with isolating transformers.

Acknowledgements Milan Stork's participation was supported by Department of Applied Electronics and Telecommunications, University of West Bohemia, Plzen, Czech Republic and by the European Regional Development Fund and the Ministry of Education, Youth and Sports of the Czech Republic under the Regional Innovation Centre for Electrical Engineering (RICE), project No. No. LO1607, the Internal Grant Agency of University of West Bohemia in Pilsen, the project SGS-2015-002 and GA15-22712S.

References

1. D.H. Boteler, et al., The Effects of Geomagnetic Disturbances on Electrical Systems at the Earth's Surface, Adv. Space Res., vol. 22, no. 1, pp. 17–27, 1998.
2. T. Makinen, Geomagnetically Induced Currents in the Finish Power Transmission System, Finnish Meteorological Institute Geophysical Publications, no. 32, pp. 101, 1993.
3. V.D. Albertson, J.N. Thorson, S.A. Miske, The Effects of Geomagnetic Storms on Electrical Power Systems, IEEE Trans., PAS-94, pp. 1931, 1974.
4. R.A. Gummow, P. Eng, GIC Effects on Pipeline Corrosion and Corrosion Control Systems, Journal of Atmospheric and Solar-Terrestrial Physics, vol. 64, pp. 1755–1764, 2002.
5. D.H. Boteler, R.J. Pirjola, H. Nevanlinna, The Effects of Geomagnetic Disturbances on Electrical Systems at the Earth Surface, Adv. Space Res., vol. 22, no. 1, pp. 17–27, 1998.
6. E.A. Eroshenko, et al., Effects of Strong Geomagnetic Storms on Northern Railways in Russia, Adv. Space Res., vol. 46, issue 9, pp. 1102–1110, 2010.
7. R.A. Gummow, GIC Effects on Pipeline Corrosion and Corrosion Control Systems, Journal of Atmospheric and Solar-Terrestrial Physics, vol. 64, pp. 1755–1764, 2002.
8. Y. Gallet, et al., Possible Impact of the Earth's Magnetic Field on the History of Ancient Civilizations, Earth Planet, Scientific Letters, vol. 246, no. 1–2, pp. 17–26, 2006.
9. V. Haak, S. Maus, M. Koret, H. Luhr, The Earth's Magnetic Field Screening and Monitoring, Physics in our Time, vol. 35, no. 5, pp. 218–224, 2003.
10. http://bdv.ig.cas.cz/index.html, http://rwcprague.ufa.cas.cz.
11. Z. Nemecek, et al., Dynamics of Magnetosphere and Ionization Processes and its Correlation with Solar Activity, Bull. GACR, vol. 14/4, no. 5–10, 2006 (in Czech).
12. J. Klezcek, Great Encyclopedia of the Universe, Academia, Prague, ISBN 80-200-0906-X, 2002 (in Czech).
13. R. Lanza, A. Meloni, The Earth's Magnetism, Springer, Berlin, ISBN 3-540-27979-2, 2006.
14. G.A. Glatsmaier, P.H. Roberts, Phys. Earth Planet, Inter., vol. 91, pp. 63, 1995.
15. E.N. Parker, Cosmically Magnetic Fields, Clarendon Press, Oxford, 1979.
16. E.N. Parker, Interplanetary Dynamical Processes, Wiley-Interscience, New York, 1963.
17. D. Stevenson, Planetary Magnetic Fields, Earth and Planetary Science Letter, 6523, pp. 1–11, 2002.
18. P. Ertepinar, et al., Archaeomagnetic Study of Five Mounds from Upper Mesopotamia between 2500 and 700 BCE, Earth and Plane. Sci. Letters, vol. 357–358, pp. 84–98, December 2012.
19. J.P. Legrand, P.A. Simon, Geomagnetic Storms and their Associated Forecasts, Solar-Terrestrial Predictions - IV, Ed. by J. Hruska, et al., NOAA, Boulder, vol. 3, pp. 191, 1993.
20. G. Verbanac, et al., Four Decades of Geomagnetic and Solar Activity: 1960–2001, Journal of Atmospheric and Solar-Terrestrial Physics, vol. 72, issue 7–8, pp. 607–616, 2010.
21. P.N. Mayaut, Analysis of Storm Sudden Comment Cements for the Years 1868–1967, Journal of Geophysics, Res. 111, 1975.
22. P. Ertepinas, et al., Archaeomatic Study of Five Mounds from Upper Meoptamia between 2500 and 700 BCE: Further Evidence for an Extremely Strong Geomagnetic field from 3000 Years Ago, Earth and Planetary Science Letters, vol. 357–358, pp. 89–98, December 2012.
23. Y. Gallet, et al., On the Use of Archeology in Geomagnetism and vice-versa: Recent Developments in Archaeomagnetism, C.R. Physic, vol. 10, no. 7, pp. 530–648, 2009.
24. T. Nagata, et al., Secular Variation of Geomagnetic Total Force during the Last 5000 Years, Journal of Geophysics Research, vol. 68, pp. 5277–5281, 1963.
25. J. Larmor, Possible Rotational Origin of Magnetic Fields of Sun and Earth, Elect. Rev. 85, 1919.
26. D. Sisan, W. Shew, D. Lathrop, Lorenz Force Effects in Magneto-Turbulence, Physics of Earth and Planetary Interiors, vol. 135, pp. 137–159, 2003.

27. B. Skala, J. Hrusak, D. Mayer, M. Stork, On Strongly Nonlinear Phenomena in Electrical Machines, New Aspects of Systems, Proceedings of 12th WSEAS International Conference on Systems, Heraklion, Greece, pp. 120–125, July 22–24, 2008.
28. R. Hide, Structural Instability of Rikitake Disc Dynamo, Geoph. Res. Letter, vol. 22, pp. 1057–1059, 1995.
29. T. Rikitake, Electromagnetism and the Earth's Interior, Elsevier, Amsterdam, 1966.
30. F. Plunian, P. Marty, A. Alemany, Chaotic Behavior of the Rikitake Dynamo with Symmetric Mechanical Friction and Azimuthal Currents, Proc. Roy. Soc. Lond., vol. 454, pp. 1835–1842, 1998.
31. D.F. Webb, Solar and Geomagnetic Disturbances during the Declining Phase of Recent Solar Cycles, Adv. Space Res., vol. 16, no. 9, pp. 57–69, 1995.
32. E. Correia, R.V. de Souza, Identification of Solar Sources of Major Geomagnatic Storms, Journal of Atmospheric and Solar-Terrestrial Physics, vol. 67, issue 17–18, pp. 1702–1705, 2005.
33. J.R. Wait, Theory of Magneto-Telluric Fields, Journal Res. National Bureau of Standards, vol. 66D, no. 5, pp. 509–541, Sept./Oct. 1962.
34. S.P. Plun Keit, et al., Solar Source Regions of Coronal Man Ejections and their Geomagnetic Efforts, Journal of Atmospheric and Solar-Terrestrial Physics, vol. 63, issue 5, pp. 398–407, March 2001.
35. P.N. Mayaud, Analysis of Storm Sudden Commencements for the Years 1868–1967, Journal Geophysics Res., vol. 80, pp. 111, 1975.
36. R. Wang, Large Geomagnetic Storms of Extreme Solar Event Periods in Solar Cycle, Adv. Space Res., vol. 40, issue 12, pp. 1835–1841, 2007.
37. J.P. Elovaara, et al., Geometrically Induced Currents in the Nordic Power Systems, Proceedings of the CIGRE, Paper 36–301, Paris, pp. 10, 30 August - 5 September 1992.
38. A.A. Viljanen, R. Pirjola, Geomagnetically Induced Currents in the Finnish High Voltage Power System - A Geophysical Review, Surv. Geophys, vol. 15, pp. 383–408, 1994.
39. R. Pirjola, M. Lehtinen, Currents Produced in the Finnish 400 kV Power Transmission Grid and in the Finnish Natural Gas Pipeline by Geomagnetically-Induced Electric Fields, Ann. Geophys., vol. 3, pp. 485–491, 1985.
40. R. Pirjola, et al., Prediction of Geomagnetically Induced Currents in Power Transmission Systems, Adv. Space Res., vol. 26, no. 1, pp. 1–14, 2000.
41. I.A. Erinmez, et al., Management of the Geomagnetically Induced Current Risks on this National Grid Company's Electric Power Transmission Systems, Journal of Atm. and Solar-Terrestrial Physics, vol. 64, pp. 743–756, March-April 2002.
42. T.S. Molinski, Why Utilities Respect Geomagnetically Induced Currents, Journal of Atmospheric and Solar-Terrestrial Physics, vol. 64, issue 16, pp. 1765–1778, 2002.
43. D. Mayer, Analysis of Electrical Circuit with Matrix Calculus, Academia, Prague, 1966 (in Czech).
44. D. Mayer, Introduction to the Theory of Electric Circuits, 2nd Ed., SNTL/ALFA, Prague, 1981 (in Czech).
45. L. Bolduc, J. Aubin, Effects of Direct Currents in Power Transformers, Part I: A General Theoretical Approach, Electric Power Systems Research, vol. 1, issue 4, pp. 291–298, 1978.
46. W.B. Gish, W.E. Feero, G.D. Rockefeller, Rotor Heating Effects from Geomagnatic Induced Currents, IEEE Trans. Power Delivery, vol. 9, no. 2, pp. 712–719, April 1994.

Index

N. Mahdavi Tabatabaei et al. (eds.), *Reactive Power Control in AC Power Systems*, Power Systems, DOI 10.1007/978-3-319-51118-4

MIX
Papier aus verantwortungsvollen Quellen
Paper from responsible sources
FSC® C105338

If you have any concerns about our products,
you can contact us on
ProductSafety@springernature.com

In case Publisher is established outside the EU,
the EU authorized representative is:
Springer Nature Customer Service Center GmbH
Europaplatz 3, 69115 Heidelberg, Germany

Printed by Libri Plureos GmbH
in Hamburg, Germany